Materials Engineering and Sciences Division 2014

Core Programming Area at the 2014 AIChE Annual Meeting

Atlanta, Georgia, USA
16-21 November 2014

ISBN: 978-1-5108-1264-2

Printed from e-media with permission by:

Curran Associates, Inc.
57 Morehouse Lane
Red Hook, NY 12571

Some format issues inherent in the e-media version may also appear in this print version.

Printed by Curran Associates, Inc. (2015)

AIChE
120 Wall Street, FL 23
New York, NY 10005-4020

Phone: (800) 242-4363
Fax: (203) 775-5177

www.aiche.org

Additional copies of this publication are available from:

Curran Associates, Inc.
57 Morehouse Lane
Red Hook, NY 12571 USA
Phone: 845-758-0400
Fax: 845-758-2634
Email: curran@proceedings.com
Web: www.proceedings.com

TABLE OF CONTENTS

386436 Casting Solvent Effects on Block Ionomer Morphology and Corresponding Property Changes

Abstract:

The role of casting solvent has been shown to play an important role in the microstructure of block copolymers that have not been, or are not able to be, annealed. Casting solvent plays a stronger influence in block ionomer ordering because these materials can possess strongly incompatible blocks that are easily templated in various solvent environments. Here, this principle is utilized to produce different morphologies from a sulfonated pentablock ionomer. The judicious selection of polar and nonpolar solvents, and their mixtures, translates to different solution structures that in turn create different solid film morphologies. Transmission electron micrographs and tomograms provide morphological information amongst samples varying in sulfonate concentration and casting solvent(s). The obtained data shows that casting solvent identity does, in fact, alter the ordering of the pentablock copolymer into vastly different morphologies. Following identification, the effect of the produced morphologies on swelling (water uptake), water vapor transmission rate (WVTR), and water-based semi-permanent deformation will be presented. Preliminary results indicate that geometric restriction of the sulfonated, hydrophilic phases reduces water uptake. Additionally, systems that exhibit more long range order have a lower apparent sulfonate-domain percolation threshold that results in maintained, high WVTR at low sulfonate concentration. Lastly, more disordered sulfonate domains translate to more easily deformed microstructures under the stress of liquid water and heat.

Phase Separation Kinetics in Pressure Sensitive Adhesive Melts Containing Styrene-Diene Based Block Copolymers

Ninad Dixit, Alicia Pape, Eugene Joseph and Stephen Martin

Virginia Polytechnic Institute and State University, Macromolecules and Interfaces Institute, Blacksburg, VA, ninad22@vt.edu.

Pressure sensitive adhesives are often processed by coating from the melt, and the final adhesive properties depend on the morphology of the resulting film. We report on the phase separation kinetics in melts of commercially available styrene-diene based block copolymers and adhesive blends containing various amounts of tackifier. A modified Percus-Yevick hard sphere model was used to determine morphological parameters from real-time small-angle X-ray scattering (SAXS) analysis. The Avrami kinetics model was used to study ordering of the polystyrene domains based on rheological analysis. The addition of tackifier in the adhesive blends led to a gradual decrease in the extent of ordering of polystyrene domains. The phase separation mechanism also shifted from a 'Nucleation and Growth' to a 'Spinodal Decomposition' type process with increasing amounts of tackifier in the blend. Isoprene chain mobility and tackifier content in the blend were found to be the most dominant factors affecting the ordering kinetics. At higher tackifier content, polystyrene domains remained in a liquid-like disordered state. We also concluded that the concentration fluctuations in the disordered melts drive the initial stage of phase separation process in adhesive blends.

379646 Impact of Conformational and Chemical Correlations on Microphase Segregation in Random Copolymers

Abstract:

Random copolymers play an important role in a range of soft materials applications and biological phenomena. For example, polymer membranes for fuel cells are frequently composed of random copolymers synthesized with hydrophobic and hydrophilic monomers. Microphase segregation of these monomers results in a solid material whose mechanical integrity arises from a space spanning random arrangement of hydrophobic domains, and ion transport occurs through the corresponding hydrophilic domains. The structure of the microphase segregated domains dictates the pathways for ion transport and ultimately the macroscopic ion mobility and material performance. An individual monomer is typically a single chemical unit whose length is comparable to a Kuhn length, resulting in a monomer segment that is structurally rigid at length scales of a segregated domain. Previous work on random copolymer phase segregation addresses the impact of correlations between the chemical identities along the chains for flexible polymers; thus, a monomer block behaves as a random walk without structural correlation associated with semiflexibility. In our work, we develop a model of semiflexible random copolymers to address the thermodynamics of microphase segregation and the structure of the segregated domains. Using a random phase approximation up to quartic order in density fluctuations, we show the onset of phase segregation and the correlation length of domains is extremely sensitive to chain rigidity. We then leverage particle-field Monte Carlo simulations to address the structure of the domains and discuss the role of density fluctuations in the thermodynamic behavior.

Sequential Infiltration Synthesis into Native –OH containing Block Copolymer

Caleb Breaux, Richard Lawson, Peter Ludovice, Clifford Henderson

cbreaux6@gatech.edumailto:Ben.Nation@gatech.edu

School of Chemical and Biomolecular Engineering

Georgia Institute of Technology, Atlanta, GA

Microphase separation of block copolymers into structures with characteristic length scales on the order of 10 nm is a promising alternative to conventional lithography. After aligning a block copolymer to a pre-patterned underlayer, one block of the copolymer must be removed, typically through an etch step. However, if the etch resistance between the two blocks are similar, it can be difficult to retain the phase separated pattern. Sequential Infiltration Synthesis (SIS) offers a route by which the etch resistance of one block may be increased by the infiltration of organometallic into one block. Extensive study of SIS has been done regarding PS-b-PMMA due to its ease of phase separation via thermal annealing, and because the PMMA block's carbonyl moieties readily act as sites for SIS of Al using the precursors trimethyl-aluminum (TMA) and water. For the SIS of other metals (i.e. Ti, W, Zn), a pretreatment of one cycle of TMA/Water is needed for the PMMA domains to have the –OH groups for their growth. Here, a high χ-block copolymer (PS-b-PHOST) with a domain that has native –OH groups is used in SIS to form perpendicular and parallel morphologies.

387881 Kinetically-Controlled Dynamics in Block Polymer Micelles

Abstract:

Amphiphilic block polymers consist of covalently-bonded hydrophobic and hydrophilic polymer segments. Akin to small molecule amphiphiles, these macromolecules self-assemble into a variety of well-defined structures in aqueous solution such as micelles and vesicles. However, the macromolecular hydrophobic block leads to very slow and seemingly nonexistent micelle dynamics. This kinetic entrapment is an attractive advantage of amphiphilic block polymers for applications in drug delivery; however, it also leads to path-dependent solution assembly. Consequently, careful optimization of preparation conditions are necessary to produce well defined, uniform, and reproducible macromolecular solution assemblies. We are interested in understanding the effects of common processing conditions on the structure, dynamics, and long-term stability of block polymer micelles. In particular, we found that organic cosolvent removal led to an unexpected micelle growth, in which the sizes evolved through a distinct bimodal distribution separated by multiple fusion events. These studies revealed that the micelle stability was critically dependent on stirring the solution, which motivated a detailed investigation into the effects of solution agitation on the chain exchange dynamics in polymer micelles. The results presented here demonstrate the influence of kinetically-controlled processes on block polymer micelle dynamics, and emphasize the importance of selecting and controlling processing conditions when preparing these assemblies.

Small Angle Scattering studies of nanoparticle packing into block copolymer micelles

Gauri Nabar[1], Mathew Souva[1], Sönke Siefert[4], Randall Winans[4], Jessica Winter[1,2], Barbara Wyslouzil[1,3]

The Ohio State University [1]William G. Lowrie Department of Chemical and Biomolecular Engineering, [2]Department of Biomedical Engineering, [3]Department of Chemistry and Biochemistry

Argonne National Laboratory [4] X-ray science division

Micellar nanocomposites are finding increasing applications in medical nanotechnology because of their ability to encapsulate a variety of nanoparticles. In our work, we synthesize micelles composed of poly(styrene-b-ethyleneoxide) (PSPEO) co-encapsulating quantum dots (QDs) and superparamagnetic iron oxide nanoparticles (SPIONS) for in-*vitro* cancer cell detection and manipulation. We have extended this approach to introduce poly(lactic-co-glycolic acid) PLGA into PSPEO micelles either by solubilization or a mixed micelle formation mechanism. Successful application of these nanocomposites requires an understanding of how their morphology evolves with changing process conditions and the nature of polymer-encapsulant interactions. To understand the multifunctional properties of the micelles, it is also important to know the number of nanoparticles encapsulated within the micelle and their distribution inside the heterogeneous micellar micro-environment. Here, we evaluate nanoparticle packing as a function of number and shape, in PSPEO micelles using Small Angle X-ray Scattering.

Our multifunctional micelles are synthesized using the interfacial instability approach, which involves combining the polymer and nanoparticles initially present in a common solvent such as chloroform with a selective solvent such as water. Micelles form, trapping the hydrophobic nanoparticles. An emulsifier, poly vinyl alcohol (PVA) is used in this process. We investigated the micelle encapsulation behavior of differently shaped nanoparticles (spheres and cylinders) at different concentrations. At sufficiently high nanoparticle concentration, crystalline behavior is exhibited. Using small angle x-ray scattering (SAXS), we observed diffraction peaks in the scattering profiles at high nanoparticle loadings. The inter-nanoparticle spacings calculated for the QDs and the SPIONS are consistent with their size and geometry. Our future efforts are directed towards studying the onset of packing behavior within micelles for different classes of nanoparticles. Further, by changing the original polymer used to a longer polymer known to form cylindrical micelles, we observed nodular loading behavior of the nanoparticles inside the micelles. The dynamic behavior of the nodes within the cylindrical micelle was studied using fluorescence microscopy.

Nanoparticle loading was also compared to that of a hydrophobic polymer, PLGA. PLGA was found to exist in two phases, solubilized in micelles or as independent microspheres, based on initial conditions. We were able to observe the scattering behavior of each of these phases to determine PLGA distribution on a volumetric basis.

Nanostructure investigation using SAXS provides an in-*situ* exploration of the sample. Combined with other analytical tools such as Transmission Electron Microscopy (TEM) and fluorescence microscopy, scattering methods yield vital structural information about the sample probed. This forms the basis for successful implementation of these nanocomposites in a wide variety fields in nanotechnology and nanomedicine.

Ionic liquids are organic salts that are liquid at or near room temperature (i.e. below 100°C) and possess unique properties that are tunable based on cation and anion composition. ILs can be attached as side chains to polymerizable monomers to create poly(ionic liquid)s, which have applications as surface coatings, anti-bacterial surfaces, and lubrication. We report the surface-initiated ring-opening metathesis polymerization (SI-ROMP) of norbornene-functionalized ionic liquids to grow poly(ionic liquid) films on gold and silicon substrates. The ionic liquids consist of the cation 1-norbornylmethylene-3-methylimidazolium with the anions bromine (Br^-), hexafluorophosphate (PF_6^-) and trifluoromethanesulfonate (TFO^-) that are easily ion exchanged into the film after SI-ROMP. The properties of the poly(ionic liquid) films depend on the selection of anion and swelling solvent, as characterized by reflectance absorption infrared spectroscopy, ellipsometry, electrochemical impedance spectroscopy, and tribometry.

391107 Synthesis of Anion-Exchange Membranes for Applications Artificial Photosynthesis and in CO2 Capture

Abstract:

Anion exchange materials are fundamental components of fuel cells and solar fuel generators, as polymer electrolyte and polymer membrane, attributed to its flexibility and ease of fabrication. Herein we report the synthesis and characterization of anion conducting membranes derived from poly(phenylene oxide) (PPO) and poly(epichlorohydrin) (PECH) precursors. A variety of ionic groups including quaternized ammonium with different alky substituents, and imidazolium were attached to the PPO and PECH backbone. A range of compositions were synthesized, and gas permeabilities, and ionic conductivities were measured as a function of ionic content. The stability of the ionomers under harsh alkaline conditions was confirmed by thermogravimetric analysis and infrared spectroscopy. The properties of the PPO- and PECH-derived ionic membranes are presented within the context of solar fuels generation, and as materials for CO_2 capture.

Polymerized ionic liquid membranes for CO_2 capture: the effect of water vapor

Luca Ansaloni[1], Jacob R. Nykaza[2], Matteo Minelli[1,3], Yossef A. Elabd[2], Marco Giacinti Baschetti[1,3]

[1]Dipartimento di Ingegneria Civile Chimica, Ambientale e dei Materiali (DICAM)
Alma Mater Studiorum - Università di Bologna, Bologna, Italy

[2]Department of Chemical and Biological Engineering,
Drexel University, Philadelphia, Pennsylvania

[3]Centro Interdipartimentale per Ricerca Industriale - Meccanica Avanzata e Materiali (CIRI-MAM)
Alma Mater Studiorum - Università di Bologna, Bologna, Italy

Polymerized ionic liquids (PILs) represent an attractive material for membrane-based gas separation technology, thanks to the remarkable properties of ionic liquids, such as the large CO_2 solubility at room temperature[1,2].

Water is a common impurity present in several gas separation processes and the investigation of its effect on the membrane transport properties is required for a reliable and comprehensive material characterization. The determination of the H_2O influence is even more important when hydrophilic moieties, such as side-chain charged ions, are present in the membrane matrix.

As previously reported[3] for perfluorosulfonated ionomers (Nafion and Aquivion), which are of interest for some gas separation applications, the presence of water enhances significantly the gas transport rate of the various gaseous species inspected, with up to a two orders of magnitude increase in the permeability and a simultaneous improvement of CO_2 selectivity versus other gases.

Accordingly, gas transport properties of imidazolium-based PIL homopolymer and block copolymer membranes are investigated under humid conditions in order to characterize the water vapor influence on the separation performance. In particular, CO_2 and CH_4 are considered as gaseous probes and their permeability is measured in two different PILs membranes: the p(MEBIm-TFSI) homopolymer and the p(MMA-b-MUBIm-TFSI) block copolymer. The synthesis of these two polymers are detailed elsewhere[2,4,5].

First, water sorption experiments are carried out by means of a quartz-spring microbalance apparatus[6] at different temperatures (25, 35 and 45°C). Experimental results show a rather low solubility of H_2O in both PILs membranes, likely due to the large hydrophobic character of the polymeric chains supporting the imidazolium moiety. A slight increase in water uptake is observed with temperature, although the solubility value remains rather limited (≤ wt. 2.5%).

The permeability of CO_2 and CH_4 in both PILs is then investigated in a humid permeometer apparatus[3], which allows the membrane characterization maintaining the polymer film at a fixed water activity, which can span in a rather wide range (0 - 75%), and at different operative temperatures (25, 35 and 45°C).

The results in terms of gas permeability shows a monotonous increasing behavior with the water activity, regardless to the low water content in the membrane, both in the homopolymer and the block copolymer, although the extent of the variation is different for the two materials. In case of p(MMA-b-MUBIm-TFSI), at 75% R.H. a 30% increase in the transport rate is observed, whereas in case of the p(MEBIm-TFSI), at the same water vapor pressure, the gas permeability is 70% larger than the values measured in dry conditions. Interestingly, the water vapor presence affects in a similar way both penetrants, and the selectivity is barely influenced by the R.H..

Differently from what is commonly observed for other membrane materials with similar water solubility (e.g. polyimides[7]), the effect of water vapor on the gas permeation process is in an overall increase of the performance of PILs membranes, making them attractive materials for gas separation applications since no dehydration step is required before the membrane module.

References:

1. Scovazzo, *J. Memb. Sci.* **343,** 199–211 (2009).
2. Chen, H *et al., Macromoleculs* **42,** 4809–4816 (2009).
3. Giacinti Baschetti *et al., Int. J. Hydrogen Energy* **38,** 11973–11982 (2013).
4. Ye *et al., Macromolecules* **45,** 7027 – 7035 (2012).
5. Nykaza *et al., Polymer* (2014). doi:10.1016/j.polymer.2014.04.003
6. Davis *et al., Macromolecules* **45,** 7486–7494 (2012).
7. Piroux *et al., J. Memb. Sci.* **232,** 115–122 (2004).

378496 Effect of the First Layer on Polyelectrolyte Multilayer Properties and Function

Abstract:

Polyelectrolyte multilayers have been attracting attention due to the ease of preparation and potential for tailorable properties. We are interested in polyelectrolyte multilayers as coatings for orthopedic implants. Polyelectrolyte multilayers were prepared with bone morphogenetic protein 2 as the first layer and control systems were prepared with a polycation as the first layer. Significantly different surface roughness was observed between the two, which impacted the proliferation and differentiation of pre-osteoblasts cultured on both coatings. Furthermore, it was observed that processing conditions such as deposition solution pH and and salt concentration impacted greatly this difference in surface roughness and also affected other physical properties. Techniques including quartz crystal microbalance with dissipation, atomic force microscopy and goniometry were employed to try to understand the observed phenomena. We will discuss our findings and present thought regarding potential mechanisms.

383798 Solution Processable Semiconductors Based on Molecular and Nanoscale Precursors

Abstract:

Dispersions of molecular or nanoscale precursors such as chalcogenidometallates and colloidal quantum dots are currently being explored as solution processable precursors to semiconducting films. Solution-based processing can provide enormous cost benefits for large-area and high-volume electronic applications such as thin-film transistors, photovoltaics, and thermoelectrics. We recently reported the synthesis of large-grained (>500 nm) CdTe thin films from colloidal quantum dot precursors that were integrated into photovoltaic devices with >12% power conversion efficiency.[1] I will summarize our recent efforts in creating semiconducting solids from solution-processable precursors and their applications in electronic devices.

1. M. G. Panthani, J. M. Kurley, R. W. Crisp, T. C. Dietz, T. Ezzyat, J. M. Luther, D. V. Talapin. High Efficiency Solution Processed Sintered CdTe Nanocrystal Solar Cells: The Role of Interfaces. *Nano Letters* 2014, 14, 670–675.

366127 Environmentally Benign Synthesis of Ultra-Thin Metal Telluride Thermoelectric Nanowires

Abstract:

Metal telluride nanowires are attractive materials for many applications, yet most synthesis recipes require hazardous reducing agents such as hydrazine or sodium borohydride. We describe a two-step large-scale production of various metal tellurides with nanowire morphology using a non-hazardous reducing agent. In the first step, Te grows one-dimensionally to form ultra-thin nanowires; in the second step, these nanowires are converted to metal telluride nanowires by adding metal precursors. Analysis of the reaction products versus time provides insights into the growth and conversion mechanisms as well as the reaction rates. The yield of each metal telluride nanowire synthesis have been calculated based on the mass of starting precursors and the mass of the washed and dried reaction products. For PbTe, CdTe, and Bi_2Te_3 the yields were approximately 83.3%, 87.5%, and 77.3%, respectively, which points to the feasibility of large-scale synthesis of these materials. For Ag_2Te and $Cu_{1.75}Te$, the yields were about 24.5% and 58.9%, respectively. These nanowires exhibit extremely narrow size distribution and a significantly enhanced thermoelectric properties.

Ultra-high Density Nanoparticles with Perfect Ordering and Controllable Dimensions Using Anodized Alumina as a Mask

Cheng Xu*, Luping Li, Yang Zhao, and Kirk J. Ziegler

Department of Chemical Engineering, University of Florida, 1006 Center Drive, Gainesville, FL 32611, US

*Presenting Author (Ph: 352-392-3122; Email: chengotopia@ufl.edu)

Abstract:

Fabrication of nanoparticles with perfect ordering on large-area substrates has gained interest for various applications in nanotechnology. These nanoparticles can be used for photoelectric devices, nanoplasmonics, data storage, and biological sensors as well as catalysts to control nanowire growth. Available techniques include combination of block copolymer template and either E-beam or focused ion beam lithography. However, it is difficult to control the thickness of the block copolymer template and subsequently use the template at high temperatures. In this study, we report a strategy of using anodized alumina template (AAO) to generate close-packed hexagonal patterns with tunable size and thickness and then applying the AAO template as a mask to perform pattern transfer from AAO to a foreign substrate. First, to demonstrate the transfer process, ultra-thin AAO templates were fabricated on aluminum using 2-step anodization. The dimensions of AAO templates could be controlled by the type of electrolyte, concentration of electrolyte, anodization voltage, temperature, and stirring rate. The close-packed hexagonal pattern provides maximum pore density that can reach10^{10}-10^{12} cm^{-2}. With the protection of a polystyrene thin film, AAO was transferred to a foreign substrate after dissolving away aluminum. Then the polystyrene was removed using ozone plasma, and gold nanoparticles were fabricated using e-beam evaporation or sputtering followed by annealing. As shown in Figure 1, the deposited gold nanoparticles have identical density and dimension as AAO template. These gold nanoparticles were used as catalysts for vapor-liquid-solid (VLS) ZnO nanowire growth and also for fabrication of nanoplasmonic devices. This technique provides a new route of fabricating large-scale, high-density nanoparticles with tunable sizes

Figure 1. Au nanoparticles with perfect ordering using AAO template as a mask.

Keywords: nanoparticles, vapor-liquid-solid nanowire growth, ZnO nanowires, anodized alumina, nanoplasmonics

Optically Abrupt Plasmonic Response of Si Nanowires via Nanoscale Spatial Control of Carrier Density

Dmitriy S. Boyuk, Li-Wei Chou and Michael A. Filler

School of Chemical & Biomolecular Engineering, Georgia Institute of Technology, Atlanta, Georgia 30332, United States

The ability to modulate carrier density in nanoscale semiconductors promises exciting new opportunities to engineer localized surface plasmon resonances (LSPRs). Near-field coupling of closely spaced resonators, which is important for controlling local field enhancement and optical energy transfer, requires that the spatial resolution of carrier density exceed that of resonator spacing. Unfortunately, for semiconductor nanowires synthesized via the vapor-liquid-solid technique, the so-called "reservoir effect" – the presence of dopant atoms in the growth catalyst even after dopant-containing precursor flow ceases – can limit profile abruptness. Here, we use *in-situ* infrared spectroscopy and discrete dipole approximation calculations to interrogate the effect of graded carrier densities on the plasmonic properties of phosphorous-doped Si nanowires. The spectral response of epitaxially aligned nanowires containing multiple axially encoded resonators (50 nm long, 20 nm spacing), each grown at 540 °C with Si_2H_6 and PCl_3 partial pressures of 4×10^{-5} and 3×10^{-5} Torr, respectively, exhibits a pronounced asymmetry characteristic of multiple, convoluted peaks. We show that this behavior arises from charge carrier densities on the order of 10^{20} cm^{-3} in the region between resonators. A reduction of nanowire growth temperature to 465 °C after fabricating each resonator dramatically decreases this effect. We use this approach to fabricate Si nanowires with spectral responses indistinguishable from those of nanowires with ideal, atomically abrupt carrier density profiles. As such, we refer to these as "optically abrupt" resonators. Our findings pave the way for exploiting Si-based plasmonic materials in applications ranging from infrared photodetection to optical interconnects.

390118 Enhanced Thermoelectric Power in Photo-Modulated Porphyrin-Bi2Te3 Nanowire Hybrid Systems

Abstract:

Solar energy is perhaps the most abundant renewable energy source, but its conversion into electric power is still challenging for photovoltaic (PV) cells because they have not achieved efficiencies and cost that can that put solar cells in a competitive advantage. Porphyrins have become an attractive choice for dye sensitized solar cell (DSSC) applications because they are relative inexpensive materials, and they can reach acceptable photon-carrier generation efficiency. Porphyrins are photoactive within the ultraviolet to visible region which represent close to 50% of the entire solar spectrum energy. Incorporating a thermoelectric element to porphyrins could increase the overall efficiency by collecting the infrared (IR) radiation. Achieving a larger efficiency by collecting from the UV to IR spectrum is the motivation for a proposed ultra-wide solar spectrum photovoltaic-thermoelectric (PV-TE) solar power cell. The understanding of the Photo-thermoelectric effects and charge transfer characteristics for the proposed PV-TE solar cell is fundamental and covered in this work. The thermoelectric characterization of porphyrin deposited onto p and n-type Bi_2Te_3 nanowires under different illumination conditions is presented. We observed charge transfer across the Bi_2Te_3 nanowire-porphyrin interface under illumination and the temperature dependence of carriers recombination within the porphyrin. Furthermore, The Seebeck coefficient is substantially increased under white light illumination indicating that residual charge in the porphyrin modulates the carrier concentration in the Bi_2Te_3nanowires, but the electrical conductivity is not appreciable altered resulting in an enhanced thermoelectric power. Photo modulation of thermoelectric power in hybrid materials provides an alternative approach to enhance the thermoelectric figure of merit.

This work was performed, in part, at the Center for Integrated Nanotechnologies, an Office of Science User Facility operated for the U.S. Department of Energy (DOE) Office of Science by Los Alamos National Laboratory (Contract DE-AC52-06NA25396) and Sandia National Laboratories (Contract DE-AC04-94AL85000).

In the field of organic photovoltaics, incorporating halogen atoms into potential molecular candidates for use in the active layer has been a topic of significant interest. This route of functionalization provides a variable to tune such properties as the molecular packing of the bulk material, single-molecule electronic properties such as HOMO and LUMO energy levels, and air stability. Like many small-molecule organic semiconductors, contorted hexabenzocoronenes (cHBCs) can be decorated with a variety of halogen atoms in various locations. Using Density Functional Theory (DFT) and other approaches, we will show how halogenation (specifically with fluorine and chlorine) might also be used to improve or worsen the electronic coupling between adjacent cHBC molecules in the absence of major differences in packing.

It is known that molecular orbitals need to be properly aligned between molecules in order to maximize the electronic coupling, thus mobility, of the bulk crystal. However, changes in molecular orbital alignment are typically the result of significant differences in molecular packing between systems. Experimentally, while multiple polymorphs of halogenated cHBCs exist, they can share the same molecular packing given the right processing conditions. This offers an excellent test-bed for this computational study. Wherever possible, x-ray diffraction data were used to deduce the structures for this study's DFT results. Where only lower quality x-ray diffraction data were available, Conformation-family Monte Carlo methods were used to predict polymorphs, and their respective simulated diffraction data were used to match with experimental results. A molecular orbital projection method was used to calculate electronic couplings and to predict mobilities. In a few cases, this mobility data will be compared to experimental data. This study will demonstrate some of the interesting electronic properties of halogenated cHBCs and may provide a different way of looking at small-molecule organic semiconductor design.

366458 Control of Point Defect Behavior in Metal Oxides Via Surface Band Bending

Abstract:

Point defects within metal oxide semiconductors such as ZnO affect the material's performance in applications for nanoelectronics, gas sensing, photonics and photocatalysis. Past work in this laboratory has shown that the presence of a nearby surface can influence the concentrations and spatial distributions of defects deep within the semiconductor – often in a controllable way. One mechanism for this influence involves band bending near the surface. The present work employs the optical modulation technique of photoreflectance to measure the magnitude V_S of band bending at polar c-axis ZnO surfaces, and demonstrates that V_S can be manipulated over a significant range (roughly 0.23-0.44 eV) through variations of both temperature and the ambient partial pressure of O_2. Separate modeling of charged oxygen interstitial motion within the ZnO indicates that the near-surface concentration of these defects scales quadratically with V_S, thereby pointing to a novel general approach by which bulk point defect concentrations can be controlled.

367928 The Fundamentals of Charge Transport at Nanoscale Oxide and Ferroelectric Interfaces

Abstract:

Here we investigate how charge transport properties at metal-semiconductor interfaces scale down to the nanoscale regime, comparing the properties to macroscopic interfaces and providing a perspective on what it means to device manufacturing. Strontium titanate - the prototypical oxide material - has been widely studied for applications in thermoelectrics, nanoelectronics, catalysis, and other uses, and behaves as an n-type semiconductor when doped. We investigated how charge transport is effected at interfaces to stronitium titanate under a wide range of conditions - by varying contact size, interface shape, dopant concentration, and surface structure and in various combinations. The results of the analysis have wide ranging implications, especially for ferroelectric oxide materials and serves as the basis for understanding and controlling switching effects - both polarization and oxygen migration based switching.

Graphene/Polyaniline Hydride Materials for Biomedical, Environmental, Electronic, and Green-Energy Applications

Yong Min and Yidong Liu

By leveraging the benefits from both functional materials of polyaniline (PANI) and structural materials of grapheme(GF), graphene/polyaniline (GF-PANI) hybrid materials were investigated in terms of selecting coupling agents, optimizing reaction conditions, studying the surface morphology, adopting different processing techniques, measuring the material performance, evaluating long term reliabilities and industrial uses. The work have been demonstrated not only on fundamental understanding the formation of GF-PANI hybrid materials, but also on industrial applications in terms of Li-ion battery, supercapacitor, water treatment, nanocomposite, anti-corrosion, transparent electrodes, conducting plastics, bio-sensor and devices, CO_2 sequestration, and printable electronics, as illustrated in Figure 1.

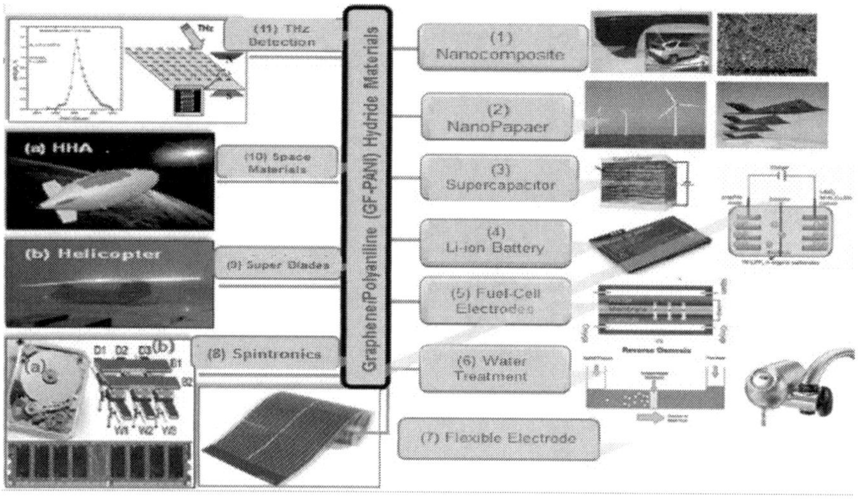

Fig. 1. Illustrated potential industrial application of GF-PANI hybrid materials.

The objective of this paper is to promote the benefits of the emerging materials and advertise their industrial usages. The detailed technical processing and material performance as well as corresponding marketing needs and customer's requests will be highlighted during the presentation. As an example, the hybrid paper shown a tensile strength of ~380 MPa, which is almost like steel, but its density is only at 1/6 of the steel. The paper also had an excellent thermal transform performance, which had been used at LED indoor lamps and street lamps. The hybrid ink/coating also presented an excellent anti-corrosion performance, which linked to billion dollars industry. An experiment indicated that < 3 wt.% of the GF-PANI addition promoted the performance >30%. The supercapacitor made from the hybrid materials displays a specific capacitance of 478 F/g at current density 0.5 A/g with 3-electrode system. After 2000 cycles, the performance could keep 88% of its initial specific capacitance.

390632 Photo-Enhanced Atmospheric Plasma-Assisted Chemical Vapor Deposition: UV-Plasma Synergies Under Ambient Conditions

Abstract:

Plasma-assisted chemical vapor deposition (PA-CVD) has been used extensively since the 1980's to deposit polymer films on substrates for applications ranging from semiconductor manufacturing to textile treatment to membrane and biomedical device processing. PA-CVD techniques are advantageous for producing ultrathin, highly-crosslinked, pinhole-free films on complex surfaces, often from starting monomers that lack the chemistry for traditional polymerization techniques, without a need for solvents or standard chemical initiators. Polymers with different material properties—e.g. surface energy, degree of crosslinking—can be produced from a single monomer by varying the geometry, electrical characteristics, and energy density (i.e. Yasuda parameter) of the plasma discharge. The polymer surface can then be decorated with a desired chemical functionality by polymerizing an appropriate monomer or by treatment with an appropriate plasma species, e.g. amination with allylamine or an ammonia plasma. Functionalization can provide a handle for further chemistry, enabling nanoparticles (antimicrobial surfaces), small molecules (microanalytical chemistry), or even biomacromolecules (tissue engineering) to be conjugated to the surface.

Recent work has sought to adapt plasma-assisted CVD and functionalization to ambient conditions with atmospheric non-thermal plasma discharge systems, for the in situ treatment of surfaces that are not amenable to processing under typical vacuum chamber CVD conditions. Many such processes have been developed, but atmospheric pressure conditions constrain the set of precursors and discharge conditions, somewhat limiting the range of achievable film properties.

In collaboration with Dr. J. Gary Eden (U. Illinois), I am evaluating the use of high-emittance photon sources—UV-A and UV-B light-emitting diodes, UV-B and UV-C microcavity plasma discharges—for enhancing plasma-assisted CVD at atmospheric pressures. I present data that suggests that photoionization of gaseous and precursor species can be utilized to extend the operating range in a dielectric barrier plasma jet by improving plasma stability and aiding in precursor fragmentation.

385018 Analysis of OH Radical Scavengers to Assess Chemical Reactions in Electrical Discharge Plasma Formed at a Gas-Liquid Interface

Abstract:

Low power pulsed electrical discharge (250 – 1000 mW) reactors that utilize water sprays with argon carrier gas have been shown to produce hydrogen peroxide (H_2O_2) from the recombination of hydroxyl radicals (OH·) which are produced by the dissociation of water. In reactors where the electrical discharge propagates at and along the interface between the gas and liquid phases, it is unknown whether the reactions to produce hydrogen peroxide occur in the gas or liquid phase. Of the many OH radical scavengers available for gas and liquid phases, there are only a few scavengers that will not alter the physical properties (e.g., conductivity, pH, water solubility, and viscosity) of the solution which in turn affect the discharge. Dyes such as methylene blue are potentially good radical scavengers for the liquid phase because the decoloration of the dye is an in-direct estimate of the amount of OH radicals scavenged, the dye is non-volatile, and the dye does not interfere with the hydrogen peroxide measurements using UV-Vis spectroscopy. However, liquid conductivity is affected by the formation of degradation products of the dye at very high dye concentrations.

Alcohols are also good candidates for OH radical scavengers. Methanol, ethanol, and propanol are highly miscible with water and the liquid conductivity is not strongly affected by the degradation products. However, lower molecular weight alcohols can diffuse into the gas phase.

In this work, methylene blue and ethanol were used as OH radical scavengers. The concentration of methylene blue was 0.1 mM which had an initial liquid conductivity of 32 µS/cm. The liquid flow rates for the dyes ranged from 0.75 to 5 ml/min. The decoloration of the methylene blue was measured at 664 nm with UV-Vis spectroscopy. The H_2O_2 was detected only in the liquid phase and was measured by a colorimetric test with $TiOSO_4$ and UV-Vis spectrometry.

Ethanol concentrations ranging from 0.1 mM to pure ethanol were introduced into the reactor at 0.75 ml/min. Proton nuclear magnetic resonance (^{1}H NMR) showed the presence of an aldehyde product in both gas and liquid streams. Ethanol and acetaldehyde were measured by gas chromatography with a flame ionization detector (GC-FID). The liquid streams leaving the reactor for the cases with ethanol were injected directly into the GC-FID. The gas stream exiting the reactor went through two cold traps and the condensed liquids were analyzed with GC-FID. Methylene blue results showed that the decoloration rate increased with increasing liquid flow rate and the addition of methylene blue did not change the H_2O_2 concentration in comparison with pure water at all flow rates. However, in alcohol solutions the amount of H_2O_2 decreased with increasing alcohol concentration. Acetaldehyde was determined to be the major product from water-ethanol solutions and it was found primarily in the gas phase. Ethanol feed concentrations at and above 1 M showed complete scavenging of the OH radical with no measurable H_2O_2. Both OH radical scavenger results suggest that the majority of the OH radical is produced in the gas phase very close to the gas-liquid interface. The OH radical production rate from acetaldehyde formation and H_2O_2 is approximately $2.72 \times 10^{-7} \pm 7.17 \times 10^{-8}$ mol/s which is within experimental error of the OH radical production rate from H_2O_2 in the pure water cases, $2.27 \times 10^{-7} \pm 1.58 \times 10^{-8}$ mol/s. This work was supported by the National Science Foundation grants CBET-0932481 and CBET-1236225.

Modeling and Analysis of Rapid Synthesis of GaAs by Hydride Vapor Phase Epitaxy Process

Min Yao, James B. Rawlings, Kevin L. Schulte and Thomas F. Kuech

Department of Chemical and Biological Engineering
University of Wisconsin-Madison

AIChE Annual Meeting
Atlanta, GA
November 16, 2014

Overview

Motivation

- Conflicts between expanding population and finite fossil fuel resources
- Renewable energy - solar power - photovoltaic (PV) systems
- Solar cells are still too expensive!
- Energy economy:
  - Increase the production rate of solar cell materials
  - Reduce total amount of materials usage
  - Increase efficiency

Introduction

Motivation

- Conflicts between expanding population and finite fossil fuel resources
- Renewable energy - solar power - photovoltaic (PV) systems
- Solar cells are still too expensive!
- Energy economy:
  - Increase the production rate of solar cell materials
  - Reduce total amount of materials usage
  - Increase efficiency
- Highly efficient materials and economical growth methods: key to the success of alternative energy

- GaAs
  - High adsorption coefficient, relative crustal abundance
  - Use of single crystal substrate, low growth rate
- Best research-cell efficiency (44.7%): multi-junction(tandem) solar cells made of III-V semiconductors

Introduction

- GaAs
  - High adsorption coefficient, relative crustal abundance
  - Use of single crystal substrate, low growth rate
- Best research-cell efficiency (44.7%): multi-junction(tandem) solar cells made of III-V semiconductors
- Typically synthesized by metalorganic vapor phase epitaxy (MOVPE): very high quality crytalline materials but very expensive!

Yao, Rawlings, *et al.* (UW-Madison) Modeling of GaAs HVPE AIChE Annual Meeting 2014 4 / 23

- GaAs
  - High adsorption coefficient, relative crustal abundance
  - Use of single crystal substrate, low growth rate
- Best research-cell efficiency (44.7%): multi-junction(tandem) solar cells made of III-V semiconductors
- Typically synthesized by metalorganic vapor phase epitaxy (MOVPE): very high quality crytalline materials but very expensive!
- Hydride vapor phase epitaxy (HVPE) ?

Introduction

- GaAs
  - High adsorption coefficient, relative crustal abundance
  - Use of single crystal substrate, low growth rate
- Best research-cell efficiency (44.7%): multi-junction(tandem) solar cells made of III-V semiconductors
- Typically synthesized by metalorganic vapor phase epitaxy (MOVPE): very high quality crytalline materials but very expensive!
- Hydride vapor phase epitaxy (HVPE) ?

MOVPE	HVPE
lower growth rate: $10\mu m/hr$	higher growth rate: $170\ \mu m/hr$
trimethyl gallium	only pure Ga and HCl (cheaper)
operate far from equilibrium	near-to-equilibrium growth

Yao, Rawlings, *et al.* (UW-Madison) Modeling of GaAs HVPE AIChE Annual Meeting 2014 4 / 23

Current GaAs HVPE System (K.L.Schulte, et al. 2012)

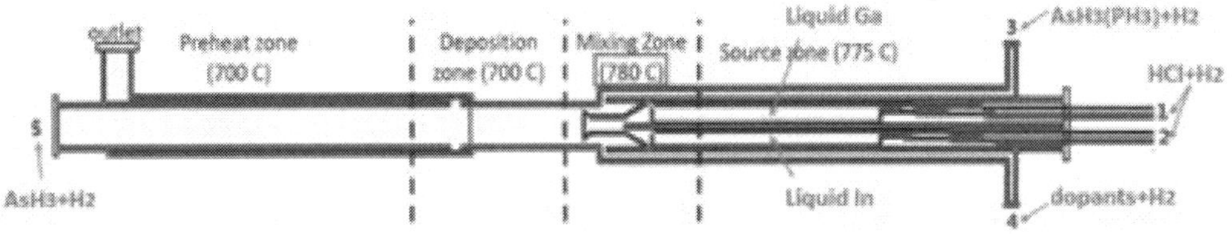

Introduction

Current GaAs HVPE System (K.L.Schulte, et al. 2012)

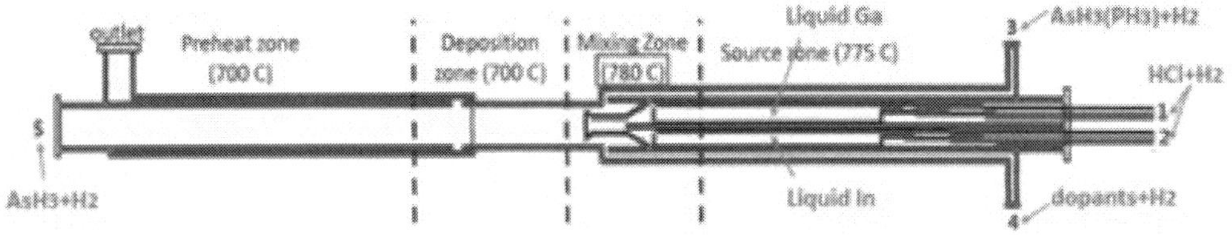

- Used to grow multilayers:
  - Continuous growth of all layers
  - Growth interruption with simultaneous equilibration of new gas flows
- Benefits of growth interruption:
  - Narrower junction transition
  - Higher device efficiencies

Yao, Rawlings, *et al.* (UW-Madison) Modeling of GaAs HVPE AIChE Annual Meeting 2014 5 / 23

Challenges and Opportunities

- Underlying **drawbacks** of manual operation: waste of reactants, poor product uniformity, long startup and switching period

Challenges and Opportunities

- Underlying **drawbacks** of manual operation: waste of reactants, poor product uniformity, long startup and switching period
- Continuous process with 'automatic' interruption needs to be developed

General Idea

- A moving belt transporting wafers between HVPE stations separated by gas curtains

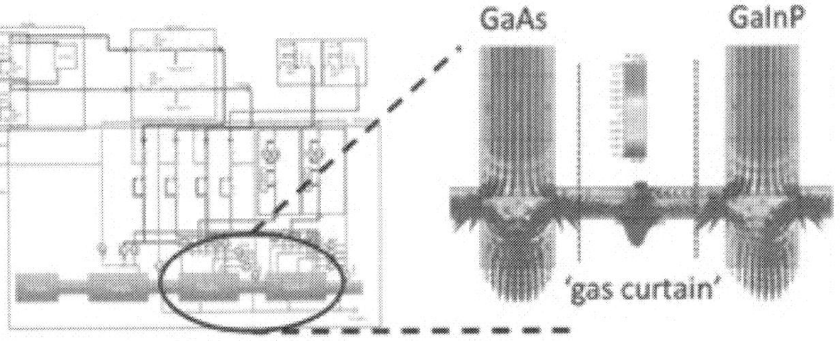

Introduction

General Idea

- A moving belt transporting wafers between HVPE stations separated by gas curtains

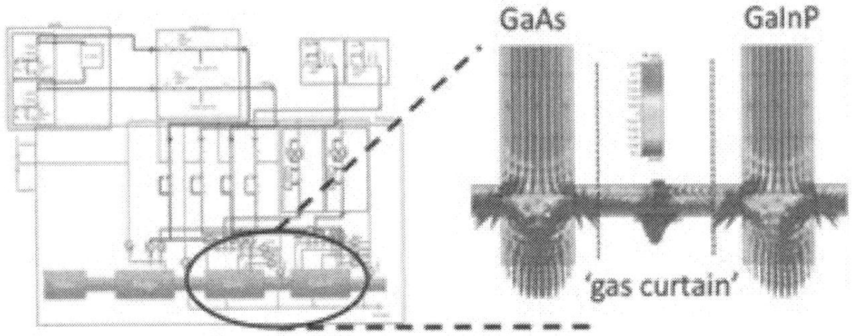

- Parameters (operating conditions, geometries) of different stations should be optimized simultaneously to meet desired materials properties and process efficacy
- **Methodology**: model-based method

Yao, Rawlings, *et al.* (UW-Madison) Modeling of GaAs HVPE AIChE Annual Meeting 2014 7 / 23

Process Analysis: General

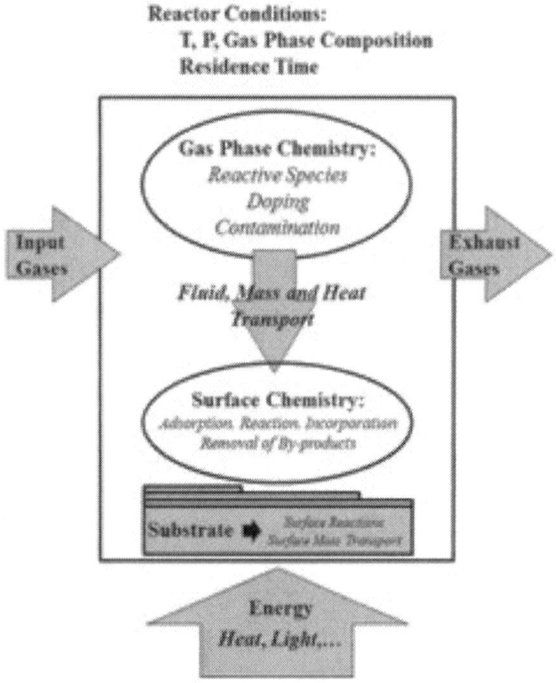

- Mass transfer
- Fluid dynamics
- Heat transport
- Chemistry
  - Gas phase chemistry
  - Surface chemistry

Process Analysis: GaAs-HVPE

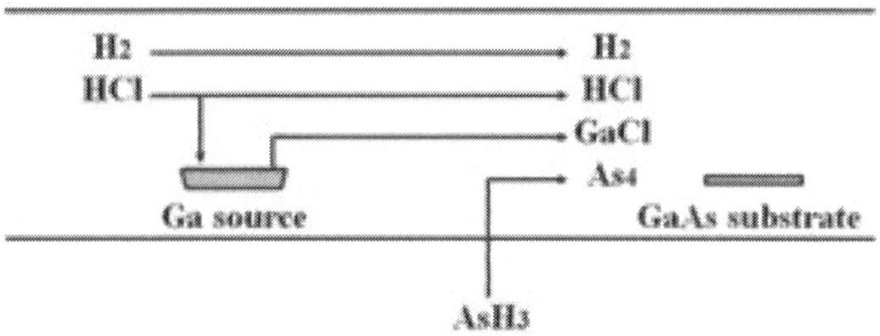

- Overall reaction:

$$\mathrm{GaCl} + \frac{1}{4}\mathrm{As_4} + \frac{1}{2}\mathrm{H_2} \rightleftharpoons \mathrm{GaAs_s} + \mathrm{HCl}$$

- Growth rate of GaAs epilayer on the substrate may be influenced by:
  - Intrinsic factors: gas phase reactions, surface adsorption, surface reactions
  - Transport effects: diffusion

Process Analysis: GaAs-HVPE

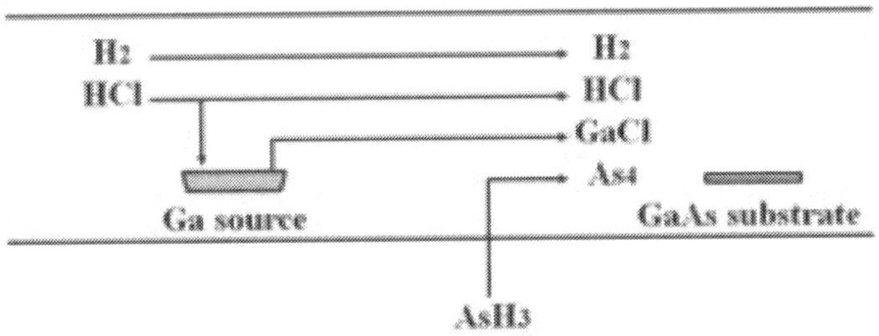

- Overall reaction:

$$\mathrm{GaCl} + \frac{1}{4}\mathrm{As_4} + \frac{1}{2}\mathrm{H_2} \rightleftharpoons \mathrm{GaAs_s} + \mathrm{HCl}$$

- Growth rate of GaAs epilayer on the substrate may be influenced by:
  - Intrinsic factors: gas phase reactions, surface adsorption, surface reactions
  - Transport effects: diffusion
  - How they work?
  - What is rate-determing?

Assumptions

Focus on the transport of species in the deposition zone:

- Ideal gas
- Isothermal: deposition temperature, typically 973 -1123 K
- Consider three principal gas species As_4, GaCl, HCl as input reactants and already well-mixed, H_2 is bulk component, large excess

Computational Tool

Apply COMSOL-Multiphysics:

- Time-dependent, 2D
- Parameters of geometries, properties and process operating conditions conform with experiments
- Module: Transport of Diluted Species, Laminar Flow

Transport Equations

- Model equations:

$$\frac{\partial \rho}{\partial t} + \nabla \cdot (\rho \mathbf{u}) = 0$$

$$\rho \frac{\partial (\mathbf{u})}{\partial t} + \rho(\mathbf{u} \cdot \nabla)\mathbf{u} = \nabla \cdot [-p\mathbf{I} + \mu(\nabla \mathbf{u} + (\nabla \mathbf{u})^T) - \frac{2}{3}\mu(\nabla \cdot \mathbf{u})\mathbf{I}] + F$$

$$\frac{\partial (c_i)}{\partial t} + \nabla \cdot (-D_i \nabla c_i) + \mathbf{u} \cdot \nabla c_i = R_i$$

- Boundary conditions:

Inlet: $\mathbf{u} = u_{in}\mathbf{n}$; $c_i = c_{i,in}$

Outlet: $\mathbf{n} \cdot \nabla(\rho \mathbf{u}) = 0$; $\mathbf{n} \cdot \nabla c_i = 0$

Walls: $\mathbf{u} = 0$; $\mathbf{n} \cdot \nabla c_i = 0$

Substrate: $\mathbf{u} = 0$; $\mathbf{n} \cdot \nabla(D_i c_i) = R_{s,i}$

Chemical Model (J.C.Hong, et al.1985)

- Proposed mechanism:

$$As_4 \underset{}{\overset{K_{As}}{\rightleftharpoons}} 2As_2$$

$$GaCl + S \overset{K_1'}{\rightleftharpoons} GaCl \cdot S$$

$$As_2 + S \overset{K_2'}{\rightleftharpoons} As_2 \cdot S$$

$$GaCl \cdot S + As_2 \cdot S \underset{k'}{\overset{H_2}{\longrightarrow}} GaAs_2(\cdot 2S) + HCl$$

$$(2S) \underset{k_e'}{\overset{HCl}{\longrightarrow}} GaCl$$

Modeling

Chemical Model (J.C.Hong, et al.1985)

- Proposed mechanism:

$$As_4 \underset{}{\overset{K_{As}}{\rightleftharpoons}} 2As_2$$

$$GaCl + S \overset{K_1'}{\rightleftharpoons} GaCl \cdot S$$

$$As_2 + S \overset{K_2'}{\rightleftharpoons} As_2 \cdot S$$

$$GaCl \cdot S + As_2 \cdot S \underset{k'}{\overset{H_2}{\longrightarrow}} GaAs_2(\cdot 2S) + HCl$$

$$(2S) \underset{k_e'}{\overset{HCl}{\longrightarrow}} GaCl$$

- Rate for deposition of GaAs:

$$G = \frac{k c_{As_4}^{1/2} c_{GaCl} - k_e c_{HCl}}{(1 + K_1 c_{GaCl} + K_2 c_{As_4}^{1/2})^2}$$

Chemical Model (J.C.Hong, et al.1985)

- Proposed mechanism:

$$\mathrm{As_4} \overset{K_{As}}{\rightleftharpoons} \mathrm{2As_2}$$

$$\mathrm{GaCl + S} \overset{K_1'}{\rightleftharpoons} \mathrm{GaCl \cdot S}$$

$$\mathrm{As_2 + S} \overset{K_2'}{\rightleftharpoons} \mathrm{As_2 \cdot S}$$

$$\mathrm{GaCl \cdot S + As_2 \cdot S} \underset{k'}{\overset{H_2}{\longrightarrow}} \mathrm{GaAs_2(\cdot 2S) + HCl}$$

$$\mathrm{(2S)} \underset{k_e'}{\overset{HCl}{\longrightarrow}} \mathrm{GaCl}$$

- Rate for deposition of GaAs:

$$G = \frac{k c_{As_4}^{1/2} c_{GaCl} - k_e c_{HCl}}{(1 + K_1 c_{GaCl} + K_2 c_{As_4}^{1/2})^2}$$

- k, k_e, K_1, K_2 calculated by fitting Shaw's experimental data (D.W.Shaw.1970)

Mesh

- Free Triangular
- Specify boundary: the active top surface of substrate
- Element size parameter: limit maximum element size under 0.25 mm

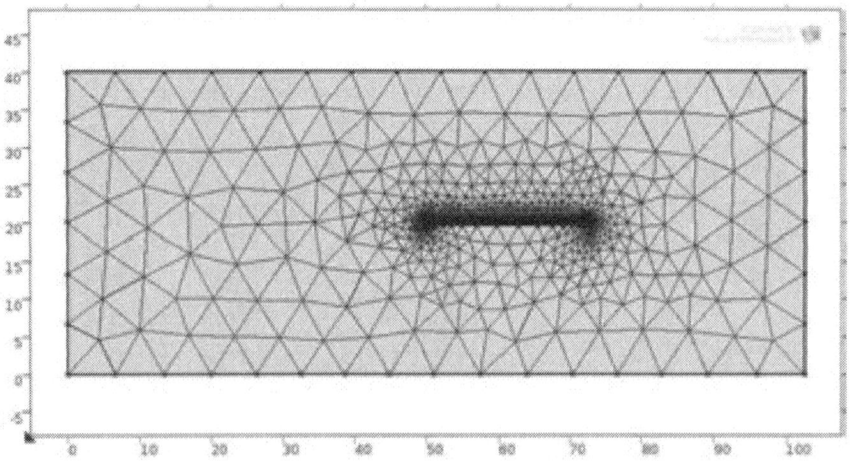

Study

- Time dependent: (0, 0.01, 10) s
- Parametric sweep: T_{dep} (973, 5, 1123) K
- Initial input concentration of each species at $T_{dep} = 973K$:

$$c_{GaCl,in} = 0.045 mol/m^3$$
$$c_{As_4,in} = 0.027 mol/m^3$$
$$c_{HCl,in} = 0.012 mol/m^3$$

Comparison with Experimental Data (D.W.Shaw.1970)

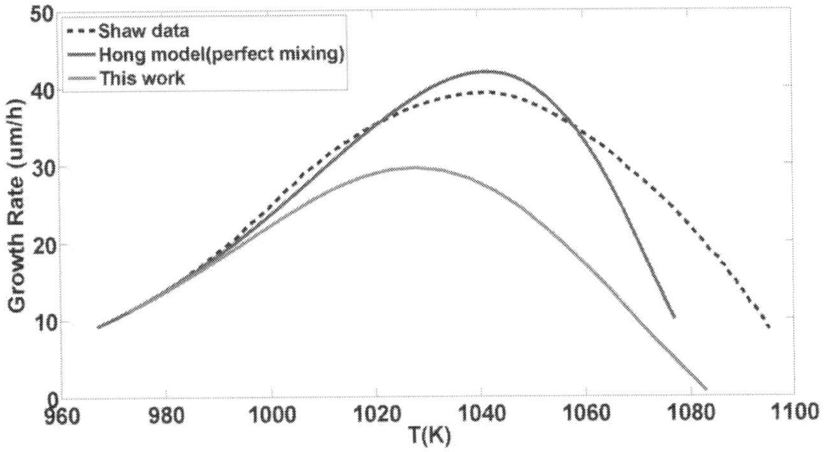

Comparison with Experimental Data (D.W.Shaw.1970)

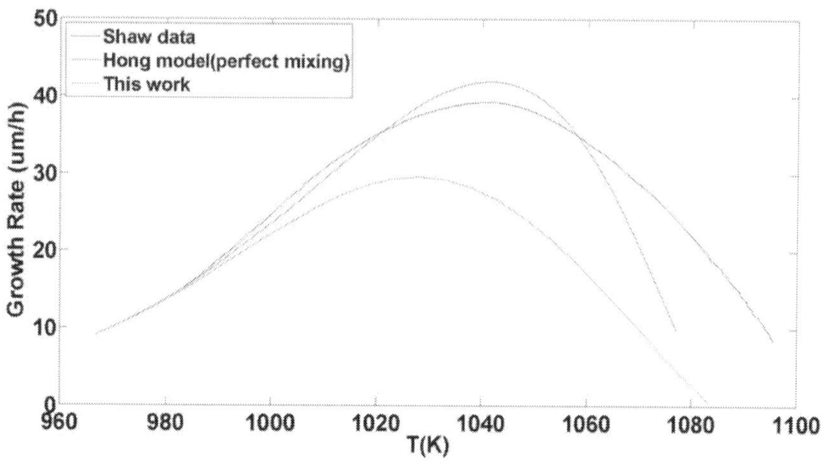

- Similar trend
- Matches satisfactorily at lower temperature
- Large discrepancies at higher temperature

GaAs HVPE Growth Kinetics

Two growth regimes (D.W.Shaw.1968-75; R.Cadoret,et al.1975)

1. Kinetically limited regime
   - Rate of growth limited by steps at the surface
   - Strongly affected by crystallographic orientation, substrate temperature
   - Flow dynamics influence is negligible

GaAs HVPE Growth Kinetics

Two growth regimes (D.W.Shaw.1968-75; R.Cadoret,et al.1975)

1. Kinetically limited regime
   - Rate of growth limited by steps at the surface
   - Strongly affected by crystallographic orientation, substrate temperature
   - Flow dynamics influence is negligible
2. Mass transfer limited regime
   - Surface reactions proceed so rapidly that equilibrium reached
   - Limited by the rate of diffusion of species from surface to the bulk
   - Equilibrium shifts towards the reactant at higher temperature

Concentration of GaCl

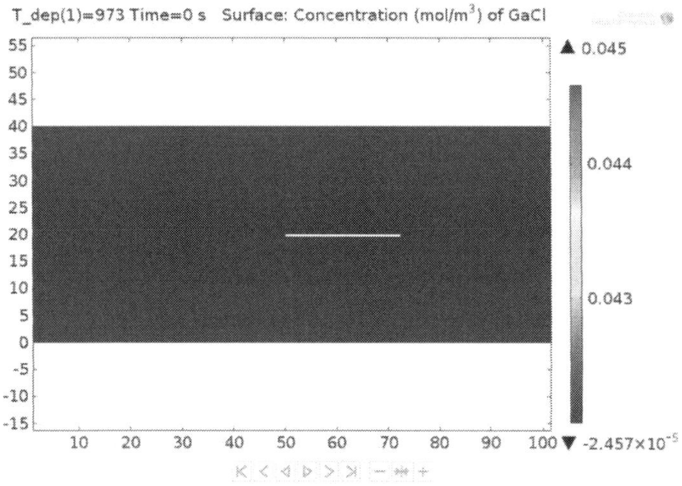

Concentration of GaCl

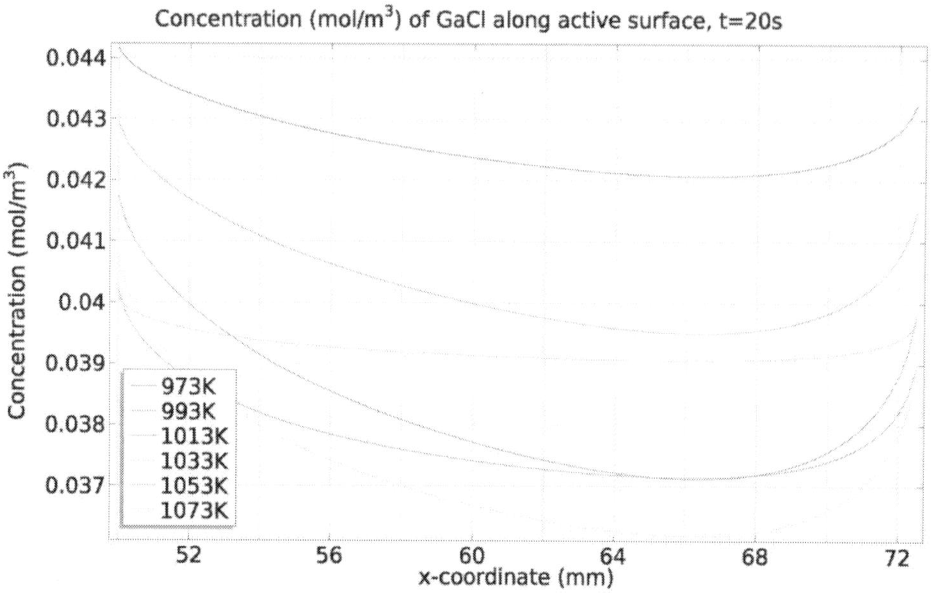

Velocity

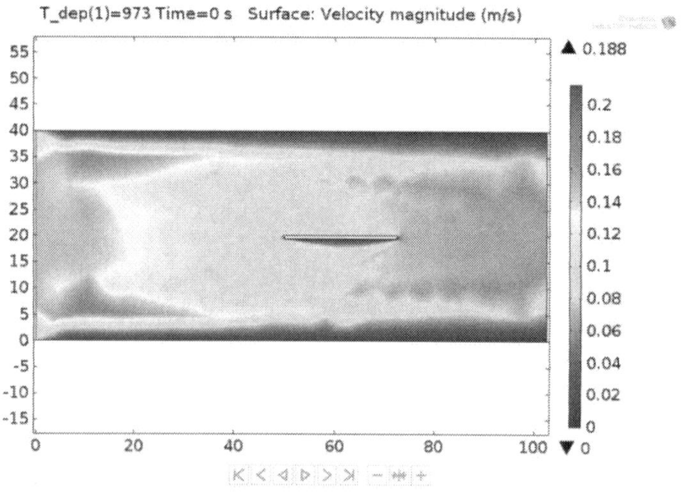

Summary

1. Implemented 2D model using COMSOL-Multiphysics packages to study the relationship between growth rate and temperature in the deposition zone

Summary

1. Implemented 2D model using COMSOL-Multiphysics packages to study the relationship between growth rate and temperature in the deposition zone
2. The model was capable of simulating the major characteristics of the epitaxial growth of GaAs

Future Work

1. Based on the current kinetic model, parameters will be optimized to match the experimental data
2. Heat transfer module will be included to model the four-zone reactor system
3. 3D model based on the validated 2D model will be built and validated by the current lab HVPE system

Conclusions

Future Work

1. Based on the current kinetic model, parameters will be optimized to match the experimental data
2. Heat transfer module will be included to model the four-zone reactor system
3. 3D model based on the validated 2D model will be built and validated by the current lab HVPE system
4. After the quantitative relationships between process inputs, states and outputs are obtained, an optimized continuous HVPE process with the design of 'gas curtain' will be developed

Acknowledgments

- Funding: NSF CMMI 1232618

387966 Large-Scale Simulation of Plasma-Facing Materials for Tokamaks and Linear Devices

Abstract:

Helium from linear plasma devices and tokamak plasmas is known to cause formation of nanometer-sized features on the surface of plasma-exposed materials after only a few hours of plasma exposure. The precise details of such surface modifications are as yet uncertain. This study examines the initial and intermediate stages of tungsten surface evolution by large-length-scale molecular dynamics (MD) simulations of helium-implanted tungsten over time scales of up to microseconds using single-crystal and polycrystalline supercell models of tungsten. The large-scale MD simulations employ state-of-the-art many-body interatomic potentials and implantation depth distributions for the insertion of helium atoms into the tungsten matrix constructed based on MD simulations of helium-atom impingement onto tungsten surfaces under prescribed thermal and implantation conditions. The large-scale MD simulations reveal features both above and below the surface, indicating that (a) the crystallographic orientation of the surface has a significant effect on retention and surface deformation, (b) the grain microstructure of the tungsten matrix (i.e., the presence or absence of grain boundaries) has strong effects on the resulting agglomeration behavior and resultant surface features, and (c) grain boundaries and surfaces are both "sinks" for helium, resulting in immobilization and/or enhanced bubble formation near such defects. Our findings have significant implications for the surface morphological evolution and the near-surface structural evolution of plasma-facing components in nuclear fusion reactors.

369804 Transport and Reactions of Mobile Helium Clusters Near Surfaces and Grain Boundaries of Plasma-Exposed Tungsten

Abstract:

The evolution of the surface morphology and the near-surface structure of plasma facing components (PFCs) in nuclear fusion reactors is impacted significantly by the implantation of helium (He) atoms. Tungsten (W) is an important PFC material due to its thermomechanical properties. In tungsten, such interstitial He atoms are very mobile and aggregate to form clusters of different sizes. The smallest of these aggregates, that is, helium clusters with up to seven helium atoms, also are mobile; the diffusional transport of such small helium clusters mediates the evolution of surface morphology and sub-surface gas bubble structure and dynamics near the surface.

In this presentation, we report the results of a systematic atomic-scale analysis of the interactions of small mobile helium clusters (He_n) with sinks, such as free surfaces and grain boundaries (GBs), and the cluster reactions near these sinks in tungsten toward development of continuum drift-diffusion-reaction models for the dynamics of mobile helium clusters in plasma-exposed tungsten. Molecular-statics (MS) simulations based on reliable many-body interatomic potentials are carried out for He_n ($1 \leq n \leq 7$) clusters near sinks (surfaces and GBs) to obtain the potential energy profiles of the He_n clusters as a function of the clusters' center-of-mass distance from a sink. Other sinks that have been investigated include regions in the vicinity of junctions where GBs intersect free surfaces. Elastic interaction potentials based on elastic inclusion theory provide an excellent description of the MS results for the cluster-sink interactions. The key parameter in the elastic models is the sink segregation strength, which is found to increase with increasing cluster size. Such cluster-sink interactions are responsible for the migration of small helium clusters by drift and for helium segregation on surfaces and grain boundaries in tungsten. As the clusters migrate toward the sinks, trap mutation and cluster dissociation reactions are activated at rates higher than in the bulk. These kinetic processes are identified and characterized based on analysis of many molecular-dynamics (MD) trajectories for each such mobile He_n cluster near surfaces and GBs. These cluster reactions are found to be responsible for important structural, morphological, and compositional features in plasma-exposed tungsten, including surface adatoms, near-surface immobile helium-vacancy complexes, and retained helium content. Such helium segregation on sinks is observed in large-scale MD simulations of helium aggregation in model polycrystalline tungsten at 933 K upon helium implantation.

379949 Invited Talk: Phase Behavior of Model Tapered Diblock Copolymers

Abstract:

Motivated by the significant recent interest in using block copolymers to create solvent-free, mechanically robust battery electrolytes, we use theory and simulation to study the microphase separated structures of model copolymers. Besides being safe and nonflammable, certain salt-doped microphase separated polymer electrolytes could save battery weight because they have a high enough shear modulus to be used in conjunction with a solid lithium electrode. This is because they are composed of AB diblock copolymers, where A is soft and facilitates ion transport while B provides mechanical strength. Depending on the strength of interactions and fraction of A, the system phase separates into various ordered phases. It would be advantages for transport and mechanical properties if regions of A and B were each continuous in all three dimensions. Unfortunately, bicontinuous phases such as the double gyroid are difficult to access at high molecular weights (desirable for good mechanical properties).

Recent experimental and theoretical work showed that linear tapered copolymers, composed of blocks of pure A and B separated by a middle "tapered" block with mixed composition that linearly varies from pure A to pure B (or from B to A for an inverse taper), can form the gyroid phase at high molecular weight. In such systems, taper length is a new adjustable parameter; we aim to guide experiments in using this parameter to control interfacial and phase behavior, with special interest in creating the gyroid phase. First, phase diagrams were created from self-consistent field theory (SCFT). The phase diagrams of normal tapered systems are similar to those of diblocks, but with a lower order-disorder transition temperature and a larger gyroid region. For more detailed information, we employ fluids density functional theory (fDFT) and molecular dynamics (MD) simulations together. These capture monomer scale packing effects and are implemented using very similar models. By comparing with the fDFT results, we can ensure the appropriate equilibrium state is formed in the MD simulations. Density profiles from SCFT, fDFT, and MD are in close agreement; tapers widen the interfacial region and large tapers decrease the maximum purity of the microphases.

370079 Charge-Based Control of Polyelectrolyte Block Copolymers

Abstract:

Polymer electrolytes based on block copolymers are a promising contender for ion transport in batteries, due to their safety compared to liquid electrolytes. However, challenges remain in designing these materials; block copolymers are sought after due to the ability to combine the mechanical stability of a glassy block with the ion transport properties of an ion-transporting block (typically poly(ethylene oxide), PEO). Upon adding a salt to mediate ion transport, this system is effectively rendered a block copolyelectrolyte with a "charged" PEO block that drastically alters the block copolymer morphology. The resulting physical system is difficult to understand due to the many levels of length scale inherent to these problems - both charge length scale as well as polymer morphology length scales are important. We have developed a theoretical method to calculate the effect of charge thermodynamics on polymer morphology, and can demonstrate how the introduction of charge along a polymer backbone can in fact be used to govern the behavior of block copolymers in a predictable and controllable fashion. Concepts such as charge size, charge cohesion, and ion entropy reveal a rich parameter space in which it is possible to use charge to tune through an array of polymer morphologies. Physical intuition elucidated by this study will hopefully guide the targeted design of new polymer-based battery electrolyte materials and more generally motivate the use of charge as a tool to govern the structure of block copolymer materials.

390168 Free Energy of Defects in Aligned Block Copolymer Systems Via Thermodynamic Integration of a Coarse Grained Block-Copolymer Model

Abstract:

Directed self-assembly (DSA) of block copolymers (BCPs) is a promising technique for producing nanoscale regular patterns. In particular, DSA of block copolymers is a leading candidate for extending current optical lithography capabilities below the 20 nm feature size scale for producing device features in semiconductor manufacturing. However, development of such techniques can be a time consuming and difficult proposition given the number of possible material and process variable combinations to explore, the limited knowledge of the fundamental behavior of such systems, and the difficulty in observing the assembled materials. Thus, the development of such DSA techniques could benefit greatly from predictive computer simulation methods that can faithfully represent the behavior of such materials and processes. One option for simulation of BCP-DSA is molecular dynamics (MD), which, when combined with realistic potentials for polymer behavior, can potentially provide more accurate simulations of the inherent polymer behavior, dynamics, and equilibrium states without oversimplifying interatomic interactions and without the need to guess modes of molecular movement. An important issue in bringing such DSA techniques to fruition is developing an understanding of defectivity in such processes and what controls these defect rates. In addition to determining equilibrium material structures and exploring the dynamics of such processes, MD models can also be used with thermodynamic integration techniques to calculate free energy differences between various states, provided a reversible path can be envisioned between them. Using such thermodynamic integration techniques, various equilibrium and metastable states can be compared. In this work, MD combined with thermodynamic integration has been used to calculate defect state free energies in self-assembled block copolymer thin films. Using such free energy data, defect densities can be calculated as functions of relevant system parameters such as the polymer and underlayer properties. In this work, defect free energies are calculated as a function of χ, N, and underlayer pattern properties (e.g. composition and multiplicity). A more complete picture of defectivity in DSA systems has evolved from this work and will be discussed. For example, though others have calculated free energies of defects as a function of χN, the independent contributions of χ and N are more informative. A high χ, low N system yields a significantly higher defect free energy than a low χ, high N system, even when χN is the same in the two systems.

Altering the Miscibility of Polymeric Blends with Electric Fields: A Study on Polystyrene / Poly(vinyl methyl ether) Blends

Annika Kriisa and Connie B. Roth
Department of Physics, Emory University, Atlanta, GA, USA

Contact information:
Annika Kriisa: akriisa@emory.edu
Connie Roth: cbroth@emory.edu

Polymeric blends exhibit limited miscibility because their macromolecular size minimizes the entropy gain on mixing. As such, the phase behavior of polymeric mixtures is extremely sensitive to enthalpic interactions and external perturbations. Methods by which the miscibility of polymeric blends could be tuned or controlled, without changing the underlying chemistry, have great practical benefits for the processing of such blends. One such possible method is the application of electric fields. Although electric fields are extensively used to orient polymer domains in blends and block copolymers, how electric fields alter the underlying miscibility of the system is poorly understood.

The application of electric fields seems experimentally simple, as they can be switched on and off instantly and effortlessly. Nevertheless, the influence of electric fields on the phase separation temperature T_s in small molecules and polymeric mixtures is not yet well known. Available theoretical calculations use thermodynamic arguments for adding an electrostatic free energy term to the total free energy of mixing and predict changes in T_s due to external electric fields that are much smaller than what most experimental results report. To date, neither theory or experiments have found a clear consensus on whether uniform electric fields enhance mixing or demixing. As only a few experimental results have been published over the past several decades with typically only small shifts in T_s, more experiments with unambiguously large shifts in T_s are needed to better understand this effect. Using a fluorescence technique we have developed for measuring the phase separation temperature T_s of polystyrene (PS) / poly(vinyl methyl ether) (PVME) blends [1], we investigate the change in T_s due to the presence of electric fields. We show that electric fields strongly enhance mixing in PS/PVME polymer blends. For example, for a 50/50 PS/PVME blend composition, T_s is increased by over 10 K for electric fields of 18 kV/mm.

[1] A. Kriisa, S.S. Park, and C.B. Roth, *J. Polym. Sci., Part B: Polym. Phys.* **2012**, *50*, 250-256.

379167 Flory-Huggins Parameter χ, from Binary Mixtures of Lennard-Jones Particles to Block Copolymer Melts

Abstract:

The Flory-Huggins parameter χ, is a quantity of paramount importance in describing the degree of segregation in block-copolymer systems. The parameter is dictated by the differences in the chemical character between the constituent monomers and the detailed molecular architecture of the copolymer. For bead-spring models with impenetrable monomer cores the calculation of χ is difficult, since the free energy of the system is not directly accessible, making additional expensive calculations necessary. We propose a novel approach for determining χ that resolves these issues by decoupling the chemical incompatibility of the monomers from the effects of molecular architecture. In particular, we have performed an extensive set of isobaric-isothermal Monte Carlo simulations of binary mixtures of Lennard-Jones particles with the same size, but with asymmetric energetic parameters. The phase behavior of these monomeric mixtures is then extended to chains with finite sizes through theoretical considerations, where we have taken the effects of molecular architecture into account by estimating an effective molecular size [1]. We demonstrate the applicability of our approach for generating parameters by reproducing the morphology diagram of a specific diblock copolymer, namely, poly(styrene-b-methyl methacrylate), which has been extensively studied in experiments.

[1] A. Chremos, A. Nikoubashman, and A. Z. Panagiotopoulos, J. Chem. Phys. **140**, 054909 (2014).

Solvent Determination in Annealing of High-X Block Copolymer PS-b-PHOST

Caleb Breaux, Richard Lawson, Peter Ludovice, Clifford Henderson

cbreaux6@gatech.edumailto:Ben.Nation@gatech.edu

School of Chemical and Biomolecular Engineering

Georgia Institute of Technology, Atlanta, GA

Microphase separation of block copolymers into structures with characteristic length scales on the order of 10 nm is a promising alternative to conventional lithography. Phase separation in block copolymers occurs when the polymer chains attain enough mobility to rearrange themselves within the film. Thermal annealing increases the mobility of polymer chains by raising the temperature of the BCP film above its Tg, while solvent annealing uses a solvent to plasticize the film to increase chain mobility. A neutral solvent is a solvent in which a vertical transition on the Flory-Huggins diagram occurs; that is, the solvent non-selective to both blocks, is desirable to prevent distortions of the pattern during the annealing step. Here, swelling of homopolymers polystyrene (PS) and polyhydroxystyrene (PHOST) were measured via QCM in search of a neutral solvent. A 50:50 volume fraction PS-PHOST BCP was then subjected to each solvent for varying times and morphology observed via ex-situ SEM and GISAXS to monitor the kinetics of the annealing process.

Relaxational behavior of glassy polymers: experimental sorption and nonequilibrium thermodynamic model

Matteo Minelli, Ferruccio Doghieri

Interdipartimental Center for Industrial Research - Advance Mechanics and Materials (CIRI-MAM)
Alma Mater Studiorum - Università di Bologna

Department of Civil, Chemical, Environmental and Materials Engineering (DICAM)
Alma Mater Studiorum - Università di Bologna

The nonequilibrium nature of glassy polymers is responsible for the very peculiar behaviors of such materials in terms of thermodynamic and transport properties, alone or in mixtures with gaseous or vapor solutes. These features are of great interest in many different applications in various industrial fields. For this reason, the study of anomalous kinetics for mass transport processes in glassy polymers has gained a great attention in the past. However, the clear understanding of the physics related as well as the development of a consistent and effective model for its description are still undisclosed.

In this work, specific vapor sorption measurements are carried out on thin polymer films samples below the T_g, in which the penetrant activity is stepwise increased following different ramps, spanning over two orders of magnitude.
To this aim, a Quartz Crystal Microbalance (QCM) system is considered and differential sorption experiments of toluene in glassy polystyrene are carried out. The specimens are obtained spin-coating a drop of 5 wt. % solution of PS (180 kDa - Versalis Spa) in toluene on the quartz crystal. After the deposition, the coated crystals are annealed in a vacuum over above T_g for at least 1 hour and then slowly cooled in vacuum for long time. The resulting film thicknesses are in the range 500-600 nm. The experiments consider controlled rates of penetrant activity increase, and the tests are performed raising the pressure in the QCM chamber at prescribed intervals, set to insure the polymeric sample reaches uniform conditions for penetrant chemical potential throughout its thickness, in each step. Different results for the apparent solubility isotherms are obtained for different rates of increase of penetrant activity, in view of the relaxation process in the glassy polymer, as induced by the sorption of the peentrant species.

The vapor sorption experiments return the kinetics of mass uptake in the thin glassy polymer samples, which are then interpreted in terms of relaxation times for volume dilation, and both models from nonequilibrium thermodynamics and from mechanics of volume relaxation are relevant. To this purpose, the well established nonequilibrium thermodynamic model for glassy polymers (NET-GP) [1] is used to the description of the nonequilibrium thermodynamic behavior of the PS/toluene mixtures below T_g. The lattice

fluid representation proposed by Sanchez and Lacombe is thus adopted [2-3], and the nonequilibrium lattice fluid (NELF) model is employed in the glassy state. This tool allows the interpretation of the mass uptake sorption data in terms of kinetics of swelling induced by the solvent in the polymer matrix.
The behaviors of volume relaxation induced in the polymer film exposed at increasing penetrant activity at the different rates are then analyzed and discussed in terms of relaxation time as function of polymer and solute density. Results are finally compared with corresponding expression retrieved for the case of pure polymer by means of mechanical pressure or temperature jumps.
The peculiar experimental procedure considered in this work, coupled with the NELF thermodynamic analysis, is indeed able to provide the characterization of sorption kinetics of a strong swelling agent in a glassy polymer. This technique is ultimately able to highlight the relaxation phenomena induced by sorption processes, alternatively and more efficiently than traditional sorption experiments, and it provides useful data for an accurate model analysis of corresponding mechanisms.

References

[1] F. Doghieri, G.C. Sarti, *Macromolecules* **29** 7885 (1996).
[2] I.C. Sanchez, R.H. Lacombe, *J. Phys. Chem.* **80** 2352 (1976).
[3] M. Minelli, F. Doghieri, *Ind. Eng. Chem. Res.* **51** 16505 (2012).

385537 Vapor-Liquid Equilibria of VOC-Loaded Shape-Memory Natural Rubber

Abstract:

The high importance of natural rubber (NR) for mankind was identified already in the beginning of the 19th century. After the discovery of the vulcanization process, NR became one of the most significant renewable materials with an annual production of about 11 billion tons at present. Recently, the so-called shape-memory natural rubber (SMNR) was discovered [1]. SMNR is a lightly cross-linked NR, which can be used to store energy and very high strains. The shape-memory properties are received due to strain-induced crystallization, which occurs by stretching the polymer during a programming procedure. Programmed SMNR samples are thus semi-crystalline and can be triggered by mechanical force, heat and even through contact with VOC (volatile organic compound) vapors.

Quitmann et al. [2] showed that constrained, lightly cross-linked SMNR even generates a reversible stress response upon exposure to VOC vapors, which depends on VOC type and concentration. This can be applied e.g. as vapor detector, as long as the mechanical reaction of the SMNR can be related to a certain VOC type and concentration. This requires the vapor-liquid-equilibrium (VLE) data of constrained SMNR/VOC systems.

In this work, VLEs of constrained and unconstrained SMNR/VOC systems were investigated at 293.15 K using a magnetic suspension balance. The measurements provide VOC equilibrium concentrations and diffusivity data. The different strains correspond also to different initial polymer crystallinities, which show a notable effect on the VLEs.

VLE modeling was performed using the Perturbed-Chain Statistical Associating Fluid Theory (PC SAFT). To account for the influence of constant strain on the VLE, a Helmholtz-energy contribution was used that accounts for network elasticity. This contribution is based on the affine network theory with a correction for the finite extensibility of the polymer chains [3].

The solubility of the VOCs in the amorphous part of the semi-crystalline SMNR was modeled based on the assumption that the crystalline phase does not absorb any solvent. The crystallinity of SMNR samples for each strain investigated was estimated based on literature data. Using the modeling approach suggested by Minelli and De Angelis [4], it was possible to correctly describe the experimental VLE data. For that purpose it was assumed that the crystallites exert an isotropic stress on the amorphous phase, which leads to reduced solubility in comparison to a non-stretched (fully amorphous) sample. This stress is quantified as an additional thermodynamic pressure inside the polymer phase and called constraint pressure p_c. p_c was considered as adjustable parameter. This allowed for quantitatively describing VLEs of SMNR/VOC systems over a broad range of strains and VOC concentrations.

References

[1] F. Katzenberg, B. Heuwers, J. C. Tiller, Adv. Mater., 2011, 23, 1909-1911.

[2] D. Quitmann, N. Gushterov, G. Sadowski, F. Katzenberg, J. C. Tiller, ACS Appl. Mater. Interfaces, 2013, 5, 3504-3507.

[3] B. Miao, T. A. Vilgis, S. Poggendorf, G. Sadowski, Macromol. Theory Simul., 2010, 19, 414-420.

[4] M. Minelli, M. G. De Angelis, Fluid Phase Equilibria, 2014, 367, 173-181.

Metal Ion Adsorption via Grafted Dendritic Surfaces

An expanding area of green technology is in the wastewater treatment for heavy metal ions. Adsorption of the fluid cations to solids has proven to be successful and recent research has investigated the improvement of this method through novel materials such as grafting dendrimers onto a solid support. We develop a reduced representation of the system via metal cations in an implicit solvent near a surface of grafted dendrimers. The system is modeled after polyamidoamine (PAMAM) dendrimers with carboxylic acid functional groups that promote the adsorption of Pb(II) ions. Martini coarse-grained molecular dynamics simulations using the LAMMPS package are utilized to predict the nanoscale behavior of the system. The goal is to observe and characterize the adsorption of the Pb(II) ions to the branched PAMAM dendrimers through the measurement of ion coordination and dendrimer dynamics. Our project entails investigating the effects of dendrimer generation and surface coverage. These simulations can identify the role of molecular scale properties on the adsorption process, thereby allowing predictions for bulk analysis including the continuum phase. These investigations have the ability to be applied to other chemical processes that make use of polymers grafted onto a support.

369062 Polypeptoids As Model Systems for Understanding Biopolymer Self-Assembly

Abstract:

Biological polymers, such as polypeptides, are responsible for many of life's most sophisticated functions due to precisely evolved hierarchical structures. Protein structures are the result of monodisperse sequences of amino acids that fold into well-defined chain shapes and tertiary structures with a degree of complexity that is very difficult to understand. Non-natural polymers offer the stability and robustness necessary for materials applications; however, our ability to control primary and secondary structure in these systems is comparatively crude. In addition, the relationship between monomer sequence and self-assembly is not well understood for biological molecules, much less synthetic polymers. Thus, there is a need to explore self-assembly phase space with sequence using a model system. Polypeptoids are non-natural, sequence specific polymers that offer the opportunity to probe the effect of sequence on self-assembly with much simpler molecular interactions and more scalable synthesis than traditional polypeptides. In this talk, I will discuss the use of this model system to understand the role of sequence on chain collapse and globule formation in solution, polymer crystallization, and block copolymer self-assembly. I will then discuss potential application as surface active agents for anti-fouling.

372269 Solute Transport in Polymer Layers

Abstract:

We study the transport of solutes in polymeric films using the device of computer simulations. While the transport properties in the rubbery state of the polymer pertain mainly to nanocomposites, the corresponding phenomena in the glassy state allow us to systematically understand the transport of gas molecules. This last aspect is particularly relevant to the use of glassy polymers as gas separation membranes. The effect of such solutes on the glass transition temperature of the polymer will also be examined.

378543 Amphiphilic Janus Particles As Solid Surfactants

Abstract:

Emulsions are multiphasic fluid mixtures in which one phase (e.g., oil) is dispersed in another (e.g., water). The stability of emulsions and emulsion-based products is crucial for their use in a variety of applications. Conventional emulsions stabilized by surfactants typically suffer from poor stability; that is, these emulsions eventually undergo phase separation via coarsening and coalescence. One alternative to surfactants is amphiphilic Janus particles, which have been proposed as excellent solid surfactants for emulsion stabilization. In fact, theoretical work shows that thermodynamically stable emulsions can be generated with amphiphilic Janus particles as stabilizing agents. An emerging idea in this area is that the shape of Janus particles, in addition to their amphiphilic nature, has a significant influence on their behavior and functionality as surface active agents. Our work is motivated by amphiphilic diblock copolymers such as poly(styrene)-block-poly(acrylic acid) that show pH-responsive surfactant properties. Such stimuli-responsive amphiphilic molecules can stabilize different types of emulsions or even induce phase inversion emulsification in response to changes in local environment. We synthesize a new class of stimuli-responsive Janus particles that change their shape and in turn amphiphilicity in response to changes in the solution pH. We show that these shape-changing Janus particles can stabilize emulsions of different types depending on the solutions pH and that the emulsions stabilized by these Janus particles can undergo phase inversion. We also demonstrate that the interaction between emulsion droplets can be tuned by controlling the solution pH. Our results present a new class of polymer particles that will further widen the functionality and properties of Janus particles as solid surfactants.

Morphology and Charge Separation in All-Conjugated Block Copolymer Photovoltaics

Recent work has demonstrated that all-conjugated block copolymers have significant potential for solution-processed photovoltaic devices, with power conversion efficiencies near 3%. However, optoelectronic properties and structure-property-processing relationships are poorly understood for this class of materials. Here, we present systematic studies on the processing, morphology, and photophysical properties of all-conjugated block copolymer systems. In contrast to coil-coil block copolymers, all-conjugated block copolymers exhibit morphology typically dominated by crystallization rather than micro-phase segregation. However, under solvent annealing, all-conjugated block copolymers self-assemble into lamellar phases with domain sizes independent of polymer block molecular weight. Block copolymers exhibit close π-π stacking of both polymer blocks in solvent annealed thin films, and lamellar ordering also appears in thermally annealed films at temperatures near the crystal melting temperatures for both blocks. Next, we explore photophysical properties in model block copolymers with different linkers between donor and acceptor polymer blocks. We analyze steady-state and transient absorbance and photoluminescence in block copolymers with a poly(3-hexylthiophene) (P3HT) block and one of two alternating copolymer blocks: poly(dioctylfluorene-*alt*-benzothiadazoledithiophene) (PFTBT) or poly(benzothiadazoledithiophene-*alt*-dioctylfluorene) (PTBTF). Block copolymers P3HT-*b*-PFTBT and P3HT-*b*-PTBTF are identical except for the chemistry of the linking group. Photoluminescence measurements reveal the formation of an additional charge-transfer state in P3HT-*b*-PTBTF which can inhibit charge separation. Device measurements show a significant drop in the photocurrent of P3HT-*b*-PTBTF compared with that that of P3HT-*b*-PFTBT. Transient absorption measurements reveal broader photo-induced absorption signal in P3HT-*b*-PFTBT, indicative of more efficient charge separation compared with P3HT-*b*-PTBTF. Both block copolymers exhibit subnanosecond geminate charge recombination. These studies show that the chemistry of the linking group has a significant impact on charge separation and recombination processes and that suppression of charge recombination is essential to further improvement of block copolymer photovoltaics.

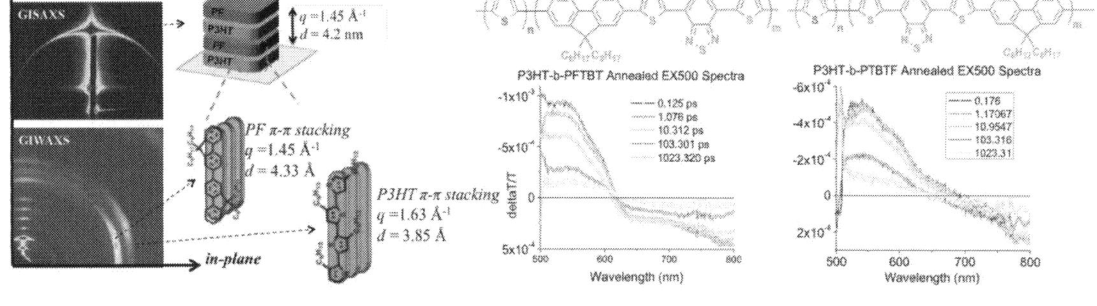

383635 A Hydrogel-Based Cell Culture Platform with Reversible Stiffening Via an Azobenzene-Containing Cross-Linker

Abstract:

The mechanical stiffness of the extracellular matrix is a dynamic property that changes during wound healing, disease progression, aging, etc. For example, tissue stiffness increases during fibrosis, and increased stiffness may contribute to a positive feedback loop that further accelerates the disease progression. Until recently, many cell culture platforms exhibited static material properties, which required replated cells to study the effects of different elastic moduli on cell phenotype. While much progress has been made in the development of substrates that can either increase or decrease in stiffness in situ, there are fewer examples of substrates that can both stiffen and soften, which may be important for simulating the effects of disease progression and resolution. Here, we present a hydrogel-based system that can reversibly stiffen and soften using an azobenzene photoisomer in the crosslinker. To form the hydrogel, poly(ethylene glycol) macromers were crosslinked with an azobenzene-containing peptide using Michael addition chemistry in phosphate buffered saline. We find that upon irradiation with cytocompatible doses of UV light (365 nm, 10 mW/cm^2, 2 min), the photoisomer induces a length change in the crosslinker and a concomitant stiffening of the hydrogel by 100-200 Pa. The resulting change in gel properties is maintained for days due to the long half-life of the photoisomer. The initial modulus of the gel can be recovered upon irradiation with visible light (400-500 nm, 10 mW/cm^2, 2 min). The developed hydrogels should enable noninvasive control of substrate modulus independent of changes in the network connectivity, allowing us to probe questions about the effect of dynamic changes in matrix stiffness on cell behavior.

388914 Engineered Protein Hydrogels to Facilitate and Respond to Neuronal Outgrowth

Abstract:

Neural regeneration within the central nervous system (CNS) is a critical unmet challenge as brain and CNS disorders continue to be the leading cause of disability nationwide. Common tissue engineering goals seek to customize cell-biomaterial interactions and guide cell behavior. Here we have developed a material that is both cell instructive and cell responsive, creating a dynamic interplay between cells and their engineered extracellular matrix (ECM). Using recombinant protein technology, we engineered a family of elastin-like protein hydrogels with multiple independently tunable properties. The concentration of adhesive ligands such as the fibronectin-derived sequence RGD can be precisely controlled without altering the mechanical properties of the hydrogel. When dorsal root ganglion neurons were cultured in 3D in these gels, RGD ligands at 1.9 $x10^7$ ligands/μm^3 promoted neuron-specific growth, and more than doubled the rate of neurite extension compared to hydrogels without the adhesive sequence. Crosslinking density was tuned to create scaffolds with elastic moduli from 0.5-2.1 kPa with constant adhesive site densities. The most compliant gels led to the greatest outgrowth from encapsulated DRGs with neurites extending over 1800 μm by day 7. In contrast, the stiffest gels permitted far fewer extensions and limited outgrowth to a maximum of 600 μm over the same time frame.

We have designed the proteins with a second set of bioactive sequences that specifically respond to changes in cell phenotype. Neural stem cells (NSCs) undergoing differentiation may change their production of the protease urokinase plasminogen activator (uPA), which has previously been found at the growth cones of extending neurites. By incorporating cell-mediated degradable subunits into the elastin-like proteins, we are able to mimic the natural remodeling of the ECM. We engineered multiple uPA target sites with different degradation kinetics into the elastin-like protein to allow neural cell-mediated control of the scaffold degradation dynamics. DRGs encapsulated in uPA degradable scaffolds extended neurites at a faster rate than in gels of similar RGD density and initial stiffness without the uPA degradable sequence.

These crosslinked scaffolds are useful for directing the growth and differentiation of multiple cell types including clinically relevant NSCs. The tunable scaffolds are responsive to neuronal cells which may be able to specifically self-modulate the release of multiple bioactive factors while undergoing differentiation. This work demonstrates the versatility and responsiveness of our modularly-designed protein hydrogels for neural cell culture and encourages continued development as a biomaterial tissue construct for treating spinal cord injury.

Nanostructured Columnar Thin Films Enhance Protein Adsorption and Cell-Material Interactions

Tadas Kasputis[1,2,5], Alex Pieper[1,4], Eva Schubert[2,4], Mathias Schubert[2,3,4], Angela K. Pannier[1,3,4]

[1]*Department of Biological Systems Engineering, University of Nebraska-Lincoln, Lincoln, NE 68588, USA*

[2]*Department of Electrical Engineering, University of Nebraska-Lincoln, Lincoln, NE 68588, USA*

[3]*Nebraska Center for Materials and Nanoscience, University of Nebraska-Lincoln, Lincoln, NE 68588, USA*

[4]*Center for Nanohybrid Functional Materials, University of Nebraska-Lincoln, Lincoln, NE 68588, USA*

Chemical and physical techniques to fabricate uniform nanotopographies have recently become an effective strategy to enhance biomolecule immobilization and cell-material interactions. Sculptured thin film (STF) substrates, fabricated by electron beam facilitated glancing angle physical vapor deposition, consist of nanocolumns ranging in size from the nano- to micro-scales with precise column orientation and spacing. In this work, nanostructured columnar features, such as nanocolumn orientation and density, of titanium (Ti) STFs were investigated for their potential to load fibronectin (FN) within the intercolumnar void spaces of the STFs and to subsequently serve as biomaterial substrates. Generalized ellipsometry was used to characterize FN adsorption to STFs and was able to discriminate between the spatial distribution of adsorbed FN either on top of or within STFs. Substrates with densely packed, vertically oriented nanocolumns were found to enhance FN adsorption two-fold relative to flat surfaces as a function of the increased three-dimensional surface area afforded by the STF nanotopography. Mouse fibroblasts or mouse mesenchymal stem cells (mMSCs) were subsequently cultured on STFs, both with and without FN, and the influences of substrate nanotopography and protein loading were investigated in the context of cell characteristics, including cell adhesion, spreading, viability, and proliferation. Studies of cell adhesion to the various Ti substrates without FN demonstrated that while flat Ti substrates do not support cell attachment, Ti substrates possessing a STF nanotopography can significantly enhance cell adhesion relative to flat Ti substrates. The addition of FN to STFs was only found to enhance cell-material interactions in mMSC cells, demonstrating a cell-dependent response to the addition of FN. The comparisons of trends between the total amount of FN adsorbed to STFs and cell-material responses between substrates containing FN indicate that the substrate nanotopography, not the enhanced amount of FN within STFs, predominantly contributes to the enhancement of cell-material interactions on nanostructured STF substrates. Upon determining the optimal nanotopography for enhancing cell-material interactions, STFs were also used to examine the influences of uniform nanotopographies on the expansion and differentiation of various stem cell lines. These findings demonstrate that nanostructured Ti-STF substrates are biocompatible substrates that enhance cell-surface interactions and the influences of STF nanotopography on cell-material interactions can be applied to the design of nanostructured substrates for the loading of biomolecules and culture of cells for biomaterial, tissue engineering, and diagnostic applications.

382851 Spatial Patterning of BMP-2 and BMP-7 on Biopolymeric Films and the Guidance of Muscle Cell Fate

Abstract:

In the cellular microenvironment, growth factor gradients are crucial in dictating cell fate. Towards developing materials that capture the native microenvironment we engineered biomimetic films that present gradients of matrix-bound bone morphogenetic proteins (BMP-2 and BMP-7). To this end layer- by-layer films composed of poly(L-lysine) and hyaluronan were combined in a simple microfluidic device enabling spatially controlled growth factor diffusion along the film. Linear long-range gradients of both BMPs induced the trans-differentiation of C2C12 myoblasts towards the osteogenic lineage in a dose dependent manner with a different signature for each BMP. The osteogenic marker alkaline phosphatase (ALP) increased in a linear manner for BMP-7 and non-linearly for BMP-2. Moreover, an increased expression of the myogenic marker troponin T was observed with decreasing matrix-bound BMP concentration, providing a substrate that it is both osteo- and myo-inductive. Lastly, dual parallel matrix-bound gradients of BMP-2 and -7 revealed a complete saturation of the ALP signal. This suggested an additive or synergistic effect of the two BMPs. This simple technology allows for determining quickly and efficiently the optimal concentration of matrix-bound growth factors, as well as for investigating the presentation of multiple growth factors in their solid-phase and in a spatially controlled manner.

368692 Platelet-like Nanoparticles As Synthetic Hemostats

Abstract:

There is significant clinical interest in synthetic hemostats that can be used in transfusion medicine for management of bleeding complications. To this end, a promising strategy is to design particulate systems with surface functionalities that allow platelet-mimetic biochemical interactions that render hemostatically important cell-cell and cell-matrix interactions. Specifically, platelets render hemostasis by marginating to the injured vascular wall from flowing blood, undergoing adhesion and activation at the injury site by binding to von Willebrand Factor (vWF) and collagen, and subsequently promoting the activation and aggregation of other neighboring platelets at this site. Inspired by these mechanisms, we have fabricated a platelet like nanoparticle (PLN) capable of performing hemostatic functions similar to their natural counterpart. Here, we will discuss the design, synthesis and applications of PLNs, fabricated via layer-by-layer (LbL) assembly, that combine both key biophysical and biochemical parameters of natural platelets on a single particle platform. PLNs were surface-functionalized with a combination of three peptides that promote platelet-mimetic hemostatic properties of vWF-adhesion, collagen-adhesion and active platelet aggregation in tandem. *In vitro* studies performed under physiologically relevant flow conditions in an artificially created wound environment confirmed the simultaneous adhesive and aggregatory functionalities of PLNs. *In vivo* studies in a mouse tail transection model demonstrated that PLNs render ~65% reduction in bleeding time and actively localize at the clot site. These results highlight the efficacy of cell-mimetic synthetic capsules in providing a platform for the biomimicry of natural cells and their applications.

387420 Fabrication of Layer By Layer Core and Hollow Particles to Study Phagocytosis

Abstract:

Fabrication of Layer by Layer Core and Hollow Particles to Study Phagocytosis
Anusha Garapaty[1], Xingjie Zan[1] and Julie A.Champion[1], (1) School of Chemical and Biomolecular Engineering, Georgia Institute of Technology, Atlanta, GA

Macrophages actively remove foreign materials introduced into the body through a process known as phagocytosis. Phagocytosis is mediated by actin-dependent processes involving adhesion, tyrosine phosphorylation, pseudopodial extension, and force generation. This cascade of events shares resemblance with cell migration, where physical and chemical parameters of the substrate play an important role. This suggests that biophysical properties of a material can also govern the process of phagocytosis. The material parameters size, shape and mechanical stiffness have been previously studied, by us and others, to establish their effect on interaction with macrophages. The two steps of phagocytosis, attachment and internalization, were shown to be affected by these parameters [1-3]. However, the combinatorial effects of these material biophysical parameters on phagocytosis have not been studied. This is in large part because current particulate materials are not able to be tuned according to all the parameters without affecting the others. To meet this challenge, we have fabricated novel particulate materials whose parameters size, shape and mechanical stiffness can be independently controlled. These materials have been fabricated by the development of a template based layer-by-layer (LbL) assembly process.

Polystyrene particles of a desired size are chosen as the template. Shape is imparted to the polystyrene particles by heat liquefaction while stretching. The polystyrene template particles are coated with polyelectrolytes poly (vinyl pyrrolidone) (PVP), poly (ethyleneimine) (PEI), and poly (acrylic acid) (PAA) by LbL self assembly. Mechanical stiffness can be varied by functionalization of PAA with thiol groups prior to LbL on particles, to achieve crosslinking between the layers. Finally, treatment with tetrahydrofuran removes the template to attain the hollow form of the particle. The hollow form combined with varying degrees of polyelectrolyte crosslinking helps achieve a range of mechanical stiffness. Size, shape and mechanical stiffness can be controlled independently by the choice of the template polystyrene size, the method of stretching, and the degree of functionalization of thiols on PAA respectively. This highlights the versatility of our system to generate a gamut of polymer carriers of controlled shape, size, and stiffness while retaining the same surface chemistry. These particles are being used to elucidate the interplay between the biophysical properties size, shape and stiffness in governing phagocytosis. This novel polymer particle platform can be functionalized with active ligands for modulation of macrophage activity and phenotype in immunological therapeutic applications.

References

1. Beningo, K.A. and Y.-l. Wang, *Fc-receptor-mediated phagocytosis is regulated by mechanical properties of the target.* Journal of cell science, 2002. 115(4): p. 849-856.

2. Champion, J.A. and S. Mitragotri, *Role of target geometry in phagocytosis.* Proceedings of the National Academy of Sciences of the United States of America, 2006. 103(13): p. 4930-4934.

3. Champion, J.A., A. Walker, and S. Mitragotri, *Role of particle size in phagocytosis of polymeric microspheres.* Pharmaceutical research, 2008. 25(8): p. 1815-1821.

388587 Facile Ways to Stabilize Nanoparticles Using Zwitterionic Polymers

Abstract:

Zwitterionic polymeric materials emerged as new generation of materials with excellent non-fouling properties. It has been demonstrated that their use as surface coatings can effectively resist binding from proteins, cells, and full blood. Nanoparticles coated with zwitterionic polymers exhibited long blood residency *in vivo*. These excellent properties were attributed to the super-hydrophilicity of these polymers providing strong hydration effects. However, this advantage disfavors their ease of use to modify hydrophobic objects, such as nanoparticles. Zwitterionic polymers are insoluble in most organic solvents where nanoparticles easily maintain their colloid stability, and no effective modification can happen in such solvents. In aqueous phase, zwitterionic polymers can be introduced to particle surface through various types of chemistry. But these syntheses are usually complex, and particles tend to aggregate during these processes. A novel facile method will be presented to effectively address the superhydrophilic zwitterionic polymer/ hydrophobic particle synthetic issue. We will demonstrate the resulting zwitterionic polymer protected nanoparticles achieve unprecedented stability comparing with established stabilizing technologies.

390344 Length Selective Delivery and Altered Subcellular Processing of Protein Stabilized Single Wall Carbon Nanotubes

Abstract:

Single wall carbon nanotubes (SWCNTs) are increasingly being investigated for biomedical imaging, sensing, and drug delivery. Selective cellular delivery and preservation of inherent SWCNT properties within cells are crucial for the realization of SWCNTs as safe and effective biomaterials. SWCNT uptake and subsequent intracellular trafficking are dictated by the cellular entry mechanism. Specialized immune cells known as macrophages are capable of two size dependent entry mechanisms, endocytosis and phagocytosis. To determine how SWCNT entry via these mechanisms affects cellular uptake and subcellular distribution, we created SWCNT dispersions with three different lengths ranging from nm to μm. In order to maintain SWCNT optical properties and promote cellular uptake into macrophages, we noncovalently dispersed SWCNTs with bovine serum albumin. Using Raman spectroscopy, we quantify cellular uptake as well as track the intracellular state of the SWCNTs as a function of SWCNT length in macrophages. We find that bulk uptake in cells increases with decreasing SWCNT length. We also observe smaller, more numerous endosomes as SWCNT length decreases, suggesting that increased bulk uptake is driven by efficient entry and processing via endocytosis. Highly concentrated intracellular regions of SWCNTs are observed in phase dense endosomes which correlate with changes in SWCNT properties observed via Raman spectroscopy. SWCNTs in these regions were found to be less isolated and more aggregated resulting in disruption of SWCNT properties including reduced NIR fluorescence intensity and suggesting the removal of the protein dispersing agent. Similar SWCNT property changes are not observed in fibroblasts treated with equal length SWCNTs at the same concentration. Longer SWCNTs forced to enter through phagocytosis remain relatively isolated with preserved SWCNT properties compared to their shorter counterparts. We hypothesize that aggregation of SWCNTs within macrophages but not fibroblasts may facilitate the retention of SWCNTs within macrophages. Further, the greater individualization of SWCNTs in fibroblasts may promote their release from the cell over time which could have useful implications for cell based delivery systems.

Functionalized Halloysite Nanotubes for Epoxy Nanocomposite Applications

Songshan Zeng[1], Christopher Reyes[2], Paul Rodgers[1], Samuel Wentworth[1], Luyi Sun[1]

1. Department of Chemical & Biomolecular Engineering and Polymer Program, Institute of Materials Science
University of Connecticut, Storrs, CT 06269 USA
Phone: 860-486-6895; Fax: 860-486-4745
Email: luyi.sun@uconn.edu

2. Department of Chemistry and Biochemistry
Texas State University, San Marcos, Texas 78666-4684

Abstract

Polymer nanocomposites have been extensively studied over the past few decades, and a wide range of applications have been developed based on their excellent performance. Halloysite nanotube (HNT), a type of naturally occurring aluminosilicate clay, has come into focus for study as a nanofiller to reinforce polymers. Herein, we report a new approach for modifying HNTs for the preparation of epoxy/HNT nanocomposites with improved stiffness and toughness. The hydroxylated HNTs (h-HNTs) from NaOH treatment was further modified by sodium dodecyl sulfate (SDS), leading to better colloidal stability as confirmed by zeta potential. With an addition of 5 phr of this SDS treated h-HNTs (HNTs-SDS) into epoxy matrix, the modulus and fracture toughness of epoxy/HNT-SDS nanocomposite was improved by 22% and 52%, respectively. The simultaneous improvements in both stiffness and toughness are believed owing to the significantly improved interface between epoxy matrix and HNTs after surface treatment.

Key Words

Halloysite, nanocomposite, epoxy, dispersibility

EFFECTS OF PROCESSING PARAMETERS ON EXPERIMENTAL ORIENTATION EVOLUTION OF GLASS FIBERS IN COMPLEX 3D FLOW OF INJECTION MOLDED COMPOSITES

Rebecca A. Minnick, Hongyu Chen, Donald G. Baird
Virginia Polytechnic Institute and State University
Department of Chemical Engineering
Randolph Hall 133
Blacksburg, VA 24061

ABSTRACT

Long Fiber-Reinforced Polymer (LFRP) composites are important due to their increased mechanical properties while still being melt processable. LFRP composites are used in industries such as automotive and aerospace. Increased mechanical properties are dependent on the orientation of fibers in the composite. Orientation is a function of the type of fiber and matrix being used, but also processing conditions such as injection speed, screw temperature, mold temperature, and mold geometry. Understanding the orientation effects of various processing parameters is essential in gaining the ability to optimize mold design so as to obtain desired mechanical properties. This work particularly focuses on understanding the effects of the injection speed processing parameter on experimental orientation of long, semi-flexible glass fibers in an end-gated plaque (EGP) exhibiting a complex 3-dimentional geometry. Three injection speeds were considered with constant mold and screw temperatures. These samples were then used to obtain experimental orientation and fiber length distribution data. Orientation data was obtained using a slightly modified version of the method of ellipses (MoE), which has been used for rigid fibers. Injection speeds of 1, 2, and 4 seconds were used with a mold temperature of 79°C. Orientation data along the center plane was obtained and compared against simulated values of orientation.

Oluwatimilehin 'Timi' Fadiran
Advisor: Dr. J. Carson Meredith

Pollen as a Renewable Reinforcing Filler for Thermosetting Polymer

Abstract

Pollen has the potential to be an effective filler in polymers in part due to its high mechanical strength and light weight. Pollen-polymer composites could form a new class of high strength, light-weight materials with biorenewable filler. Due to pollen's unique architectures, these composites also provide insights to how filler microstructure may affect wetting and adhesion between phases. In this work, the effectiveness of pollen fillers in a thermosetting polymer is characterized as a function of pollen loading and surface treatment. We have found that an acid-base hydrolysis treatment of pollen can be used to tune mechanical, interfacial, and thermal properties in the thermoset. The acid-base treatment exposes and generates additional surface functional groups improving adhesion between pollen and the polymer matrix. Additionally, it reduces the pollen density versus native pollen to approximately 1.16 g/cc much lower than conventional fillers (talc, silica, calcium carbonate) and even many advanced fillers (carbon nanotubes, cellulose, wood flour). This is due to the elimination of intracellular material which makes the pollen hollow. Finally, the treated pollen is able to simultaneously stiffen and strengthen the thermosetting polymer due to its higher mechanical stiffness and the high degree of adhesion between the phases. Thus, pollen is an attractive filler for creating high strength, light-weight materials with a sustainable plant sourced filler.

387347 High Performance Multifunctional Polystyrene Nanocomposites

Abstract:

In this paper, the high performance polystyrene (PS) [1] polymer nanocomposites (PNCs) have been fabricated by a solvent extraction method [2] filled with various Fe_3O_4nanoparticles (NPs) and carbon nanofibers (CNFs) loading levels. The effects of nanofillers loading levels on the morphology, thermal stability, crystalline phase, thermomechanical properties, melt rheological behaviors, flammability, magnetic properties and dielectric properties were systematically studied. The good nanofillers dispersion and strong interfacial interaction between nanofillers and PS matrix are observed by scanning electron microscope (SEM). The increased thermal stability of PNCs is observed by thermogravimetric analysis (TGA) with increasing the CNFs loading. Furthermore, nanofillers have an affirmative influence on the thermomechanical and flammable properties of PNCs based on the strong interaction between the nanofillers and PS matrix. Meanwhile, the nanofillers have effectively improved the melt rheological behaviors and dielectric property of the PNCs. This work demonstrates a facile and cost-effective approach to construct the good dispersion and strong interfacial interaction between nanofillers and polymer matrix, which enhances the properties of PS matrix.

[1] X. Yan, Q. He, X. Zhang, H. Gu, H. Chen, Q. Wang, L. Sun, S. Wei and Z. Guo; Magnetic Polystyrene Nanocomposites Reinforced with Magnetite Nanoparticles; Macromolecular Materials and Engineering, 299(4), 485-494 (2014)

[2] X. Chen, S. Wei, Y. Atarsingh, R. Patil, J. Zhu, R. Ximenes, L. Sun and Z. Guo; Poly(propylene)/Carbon Nanofiber Nanocomposites: Ex Situ Solvent-Assisted Preparation and Analysis of Electrical and Electronic Properties; Macromolecular Materials and Engineering, 296 (5), 434-443 (2011)

385075 Preparation of Novel Composite Materials Via CO-Coagulation of Nanoparticles

Abstract:

Since the term 'nanocomposites' has been proposed for the first time by Theng in 1970, the science dealing with these materials has arisen at the border of different areas of knowledge.

Numerous procedures for the preparation of nanocomposite materials have been investigated in order to generate materials with controlled features, which is still one of the great challenges in nanotechnology.
The purpose of this work is to use the extended and large knowledge of self-assembly behavior of spherical nanoparticles in aqueous solutions and its dependence on interparticle interactions for the preparation of various polymeric nanoparticles and the investigation of their self-assembly, to generate hybrid nanostructured composites. The nanocomposites are prepared by blending together two colloidal suspensions made of charged particles. The first one is a suspension of hard silica nanoparticles (20 nm diameter) at different concentrations, while the second one consists of soft poly(butyl acrylate/methyl methacrylate) copolymer particles (80 nm diameter). The advantage of working in aqueous suspensions is that interparticles interactions of charged nanoparticles can be controlled. In order to investigate the specific role of such interactions on the final structure and mechanical properties of the materials, the nanocomposites have been synthetized following three different assembly strategies. The first one consists of simply mixing of stable particle suspensions with same surface charge. In the second case, instead, the filler particles were subject to aggregation and gelation while the matrix particles remained in stable suspension. The last case is that of hetero-aggregation of silica and rubber particles suspensions bearing opposite surface charges.
The solvent (i.e., water) was removed from all samples, which have been annealed, giving rise to nanostructured polymeric composites. All samples structure and mechanical properties have been characterized using different techniques (SAXS, AFM, SEM, tensile tests).

Cross-Linking Graphene Oxide-Polyethyleneimine Hybrid Hydrogel Film Containing Ciprofloxacin: One-Pot Synthesis, Controlled Drug Release and Anti-Microbial Performance

Tiefan Huang,[1] *Lin Zhang,*[1] *Huanlin Chen,*[1] *Congjie Gao*[1,2]

1.Key Laboratory of Biomass Chemical Engineering of MOE, Department of Chemical and Biological Engineering, Zhejiang University, Hangzhou 310027, China

2. National Engineering Research Center for Liquid Separation Membrane, Hangzhou 310012, China

Abstract:
A novel drug delivery system based on graphene oxide-polyethyleneimine hybrid hydrogel film, in which graphene oxide nanosheets in the film were cross-linked by PEI, was prepared via one-pot synthesis technique. Due to cross-linking, mechanic stability of the film was significantly improved compared with uncross-linking graphene oxide film. The release behaviour of ciprofloxacin was investigated under various pH in vitro, which showed slow drug release without initial burst effect. By applying release kinetic models to the drug release processes, the mechanism of release of CF from the graphene oxide-polyethyleneimine hybrid hydrogel film was found to follow the near zero-order kinetics. E. coli was selected as a model bacterium to evaluate the antibacterial activity of this system. Compared with graphene oxide film and graphene oxide-polyethyleneimine hybrid hydrogel film, the ciprofloxacin loaded hybrid hydrogel film exhibited a significant antibacterial effect, which can be attributed to the ciprofloxacin releasing and diffusing-out from the film. We propose that this study opens the avenue to designing suitable crossing-linking GO hybrid films for biomedical area as well as many other application.

Ultra-thin anti-corrosion coating: a 0.5 nm thick aluminosilicate film (two-dimensional zeolite) protects Ru(0001) from oxidation

J.A. Boscoboinik

Center for Functional Nanomaterials, Brookhaven National Laboratory. Bldg. 735, Upton, NY 11973 (USA).

The oxidation of a Ru(0001) was explored for a clean and aluminosilicate-coated surfaces using ambient-pressure X-ray photoelectron spectroscopy (AP-XPS) in O_2 ambient at temperatures ranging from 300 K to 923 K. The coating consists of an ultrathin (0.5 nm thick) well-defined aluminosilicate framework, which has been recently proposed as a model system for surface science studies of zeolites. [1-3] Heating Ru(0001) to high temperatures in the presence of oxygen results in the formation of a ruthenium surface oxides. For the case of an O_2 pressure of 1×10^{-4} mbar, three different oxide thicknesses are detected, depending on the annealing temperature. However, when the ultrathin aluminosilicate framework (2D-zeolite) is prepared on top of the Ru(0001) surface, only chemisorbed oxygen is detected on the surface when the crystal is exposed to the same conditions that lead to the formation of the oxide.

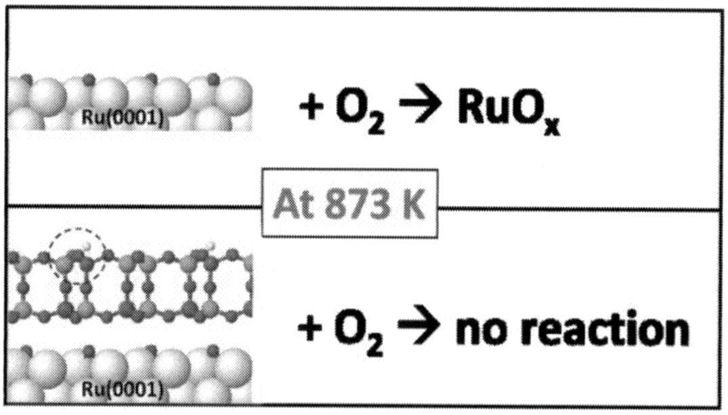

References

[1] J.A. Boscoboinik, X. Yu, B. Yang, F.D. Fischer, R. Wlodarczyk, M. Sierka, S. Shaikhutdinov, J. Sauer, H.-J. Freund, *Angew. Chem. Int. Ed.* 51 (2012) 24, 6005-6008. *Angew. Chem.* 124 (2012) 6107-6111.

[2] J.A. Boscoboinik, X. Yu, B. Yang, S. Shaikhutdinov, H.-J. Freund. *Micropor. Mesopor. Mater.* (2013) 165, 158-162.

[3] J.A. Boscoboinik, X. Yu, S. Shaikhutdinov, H.-J. Freund. *Micropor. Mesopor. Mater.* (2014) 189, 91-96.

Research carried out in part at the Center for Functional Nanomaterials, Brookhaven National Laboratory, which is supported by the U.S. Department of Energy, Office of Basic Energy Sciences, under Contract No. DE-AC02-98CH10886. JAB acknowledges the Alexander von Humboldt Foundation for support during a stay at the Fritz Haber Institute of the Max-Planck Society in Berlin, Germany.

Precursor utilization in atomic layer deposition on nanostructured materials in fluidized bed reactors

Fabio Grillo, Michiel T. Kreutzer, J. Ruud van Ommen*

*Delft University of Technology, Department of Chemical Engineering, 2628 BL Delft, The Netherlands

Abstract

Atomic layer deposition (ALD) in fluidized bed reactors (FBRs) is a promising technology for coating and functionalizing nano-structured materials, such as, nanoparticles and micron-sized nano-porous particles [1, 2, 3, 4, 5, 6]. One of the greatest potentiality of such technique lies in the inherent scalability of FBRs, which may have the latter play a major role in scaling up the nanotechnology sector by enabling the processing of the large quantities of material needed for several practical applications [7, 8]. Nevertheless, the economic and environmental feasibility of the scale-up of such process strongly depends on the efficiency with which the ALD precursors are used, as they represent a major operating cost.

In the present work, a multi-scale dynamic model is presented as a means to understand and optimize the precursor utilization during ALD on nanoparticles and nanoporous micron-sized particles in FBRs. As a case study, the model is used for studying the dynamics of the deposition of alumina through trimethylaluminum and water on both titania nanoparticles and micron-sized nano-porous -alumina particles at low (~1 mbar) and atmospheric pressure. In doing so, we investigate the influence of operating parameters, such as, pressure, reactor height and internal structure of both nanoparticle agglomerates and nano-porous micron-sized particles, on the precursor utilization efficiency.

The major result is that, at proper operating conditions, fast ALD reaction kinetics enables the saturation of the particles surface area with hardly any loss of precursors. In particular, our results suggest that the internal structure of nanoparticles agglomerates has little to no influence on the precursor utilization efficiency. On the other hand, the internal structure of nanoporous micron-sized particles introduces considerable internal mass transfer limitations, which might affect the precursor utilization efficiency, especially in small scale units. Furthermore, at low pressures, the reduction in the gas-solid contact efficiency, introduced by gas owing through the FBR as bubbles, has negligible effect on precursor losses, whereas at atmospheric pressure the reactor height should be designed according to bubble properties and bubble rising velocity to avoid such losses. Finally, simple scaling rules for the optimization of the precursor utilization are proposed.

References

[1] D. M. King, X. Liang, A. W. Weimer, Functionalization of fine particles using atomic and molecular layer deposition, Powder Technology 221.

[2] X. Liang, Y. Zhou, J. Li, A. W. Weimer, Reaction mechanism studies for platinum nanoparticle growth by atomic layer deposition, Journal of Nanoparticle Research 13 (2011) 3781-3788.
[3] J. Li, X. Liang, D. M. King, Y.-B. Jiang, A. W. Weimer, Highly dispersed Pt nanoparticle catalyst prepared by atomic layer deposition, Applied Catalysis B: Environmental 97 (2010) 220-226.
[4] A. Goulas, J. R. van Ommen, Atomic layer deposition of platinum clusters on titania nanoparticles at atmospheric pressure, Journal of Materials Chemistry 15 (2013) 46474650.
[5] R. Beetstra, U. Lafont, J. Nijenhuis, E. Kelder, J. van Ommen, Atmospheric pressure process for coating particles using atomic layer deposition, Chemical Vapor Deposition 15 (2009) 227-233.
[6] Y. Zhou, D. M. King, X. Liang, J. Li, A. W. Weimer, Optimal preparation of Pt/TiO2 photocatalysts using atomic layer deposition, Applied Catalysis B: Environmental 101 (2010) 54-60.
[7] D. M. King, J. A. Spencer II, X. Liang, L. F. Hakim, A. W. Weimer, Atomic layer deposition on particles using a fluidized bed reactor with in situ mass spectrometry, Surface and Coatings Technology 201 (2007) 9163–9171.
[8] J. R. Ommen, J. M. Valverde, R. Pfeffer, Fluidization of nanopowders: a review, Journal of Nanoparticle Research 14 (2012) 1-29.

389590 Engineering Lithium-Containing Ionic Conductive Thin Films By Atomic Layer Deposition for Lithium-Ion Battery Applications

Abstract:

Lithium (Li)-ion batteries have drawn much attention for their outstanding performance in portable electronic applications. These batteries have the potentials to function as miniaturized power source for microelectromechanical (MEMS) devices through the fabrication of 3-dimensional configurations. To fabricate a fully functional 3D Li-ion microbattery, however, an ultra-thin and highly conformal electrolyte layer is required to coat the 3D electrodes. The solid oxide Li-ion conductor, lithium aluminosilicate ($Li_xAl_ySi_zO$, LASO), synthesized by atomic layer deposition (ALD) is a promising electrolyte material for 3D battery applications owing its adequate ionic conductivity as well as improved electrode stability.

The self-limiting characteristic of ALD allows for precise control of thickness and composition of complex oxides and results in a highly conformal and pinhole-free coating even on highly complex structures such as high aspect ratio 3D electrodes. The metal precursors, lithium t-butoxide (LTB), trimethylaluminum (TMA), tris(tert-butoxy)silanol (TTBS), and tetraethylorthosilicate (TEOS) were used to form $Li_xAl_ySi_zO$ via ALD. *In-situ* FTIR was implemented to study the incubation time and growth mechanisms of each constituent oxide on the other to improve the controllability of deposited films. *In-situ* FTIR studies revealed that the growth mechanism of silicon oxide is strongly affected by the underlying oxide layer, exhibiting different surface reaction mechanisms during the incubation stage.

Li-ion conductivities and the activation energy of as-deposited LASO/LAO/LSO films with respect to lithium contents and the film thickness were studied. The LASO ALD coating on 3D carbon array posts were confirmed to be conformal and uniform using transmission electron microscopy (TEM) imaging. A Li-ion half-cell consisting of LASO coated on 3D carbon array electrode has shown promising improvement in efficiency when compared to bare electrode.

Lithiation cycling tests of thin LASO/LAO/LSO films were found to be functions of both composition and thickness. The reversibility and kinetics of insertion as well as the effect on the cycling stability from the direct deposition of LASO/LAO/LSO on potential anode materials, SiNWs were also investigated using *in-situ* TEM observations during lithiation.

Symmetry-Breaking in Light-Trapping Nanostructures on Silicon for Solar Photovoltaics

Sang Eon Han[1,4*], Swapnadip Ghosh[2,3], Tianhao Cai[4], Brittany Hoard[1], and Sang M. Han[1,2,3,4]
[1]Nanoscience and Microsystems Engineering, University of New Mexico
[2]Electrical Engineering & Computer Engineering, University of New Mexico
[3]Center for High Technology Materials, University of New Mexico
[4]Chemical & Nuclear Engineering, University of New Mexico

Current c-Si solar cells, which comprise 80-90% of the solar cell market, typically use 100-300 μm thick silicon wafers to absorb sunlight adequately and these wafers account for about a half of the total module cost. To reduce the cost by using thin c-Si films, highly efficient light trapping is required. For thin-film photovoltaics, nanoscale structures are typically involved for light trapping because the film thickness becomes comparable to the wavelength of sun light. While diverse nanostructures have been studied to break the light-trapping limit of geometric optics, known as the Lambertian limit, highly efficient nanostructures that can be easily manufactured have not been demonstrated. We have previously predicted that symmetry-breaking in light-trapping periodic nanostructures on thin films can approach the Lambertian limit very closely. Herein, we will present how the systematic symmetry-lowering increases light-trapping in thin-film photovoltaics. Specifically, we will show that, as the symmetry is systematically lowered, the number of optical absorption peaks increases in a manner consistent with group theory. When the nanostructures were optimized to maximize the integrated absorptance over a spectral band of interest, the integrated absorptance increased by symmetry-breaking. Further, we will demonstrate the experimental realization of such low-symmetry structures using simple wet etching methods on c-Si(100) wafers. Our method of symmetry breaking would provide an ideal experimental platform to study the effect of symmetry breaking on diverse phenomena beyond light trapping.

* Author to whom all correspondence should be made.

389052 Modeling of Silicon Dopant Activation in Ingaas

Abstract:

Driven by interest from both the semiconductor and energy industries, III-V materials, especially ternary compounds such as InGaAs (on Si), have attracted a lot of attention because of their direct electronic band gap for most of these materials and their outstanding electrical conductivity. More practically, they offer a potential solution to the increasing problem of over-heating in current transistors. Compared to Si, III-V compound semiconductors (e.g., GaAs or InP) have lower conduction band effective mass and higher electron mobility. In particular, a ternary compound such as $In_{0.53}Ga_{0.47}As$, which is lattice-matched to InP, can be integrated on Si substrates to support device fabrication. However, preliminary experimental investigations show only poor activation of Si dopants in InGaAs (~15%), which limits their usefulness. Our atomic-scale simulations are intended to unravel the reasons behind the poor activation. We have used first-principles density-functional theory and semi-empirical molecular modeling using Tersoff potentials for the multi-scale modeling. We have parameterized Tersoff potentials for Si-X (X = In, Ga, and As) interactions. This parameterization allowed us to perform diffusion calculations on large systems. Calculation of substitutional and interstitial defect formation energies provides a picture of the preferences of Si atoms to occupy substitutional and interstitial cation sites in InGaAs. We have evaluated the formation energy dependence up to the second nearest neighbors. The nudged elastic band (NEB) method is used to calculate migration energies, which ultimately leads us to activation energies.

Tunable photoluminescence property of SiC nanocrystals fabricated by thermal plasma enhanced CVD process

Tengfei Cao, Haibao Zhang, Binhang Yan, Yi Cheng*
Department of Chemical Engineering, Tsinghua University, Beijing 100084, P.R.China
Email: yicheng@tsinghua.edu.cn

Silicon carbide (SiC) is a well-known wide-bandgap semiconductor for excellent thermal, mechanical and electrical properties, which has been the material of choice for high voltage and high current devices operated at elevated temperatures.[1, 2] SiC is also biocompatible with blood which leads to its biological application.[3] SiC nanocrystal is a new form of SiC with nano-sized SiC grains. It has a photoluminescence property at room temperature besides the outstanding properties of bulk SiC material. Together with its chemical inertness and biocompatibility, SiC nanocrystal has been ultilized in *in vivo* biological imaging as biomarkers.[4]

Wet chemical etching[5] and plasma enhanced chemical vapor deposition (PECVD)[6] are the main methods for SiC nanocrystal fabrication. The two methods both have the feature of low production capability which hampers the production of SiC nanocrystal in large scale. Thermal plasma intensification has been adopted in high-rate CVD of nanocrystalline silicon with room temperature visible photoluminescence property.[7] Thermal plasma offers higher temperature and higher reactive species density compared with cold plasma and thus provides much higher fluxes of energy and matter in nano material fabrication.[8] In this work, thermal plasma enhancement was adopted for the CVD of SiC nanocrystals with $SiCl_4$ as Si source. And optical emission spectroscopy (OES) diagnostic[9] was carried out to analyze the gas phase composition in plasma and better understand the deposition process.

The crystal structure of the samples was characterized by X-ray diffraction (XRD). Diffraction peaks of (111), (200), (220) and (311) crystal planes of β-SiC were observed. Surface chemical composition of the samples was analyzed by X-ray photoelectron spectroscopy (XPS) and C atomic fraction and SiC mole fraction of the surface was calculated. Visible photoluminescence of the samples were observed with the excitation of ultraviolet (UV) light at room temperature. And this photoluminescence property can be tunable with the control of grain size of SiC nanocrystals. In this presentation, we will provide the experimental results on the high-rate fabrication of SiC nanocrystals and the tunable photoluminescence property of the products.

Keywords: Silicon carbide; Nanocrystal; Photoluminescence; Thermal plasma; PECVD; High deposition rate

References:

1. D. Nakamura, I. Gunjishima, S. Yamaguchi, T. Ito, A. Okamoto, H. Kondo, S. Onda and K. Takatori, *Nature*, 2004, **430**, 1009-1012.
2. C. R. Eddy and D. K. Gaskill, *Science*, 2009, **324**, 1398-1400.

3. P. Aspenberg, A. Anttila, Y. T. Konttinen, R. Lappalainen, S. B. Goodman, L. Nordsletten and S. Santavirta, *Biomaterials*, 1996, **17**, 807-812.
4. J. Y. Fan, H. X. Li, J. Iiang, L. So, Y. W. Lam and P. K. Chu, *SMALL*, 2008, **4**, 1058-1062.
5. J. Zhu, Z. Liu, X. L. Wu, L. L. Xu, W. C. Zhang and P. K. Chu, *Nanotechnology*, 2007, **18**.
6. T. TAKESHITA, K. ICHIGE, Y. KURATA and S. HASEGAWA, *J Appl Phys*, 1991, **69**, 7945-7947.
7. T. F. Cao, H. B. Zhang, B. H. Yan and Y. Cheng, *RSC ADVANCES*, 2013, **3**, 20157-20162.
8. K. K. Ostrikov, U. Cvelbar and A. B. Murphy, *J Phys D Appl Phys*, 2011, **44**.
9. T. F. Cao, H. B. Zhang, B. H. Yan, W. Lu and Y. Cheng, *RSC Advances*, 2014, **4**, 15131-15137.

381314 Role of Light Scattering in Hybrid Solar Cells

Abstract:

Hybrid solar cells, with an inorganic/organic interface for charge separation, have been extensively investigated in the past decade in order to replace the expensive Si based technology with an inexpensive alternative. Typically, these devices incorporate a mesoporous TiO_2 film which is decorated with dye molecules and filled with a hole transport material, to conduct the electrons and holes, respectively. Recently, devices with an liquid electrolyte have been able to reach up to 13% conversion efficiency. However, the TiO_2 mesoporous films used for solid state dye sensitized solar cells and hybrid solar cells have a limited light absorption due to thickness of the film (500 nm – 2 μm) required for efficient charge transportation. An elegant approach to increase the light absorption is to induce "defects" within the mesoporous film, causing light to scatter within the active layer of the device. In this work, we combine the commonly used 25 nm particles with other nanostructures, including 200 nm TiO2 particles, TiO_2 nanowires, and Sn doped nanowires, all of which produce light scattering due to their dimensions and disorder within the active layer. Through this approach, we are able to correlate an improvement in conversion efficiencies of around 25 % to the light scattering. Furthermore, the incorporation of these nanowire structures increases the mobility of the electrons, allowing for increased charge extraction and reduced recombination. These two phenomena can be simultaneously engineering due to the crystallinity of the "defects" within the films and the cascading conduction bands produced with the incorporation of the doped TiO_2 wires. The extent of the reduction in recombination is quantified through photovoltage decay and impedance spectroscopy measurements and compared to the standard mesoporous TiO_2 devices.

380985 Supramolecule-Directed Immobilization of Nanoparticles on Solid Surfaces

Abstract:

We have developed an in situ synthesis route for preparation of a layer of immobilized nanoparticles on a solid surface by employing covalently tethered supramolecular layer as a directing agent. The nanoparticles were immobilized and stabilized by a calix[7]hydroquinone monolayer which in turn was immobilized on the substrate surface by covalent layer-by-layer assembly. The calix[7]hydroquinone moiety served as a nano-reactor, container and template in the in situ synthesis process. The nanoparticles were formed by reduction of the metallic ions by the hydroquinone form of the calix[7]arene. The size of the nanoparticles can be tuned continuously up to islands of 100 nm by repeated application of the redox properties of calix[7]arene. A variety of metals can be formed as nanoparticles by this process, including Ag, Pt, Pd, Cu and Au. Metallic ions with less charges, lower oxidation state in the reaction and higher standard redox potential provided better yield. Bimetallic nanoparticles with a core-shell structure were synthesized by a sequential deposition of different metals on the same monolayer. The process is applicable to flat surfaces as well as to those with irregular topography. A significant advantage derived by this method of formation and assembly is its surfactant-free nature that enables access to the active surface of the nanoparticle. The process of in situ synthesis under controlled conditions, together with calibrated variation of particle size has the potential to serve as building blocks for electronic nano-devices, nano-catalysts, etc.

377913 A Retinoic Acid-Enhanced Human Blood-Brain Barrier Co-Culture Model Constructed from Scalable Cell Sources

Abstract:

The blood-brain barrier (BBB), which consists of the endothelial cells that line brain capillaries, is a highly impermeable interface between the bloodstream and the brain parenchyma. While the physiological purpose of the BBB is to maintain brain health and homeostasis, its restrictive properties also present a challenge for drug delivery. Over the last several decades, researchers have developed many permutations of *in vitro* BBB models to create a suitable system that mimics the *in vivo*BBB for drug screening applications. These models are most often comprised of primary brain microvascular endothelial cells (BMECs) and support cells of the neurovascular unit such as pericytes, astrocytes, and neurons. However, due to yield and availability constraints, the cells in these models are typically derived from various animal species which may not appropriately represent the properties of the human BBB. In an effort to overcome these limitations, we previously introduced a method for deriving BBB endothelium from human pluripotent stem cells (hPSCs). Here, we present improvements made to this system, including the identification of retinoic acid (RA) as a substantial contributor to BBB properties and the incorporation of primary human brain pericytes and human astrocytes and neurons derived from neural progenitor cells (NPCs).

The addition of RA to differentiating hPSC cultures consisting of neural cells and immature BMECs induced VE-cadherin expression in the PECAM-1^+/GLUT-1^+ BMECs, indicating their maturation along the endothelial lineage. RA did not increase BMEC differentiation efficiency but did promote BMEC proliferation to increase yield by approximately 2-fold. Upon purification, RA-treated BMECs exhibited greater organization of tight junction strands, possessed transendothelial electrical resistance (TEER) that was substantially greater than untreated controls, and demonstrated increased multidrug resistance protein (MRP) efflux activity. When co-cultured with primary human brain pericytes and NPC-derived astrocyte/neuron mixtures, the hPSC-derived BMECs exhibited TEER reproducibly exceeding 5000 Ωxcm^2. This model possesses barrier properties approaching *in vivo* measurements in rats (5900 Ωxcm^2) and the theoretical maximum of BBB resistance (8000 Ωxcm^2), whereas no previous human model has exceeded 500 Ωxcm^2 and most animal models fall below 2000 Ωxcm^2.

This fully human BBB model should have significant utility for drug screening applications given its substantial barrier properties. Because the human cells in this model are obtained from renewable sources, the overall system can be readily scaled to handle high throughput workloads. Moreover, in conjunction with advancements in cellular reprogramming and genome editing, this system could be highly attractive for disease modeling applications and assist in understanding how an individual's genetic profile may dictate drug distribution in the central nervous system.

379286 Scalable Generation of Functional Pancreatic β Cells from Human Pluripotent Stem Cells for Tissue Engineering and Drug Screening Applications

Abstract:

The generation of functional, insulin-secreting pancreatic β cells from stem cells *in vitro* would provide an unprecedented cell source for cell replacement therapy and drug screening in diabetes. However, insulin-producing cells generated *in vitro* from human pluripotent stem cells, including embryonic stem (ES) and induced pluripotent stem (iPS) cells, lack many characteristics of bona fide β cells, including glucose-stimulated insulin secretion, appropriate marker expression, and rapid function after transplantation. Furthermore, production at a scale for practical use is highly desirable and technically challenging.

We have developed an *in vitro*, suspension-based protocol that from both ES and iPS cells produces β cells, termed stem cell-β cells (SC-β), in a 300-mL spinner flask using defined factors. SC-β cells increased intracellular [Ca^{2+}] and secreted insulin in response to multiple sequential high glucose challenges at amounts comparable to cadaveric β cells. Insulin was packaged into secretary granules in a manner that was indistinguishable from cadaveric β cells when visualized by electron microscopy. In addition to insulin, SC-β cells expressed the β cell markers PDX1, MAFA, and NKX6-1 and did not express other non-β cell markers, including glucagon, somatostatin, ghrelin, and pancreatic polypeptide. This scalable protocol produced approximately $3x10^8$ cells, of which 30-50% are SC-β cells.

As soon as two weeks after transplantation into SCID/Beige mice, high amounts of human insulin was detected in the serum of mice that received SC-β cells, similar to that found in mice transplanted with cadaveric β cells. The amount of human insulin increased after intraperitoneal injection of glucose, demonstrating the cells were functional *in vivo*. No human insulin was detected in mice transplanted with hESC-derived pancreatic progenitors and non-functional insulin-producing endocrine, consistent with previous findings. When transplanted into diabetic NRG-Akita mice, SC-β cells rapidly reversed hyperglycemia and maintained normoglycemia for 4 months.

To further demonstrate the utility of these SC-β cells, we treated SC-β cells with different categories of known anti-diabetic drugs that increase insulin secretion and with prolactin, which increases β cell proliferation. Increasing β cell function and mass are highly sought after drug targets for treating diabetes. Treatment with anti-diabetic drugs increased insulin release from SC-β cells. In addition, treatment with prolactin increased proliferation of SC-β cells, as indicated by an increase in Ki67 staining.

We have developed a scalable process to generate functional SC-β cells *in vitro* that are virtually indistinguishable from cadaveric β cells. This protocol is able to produce $>10^8$ SC-β cells per batch, meaning it is the first scalable human source of functional β cells for therapy and drug discovery. Given that the only alternative is the very limited and unreliable supply of cadaveric β cells, this discovery represents a major advance in stem cell engineering for treating diabetes.

The use of human pluripotent stem cells (hPSCs) provide great potential to produce large quantities of cardiomyocytes *in vitro* which can potentially be used for high-throughput pharmaceutical drug testing and myocardial repair. For either application, highly reproducible 3D tissues, that can mimic the native myocardium, are desirable to better mimic the heart's natural physiology. Initially, stem cell differentiation protocols used a 3D embryoid body (EB) aggregation approach, resembling developmental steps primarily in their structural nature. Due to several obstacles, i.e. insufficient cardiomyocyte production and reproducibility, state-of-the-art cardiac differentiation protocols transitioned into 2D sheets and a fully defined media including small molecules. With the ability to successfully produce stem cell derived cardiomyocytes, improvements in total cardiomyocyte number per area, cardiomyocyte maturation, and fully functional 3D cardiac tissues with minimum cell handling steps by overcoming 2D cardiomyocyte dissociation, is desired. Thus, the objective of this study was to create 3D hydrogels to culture and directly differentiate human induced pluripotent stem cells (hiPSCs) into mature, aligned, and synchronously contracting engineered cardiac tissues.

IMR90-1 hiPSCs were cultured and expanded on hESC qualified Matrigel in mTeSR-1 media. HiPSCs were dissociated into clumps or single cells, counted, centrifuged, and resuspended in PEG-fibrinogen precursor solution at 60×10^6 cells/mL, which consisted of PEG-fibrinogen polymer precursor, triethanolamine (TEOA), N-vinylpyrrolidone (NVP), and the photoinitiator Eosin Y. The cell-PEG-fibrinogen mixture was transferred to a polydimethylsiloxane (PDMS) mold on acrylated glass and photocrosslinked using a visible light source to form 3D microislands. Microislands were cultured in mTeSR-1 media for three days followed by initiation of differentiation (day 0). Stem cell viability 24 hours post-encapsulation was assessed using LIVE/DEAD assay, which showed that clump and single hiPSCs remained viable in their new, 3D microenvironment. Starting on day 0, cardiac differentiation timeline follows previously published differentiation protocols (Lian, X. et al, 2012). Temporal changes throughout cardiac differentiation were assessed by phase contrast microscopy, immunohistochemistry, RT-qPCR, transmission electron microscopy (TEM), and electrophysiology. First isolated areas of contractions within microislands were visible on day 7 of differentiation. Spontaneously contracting areas and force of contraction increased over time, resulting in fully and synchronously contracting 3D cardiac microislands which remained their function for over three months. Proliferating cell nuclear antigen (PCNA) immunostaining of microislands on day 10 of differentiation showed that proliferating cells within microislands were primarily located on tissue edges. Cardiac localization, alignment, and maturation were assessed by staining engineered cardiac microislands with α-sarcomeric actinin (aSA, cardiac marker) and connexin 43 (Cx43, gap junction protein) on day 10, day 20, day 30, and three months. Sarcomeres were visible at all four time points but developed a more defined and highly aligned structure within microislands over time and showed mature gap junctions at three months. Pluripotency gene Oct4 and cardiac genes myosin heavy chain 6 and 7 (MYH6, MYH7) expression of hiPSCs, day 0, day 10, and day 20 microislands were compared to time-matched 2D cardiac sheets. RT-qPCR revealed a significant decrease in Oct4 gene expression in day 0 3D microislands and 2D sheets compared to hiPSCs. The MYH7/MYH6 ratio, which was used to determine cardiac maturation, showed an increase in 2D sheets and 3D microislands from day 10 to day 20. Although ratios increased in both culture systems, 3D cultured cells matured at a slower rate compared to 2D cardiomyocytes initially, but showed features of mature cardiomyocytes after three months using TEM. Lastly, contracting microislands were dissociated to obtain electrophysiological characteristics. Recording of calcium transients in human cardiomyocytes were obtained using an IonOptix Myocyte Calcium and Contractility Recording System. Samples were loaded with Fura-2AM dye and perfused with Tyrode's solution. Cardiomyocytes were paced from 0.5-1.5 Hz and stimulated at 30 V. Both 2D and 3D cultured cardiomyocytes showed 1:1 correspondence to electrical stimuli of up to 1.5 Hz.

In this study we showed that clump and single hiPSCs can be cultured and directly differentiated into contracting 3D cardiac tissues using the biomaterial PEG-fibrinogen. In this new, 3D microenvironment hiPSCs remained viable, proliferated, and formed highly reproducible synchronously contracting cardiac tissues. Although cardiac differentiation in 3D progresses slower initially compared to 2D cultured cardiomyocytes, 3D cultured cells responded to pacing frequencies up to 1.5 Hz, showed fully aligned and well defined sarcomeres, and remained contractile functions over several months that led to fully mature cardiomyocytes *in vitro*, which mimics features of native human heart development. This one-step encapsulation procedure of hiPSCs into PEG-fibrinogen produces highly reproducible, physiologically relevant, aligned, and mature 3D engineered cardiac tissues *in vitro* for possible uses in drug-testing or myocardial injections.

385823 The Effect of Alginate Capsule Composition on Pancreatic Differentiation of Human Embryonic Stem Cells

Abstract:

Introduction: Type 1 diabetes (T1D) is an autoimmune disease affecting millions of people worldwide wherein the beta cells of the pancreas are destroyed resulting in insulin dependence. Shortage of donor islets combined with immune rejection limits islet transplantation from becoming a viable therapy. We have shown previously that calcium alginate encapsulation for differentiation of human embryonic stem cells (hESCs) results in efficient differentiation to insulin producing cells. Due to the unlimited supply of hESCs and the immunoisolation capability of alginate capsules, this could be used as an alternative treatment option for T1D. Previous studies for islet encapsulation have shown that alginate capsule composition has a significant effect on the materials capacity to immunoisolate. On the other hand, such capsule composition can modify the differentiation of the encapsulated cells. Hence in this study we are evaluating the effect of such material properties, namely cation concentration and type, on pancreatic differentiation of encapsulated hESCs.

Materials and Methods: We developed a stage wise directed differentiation protocol to derive pancreatic cells from hESCs. Definitive endoderm was induced with ActivinA and Wnt3Afor 4 days followed by pancreatic progenitor induction with Cyclopamine and Cyclopamine with Retinoic Acid for 2 days each, respectively. A single cell suspension of ROCK inhibitor treated hESCs were suspended in 1.1% alginate and 0.2% gelatin followed by encapsulation by drop wise addition into a divalent cation bath. hESC were encapsulated using either $CaCl_2$ or $BaCl_2$ as the crosslinking solution with alginate composed of low guluronic acid content. The hESCs were differentiated entirely under encapsulation and characterized after completion of pancreatic differentiation. Viability and proliferation was analyzed for each encapsulation configuration. Pancreatic commitment was analyzed by PDX1 gene and protein expression.

Results: Our results to date show that encapsulation and pancreatic differentiation of hESCs using standard literature concentrations for $CaCl_2$ (100 mM) or $BaCl_2$ (20 mM) show little difference in viability by the LIVE/DEAD assay. However, upon completion of differentiation, calcium alginate (CAlg) gels contained a small number of cell aggregates which were large and heterogeneous in shape. The barium alginate (BAlg) gels contained many small cell aggregates which were homogenous. This indicates that the CAlg gel is more compliant to cell migration and proliferation, while the BAlg gel is more restrictive. Gel stiffness measurements by atomic force microscopy show the BAlg gel to be approximately 2 fold stiffer than the CAlg gel. Gene expression analysis for *PDX1* was significantly higher when hESCs were differentiated in CAlg gels. Lowering the $BaCl_2$ concentration to 10 mM yielded a gel stiffness similar to the $CaCl_2$ gel, and subsequently resulted in increased differentiation.

Discussion: Our results indicate that even in the presence of chemical factors, substrate stiffness greatly affects the efficiency of pancreatic differentiation. Many protocols in the literature use barium alginate for islet encapsulation due to its increased immunoisolation. However, using this gel for pancreatic differentiation of hESC hinders differentiation. Further studies are currently underway to evaluate different alginate types and cations, as well as using finite element modeling to characterize the local stresses felt by the cell aggregates and how gel stiffness (or composition) affects this.

388851 Hydrogels for Controlling Neural Stem Cell Fate through Intracellular Redox State

Abstract:

Neurological injuries and diseases often result in an inflammatory environment hostile toward regeneration. This causes an acute release of reactive oxygen species (ROS) that can result in a secondary insult and toxicity for transplanted cells. Even at sub-lethal levels, an increase in extracellular ROS can change the intracellular redox state and affect key intracellular pathways. Here we focus on the redox capacity of the hydrolytically degradable poly(lactic acid)-b-PEG-b-poly(lactic acid) dimethacrylate block copolymer (PEG-PLA) to influence encapsulated NSC survival and function.

As PEG-PLA hydrogels degrade they release lactic acid, which we have previously shown to scavenge free radicals and rescues NSCs from free radical-induced death in a concentration dependent manner. We systematically synthesized a family of PEG-PLA hydrogels ranging from 0-100% degradable content to quantify the release of lactic acid into the medium and its neuroprotective capabilities. While keeping the initial (< 24 hours) mechanical properties constant, the degradable macromer content had an immediate (30 minutes) and prolonged (7 days) effect on intracellular redox state indicating fewer intracellular ROS. Cells encapsulated in hydrogels with increasing degradable content were more viable and more proliferative, with the 100% degradable material supporting a three-fold increase in metabolic ATP content and a more than two-fold increase in DNA content compared to the 0% degradable material after 7 days. In these highly degradable 3D hydrogels, we further showed that NSCs had greater concentrations of total glutathione. Our current work extends these findings to create permissive microenvironments for oligodendrocyte precursor cell proliferation and differentiation to facilitate function myelination of regrowing axons. These hydrogels may be beneficial for cell transplantation therapy, where injured host cells as well as transplanted cells may be stimulated to survive via ROS-scavenging properties of the hydrogel.

390556 Direct Reprogramming of Skin Derived Stem Cells into Functional Neural Crest Stem Cell Fate

Abstract:

Introduction: Neural crest stem cells (NCSC) are multipotent cells that arise during embryonic development at the junction of neural and non-neural ectoderm and migrate laterally along the length of developing notochord (Sauka-Spengler & Bronner-Fraser, 2008), in process, giving rise to diverse cell types including peripheral neurons, Schwann cells, melanocytes, corneal endothelium, cartilage, bone, smooth muscle etc. In the present work, we provide first evidence that skin stem cells (SSC) can be coaxed to acquire NCSC fate under chemically defined conditions (without any genetic modification of the SSC) and resulting NCSC efficiently differentiate into functional neural crest (NC) lineages i.e. neurons, smooth muscle cells (SMC), osteocytes, chondrocytes, adipocytes, and endothelium.

Materials & Methods: Human neonatal foreskin SSC and adult skin SSC were isolated by separating epidermis from the dermis by enzymatic treatment and grown either on 3T3-J2 murine fribroblasts feeder (Rheinwald & Green, 1975) as colonies or in SSC serum free medium as feeder-free culture. SSC specific promoter driven puromycin cassette was cloned into a lentiviral vector and SSC were transduced and puromycin selected to isolate bona fide SSC. To induce SSC into neural crest fate, SSC were cultured in type1 collagen coated dishes in presence NC induction medium (NCIM) for 8-12 days. SSC derived neural crest stem cells (SSC-NCSC) were analysed transcriptionally (through global RNA sequencing (RNASeq) and quantitative real time PCR (qRT-PCR) as well as translationally (through flow cytometry, immunocytochemistry, and immunoblotting). To demonstrate that SSC possess the ability to undergo NC transformation at single cell level, clonal cultures of SSC were established and induced to acquire NC fate by NCIM treatment. To establish the multipotentiality of SSC-NCSC, they were further induced to differentiate into neuronal, myogenic, osteogenic, chondrogenic and endothelium fate by lineage specific differentiation medium (DM)/condition and characterized molecularly as well as functionally.

Results and Discussion: SSC promoter-Puro selected homogenous population of SSC (Fig1.A) grew as epithelial colonies and after 8-12 days of induction treatment small spindle shaped cells appeared to be delaminating from SSC colonies (Fig1. A). These emigrating cells were termed as SSC-NCSC due to their similarity with NCSC in terms of transcriptional and translational profile (Fig. 1B, 1C). The key NC markers i.e. SOX10, PAX3, NES, FOXD3, MSX1, IRX3, B3GAT1 and, NGFR were upregulated in SSC-NCSC while SSC markers were downregulated (Fig. 1B, 1C). Global RNASeq analysis shows SSC-NCSC clustered close to human embryonic stem cell (hSSC) derived NCSC. SSC-NCSC differentiated into mesenchymal lineages i.e. osteocytes, chondrocytes and adipocytes as evidenced by RT-PCR analysis and functional stains von kossa, oil red O and alcian blue, respectively (Fig1D, non-induced controls are in inset). SSC-NCSC also differentiated towards functional SMC as they expressed ACTA2, CNN1 and MYH11 (Fig1E, ACTA2) and fibrin based vascular media engineered using SSC-NCSC derived SMC displayed contractility in response to receptor mediated and non-receptor mediated vasoagonists. Approximately, 23% SSC-NCSC expressed CD31 and could uptake LDL and formed tube on Matrigel suggestive of endothelial potential of SSC-NC (Fig.1E). Finally, SSC-NCSC were successfully differentiated towards peripheral neurons as suggested by immunostaining (TUBB3, TH, NeuN, Peripherin) (Fig1.F, TUBB3). Electrophysiological analysis showed that SSC-NCSC derived neurons could generate action potentials upon stimulation indicating their functional maturity (Fig1.F). SSC colony derived from single cell could also be induced into NC fate suggesting clonal transdifferentiation potential of SSC. In addition, adult SSC from (45 year old, male) also gave rise to SSC-NCSC suggesting limited effect of aging on NC potential of SSC.

Conclusions: This is first report of direct reprogramming of SSC into functional neural crest fate which has implications in stem cell biology, direct reprogramming and regenerative medicine.

383573 Resolving Proliferation-Potency Relationship Identifies Survival Marker for Multipotent Mesenchymal Stem Cells

Abstract:

Heterogeneity in cell survival is a major impediment to realizing the therapeutic potential of mesenchymal stem cells (MSCs). The present study resolves the proliferation-potency relationship for MSCs at the clonal level and employs this knowledge to identify a novel cell-surface marker to enrich multipotent MSCs with enhanced survival. Clonal analysis of the proliferation-potency relationship was performed with a unique high-capacity assay that generates single-cell derived MSC colonies that are classified according to their trilineage potential to exhibit adipo- (A), chondro- (C), and osteogenesis (O) as a measure of potency. The proliferation potential of MSC clones was analyzed from complementary approaches of colony-forming efficiency, ex vivo expansion, senescence and cell death. These different approaches yielded similar results, which makes a compelling argument for the following proliferation-potency relationship that was detected. There was a broad range of proliferation potentials of clones isolated from early passage MSCs. For example, the median specific growth rate was 0.85 per day (20 h doubling time) for OAC clones and was 5-fold less for O clones ($P<0.01$). As multipotent MSCs progressed through a complex hierarchy of lineage commitment, proliferation potential correlated to overall potency, but it was independent of the specific lineage pathway of commitment in that OA and OC clones had similar proliferation potentials. The gradient in this correlation was pronounced with highly clonogenic OAC clones at one extreme and senescent O clones at the other. In fact, more than 75% of cells in O clonal cultures stained positive for senescence-associated beta-galactosidase activity vs. less than 10% for OAC clones ($P<0.001$). The proliferation-potency relationship served as a template to identify neuron-glial antigen 2 (NG2) as a cell-surface marker for MSC survival. Fluorescence-activated cell sorting revealed that MSCs selected for low scatter and high NG2 surface expression had a 50% improvement in clonogenic survival in vitro relative to low scatter MSCs and a 100% improvement relative to parental MSCs. This is the first report of a role for NG2 in the survival of normal diploid cells. The relationship between NG2 and MSC survival is particularly significant because cell survival is a basic prerequisite for any MSC therapy to be effective. As such, there are numerous basic research and clinical applications for NG2 as a MSC survival marker. This research was funded by NIH (R03EB007281) and NSF (CBET-1066167).

Metabolic and Phenotypic Heterogeneity of Human Mesenchymal Stem Cells in Tissue Engineering

Yijun Liu, [1] Nathalie Muñoz, [2]
Timothy M. Logan[2], Teng Ma[1],
[1] Department of Chemical and Biomedical Engineering [2] Department of Chemistry and Biochemistry,
Florida State University

Introduction

Stem cells have unique metabolic characteristics that play an essential role in maintaining their self-renewal and multi-potentiality. Human mesenchymal stem cells (hMSCs) are a primary cell source in cell therapy for a wide range of diseases. However, culture induced changes in hMSC phenotype have been well documented and the suitability of culture-expanded hMSCs for eventual therapeutic application is a significant barrier in hMSC translation to clinical application. hMSCs are intrinsically heterogeneous and composed of slow and rapidly dividing subpopulations that exhibit varying differentiation and functional properties. Immediately after isolation and upon culture expansion, BM-MSCs acquire and accumulate genetic and phenotypic changes and that many proliferating cells progressively enlarge, exit cell cycle, and become senescent. The content of the clonogenic and multipotent subset in heterogeneous hMSC culture is not only correlated with cell proliferation but also with therapeutic outcome. Thus, understanding the metabolic characteristics of hMSC subsets has important implication in optimizing hMSC properties. Our prior studies have shown that oxygen tension significantly regulates hMSC metabolism and influences expansion, secretion of trophic factors, and differentiation [1,2,3]. The objective of present study is to investigate the relation between hMSC phenotypic and metabolic properties to develop an metabolic strategy for hMSC expansion while maintaining their multilineage potential and therapeutic potency.

Materials and Methods

Low density culture was used to enrich the clonogenic subset whereas standard seeding density was used as control. A GC-MS-based metabolic profiling approach was employed to analyze the fate of the ^{13}C labeled glucose and the label incorporation in the major intracellular metabolites involved in glycolysis and TCA cycle in (1) hMSC clonal population, (2) hMSC standard culture. The isotope enrichments in the metabolites were calculated and used to infer activities of specific metabolic enzymes and the associated pathways. In addition, metabolic modulator such as dichloroacetic acid (DCA), 2-deoxyglucose (2-DG), and antimycin A (AmA), were used to selectively modulate the specific metabolic pathways and evaluate their regulatory roles in regulating hMSC phenotypic, secretory, and multi-lineage properties.

Results

The results revealed that clonogenic hMSC subpopulation can be selectively enriched by plating cells in clonal and low density (LD) culture (10 and 100 cells/cm^2, respectively), indicated by elevated expression of CD146, expression of stem cell marker genes, and CFU-F formation compared to cells plated at medium (MD)- or high (HD)-density (1000 and 3000 cells/cm^2, respectively). Global metabolic profiles revealed by gas chromatography-mass spectrometry (GC-MS) of cell extracts showed clear distinction between LD and HD cultures, and exhibited density-dependent differences in coupling of glycolysis to the TCA cycle. Metabolic inhibitors revealed density-dependent differences in the dependence on glycolysis versus oxidative phosphorylation (OXPHOS) for ATP generation, in glutamine metabolism, and in the dependence on the pentose phosphate pathway (PPP) for maintaining cellular redox state and density-dependent sensitivity exogenous reactive oxygen species. We also show that active OXPHOS is not required for proliferation of the clonogenic subset and contributes to cellular senescence during expansion. Together, the results revealed the mechanistic connection between hMSC phenotypic and metabolic heterogeneity and demonstrate that regulation of hMSC metabolism as a strategy in preserving hMSC properties during expansion.

Conclusions

The unique metabolic characteristics of the clonogenic subpopulation suggest a novel approach for optimizing in vitro expansion of hMSCs. hMSCs population heterogeneity is associated with metabolic heterogeneity, with aerobic glycolysis supports expansion of the clonogenic population whereas active OXPHO contributes to cellular senescence during expansion. The identification of metabolic phenotype of the clonogenic subset in the heterogeneous MSC population suggests a metabolic strategy to preserve and amplify the clonogenic subset in hMSC expansion and application.

References

1. Grayson, W.L., Zhao, F., Bunnell, B., Ma, T., Hypoxia enhances proliferation and tissue formation of human mesenchymal stem cells. Biochemical and Biophysical Research Communications, 2007. 358(3): p. 948-953.3.

2. Kim, J., Ma, T., Autocrine fibroblast growth factor 2-mediated interactions between human mesenchymal stem cells and the extracellular matrix under varying oxygen tension. Journal of Cellular Biochemistry, 2013. 114(3): p. 716-727

3. Munoz, N., Kim, J., Liu, Y., Logan, T.M., Ma, T., Gas chromatography-mass spectrometry analysis of human mesenchymal stem cell metabolism during proliferation and osteogenic differentiation under different oxygen tensions. Journal of Biotechnology, 2014. 169: p. 95-102.

377753 Nanoscale Organic Hybrid Materials - Multiscale Structure and Dynamics

Abstract:

Soft colloids have attracted significant interest as model systems for studying glassy fluid structure and dynamics. This attraction arises in part from their less fragile behavior when compared to hard sphere colloids and their ability to reach equilibrium even when in a jammed state. This talk focuses on structure and dynamics of a class of soft colloids created by densely grafting oligomers or ionic liquid molecules to inorganic nanoparticles. Termed Nanoscale organic hybrid materials (NOHMs), these materials form stable, self-suspended suspensions, which allow glassy fluid physics, rheology, and structure to be studied without complications from enthalpic interactions between a solvent and the suspended phase. The effect of "softness" and confinement on the structure and dynamics of the nanoparticle cores and polymeric corona in NOHMs are explored in the talk. In particular, by varying the molecular weight, grafting density, and chemistry of the tethered polymer chains it is shown that jammed materials can be designed with tunable softness, elastic moduli, and ion transport properties that are all strong functions of particle curvature. By manipulating the size and chemistry of the tethered molecules, we further show that the materials can be used as model nanoparticle additives in polymeric hosts to study phase behavior and dynamics.

383070 Surface Properties of Biomimetically Designed Polyelectrolyte Coatings

Abstract:

Controlling interactions and adhesion between surfaces and their environments remains an active challenge in materials science. Desirable properties based on these interactions include surfaces that either demonstrate stickiness to water droplets or ease of shedding droplets, surfaces that resist ice formation or steam condensation are all desirable for various applications. Nature has created numerous materials that might be used as the basis of coatings for these various applications in the surfaces of a number of plants and animals. In the majority of these cases, these superior properties are achieved not with exotic chemistries, but by using a range of textures that enhance the surface properties. For example, the hierarchical nano and microscaled texture used by the lotus leaf to create a self-cleaning, superhydrophobic surface is often cited as an inspiration for artificial surfaces. The wide range of facile techniques available for processing polymers presents an opportunity for the creation of arbitrary surface textures and morphologies. One such example is the "layer-by-layer" (LbL) assembly technique, in which the complexation of oppositely charged polyelectrolytes is directed onto a surface. Using the LbL method films with can be created with a great deal of control as to the placement of film components, which can include nanoparticles or other multivalent materials in addition to polyelectrolytes. Discussed here will be different strategies to create biomimetic polyelectrolyte based coatings that have properties ranging from sticky to slippery with respect to water and other liquids. Our sticky surfaces are a special case of hierarchical texture while our slippery surfaces are created by infusing liquids into a textured surface, replacing the air pockets used in the lotus leaf structure.

371371 Polymerized Ionic Liquid Block Copolymers: Highly Versatile Ion Conductors

Abstract:

Polymerized ionic liquid (PIL) block copolymers constitute a distinct set of block copolymers that are a relatively new type of polymer electrolyte. These new materials have recently been synthesized by several research groups for applications ranging from energy storage and production to gas and water purification. PIL block copolymers combine the advantageous properties of both PILs and block copolymers and are synthetically highly versatile with numerous cation and anion chemistries available. Specifically, the unique physiochemical properties of PILs, such as high solid-state ionic conductivity, high chemical, thermal and electrochemical stability, and widely tunable physical properties (e.g., via anion exchange), are incorporated in the block copolymer architecture, which allows for self-assembly into a range of nanostructures, where morphology type and domain size are tunable. In this talk, a variety of PIL block copolymers synthesized in our research group will be discussed along with the impact of various chemistries and chain architectures on the solid-state morphology and ion conductivity. The application of these materials to fuel cells and batteries will also be discussed.

386793 Nanoscale Confinement Effects on Polymer Properties: Recent Advances in Characterization and Understanding and Important Questions to be Addressed

Abstract:

Polymeric materials are employed in confined states in many important technological applications. These include nanoscale and nanostructured films used in lithographic applications for manufacturing microelectronic chips as well as well-dispersed and/or exfoliated nanocomposites where all polymer chains are less than ~100 nm from a nanofiller interface. The study of how nanoscale confinement affects the properties of polymeric materials is now 20 years old, dating back to a 1994 publication by Keddie, Jones and Cory in which ellipsometry was used to demonstrate a dramatic reduction in the glass transition temperature (Tg) of polystyrene films supported on silica when thickness decreased below ~40 nm. Since that time, major nanoscale confinement effects have been demonstrated in a variety of polymer properties, including Tg, physical aging and modulus. Furthermore, a variety of studies have contributed to demonstrating the tunability of such effects based on changes in polymer repeat unit structure, the presence (and strength) or absence of attractive polymer/substrate or nanofiller interactions, and the addition of small molecule diluent. Nevertheless, major questions have remained unanswered in the field, including why different polymer molecules that lack attractive interactions with a substrate (or have no substrate, as with a free-standing film) exhibit starkly different magnitudes of the Tg-confinement effects. As well, major stumbling blocks to certain types of characterization, such as the use of conventional differential scanning calorimetry (DSC) for quantifying the effects of confinement on Tg and physical aging, have hampered progress in the field.

Here, I shall present recent progress from my research group that has positively addressed such questions and overcome such stumbling blocks. In particular, I shall describe work that demonstrates the important role of the fragility of a polymer in influencing both the magnitude of the Tg-confinement effect in supported polymers and the film thickness at which such effects first become noticeable. The observed effect is that the more fragile the polymer, the greater the Tg-confinement effect. I shall explain the underlying cause behind this key role of fragility on confinement effects and also demonstrate that some polymer system systems with very low fragilities exhibit no confinement effect for films as thin as 20 nm. I shall also describe how we have developed a new method to characterize the effect of confinement on polymer fragility via ellipsometry. With this new approach, we have demonstrated a dramatic decrease in fragility with nanoscale film thickness and that limiting values are approximately achieved in sufficiently thin films.

I shall also show how differential scanning calorimetry can be easily employed to characterize Tg-confinement effects in polymeric nanotubes made using anodized alumina templates. The Tg-confinement effects measured by DSC for polystyrene nanotubes made by melt infiltration show excellent correspondence with Tg-confinement effects determined from fluorescence measurements on the same nanotubes as well as with Tg-confinement effects of supported polystyrene films. With this advance, it will be relatively easy in the future for any research team having access to DSC to study and characterize confinement effects. Additionally, we shall demonstrate how changing the substrate can result in Tg-confinement effects for polystyrene films changing sign. For silica substrates which have no attractive interactions with polystyrene, Tg decreases with decreasing thickness below ~40 nm. We have found that carbon-based substrates can result in increases in Tg of polystyrene films of the same thickness that show decreases in Tg when placed on silica. This is because of pi-pi bonding interactions between polymer and the carbon-based substrate, a result that has important implications in polymer/carbon-black based composites found in the tire industry as well as novel graphene-based nanocomposites.

Although significant progress has been made very recently in this field, major new questions have emerged. Possibly most important among these is why thermodynamic and pseudo-thermodynamic methods of Tg measurement, including ellipsometry, x-ray and neutron reflectivity, fluorescence, capacitive dilatometry, and DSC, report strong effects of confinement on polymer properties such as Tg while measures of dynamics related to the alpha-relaxation process show suppressed effects. In bulk materials, there is a strong correlation between the measured Tg and dynamics that is apparently less robust or missing at the nanoscale. I shall highlight this issue and other questions that will be important for researchers to explore in the coming decade.

383163 Accessing Polymorphism and Tuning Molecular Orientation of Molecular Semiconductors through Post-Deposition Processing

Abstract:

Though both the molecular orientation and crystal structure of molecular semiconductors are known to impact charge transport in thin-film devices, controllably accessing different polymorphs and tuning both the in-plane and out-of-plane molecular orientations remain challenging, requiring careful selection of film deposition conditions, film thickness, and substrate chemistry. Without necessitating any specification of these parameters, we have been able to impose specific polymorphs and/or preferential molecular orientation through post-deposition processing.

In the case of contorted hexabenzocoronene (HBC), we can access two distinct polymorphs and independently vary the distributions of its out-of plane molecular orientation. Starting with a thermally evaporated amorphous HBC film, thermal and solvent-vapor annealing induces the thin film to crystallize into distinct polymorphs; we can further transform between the polymorphs with subsequent rounds of processing. Because the molecular orientation distribution is primarily determined during the first processing step, subjecting already crystalline thin films to different post-deposition processing conditions allows us to access different polymorphs without altering the molecular orientation distribution, thus enabling us to decouple the influence of each on charge transport.

In the case of solution-processable triethylsilylethnyl anthradithiophene (TES ADT), we have been able to exploit its differential crystallization rates on substrates having different surface energies to guide the in-plane growth direction of neighboring crystals. The systematic engineering of solely low- and high-angle grain boundaries between neighboring crystals have allowed us to elucidate their relative contribution to charge transport.

392048 Exciton Engineering - Using Transport and Reaction Engineering to Solve Problems in Solar Energy

Abstract:

The commercial successes of conventional solar cells (SCs) have emboldened researchers to raise efficiency, reduce cost, and develop new functionalities for next-generation photovoltaics (PVs). Several parallel fields, *e.g.*, bulk crystal SCs, dye-sensitized SCs, organic PVs, are approaching these challenges from different angles. Nanostructured PVs, such as nanowire or semiconducting single-walled nanotube (SWNT), offer tremendous potential through reduced materials costs, scale-able fabrication, and unique and tunable properties. In this presentation, I will argue that chemical engineers are uniquely positioned to use their foundational skills in understanding population balances, transport, and reactivity to analyze solar energy harvesting systems. The exciton is a quasiparticle electron-hole pair whose diffusion and "reaction" to other entities is at the heart of photocatalysis and photoconduction in many systems. Excitons are produced when photons interact with some forms of matter, and their diffusion can be described using recognizable variants of Ficks law as they seek to minimize their chemical potential gradient. Excitons "chemically react" at surfaces to produce electrons and holes in a process that is central to photoconduction. They decay to phonons as they heat the surrounding lattice and radiative recombine to often emit a detectable photon, allowing precision measurement of their kinetics. This presentation will introduce the concept of "Exciton Engineering" – a term that we have used in our laboratory for the past 5 years. I will show that excitons have a Thiele modulus for the first time, and that reaction engineering concepts can be utilized to optimize the design of solar energy devices. Validation of this approach with several experimental systems will be presented, including an exciton antenna functioning as a solar funnel for photon concentration. I will also present our work on the all carbon photovoltaic cell using single walled carbon nanotubes and no other matrix as the photoabsorber for the first time. We have mathematically analyzed current and future nanotube based photovoltatic devices to produce important scaling laws that point towards their optimal design for maximal efficiency. Lastly, I will conclude with a perspective on how Exciton Engineering may lead to next generation solar harvesting devices.

382247 Copper Zinc Tin Sulfide Thin Films for Solar Cells

Abstract:

The photovoltaic market is dominated by silicon solar cells but low absorption coefficient of silicon necessitates the use of relatively thick (0.1-0.3 mm), high electronic quality and expensive wafers, which increases the cost of solar-to-electric energy conversion. Use of solar cells made from thin semiconductor films is growing rapidly because they are typically a factor of 100 thinner than Si devices and cost less. Amongst the thin film solar cells, those based on the copper indium gallium selenide ($CuIn_{1-x}Ga_xSe_2$ or CIGS) and CdTe absorbers have emerged as highest efficiency and lowest cost alternatives to silicon. However, terawatt-level production of these thin film solar cells may be limited by the scarcity of Te and In. Thus, an important problem in developing the next generation thin film solar cells is the synthesis of direct band gap semiconductors that are made of nontoxic and abundant elements. The rapid rise in power conversion efficiencies of thin-film solar cells based on Cu_2ZnSnS_4 (CZTS), $Cu_2ZnSnSe_4$ (CZTSe) and their alloys $Cu_2ZnSn(S_{1-x}Se_x)_4$ (CZTSSe) attests to their potential as low-cost, earth-abundant alternatives to CIGS and CdTe. To date, many thin CZTS and CZTSe film formation methods have been developed and solar cells with efficiencies exceeding 12% have been demonstrated. Increasing solar cell efficiencies relies heavily upon trial-and-error optimization of multiple film properties during the film's synthesis. Ideally, a monolayer of single crystal grains with sizes on the order of the film thickness is required to reduce charge recombination at grain boundaries. This talk will focus on elucidating the factors that affect the microstructure and composition of thin CZTS and CZTSSe films formed by annealing, in a sulfur or selenium containing atmosphere, of thin films deposited from colloidal CZTS nanocrystal dispersions. In this method, 20 to 30 nm diameter CZTS nanocrystals are synthesized in hot (150-300 °C) oleylamine from copper, zinc, and tin diethyldithiocarbamates. Following, thin (2-3 mm) coatings are drop cast from colloidal dispersions of these nanocrystals, which are then annealed in a well-controlled closed system to form large grain microcrystalline films. Abnormal and normal grain growth compete with each other and lead to bimodal grain size distributions. The factors that affect the abnormal and normal grain growth rates include, the temperature, the substrate, and the sulfur (selenium) vapor pressures. Substrates containing alkali impurities lead to larger grains at lower temperatures. Annealing films in selenium leads to the formation of a layer of 2-5 μm size CZTSSe grains on top of a nanocrystalline layer that is rich in carbon: the carbon originates from the dispersion-stabilizing ligands on NC surfaces. Tendency of CSe_2 to polymerize leads to carbon segregation at the CZTSSe-substrate interface. In contrast, films annealed with sulfur do not show such distinct carbon-rich layers and most of the carbon volatilizes from the film during annealing. Careful balancing of these mechanisms is required to obtain films with microstructures suitable for solar cells.

Synthesis and Integration of
Functionally Enhanced Complex Material Systems

Jane P. Chang*

Department of Chemical and Biomolecular Engineering
University of California, Los Angeles (UCLA)
Los Angeles, CA 90095

The introduction of novel and functionally improved materials into silicon based integrated circuits is a major driver to enable the continued down-scaling of circuit density and performance enhancement in analog, logic, memory and sensing devices. Many novel materials, such as complex metal oxides, magnetic materials and phase change materials, are introducing new and unique functionalities that can enable continued technological advancement. The demand of engineering and precision control of these multi-functional materials and their interfaces at an atomic level has grown immensely due to their versatile applications. However, many of these materials are much harder to pattern, thereby posing significant challenges to their large scale integration. In this talk, I will discuss current research advances in atomic layer deposition for synthesizing multifunction and complex metal oxides with tailored electronic, chemical, interfacial, thermal properties and microstructures. Specifically, I will highlight our most recent research on the engineering of multi-ferroic materials for applications in memory, antenna and motors. For integration, I will discuss a generalized approach based on combined thermodynamic assessment and kinetic validation to aid the selection of viable etch chemistry leading to enable the high fidelity pattern transfer for large scale integration.

* corresponding author e-mail: jpchang@ucla.edu

Poly-L-Arginine Based Materials As Instructive Substrates for Fibroblasts Synthesis of Collagen

Kaitlin M. Bratlie[1,2]

[1]Department of Material Science & Engineering, Iowa State University, Ames, IA, USA
[2] Department of Chemical & Biological Engineering, Iowa State University, Ames, IA, USA

The properties of substrates in the extracellular microenvironment affect cells and are key factors in controlling cellular behavior and responses.[1] Implanted biomaterials, tissue engineering constructs, and artificial organs all interact to some extent with surrounding tissues and cells. Ultimately, their *in vivo* fate depends on the outcome of this interaction. The impact of biomaterials on cells has been widely studied;[2–5] however, the ability to tune the response of fibroblasts for applications in wound healing and tissue engineering, in particular the substrates ability to alter the angle of collagen deposition through changing chemical functional group alone has not been explored.

The library of materials used here is based on the basic amino acid arginine. Arginine-rich peptides have been extensively explored as intracellular delivery vehicles as their guanidine functionality allow them to interact with the phosphate groups in cell membranes.[16–18] Thus, we have developed a library based on arginine on the premise that interactions between the arginine moieties and the cell membrane may result in altered cell morphology, cytokine production, and collagen fiber deposition angles by fibroblasts.

In this work, the ability to tune the fibroblast cell response in particular was examined by measuring cell morphology, cytokine production, and collagen fiber deposition angles in response to a library of arginine-based materials. The data presented here shows large variation of VEGF secretion (~600 ng/mL to 3200 ng/mL), collagen fiber orientation (0.003 < polarization anisotropy <0.3357), cell migration (~15 < persistence time < 120 min, 0.11 < speed < 0.23 μm/min), and cell morphology (0.0387 < form factor < 0.1072). These findings demonstrate important progress in the ability to control cellular responses through biomaterials and the potential to attain the desired outcome with exposure to these materials in wound healing and tissue engineering.

1. Lord, M. S., Foss, M. & Besenbacher, F. Influence of nanoscale surface topography on protein adsorption and cellular response. *Nano Today* **5,** 66–78 (2010).

2. Acharya, A. P., Dolgova, N. V, Clare-Salzler, M. J. & Keselowsky, B. G. Adhesive substrate-modulation of adaptive immune responses. *Biomaterials* **29,** 4736–50 (2008).

3. Schutte, R. J., Xie, L., Klitzman, B. & Reichert, W. M. In vivo cytokine-associated responses to biomaterials. *Biomaterials* **30,** 160–168 (2009).

4. Anderson, J. M., Rodriguez, A. & Chang, D. T. Foreign body reaction to biomaterials. *Semin. Immunol.* **20,** 86–100 (2008).

5. Nilsson, B., Ekdahl, K. N., Mollnes, T. E. & Lambris, J. D. The role of complement in biomaterial-induced inflammation. *Mol. Immunol.* **44,** 82–94 (2007).

(112b) Role of 3D Matrix Stiffness and Matrix Adhesions in Regulating the Response of Human Cancer Cells to Cytotoxic Compounds

Kalpith Ramamoorthi[1], Yasaman Chehreghanianzabi[2], Silviya Petrova Zustiak[2,], and Prashanth Asuri[1,]*, [1]Department of Bioengineering, Santa Clara University, Santa Clara, CA, [2]Department of Biomedical Engineering, Saint Louis University, St. Louis, MO*

Abstract

There is an immediate need to develop highly predictive *in vitro* cell-based assays that provide reliable information on cancer drug efficacy and toxicity. Existing *in vitro* screens are typically performed under two-dimensional (2D) cell culture conditions that do not adequately capture the complexity of *in vivo* tissue, thus impacting their capacity to accurately and reproducibly predict human response to drug candidates. Development and use of biomaterial-based three-dimensional (3D) culture platforms has recently gained much scientific interest as research indicates that 3D cultures of cancer cells can adequately mimic the *in vivo* tumor conditions. Moreover, it has been recognized that the biophysical and biochemical properties of 3D culture matrices can play key roles in regulating various cancer cell fates, including invasion and proliferation as well as their response to chemicals. Therefore, using alginate/collagen-based scaffolds of varying mechanical stiffness and presenting different densities of cell-binding ligands, we explored the role of microenvironmental biophysical cues on human glioblastoma cell response to cytotoxic compounds. These experiments suggested the ability of both matrix stiffness and cell matrix adhesions to strongly influence cell response to cytotoxins. To confirm the roles of matrix stiffness and integrin-mediated cell adhesions, and not drug diffusion on the altered drug responses, we performed fluorescence correlation spectroscopy experiments that showed nearly identical drug diffusion rates for the various alginate/collagen scaffolds used in the study. Additionally, our preliminary mechanistic experiments using inhibitors of integrin-mediated adhesion implied a strong role of integrin-mediated adhesions in dictating the toxic response of human glioblastoma cells. This study therefore indicates that 3D matrix stiffness and cell matrix adhesions are important parameters in the design of more predictive *in vitro* cell-based platforms for drug development and toxicity screening.

Introduction

Advances in target-oriented, combinatorial, and diversity-oriented synthesis has revolutionized the pharmaceutical industry by rapidly expanding the chemical space for the discovery of new cancer drugs [1,2]; however, we have yet to see a commensurate increase in the number of FDA approved drugs. Current drug development pipelines require 12-15 years and billions of dollars to bring a drug to market [3]. Additionally, drugs might fail at later stages of development as a direct result from the poor predictive capacity of the current drug efficacy and toxicity screening methods. Central to this problem is the use of *in vivo* animal models that are time-consuming, expensive, and low-throughput, and not entirely reliable in predicting drug efficacy and toxic responses in humans [4]. For these reasons, development of alternate biomaterial-based platforms that adequately mimic *in vivo* tumor microenvironments and can provide high concordance between *in vitro* assays and *in vivo* human responses in a reproducible and reliable manner is needed. In order to develop relevant *in vitro* testing models that provide reliable information on compound efficacy and safety, it is critical that these models sufficiently capture the properties of the *in vivo* tumor microenvironment. Over the last decade or so, it has become clear that existing *in vitro* cell-based screens typically performed under 2D culture conditions do not adequately capture the complexity of *in vivo* tissue, and thereby fail to reproducibly predict human response to chemical drugs [5].

Biomaterial-based 3D cultures of cancer cells have recently been proposed as relevant models to study tumor progression and metastasis owing to their ability to adequately mimic the *in vivo* tumor conditions [6,7]. It has also been suggested that such 3D culture models may also better identify oncogenic pathways for the discovery of new chemotherapeutics [8,9]. 3D microenvironmental cues in guiding cell responses to drugs has also gained much scientific interest as research indicates that biophysical and biochemical properties of the cellular microenvironment can significantly alter cell response to drugs and toxins. In addition to the already well-established role of drug diffusion on attenuated drug response [10,11], differential drug responses in 3D culture conditions relative to 2D have also been attributed to presentation of integrin binding sites and matrix stiffness [12-15]. However, it is not clear if these different matrix properties act independently or synergistically to regulate cell responses. Furthermore, the mechanisms behind altered responses are not fully established. Addressing these gaps in the knowledge base will allow development of reliable cell culture assay platforms for drug discovery and toxicity testing.

Here, we used 3D alginate/collagen hybrid hydrogel scaffolds that allow independent variation of matrix stiffness and densities of integrin-binding ligands as a flexible platform to explore the role of 3D matrix properties on cell responses to drugs. Human glioblastoma cells were chosen as the model cell type to investigate the role of 3D matrix properties on the toxic response. These studies indicated that both matrix stiffness and cell matrix adhesions strongly influenced glioblastoma cell response to cytotoxic compounds, with higher IC_{50} values observed for cells encapsulated in stiff hydrogel matrices as well as hydrogel matrices presenting ligands for cell adhesion. We also performed diffusion assays that confirmed that the rates of chemical diffusion in the various hydrogel preparations were not different and did not contribute to the differential response to toxins. Finally, inhibition of integrin attachment using RGD peptide led to a decrease in the tolerance of glioblastoma cells for the toxic compounds, thereby confirming the role of cell matrix adhesions on the toxic response.

Materials and methods

Materials

Human U-251 and U-87 glioblastoma cells were obtained from ATCC (Manassas, VA). Dulbecco's modified Eagle medium (DMEM) was purchased from Mediatech (Manassas, VA); clear DMEM (no phenol red) from Sigma Aldrich (Saint Louis, MO); fetal bovine serum and penicillin–streptomycin from Invitrogen (Carlsbad, CA). Medium viscosity alginate was purchased from Sigma Aldrich (St. Louis, MO) and rat tail Collagen I from Life Technologies (Grand Island, NY). Synthetic RGD peptide (inhibits integrin-mediated cell binding) was purchased from Selleck Chemicals (Houston, TX). The test chemical compounds, acrylamide and cadmium chloride for the toxicity assays and Rhodamine 6G (R6G) for FCS were all obtained from Sigma Aldrich (St. Louis, MO) and the MTT cell proliferation assay kit was from ATCC (Manassas, VA).

Preparation and characterization of hydrogel scaffolds

Ice-cold stock solutions of alginate (3% w/v) and collagen (3 mg/mL) were first mixed gently on ice to yield 0.25 or 1% w/v alginate containing 0, 0.5, or 1 mg/mL collagen, before the addition of 1M NaOH to adjust the mixture to neutral pH. The alginate/collagen mixtures were pipetted into molds, followed by the addition of ice-cold 100 mM $CaCl_2$ solution to initiate alginate gelation. After 10 min, $CaCl_2$ was replaced with 200 μL of PBS. The hydrogel mixtures were then incubated for 24 h in a 37 °C 5% CO_2 incubator for collagen gelation. Rheological measurements of the alginate/collagen hydrogels were carried out in an Anton Paar MCR302 rotational rheometer at 37 °C using parallel plate geometry. The hydrogel samples were carefully loaded onto the rheometer using a spatula and excess water was removed by gently pressing with tissue paper. The samples were allowed to equilibrate for 5 min before starting the measurements. First, amplitude sweeps at a constant frequency of 1 Hz were performed to ensure that the measurements were carried out in the linear viscoelastic regimes for the alginate samples. Dynamic sweep tests over frequencies ranging from 0.1 to 100 Hz were then recorded in the linear viscoelastic regimes (strain amplitude = 0.01) to determine the shear storage modulus, G' that relates to the elastic behavior of a viscoelastic material and is a measure of its stiffness.

Cell encapsulation in hydrogel scaffolds

The glioblastoma cells were maintained in Dulbecco's modified Eagle medium supplemented with 10% fetal bovine serum and 1% penicillin–streptomycin at 37 °C in a 5% CO_2 humidified environment. The cells were grown in 100 mm tissue culture dishes to 60–80% confluency and subcultured at a 1:4 ratio with 0.25% trypsin. For cell encapsulation in hydrogel scaffolds, equal volumes of at 600,000 cells/mL in culture media and alginate/collagen solutions were first mixed gently on ice. 50 μL of the cell-hydrogel mixture was pipetted into wells of a 96-well plate, followed by the addition of 100 μL of ice-cold 100 mM $CaCl_2$ solution. After 10 min, $CaCl_2$ was replaced with 100 μL of fresh media, before transferring the well plates to a 37 °C 5% CO_2 incubator for collagen gelation. The cells were cultured for at least 48 h in the hydrogel scaffolds to acclimate to the 3D environment prior to further testing.

Cell toxicity assays

Cell toxic response to test chemicals was measured using the formazan-based MTT assay as per the manufacturer's instructions, and the formazan absorbance was measured at 570 nm on a Tecan Infinite 200 PRO spectrophotometer. The test chemical concentrations were selected based on previous literature: acrylamide (0.01–100 mM) and cadmium chloride (0.1–1000 μM). Glioblastoma cells encapsulated in hydrogel scaffolds were exposed to the chemicals for 24 h before being assayed for viability; all measurements were performed using six different concentrations in triplicate and reported relative to the value for cells in the hydrogel scaffolds not exposed to the test chemicals.

Preparation of hydrogels for fluorescence correlation spectroscopy measurements

Alginate solutions of concentrations 0.25% and 1% w/v were prepared by adding calculated amount of alginate, 2.5 mg and 10 mg, respectively, to 1 mL of clear media (without phenol red) and placed in a 37 °C water bath overnight to obtain uniform solutions. 200 μL of each solution was pipetted in chambered cover glass (Thermo Scientific); 50 μL of $CaCl_2$ was pipetted on top of the gel solutions and left for 10 minutes. The $CaCl_2$ solution was then aspirated and the gels were soaked in supplemented clear media (10% FBS and 1% penicillin–streptomycin) containing 10 nM Rhodamine 6G for 24 hours before measurements. We also prepared 0.25% and 1% alginate gels containing 0.5 mg/mL collagen. The gels were prepared as described above except that 0.5 mg/mL ice-cold collagen was added to the gel precursor solution prior to the addition of $CaCl_2$.

FCS theory and instrumentation

The FCS technique has been discussed in detail elsewhere [16-19]. FCS has a confocal set-up where the excitation radiation is provided by a laser beam (excitation at 514 nm) and directed into a microscope objective via a dichroic mirror and focused on the sample. As the sample molecules are usually dissolved in aqueous solution, water immersion objectives with a high numerical aperture (ideally > 0.9) are used. The fluorescence light from the sample is collected by the same objective and passed through the dichroic and the emission filter. The pinhole in the image plane (field aperture) blocks any fluorescence light not originating from the focal region, thus providing axial resolution. Afterwards, the light is focused onto the detector, preferably an avalanche photodiode or a photomultiplier with single photon sensitivity. Recorded intensities of fluctuations from fluorescent molecules are correlated by time to give probe translational diffusivity. Analysis of the autocorrelation function, $G(\tau)$, provides information about the mechanisms underlying the intensity fluctuations [16]:

$$G(\tau) = 1 + \frac{1}{N}\frac{1}{\left(1+\frac{\tau}{\tau_d}\right)}\frac{1}{\left(1+p\frac{\tau}{\tau_d}\right)^{0.5}} \qquad (1)$$

where N is the average particle number in the detection volume, τ is the delay time, τ_d is the characteristic diffusion time, and $p=(r_0/z_0)^2$ is an instrumental constant (where r_0 and z_0 are the radius and axial length of the focused laser beam spot, respectively). Assuming a three-dimensional Gaussian profile of the excitation beam, τ_d can be related to diffusivity by the following equation:

$$\tau_d = (r_0)^2 / 4D \qquad (2)$$

FCS measurement and data analysis

Measurements of R6G diffusivity in alginate and alginate/collagen hydrogels were performed on an FCS unit equipped with a 514 nm laser and a Zeiss LSM510 confocal microscope. Measurements were performed at 22 °C in FBS supplemented DMEM medium without phenol red since phenol red interferes with the fluorescent signal from R6G. Acquisition times of 215 s were used to optimize signal-to-noise ratio. Low laser intensity was used throughout to avoid activation of the fluorophore triplet states as well as photobleaching. The instrument was calibrated with R6G (diffusivity for R6G = 2.8×10^{-10} m^2s^{-1} in water). The detection volume was estimated to be 13 fl. The resultant autocorrelation function was fitted with the model depicted in Eq. 1, which describes normal diffusion of a single monodisperse fluorophore, i.e. R6G used in the experiments. Minimum of 5 different spots at various locations (i.e. illuminated volumes) were measured per single hydrogel to obtain an average characteristic diffusion time, τ_d, for each sample.

Results

Preparation and characterization of alginate/collagen scaffolds for glioblastoma culture

To evaluate the role of 3D matrix properties on cell responses to toxic chemicals, we first performed mechanical and biochemical characterization of the hydrogel scaffolds. We developed hybrid scaffolds by using alginate that lacks specific recognition sites for cell adhesion [20] and collagen that facilitates cell matrix adhesions via integrins. Elastic modulus of the alginate/collagen hydrogels was characterized using rotational rheometry. Table 1 shows the values of the storage modulus (G'), which is a measure of the elastic properties of a viscoelastic material.

Table 1. *Storage modulus of alginate/collagen hydrogels as characterized by rotational rheometry.*

Final concentration in the scaffold		**Storage modulus, *G'* (Pa)**
Alginate (w/v)	Collagen (mg/mL)	
0.25%	0	398 ± 22
	0.5	413 ± 53
	1	1036 ± 69
1%	0	3214 ± 158
	0.5	3383 ± 671
	1	4802 ± 442

The elastic moduli of the alginate/collagen hydrogels were clearly dependent on both alginate and collagen concentrations. Interestingly, addition of collagen had a greater effect on the mechanical properties of the 0.25% alginate relative to the 1% alginate hydrogel, at least for 1 mg/mL collagen. However, the stiffness of hybrid scaffolds prepared using 0.5 mg/mL collagen was comparable to the stiffness of pure alginate scaffolds. Since we were interested in understanding the individual and synergistic roles of both matrix stiffness and cell matrix adhesions on cell responses, we elected to pursue further studies with alginate/collagen hydrogel scaffolds prepared using 0 and 0.5 mg/mL collagen.

Next, we assessed the ability of the hydrogel scaffolds to support cell culture using human U-251 and U-87 glioblastoma cells as the model under defined growth conditions, using serum supplemented culture media at 37 °C in a 5% CO_2 humidified environment. Under these conditions, consistent with previous studies, we observed >80% cell viability after 1 week of culture in the alginate/collagen scaffolds. Further, we also established the ability of alginate/collagen scaffolds to support cell adhesions with the hydrogel matrix and subsequently greater cell spreading. Specifically, as seen in Figure 1, we observed extension of astrocytic processes for glioblastoma cells in alginate/collagen scaffolds, but not in pure alginate scaffolds.

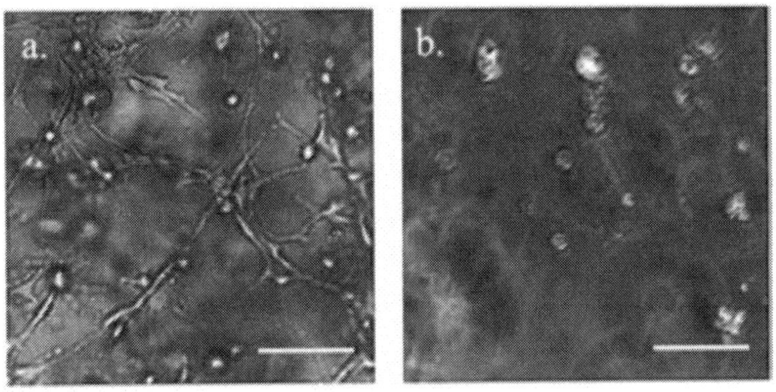

Figure 1. *Phase contrast images taken on day 3 demonstrating that U-87 cells extend processes in (a) 0.25% alginate/0.5 mg/mL collagen hybrid scaffolds, but not (b) pure 0.25% alginate gels. The scale bar depicts 200 μm.*

3D matrix stiffness and cell matrix adhesions influence cell response to chemicals

After establishing the ability of the alginate/collagen scaffolds hybrid hydrogel to support culture of glioblastoma cells, we then proceeded to test the influence of 3D matrix properties on cell response to neurotoxins, namely acrylamide and cadmium chloride. In a previous study, we had demonstrated that glioblastoma cells encapsulated in soft alginate scaffolds were more sensitive to toxic chemicals relative to cells cultured in stiffer alginate scaffolds [14]. Here, we further investigated the effect of cell matrix adhesions facilitated by collagen on the response of glioblastoma cells to toxins. As seen in Figure 2 and consistent with our previous investigations, U-87 and U-251 cells encapsulated in 0.25% alginate scaffolds showed statistically higher sensitivity (lower IC_{50} values) to the toxins relative to those cultured in 1% alginate. Furthermore, experiments using the hybrid scaffolds revealed that 3D cell matrix adhesions significantly enhanced the toxin tolerance of both U-87 and U-251 cells in soft conditions, but did not significantly affect the response of cells in stiff conditions. These results are not entirely surprising, as earlier studies have already demonstrated role of integrin-mediated cell adhesion and integrin signaling pathways in ECM-mediated resistance to chemotherapeutic agents [21,22].

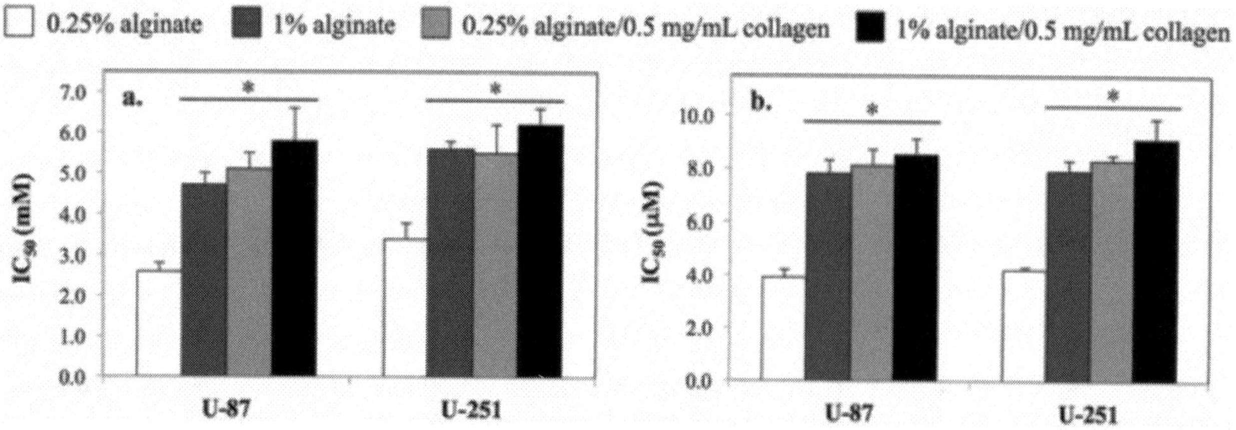

Figure 2. *Half-maximal inhibitory concentration (IC_{50}) of (a) acrylamide and (b) cadmium chloride for U-87 and U-251 cells cultured in 0.25% alginate (white bars), 1% alginate (dark grey bars), 0.25% alginate/0.5 mg/mL collagen (light grey bars), and in 1% alginate/0.5 mg/mL collagen (black bars) scaffolds. *$P<0.01$ for IC_{50} values for cells cultured in 1% alginate and 0.25% or 1% alginate/0.5 mg/mL collagen compared with those cultured in 0.25% alginate, as determined by Student's unpaired two-tailed t test. Error bars represent standard deviation of three samples.*

Taken together, while our data strongly suggested the ability of 3D matrix stiffness and cell matrix adhesions to regulate cell responses to toxins, the role of impeded drug diffusion in 3D scaffolds on reduced drug response has also been well established. Therefore, we performed FCS measurements of a fluorescent dye, namely R6G, diffusing in the various hydrogel preparations described above. R6G is a small molecule with a molecular weight of 479.02 Da (Figure 3) and, thus, is used as a model fluorophore to approximate the diffusion of the small molecular toxins used in this study.

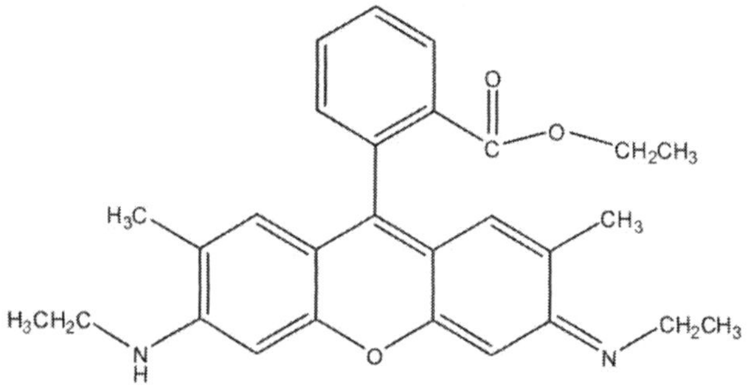

Figure 3. *Chemical structure of Rhodamine 6G used for FCS measurements.*

Normalized autocorrelation functions for R6G in media alone and in the pure alginate hydrogels of different stiffness is presented in Figure 4a. A shift towards longer diffusion times is evident for the autocorrelation functions for both alginate gels as compared to R6G diffusivity in media alone. The residuals on Figure 4 b-d demonstrate that all data is well fitted by a single component fit – Eq. 1. Even though Eq. 1 has been established for fluorophores diffusing in solutions, it is common for the equation to be used in hydrogel systems for the regime where the solute hydrodynamic radius (R_h = 0.8 nm for R6G) is smaller than the mesh size of the hydrogel (alginate is a macroporous hydrogel) [17].

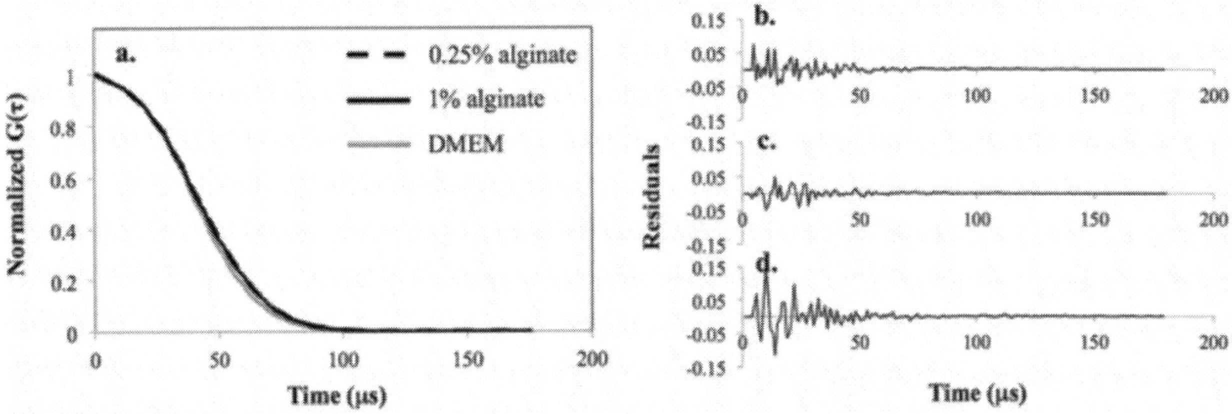

Figure 4: *(a) Normalized autocorrelation functions for R6G diffusion in 0.25% alginate (black dotted line), 1% alginate (black line), or media alone (light grey line); **(b)** 0.25% w/v alginate gel, **(c)** 1% w/v alginate gel, and **(d)** R6G in clear media alone indicate the goodness of the fit.*

The resultant normalized diffusion times, τ_0/τ_d, where τ_0/τ_d is equivalent to the normalized effective diffusion coefficient D_e/D_0, are given on Figure 5. The data indicates that there is no significant difference in the diffusivity of R6G in the different alginate hydrogels, either pure or with collagen, but that the diffusivity is overall attenuated in the hydrogel matrix by ~30% as compared to the diffusivity in media alone.

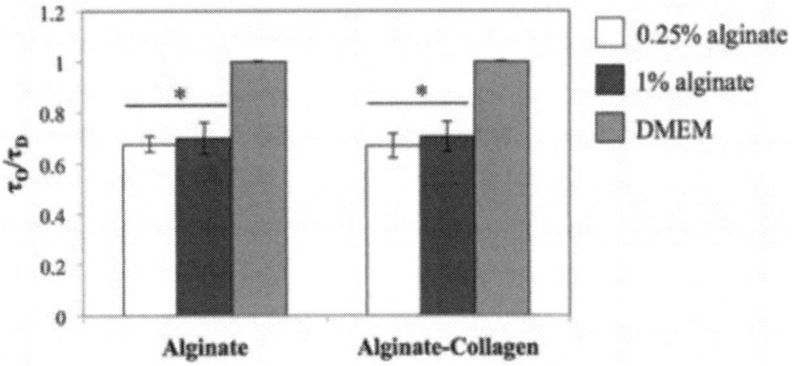

Figure 5. *Normalized diffusivities for R6G in clear media and pure alginate and alginate/collagen hydrogels of 0.25% and 1% w/v as measured by fluorescence correlation spectroscopy. *P<0.01 for diffusivity values for R6G diffusing in 0.25% or 1% alginate and 0.25% or 1% alginate/0.5 mg/mL collagen compared with R6G diffusing in media alone, as determined by Student's unpaired two-tailed t test. Error bars represent standard deviation of three samples with minimum of five measurements per sample.*

Addition of integrin inhibitor reverts acquired chemoresistance due to adhesion to collagen

We performed additional experiments to confirm the role of cell matrix adhesions on cell response to toxins. For this, we exploited the ability of synthetic peptides to compete for integrin-binding sites, thereby interfering with cell matrix adhesions. We repeated the toxicity assays in the presence of the commercially available RGD peptide that has been previously shown to inhibit integrin-mediated cell adhesions. Under these conditions, as seen in Figure 6, addition of RGD significantly impacted toxic responses for cells encapsulated in alginate/collagen gels made using 0.25% alginate, but not 1% alginate. Moreover, addition of RGD did not impact the IC_{50} values for U-87 cells cultured in pure alginate gels. Thus, in combination, these experiments confirmed the observations of cell matrix adhesion-dependent increase in tolerance of U-87 cells cultured in soft hybrid hydrogels.

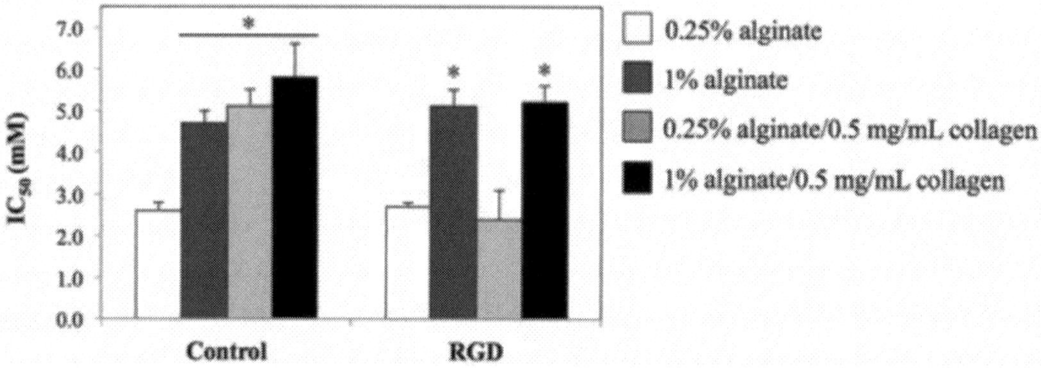

Figure 6. *Half maximal inhibitory concentration (IC_{50}) of acrylamide for U-87 cells cultured in 0.25% alginate (white bars), 1% alginate (dark grey bars), 0.25% alginate/0.5 mg/mL collagen (light grey bars), and in 1% alginate/0.5 mg/mL collagen (black bars) scaffolds in the presence and absence of the integrin inhibitor – RGD. For control experiments, *$P<0.01$ for IC_{50} values for cells cultured in 1% alginate and 0.25% or 1% alginate/0.5 mg/mL collagen compared with those cultured in 0.25% alginate, and for RGD experiments, *$P<0.01$ for IC_{50} values for cells cultured in 1%/0 or 0.5 mg/mL collagen compared with those cultured in 0.25% alginate/0 or 0.5 mg/mL collagen, as determined by Student's unpaired two-tailed t test. Error bars represent standard deviation of three samples.*

Discussion

Recent years have witnessed advances in the development of 3D biomaterial-based *in vitro* cell culture models that mimic the *in vivo* cellular microenvironment. Such organotypic tissue mimics are emerging as useful tools for addressing a key bottleneck in drug development – low clinical efficacy and safety. It is well recognized that cancer cells cultured in 3D conditions more closely resemble *in vivo* tumors and are more malignant and invasive than those cultured in conventional 2D monolayer culture conditions. Moreover, cancer cells cultured in 3D exhibit increased resistance to chemotherapy drugs relative to 2D cultures. Although this was previously attributed to limited drug diffusion in 3D matrices, it is now appreciated that other matrix properties, including mechanical stiffness and adhesion ligand density, also play a significant role in regulating cell responses to chemicals. However, it is still not clear if these different biochemical and biophysical matrix properties act independently or synergistically to regulate cell responses to chemicals.

To that end, in this study we used alginate/collagen-based hybrid hydrogel scaffolds that allow independent variation of stiffness and ligand densities as a flexible platform to explore the role of 3D matrix properties on cell responses to drugs. Glioblastoma multiforme tumor cells were used as the model system; GBM is among the most common, aggressive, and lethal forms of human cancer, affecting nearly 20,000 people a year in the US [23]. Moreover, it has been established that biochemical and biophysical matrix cues play key roles in regulating GBM cell invasion and proliferation [24,25]. Our results indicated the ability of the biochemical and biophysical matrix properties to also influence cell response to toxic chemicals. Specifically, we observed that both higher matrix stiffness and cell matrix adhesions increase the tolerance of human glioblastoma cells to the tested neurotoxins. These observations were validated using FCS experiments that demonstrated statistically similar drug diffusion rates for all hydrogel preparations and integrin inhibition experiments that demonstrated reversal of U-87 cell tolerance to toxins in soft conditions.

Future experiments will further explore the individual and synergistic roles of various biophysical and biochemical cues of the 3D microenvironment on cell responses to chemicals and establish the key matrix parameters responsible for *in vivo*-like drug responses. We believe that these studies will significantly contribute to the development of cancer cell-based *in vitro* assay platforms that efficiently mimic *in vivo* human responses to anti-tumor drugs and consequently decrease the reliance of drug efficacy data solely from animal testing.

Acknowledgements

This work was funded by start-up funds provided to Dr. Asuri and Dr. Zustiak by Santa Clara University and Saint Louis University, respectively as well as by a President's Research Fund (PRF) grant awarded to Dr. Zustiak by Saint Louis University.

References

1. Schreiber SL (2000) Target-oriented and diversity-oriented organic synthesis in drug discovery. *Science* 287: 1964-1969.
2. Al-Lazikani B, Banerji U, Workman P (2012) Combinatorial drug therapy for cancer in the post-genomic era. *Nat. Biotechnol.* 30: 679-692.
3. Siddiqui M, Rajkumar SV (2012) The high cost of cancer drugs and what we can do about it. *Mayo Clin. Proc.* 87: 935-943.
4. Mak IW, Evaniew N, Ghert M (2014) Lost in translation: animal models and clinical trials in cancer treatment. *Am. J. Transl. Res.* 6: 114-118.
5. Smalley KS, Lioni M, Herlyn M (2006) Life isn't flat: taking cancer biology to the next dimension. *In Vitro Cell Dev. Biol. Anim.* 42: 242-247.
6. Yamada KM, Cukierman E (2007) Modeling tissue morphogenesis and cancer in 3D. *Cell* 130: 601-610.
7. Provenzano PP, Eliceiri KW, Keely PJ (2009) Shining new light on 3D cell motility and the metastatic process. *Trends Cell Biol.* 19: 638-648.
8. Li Q, Chow AB, Mattingly RR (2010) Three-dimensional overlay culture models of human breast cancer reveal a critical sensitivity to mitogen-activated protein kinase kinase inhibitors. *J. Pharmacol. Exp. Ther.* 332: 821-828.
9. Breslin S, O'Driscoll L (2013) Three-dimensional cell culture: the missing link in drug discovery. *Drug Discov. Today* 18: 240-249.
10. Minchinton AI, Tannock IF (2006) Drug penetration in solid tumours. *Nat. Rev. Cancer* 6: 583-592.
11. Mitra M, Mohanty C, Harilal A, Maheswari UK, Sahoo SK, et al. (2012) A novel in vitro three-dimensional retinoblastoma model for evaluating chemotherapeutic drugs. *Mol. Vis.* 18: 1361-1378.
12. Hehlgans S, Haase M, Cordes N (2007) Signalling via integrins: implications for cell survival and anticancer strategies. *Biochim. Biophys. Acta* 1775: 163-180.
13. Aoudjit F, Vuori K (2012) Integrin signaling in cancer cell survival and chemoresistance. *Chemother. Res. Pract.* 2012: 283181.
14. Ramamoorthi K, Hara J, Ito C, Asuri P (2014) Role of Three-Dimensional Matrix Stiffness in Regulating the Response of Human Neural Cells to Toxins. *Cellular and Mol. Bioeng.* 7: 278-284.
15. Zustiak S, Nossal R, Sackett DL (2014) Multiwell stiffness assay for the study of cell responsiveness to cytotoxic drugs. *Biotechnol Bioeng.* 111: 396-403.
16. Magde D, Elson EL, Webb WW (1974) Fluorescence correlation spectroscopy. II. An experimental realization. *Biopolymers* 13: 29-61.
17. Zustiak SP, Boukari H, Leach JB (2010) Solute diffusion and interactions in cross-linked poly (ethylene glycol) hydrogels studied by fluorescence correlation spectroscopy. *Soft Matter* 6: 3609-3618.

18. Zustiak S, Riley J, Boukari H, Gandjbakhche A, Nossal R (2012) Effects of multiple scattering on fluorescence correlation spectroscopy measurements of particles moving within optically dense media. J. *Biomed. Opt.* 17: 125004-125004.
19. Zustiak SP, Nossal R, Sackett DL (2011) Hindered diffusion in polymeric solutions studied by fluorescence correlation spectroscopy. *Biophys. J.* 101: 255-264.
20. Rowley JA, Madlambayan G, Mooney DJ (1999) Alginate hydrogels as synthetic extracellular matrix materials. *Biomaterials* 20: 45-53.
21. Sethi T, Rintoul RC, Moore SM, MacKinnon AC, Salter D, et al. (1999) Extracellular matrix proteins protect small cell lung cancer cells against apoptosis: a mechanism for small cell lung cancer growth and drug resistance in vivo. *Nat. Med.* 5: 662-668.
22. Serebriiskii I, Castello-Cros R, Lamb A, Golemis EA, Cukierman E (2008) Fibroblast-derived 3D matrix differentially regulates the growth and drug-responsiveness of human cancer cells. *Matrix Biol.* 27: 573-585.
23. Holland EC (2000) Glioblastoma multiforme: the terminator. *Proc. Natl. Acad. Sci. USA* 97: 6242-6244.
24. Reyes SB, Narayanan AS, Lee HS, Tchaicha JH, Aldape KD, et al. (2013) alphavbeta8 integrin interacts with RhoGDI1 to regulate Rac1 and Cdc42 activation and drive glioblastoma cell invasion. *Mol. Biol. Cell* 24: 474-482.
25. Ananthanarayanan B, Kim Y, Kumar S (2011) Elucidating the mechanobiology of malignant brain tumors using a brain matrix-mimetic hyaluronic acid hydrogel platform. *Biomaterials* 32: 7913-7923.

387926 Increasing the Polymer Backbone Length of Fumarate-Based Polymer Scaffolds Does Not Alter MSC Osteoblastic Differentiation

Abstract:

Poly(propylene fumerate) (PPF) has been developed and used widely in the field of tissue engineering, specifically bone tissue engineering. Unfortunately, the rate of degradation of PPF is less than the rate of bone formation. Due to this limitation, work in our laboratory focuses on new synthetic routes to increase the ester hydrolysis of fumarate-based polymers over PPF. Poly(butylene fumarate) (PBF), an unsaturated linear polyester which differs from PPF by a methylene group, has been synthesized through a ring opening polycondensation reation from maleic anhydride (MA) and butylene glycol (BG). Previously we have shown that after UV initiated crosslinking, PBF samples have a faster rate of degradation than PPF samples as determined through mass loss and changes in compressive moduli. Our latest efforts have focused on the cytocompatibility and osteoblastic differentiation of NIH-3T3s (a pre-osteoblastic cell line) and primary rat bone marrow derived messenchymal stem cells (MSC) on the new polymer. We will present results demonstrating that PBF supports attachment, viability, proliferation, and osteoblastic-differentiation as well, if not better, than PPF.

Safety and biodistribution of pathogen mimicking polyanhydride nanovaccines

Julia Vela Ramirez[1], Jonathan Goodman[1], Rajarshi Roychoudhury[2], Paola Boggiato[3], Paul Lueth[3], Bryan Bellaire[3], Jesse Hostetter[4], Nicola Pohl[2], Michael Wannemuehler[3] and Balaji Narasimhan[1]

[1]Department of Chemical and Biological Engineering, Iowa State University, Ames, Iowa 50011
[2]Department of Chemistry, Indiana University, Bloomington, IN 47405
[3]Department of Veterinary Microbiology and Preventive Medicine, Iowa State University, Ames, Iowa 50011
[4]Department of Veterinary Pathology, Iowa State University, Ames, Iowa 50011

The development of novel strategies to improve adjuvant formulations that can be targeted to immune cells to enhance the delivery of antigen by activation of specific receptors is an important area of interest in the design of vaccine formulations. The use of biodegradable particles has shown promising characteristics in this regard, with the particles playing dual roles, as adjuvants and delivery vehicles. These particles have demonstrated enhanced antigen presentation and stimulation of antigen presenting cells (APCs), which are fundamental in initiating an immune response. The particle surface can be functionalized to target them to cellular populations or receptors that are important in the initial immune response. In particular, C-type lectin receptors are pattern recognition receptors that recognize carbohydrate moieties (which are present in viruses and bacteria). In these studies, the safety and cellular distribution of pathogen mimicking carbohydrate-functionalized polyanhydride nanoparticles was analyzed.

After *in vivo* administration, nanoparticle formulations interact with a variety of cells, including APCs. The first step before analyzing the effects of active targeting strategies upon the efficacy of carbohydrate-functionalized nanoparticles is an assessment of their safety. In addition to the toxicity studies, an understanding of the distribution and uptake of these particles by cellular populations at the site of immunization is required to rationally select formulations that enable tailored immune responses.

In these studies, the effect of carbohydrate functionalization on the safety and cellular biodistribution of polyanhydride nanoparticles following intranasal administration in Swiss Webster mice was investigated. Polyanhydride nanoparticles based on 1,6-bis-(*p*-carboxyphenoxy)hexane (CPH), and 1,8-bis-(*p*-carboxyphenoxy)-3,6-dioxaoctane (CPTEG) were used in this study. Surface functionalization with di-mannose, galactose and glycolic acid (as a control) was performed using an amine-carboxylic acid coupling reaction. Carbohydrate-functionalized polyanhydride nanoparticles showed no adverse consequences after intranasal administration in mice. Histopathology of liver, kidneys and lungs was analyzed. In addition, analysis of serum and urine biomarkers from mice that received a 5 mg dose of nanoparticle formulations subcutaneously showed no significant differences when compared to the corresponding markers from animals that received saline. The profile of cellular uptake and biodistribution was found to be dependent on the surface functionalization of the nanoparticles, however, all the treatment groups displayed enhanced internalization over 96 hours. We also developed a novel methodology to track the nanoparticles in lung tissue after paraffin-embedding using immunohistochemistry. Based on these analyses, carbohydrate-functionalized nanoparticles are safe for *in vivo* applications, however, there is a direct effect of the surface chemistry on their cellular biodistribution post-administration, which may have an important role in their performance and efficacy. This work has important implications for enhancing the *in vivo* performance of nanoparticle-based vaccine delivery systems.

POSS-Containing Biodegradable Polymer Networks for Bone Tissue Engineering Applications

Shanfeng Wang

Department of Materials Science and Engineering, The University of Tennessee,
Knoxville, TN 37996

Here I present three groups of injectable and photo-crosslinkable polymeric biomaterials containing polyhedral oligomeric silsesquioxane (POSS) prepared in our research group for bone tissue engineering applications. In the first study, photo-crosslinkable hybrid organic-inorganic composites were developed from poly(propylene fumarate) (PPF) and methacryl-polyhedral oligomeric silsesquioxane (mPOSS), which has eight methacryl groups tethered with a cage-like hybrid inorganic-organic nanostructure. Blending mPOSS with PPF was found to decrease the viscosity of PPF and expedite the photo-crosslinking process. Mouse pre-osteoblastic MC3T3-E1 cell adhesion, spreading, proliferation, differentiation, and gene expression were examined on crosslinked PPF/mPOSS disks. In the second study, a series of novel injectable and photo-crosslinkable PPF-*co*-POSS copolymers were synthesized via two-step polycondensation to improve both stiffness and toughness simultaneously, and to promote biological performance of bone tissue engineering scaffolds. Tensile and compressive moduli and facture toughness were enhanced for crosslinked PPF-*co*-POSS when ϕ_{POSS} increased up to 10%, where POSS nanocages were well distributed and their crystallinity were completely confined in the networks. Decreases in these mechanical properties were observed at higher ϕ_{POSS} because of decreased crosslinking density and larger POSS aggregates. The mechanical properties with the non-monotonic dependence on ϕ_{POSS} were correlated with MC3T3-E1 cell attachment, spreading, proliferation, differentiation, and gene expression, which all maximized at ϕ_{POSS} of 10%. In the third study, poly(ε-caprolactone) triacrylate (PCLTA) networks were incorporated with a photo-reactive methacryl isobutyl polyhedral oligomeric silsesquioxane. POSS nanocages tethered in the PCLTA networks greatly enhanced the mechanical and rheological properties, but did not significantly alter the surface wettability and the capability of adsorbing serum proteins from cell culture media. Better MC3T3-E1 cell attachment, spreading, and proliferation were found on the networks with ϕ_{POSS} of 10-20% than the networks containing no POSS. In summary, nanohybrid polymer/POSS networks with improved mechanical properties and osteoconductivity have great potential as scaffolding materials for bone repair and regeneration.

387939 Failure Analysis of Transvaginal Mesh Products – a Biomaterials Perspective Using Materials Science Fundamentals

Abstract:

Transvaginal mesh products have been used in recent years to treat stress urinary incontinence (SUI) and pelvic organ prolapse (POP) in female patients. The meshes used as the support structure of the SUI and POP products are typically manufactured using monofilaments of various forms of polypropylene that are woven in a variety of patterns and shapes. The properties of the polypropylene mesh component of the SUI and POP products along with an overview of prior studies on the effectiveness of this material selection will be discussed. In particular, the biocompatibility of polypropylene used in these specific applications will be addressed. Finally, a failure analysis mechanism for the mesh component of the SUI and POP products will be proposed.

386986 Liposome Coated Silk Microspheres Are Effective Lubricants

Abstract:

We present results of our work to develop a liposome based coating on silk microspheres to couple the lubrication properties of highly hydrated phospholipid head groups with the cushioning effects and wear resistance of silk. We show that this liposome coated silk microsphere suspension is an effective lubricant, capable of providing a low coefficient of friction on the order of 0.02, minimizing surface wear and avoiding degradation even over relatively prolonged cycling conditions. We propose that the lubricating properties of the liposome coated silk microspheres are a result of an efficient rolling mechanism. Details of the nanostructure of liposome coated silk microspheres were characterized by confocal laser scanning microscopy and scanning electron microscopy, the lubrication properties and nano-scaled surface wear were characterized using a tribometer and an atomic force microscope. With current therapeutic options for improving joint comfort being limited, liposome coated silk microspheres, which are biocompatible and slowly degradable, plus the appropriate tribological properties, offer an intriguing opportunity. This work was supported by the National Science Foundation.

Microcontact Printing of Polyelectrolytes on PEG for Micropatterning Nanoparticles, DNA, Proteins and Cells

Zhibin Wang

11/17/2014

Department of Chemical and Biomedical Engineering

FAMU-FSU College of Engineering

Florida State University

1

Microcontact printing of polyelectrolytes for micropatterning

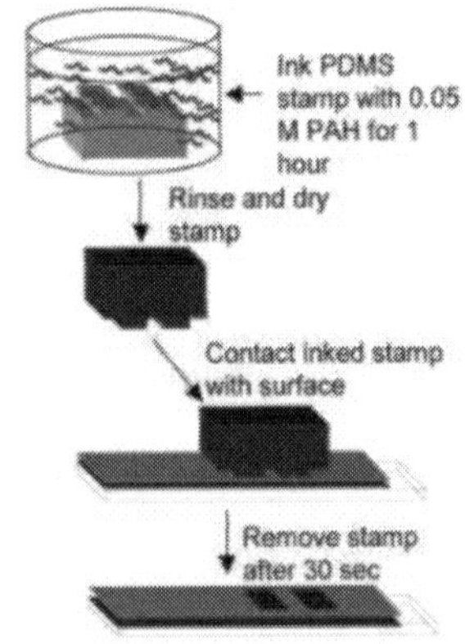

AFM image of PAH stripes on PEM surface

P. T. Hammond et al. *Langmuir* 2003

Fluorescence image of protein microarray

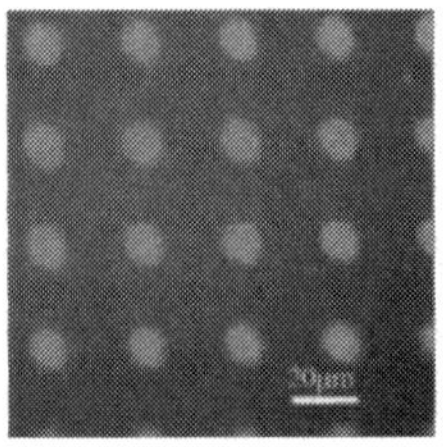

P. T. Hammond et al. *Biomacro.* 2004

AFM image of DNA nano-array

Cha et al. *ACS Nano.* 2009

PAH: Poly(allylamine hydrochloride)

PEM: Poly electrolyte multilayer

Polyethylene glycol grafted surface for micropatterning

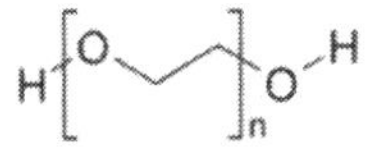

Polyethylene glycol (PEG)

- Hydrophilicity
- Flexibility
- Nontoxicity
- Nonimmunogenecity

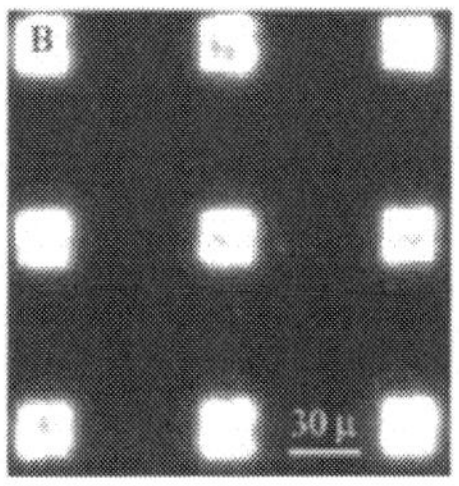

Zhang et al. *Langmuir* 2002

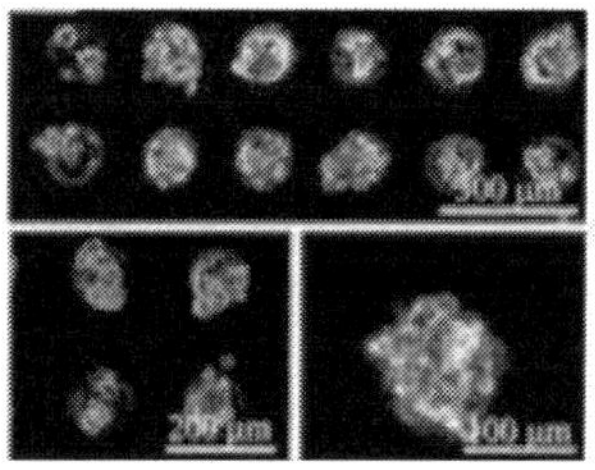

Yoshimoto et al. Lab Chip 2009

Steric repulsion (against van der Walls and hydrophobic interaction)

- **Osmotic (solvation of PEG chains)**
- **Elastic (conformational entropy of PEG chains)**

Fabrication of PEG grafted surface for micropatterning

Self-assembly monolayer of short chain PEG

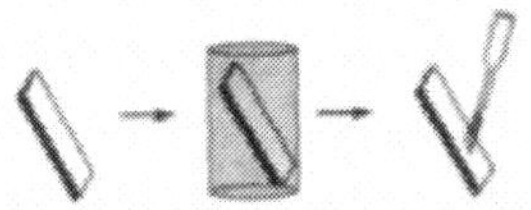

Freshly prepared gold slide → Organic thiol oligo (ethylene oxide) solution overight → Rinse and dry

M. Whitesides *et al. J. Am. Chem. Soc.* 1993

Photolithography

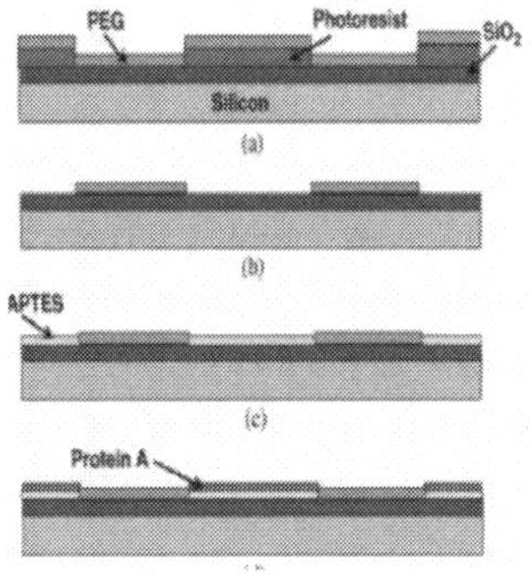

Zohar et al. *Nanotech.* 2005

Soft lithography

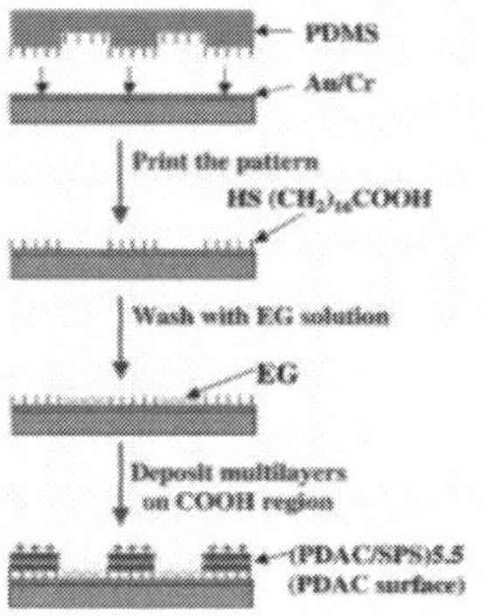

P. T. Hammond et al. *Adv. Mater.* 2002

Drawbacks: Expensive facilities; Harsh condition; Multiple processing steps

Microcontact printing of polyelectrolyte on PEG

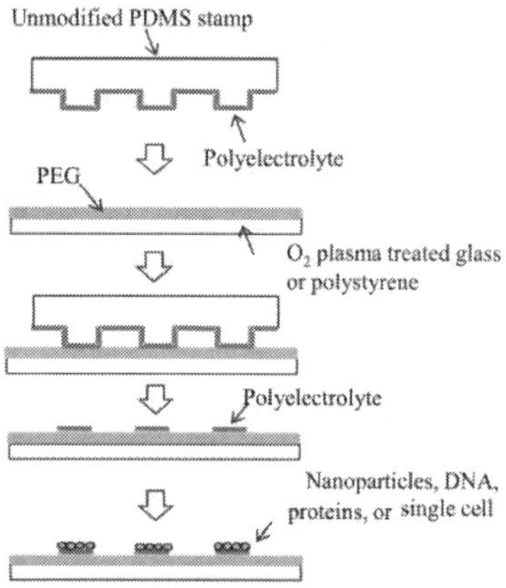

Schematic of microcontact printing of polyelectrolyte on PEG

Characterization of PEG grafted surface

Short chain of PEG silane

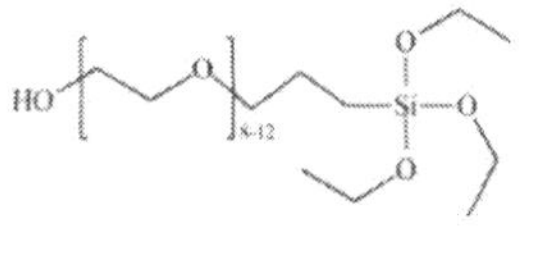

Contact angle measurement of PEG grafted surface

AFM image of PEG grafted surface

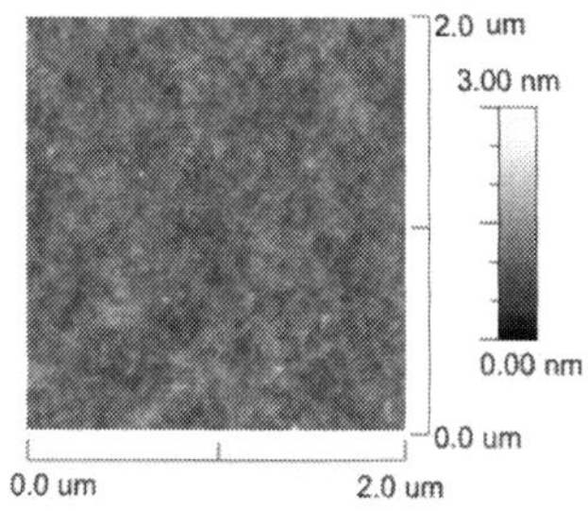

Microcontact printing of PAH on PEG

Fluorescence and AFM images of poly(allylamine chloride) (PAH) dots on PEG

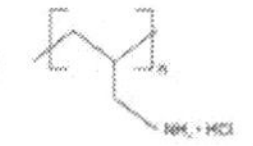

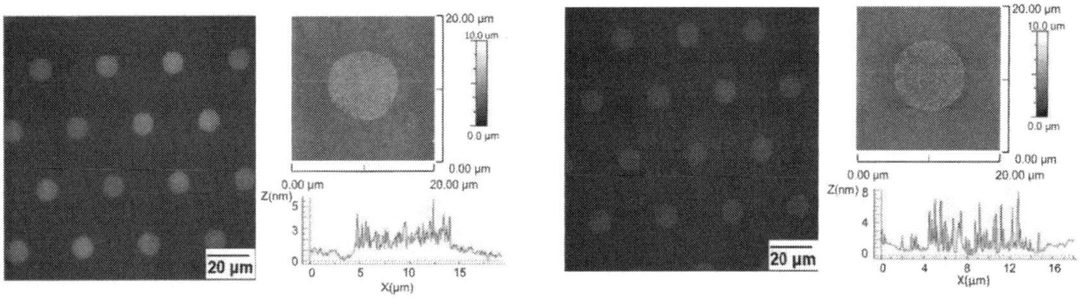

Before (left) and after (right) 1 day incubation in PBS

Z Wang et al. *Soft Matter* 2012

Microcontact printing of polyelectrolyte on PEG

Branched poly(ethylene imine)-RITC

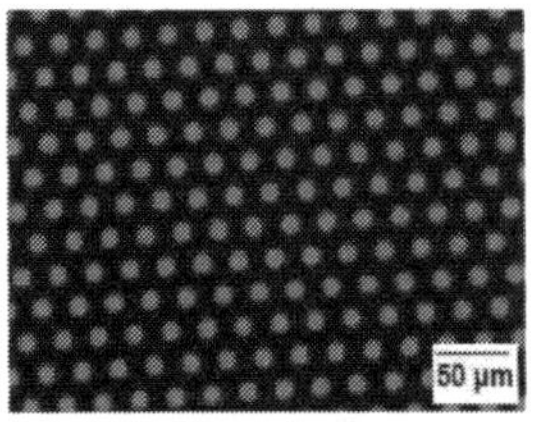

Linear poly(ethylene imine)-RITC

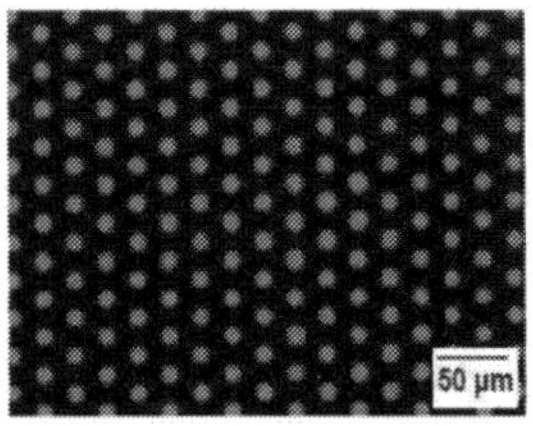

Chitosan (red) and poly(lysine) (green)

Stretched DNA bundles (YOYO-1)

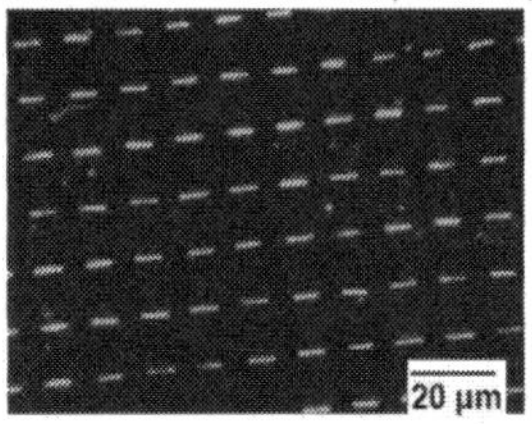

Bilayer of Poly(sodium 4-styrene sulfonate)/PAH-RITC

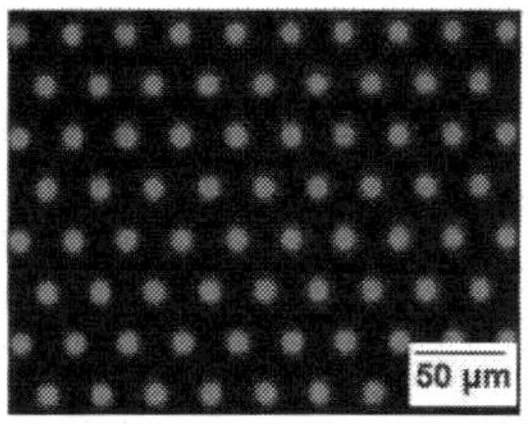

Printed polyelectrolyte on PEG for micropatterning nanoparticles and biomaterials

Quantum dots (red)

μCP of PDAC Poly(diallydimethylammonium chloride)

DNA molecule (YOYO-1)

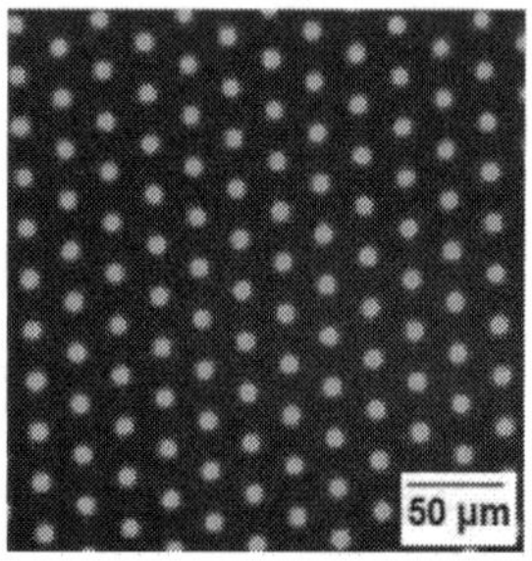

μCP of PAH

Neutravidin (red) and biotin (green)

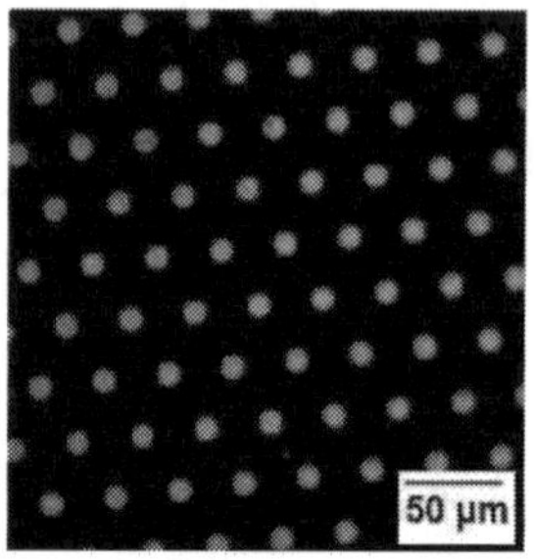

μCP of PAH and modified by glutaraldehyde

Printed polyelectrolyte on PEG for micropatterning live cells

K562 cell patterned by μCP of PAH on PEG grafted glass

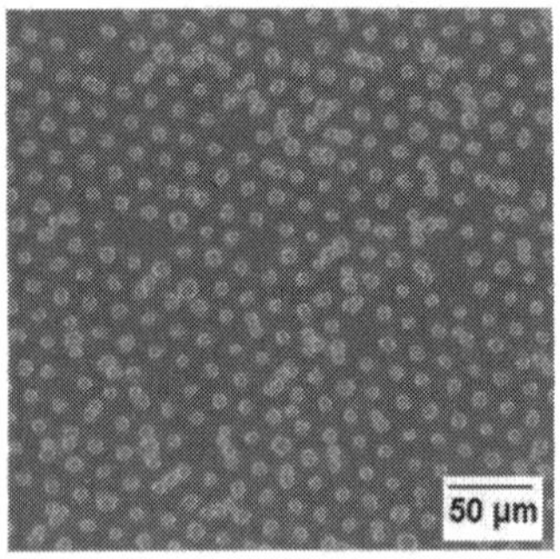

K562 cell patterned by μCP of PAH on PEG grafted polystyrene

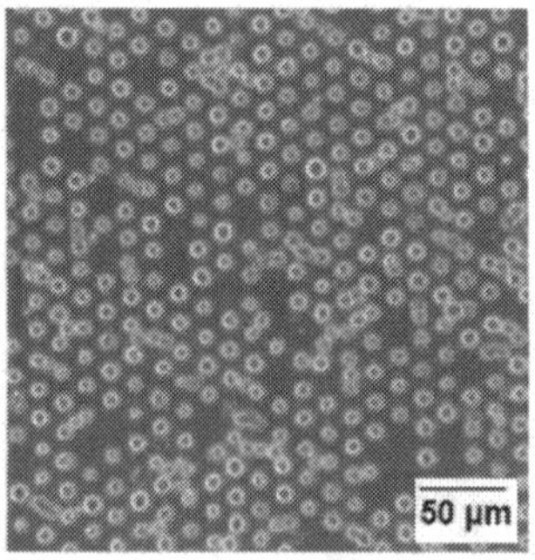

Patterned cells viability (24 h)

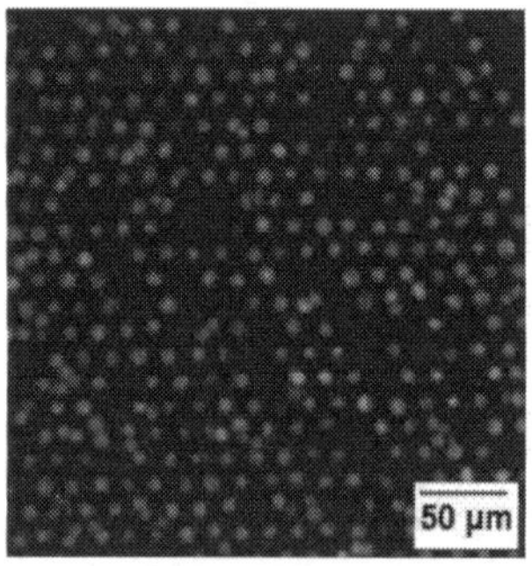

Green: Acridine orange
Red: Propidium iodide

Summary

- A novel μCP technique have been developed based on printing polyelectrolytes directly on a PEG-silane SAM using unmodified PDMS stamps.

- Monolayer-like structures of PAH, PLL, chitosan, linear PEI, branched PEI and PDAC, PAH/PSS bilayer, and stretched DNA molecular bundles have been printed.

- The printed structures are stable under aqueous conditions and allow for micropatterning of nanoparticles, DNA, protein, and live cells.

- This method is potentially useful in fields of tissue engineering and biomedical devices.

Acknowledgement

Dr. Guan's group:

Dr. Jingjiao Guan (Chemical & Biomedical Engineering, FSU)

Dr. Peipei Zhang (University of Maryland)

Dr. Brett Kirland (FDLE)

Dr. Yingru Liu

Dr. Eric Lochner (Physics, FSU)

Dr. Ken Knappenberger (Chemistry, FSU)

Florida HEALTH

The Florida Department of Health

Thank you

Structure & Synthesis of Cage-Based Small-Pore Zeolites

Tracy Davis, Dan Xie, Christopher Lew, Stacey Zones, Robert Saxton
Chevron, Richmond, CA, USA.

Cage-based small pore zeolites (e.g., CHA, AEI, AFX) serve as catalysts and adsorbents in a number of commercial applications. Specifically, CHA and AEI have been shown to perform very well as deNOx catalysts [1, 2], and CHA is used commercially in the conversion of methanol-to-olefins [3]. The high silica forms of CHA, AEI and AFX were first discovered at Chevron [4,5,6], and research surrounding their synthesis continues. Additionally, work is ongoing to determine the potential of SSZ-52, the structure of which was recently shown to bear important similarities to CHA [7].

In this paper, the structures of CHA, AEI, AFX and SSZ-52 will be considered and compared in detail. While all of these zeolites are characterized by multi-dimensional small-pore structures and belong to the ABC-6 family, differences in the stacking sequence of the double 6-rings result in cages of varying dimensions. Such differences are a direct result of both the inorganic and organic species present during synthesis. This relationship between synthesis conditions and structure formation will be highlighted and specific attention will be paid to SDA-zeolite interaction energies. Following on a careful analysis of these zeolite structures, their performance as deNOx catalysts will be presented.

1. M. Jacoby, *Chem. Eng. News* 90 (2012) 10-16.
2. M. Moliner, C. Franch, E. Palomares, M. Grill, A. Corma, *Chem. Commun.* 48 (2012) 8264-8266.
3. P. Barger, B. Vora, U.S. Patent 6531692, 2003.
4. S. Zones, U.S. Patent 4544538, 1985.
5. S. Zones, Y. Nakagawa, S. Evans, G. Lee, U.S. Patent 5958370, 1999.
6. S. Zones, U.S. Patent 5194235, 1993.
7. D. Xie, L. McCusker, C. Baerlocher, S. Zones, W. Wan, X. Zou, *JACS* 135 (2013) 10519-10524.

378025 One-Step Dual Template Synthesis of Hierarchical Layered Zeolite with Tailored Textural and Catalytic Properties

Abstract:

A one-step dual template strategy was employed to achieve facile synthesis of meso-/microporous layered zeolites with tailored morphology, mesoporosity and catalytic activity. Tetrapropyl ammonium hydroxide (TPAOH) and diquaternary ammonium surfactant ($[C_{22}H_{45}\text{-}N^{+}(CH_3)_2\text{-}C_6H_{12}\text{-}N^{+}(CH_3)_2\text{-}C_6H_{13}]Br_2$, $C_{22\text{-}6\text{-}6}$) with a composition of $100SiO_2/xC_{22\text{-}6\text{-}6}/y$TPAOH/$4000H_2O$ were used as the templates in meso-/microporous layered zeolite synthesis. The influences of several parameters on the zeolite formation were investigated, including concentration of TPAOH and $C_{22\text{-}6\text{-}6}$, hydrothermal synthesis time, and ageing temperature and time prior to hydrothermal crystallization. The sizes of mesoporous and microporous domains of the zeolite and thus its meso- and microporosity can be modulated by simply tuning these parameters. The catalytic studies showed that the catalytic activity and selectivity can be systematically modulated by varying these parameters in the synthesis. The ability to form hierarchical zeolite particles with tunable meso-/microporosity is important to keep their potential applications as broad as possible for a specific zeolite structure. The dual template synthesis approach presented is potentially applicable to form other types of meso-/microporous zeolite materials.

386832 From Self-Assembled Mesoporous Thin Films to Kesterite Solar Cells

Abstract:

Given the terawatt-scale of future energy needs, the most promising future photovoltaic materials should be Earth abundant with their primary mineral resources distributed across several geographic regions and their supply chains robust to reduce concerns of price volatility. In addition, the process of forming the solar cell should be scalable, low-cost, and not utilize dangerous or toxic materials. One of the strongest candidates appears to solution processed kesterite structures of $Cu_2ZnSn(S,Se)_4$ or similar alloys. Our most recent result is the development of benign-solvent molecular-inks that yield ~11% efficient solar cells from these kesterites. This presentation will focus on the materials science and physics that led from a starting point focused on the development of nanostructured thin film photovoltaics based on self-assembled mesoporous ceramic films to the current state-of-the-art. We will show the results of composition spread experiments coupled with spatially resolved photoluminescence and a new theoretical framework to analyze photoluminescence that reveal the effects of native point defects and doping on material quality and device performance.

368256 Modeling the Formation of Ordered Nanoporous Silica Materials

Abstract:

Porous materials are ubiquitous in modern technology. Traditional applications include catalysis and adsorption separations, while more recent applications include membrane separations, biosensors, drug delivery and low-k dielectrics for microelectronics. The development of new porous materials with controlled pore structure that can be tailored for specific applications is a very active area of materials research. Understanding how such materials form and how to control the pore geometries will help facilitate advanced applications of these materials. Molecular modeling of syntheses of porous materials is challenging because large system sizes are needed to have representative sampling of the processes, and long time scales are required to observe the formation of ordered structures. In this presentation we describe the development and application of models of porous material assembly, focusing on silica materials. We show how quite simple models based on the assembly of corner-sharing silica tetrahedra can describe the self-assembly of a wide range of ordered nanoporous materials, including ordered mesoporous silicas and all-silica zeolite frameworks.

363545 Site-Specific Placement of Phases in Porous Nanocomposites

Abstract:

Templating methods developed over the past two decades have enabled syntheses of nanoporous materials with well-controlled architecture, suitable for a variety of applications that include energy storage and conversion, catalysis, opto-electronics and photonics, sensing, sorption and separation, controlled storage and release, and more. Control over dimensions and morphology of both the pore space and the solid wall components is achieved by hard templating and/or soft templating. Hard templating involves confinement of precursors in a solid mold, such as colloidal crystals, anodic alumina membranes, track-etch membranes, lithographic patterns, and other porous solids. Soft templating relies on dynamic structures formed from emulsion droplets, surfactants, block-copolymers, ionic liquids and other micelle-forming substances to direct the organization of precursor components. Multiple templating methods can be combined to achieve hierarchical pore structures with defined pore size distributions and pore interconnectedness.

Next-generation nanostructured materials will require multiple components, including multiple active phases (catalytic, electroactive, conductive, magnetic, etc.) and phases that provide a structural backbone, stabilization, and perhaps transport pathways. It will be critical to have reliable control over component placement, ideally through self organization and in a single step. Site-specific placement of component phases extends the level of achievable complexity in nanocomposite materials and governs the performance of porous nanocomposites, because it determines component spacing (dispersion, interactions), accessibility, and grain growth/aggregation. Positioning of phases also affects relative dimensions of components and relative amounts that are in contact with interactive species. For example, in catalyst materials, the location of active components on a supporting matrix impacts their accessibility and long-term stability. In electrode materials for electrical energy storage devices, such as rechargeable batteries and supercapacitors, it is often necessary to disperse active material throughout a conductive phase or coat it with a protective phase. The relative location of such multiple phases affects conductivity, accessibility to electrolyte, and the ability of active material to maintain nanosized dimensions during insertion or deinsertion of charged species.

This presentation will provide on overview of some methods that allow placing of individual components in porous nanocomposites at specific sites, such as the pore surfaces, within the pore walls, at specific nodal points of a porous skeleton, or uniformly distributed throughout the structure. These methods include capillary condensation of volatile precursors between closely spaced surfaces of the hard template, or modulating the interactions between precursors and a hard template. The interactions are tuned through precursor selection, choice of ligands that complex with the precursor and may influence polymerization of sol-gel precursors, and through surface modification of the hard template. Interactions between multiple components confined in a template can also be exploited to control the morphology of the porous microparticles and hierarchically structured extended porous solids.

369556 Structure-Reaction Property Relationships for Lewis Acid Molecular Sieves

Abstract:

Lewis acidic tetravalent heteroatoms (Sn^{4+}, Ti^{4+}, Zr^{4+}) isolated within silica-based molecular sieves catalyze sugar isomerization and epimerization reactions as well as Diels Alder reactions of ethylene with substituted furans. Here, I will discuss the atomic arrangements and local environments of the Lewis acid centers and their reaction properties.

369125 Multiscale Ab Initio Modeling of Catalysts

Abstract:

In this talk, multiscale simulation will briefly be introduced as an enabling technology that bridges the gap between scales. Understanding the emergent properties of heteroepitaxial catalysts and the in silico prediction of suitable bimetallic catalysts will be elucidated as an application domain. We will demonstrate this methodology for the specific example of ammonia decomposition on single metals and bimetallics. It will be shown that incomplete monolayer bimetallics form an important class of catalytic materials.

Modeling CO_2 adsorption in metal organic frameworks

Laura Gagliardi

Department of Chemistry, Supercomputing Institute, and Chemical Theory Center, University of Minnesota, 207 Pleasant Street SE, Minneapolis, Minnesota 55455-0431, USA

During the formation of some metal organic frameworks (MOFs), the metal centers coordinate with both linker and solvent molecules. Subsequent activation removes the solvent molecules and creates open metal sites, which have strong affinity for CO_2. At present the interactions of CO_2 with these open metal sites are so poorly understood that commonly used force fields underestimate the adsorption of CO_2 in a recently synthesized MOF, Mg-MOF-74 at carbon capture conditions by as much as two orders of magnitude. Here we present a systematic procedure to develop a force field for Mg-MOF-74 and related materials using high-level quantum chemical calculations.[1] This transferable force field accurately takes into account the chemistry of the open metal sites. Recently we have employed a similar procedure to study also Fe-MOF-74.[2]

In the second part, results of a recent quantum chemical study[3] on the mechanism of carbon dioxide adsorption in an alkylamine functionalized metal-organic framework will be presented. Our calculations demonstrate that in these materials an ammonium carbamate species is formed via the insertion of CO_2 into the Mg-N_{amine} bonds.[4]

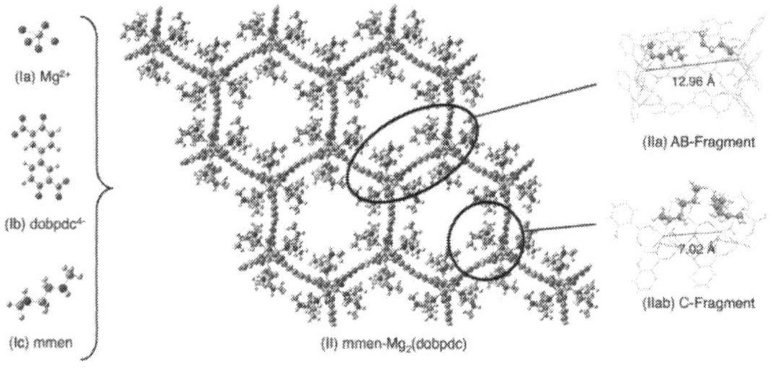

[1] A. L. Dzubak, L.-C. Lin, J. Kim, J. A. Swisher, R. Poloni, S. N. Maximoff, B. Smit, and L. Gagliardi Ab initio Carbon Capture in Open-Site Metal Organic Frameworks Nature Chemistry 4, 810-816 (2012)

[2] J. Borycz, L.-C. Lin, E. Bloch, J. Kim, A. L. Dzubak, R. Maurice, D. Semrouni, B. Smit, and L. Gagliardi; CO_2 Adsorption in Fe_2(dobdc): A Classical Force Field Parameterized from Quantum Mechanical Calculations J. Phys. Chem. C. **DOI:** 10.1021/jp500313j (2014)

[3] N. Planas, A. L. Dzubak, R. Poloni, L.-C. Lin, A. McManus, T. McDonald, J. Neaton, J. Long, B. Smit, L Gagliardi The mechanism of carbon dioxide adsorption in an alkylamine functionalized metal-organic framework *J. Am. Chem. Soc.*, *135* (20), 7402–7405 (2013)

[4] S. O. Odoh, A. L. Dzubak, B. Vlaisavljevich, S. Schnell, N. Planas, K. Lee, J. Neaton, B. Smit, and L. Gagliardi, submitted (2014)

Controlling Solar Cell Active Layers via Surface Modification and Gas Expanded Polymer Annealing

AIChE Abstract for November 2014 Conference: Atlanta, GA

Sarah Russell[†‡], *Holly Stretz* [*†], *Mark Dadmun* [□Δ], *S. Michael Kilbey II* [□§], *and Zach Seibers*[J]

[†]Department of Chemical Engineering, Tennessee Technological University, Cookeville, Tennessee 38505, United States

[‡]Center for Manufacturing Research, Tennessee Technological University, Cookeville, Tennessee 38505, United States

[□]Department of Chemistry, University of Tennessee, Knoxville, Tennessee 37996, United States

[J]Department of Energy Science and Engineering, University of Tennessee, Knoxville, TN 37996, United States

[Δ]Chemical Sciences Division, Oak Ridge National Laboratory, Oak Ridge, Tennessee 37831, United States

[§]Department of Chemical and Bimolecular Engineering, University of Tennessee, Knoxville, TN 37996, United States

Controlling and characterizing the morphology of the active layer in organic photovoltaics (OPVs) is essential for improving sunlight-to-energy conversion. While the most widely used active layer is based on a blend of the donor-type polymer, poly(3-hexylthiophene) (P3HT) and the fullerene [6,6]-Phenyl-C_{61}-butyric acid methyl ester (PCBM), various thermal or solvent annealing protocols are required to achieve modest power conversions efficiencies of ~5%. Here the use of gas expanded polymer (GXP) annealing, which relies on varying pressure conditions by introducing and removing CO_2 gas at a constant flow rate, is investigated as a process to improve the dispersion of PCBM within a P3HT film. GXP annealing causes the thin film to swell and deswell which leads to a change in crystallinity of the polymer P3HT, as observed by X-ray diffraction, UV-Vis spectroscopy and contact angle measurements. Neutron reflectivity, which is sensitive to the PCBM depth profile suggests that pressure oscillation range of 875-900 psi causes the PCBM to distribute more uniformly throughout the film compared to an as-cast sample. Modification of the film/substrate interface using a surface-grafted P3HT brush followed by GXP annealing can also alter the active layer morphology. The efficacy of this surface modification will be described.

376868 Polymer Functionalized Graphene Oxide As Thermally Responsive Ion Permeable Membrane

Abstract:

A UCST polymer, poly(sulfobetaine) (PMABS), which exhibits upper critical solubility temperature (UCST) behavior, was successfully grown onto graphene oxide sheets with an atom-transfer polymerization process. The modified graphene oxide sheets could be filtered to form free-standing papers that were then fabricated into a thermally-responsive membrane usable in non-aqueous solutions. It was tested as a separator in a electrochemical cell, separating a Li foil and a reduced graphene oxide paper as the two electrodes in a configuration that was similar to a half-cell in a Li ion battery. In the absence of the modified membrane, the specific storage capacitive for Li of the reduced graphene oxide paper increased ~30% when heated from 20°C to 80°C. In contrast, a PMABS-modified separator caused the storage capacity to decrease by over 50%. The membrane were characterized using SEM, AFM, UV spectroscopy an quartz crystal microbalance. This reversible thermal reduction in ion permeability can be used to mitigate thermal runaway of a Li ion battery, or tune the thermal response of an ion sensor or an electrode, among other applications.

387200 Epoxy/Magnetite/Carbon Nanofibers Nanohybrids for Anticorrosion Coatings

Abstract:

Corrosion protection has become an issue of prime importance in modern society.[1, 2] Epoxy/magnetite/carbon nanofibers nanohybrid coatings have been studied for anticorrosion purposes of stainless steel (SS). The thickness effect on the anticorrosion properties of the coatings were investigated in 3.5 wt% NaCl aqueous solution by monitoring the open circuit potential (E_{ocp}) and tracing quasi-stationary polarization (tafel) curves of the coatings-coated stainless steel electrode. A significantly positive shift in the open current potential, Eocp, was observed from the Eocp-time curves of the nanocomposite coated stainless steel electrodes. Consistently, extrapolation of Tafel plots gave a much more positive corrosion potential (E_{corr}) and much lower corrosion current (I_{corr}). The thinner nanocomposite coating lost its protectiveness after 40 hrs whereas the thicker nanocomposite coating retains its anticorrosion properties during the study. Fourier transform infrared spectroscopy shows no chemical changes in the nanocomposite coatings, suggesting water infiltration induced coating failure mechanism.

References

1. H. Wei, D. Ding, S. Wei and Z. Guo, *Journal of Materials Chemistry A*, 2013, **1**, 10805-10813.

2. J. Zhu, S. Wei, I. Y. Lee, S. Park, J. Willis, N. Haldolaarachchige, D. P. Young, Z. Luo and Z. Guo, *Rsc Advances*, 2012, **2**, 1136-1143.

387863 Multifunctional Epoxy Nanocomposites Reinforced with Nanofiller in Various Types

Abstract:

Epoxy polymer nanocomposite (PNCs) were systematically studied with different types of nanofillers, including polyaniline nanofibers and nanospheres, polypyrrole nanofibers and nanospheres, magnetite nanospheres, carbon coated iron (Fe@C) nanospheres, graphene nanosheets and graphene nanosheets decorated with core-shell nanoparticles have been fabricated. The curing process of the PNCs was monitored by fourier transform infrared spectroscopy test and the curing extent of the PNCs were calculated based on the residual heat obtained in the DSC test. Reduced viscosity observed in the uncured liquid samples was attributed to the interaction between the nanofillers and the polymer matrix, and the nanofillers were found to favor the disentanglement of the polymer chain. The mechanical property study of the cured solid samples indicated that at lower loading nanofiller displayed a stress releasing function and led to increased tensile strength of the PNCs. In addition, due to the enhanced interaction between the functionalized magnetite nanospheres and the epoxy resin, the tensile strength was also increased. Meanwhile, thermo-mechanical properties including storage and loss modulii, glass transition temperature were evaluated with the free volume change in the PNCs. The dielectric property study indicated that the interfacial polarization in the PNCs depends on the specific surface area of the nanofiller. Electrical conductivity studied indicated that the nanofiller with high aspect ratio can form network structure more easily and lead to the percolation behavior show in lower loading. Finally, the thermal stability and heat release rate study of the PNCs revealed that all the nanofillers applied would promote the char formation of the epoxy matrix and led to reduced flammability.

389204 Flame Retardant Polypropylene Nanocomposites Using a Variety of Nanofillers

Abstract:

Polypropylene (PP) nanocomposites reinforced with a variety of nanofillers such as transitional metal@metal oxide core-shell nanoparticles (NPs),[1,2] magnetite NPs, carbon nanotubes, graphene nanosheets, magnesium hydroxide, aluminum hydroxide, and silicon dioxide were prepared through in- or ex- situ solution based method for flame retardancy application. In order to improve the filler dispersion quality, polymer-filler interfacial bonding and the flame retardant efficiency, the filler surface functionalization/treatment and the addition of surfactant were integrated into the fabrication of PP nanocomposites. The filler surface modification and dispersion were investigated through transmission electron microscopy (TEM) and scanning electron microscopy (SEM). The physicochemical properties of thus prepared nanocomposites including thermal stability, flammability, melting rheological behaviors were studied using thermogravimetric analysis, micro-scale combustion calorimetry, UL-94 vertical burning, and limiting oxygen index. The flame retardancy mechanism for each composition will be discussed in detail and the advantages as well as the disadvantages will also be compared.

References

1. **Q. He,** T. Yuan, X. Zhang, Z. Luo, N. Haldolaarachchige, L. Sun, D. P. Young, S. Wei, Z. Guo, Magnetically Soft and Hard Polypropylene/Cobalt Nanocomposites: Role of Maleic Anhydride Grafted Polypropylene, ***Macromolecules*** 46 (**2013**): 2357-2368

2. **Q. He**, T. Yuan, S. Wei, Z. Guo, Catalytic and Synergistic Effects on Thermal Stability and Combustion Behavior of Polypropylene: Influence of Maleic Anhydride Grafted Polypropylene Stabilized Cobalt Nanoparticles, ***Journal of Materials Chemistry A,*** 1 (**2013**): 13064–13075

389310 Multifunctional Polymer Nanocomposites Toward Electromagnetic Interference Shielding

Abstract:

Multifunctional polymer nanocomposites including polyurethane nanocomposites reinforced with iron-silica core-shell nanoparticles, polypropylene nanocomposites reinforced with magnetic (magnetite, iron) nanoparticles and carbon nanotubes are fabricated for electromagnetic interference shielding applications. Parameters including filler surface treatment (controlled shell coating thickness), filler composition, and the effect of surfactant were systematically studied. Filler dispersion quality, morphology and the magnetic properties (coercivity, saturation magnetization) were studied. In addition, relative complex permittivity and relative complex permeability are investigated and the corresponding microwave absorption performances were calculated within the tested frequency ranging from 2 - 18 GHz. Results indicate that the polypropylene nanocomposites reinforced with the combination of iron nanoparticles and carbon nanotubes exhibited a reflection loss in a wider frequency ranges with the maximum reflection loss ranged from -22 to around -44 dB under different thickness. The results indicated that the relatively match permeability and permittivity was responsible for the high efficiency in the microwave absorptions. Also, the eddy current loss for the as-prepared nanocomposites was also studied and its influence on the microwave absorption performance was discussed in detail.

Graphene is as an attractive filler material for polymers because of its excellent electrical, mechanical and thermal properties. In this work, we report a massive increase in the electrical conductivity of a graphene nanoplatelet (GNP)/polystyrene composite by the addition of non-conducting particles. The non-conducting filler acts as a highly effective dispersion aid, preventing the sheet-like GNP from agglomerating during the solvent casting process used to fabricate the composite. The enhanced dispersion of the GNP leads to orders of magnitude enhancement in electrical conductivity compared to samples without this filler. Also adding rubber balls as filler materials along with graphene will help in fabricating multi-functional polymer composites with improved mechanical and as well as electrical properties. The rubber balls will help in keeping graphene dispersed to improve electrical property, adding enhanced mechanical property in the polymer to fabricate multi-functional polymer composite materials.

394222 Slow Growth Increases Myo-Inositol Availability for Glucaric Acid Production in S. Cerevisiae

Abstract:

Glucaric acid has been classified as a top value-added chemical from biomass for potential uses ranging from cancer therapies to sustainable nylons by the Department of Energy. A biosynthetic pathway was constructed in *E. coli* to convert glucose to glucaric acid but activity of the *Mus musculus* myo-inositol oxygenase (MIOX) enzyme was rate-limiting. New expression constructs enabled a one-step knock-in of the four pathway genes to *S. cerevisiae.* Protein sequence similarity mapping paired with simple structure-function hypotheses enabled efficient sampling of over 300 MIOX enzymes. Substituting a plant MIOX in the pathway increased glucaric acid titers detected in the YPD culture broth at 72hrs. Supplementation with 60mM myo-inositol tripled glucaric acid titers for yeast expressing either the plant MIOX (1.5 g/L) or the mouse MIOX (0.94g/L). However, myo-inositol supplementation did not affect glucaric acid titers in yeast expressing the mouse MIOX during slower growth on non-fermentable carbon sources such as glycerol or ethanol. Myo-inositol is a component of yeast membrane phospholipids and carbon-limited chemostat experiments demonstrated that slow growth rates lead to increased myo-inositol availability and glucaric acid titers. We pursued a fed-batch strategy in YPD with yeast expressing the plant MIOX and achieved final titers of 1.2g/L.

Development of Tissue Phantoms for a New Breast Cancer Detection Technique

Matthew A. Conrad
Caitlin B. Douglas
Adam J. Nolte, PhD

Rose-Hulman Institute of Technology | Department of Chemical Engineering

Background Information

- Healthy breast tissue has a Young's modulus of ~3 kPa and tumorous tissue has a Young's modulus of ~20 kPa

⇨ Gelatin hydrogels seem to be acceptably soft

- 250 Bloom B Gelatin was utilized

Rose-Hulman Institute of Technology | Department of Chemical Engineering

Research Goals

- To prepare gelatin tissue phantoms of a Young's modulus of healthy tissue with tumors of various sizes and geometries
- To characterize the mechanical properties such as the effects of aging, multiple tests, strain rate, and sodium alginate as an additive

Rose-Hulman Institute of Technology | Department of Chemical Engineering

Tissue Phantom Preparation

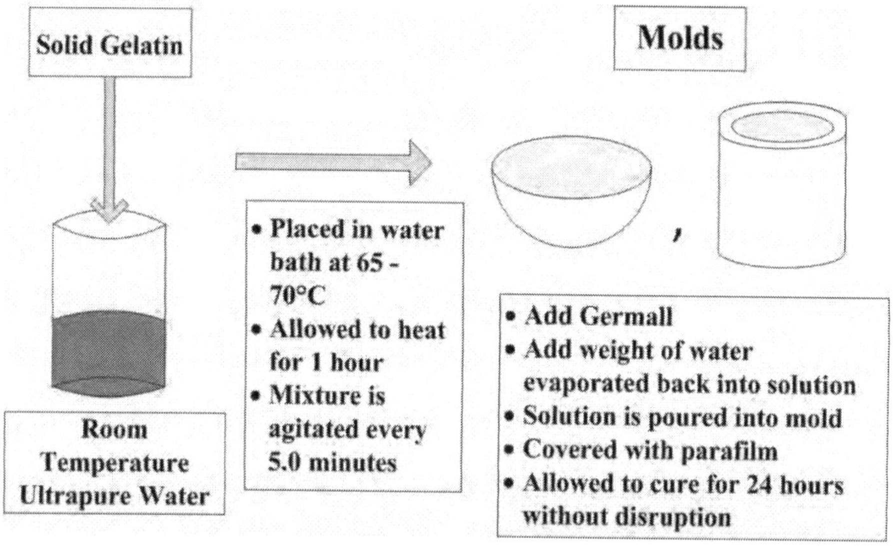

Rose-Hulman Institute of Technology | Department of Chemical Engineering

Tissue Phantom Preparation

Rose-Hulman Institute of Technology | Department of Chemical Engineering

Tumor Inclusion

Two proposed methods:

Liquid Injection via syringe

- 12.0% w/w gelatin liquid solution syringe-injected into congealing 4.0% w/w gelatin solution

⟹ • Not effective

Pre-made inclusions

- 12.0% w/w solid gelatin inclusion placed in congealing 4.0% w/w gelatin solution (T ~ 23.5°C)

⟹ • More effective approach

Rose-Hulman Institute of Technology | Department of Chemical Engineering

Mechanical Testing

Mechanical properties measured:

Young's Modulus
Fracture Strength
Toughness

- Samples tested in compression with an Instron with a 10 kN LC
  - Young's Modulus from strains less than 10%
  - Some samples tested to failure

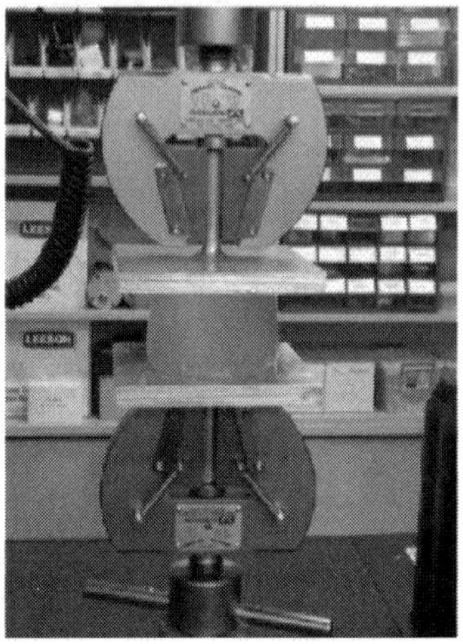

Rose-Hulman Institute of Technology | Department of Chemical Engineering

Concentration Dependence

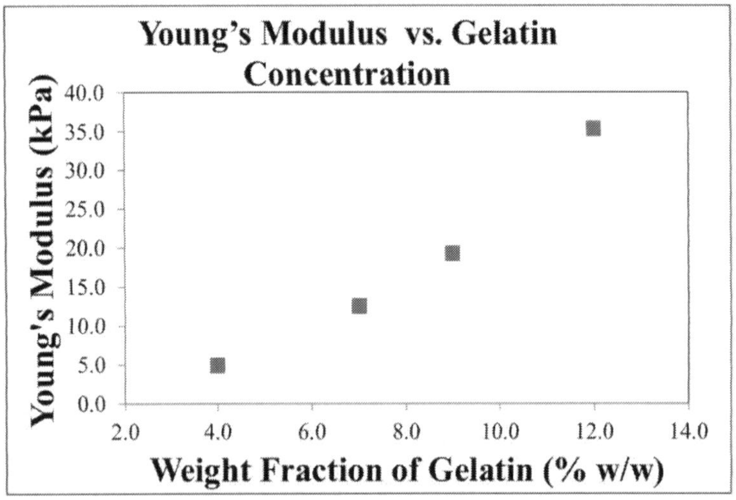

Rose-Hulman Institute of Technology | Department of Chemical Engineering

Aging Effects

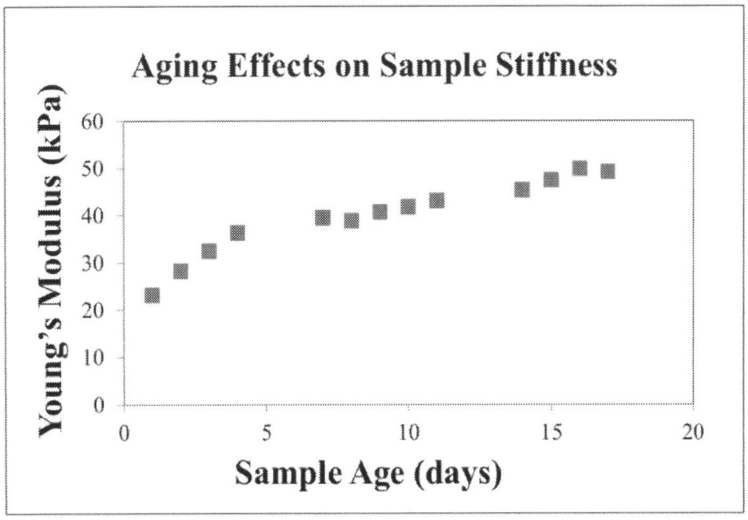

Rose-Hulman Institute of Technology | Department of Chemical Engineering

Testing Speed

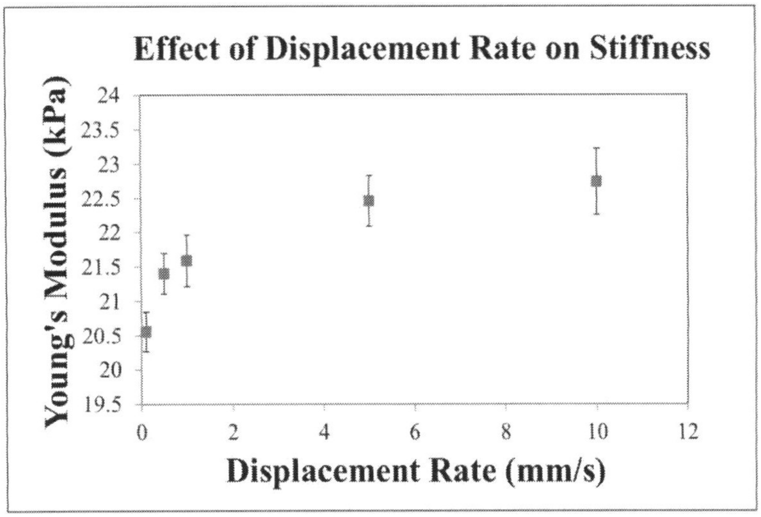

Rose-Hulman Institute of Technology | Department of Chemical Engineering

Multiple Test Effect on Stiffness

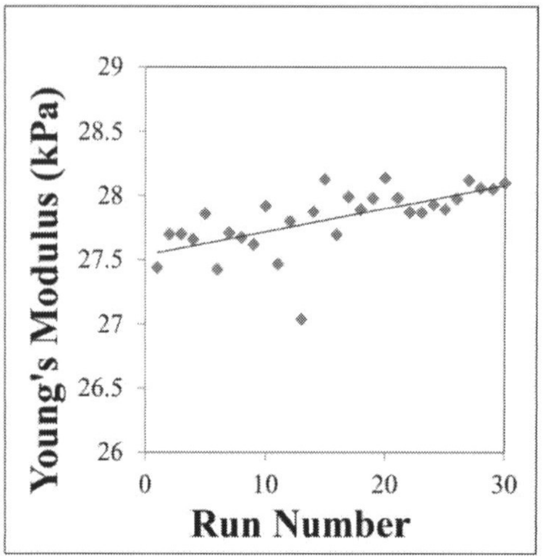

Uncovered during testing

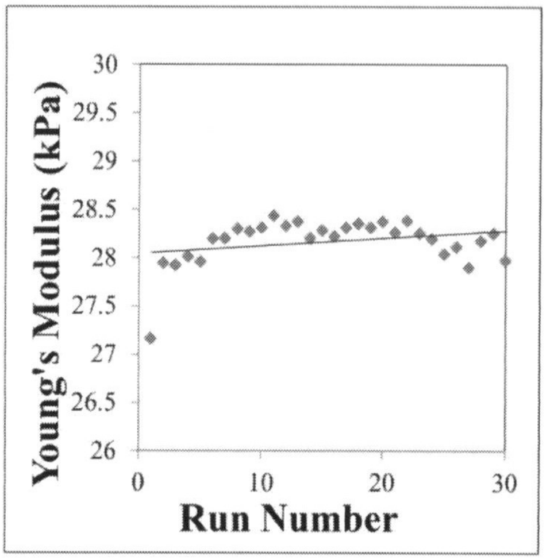

Covered during testing

Rose-Hulman Institute of Technology | Department of Chemical Engineering

Additive Effects: On Stiffness

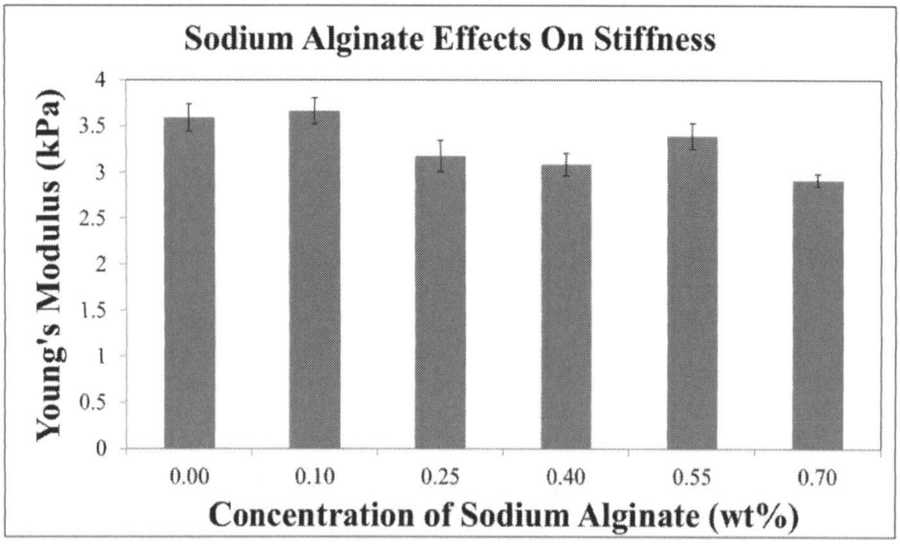

Rose-Hulman Institute of Technology | Department of Chemical Engineering

Additive Effects: On Fracture Strength

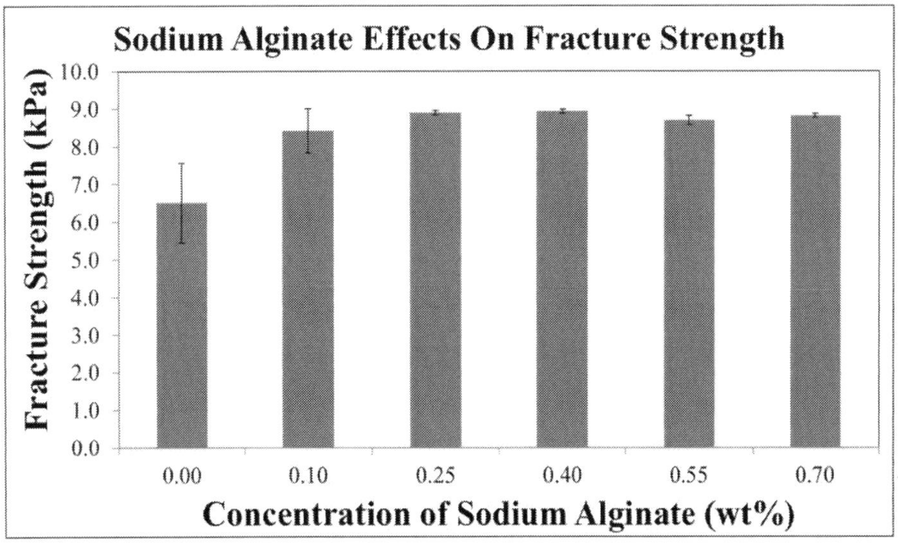

Rose-Hulman Institute of Technology | Department of Chemical Engineering

Additive Effects: On Toughness

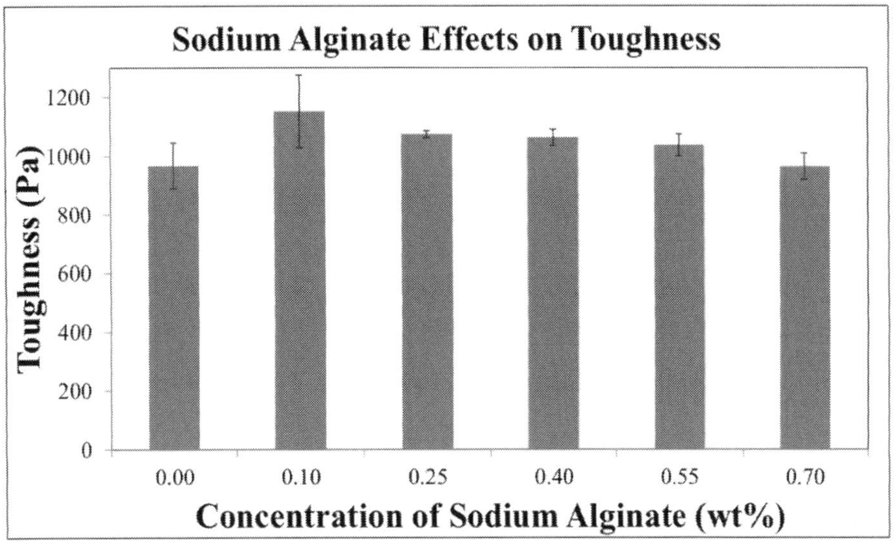

Rose-Hulman Institute of Technology | Department of Chemical Engineering

Conclusion

- Gelatin hemispheres can be made to mimic healthy breast tissue with tumor inclusions by inserting a pre-made tumor into a congealing gelatin solution

- Increasing the sodium alginate concentration appears to improve the fracture strength of gelatin hydrogels while decreasing the stiffness

Rose-Hulman Institute of Technology | Department of Chemical Engineering

Future

- Confirmation of results with texture analyzer

- More accurate relaxation tests

- Optimizing concentrations of additives

Rose-Hulman Institute of Technology | Department of Chemical Engineering

Acknowledgements

- ***This research was funded by the National Science Foundation under Grant No. ECSS-1306808***
- ***Professor Michael Insana and research group members (UIUC)***
- ***Professors Olson and Throne (RHIT)***

Rose-Hulman Institute of Technology | Department of Chemical Engineering

Relaxation Effects

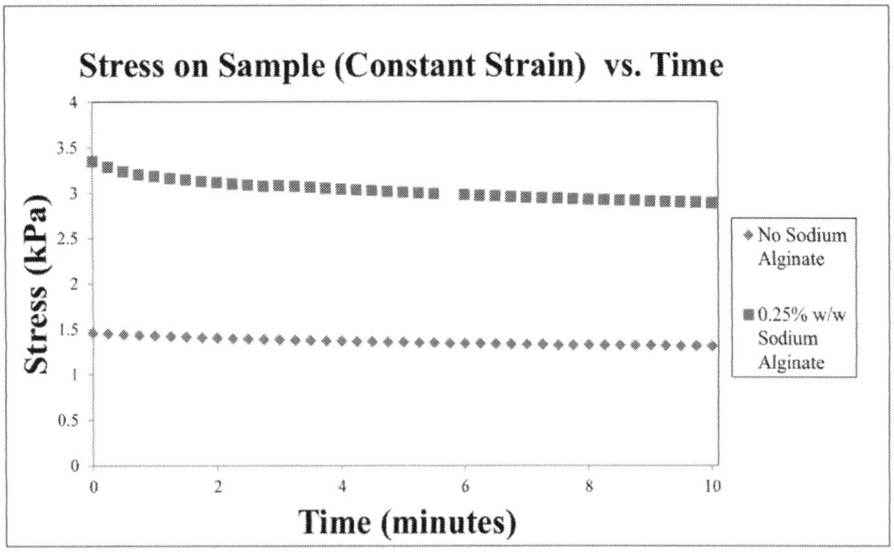

Rose-Hulman Institute of Technology | Department of Chemical Engineering

394192 Effective Antisense Design Using Ensemble of Energetically Sub-Optimal Secondary mRNA Structures

Abstract:

There is a deficit of effective therapies in the current research against bacterial infection. Many therapeutic strategies seek to use small molecules to target the infectious pathogen. One approach to disease inhibition involves direct manipulation of the pathogen at the RNA level. Part of the central dogma, messenger RNA (mRNA) is a transcript of genetic information that encodes the fundamental instruction for protein production. Inhibiting translation of mRNA effectively prevents synthesis of proteins.

In order to effectively block the RNA's message from being read, it must be physically accessed by the therapeutic agent. A field of study has arisen in which a complementary nucleic acid binding strand, called antisense[1], is generated to impede protein synthesis. One of the issues that arises in creating effective antisense RNA is the ability to find appropriate target sites on the mRNA to be inhibited. This problem is in part due to fluctuating secondary structure sequestering target sites. In fact, the kinetics of physical accession of the antisense to the target strand are suggested to be the rate-limiting factor[2] and thus the cause of inefficiency of this method. However, the fluctuations among suboptimal conformations are energetically inherent to the nucleic acid sequence and can be predicted. A software package, GenAVERT, has been developed to thermodynamically determine effective target sites by identifying the ensemble of the most probable energetically suboptimal states and determine which regions of mRNA are most accessible[3]. Once properly identified, a successful and effective antisense can, in theory, be synthesized.

In this work, a proof of principle will be demonstrated in *E. coli* with the downregulation of a short-lived variant of the green fluorescent protein through construction of an inducible antisense molecule. Statistical reduction in the fluorescence upon antisense induction will affirm the efficacy of this model.
Supported by fundamental energetic principles, this method of producing an antisense sequence complement for the most accessible mRNA target region has the potential to greatly reduce the strife of pathogenic infection.

---

[1] Lee, L. and Roth, C. (2003). Antisense technology in molecular and cellular bioengineering. *Current Opinion in Biotechnology*, [online] 14(5), pp.505-511.

[2] Patzel, V. and Sczakiel, G. (2000). In vitro selection supports the view of a kinetic control of antisense RNA-mediated inhibition of gene expression in mammalian cells. *Nucleic Acids Research*, [online] 28(13), pp.2462-2466.

[3] Johnson, E. and Srivastava, R. (2013). Volatility in mRNA secondary structure as a design principle for antisense. *Nucleic Acids Research*, [online] 41(3), pp.e43-e43.

Nanotechnology REU Summer 2014

Raghav Malik
AIChE National Conference: Undergraduate Research Forum II: Bioengineering, November 17th
Supervising Graduate Student: Giulia Suarato
Principal Investigator: Dr. Yizhi Meng
Department of Materials Science and Engineering

NSF

Stony Brook University

Proanthrocyanidin (PAC)

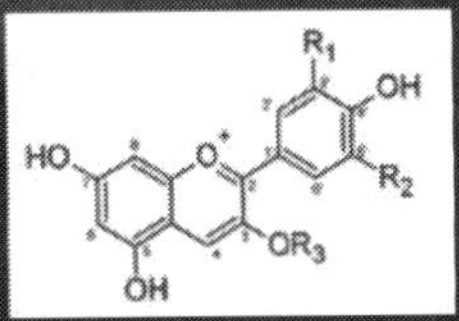

- Organically derived (from fruits and wines)
- Digests in saliva
- Promising chemotherapeutic effects, novel cancer cytotoxicity
- Encapsulation to improve bioavailability

NSF

Stony Brook University

Kom Kamonpatana, et. al. Susceptibility of anthocyanins to ex vivo degradation in human saliva, Food Chemistry. 2012: Vol. 135, 738-747

Chemotherapeutic Effects of PAC

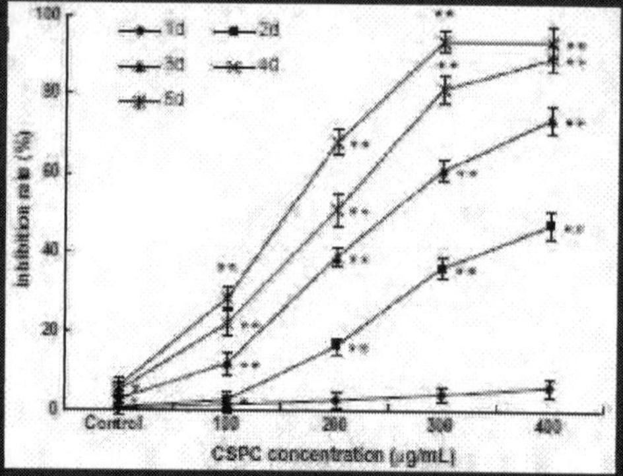

- Human Hepatoma G2 Cells
- Inhibition Rate: (Control – Treated / Control)*100
- Inhibition Rates Dose-Dependent
- Inhibition Time Dependent

NSF

Stony Brook University

H. Zhang, J. Ke, T. Shao, J. Li, Y. Duan, Y. He, C. Zhang, G. Chen, G. Sun, X. Sun. Food and Chemical Toxicology. 2014: Vol. 64, 166-176

Objective

Encapsulate PAC using chitosan and characterize the resulting nanoparticle as well as study the effect on MG-63 Osteosarcoma cells *in vitro*.

NSF

Stony Brook University

5

Why Chitosan?

- High Bioavailability
- Organic, derived from Chitin
- Biodegradable
- Hydrophilic

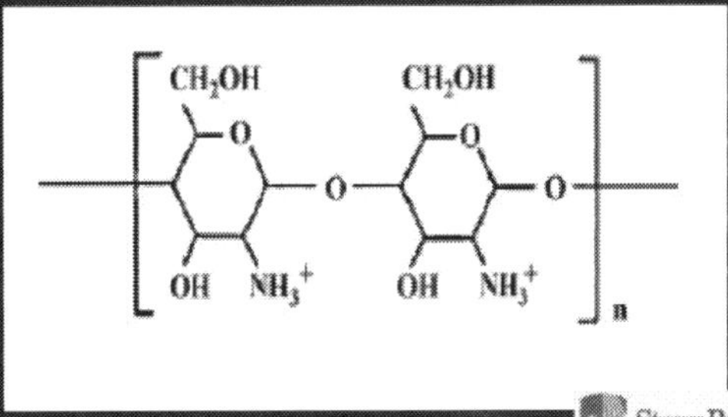

Chitosan

Stony Brook University

6

Encapsulation Potential

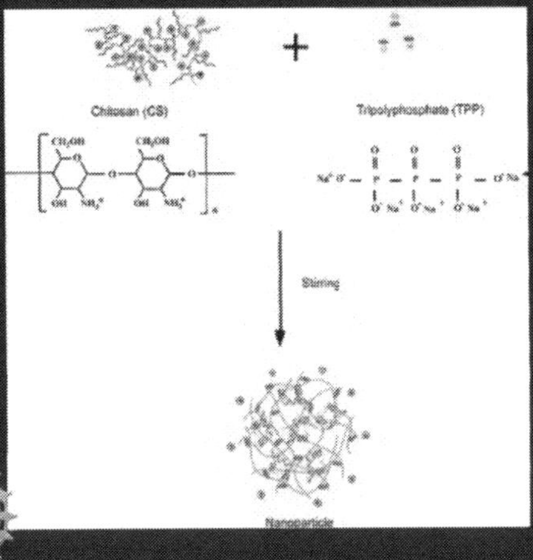

- Ionic Gelation Method
  - Pluronic F68 (detergent)
  - GCS:TPP 8.7:1 and 2.9:1
- Sizes are typically 200-300 nm

Stony Brook University

Adriana Trapani, Johannes Sitterberg, Udo Bakowsky, Thomas Kissel. International Journal of Pharmaceutics, 2009: Vol. 375, 97-106

7

Structure of Glycol Chitosan

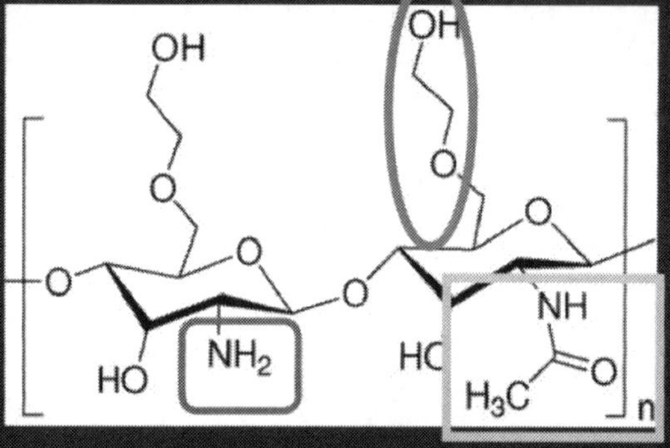

NSF

Stony Brook University

8

Approach

- Synthesis
- Characterization
- Delivery *in vitro*

NSF

Stony Brook University

9

Fabrication and Characterization of NPs

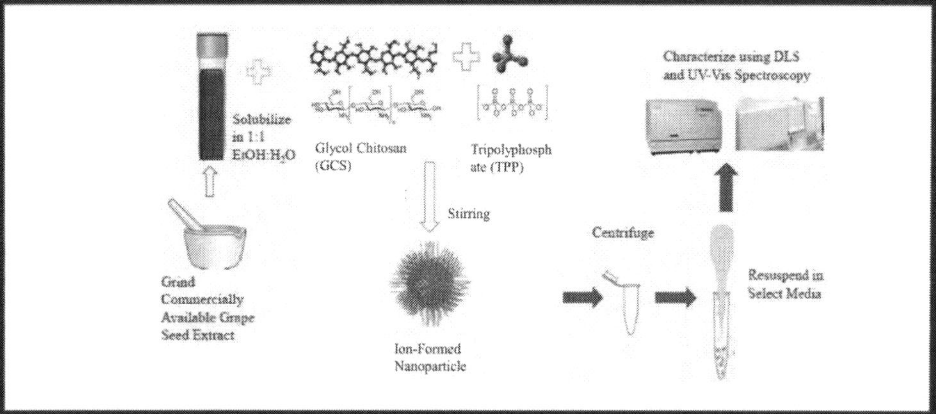

Stony Brook University

10

Nanoparticle Characterization

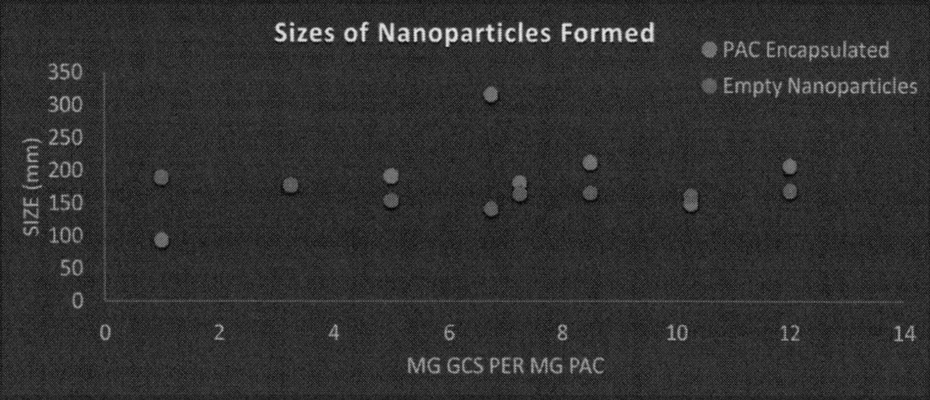

Stony Brook University

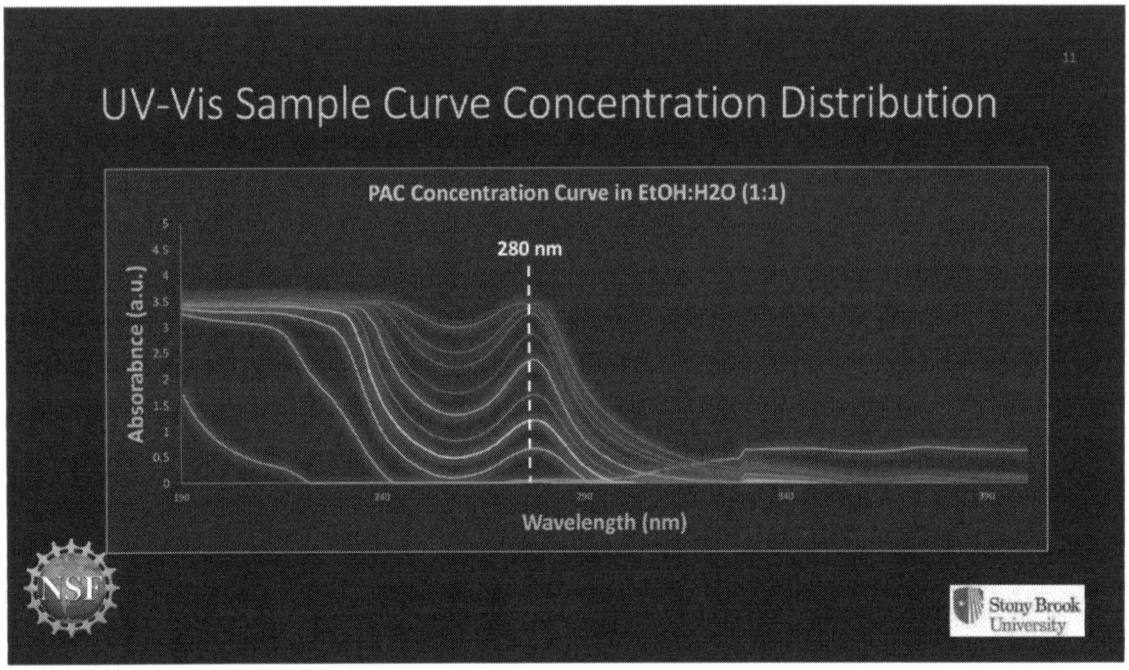
11
UV-Vis Sample Curve Concentration Distribution
PAC Concentration Curve in EtOH:H2O (1:1)
280 nm
Absorabnce (a.u.)
Wavelength (nm)
5
4.5
4
3.5
3
2.5
2
1.5
1
0.5
0
190
240
290
340
390
NSF
Stony Brook University

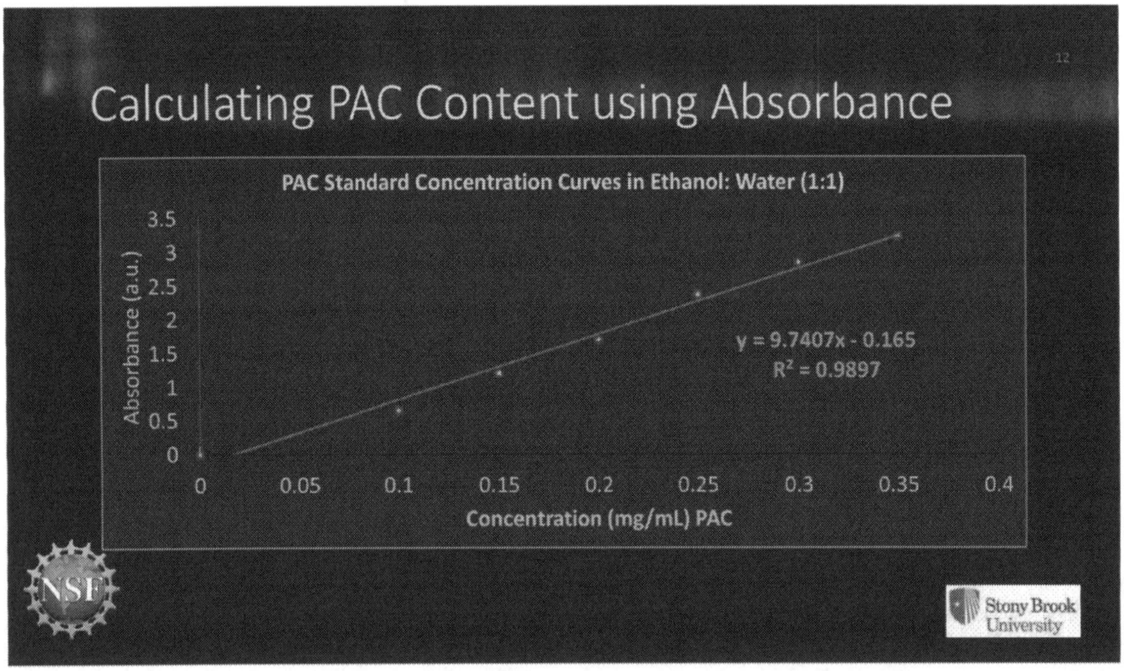
12
Calculating PAC Content using Absorbance
PAC Standard Concentration Curves in Ethanol: Water (1:1)
Absorbance (a.u.)
3.5
3
2.5
2
1.5
1
0.5
0
y = 9.7407x - 0.165
R² = 0.9897
0
0.05
0.1
0.15
0.2
0.25
0.3
0.35
0.4
Concentration (mg/mL) PAC
NSF
Stony Brook University

13

Encapsulation Efficiencies of GCS NPs

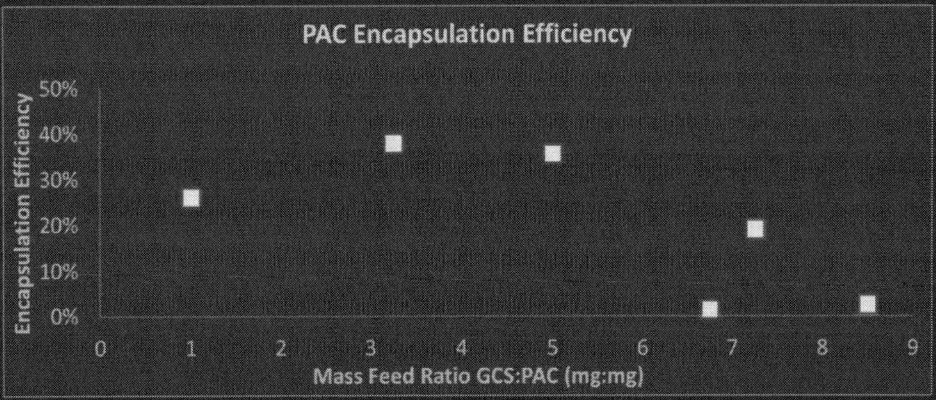

NSF

Stony Brook University

14

Delivery

- Cells used: MG-63 Human Osteosarcoma Cells
- Morphological study of cells after application
  - Comparing free and encapsulated PAC
- Cell Viability Assay done using MTS Assay
  - Comparing free and encapsulated PAC

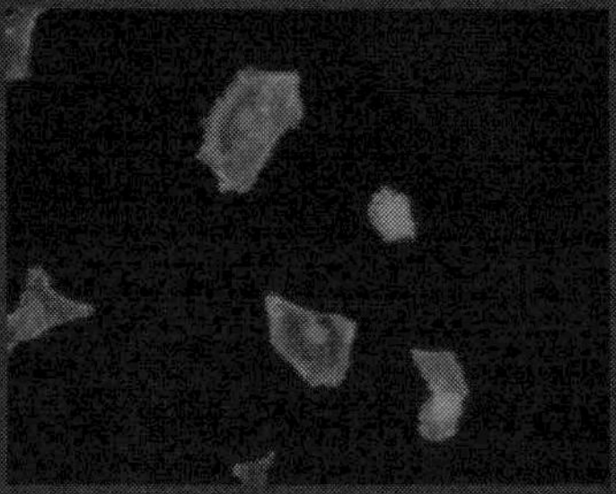

NSF

Stony Brook University

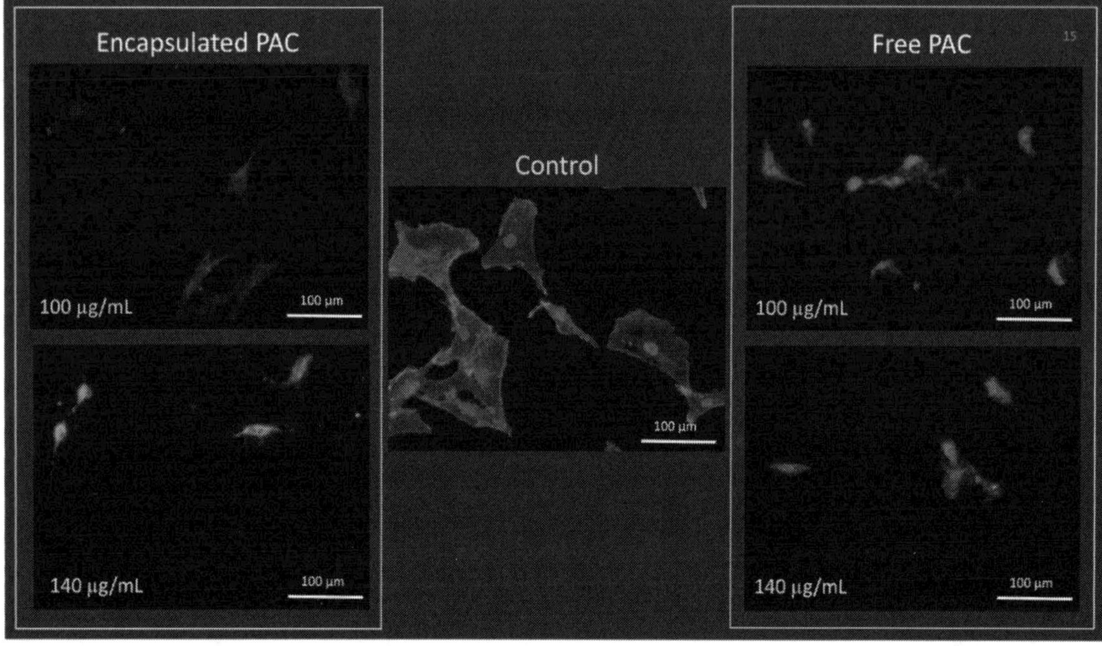

16

Cell Viability after 72 Hours

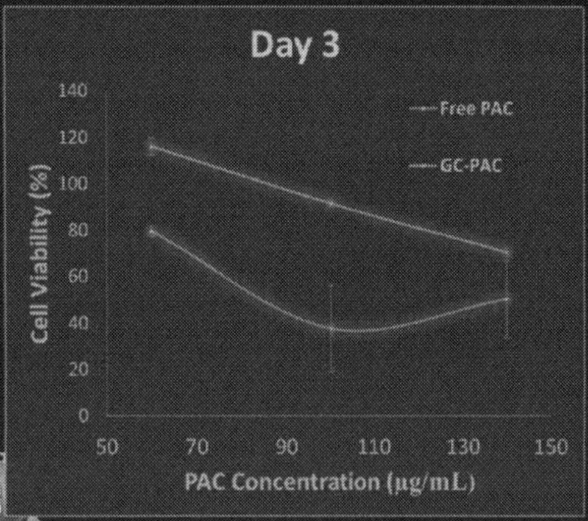

- MG-63 Osteosarcoma cells doubling time <72 hours
- Cell Viability is dose dependent
  - Can be adjusted for desired effect

Stony Brook University

17

Percent Cell Viability

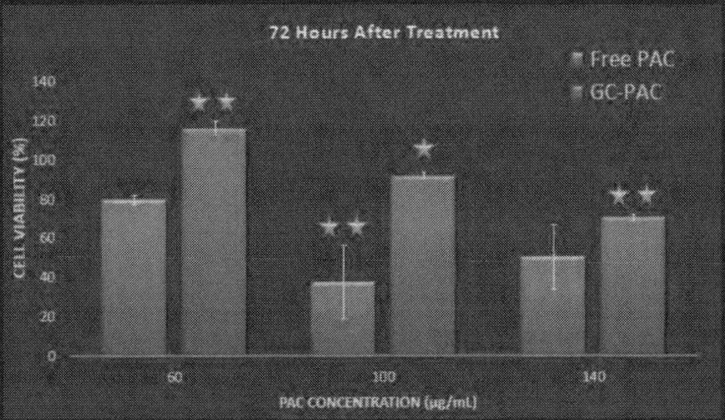

★ = $P < 0.05$

★★ = $P < 0.01$

18

Conclusions

- Most Efficient Mass Feed Ratios: 5 mg : 1 mg GCS:PAC
- Morphological analysis shows similar effects by both free and encapsulated PAC
- MTS Analysis indicates encapsulated PAC is advantageous over free PAC in its dose-dependency as well as effectiveness with greater amounts of PAC

Future Studies

- Release Kinetics of GCS-PAC NPs
- Amount of GCS in formed NPs

19
Acknowledgements
I would like to thank the National Science Foundation, Giulia Suarato, Dr. Yizhi Meng, the Center for Inclusive Education, and all other students and mentors involved in the REU for making this research possible by providing the resources necessary
NSF
Stony Brook University

394191 Single Cell Isolation Via Microfluidic-Based Droplet Delivery

Abstract:

Commercial technologies such as fluorescent-activated cell sorting (FACS) for isolating and delivering single cell samples are often inappropriate for smaller research facilities due to their size and cost. Researcher labs conducting experiments on single cells require an inexpensive and accessible method for creating samples. We demonstrate the practicability of using a gravity-driven microfluidic device under a bench-top optical trap to generate droplets containing isolated single cells for delivery onto a 96 well plate. The microfluidic device dispenses droplets of constant volume at fixed intervals. We establish functionality by delivering droplets containing glutaraldehyde fixed red blood cells (RBCs) onto glass slides. A fluorescent confocal microscope capable of imaging the glutaraldehyde-induced fluorescence of the RBCs is employed to count the number of cells contained per droplet. We attained 64% fidelity of single-cell samples, indicating that delivery from our device is not stochastic. Biomedical researchers often expect droplet delivery methods to be compatible with a 96 well plate standard; to this end we achieved automated delivery into each well of a 96 well plate using a Teensy 3.0 encoded motorized stage. Additionally, we implemented a side and forward scatter detection scheme around the optical trap to show the feasibility of fully automating the cell isolation and droplet delivery process. The entire apparatus is compact and portable enough to fit within a sterile environment to prevent contamination of samples. We propose that our setup is an inexpensive and compact alternative to flow cytometers that is accessible to most labs and can reliably sort single cells into droplets for sterile delivery.

394190 Self-Assembling Zwitterionic Nanogels As Immune Isolating Coatings for Stem Cell Derived Pancreatic Islet Transplantation

Abstract:

The concept of pancreatic islet transplantation to patients with diabetes to restore normoglycemia is a viable clinical strategy. However, widespread clinical application of islet transplantation remains limited because of the harmful side effects of the necessary immunosuppressive therapy to prevent host rejection of transplanted cells. Here we report the fabrication of a superbiocompatible nanogel and a novel coating strategy for protection of stem cell derived pancreatic islets against host rejection in vivo. Zwitterionic nanogels (100-300nm in size) were fabricated using an inverse emulsion polymerization synthesis scheme and further surface modified with CLICK reactive functional groups. Stem cell derived pancreatic islet cells were engineered to express azide functionalities on their cell membranes. Confocal fluorescence microscopy was used to verify covalent reaction and assembly of dual-layered nanogel coatings on engineered islet cell clusters. Cytocompatibility of coating and islet cell functionality were examined using a combination of in vitro glucose tolerance tests, live/dead staining, and in vivo studies on mice treated with streptozocin, a small molecule drug that destroys the beta cells. Our results indicate that the developed nanogel formulation and encapsulation strategy provides a viable strategy for prolonged protection and isolation of transplanted islets against host rejection

394187 Modeling of Selenium Nanoparticle Deposition for Optimized Production of Antibacterial Surfaces

Abstract:

The Webster Nanomedicine lab designs and explores applications of nanomaterials to inhibit bacterial growth and cancer cell responses. Due to its beneficial effects on the body, and known chemopreventive and chemotherapeutic properties[1], selenium has been used in its nano-form to impart nano-roughness and chemical effects in these experiments. To ensure precise and consistent selenium nanoparticle (SeNP) coverage, we aim to tabulate several relevant aspects of the particle. By adjusting the parameters of the reaction producing SeNP, a variety of data can be collected and analyzed. Characteristics that are pertinent include the nanoparticle size, coverage, surface energy, and roughness. Eventually, a cubic response surface model will be developed using a central composite design (CCD), which will be able to predict the interaction of the particles with bacteria and cells.

SeNP are produced through a reaction of glutathione (GSH) with sodium selenite (Na_2SeO_3). The reactants are introduced in a 4:1 molar ratio, respectively. Thin discs of poly(l-lactic acid) (PLLA) are added to the reaction, providing a surface for the cells to grow. In order to precipitate out the SeNP, 2M NaOH is added to the mixture. The reaction is then halted by the addition of DiH_2O. The experimental design used four factors: time in GSH/ Na_2SeO_3, time in NaOH, reaction volume, and the number of samples in the volume. Three results are compared: nanoparticle size, nanoparticle coverage, and surface energy of each sample. Both the size and the coverage of the produced nanoparticles can be measured using scanning electron microscopy. The surface roughness is portrayed through atomic force microscopy. Surface energy can be calculated utilizing the contact angle, which is defined using a Goniometer.

Previous studies show longer time increments (in both cases) increase the deposition of SeNP on the substrate. However, as the increments reach larger values of time (30-60 seconds), there is less of a change in coverage. This suggests the existence of a saturation point. On the PLLA, there is a finite number of binding sites for the nanoparticles. The size of the nanoparticle, with time being the only deviation, remains consistent. It is predicted that development time and reaction volume directly correlate. Recent observations of previous studies have shown that increased coverage correlates with development time in both steps, and therefore, roughness.

Through characterization of the SeNP product, a cubic response surface will be calculated relating the input parameters to expected results, optimizing the coverage of nanoparticles on substrates. These efforts conserve both time and resources because reactions can be designed with focused specifications.

371020 Engineering Modular Delivery Vehicles Using Biomimetic Polyelectrolytes

Abstract:

Electrostatically driven polymer self-assembly mechanisms are vastly underexplored compared to that of amphiphilic based assemblies, and yet offer unique opportunities for both encapsulation of charged therapeutics and controlled delivery via the tailoring of intermolecular interactions using pH and salt. Additionally, using polypeptide-based materials we have recently shown that oppositely charged polyelectrolytes self-assemble into either liquid or solid complexes because of their secondary structure, providing another avenue to tune material properties. Using molecular engineering, nanoscale stabilization of polyelectrolyte complex formation can be achieved by coupling the polyelectrolyte to a neutral yet hydrophilic block, forming nanometer sized micelles with a polyelectrolyte complex core of either solid or liquid nature and a hydrophilic corona. In this work, we characterize the structure and stability of these polypeptide based model micellar systems using static and dynamic light scattering, electron microscopy, circular dichroism and small angle x-ray scattering. Additionally, we create polyelectrolyte complex micelles that contain therapeutically relevant charged molecules such as miRNA and peptides, specifically for the treatment of atherosclerotic lesions and cancer. The modular nature of these assemblies enables the addition of a targeting ligand to increase the efficacy of delivering miRNA, creating polyelectrolyte complex micelles with a targeting ligand outside the corona of the micelle. Initial cell studies conducted with the different miRNA containing micelles will be discussed.

This work is supported by the University of Chicago and the U.S. Department of Energy Office of Science program in Basic Energy Sciences and the Materials Sciences and Engineering Division.

ID: 378044
Password: 248756

AIChE – Biomaterials – 08B15 Biomaterials: Faculty Candidates I
Title: Self-Assembled Hydrogels Utilizing Polymer-Nanoparticle Interactions
Authors: Mark W. Tibbitt, Eric A. Appel, and Robert Langer

Hydrogels are an important class of biomaterial that have received significant attention for controlled drug-delivery applications on account of their similarity to soft biological tissue and highly tunable mechanical properties.[1] Shear-thinning hydrogels are water-swollen dynamically cross-linked polymer networks exhibiting viscous flow under shear stress (shear-thinning) and rapid recovery of mechanical properties when the applied stress is relaxed (self-healing).[2] These properties afford minimally invasive implantation in vivo though direct injection or catheter delivery to tissues, contributing to a rapid gain in interest in their application for drug delivery and tissue engineering.[3] Herein, we report the preparation and application of shear-thinning, injectable hydrogels driven by non-covalent interactions between hydrophobically-modified biopolymers (BPs) and drug-loaded biodegradable nanoparticles (NPs) comprised of poly(ethylene glycol)-block-poly(lactic acid) (PEG-block-PLA). This materials approach builds upon the recent paradigm of NPs as molecular binders in polymeric systems.[4] The transient and reversible interactions between BPs and NPs enable flow under applied shear stress, followed by rapid self-healing when the stress is relaxed, demonstrating the shear-thinning and injectable nature of the material. The inclusion of PEG-block-PLA NPs facilitates a high-loading content of hydrophobic molecules within the hydrogel, while a second, hydrophilic molecule, can be loaded simultaneously into the aqueous phase of the gel. Delivery of the hydrophilic molecule is controlled by Fickian diffusion through the gel whereas erosion-based release from the gel surface governs delivery of the hydrophobic molecule. In this manner, multiple therapeutic compounds can be delivered with differential release curves. Additionally, the biocompatible, shear-thinning hydrogels can be implanted into the subcutaneous space of mice for controlled release of therapeutics *in vivo* or as a cell carrier for regenerative medicine applications. Overall, this presentation will demonstrate the facile synthesis of an injectable hydrogel affording controlled delivery of therapeutics with minimally invasive application.

References
1. Hoffman AS *Adv Drug Delivery Rev* **54** (2002) 3-12
2. Appel EA *et al. Chem Soc Rev* **107** (2010) 18392-18397
3. Guvendiren M *et al. Soft Matter* **8** (2012) 260-272
4. Rose S *et al. Nature* **505** (2014) 382-385

378549 Injectable Hydrogels with in Situ Double-Network Formation for Cell Transplantation

Abstract:

Cell transplantation via direct cell injection at the target site is a minimally invasive strategy for treating various injuries and degenerative diseases. However, cell viability is typically only 5% in these procedures, and the therapeutic success critically hinges on the survival and subsequent maintenance of the transplanted cells. To protect these cells, we designed an injectable hydrogel that undergoes two different physical crosslinking mechanisms. The first crosslinking step occurs *ex vivo* through peptide-based molecular recognition to encapsulate cells within a physical hydrogel, while the second crosslinking step occurs *in situ* through a thermal phase transition to form a reinforcing network. At room temperature, the physical hydrogel is shear-thinning and self-healing to facilitate gentle cell encapsulation and transplantation by syringe injection. At body temperature, the hydrogel forms a secondary network resulting in a 10-fold increase in shear modulus and significantly reduced erosion rates and prolonged retention time. Human adipose-derived stromal cells (hASCs) injected through a 28-gauge syringe needle were significantly protected from disruptive mechanical forces when encapsulated within the hydrogel compared to medium alone. *In vivo* subcutaneous injection of hASCs in a murine model demonstrated significantly improved cell retention when delivered in a double-network hydrogel compared to a single-network gel or medium alone. These results suggest that in situ formation of a reinforcing network within an already existing hydrogel can enhance mechanical properties and retention time, and thereby improve long-term cell viability. These cell-delivery materials are being further investigated for the ability to support functional tissue regeneration at sites of ischemia.

Introduction: Collagen hydrogels have been widely investigated as scaffolds for vascular tissue engineering due in part to the abundance of collagen in the vessel wall and the capacity of collagen networks to support cell spreading. However, collagen hydrogels display relatively low stiffness and strength, have a tendency to undergo rapid degradation, and are highly susceptible to cell-mediated contraction. In the current work, we develop and characterize a hybrid natural-synthetic interpenetrating network (IPN) that retains the benefits of collagen in terms of enabling robust cell elongation, but which displays enhanced stiffness and strength as well as increased resistance to cell-mediated contraction and degradation. Toward this end, we first formed a collagen-based hydrogel, infused this hydrogel with a solution of poly(ethylene glycol) diacrylate (PEGDA) plus photoinitiator, and subsequently crosslinked the infusate by exposure to longwave UV light. This approach allowed for the tailoring of scaffold mechanical performance by varying the molecular weight of the infusing PEGDA, and resulted in networks which displayed the resistance of PEGDA to degradation and contraction. Furthermore, we demonstrated the ability of these collagen I-PEGDA (coll-PEGDA) IPNs to support the differentiation of encapsulated human mesenchymal stem cells (MSCs) toward a vascular smooth muscle cell (VSMC) fate.

Materials and Methods: *Fabrication of Coll-PEGDA IPN.* Collagen hydrogels (3 mg/mL) were formed at 37 °C. These collagen hydrogels were then infused with a solution of 11 wt% PEGDA (3.4 kDa, 6.0 kDa and 10.0 kDa) and Irgacure 2959 as photoinitiator, and subsequently crosslinked by exposure to longwave UV light for 6 min. Rheological and tensile properties were evaluated for each IPN formulation. The micro- and nano-structures of the IPNs were also characterized by laser reflection confocal and scanning electron microscopy (SEM). *Construct Culture and Analysis.* Human bone-marrow derived MSCs (Lonza) were encapsulated at $1x10^6$ cells/cm^3 within the collagen hydrogel prior to 6.0 kDa PEGDA network formation. Resulting constructs were cultured for 2 wks in DMEM with 10% MSC-qualified FBS supplemented with 10 ng/mL of TGF-β_3 and 1% antibiotic/antimycotic. Gene expression was then analyzed relative to GAPDH using qRT-PCR.

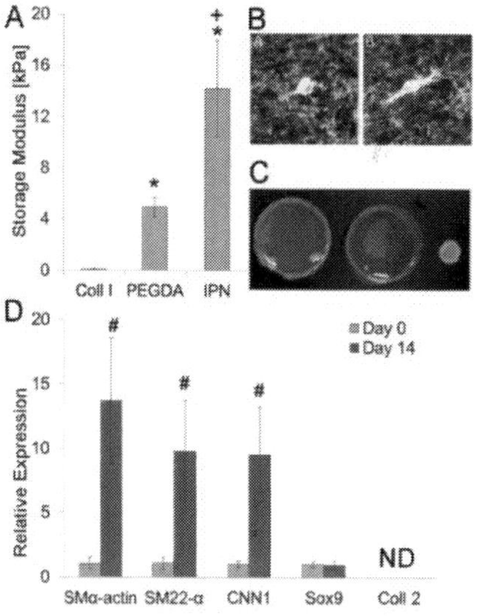

Fig. 1. (A) Storage modulus of Coll-PEGDA IPNs relative to pure PEGDA and pure collagen counterparts. (B) Control over extent of cell spreading within the IPN networks (collagen fibers - green, cells - yellow). (C) Size of Coll-PEGDA IPN, PEGDA, and collagen hydrogels following 2 wks of culture. (D) Relative gene expression of VSMC and chondrogenic markers in Coll-PEGDA IPNs in media supplemented with TGFβ_3.. *, $^+$, significantly different from pure collagen and PEGDA; # significantly different from day zero

Results and Discussion: For a 11 wt% 3.4 kDa PEGDA infusate, the formation of a coll-PEGDA IPN resulted in 140-fold increase in elastic modulus relative to the pure collagen counterpart and a nearly 10-fold increase in modulus relative to the corresponding pure PEGDA gel (**Fig. 1A**). The coll-PEGDA IPNs also exhibited tensile strengths ranging from 50-110 kPa, depending on the MW of PEGDA utilized (data not shown). These tensile strengths were approximately 2-fold greater than the corresponding pure PEGDA gels and markedly greater than that of pure collagen gels (< 1 kPa tensile strength). In addition, these IPN systems allowed for high viability (> 90%) cell encapsulation via the inclusion of cells within the collagen hydrogel precursor solution (data not shown). By varying the time between the formation of the cell-laden collagen hydrogel and the formation of the PEGDA secondary network, the degree of cell elongation within the IPN could be controlled from round to highly elongated (**Fig. 1B**). Finally, the capacity of Coll-PEGDA IPNs to support MSC smooth muscle progression was analyzed following 2 wks of cell culture in the presence of TGF-β_3, a growth factor which can stimulate VSMC or chondrogenic differentiation. Importantly, elongated MSCs within the coll-PEGDA networks showed up to a 14-fold increase in VSMC marker expression relative to day 0, whereas no increase in chondrogenic markers was observed ($p < 0.05$; **Fig 1D**).

Conclusions: Coll-PEGDA IPNs fabricated via sequential infiltration have significant potential in vascular graft applications due their ability to support smooth muscle lineage progression and their enhanced stiffness and strength and increased resistance to cell-mediated contraction (**Fig. 1C**) relative to pure collagen.

Acknowledgements: We gratefully acknowledge funding from the NIBIB.

387143 Delivery of SDF-1α Lowers the Effective BMP-2 Dose for in Vivo Mineralization

Abstract:

The use of bone morphogenic proteins (BMPs) shows promise in therapies for improving bone regeneration; however, high supraphysiological concentrations required for a desired osteoinductive effect, costs, and patient variability have prevented the full advantages of BMP-based therapeutics from being realized. Thus, one strategy is to deliver synergistic molecules with BMP to enhance efficacy and lower doses. In this work, a matrix metalloprotease (MMP)-sensitive hyaluronic acid (HA)-based hydrogel was used for the delivery of both stromal cell-derived factor-1 alpha (SDF-1α) and BMP-2 towards improving BMP-induced osteogenesis. SDF-1α has been shown to play an important role in stem cell trafficking and HA hydrogels are known to increase extracellular matrix production. Additionally, our lab has previously shown that HA is capable of inducing cell chemotaxis through the CD44 receptor.

Maleimide-modified HA (MaHA) was cross-linked using a difunctional MMP-sensitive peptide according to an addition reaction, allowing for protease-mediated hydrogel degradation and growth factor release. The compressive and rheological properties of MaHA hydrogels were characterized as a function of hydrogel polymer content (1-5 wt%) and functionalization (10-40%f), where mechanical properties and gelation time were strongly dependent on cross-linking density. To determine growth factor release, MaHA hydrogels loaded with either 100 ng/scaffold of SDF-1α or BMP-2 were degraded in 10, 2, or 1 U/ml type II collagenase (non-specific MMP degradation) and compared to degradation in buffer alone. In general, BMP-2 and SDF-1α release corresponded closely with hydrogel degradation profiles, where limited release occurred through diffusion in the absence of MMPs. In the presence of MMPs, hydrogel degradation and growth factor release was controlled as a function of both collagenase concentration and hydrogel formulation, where faster growth factor release occurred at higher collagenase concentrations and for hydrogels with lower cross-link densities. In vitro SDF-1α bioactivity was determined by quantifying cellular invasion or outgrowth from a hydrogel-encapsulated cell pellet into the surrounding hydrogel over time. After four days, significant invasion from the cell pellet was observed for hydrogels loaded with 250 ng of SDF-1α and limited invasion was seen without any SDF-1α loaded into the hydrogels. Lastly, a critical-sized cranial defect was used to determine the effect of growth factor delivery on bone formation in vivo. The animal group with combined SDF-1α and BMP-2 delivery showed significantly higher bone formation when compared to groups loaded with the same BMP-2 or SDF-1α concentration delivered separately or the empty defect group. Furthermore, histology indicated increased in vivo cellular invasion for hydrogels loaded with SDF-1α.

Mechanically Dynamic Hydrogels for Probing Hepatic Stellate Cell Response in Fibrosis

Steven R. Caliari[1], Maryna Perepelyuk[2], Rebecca G. Wells[2], Jason A. Burdick[1]
[1]Department of Bioengineering, University of Pennsylvania, [2]Department of Medicine, University of Pennsylvania

Introduction: Chronic liver injuries drive hepatic stellate cell (HSC) differentiation into myofibroblasts. This activation process leads to excessive extracellular matrix (ECM) deposition, resulting in fibrosis and loss of liver function. While static model systems have provided insight into the role of mechanics in the development of fibrosis, there is a paucity of engineered model systems that can recapitulate the temporal dynamics of disease states in order to understand relevant cellular signaling events. We are addressing this issue by developing a suite of static and dynamic hyaluronic acid (HA) hydrogel substrates to model the mechanical progression and regression of liver fibrosis.

Results: Stiffening and softening methacrylated HA (MeHA) systems were developed to mimic the progression and regression of liver fibrosis respectively. Soft-to-stiff MeHA gels were fabricated by an initial Michael-type addition reaction followed by visible light-mediated secondary radical crosslinking at user-defined time point(s) with no negative impact on HSC viability observed. HSC myofibroblast differentiation was correlated with substrate stiffness, not the generation of radicals during secondary crosslinking. The timing of stiffening significantly affected HSC phenotype with later stiffening (day 7) resulting in more rapid cytoskeletal re-organization as measured by cell spread area and α-smooth muscle actin (α-SMA) expression compared to early stiffening (day 1). After 14 days, ~ 95% of HSCs cultured on stiffened gels expressed α-SMA regardless of early (day 1) or late (day 7) stiffening. Stiff-to-soft gels mimicking the regression of fibrosis were formed by incorporating ester-containing hydrolytically-degradable crosslinkers. Mechanics could also be cycled to model repeated injury. Stiff-to-soft gels (Day 0 stiffness: 17.3 ± 4.5 kPa, Day 14 stiffness: 3.4 ± 1.0 kPa) could be re-stiffened by radical polymerization of the remaining gel methacrylates (re-stiffened modulus: 30.0 ± 3.7 kPa).

Conclusions: We describe a series of stiffening and softening hydrogel constructs to explore the influence of dynamic changes in mechanical properties on HSC phenotype during fibrosis activation and regression. Critically, this platform enables nuanced interrogation of HSC phenotype in response to an evolving set of microenvironmental cues. Ongoing work is tracking *in vivo* liver mechanics during fibrosis induction and recovery to guide rational biomaterial design. Together, the dynamic substrates developed here will be useful as streamlined models of other disease states in the wound healing-fibrosis-cancer triad and as *ex vivo* tools to identify potential targets for therapeutic intervention.

387816 Engineering a Bioartifical Pancreas: Tuning Device Geometry Prevents Foreign Body Immune Responses and Fibrosis to Enable Prolonged Efficacy in Diabetic Rodents

Abstract:

Introduction: Transplantation of donor pancreatic islet cells into diabetics to restore normoglycemia has been common practice for decades. However, major hurdles still exist that drastically limit the efficacy of this treatment for patients. One major problem is the need for immunosuppressive therapies to prevent rejection of transplanted cells. To overcome this barrier efforts have been placed in developing a bioartificial pancreas, whereby islets are encapsulated within a semi-permeable immunoprotective barrier and then transplanted into diabetic patients. The long-term function of the bioartificial pancreas is dependent on its interaction with the host immune system. Immune recognition initiates a cascade of cellular processes leading to foreign body reactions, which include persistent inflammation, fibrosis (walling-off), and damage to the surrounding tissue. These unwanted effects are deleterious to the function and viability of the encapsulated islet cells. Towards the development of more biocompatible hydrogels we investigated the influence of hydrogel geometry on host recognition and fibrosis.

Materials and Methods: To investigate our hypothesis we fabricated a series of Ba cross-linked SLG20 alginate microcapsules of varying geometries with precise dimensions. Using wild type C57BL/6 mice, a robust model for fibrosis we interrogated the effects of fibrosis formation on our capsules to identify a unique geometry that proved to be most efficacious at resisting fibrosis. Rat and human derived pancreatic islet cells where then encapsulated lead capsule formulations and transplanted to Streptozocin (STZ) treatment diabetes induced C57BL/6 mice and monitored ability to restore normoglycemia over time.

Results and Discussion: Alginate based hydrogels of tuned geometry significantly abrogated foreign body reactions and fibrosis when compared to conventional alginate microcapsules. Tuned hydrogel capsules resisted fibrosis for at least 6 months when transplanted in the intraperitoneal space, a common site for bioartifical pancreas transplantation. Our study of probing immune cells recruited to transplanted materials implicates the limited activation of macrophages, a key driver of host recognition, as central to these truncated foreign body responses. Most significantly, diabetic mice treated with tuned format bioartificial pancreas maintained normoglycemic for over 6 months, producing a 3-fold improvement over mice treated with conventional format alginate based bioartificial pancreas.

Conclusions: We have demonstrated that by simply tuning the geometry of transplanted materials we can drastically influence their host recognition and propagation of foreign body reactions. Using these findings we fabricated a tuned format alginate based bioartificial pancreas, which was demonstrated to be significantly longer lasting and more efficacious than conventional format bioartificial pancreas systems. As such, we believe tuned format capsules could lead to a much improved cell encapsulation therapy strategy for type 1 diabetes. Furthermore, we believe that our findings have significant implications on the design of in vivo-transplanted biomedical devices for a range of applications including cell transplantation, controlled drug release, implantable sensors, and prosthesis for tissue engineering.

Candidate: **Natesh Parashurama MD PhD**

Title: Liver Regenerative Medicine and In Vivo Molecular Imaging for the Study of In Vivo Liver Organogenesis, Liver Disease and Development of New Diagnostics and Therapeutics

The emerging field of liver (hepatic) regenerative medicine encompasses understanding hepatic stem cell biology, organogenesis, tissue regeneration, and tissue homeostasis for the study of liver disease and development of new diagnostic and therapeutic strategies. The liver, the largest internal organ in the body, performs critical metabolic, endocrine, and synthetic physiological functions. As such, it displays elaborate cellular phenotypes and intricate tissue architecture, including hexagonal, lobular structure, complex vascular spaces with specialized endothelial cells, tissue specific fibroblasts and macrophages, complex blood supply (hepatic artery and portal vein) and heterogeneous billiary duct cells which line billiary duct drainage systems. With respect to capacity for organ regeneration, the liver is unmatched. For example, preclinical, rodent studies have indicated acute hepatocellular loss results in compensatory replication of neighboring adult hepatocytes. In contrast, chronic hepatocellular injury, which often leads to end-stage disease, is compensated for by small, liver stem/progenitor cells lying at the junction of intrahepatic bile ducts and the Canals of Hering. Furthermore, circulating and immune cells, like bone marrow mononuclear cells, mesenchymal stem cells, and tissue macrophages, all have major roles in orchestrating repair in liver injury. These repair mechanisms have been translated to established treatments, like human liver transplantation, and promising therapies, such as human hepatocyte, human bone marrow-derived, and human liver stem/progenitor cell transplantation. However, current and new therapies require harvesting of human tissue, underlying mechanisms of tissue homeostasis remain elusive and hard to decipher in vitro, in animals and humans, and new approaches and techniques are needed answer critical scientific questions and take steps towards new diagnostics and therapeutics aimed towards regeneration.

A deeper understanding of human, three-dimensional (3D) liver organogenesis would naturally advance liver regenerative medicine. Unfortunately, crucial steps of liver organogenesis take place in utero, and most research has been focused on making functional hepatocytes rather than re-creating liver organogenesis. Nevertheless, this would lead to more accurate studies of acute and chronic regenerative capacity, metabolic, detoxification, and synthetic functions, and disease modeling. A second point is the need for being able to assess 3D liver organogenesis while it is occurring. Current techniques rely on removing tissue, treating the tissue and analyzing it in a static fashion. However, techniques are also needed to assess dynamic molecular events in real time, in the intact living subject, during organogenesis or regeneration. These needs are addressed by techniques in the field of molecular imaging, which aims to study biological events and pathways in the intact living subjects. Thus the combination of studying 3D liver organogenesis, and applying current techniques and developing new techniques in molecular imaging, will likely further the field of hepatic regenerative medicine.

393651 Implantable Scaffolds to Engineer a Pre-Metastatic Niche for Early Detection and Treatment of Breast Cancer Metastasis

Abstract:

Breast cancer metastasis continues to remain the highest contributor to breast cancer-related deaths worldwide. A major limitation in developing life-preserving timely interventions is the scarcity of robust technologies capable of early detection of metastatic disease as well as model systems that allow systematic screening and examination of factors contributing to tumor metastasis in a controlled setting. Prior to colonization of a metastatic site, a "pre-metastatic niche" that consists of cells, soluble factors, and extracellular matrix components is established, acting as a supportive microenvironment for tumor cell homing and colonization. Importantly, the existence of this niche indicates that metastasis to a particular site is not random but is pre-determined. This view indicates that an environment could be engineered to attract these metastatic cells. To this end, we previously developed microporous poly(lactide-co-glycolide) (PLG) scaffolds that induce metastatic tumor cell homing to a defined site (e.g., subcutaneous space or peritoneal fat pad) *in vivo* for early detection of metastatic disease and subsequently reduce tumor burden in major organs (e.g., lung). However, translation of this technology to the clinical setting necessitates biomaterials with greater long term stability. To address this need, we developed microporous poly(ε-caprolactone) (PCL) scaffold implants that exhibit little to no degradation over a period of 14 weeks when implanted subcutaneously in tumor free mice, demonstrating stability of the implant for long term studies and eventual translation. Moreover, we have demonstrated that PCL scaffolds recruit metastatic tumor cells and reduce tumor burden in major organs (e.g., lung) when implanted in the subcutaneous space of tumor bearing mice. We are investigating the utility of PCL implants in mouse models that recapitulate the application of this technology in the clinical setting. For example, we have demonstrated that PCL scaffolds recruit metastatic tumor cells when implanted subcutaneously for longer time periods *in vivo* (e.g., ~ 1 month) prior to tumor induction - a situation that mimics the chronic response to a scaffold implant. Further, to mimic clinical disease progression, we have established a post-surgical model of breast cancer metastasis incorporating PCL scaffolds implants and are currently examining their ability to influence survival. In sum, these implants with early detection and therapeutic potential could be easily integrated into current breast cancer disease management plans and hold significant promise to lead to a reduction in breast cancer mortality in the near future.

384800 Structural Analysis of Lithiated Lignin-Derived Carbon Composite Anodes

Abstract:

Novel lignin-based carbon composite materials have been developed for use as anodes in lithium-ion batteries. The structure of the uncharged anodes has been previously characterized via pair distribution functions (PDFs) obtained from neutron scattering experiments. Computational models of the uncharged composite systems were developed and studied using classical molecular dynamics simulations. The PDFs of the model systems successfully reproduced the experimental PDFs, and the model PDFs were then decomposed into component functions to understand the origin of each characteristic feature of the overall PDF. In this presentation, we extend our work to include an understanding of the charged and discharged anodes via the use of reactive lithium-ion simulations. The resulting change in the uncharged composite PDF due to lithiation and subsequent de-lithiation is explained by developing a relationship between the properties of the composite model (volume fraction of crystalline material, characteristic size of the crystalline domains, composite density, and ion loading) and the unique features of the PDFs, including peak height, peak location, and long-range oscillations.

N.M. was supported by a grant from the Oak Ridge Associated Universities High Performance Computing Program, by a grant from the Sustainable Energy Education and Research Center of the University of Tennessee and by a grant from the National Science Foundation (DGE-0801470). This research project used resources of the National Institute for Computational Sciences (NICS) supported by NSF under agreement number: OCI 07-11134.5. This research at Oak Ridge National Laboratory's Spallation Neutron Source was sponsored by the U.S. Department of Energy, Office of Basic Energy Sciences.

357725 Composite Characterization By a New Non Destructive Technology (NDT): Mobile Fourier Transform Infrared Spectroscopy (FTIR)

Abstract:

The prospect of greater fuel economy and lower operating costs linked to sturdier and lighter airframe, has led to rapid adoption of composite materials in commercial aircraft. The increased use of composites presents new challenges in maintaining safe flying conditions and already, calls have been made for improving inspection of composite materials. All aircrafts are subject to environmental damages; however, composite materials are susceptible to environmental damages that are different than that for traditional aluminum materials. The resin component of the composite consists of organic molecules (most often epoxy based), which are susceptible to the affect of a number of external stressors such as heat, cold, moisture, ultraviolet light and exposure to chemicals. Analyzing the affect of these environmental stressors on in-service composite based aircraft requires new techniques and strategies.

Oak Ridge National Labs studied several spectroscopic techniques for the determination of extent of heat damage on composite panels. The study found that diffuse reflectance FTIR (Fourier Transform Infrared Spectroscopy) or DRIFT spectroscopy and laser induced fluorescence had the highest degrees of success, but both techniques required development of a field unit that could be used for NDI (Non Destructive Inspection).

With recent progresses in optoelectronics, FTIR instruments have now reached a critical form factor to be used in places like aircraft hangars. Combined with easy to use interface, portable FTIR can be operated by non-technical users for composites repair of in-service parts, manufacturing QC and also composite coating QC for identification among other things. In this talk, we will focus on applications of handheld FTIR for analysis of polymers, composites/coatings and the potential to use FTIR as a quality control tool for composite bonding. First, we will show how FTIR can be correlated to heat damage of BMI composites for aerospace applications such as the maintenance of the Boeing 787 fuselage. Subsequently, we will cover how FTIR can be used to determine composites and/or coatings curing time on large composite parts such as wind turbine blades. Finally, we will show how FTIR can be used to detect manufacturing contaminants such hydrocarbon or silicon based oil on composites parts, assess the level of preparedness of composites from work developed by the University of Washington (AMTAS) group.

Effect of Amphiphilic Surfactants on the Properties of Polyethylene – Graphene Nanocomposites

Vikas Mittal, Ali U Chaudhry

Chemical Engineering Department,
The Petroleum Institute, Abu Dhabi, UAE

Nanocomposites

- Heterogenous materials with at least one dimension of filler phase in nm
- Aluminosilicates, nanotubes, graphene etc. have been extensively used as fillers
- Enhanced properties, e.g. elastic modulus, tensile strength, heat deflection, gas permeability even at low filler content
- Low density and high transparency of composites

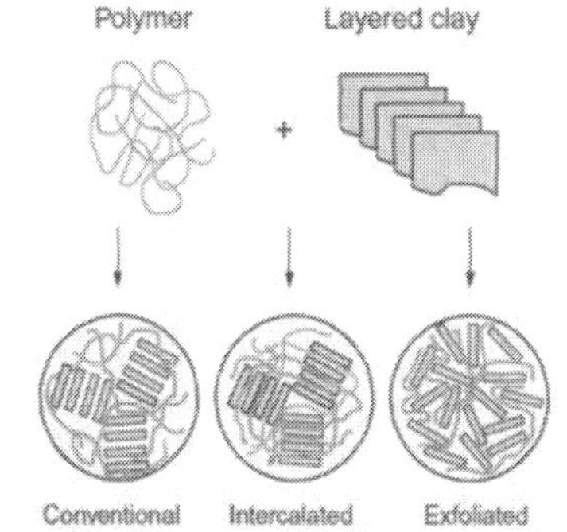

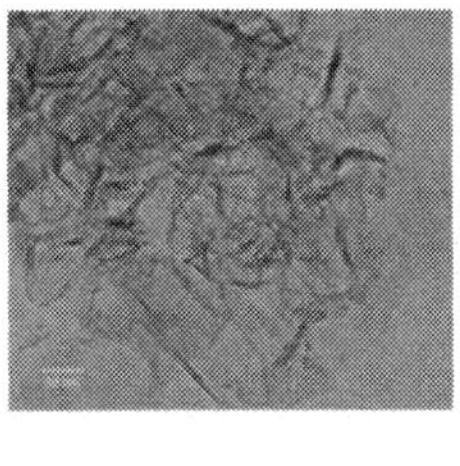

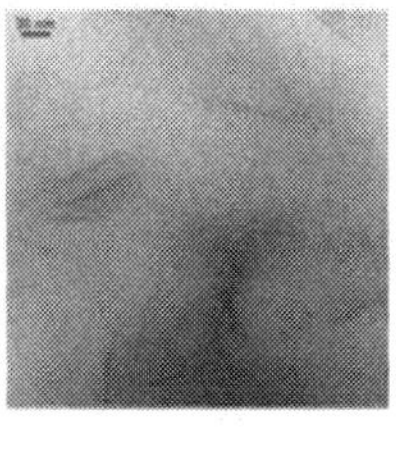

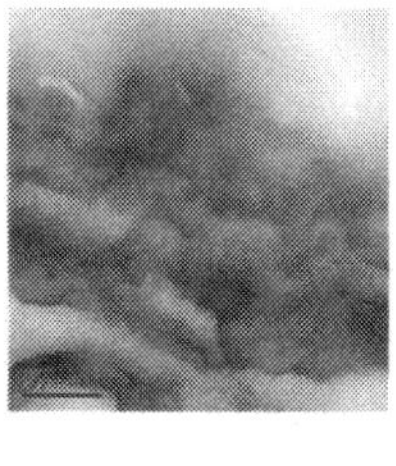

Approach 1: Surface Treatment

- exchange of surface inorganic cations with organic ions
- exchanging the inorganic cations with a monomer/initiator and subsequent polymerization
- physical adsorption on the surface
- other chemical reactions with reactive surface modifications

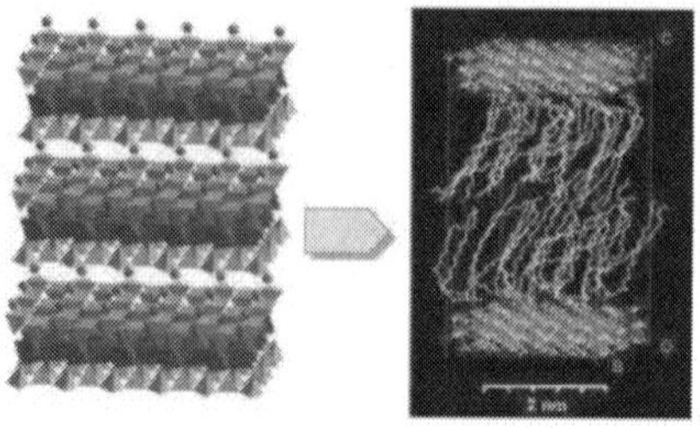

Courtesy: Dr. Heinz, University of Akron

Approach 2: Polarization of the Polymer Matrix

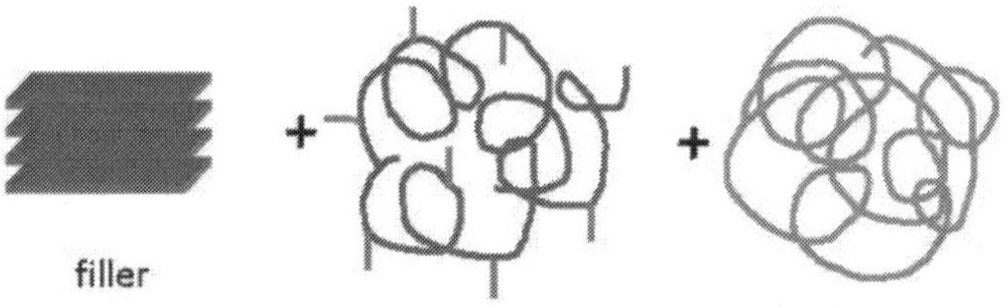

Aim/Objective

- To develop polyethylene-graphene oxide nanocomposites using compatibilizers.
- Different compatibilizers have been reported in the literature to enhance graphene dispersion in the non-polar polymer matrices, but the use of various graphene types, different surface modifications and polymer grades as well as different synthesis routes to generate nanocomposites makes the comparison of the resulting nanocomposite properties difficult.
- A systematic study using a wide variety of compatibilizers with different functionalities is required, while keeping the filler type and nanocomposite generation route constant.
- Thus, the specific goal of the current study was to explore the effect of a variety of different compatibilizers on the mechanical, rheological, thermal and morphological characteristics of nanocomposites generated by melt mixing with high density polyethylene as matrix and graphene oxide (of low polarity) as filler.

4

Polymer + Compatibilizers

Compatibilizer name	Polymer/ Compatibilizer	Supplier and trade name	Co-monomer content, %	MFI, g/10 min (2.16 kg, 190°C)	Density, g/cm^3	Peak melting temperature, °C
-	PE	Borouge BB2581	--	0.35	0.958	147
Ethylene acrylic acid copolymer	EAA	Exxon Mobil ESCOR 5050	9% AA	8.4	0.936	97
Ethylene vinyl acetate-g-maleic anhydride	EVA-MA	Dupont Fusabond C190	--	16	0.950	71
Enhanced polyethylene resin	EPE	Dow ELITE 5230F	--	4	0.916	122
Ethylene methyl acrylate copolymer	EMA1	Dupont Elvaloy AC 12024S	24% MA	20	0.944	88
Ethylene methyl acrylate copolymer	EMA2	Dupont Elvaloy AC 15024S	24% MA	50	0.944	88
Ethylene methacrylic acid copolymer with zinc ion	EMAZ	Dupont Surlyn 9020	--	1	0.96	85
Ethylene-α-olefin copolymer	E-AO	Dow Affinity EG8150G	--	0.5	0.868	56
Chlorinated polyethylene 25%	CPE25	Lianda Corporation, Weipren 6025	25% chlorination	1.8	1.1-1.3	140
Chlorinated polyethylene 35%	CPE35	Xuran Chemicals	35% chlorination	1.9	1.1-1.16	amorphous
Ethylene butyl acrylate copolymer	EBA	Dupont Elvaloy AC34035	35% BA	40	0.93	90
Ethylene-g-maleic anhydride copolymer	PE-MA	Aldrich PE-g-MA	--	500 cP viscosity (140°C)	--	--

Nanocomposites

Compatibilizer name	Composite code
-	PE/G
Ethylene acrylic acid copolymer	PE/G/EAA
Ethylene vinyl acetate-g-maleic anhydride	PE/G/EVA-MA
Enhanced polyethylene resin	PE/G/EPE
Ethylene methyl acrylate copolymer	PE/G/EMA1
Ethylene methyl acrylate copolymer	PE/G/EMA2
Ethylene methacrylic acid copolymer with zinc ion	PE/G/EMAZ
Ethylene-α-olefin copolymer	PE/G/E-AO
Chlorinated polyethylene 25%	PE/G/CPE25
Chlorinated polyethylene 35%	PE/G/CPE35
Ethylene butyl acrylate copolymer	PE/G/EBA
Ethylene-g-maleic anhydride copolymer	PE/G/PE-MA

6

Tensile Properties

Composite	Tensile modulus, MPa	Yield strain, %	Peak stress, MPa	Total elongation, mm
PE	1063	9.0	62	4.4
PE/G	1203	8.5	61	4.0
PE/G/EAA	1489	8.5	70	4.6
PE/G/EVA-MA	1067	8.2	57	4.4
PE/G/EPE	1370	8.4	70	4.5
PE/G/EMA1	1222	8.3	51	4.7
PE/G/EMA2	1412	8.3	66	4.4
PE/G/EMAZ	1327	8.9	60	4.7
PE/G/E-AO	1546	8.5	64	4.6
PE/G/CPE25	1557	7.6	68	4.1
PE/G/CPE35	1268	8.5	64	4.4
PE/G/EBA	1420	7.9	64	4.5
PE/G/PE-MA	1234	8.4	69	4.5

7

Rheological Properties

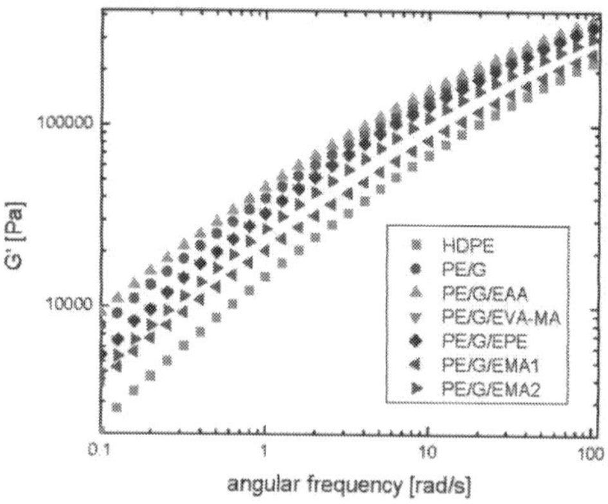

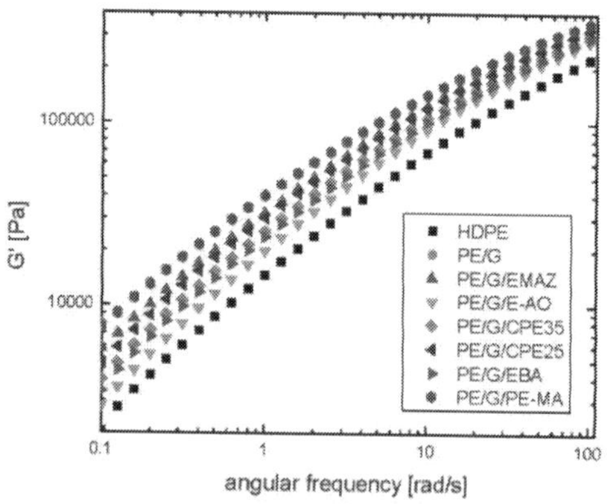

8

Rheological Properties

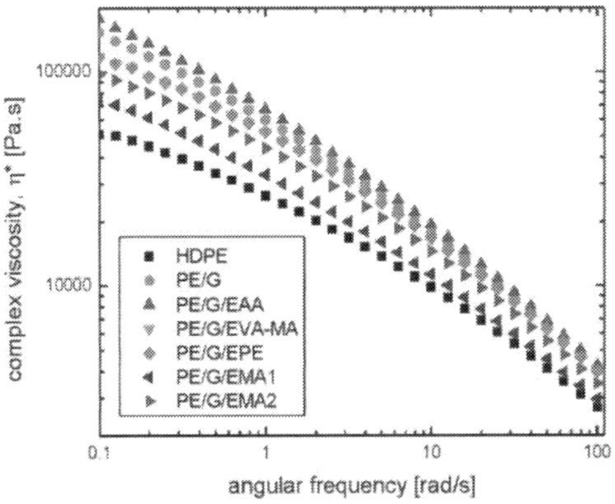

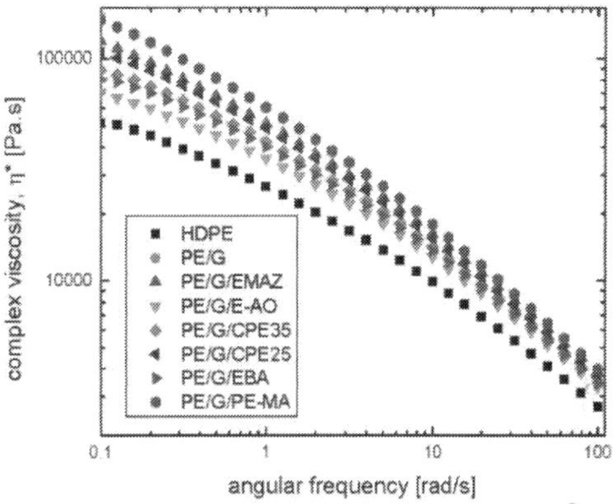

9

Thermal Degradation

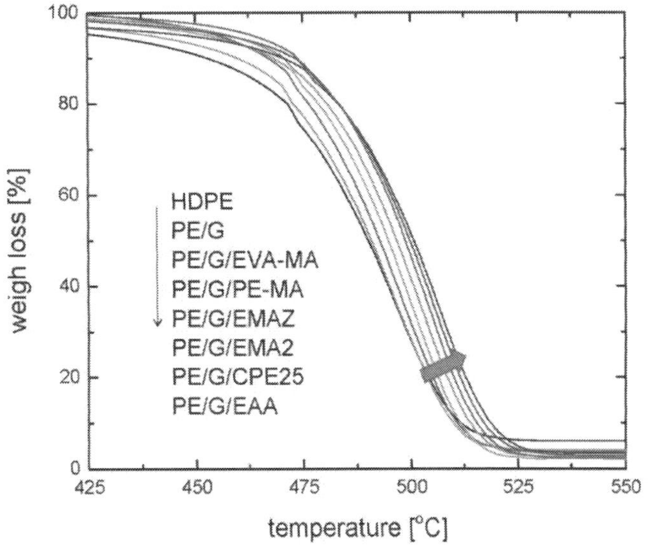

10

Calorimetric Properties

Composite	ΔH, J/g	Peak meting point, °C	Peak crystallization point, °C	Crystallinity, %
PE	151	135.9	115.5	52
PE/G	146	135.8	115.0	50
PE/G/EAA	148	137.8	115.3	51
PE/G/EVA-MA	139	139.6	114.7	48
PE/G/EPE	138	137.4	114.9	48
PE/G/EMA1	147	137.3	115.0	51
PE/G/EMA2	128	137.1	115.4	44
PE/G/EMAZ	121	134.3	115.6	42
PE/G/E-AO	145	137.3	115.6	50
PE/G/CPE25	153	138.1	115.4	52
PE/G/CPE35	144	133.0	116.1	50
PE/G/EBA	121	138.6	115.5	42
PE/G/PE-MA	145	140.0	114.4	50

11

Calorimetric Properties

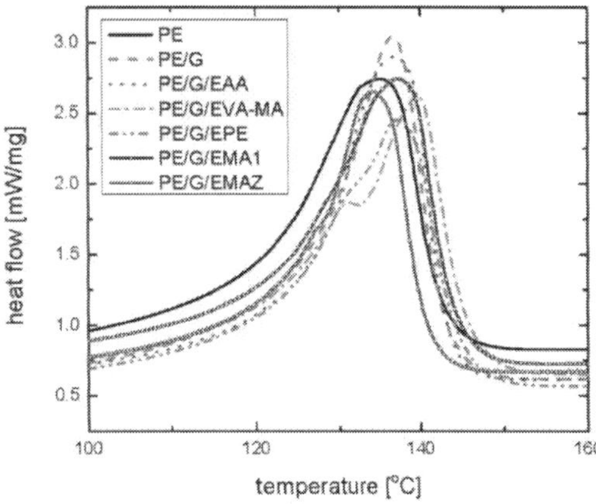

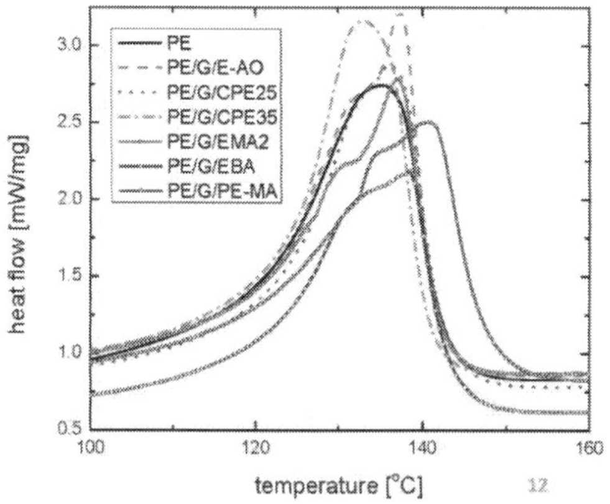

12

Morphology

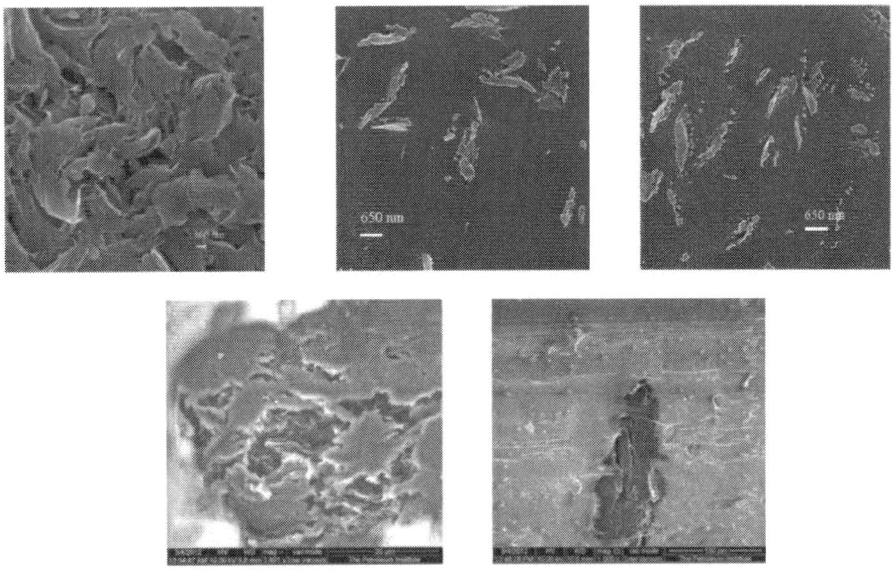

SEM micrographs of (a) pristine graphene oxide; (b) PE/G/CPE25;
(c) PE/G/EAA; (d) PE/G/EVA-MA and (e) PE/G/PE-MA nanocomposites.

13

Morphology

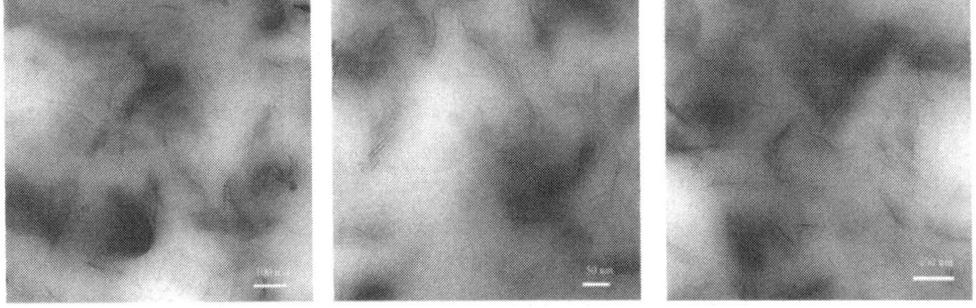

TEM micrographs of PE/G/5%EAA, PE/G/10% EAA and PE/G/15%EAA nanocomposites

14

Mechanical Properties

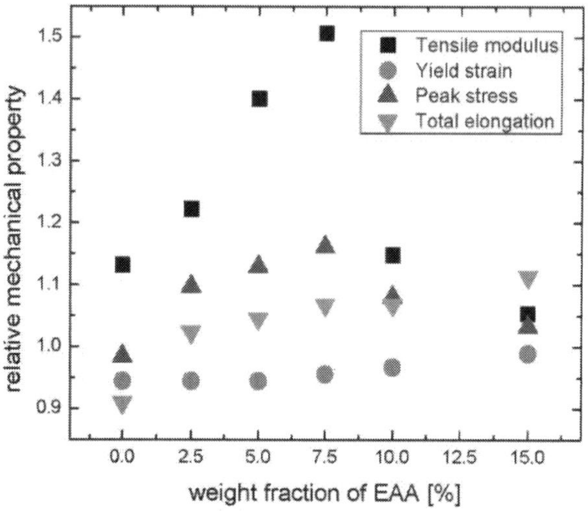

15

Thermal Properties

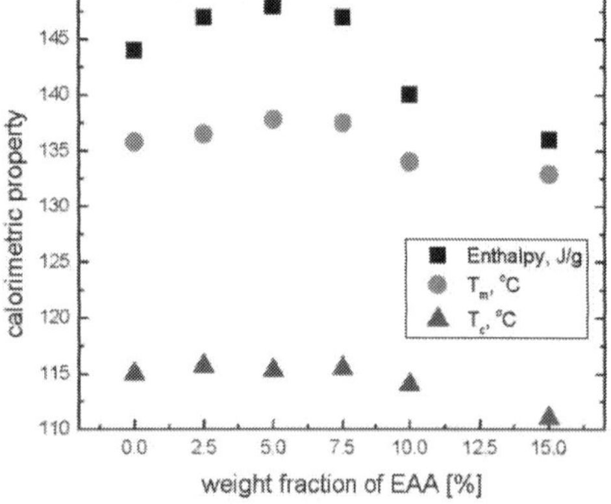

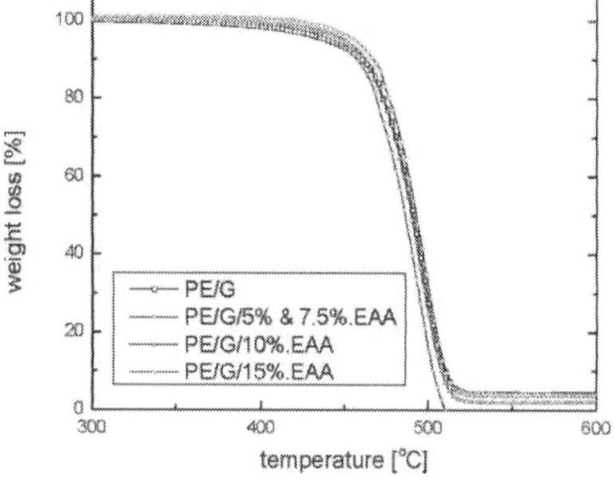

16

Rheological Properties

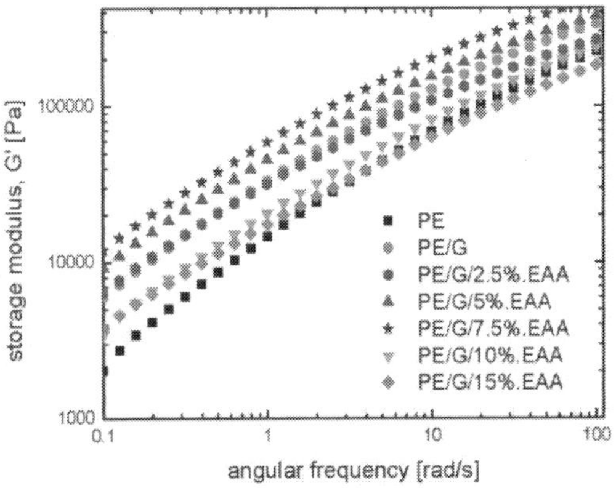

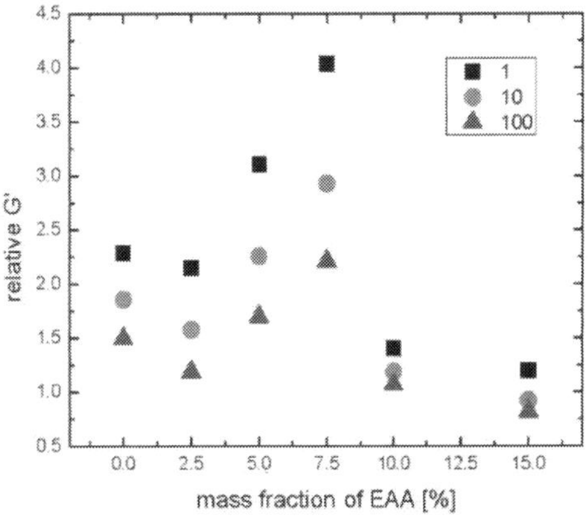

17

Conclusions

- The molecular weight, extent of grafting and polarity in the compatibilizers were the factors which determined their interactions with the filler surface.
- An increment of 40-45% in tensile modulus was observed for nanocomposites with CPE25, E-AO and EAA compatibilizers.
- The rheological and thermal degradation properties also followed the trends observed for tensile properties.
- The SEM investigations also confirmed the findings from mechanical and rheological analysis with composites with EAA, CPE25 specifically demonstrating extensively exfoliated filler platelets.
- Observed properties in the composites with varying amounts of EAA revealed that 7.5 wt% was the optimum amount of EAA in the composites as the mechanical and rheological properties exhibited improvement due to filler exfoliation. Beyond 7.5 wt% content of EAA in the composites, the extent of filler exfoliation though was observed to increase, the performance of the nanocomposites deteriorated due to matrix plasticization.

18

THE PETROLEUM INSTITUTE

Thank You

19

Synthesis and Morphology

Routes for Preparation

In-situ polymerization, melt intercalation and solvent method

Influence on Properties

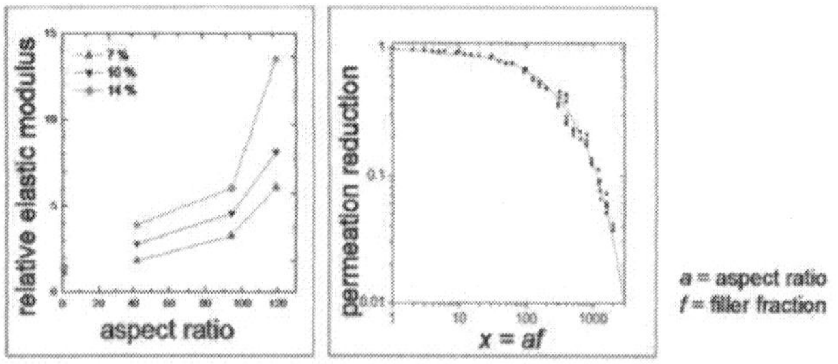

Osman MA, Atallah A, Müller M, Suter UW. Polymer 2001, 42, 6545
Gusev AA, Lusti HR. Advanced Materials 2001, 13, 1641

QATAR UNIVERSITY

Qatar National Research Fund
Member of Qatar Foundation

"Synthesis and Characterization of Novel Ni-B-Y_2O_3 Composite Coatings"

Funded by QF through NPRP4-662-249

Presented by:
Prof. Ramazan Kahraman
Department of Chemical Engineering
Qatar University, Doha, Qatar

24-Nov-14 1

Sequence of Presentation

- Introduction
- Alloy and Composite Coatings
- An Overview of Ni-B Coatings
- Experimental
  - Process flow chart
  - Experimental facilities
  - Solution bath composition
- Results
  - Chemical and Structural analyses
  - Surface morphology
  - Mechanical properties
  - Electrochemical properties
- Summary
- Future Work
- Acknowledgements

24-Nov-14 2

Concept of Corrosion

Electrochemical degradation of materials is called corrosion. In other words, it is eating away of materials by electrochemical reactions.

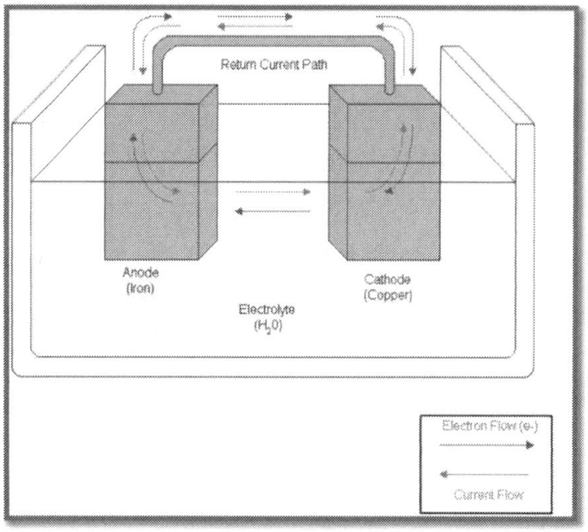

Source: webapps.cee.vt.edu

Corrosion Damages

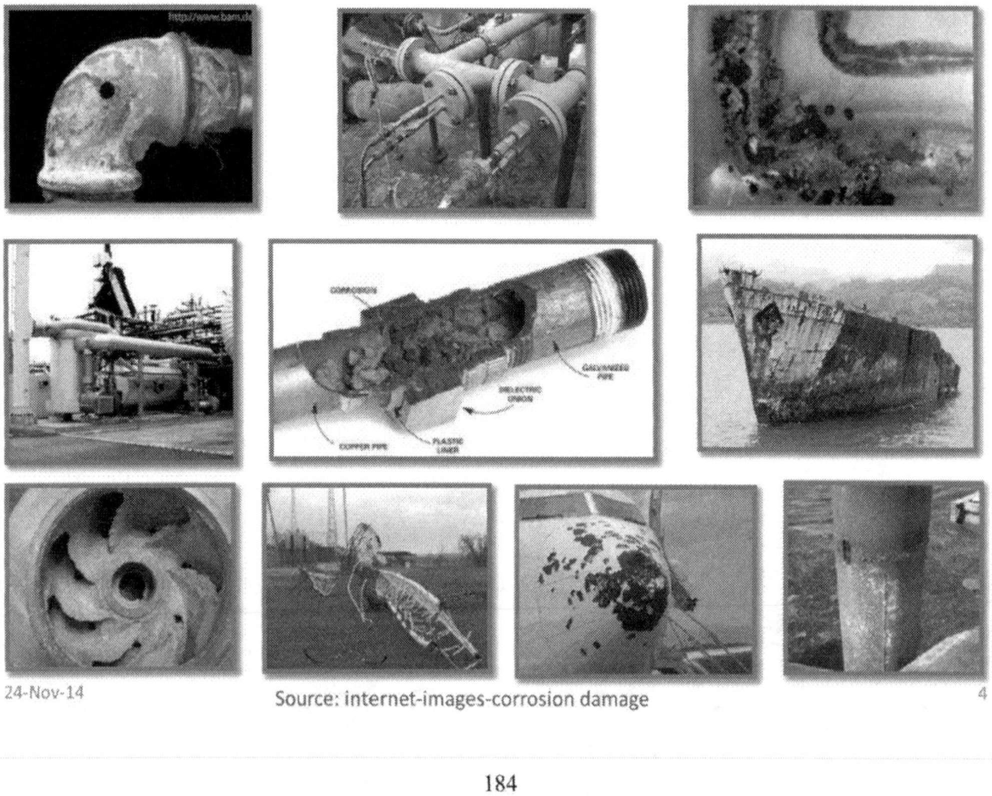

Source: internet-images-corrosion damage

Cost of Corrosion

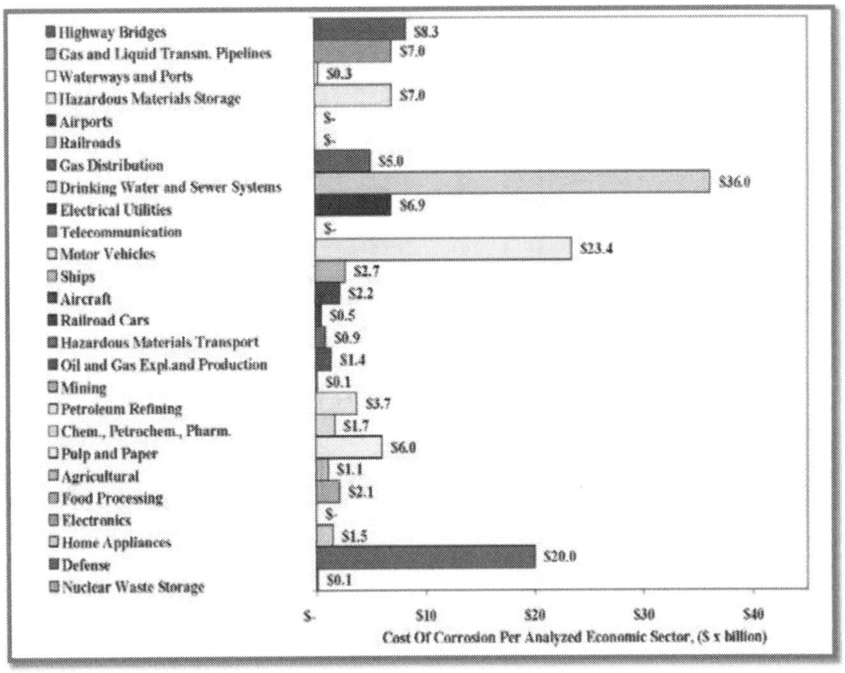

24-Nov-14 Source: NACE international 5

Why Coatings?

- Coatings are applied to control deterioration in metallic surfaces due to **"corrosion"** and to improve mechanical properties like **"wear"**.
- Wear and corrosion damage result in:
  - **Loss of plant efficiency and at worst a shutdown.**
  - **Raises safety concerns.**
  - **Costs billions of dollars annually**
- About 40% of the total cost could be avoided by proper corrosion prevention methods.
- The combined effects of wear and corrosion can result in much greater material loss than each process taken alone.
- **The research on coating technology is critical all over the world especially in Qatar, where every minute of natural gas and oil production and transport counts.**

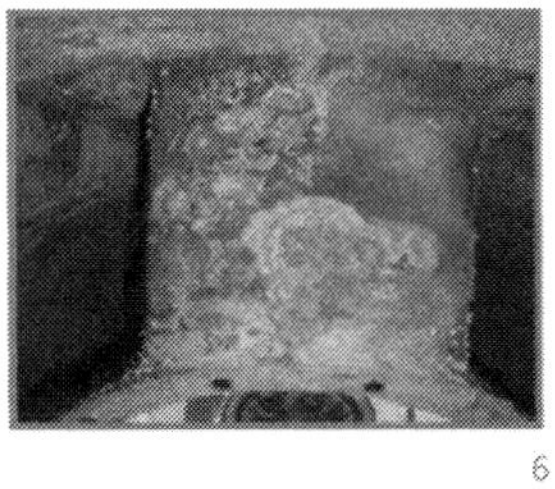

24-Nov-14 http://www.marineengineering.org.uk/Repairs/bt4.jpg 6

Coating Applications

- Coatings are applied on many working parts, components and machineries to improve their mechanical and corrosion properties.
- They are widely used in many industries to improve the performance of working components.
  - **Oil and gas**
  - **Automotive**
  - **Aerospace**
  - **Textile**
  - **Manufacturing**
  - **Electrical and electronic**
  - **Sea water desalination ... etc**

24-Nov-14

Source: internet

7

Alloy Coatings

- An alloy coating is a coating system which contains some metal as an additive to improve its properties.
- Ni-P-X (X=Co, W, Cr, Cu, Zn, ... etc)
- Every metal added to the coating has a specific purpose.

  Alloying coatings have superior properties such as;
  - High hardness
  - High wear properties
  - High corrosion resistance
  - High temperature stability
  - Good magnetic properties
  - Improved surface topography

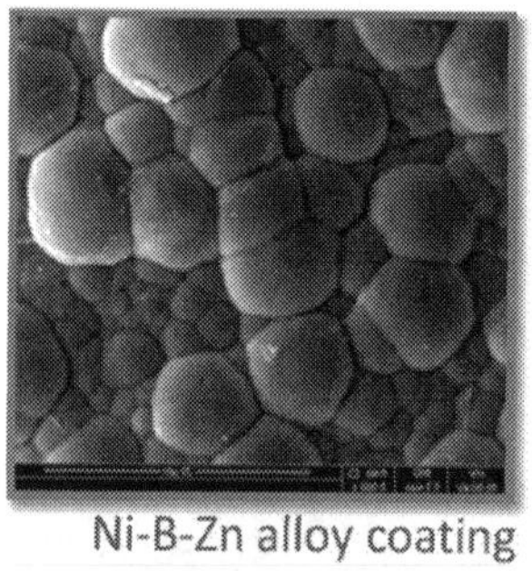

Ni-B-Zn alloy coating

24-Nov-14

Shakoor et al , International J. of Electrochemical Sci. *(2014)*, **Accepted.**

8

Composite Coatings

- Primarily, a composite coating comprises of two physically and chemically distinct phases at least one is on nano scale.
- A nanocomposite coating comprises of at least two phases: a nano crystalline phase and an amorphous phase, or two nano crystalline phases.
- Ni-P-X (X=TiO_2, SiO_2, ZrO_2, CeO_2, Al_2O_3 ... etc)
- Nanocomposite coatings have improved wear and corrosion resistant properties.

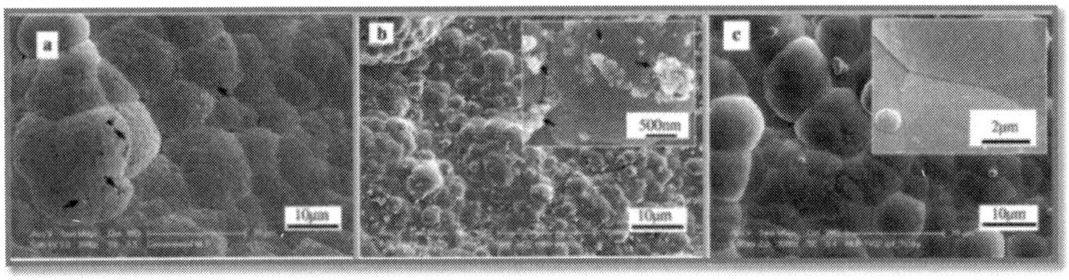

24-Nov-14 W. Chen, W. Gao ,et al,J. Surface & Coatings Technology 204 (2010) 2493–2498. 9

Ni-B Coatings

- Ni-B coatings are attractive due to:
  - High hardness
  - High wear resistance
  - Cost effectiveness
  - Thickness uniformity
  - Good lubricity and ductility
- These tempting properties make them suitable for many industries like:
  - Automobile, oil and gas, aerospace, petrochemical
  - Electronics, paper, food, computer, nuclear, ... etc.
- Further improvement in properties is essential so that new developments and more challenging requirements can be fully addressed.

24-Nov-14 10

An Overview of Ni-B Coatings

- Coating systems:
  - Ni-P-X (X=Zn, Cu, W, Cr, Co, PTFE, ... etc)
  - Ni-B-X (X=Co, Zn)
  - Ni-P-X (X=TiO_2, ZrO_2,Y_2O_3,Al_2O_3 , Si_3N_4, ... etc)
  - Ni-B-X (X= Si_3N_4, Al_2O_3)
- Deposition Processes:
  - Conventional electrodeposition process
  - Conventional electroless plating process
  - Pulsed electrodeposition process
  - Sol. enhanced electroplating process
  - Sol. enhanced electroless plating process
- Substrates :
  - Mild steel
  - Stainless steel
  - Copper
  - Brass
  - Magnesium alloys and aluminum alloys

24-Nov-14 11

Process Flow Chart

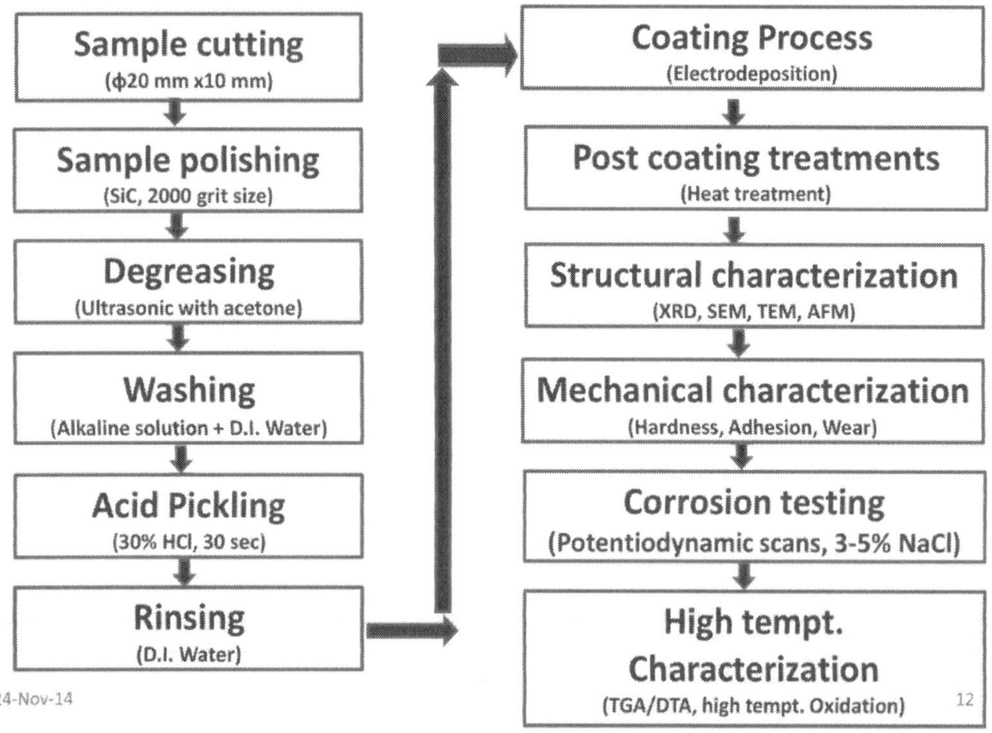

24-Nov-14 12

Experimental Set up

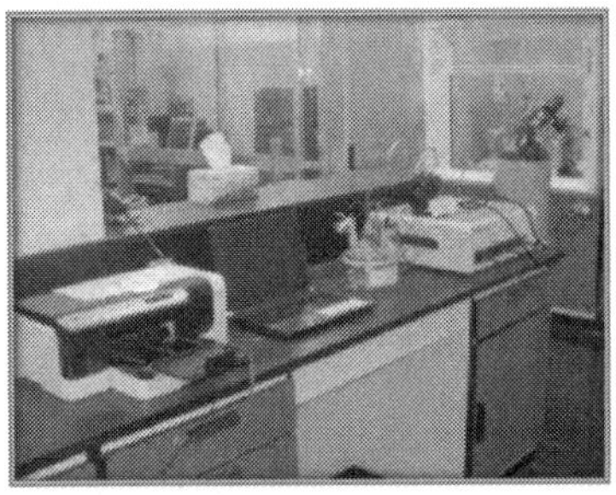
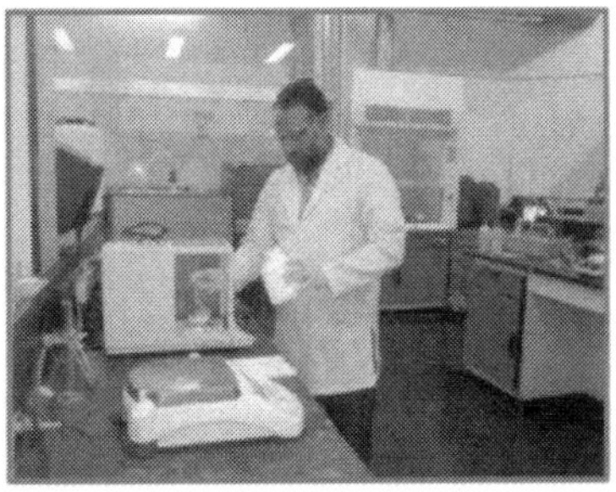
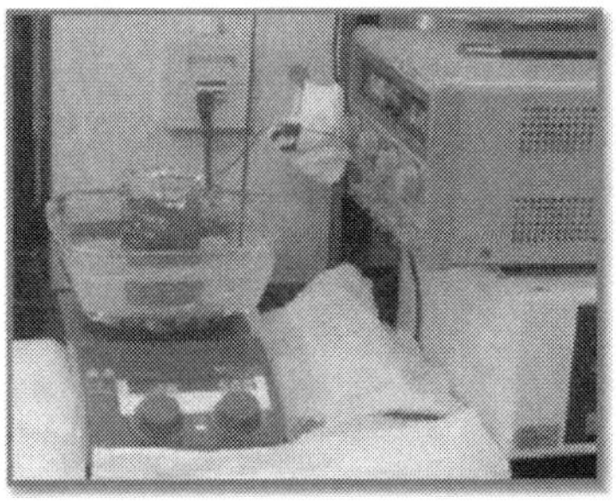
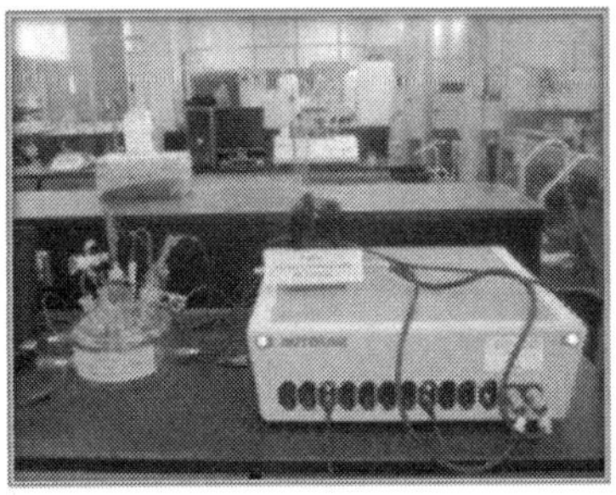

Ni-B Solution Bath Composition

Sr.#	Chemicals	Composition
1	Nickel sulphate hexa hydrate	240g/l
2	Nickel chloride hexa hydrate	45g/l
3	Boric acid	30g/l
4	Dimethylamine borane complex	3g/l
5	Yttrium oxide powder	0 and15 g/l
	Operating conditions	
1	pH	4
2	Temperature	43±°C
3	Deposition time	30 minutes
4	Current density	50 mA/cm^2
5	Bath agitation	300 rpm

Compositional analysis

EDX analysis:

Coatings	Ni (Wt.%)	B (Wt.%)	Y_2O_3 (Wt.%)
Ni-B	92.339	7.661	-
Ni-B-Y_2O_3	78.59	12.95	8.46

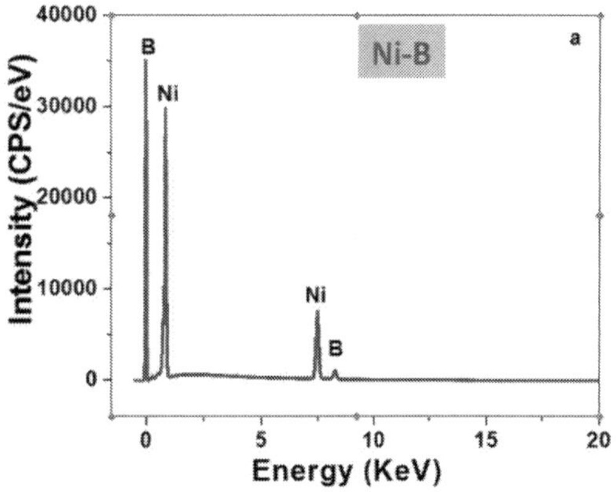

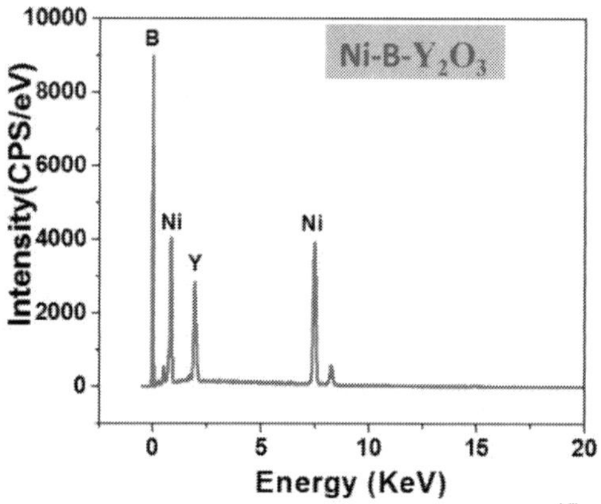

Structural Analysis....XRD

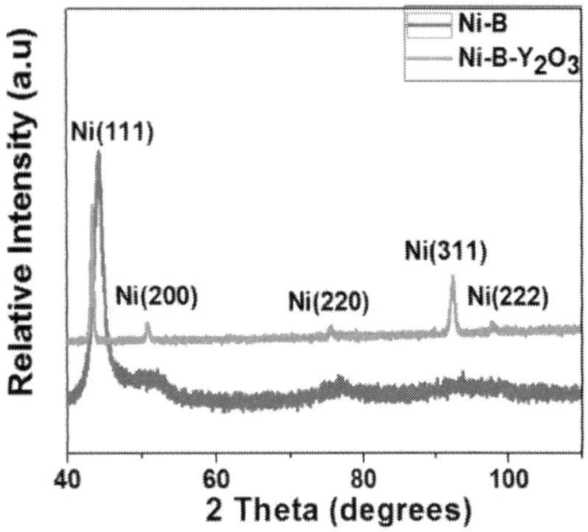

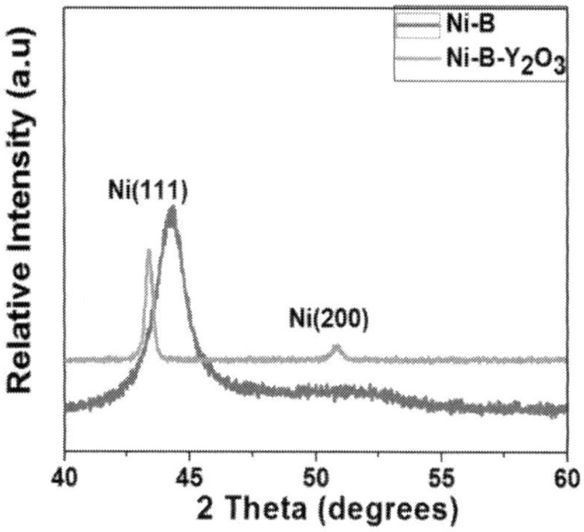

Surface Morphology....SEM/AFM

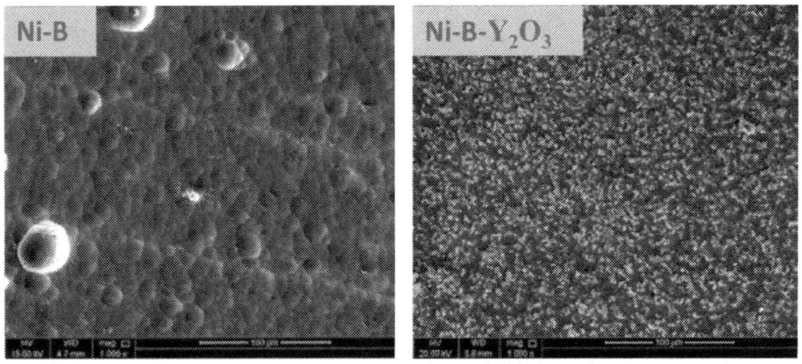

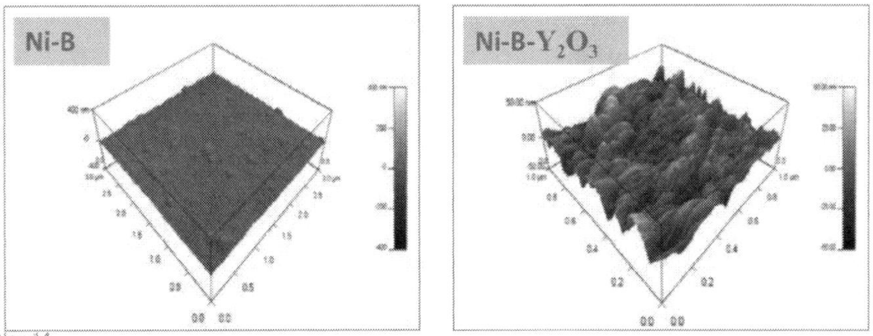

24-Nov-14 17

Thermal Transformation....DSC

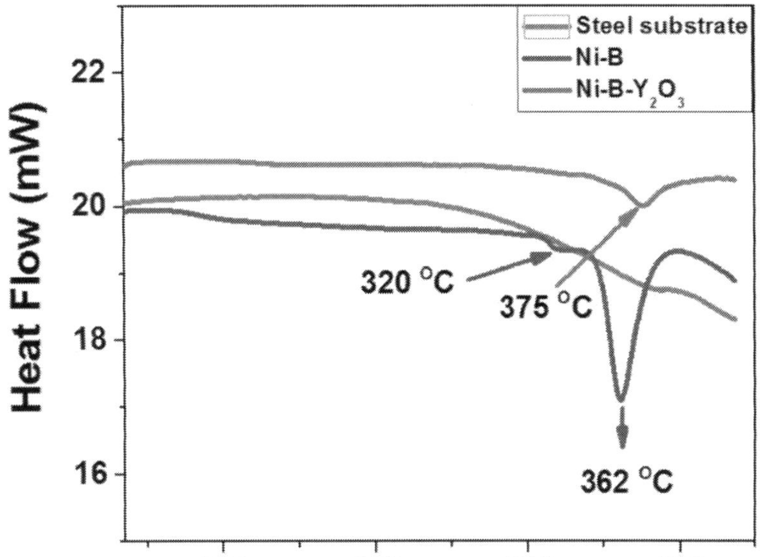

24-Nov-14 18

Nanomechanical Properties

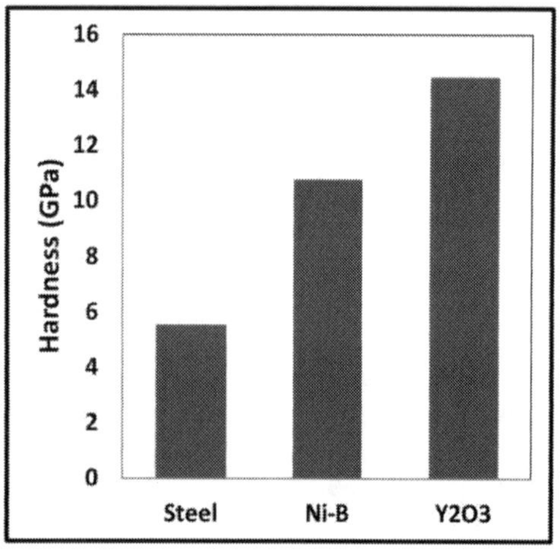

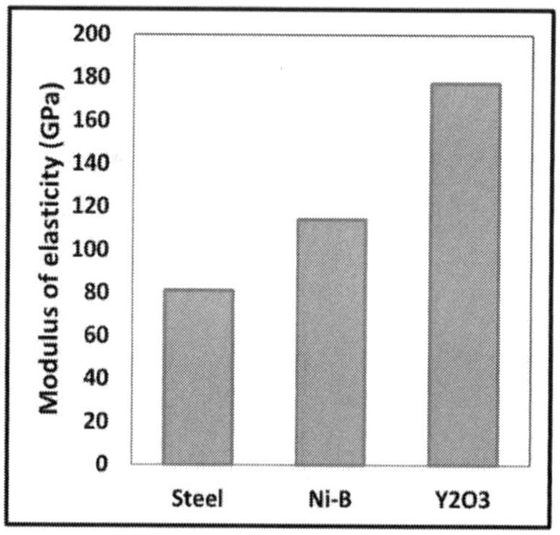

24-Nov-14

19

Electrochemical Properties

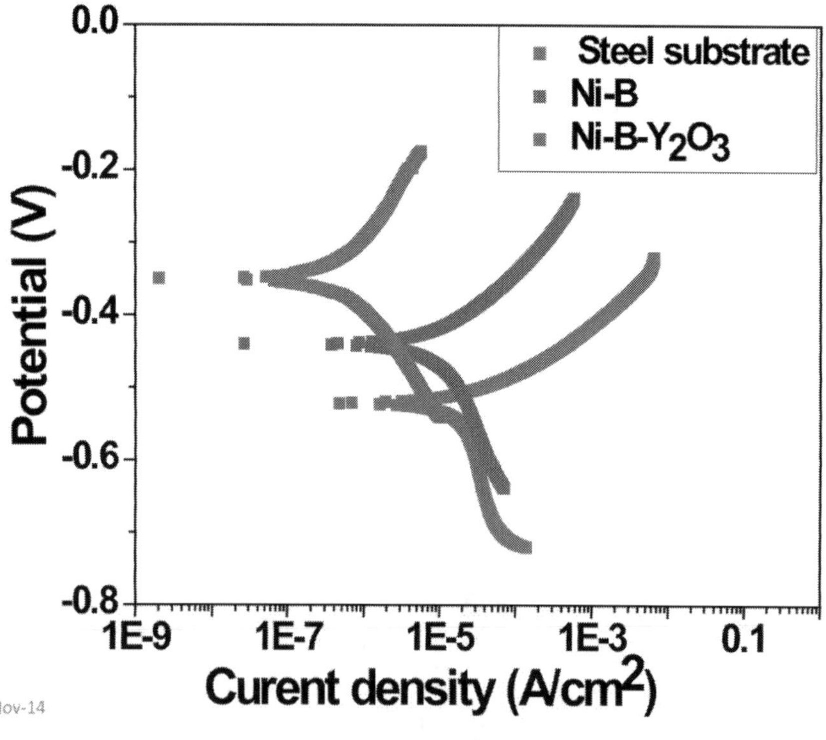

24-Nov-14

20

Summary

- Corrosion damages and the importance of its prevention has been presented.
- An overview of previous research on Ni-B (and Ni-P) coatings has been presented.
- Synthesis and characterization of electrodeposited Ni-B and Ni-B-Y_2O_3 coatings has been presented.
- Addition of Y_2O_3 to Ni-B matrix results in
  - **Improvement in crystallinity.**
  - **Increase in surface roughness.**
  - **Improvement in mechanical properties (hardness and elastic modulus).**
  - **Improvement of corrosion resistance.**
- Conclusion: Ni-B-Y_2O_3 composite coatings are useful for high wear and corrosive applications.

Future Work

- Effect of heat treatment on properties of Ni-B-Y_2O_3 composite coatings.
- Wear properties of Ni-B and Ni-B-Y_2O_3 composite coatings.
- High temperature oxidation behavior of developed coatings.
- Improvement of properties of Ni-B-Y_2O_3 composite coatings by pulse electrodeposition technique.

Acknowledgements

- Prof. Ramazan Kahraman(QU), Prof. Wei Gao (UoA), Dr. Shakoor (QU), Dr. Yuxin Wang (UoA) and Dr. Umesh (QU) are Obliged to:
    - **QNRF** for providing sufficient financial support through its **NPRP-4-662-2-249** research grant.
    - **Qatar University** (QU) for providing an opportunity to conduct this research work in the Dept. of Chemical Engineering.
    - **Dept. of Mechanical and Industrial Engineering,** QU for utilization their experimental facilities.
    - **Center for Advanced Materials (CAM),** QU for utilization their AFM facility.

24-Nov-14 23

24-Nov-14 24

Structural Characterization of Carbon and Silicon Oxycarbide Nanodomains in Silicon Oxycarbide Polymer Derived Ceramics

Scarlett Widgeon,[1] Wenruo Bai,[2] Gabriela Mera,[3] Sabyasachi Sen,[4] Ralf Riedel,[3] Alexandra Navrotsky[5]

[1]Department of Chemical Engineering, University of California, Santa Barbara, Santa Barbara, CA 93106 USA
[2]Department of Electrical Engineering, University of Washington, Seattle, WA 98195 USA
[3]Institut für Materialwissenschaft, Technische Universität Darmstadt, Darmstadt, Hessen D-64287 Germany
[4]Department of Chemical Engineering and Materials Science, University of California, Davis, Davis, CA 95616 USA
[5]Peter A. Rock Thermochemistry Laboratory and NEAT ORU, University of California, Davis, Davis, CA 95616 USA

Silicon oxycarbide polymer-derived ceramics (SiOC-PDCs) are nanocomposite materials that exhibit unusual properties, such as oxidation resistance, corrosion resistance, and low crystallization rates at elevated temperatures. These properties make PDCs good candidates for a wide range of high temperature applications and the properties can be altered by adjusting the Si:O:C ratios. Previous studies show that the SiOC-PDCs consist of domains of amorphous carbon and of amorphous SiOC that consists of SiC_xO_{4-x} tetrahedral units. However, it remained unclear whether the nano-sized carbon clusters are embedded in a continuous SiOC network or if the SiOC phase forms nano-sized clusters that are surrounded by a continuous carbon network. Furthermore, the nature of the spatial distribution and connectivity of the SiC_xO_{4-x} tetrahedral units has remained controversial in the literature. Solid-state ^{29}Si NMR spectroscopy was used to reveal the morphology, spatial distribution, and connectivity of the SiC_xO_{4-x} tetrahedral units in the SiOC network. The ^{29}Si NMR data suggest that the structure of the SiOC-PDCs consists of a continuous mass-fractal backbone of corner-shared SiC_xO_{4-x} tetrahedral units with voids occupied by amorphous graphitic carbon. The morphology of the SiOC network and spatial distribution of the SiC_xO_{4-x} tetrahedral units were also studied using Reverse Monte-Carlo simulations (RMC). The resulting simulated structure shows oxygen-rich and carbon-rich regions, consistent with results derived from ^{29}Si NMR spectra. Additionally, the results suggest that SiOC-PDCs containing higher carbon contents may result in a bicontinuous network consisting of the amorphous carbon and SiOC networks. This information leads to understanding of the Si:O:C ratios necessary to form percolation networks that can give rise to transport properties, such as electrical conductivity.

385285 Microstructure Characterization of Magnesium Oxide-Based Wallboard Composites for Enhanced Structural Properties and Fire/Water Resistance

Abstract:

Despite being used on a large scale since antiquity in ancient China, magnesium oxide (MgO) building products are not used in the United States. This largely stems from the inability of prior manufacturers, primarily Chinese based, to produce consistent and quality materials that meet quality control requirements for US building standards. MgO building products have the potential to displace current products simply based on their superior properties that can be achieved with strict production quality control. The focus of this work is a collaboration between JetProducts LLC and Clemson University to develop correlations between the MgO cement molecular / nanoscale properties (crystalline structure, composition, microstructure) and the macroscale properties (strength, flexibility, and fire/moisture barrier properties), which are significantly influenced by the manufacture parameters(raw material, additives, and curing process).

Current residential and commercial construction materials are dominated by oriented strand board (OSB), plywood, gypsum board (sheetrock), and Portland cement-based boards. Each of these options has significant drawbacks from a sustainability standpoint. OSB and plywood are moisture impermeable, highly flammable, and contain toxic formaldehyde-based resins. Sheetrock and Portland cement-based materials are not water resistant and have a very high CO_2 footprint, unless they contain a high fraction of fly ash, which then imparts potential toxicity issues due to the significant heavy metal contamination. Furthermore, the presence of crystalline silica in these materials can lead to occupational Silicosis. MgO construction materials have significant potential in the US residential and commercial markets with answers to fire resistance, contractor health/safety issues and manufacturing carbon footprint concerns. JetProducts is a startup small business that began in 2005 and has become a primary producer of MgO-based construction materials in the US, which is a rapidly expanding market. This work has the potential to transform the building products industry by providing a fundamental understanding of the underlying phenomena that influence the properties of MgO cement boards, which will enable the broad emergence of a greener and more sustainable product line.

In this project, we have obtained several wallboard samples of varying compositions and cured under different conditions to impart a range of mechanical properties. We also obtained samples from Chinese manufacturers for comparison. We have characterized these material using x-ray diffraction, thermal analysis, energy-dispersive X-ray spectroscopy, and high-resolution environmental SEM. Based on these results, we can draw correlations that connect the macroscale mechanical properties with the material chemical composition, MgO crystalline phase, grain size, and composite morphology.

2014 AIChE Abstract

Controlling the Composition and Structure of Polymer Films Formed on Liquid Substrates via Initiated Chemical Vapor Deposition

Laura C. Bradley and Malancha Gupta

The initiated chemical vapor deposition (iCVD) technique is a solventless polymerization process typically used to deposit thin polymer coatings onto solid substrates. We were recently the first group to introduce low-vapor pressure liquids, such as ionic liquids and silicone oil, into the iCVD process. We found that the deposition onto liquids can result in the formation of polymer films or polymer particles. Using liquid substrates introduces surface tension and solubility effects that are not relevant for solid substrates, and our work aims to understand how these effects can be controlled to tune the polymer morphology and composition. This talk will focus on demonstrating how film formation on ionic liquid substrates can be controlled to fabricate layered polymer films and polymer-ionic liquid composite materials. We have found that the most significant variable influencing the composition and structure of polymer films is the monomer solubility in the liquid. For systems in which the monomer is insoluble polymerization only occurs at the liquid-vapor interface, whereas when the monomer is soluble in the liquid polymerization can also occur within the bulk. We have shown that soluble monomers can polymerize within the bulk by monomer absorption into liquid or diffusion through polymer films at the liquid surface, and varying the order in which the soluble and insoluble monomers are introduced into the system leads to different film structures. The iCVD technique also enables process parameters such as reaction time and pressure to be easily tuned to control the amount and molecular weight of polymer formed in the bulk to make polymer-ionic liquid gels. The molecular weight of polymer formed in the bulk is orders of magnitude higher than polymer formed at the liquid-vapor interface and therefore we observe a transition of thin liquid layers from a viscous liquid to a gel at low polymer concentrations. We are currently studying the copolymerization of reactive ionic liquid substrates with vapor phase precursors to covalently attach the ionic liquid functionality to polymer films. Our work provides an understanding of the mechanism of film formation using liquid substrates in order to control the structure and composition of new composite materials for sensing, separation, and electrochemical applications.

The objective of this presentation is to illustrate the roles of elastocapillary instability in the pattern formation and adhesion in soft confined films. Such an adhesively stressed thin film confined between two rigid flat substrates auto-roughens with its dominant wavelength (λ) exhibiting pronounced dependence on the film thickness (H). A linear stability analysis of the elastic field equations shows that the instability has two different regimes, the scaling $\lambda \sim H$ is observed when the role of the surface tension is negligible. However, a long wave instability [] ensues when the surface tension force is significant. These two different regimes, indeed, are the limiting cases of a more general scaling relation:

, which has been confirmed by experiments performed with thin hydrogel films of different thickness and elasticity. A potentially interesting consequence of the role of capillarity in selection of wavelength in soft thin films is that the conventional Griffith's debonding stress is to be modified by an elastocapillary number. This new relationship has also been verified experimentally by measuring pull-off stresses of rigid objects from hydrogel films of different thickness and elastic moduli. This new result is not only important in understanding the adhesion of soft films, but it is significant even in high modulus films as long as it is very thin.

Air Stable, High Performance N-Type Polymer Semiconductor Field-Effect Transistors Processed From Non-Halogenated Solvents

Boyi Fu and Elsa Reichmanis

School of Chemical & Biomolecular Engineering, Georgia Institute of Technology, Atlanta, GA

The development of air stable, solution processable electron transport polymer semiconductors continues to be a challenge, due to issues associated with stabilization of the LUMO energy level. This level could be controlled via incorporation of electron deficient units which serve to reduce the electron density of the parent polymers. Herein, we replaced the electron rich bithiophene units within a hole transport polymer semiconductor, poly(tetrathiophene-diketopyrrolopyrrole), by electron deficient bithizole units, to develop an acceptor-donor-acceptor (A-D-A) polymer, poly(bithiazole-bithiophene-diketopyrrolopyrrole), PDBTz, showing electron transport characteristics. The 5-decylheptadecyl side chain, which has been demonstrated to benefit both polymer solubility and effective π-π intermolecular interactions, was incorporated into PDBTz, allowing for solution-processing onto organic electronic and optoelectronic devices from a wide range of solvents. For instance, non-halogenated and lower-toxicity solvents such as o-xylene, p-xylene, and tetralin proved to be effective. PDBTz exhibits field-effect electron mobilities in excess of 0.1 $cm^2V^{-1}s^{-1}$, depending on choice of gate dielectric, source-drain electrodes, and transistor architecture. PDBTz based field-effect transistors exhibit no change in electron mobility, current on/off ratio and threshold voltage after exposure to ambient conditions for more than 3 weeks. The PDBTz platform represents a viable acceptor material for field-effect transistors, solar cells and thermoelectric applications.

Molecular Scale Behavior of Superspreading Agents

Rolf E. Isele-Holder[1], Ahmed E. Ismail[1]

[1]AICES Graduate School and Aachener Verfahrenstechnik: Molecular Simulations and Transformation, RWTH Aachen University, Germany

Abstract

The capability of trisiloxane surfactants—copolymers composed of short blocks of poly(dimethyl siloxane) and poly(ethylene oxide)—to facilitate superspreading, the ultra-rapid spreading of water droplets on hydrophobic polymers (and other substrates), is a unique phenomenon exploited in various industrial products, such as herbicides, cosmetica, and inks. In spite of numerous studies in the last 20 years, Superspreading is still poorly understood and subject to intensive debate, even though a full understanding of the underlying mechanism is needed to identify and develop improved superspreading agents. A central problem is that relevant processes occur on length scales difficult to access either experimentally or via continuum simulation methods. The ability of molecular dynamics simulations to resolve such nanoscale details make it a promising tool to study superspreading, providing that technical challenges in the simulations are adequately addressed.

One such challenge is that the force fields characterizing the polymers have been developed for bulk-phase simulations. We address this problem with the development of a fast long-range dispersion solver, applicable in large-scale simulations, that resolves problems related to the truncation of dispersion potentials, a major source of inaccuracies in interfacial simulations [1,2]. Moreover, we have developed and optimized models for the description of several polymers, namely polyolefines, poly(dimethyl siloxane), poly(ethylene oxide), and poly(tetrafluoroethylene), for use in simulations with water with an emphasize on interfacial properties. These models have been applied in a series of simulations of surfactants based on these polymers at their interface with water, and in large-scale simulations of spreading droplets of aqueous surfactant solutions on various polymer substrates [3,4].

We point out important differences between the different surfactants and show how these differences relate to the superspreading phenomenon. We focus in detail on the mechanisms by which the different surfactants lower the surface tension of water. Moreover, we investigate the formation and stability of clusters at the interfaces and analyze the dynamics of the different surfactants. Our spreading droplet simulations show that our developed model reproduces features characteristic of superspreading, and suggest that the mechanism that drives superspreading is simpler than the "unzipping" or "bilayer" mechanisms currently advocated in the scientific literature [5,6].

References

[1] R.E. Isele-Holder, W. Mitchell, A.E. Ismail, J. Chem. Phys., 2012, 137, 174107.

[2] R.E. Isele-Holder, W. Mitchell, J. Hammond, A. Kohlmeyer, A.E. Ismail, J. Chem. Theory Comput., 2013, 9, 5412-5420.

[3] R.E. Isele-Holder, A.E. Ismail, J. Phys. Chem. B (submitted)

[4] R.E. Isele-Holder, A.E. Ismail, Langmuir (in preparation)

[5] J. Venzmer, Curr. Opin. Colloid Interf. Sci., 2011, 16, 335-343.

[6] E. Ruckenstein, Colloid Surface A, 2012, 412, 36-37.

Solid-State Self-Assembly: Fundamentals and Applications

***Yoonseob Kim*[1], *Nicholas A. Kotov*[1,*]**
[1]Department of Chemical Engineering, University of Michigan, Ann Arbor, MI, 48109-2136.
[*]To whom correspondence should be addressed. E-mail:kotov@umich.edu, Fax: (734) 764-7453

Nanoscale science and technologies has been developed tremendously during the last two decades, introducing a variety of nanomaterials with unique properties. However, incorporation of the properties into macroscale functional applications has been limited. An essential challenge is the integration of such unique properties into assemblies for micro- and macroscale devices and discovering appropriate conditions under which this can be accomplished. Here we explore the self-assembly of nanoparticles (NPs) in solid-state for discovering fundamental understandings of mechanisms and dynamics for various engineering applications. 1. Yoonseob Kim et al. ***Nature***, 2013, 500, 59-63 first demonstrated an example of excellent stretchable conductors from self-assembly of spherical NPs in solid-state. Free-standing stretchable conductors were assembled by layer-by-layer assembly. High conductivity and stretchability were observed and the properties originated from dynamic self-organization of NPs under stress. Modified percolation theory to incorporate the self-assembly of NPs gave excellent match with experimental data. 2. The current study demonstrated the chiroptical activities of NPs in solid-state for the applications of biosensing devices and optoelectronics. Unconventional transfer of macroscale stresses shaped nanoscale assemblies with handedness. Chiroptical responses were reversibly tunable by controlled stresses. Computational simulation supported the chiroptical properties originated from self-assembly of NPs in solid-state.

The works were funded, in part, by Rackham Predoctoral Fellowship, University of Michigan, STX foundation, Seoul, Korea, and US Air Force Office of Scientific Research.

Title: Zwitter-wettability and its application to antifogging coatings

Division: 8A (Polymers) Graduate Student Awards (GSA) Session

Author: Hyomin Lee, Jonathan B. Gilbert, Francesco E. Angile, Daeyeon Lee, Michael F. Rubner and Robert E. Cohen

Abstract: An optically transparent, ultrathin lamellar polymer system, consisting of a hydrophilic reservoir of chitosan/carboxymethyl cellulose suitably capped with various hydrophobic layers, was designed to produce remarkable antifogging coatings. Layer-by-layer assembly methods were used to construct the conformal coatings on a variety of smooth and rough substrates; detailed morphological and wettability characterization was accomplished using a combination of quartz crystal microbalance experiments (QCM-D), depth-profiling X-ray photoelectron spectroscopy (C_{60} cluster-ion), and goniometric/tensiometric measurements of advancing and receding contact angles. The successful performance of the coatings relies on their Zwitter-wettable character, defined as the simultaneous capability to resist wetting by water in the condensed liquid state (hydrophobicity), while remaining highly hydrophilic to molecular water that diffuses rapidly into the film from the vapor phase. These mechanistic concepts and the quantitative morphological characterizations enabled us to design Zwitter-wettable films that retain optical clarity under a wide range of environmental challenges. Frost-resistant glass doors for supermarket freezer units, antifogging lenses for personal, sports and military eye-ware, and enhanced image clarity in endoscopic surgical procedures are among the applications that may benefit from these novel Zwetter-wettable polymer systems.

Comparison of the single molecular dynamics of linear and circular DNAs in microfluidic planar extensional flows

Yanfei Li[1], Christopher A. Brockman[2], Daniel Y. Yates[3], Kai-Wen Hsiao[2], Gregory B. McKenna[1], Charles M. Schroeder[2], Michael J. San Francisco[3], Julia A. Kornfield[4], Rae M. Anderson[5]

[1] *Department of Chemical Engineering, Texas Tech University, Lubbock, TX 79409, United States*

[2] *Department of Chemical & Biomolecular Engineering, University of Illinois at Urbana-Champaign, Urbana, IL 61801, United States*

[3] *Department of Biological Sciences, Texas Tech University, Lubbock, TX 79409, United States*

[4] *Division of Chemistry & Chemical Engineering, California Institute of Technology, Pasadena, CA 91125, United States*

[5] *Department of Physics, University of San Diego, San Diego, CA 92110, United States*

ABSTRACT

Understanding the elongational flow behavior of macromolecules is of great interest due to its wide application in turbulent drag reduction [1], fiber spinning [2], injection molding [3], and DNA/polymer purification techniques [4-6]. It was predicted by P. G. de Gennes in 1974 [7] that above a critical extension rate, a coiled linear polymer experiences a steady state coil-stretch transition (CST) to adopt a relatively open configuration, and conformation hysteresis occurs when the molecular weight becomes sufficiently high. This remarkable prediction has been proven to be true for linear polymers by many techniques, such as flow birefringence [8], microfluidic trap [9], and Brownian dynamics simulation [10]. However, due to the experimental difficulty in synthesizing high molecular weight circular polymers [11], it remains unknown whether ring-like polymers exhibit a similar CST, and, if so, how their steady state and transient elongational flow behavior compares with that of linear counterparts.

DNA, due to its biological significance [4], easiness to be fluorescently labeled [9], and uniformity in molecular structure, is used in the present study as a model polymer. We, for the first time, examine the dynamics of large circular and linear DNA molecules with the goal of providing fundamental data in order to compare the CST [7, 9], conformation hysteresis [12], and molecular individualism (MI) [9] for the two types of macromolecules. To do this, we combined the biological synthesis of DNA molecules following the procedures of Douglas E. Smith and coworkers [13] with direct imaging of fluorescently labeled DNA at the stagnation point in a novel microfluidic device that has a cross-slot design to develop a stagnation point extensional flow that permits direct imaging of the fluorescently labeled DNA [14].

We found that the relaxation time of linear DNA is about 3.2 times of that of circular counterpart of the same molecular weight, and this is true for both 45 and 115 kbp (kilo base pair). Also, as shown in Figure 1, both linear and circular DNAs exhibit a sharp transition in molecular configurations at a Weissenberg number Wi of about 0.4 – 0.5 (Wi = flow rate multiplied by relaxation time). However, for both 45 and 115 kbp circular DNA, the transition is shallower compared with the linear DNA. Unlike linear DNA which tends to adopt folded configurations when MI occurs during a transient stretch [9], the 115 kbp circular DNA prefers forming a kink near the middle of the chain; however, 45 kbp circular DNA always shows a dumbbell-like configuration without any molecular individualism. Therefore, molecular individualism not only is a function of molecular weight, but also closely relates to molecular topology. Within the range of molecular weight investigated in this work (45 and 115 kbp), conformation hysteresis is not observed for circular DNAs.

To summarize, circular DNAs, due to their unique molecular structure, possess shorter relaxation times, shallower slopes in steady state CST curves, and different transient MI behaviors, comparing with linear DNAs. This implies that quite different physics [15], such as entropic elasticity, viscous drag, hydrodynamic interactions, excluded volume effects, internal viscosity, and self-entanglement, could be involved in the flows of circular and linear DNAs. Therefore, this comparison of circular and linear macromolecules serves as an ideal platform for scientists to explore the structure-property relationship of macromolecules, both experimentally and in a simulation sense.

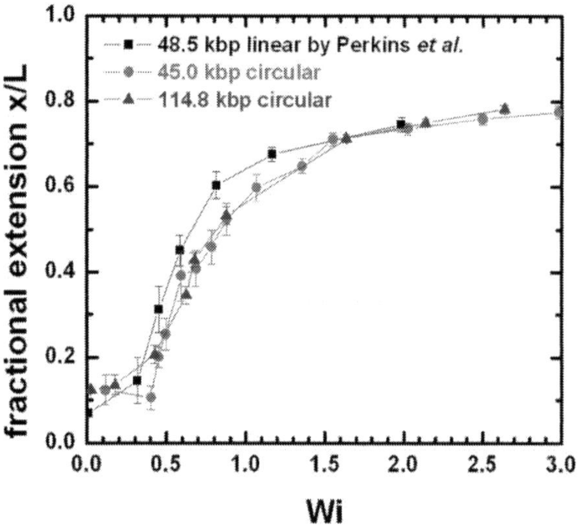

Figure 1. Steady state fractional extension x/L as a function of Weissenberg number Wi for three DNA samples: 48.5 kbp linear, 45 and 115 kbp circular.

References

1. J.L. Lumley, "Drag Reduction by Additives", Annual Review of Fluid Mechanics **1**, 367–384 (1969).
2. A. Ziabicki, "Fundamentals of fiber formation", John Wiley and Sons, London, (1976).
3. A. I. Isayev and M. Modic, "Self-Reinforced melt processible polymer composites: Extrusion, compression, and injection molding", Polymer Composites **8**, 158–175 (1987).
4. N.A. Horn, J.A. Meek, G. Budahazi, and M. Marquet, "Cancer gene therapy using plasmid DNA: purification of DNA for human clinical trials", Human Gene Therapy **6**, 565–573 (1995).

5. J. Han, H. G. Craighead, "Separation of Long DNA Molecules in a Microfabricated Entropic Trap Array", Science **288**, 1026-1029 (2000).
6. E. E. Borujeni and A. L. Zydney, "Separation of plasmid DNA isoforms using centrifugal ultrafiltration", BioTechniques **53**, 49-56 (2012).
7. P.G. de Gennes, "Coil-stretch transition of dilute flexible polymers under ultrahigh velocity gradients", Journal of Chemical Physics **60**, 5030–5042 (1974).
8. G.G. Fuller, and L.G. Leal, "Flow birefringence of dilute polymer solutions in two-dimensional flows", Rheologica Acta **19**, 580–600 (1980).
9. T.T. Perkins, D.E. Smith, and S. Chu, "Single Polymer Dynamics in an Elongational Flow", Science **276**, 2016–2021 (1997).
10. R. G. Larson, Hua Hu, D. E. Smith and S. Chu, "Brownian dynamics simulations of a DNA molecule in an extensional flow field", Journal of Rheology **43**, 267–304 (1999).
11. M. Kapnistos, M. Lang, D. Vlassopoulos, W. Pyckhout-Hintzen, D. Richter, D. Cho, T. Chang, and M. Rubinstein, "Unexpected power-law stress relaxation of entangled ring polymers", Nature Materials **7**, 997–1002 (2008)
12. C.M. Schroeder, H.P. Babcock, E.S.G. Shaqfeh, and S. Chu, "Observation of Polymer Conformation Hysteresis in Extensional Flow", Science **301**, 1515–1519 (2003).
13. S. Laib, R.M. Robertson, and D.E. Smith, "Preparation and Characterization of a Set of Linear DNA Molecules for Polymer Physics and Rheology Studies", Macromolecules **39**, 4115–4119 (2006).
14. M. Tanyeri, M. Ranka, N. Sittipolkul, and C.M. Schroeder, "A microfluidic-based hydrodynamic trap: design and implementation", Lab on a Chip **11**, 1786–1794 (2011).
15. R. G. Larson, "The rheology of dilute solutions of flexible polymers: Progress and problems", Journal of Rheology **49**, 1–70 (2005).

378165 Electric-Field Induced Assembly of Anisotropic Polymeric Particles

Abstract:

Physical interactions can be exploited to assemble polymeric particles into complex structures with exotic properties. Here we report our work on the directed-assembly of polymeric particles with anisotropic interactions induced by AC electric fields. The key interactions such as the double layer, dipolar, dielectrophoretic, and electrohydrodynamic interactions can all be controlled precisely by tuning experimental parameters which include salt concentrations, frequency, and field strength. Moreover, both shape and surface functionality on those particles can also be modulated, which significantly affect the particle interactions too.

In the first example, we report the formation of well defined zig-zag oligomers from polymeric spheres. Those oligomers can also connect and form two-dimensional hierarchical structures. Such an unique assembly can be attributed to the competition and balance among dipolar, dielectrophoretic, and double layer interactions between spheres at different planes. In the second example, we will talk about the assembly of polymeric dimers. By systematically changing the size ratios between two lobes in the dimer, we find the colloidal interactions are strongly dependent on the relative orientation between neighboring dimers. At low frequencies, one standing dimer associate with several lying dimers to form a family of clusters with right- and left-handedness. Interestingly, the broken symmetry also induces an unbalanced electrohydrodynamic flow surrounding the chiral particles and causes them to rotate in opposite directions according to their handedness. At high frequencies, the strong electrostatic and electrohydrodynamic attraction between oppositely oriented dimers induces the formation of one- and two-dimensional structures that closely resemble the Ising lattices. By combing the anisotropy in both particle property and field-induced interactions, we demonstrate that the assembled structures of polymeric particles can be significantly enriched.

Characterizing the nanoscale building blocks of a stable polymer glass

Kimberly B. Shepard, Craig B. Arnold and Rodney D. Priestley

Glassy materials are amorphous solids that are technologically ubiquitous, with applications in diverse areas, including pharmaceuticals, electronic devices, membranes, multifunctional coatings and engineering materials. The glass transition temperature, T_g, is the temperature at which the material transforms from the rubbery state into the glassy state upon cooling, and is the most important property of a glass, as it determines the temperature at which the material can be processed, used and stored. While it is technologically desirable to control the T_g and related properties, *e.g.* density, of a glass-forming material, the techniques for doing so are limited. Varying the cooling rate during glass formation can induce only modest changes in T_g. Recently, physical vapor deposition[1] and magnetron sputtering[2] have been employed to fabricate small molecule and metallic glasses with greatly enhanced T_g, however, these techniques are not applicable to polymer glasses.

In this work, we focus on a laser-deposition technique, Matrix Assisted Pulsed Laser Evaporation (MAPLE),[3] by which we fabricate poly(methyl methacrylate) glasses with exceptional properties. Polymer glasses deposited via MAPLE at low growth rates and controlled temperatures exhibit 40K increased T_g, 40% reduced density, enhanced kinetic stability and a spherical-like nanostructure.[4] This unique combination of properties is attributed to the physics of the MAPLE process, in which nano- and micro-size clusters of polymer and solvent undergo rapid cooling and solvent stripping. Upon evaporation of the solvent, each cluster forms a polymer nanoglobule, which is the nanoscale building block of the nanostructured film. Here, we report an experimental measurement of the velocity of the depositing nanoglobules, which proves the average velocity is sufficiently low that the nanoglobules remain intact upon impact with the substrate, where they build up to form a nanostructured film.[5]

In our previous work, experiments which indirectly probed the thermal properties of the nanoglobules indicated that the enhanced T_g and stability may be a result of the film's nanostructure.[6,7] Bulk MAPLE films which do not exhibit a nanostructure also lack enhanced stability. We directly measured the thermal properties of isolated nanoglobules via a nanoscale dilatometry method using heated atomic force microscopy and connected these results to the properties of the bulk films. MAPLE-deposited nanoglobules exhibit enhanced thermal stability and large excess volumes compared with control nanodroplets of similar size, in good agreement with the properties observed in the bulk.[5] Thus, we relate the nanoscale and macroscale properties of MAPLE-deposited polymer glasses in the context of the physics of the MAPLE process. The effect of MAPLE-deposition parameters on the glass' modulus measured via Brillouin Light Scattering will also be discussed.

References

1. Swallen, S. F. *et al.* Organic Glasses with Exceptional Thermodynamic and Kinetic Stability. *Science* **315,** 353–356 (2007).
2. Yu, H.-B., Luo, Y. & Samwer, K. Ultrastable Metallic Glass. *Adv. Mater.* **25,** 5904–5908 (2013).
3. Shepard, K. B. & Priestley, R. D. MAPLE Deposition of Macromolecules. *Macromol. Chem. Phys.* **214,** 862–872 (2013).
4. Guo, Y. *et al.* Ultrastable nanostructured polymer glasses. *Nat. Mater.* **11,** 337–343 (2012).
5. Shepard, K. B., Arnold, C. B. & Priestley, R. D. Transport and Stability of Laser-Deposited Amorphous Polymer Nanoglobules. Submitted.
6. Shepard, K. B., Guo, Y., Arnold, C. B. & Priestley, R. D. Nanostructured morphology of polymer films prepared by matrix assisted pulsed laser evaporation. *Appl. Phys. A* **110,** 771–777 (2013).
7. Shepard, K. B., Arnold, C. B. & Priestley, R. D. Origins of nanostructure in amorphous polymer coatings via matrix assisted pulsed laser evaporation. *Appl. Phys. Lett.* **103,** 123105 (2013).

367069 Design of Tunable Hpmcas-Inspired Polymers As Modular Oral Excipients to Deliver Poorly Water-Soluble Drugs

Abstract:

Spray-dried dispersions (SDDs) have emerged as remarkable drug delivery materials for improving limited oral bioavailability and providing therapeutic safety with high patient compliance. In this approach, synthetic and natural polymers hold tremendous potential as excipients to solubilize potent but highly lipophilic drugs in pharmaceutical formulation by maintaining drug molecules in an amorphous state. Unfortunately, excipient selection currently takes an Edisonian trial-and-error approach with little to no potential of accurately predicting key thermodynamic and kinetic phenomena. Herein, we explore fundamental polymer-drug interactions through the controlled synthesis and study of a multicomponent acrylic polymer, inspired by hydroxypropyl methylcellulose acetate succinate (HPMCAS). By strategically varying the precise monomeric incorporation and chemistry of these modular HPMCAS analogs, we comprehensively examined the role of polymers in producing stable amorphous SDDs with a model hydrophobic drug (probucol) at 10, 25, and 50 wt % drug loadings. To assess the overall amorphicity and miscibility, a combination of scanning electron microscopy imaging, powder X-ray diffraction, modulated differential scanning calorimetry, and Fourier transform infrared spectroscopy was utilized. Dissolution experiments of these HPMCAS-based SDDs in solution revealed dramatic difference in drug release kinetics, supersaturation generation, and bioavailability as a function of polymeric chemical specificity. The solubility of unprocessed probucol was increased from 0.5% to ~100% drug release in the presence of engineered polymers, comparable to HPMCAS. We identify chemically-driven interactions as crucial ingredients for facilitating high initial release and supersaturation maintenance. In particular, modulating polymer swelling in aqueous media by (1) increasing the ionizable carboxylic acid moieties and (2) applying selective chemical deprotection of hydrophobic acetate groups provided electrostatic stability and access towards polymer-drug hydrogen bonding with complementary hydrophobic interactions. Our results show the practical utility of tunable polymeric excipients for elucidating important structure-property relationships to effectively solubilize and deliver otherwise intractable drugs.

The material properties and performance of crystalline inorganic solids are often determined by the planes that are exposed on the crystal surface, e.g., catalysts, PV materials, nonlinear optical materials, etc. Currently, the protocols for synthesizing (and hence manufacturing) these materials are invented by chemical intuition coupled with trial-and-error experiments. It would be helpful to have a computer-assisted method whereby "experiments" could be carried out rapidly in a computer so that a short list of potentially successful experiments could be identified and tested in the laboratory. As a step towards creating such an approach, we have developed a mechanistic model for predicting the growth of inorganic crystals grown from solution. The model is based on the classical spiral growth mechanism of Burton, Cabrera and Frank [1].

The concept of a building unit of a periodic bond chain (PBC) is applied to simplify the identification of the PBC directions in the bulk crystal using the Hartman-Perdok theory [2]. The long-range nature of the electrostatic interactions necessitates a Madelung-type approach for the calculation of the interaction energies along the PBC directions. The variation in the electronic structure of the atoms in the surface layer is captured using the bond valence method [3], that can predict the partial charges of surface atoms, provided the local surface and solvent structure around the atoms is known. A space partitioning method is applied to sum up the 3D electrostatic interactions that involve non-bulk partial charges of the growth units situated in the kink sites on a spiral edge.

The step velocity of a spiral edge depends on the density of kink sites present along the edge and the net rate of incorporation in those kink sites [4]. A general expression is presented here to calculate the kink incorporation rate along a spiral edge on an AB-type crystal surface containing cations and anions in various orientations. The attachment and detachment fluxes from a kink site depend on experimental/engineering deign parameters (e.g., supersaturation, ionic activity ratio) and potential energy barriers, that are calculated from kink site potential energies.

The growth model successfully predicts the dependence of the step velocity on supersaturation and ionic activity ratio for the spiral edges on the (10-14) surface of calcite crystals, as compared to the *in situ* AFM measurements reported in the literature [5,6]. This high-fidelity crystal growth model can be applied to a wide class of inorganic crystals grown from solution, such as barium sulfate, titanium dioxide, potassium dihydrogen phosphate (KDP), etc. Ultimately, the model will provide a tool for predicting crystal morphology so that suitable growth media can be designed (e.g., solvent(s), additives, supersaturation) to deliver inorganic crystals of desired shapes depending on their applications.

References

1. Burton, W. K., Cabrera, N., and Frank, F. C. *Phil. Trans. Roy. Soc. A* **243**(866), 299-358 (1951).

2. Hartman, P. and Perdok, W. G. *Acta Crystallogr.* **8**(1), 49-52 (1955).

3. Brown, I. D. *The Chemical Bond in Inorganic Chemistry: The Bond Valence Model.* Oxford University Press, (2002).

4. Kuvadia, Z. B. and Doherty, M. F. *Cryst. Growth Des.* **11**, 2780-2802 (2011).

5. Teng, H., Dove, P. M., and De Yoreo, J. J. *Geochim. Cosmochim. Acta* **64**(13), 2255-2266 (2000).

6. Stack, A. G. and Grantham, M. C. *Cryst. Growth Des.* **10**(3), 1409-1413 (2010).

391766 Structure and Properties of Aqueous Methylcellulose Solutions and Gels

Abstract:

Cellulose represents the most abundant polymer across the planet. This polysaccharide can be chemically modified to impart solubility in water resulting in a plethora of commercially important products ranging from construction materials to delivery vehicles for pharmaceuticals. Substitution of a fraction of the hydroxyl groups on the native glucose rings with alkoxy moieties leads to the versatile class of polymeric materials known as cellulose ethers. Methylcellulose is especially important owing to a lower critical solution temperature (LCST) when mixed with water leading to gelation at elevated temperatures at relatively low concentrations. This nontoxic, biodegradable material, generally recognized as safe by the US Food and Drug Administration, finds widespread uses in food applications and other commodity products. Only recently has the mechanism of gelation been elucidated, based on cryogenic transmission electron microscopy (cryo-TEM), small-angle neutron scattering (SANS), and dynamic mechanical spectroscopy (DMS) measurements. This lecture will describe the discovery of a remarkable fibrillar structure that accompanies gelation and the associated temperature dependent viscoelastic properties that make methylcellulose so useful.

371668 Energy Efficient Gas Separations: Engineering Problems and Solutions

Abstract:

In principle, advanced membrane approaches that minimize energy intensive phase change-driven separations offer an order of magnitude reduction in energy intensity per unit of product purified. Implementing such approaches, however, requires coupling advanced materials and advanced processing to control costs of the final devices. Polymer-derived membranes based on crosslinked glassy polymers, polymer-inorganic nanoparticle hybrids and carbon molecular sieves are particularly attractive materials that cover different areas in the broad separation landscape. Examples of these new materials, devices based on them, as well as their manufacturing and the savings they enable will be discussed for high impact separations cases.

391911 Confinement, Electron Transfer and Lewis Acidity: Growing Facets of Zeolite Catalysis

Abstract:

Most of the zeolite research in the area of catalysis centers on the acid/base properties of zeolites and on the catalytic properties of extra-framework cations for redox and hydrogenation/dehydrogenation reactions on zeolites with, for example, Fe/Cu or Ga/Zn in extra-framework positions. Yet there are other zeolite properties that can be exploited to carry out catalytic reactions. This talk will describe recent investigations in our laboratory that highlight specific instances of catalysis by confinement and electron transfer in reactions such as NO oxidation and alkane activation. We will also discuss the catalytic properties of beta zeolites containing titanium, tin, zirconium and hafnium coordinated to the zeolite framework. This class of materials, having Lewis acidic properties but no Bronsted acids, has been investigated for the *Meerwein–Pondorf–Verley* reduction of carbonyl compounds and for the Oppenauer oxidation of alcohols. We will show that these materials are also effective catalysts for other 'classical' Lewis-catalyzed reactions such as the Diels-Alder reaction.

370231 Double Membrane Redox Flow Batteries for Economical and Efficient Renewable Electricity Storage

Abstract:

Redox flow batteries (RFBs) are considered as the most promising technology for economical and scalable electricity storage for wind and solar energy. By decoupling the energy storage and the power delivery from each other, RFBs hold a level of design flexibility and system scalability unavailable to traditional rechargeable batteries. Since their invention, RFBs have stayed with single ion-exchange membrane (single-IEM) configuration that cannot effectively use redox pairs with mixed charges and/or supporting electrolytes with different pHs. In this work, we show a universal design concept based on a double-IEM configuration that brings unprecedented freedom in choosing both redox pairs and supporting electrolytes. Two aqueous RFB examples are featured: (1) ultra-high voltage zinc-cerium RFB with a standard cell voltage of 3.08 V, the highest among all known aqueous RFBs; and (2) ultra-low cost sulfur-iron RFB with a standard cell voltage of 1.22 V, based on two highly available electroactive elements (iron and sulfur being the 1st and 5th most produced elements worldwide, respectively). The double-IEM concept is general and can be applied to non-aqueous RFBs.

Multiscale Systems Engineering for the Discovery of Cost-effective Zeolites for Natural Gas and Biogas Purification

Christodoulos A. Floudas, Eric L. First, and M. M. Faruque Hasan

Department of Chemical and Biological Engineering, Princeton University, Princeton, NJ 08544, USA

Significant methane sources in the U.S. are undeveloped due to the high cost of separating CO_2. These sources include natural gas, with 10% of U.S. reserves [1], or 30 trillion cubic feet [2], requiring purification to reduce CO_2 content to meet pipeline specifications. Shale gas and coalbed methane are other major fixed sources of methane that often require CO_2 removal. Biogas, which is released by landfills and animal waste, is primarily a mixture of methane and CO_2. In addition to being a major source of greenhouse gas emissions, biogas is a continuously-generated, untapped energy resource.

To take advantage of these opportunities for enhanced domestic energy production, cost-effective technologies are needed for methane/CO_2 separation. Pressure swing adsorption (PSA) is a promising option that utilizes a microporous sorbent, such as a zeolite or metal-organic framework, to selectively capture CO_2 from methane. We have developed a novel computational framework to identify the best zeolites, optimal PSA process topology, and operating conditions to minimize the total separation cost, including investment, operating, and material costs [3]. The importance of a combined material screening and process optimization approach was demonstrated previously for CO_2/N_2 separation [4].

Our multiscale methodology utilizes zeolite pore characterizations from ZEOMICS [5] to calculate metrics including shape, size, pore, and adsorption selectivities [6–9]. A detailed nonlinear algebraic and partial differential equation (NAPDE) model [10] is used to simulate the temperature, pressure, and composition profiles throughout the three-step PSA cycle. We employ an efficient Kriging-based grey-box constrained optimization approach [3,4,10,11] to find the minimum-cost solution to the complex NAPDE model.

We have considered a variety of feed compositions representative of methane sources with CO_2 content ranging from 5–50%. Constraints for 97% purity and 95% recovery are imposed to meet pipeline specifications and minimize losses. Eight novel zeolites are identified for economically separating CO_2 from methane, including the top zeolite WEI, which has a total separation cost ranging from \$0.15/MMBTU for 5% CO_2 sources to \$1.44/MMBTU for 50% CO_2 sources. Given the latest natural gas Henry Hub price of \$4.90 (March 2014) [12], the proposed technology may be highly desirable for developing previously uneconomical energy sources.

References:

[1] Baker RW, Lokhandwala K. Natural gas processing with membranes: an overview. Ind Eng Chem Res. 2008;47(7):2109–2121.

[2] U.S. Energy Information Administration. Proved Reserves, Reserves Changes, and Production. 2013. http://www.eia.gov/naturalgas/data.cfm. Accessed on December 20, 2013.

[3] First EL, Hasan MMF, Floudas CA. Discovery of novel zeolites for natural gas purification through combined material screening and process optimization. AIChE J. 2014;60(5):1767–1785.

[4] Hasan MMF, First EL, Floudas CA. Cost-effective CO_2 capture based on in silico screening of zeolites and process optimization. Phys Chem Chem Phys. 2013;15(40):17601–17618.

[5] First EL, Gounaris CE, Wei J, Floudas CA. Computational characterization of zeolite porous networks: an automated approach. Phys. Chem. Chem. Phys. 2011;13(38):17339–17358.

[6] Gounaris CE, Floudas CA, Wei J. Rational design of shape selective separation and catalysis–I: Concepts and analysis. Chem. Eng. Sci. 2006;61(24):7933–7948.

[7] Gounaris CE, Wei, J, Floudas CA. Rational design of shape selective separation and catalysis–II: Mathematical model and computational studies. Chem. Eng. Sci. 2006;61(24):7949–7962.

[8] Gounaris CE, Wei J, Floudas CA, Ranjan R, Tsapatsis M. Rational design of shape selective separations and catalysis: Lattice relaxation and effective aperture size. AIChE J. 2009;56(3):611–632.

[9] First EL, Gounaris CE, Floudas CA. Predictive framework for shape-selective separations in three-dimensional zeolites and metal-organic frameworks. Langmuir. 2013;29(18):5599–5608.

[10] Hasan MMF, Baliban RC, Elia JA, Floudas CA. Modeling, simulation, and optimization of postcombustion CO_2 capture for variable feed concentration and flow rate. 2. Pressure swing adsorption and vacuum swing adsorption processes. Ind. Eng. Chem. Res. 2012;51(48):15665–15682.

[11] Hasan MMF, Boukouvala F, First EL, Floudas, CA. Nationwide, Regional, and Statewide CO_2 Capture, Utilization, and Sequestration Supply Chain Network Optimization. Ind. Eng. Chem. Res. 2014;53(18):7489–7506.

[12] U.S. Energy Information Administration. Natural Gas Spot and Futures Prices. 2014. http://www.eia.gov/dnav/ng/ng_pri_fut_s1_m.htm. Accessed on May 16, 2014.

In 2003, we introduced the idea of using exfoliated zeolite layers for thin film and nanocomposite applications.[1] For such applications, a high-aspect ratio, preservation of crystal (and pore) structure, thickness and aspect ratio uniformity, and elimination of amorphous and non-exfoliated contaminants are of paramount importance. These properties could not be fulfilled by the previously reported, solution-based, exfoliation methods and require an exfoliation methodology based on melt blending with a polymer, which ensured structure preservation of the exfoliated layers.[2] Removal of the polymer and purification of the 2-D zeolites from non-exfoliated particles turned out to be a challenging problem, which was solved by the development of gradient centrifugation methods combined with selection of appropriate solvents. With this approach, single-unit-cell thick zeolite nanosheets with intact crystal and micropore structure were shown to be possible.[3, 4, 5, 6] The structural integrity and unprecedented purity and uniformity of these microporous nanosheets, open exciting possibilities for technological breakthroughs in molecular sieve membrane fabrication, synthesis of hierarchical catalysts and polymer-zeolite nanocomposites. Moreover, zeolite nanosheets enable for the first time zeolite pore mouth adsorption and catalysis to be studied by traditional uptake methods as well as surface science techniques. However, along with the exciting possibilities, challenges abound. For example, the in-plane dimensions of the existing nanosheets are in the sub-micrometer range limiting potential applications and processability as thin films. Moreover, the two exfoliated zeolites currently available are only a small fraction of zeolite topologies one would like to have available for a representative set. Earlier attempts to exfoliate other layered zeolites, including certain layered silicates and aluminophosphates with microporous layers, did not preserve the crystallographic order of the layers. Synthesis of high aspect ratio zeolite and other crystalline nanoporous nanosheets, methods to characterize their structure and properties, along with their processing and assembly to create functional devices will be the focus of this talk.

---

[1] Jeong H, Nair S, Vogt T, Dickinson L, Tsapatsis M. A highly crystalline layered silicate with three-dimensionally microporous layers. *Nature Materials*. 2003;2(1):53-58.
http://dx.doi.org/10.1038/nmat795

[2] Maheshwari S, Jordan E, Kumar S, Bates FS, Penn RL, Shantz DF, Tsapatsis M. Layer structure preservation during swelling, pillaring, and exfoliation of a zeolite precursor. *J Am Chem Soc*. 2008;130(4):1507-1516.
http://dx.doi.org/10.1021/ja077711i

[3] Xu D, Swindlehurst GR, Wu H, Olson DH, Zhang X, Tsapatsis M. On the synthesis and adsorption properties of single-unit-cell hierarchical zeolites made by rotational intergrowths. *Advanced Functional Materials*. 2014;24(2):201-208.
http://dx.doi.org/10.1002/adfm.201301975

[4] Zhang X, Liu D, Xu D, Asahina S, Cychosz KA, Agrawal KV, Al Wahedi Y, Bhan A, Al Hashimi S, Terasaki O, Thommes M, Tsapatsis M. Synthesis of self-pillared zeolite nanosheets by repetitive branching. *Science*. 2012;336(6089):1684-1687.
http://dx.doi.org/10.1126/science.1221111

[5] Agrawal KV, Topuz B, Jiang Z, Nguenkam K, Elyassi B, Francis LF, Tsapatsis M, Navarro M. Solution-processable exfoliated zeolite nanosheets purified by density gradient centrifugation. *AIChE J*. 2013;59(9):3459-3467.
http://dx.doi.org/10.1002/aic.14099

[6] Varoon K, Zhang X, Elyassi B, Brewer DD, Gettel M, Kumar S, Lee JA, Maheshwari S, Mittal A, Sung C, Cococcioni M, Francis LF, McCormick AV, Mkhoyan KA, Tsapatsis M. Dispersible exfoliated zeolite nanosheets and their application as a selective membrane. *Science*. 2011;334(6052):72-75.
http://dx.doi.org/10.1126/science.1208891

391281 Multilamellar MFI-Type Zeolites: The Effect Silica Sources and of Heteroatoms on Their Properties

Abstract:

Hierarchical zeolites do not only offer the possibility of reducing mass transfer limitations, but they also allow the catalytic conversion of large molecules over zeolites. However, the extra porosity in hierarchical zeolites often leads to low selectivity of the product. Thus, introducing the mesopores with optimum size is very crucial in the synthesis of hierarchical zeolites, because this will minimize the mass transport limitations and simultaneously prevent the formation of unwanted bulky products. Therefore, novel synthesis routes are required to prepare hierarchical zeolites with tunable porosity, *e.g.,* the preparation of multilamellar MFI-type zeolites [1]. In this contribution, recent advances in the synthesis of MFI nanosheets under different experimental conditions will be discussed. In addition, the influence of silica and aluminium source as well as the kind and the content of heteroatoms like aluminium in tuning the textural and the catalytic properties of this material will be in the focus of this work.

Nanosheets of MFI zeolite with Si/Al of 25, 100 and infinity were synthesized under static conditions and tumbling conditions. The influence of different synthesis parameters like content of the heteroatoms, alkalinity and aluminium source were studied. The catalytic cracking of cumene and di-isopropylbenzene were chosen as test reactions to characterize the effect of the incorporation of the heteroatoms.

References:

[1] M. Choi, K. Na, J. Kim, Y. Sakamoto, O. Terasaki, R. Ryoo, *Nature*, 461 (**2009**) 246.

379240 Hollow, Mesoporous Titano-Silicate Micro-Spheres As Heterogeneous Catalysts

Abstract:

A major challenge in the synthesis of titano-silicates is chemical non-uniformity in the framework and irregularity of particle size. A high concentration of isolated Ti sites is desired for applications as heterogeneous catalysts for the selective oxidation of hydrocarbons. However, at high Ti doping concentrations, Ti-O-Ti domains in the framework are often observed, instead of Ti-O-Si. Also, the formation of hollow microspheres instead of solid particles is of interest to eliminate poorly accessible material inside the particles, which could cause low catalytic effectiveness.

Li and Coppens have synthesized stable, hollow, mesoporous titanium-silica micro-spheres, in which the mesopores are templated using a surfactant, and the hollow core is formed by carrying out the synthesis in a rapidly stirred oil/water/oil emulsion.[1] However, the particles had a broad size distribution, and shell thickness was only controlled within a narrow range. Also, while Ti sites were quite well isolated at low concentrations[2], some of the Ti was still present as Ti-O-Ti microdomains at higher Ti concentrations.

Further experimental work is required to improve the structure, and associated catalytic performance, by altering the synthesis conditions. We use a Doehlert matrix approach, which is a second-order statistical technique, widely utilized as a chemometric tool in analytical sciences, to discover the most informative experimental conditions. Using such a rigorous statistical approach drastically reduces the required number of experiments. The method allows experimenters to achieve uniformity in space filling and flexibility in defining the reaction boundaries. Nitrogen adsorption/desorption, SEM, TEM, FTIR, Raman and UV-Vis absorption are used to characterize the structure of the materials.

Subsequently, catalytic characterization of the microspheres is carried out by extending the approach of Baca *et al.*, which involves the epoxidation of cyclohexene with *tert*-butylhydroperoxide.[2] This reaction is employed to validate the improved materials properties obtained via the statistically guided synthesis optimization. The relationship between the controlled hierarchical structure and chemical composition, on the one hand, and catalytic efficiency, on the other hand, is discussed.

[1] W. Li and M-O. Coppens, "Synthesis and characterization of stable hollow Ti-silica microspheres with a mesoporous shell", *Chem. Mater.* **2005,** *17,* 2241-2246

[2] M. Baca, W. J. Li, P. Du, G. Mul, J.A. Moulijn, and M.-O. Coppens, "Catalytic characterization of mesoporous Ti–Silica hollow spheres", *Catalysis Letters* **2006**, *109*, 207-210

382403 Control of the Location of Framework Al Atoms in ZSM-5 Zeolite

Abstract:

The acidic properties of aluminosilicate-type zeolites originate from the presence of protons balancing the negative charge induced by the framework Al in tetrahedral sites (T-sites). The catalytic properties of zeolites depend on many factors such as pore structure, acid strength and acid amounts. In addition these factors, recently, the location and distribution of Al atoms in the zeolite framework has been recognized as an important factor for activity and selectivity, because they would profoundly affect the accessibility of molecules to acid sites and the spatial constraints of the reaction field in the pores; however, their control is difficult to achieve. Furthermore, the relationship between Al distribution and the acid strength has not been fully understood. Thus, many zeolite researchers have seriously tackled the estimation of the distribution of framework Al atoms as well as the control of the location of acid sites in the pores.

ZSM-5 zeolite with the **MFI** topology, which is one of the most widely studied and commercially important zeolites, has twelve distinct T-sites and consists of parallel and straight 10-membered ring (MR) channels intersected by sinusoidal 10-MR channels. Here we report the synthesis of the ZSM-5 catalysts with the Al distribution in the framework controlled and their catalytic performance in the MTO reaction.

The ZSM-5 zeolites were hydrothermally synthesized with tetrapropylammonium cation (TPA^+) in the presence or absence of Na^+. Thus prepared ZSM-5 samples were denoted by [TPA,Na]- and [TPA]-HZSM-5, respectively. Considering the size of TPA^+, it would only be located at the intersections of the **MFI** structure. Hence, we assumed that the acid sites in ZSM-5 synthesized with TPA^+ in the absence of Na^+ are selectively located solely at the intersections, and that the acid sites in ZSM-5 synthesized with TPA^+ and Na^+ are located not only at the intersections but also in narrow straight and/or sinusoidal channels.

381430 Rational Design of Organic Structure-Directing Agents for Synthesis of Silica-Based Microporous Materials

Abstract:

Representative silica-based microporous materials include zeolite and sol-gel derived silica. Currently, zeolites are classified into more than 210 types based on their framework structures by the International Zeolite Association (IZA), although it is theoretically predicted that there are over a million possible frameworks. Discoveries of zeolites with new framework structures and/or properties, however, have mainly relied upon (pseudo)serendipity and Edisonian trial-and-error approach. Designed synthesis of zeolites with tailored structures and properties is still very difficult, mainly because of a lack of understanding in their formation, conventionally occurred through a complex series of "black-box" hydrothermal reactions. This limits their uses in emerging applications needing more finely tuned architectures.

Here we report our attempts to rationally design organic structure-directing agents for synthesis of microporous silica by a computational calculation. Our calculation is performed in two steps: i) screening by the semi-empirical quantum mechanics (QM) or molecular mechanics (MM) using the GULP program, followed by ii) the more accurate calculation by the QM/MM hybrid method. This talk contains i) design of novel anionic organic structure-directing agents for synthesis of microporous silica and ii) design of cationic organic structure-directing agents for tuning the frameworks and properties of zeolites.

Controlling Polymorphism in Organic-Free Syntheses of Zeolites

Matthew D. Oleksiak, Marlon Conato, and Jeffrey D. Rimer
University of Houston, Department of Chemical and Biomolecular Engineering
4800 Calhoun Rd, Houston, TX 77204

The unique acidity, thermal stability, and shape-selectivity of nanoporous zeolites are utilized in industrial applications spanning catalysis to ion exchange. The use of organics in zeolite synthesis is cost prohibitive, and avoided when possible; however, difficulties in the formation of zeolites via organic-free synthesis routes are prevalent [1]. The most common challenge is the formation of undesirable crystal phase(s), i.e. impurities. Oftentimes, organic structure directing agents (OSDAs) are used to form phase-pure crystals, though these additives are expensive and must be removed by post-synthesis calcination. In order to both understand and control organic-free polymorphism, we constructed kinetic phase diagrams based on varying synthesis conditions using only inorganic ions (e.g., alkali metals) as the structure directing agents [2]. These kinetic phase diagrams revealed regions of pure and mixed phases of eight zeolite framework types synthesized from colloidal silica and sodium aluminate.

Our systematic study has elucidated the phase behavior of many crystal structures, including for the first time the GIS framework type, a small-pore zeolite that exhibits two polymorphs: zeolites P1 and P2. This small-pore crystal structure, along with others (e.g., ANA type), have recently been shown to be catalytically active when noble metals are incorporated as extra-framework cations during crystallization. [3] By generating phase diagrams over a range of synthesis parameters, we have determined the conditions required to produce pure P1 and P2, as well as their structural analogues MER and PHI. These polymorphs differ by slight changes in the connectivity of their secondary building units – an outcome that can be controlled by judiciously selecting the silicon-to-aluminum and/or silicon-to-hydroxide ratios of the growth solution. We have shown that crystal phase transformations are governed by both thermodynamic and kinetic factors; and we have used a combination of experimental techniques to explore whether the transformation from one polymorph to another occurs by a ubiquitous pathway.

Additionally, we have investigated the growth mechanisms of two common zeolites, FAU and LTA. From time-elapsed studies of crystallization, we have shown that precursor particles are comprised of core-shell structures with a Si-rich core and an Al-rich shell wherein the nucleation of crystals on the exterior surfaces of core-shell particles has been observed. Collectively, these studies seek to identify the relative roles of classical and nonclassical pathways governing zeolite nucleation and crystal growth as a means of more effectively controlling phase purity, and ultimately their physicochemical properties for a wide range of commercial applications.

References:

[1] Oleksiak, M.D. and Rimer, J.D., Synthesis of Zeolites in the Absence of Organic Structure-Directing Agents: Factors Governing Crystal Selection and Polymorphism, *Rev. Chem. Eng.* 30 (2014) 1-49

[2] Maldonado, M., Oleksiak, M.D., Chinta, S., Rimer, J.D., Controlling Crystal Polymorphism in Organic-Free Synthesis of Na-Zeolites, *J. Am. Chem. Soc.* 135 (2013) 2641-2652

[3] Goel, S., Wu, Z., Zones, S.I., and Iglesia, E., Synthesis and Catalytic Properties of Metal Clusters Encapsulated within Small-Pore (SOD, GIS, ANA) Zeolites, *J. Am. Chem. Soc.* 134 (2012) 17688-17695

387081 Tuning Morphology and Activity of Three-Dimensionally Ordered Mesoporous Oxide Catalysts By Templated Solvothermal Synthesis

Abstract:

Realization of efficient and highly-selective liquid-phase catalytic conversion of bulky organic molecules (e.g., biorefinery, environmental remediation) demands the development of next-generation catalysts with ordered and tunable meso- or hierarchical-porosity in order to simultaneously meet demands on stability in the face of possible high-temperature aqueous phase processing and reduction of mass transport limitations. In this talk, we will discuss a hierarchical templating strategy enabling realization of high-surface area, large-pore volume metal oxide materials (e.g., TiO_2, ZrO_2) with three-dimensionally ordered mesoporosity. The approach relies on the facile synthesis and template-free assembly of monodisperse, size-tunable (ca. $\geq$ 10 nm) silica nanoparticles into 3D-ordered colloidal crystals, and their replication via solvothermal nucleation and crystal growth of the desired metal oxide within the 3D-interconnected template pore topology. We will demonstrate how this solvothermal processing results in sub-micron, spherical 3DOm titania materials.

In an effort to establish critical synthesis-structure-function relations for these materials, we will discuss a comprehensive study that elucidates the role that the template and the solvent (e.g., primary, secondary, tertiary alcohols) employed during material synthesis play in modulating particle size, crystal nucleation and growth, pore volume, and the dominant crystal faceting of the synthetic 3DOm metal oxides. Studies comparing the intrinsic kinetics associated with the synthetic 3DOm titania materials for biomass derivative conversion (e.g., glucose-to-fructose isomerization as well as fructose dehydration to HMF) and photocatalytic decomposition of 'bulky' surrogate molecules (e.g., methylene blue) help establish structure-function relations enabling tunability and optimization of catalytic activity with judicious choice of the solvent employed during the materials synthesis process.

An Alternative Strategy for Synthesizing Amine-Functionalized Silica Materials and Its Implications for Catalysis and Adsorptive Separations

Eric G. Moschetta, Adam P. Holewinski, and Christopher W. Jones*

School of Chemical & Biomolecular Engineering

Georgia Institute of Technology, 311 Ferst Drive NW, Atlanta, GA 30332

*cjones@chbe.gatech.edu

Amine-functionalized materials such as mesoporous silica are frequently used in catalytic and adsorption applications due to the facile surface customization of the silica from inexpensive starting materials. Amines grafted to the surface of mesoporous silica act as Lewis bases, making them excellent functional groups for capturing acidic CO_2 in adsorption processes and effective base catalysts, either as lone basic sites or as partners for the weakly Brønsted acidic silanols native to the surface of the silica for bifunctional (cooperative) catalysis. However, during functionalization via aminosilane grafting, amines may cluster on the surface,[1] reducing the number of silanols the amines can interact with to facilitate cooperative catalysis. In adsorption processes, the molecular understanding of the exact mechanism of CO_2 adsorption by amine adsorbents is still unclear, and both amine-amine and amine-silanol interactions have been implicated as playing a role in CO_2 capture. As such, there is interest in new synthetic methods to allow additional methods controlling amine density and distribution on the surface of amine-functionalized silica materials.

In this work, we discuss a new method for synthesizing amine-functionalized silica materials, in which alkylhalide-functionalized SBA-15 silica is converted to alkylamine-functionalized silica. Silica samples are placed in a Teflon sample holder containing six separate chambers to allow as many different samples to be aminated simultaneously. We prepared alkylhalide-functionalized silicas at similar alkylhalide loadings using Cl, Br, and I as the halides and aminated them under identical conditions to evaluate the effect of the halide on the conversion to primary amines. Then, we examined the effect of the overall alkylhalide loading on the conversion to primary amine, using the halide that led to the highest conversion from the first study. We used elemental analysis to determine the conversions in each case and other material characterizations (e.g. TGA, XRD, nitrogen physisorption) to ensure the physical properties of the silica samples were unchanged after amination. This versatile synthesis method is expected to allow for enhanced characterization of these materials by relatively infrequently used yet powerful NMR techniques.

[1] J. C. Hicks, R. Dabestani, A. C. Buchanan, C. W. Jones, *Chem. Mater.* **2006**, *18*, 5022–5032.

The Triplet "Structure-Property-Synthesis Methodology" Multi-Scale Study based on Stimuli-Responsive Brush Copolymers

Jin-Jin Li, Zheng-Hong Luo*

Department of Chemical Engineering, Shanghai Jiao Tong University, Shanghai 200240, P. R. China

Introduction

▪ ***Stimuli-responsive brush copolymers*** have aroused great interest over the past several decades, however their applied research is currently very little.[1]

▪ Facing challenges from dynamic and demanding markets, the ongoing paradigm of ***chemical engineering*** begins to shift from ***process design to product design***.[2] It will reemphasize the research on molecular structure-property relationship in the chemistry field including organic, inorganic, and biochemistry.[3]

▪ Chemical engineers should do some contribution to ***bridge the gap*** between chemistry and chemical product engineering.[3]

Structure-Synthesis Methodology Relation[4,5]

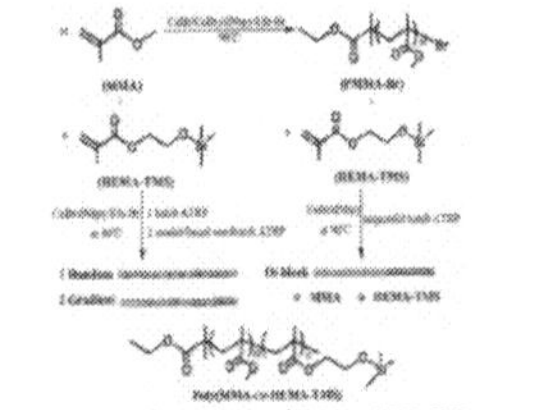

Scheme 1. Synthetic Scheme of P(MMA-*co*-HEMA-TMS)

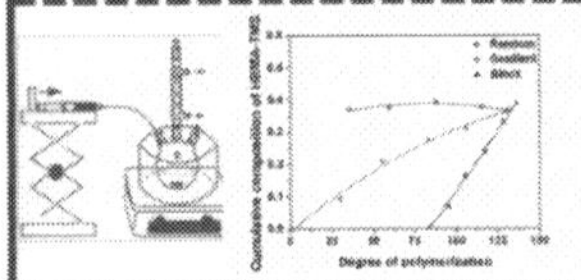

Figure 1. Cumulative HEMA-TMS composition in poly(MMA-*co*-HEMA-TMS) as a function of the number-average chain length

Random: Batch ATRP
Di-Block: Sequential, batch ATRP
Gradient: Model-based semibatch ATRcoP

$$F_{ins}(M1)_i = \frac{\Delta DP(M1)}{\Delta DP(M1)+\Delta DP(M2)} = \frac{\Delta DP(M1)}{\Delta DP(total)} = \frac{conv_i \cdot F_{cum}(M1)_i - conv_{i-1} \cdot F_{cum}(M1)_{i-1}}{conv_i - conv_{i-1}}$$

$F_{ins}(M1)_i$: the instantaneous composition of M1
$conv_i$: the total conversion of both monomers
$F_{cum}(M1)_i$: the cumulative composition of M1

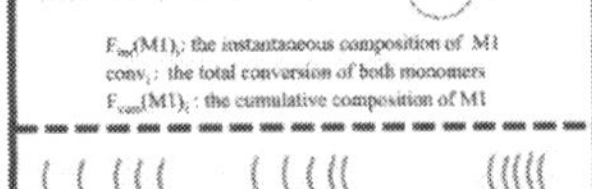

- Incorporating stimuli-responsive side chains
  1. Poly acrylic acid (PAA): pH responsive polymer.
  2. Poly(N-isopropylacrylamide) (PNIPAAm): temperature responsive polymer.
- Stimuli-responsive brush with varied grafting densities were obtained through adjusting backbone structure

Case 1. Structure-Ca^{2+} Responsivity[6]

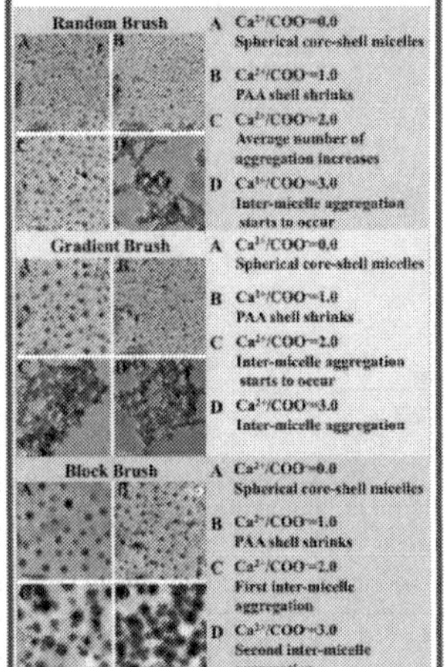

Figure 2. TEM images of micelles formed in poly(MMA-*co*-BIEM-*graft*-AA) solutions at different Ca^{2+}/COO^- values

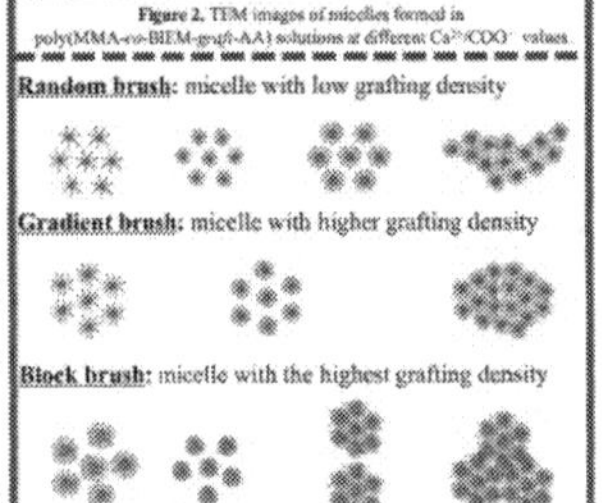

Scheme 2. Schematic illustrations of Ca^{2+}/COO^- complexation induced micelle aggregation of poly(MMA-*co*-BIEM-*graft*-AA)

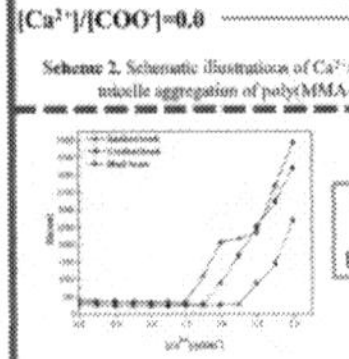

- The onset aggregation Ca^{2+} concentration Block<Gradient<Random

Figure 3. Ca^{2+} concentration dependence of the light transmittance through 0.1 mg/ml poly(MMA-*co*-BIEM-*graft*-AA) solutions.

- Degree of aggregation Block>Gradient>Random
- Block brush copolymer undergoes twice aggregation

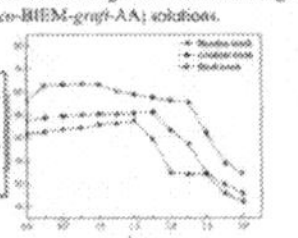

Figure 4. Ca^{2+} concentration dependence of the light transmittance through 0.1 mg/ml poly(MMA-*co*-BIEM-*graft*-AA) solutions.

Case 2. Structure-Temperature Responsivity[7]

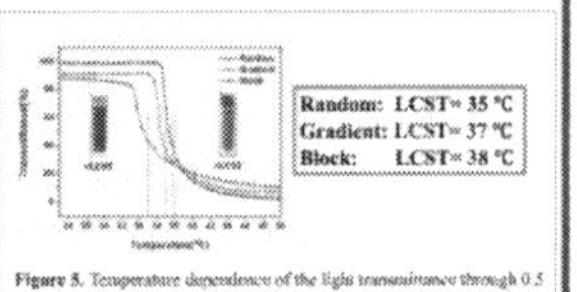

Figure 5. Temperature dependence of the light transmittance through 0.5 mg/ml poly(MMA-co-BIEM-*graft*-NIPAAm) solutions.

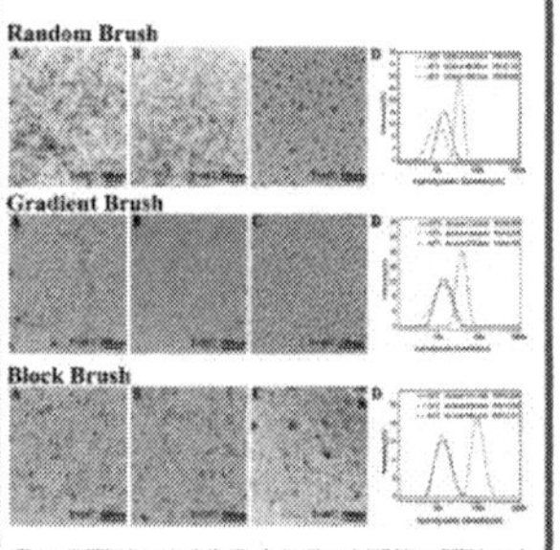

Figure 6. TEM images of micelles formed in poly(MMA-*co*-BIEM-*graft*-NIPAAm) solutions (A, B, C) and DLS results (D) at different temperature.

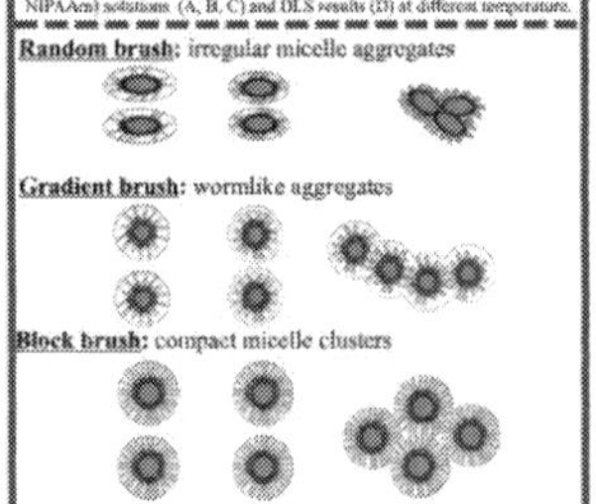

Scheme 3. Schematic illustrations of the thermo-responsive micellar behaviors of poly(MMA-*co*-BIEM-*graft*-NIPAAm).

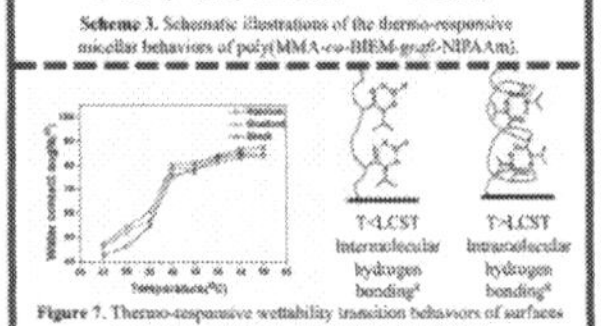

Figure 7. Thermo-responsive wettability transition behaviors of surfaces modified by poly(MMA-*co*-BIEM-*graft*-NIPAAm).

- All modified surfaces have a temperature tunable wettability
- The microcosmic structure of the molecules does not much affect the macroscopic phase transition behaviors on the surface.

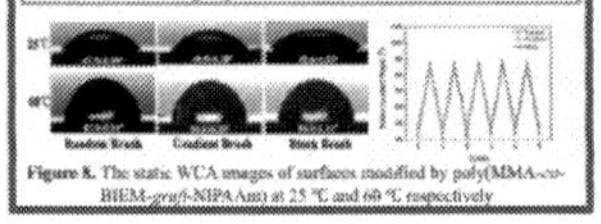

Figure 8. The static WCA images of surfaces modified by poly(MMA-*co*-BIEM-*graft*-NIPAAm) at 25 °C and 60 °C respectively

Chemical Product Engineering Perspective

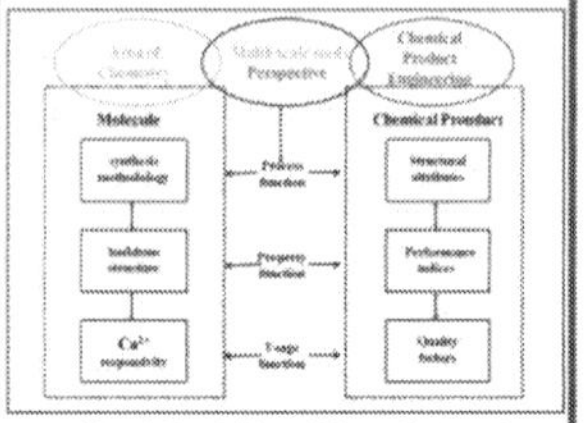

Scheme 4. Schematic Relationship between Chemistry and Chemical Product Engineering

- Chemical product engineering perspective was introduced to develop the as-prepared Ca^{2+}-responsive and temperature-responsive brush copolymers into new chemical products, which can be used as water treatment agent and thermal responsive coating.

Conclusion

▪ Established the ***relationship*** between the ***synthesis methodology*** and the ***backbone molecular structure***.

▪ Prepared a series of ***stimuli-responsive brush copolymers*** with ***random, gradient and block*** backbone structures respectively.

▪ Completed the ***triplet*** "backbone molecular structure- stimuli responsivity-synthesis methodology" ***multi-scale study*** in the field of chemistry.

▪ ***Chemical product engineering perspective*** was introduced to develop stimuli-responsive brush copolymers into new chemical products.

Acknowledgements

The authors thank the National Natural Science Foundation of China (No. 21276213), the Research Fund for the Doctoral Program of Higher Education (No. 20130073110077) and the National High Technology Research and Development Program of China (No. 2013AA032302) for supporting this work.

References

1. Lee H, Pietrasik J, Sheiko S S, Matyjaszewski K. Stimuli-responsive molecular brushes. *Prog. Polym. Sci.* **2010**, 35, 24-44.
2. Cussler, E L, Wei J. Chemical product engineering. *AIChE J.* **2003**, 49, 1072~1075.
3. Costa R, Moggridge G D, Saraiva P M. Chemical product engineering: an emerging paradigm within chemical engineering. *AIChE J.* **2006**, 52, 1976-1986.
4. Zhou Y-N, Li J-J, Luo Z-H. Synthesis of gradient copolymers with simultaneously tailor-made chain composition distribution and glass transition temperature by semibatch ATRP: from modeling to application. *J Polym. Sci. Part A: Polym. Chem.* **2012**, 50, 3052-3066.
5. Li J-J, Zhou Y-N, Luo Z-H. Synthesis and pH-responsive micellization of brush copolymers poly(methyl methacrylate-co-2-(2-bromoisobutyryloxy)ethyl methacrylate-graft-acrylic acid): role of composition profile. *Soft Matter* **2012**, 8, 11051-11061.
6. Li J-J, Luo Z-H. Case study to bridge the gap between chemistry and chemical product engineering: from molecules to products based on brush copolymers having different backbone composition profiles. *Ind. Eng. Chem. Res.* **2014**, 53, 1900-1908.
7. Li J-J, Zhou Y-N, Luo Z-H. Thermo-responsive brush copolymers with structure-tunable LCST and switchable surface wettability. *Polymer* **2014**. http://dx.doi.org/10.1016/j.polymer.2014.10.025
8. Sun T, Wang G, Feng L, Liu B, Ma Y, Jiang L, Zhu D. Reversible switching between superhydrophilicity and superhydrophobicity. *Angew. Chem. Int. Ed.* **2004**, 43, 357-360.

373811 Temperature-Triggered Shape Transformations in Layer-By-Layer Microtubes

Abstract:

Nano- and microstructured layer-by-layer (LbL) assemblies have been extensively studied for various applications. In particular, one-dimensional LbL microtubes have gained interest for their ability to shrink or swell in response to changes in pH. For LbL microtubes, surface interactions dominate over bulk behavior, leading to a variety of interesting responses to temperature. Here, it is shown that temperature can also be utilized as a trigger for shape transformations, and the results are markedly different from the case where pH is used. We report a temperature-triggered transformation of LbL microtubes of poly(allylamine hydrochloride) (PAH) and poly(acrylic acid) (PAA). PAH/PAA nanotubes were prepared using the polycarbonate (PC) membranes as sacrificial porous templates. Transformation of both freely released nanotubes and unreleased nanotubes (remain within the template) was investigated in aqueous media as a function of temperature and time using confocal laser scanning microscopy and scanning electron microscopy. When released microtube suspensions were incubated at high temperatures, the microtubes became shorter, fatter, and ellipsoid in shape. In contrast, unreleased microtubes showed periodic perforations on the surface following thermal treatment, which was suggestive of Rayleigh instabilities. In both cases, pronounced transformation occurred above the hydrated glass transition temperature of the PAH/PAA multilayer.

360208 Development and Characterization of Polyurethane Coatings for Enhanced Anti-Corrosive Properties

Abstract:

Abstract:

Around 90% of the polyols used for in production of polyurethanes currently are based on polyethers obtained from ethylene and propylene oxides. Vegetable oils also produce green polyols.

Polyols from palm oil and palm kernel oil were explored for functional polymeric materials such as epoxy, polyesteramide, alkyd and polyurethane.

Polyurethane coatings were extensively investigated for comparative functional properties. EIS analysis of the polyurethane coated steel samples was performed for corrosion resistance analysis. Nyquist and Bode plots were estimated for degradation. Adhesion of the coatings to the metallic substrates was investigated.

Also, few interesting observations regarding the comparative coating behavior on carbon steel and galvanized steel substrates are reported in this paper.

LAMAR UNIVERSITY
A Member of The Texas State University System

Anticorrosive Conductive Polyurethane Multiwalled Carbon Nanotubes NanoComposites

Daowei Ding,[1] Huige Wei,[1] Suying Wei[2], and Zhanhu Guo[1]

[1]Integrated Composites Laboratory (ICL) Dan F. Smith Department of Chemical Engineering Lamar University, Beaumont, TX 77710 USA
[2]Department of Chemistry and Biochemistry Lamar University, Beaumont, TX 77710 USA

INTRODUCTION

Corrosion, which usually involves the oxidation of metals and the reduction of oxygen, protons (H^+), and/or water,[1] imposes a major impact on the economics of industrial nations. Therefore, corrosion protection has become an issue of prime importance especially in the modern metallic finishing industry.[2]

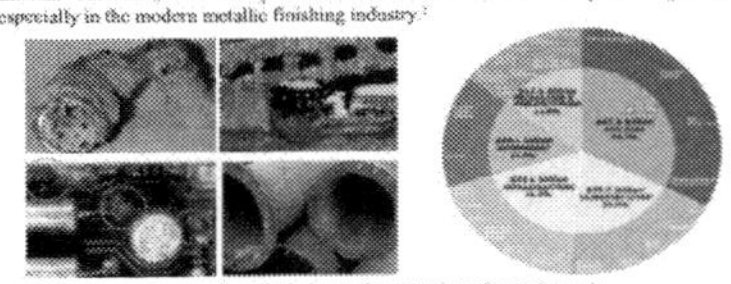

Fig. 1 Typical corrosion examples and corrosion costs.

Varieties of strategy have been developed to control the dynamics of corrosion (slowing down the kinetics and/or altering the corrosion mechanism). Conventional techniques including cathodic protection and anodic protection have been employed to decrease the oxidation or reduction rate in the corrosion process, respectively.[3]

Recently, intrinsically conductive polymers (ICPs) such as polyaniline, polypyrrole, polythiophene and their derivatives, have become a topic of current interest for corrosion prevention of metals.

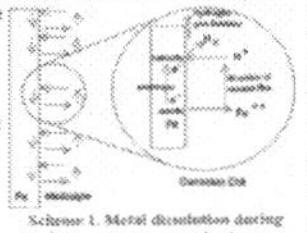

Scheme 1. Metal dissolution during the ICP coatings on metal substrates.

The problem existing for these ICPs prepared by chemical methods is that they are not easily melted and only slightly soluble in industrial solvents, and thus impose lethal limitations on their applications for anticorrosive coatings. To avoid this problem, direct electropolymerization of monomers on metal substrates to obtain thin conductive polymer films is preferred. Unfortunately, the dissolution of the metal substrate (or its passivation layer) occurring simultaneously during the polymerization or the passivation layer that acts as a barrier and inhibits the further electron transfer and the monomer oxidation in the acidic solutions remains the challenge.

Conventional organic coatings or paints, including epoxy, alkyd primers, polyvinyl acetate, and polyvinyl butyral have also been used to protect metal substrates from corrosion due to their low cost, easy application and aesthetic.

Fig. 2 Paints for anticorrosion protection.

Polyurethane (PU) has been considered as one of the most versatile polymeric materials that are widely utilized as furniture, coatings, adhesives, constructional materials, fibers, paddings, paints, elastomers and synthetic skins in the medical, automotive and industrial fields. More research efforts have been focused on the study of magnetic properties, or giant magnetoresistance behaviors of PU nanocomposites. On the other hand, carbon nanotubes (CNTs) have been intensely studied for a variety of scientific and technological applications due to their highly unique electronic, mechanical, catalytic, adsorption, and transport properties. Here, conductive PU nanocomposites filled with 5.0 wt% multiwalled carbon nanotubes (MWNTs) were fabricated using an in situ surface-initiated-polymerization (SIP) method for anticorrosion of stainless steel purpose.

EXPERIMENTAL

▪ Materials

Multiwalled Nanotubes (MWNTs)

Polyurethante (PU) two-part monomers

HO-R₁-OH — **diol**

O=C=N-R₂-N=C=O — **diisocyanate**

▪ PU/MWNTs Nanocomposite Preparation

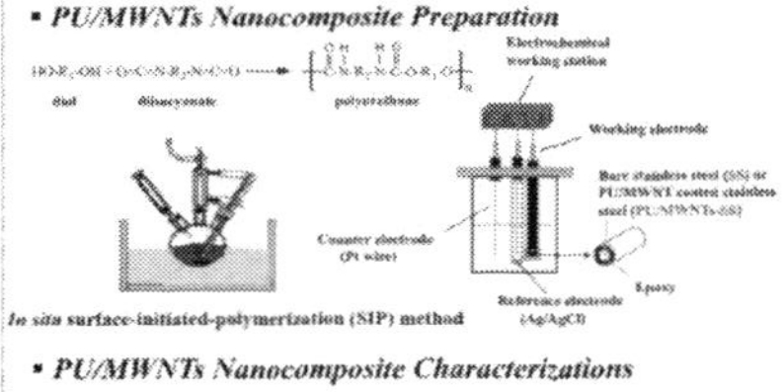

In situ surface-initiated-polymerization (SIP) method

▪ PU/MWNTs Nanocomposite Characterizations

- ➢ Chemical Structure
  - FT-IR spectra (Bruker Inc. Vector 22)
- ➢ Thermal Stability
  - TGA, TA Instruments Q-500
- ➢ In situ Conductivity
  - Cyclic Strain
- ➢ Anticorrosive Performances Evaluation
  - Open Circuit Potential (E_{ocp}) Evolution
  - Tafel Polarization Analysis
  - Electrochemical Impedance Spectroscopy (EIS)

RESULTS

▪ ATR and SEM

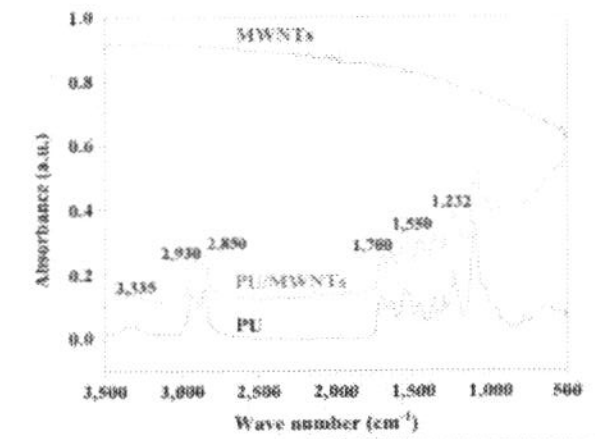

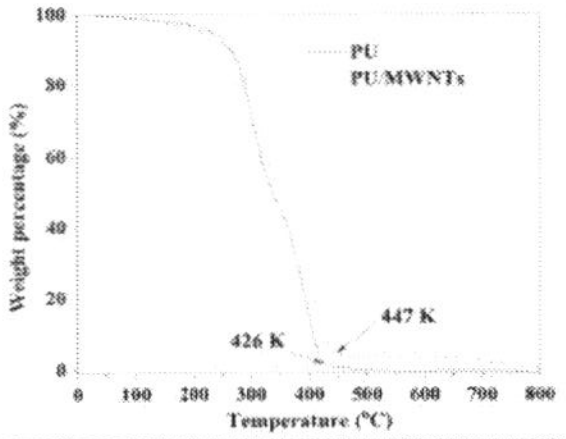

▪ In situ Electricity of PU/MWNTs in the Cyclic Strain Test

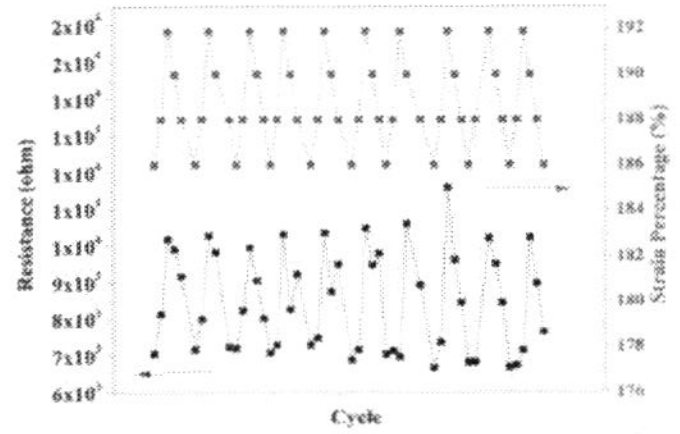

▪ E_{ocp} Evolution of SS and PU/MWNTs-SS Electrode

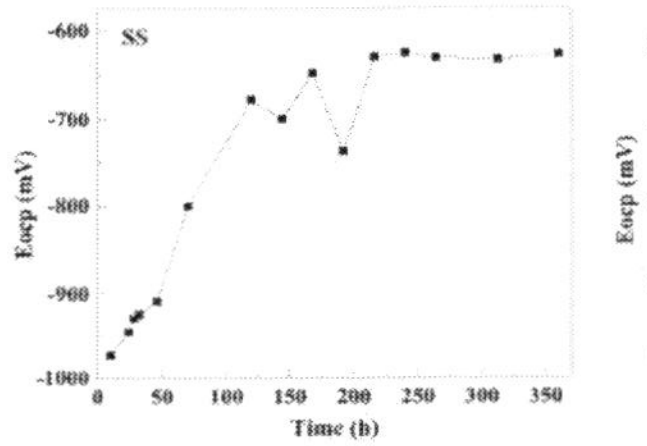

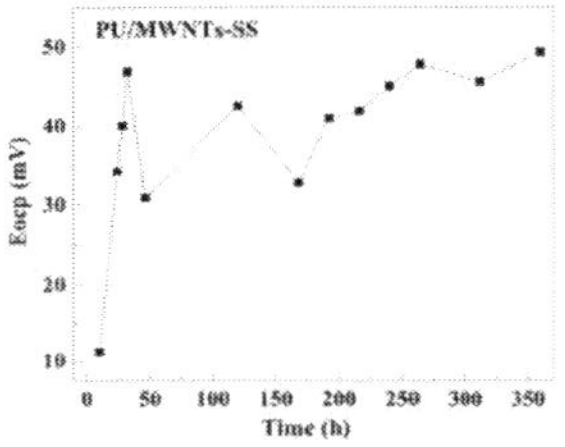

▪ Tafel Polarization Analysis

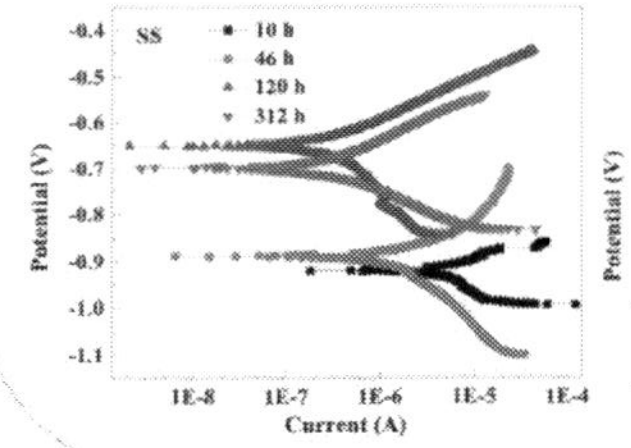

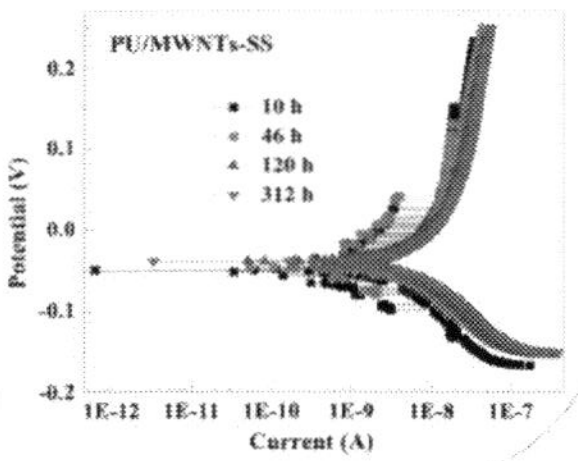

RESULTS

E_{corr} and I_{corr} values extrapolated from Tafel curves

Electrodes	E_{corr} (mV)	i_{corr} (A/cm²)	b_a (mV)	b_c (mV)	R_p (Ω)	P_{EF}%
SS	-698.0	1.62×10^{-4}	105.97	82.64	7.06×10^{3}	—
PU/MWNTs-SS	-43.6	7.63×10^{-6}	318.16	134.45	3.04×10^{5}	97.70

▪ EIS for Anticorrosion Mechanism Study

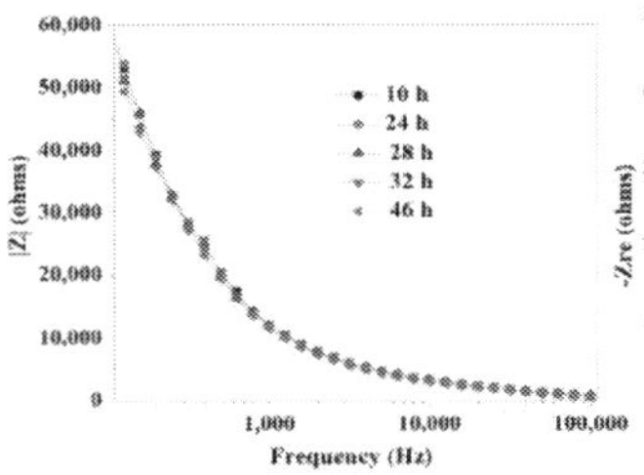

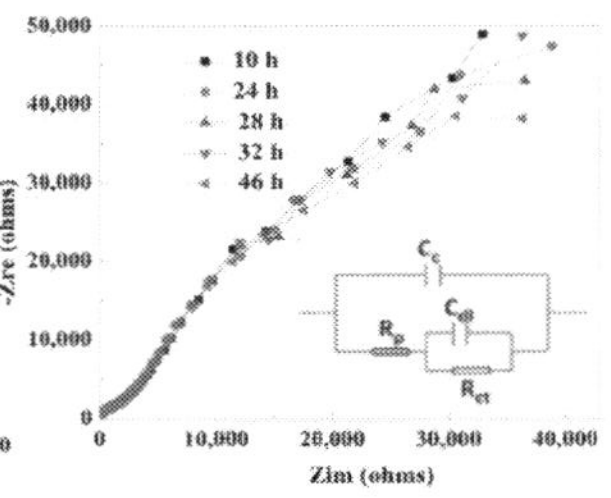

Fitting parameter values from the equivalent circuit

time (h)	R_p (Ω)	C_c (F)	C_{dl} (F)	R_{ct} (Ω)
10	4974	2.234×10^[illegible]	1.346×10^[illegible]	8.22×10^{4}
24	4709	2.352×10^[illegible]	1.332×10^[illegible]	7.751×10^{4}
28	4786	2.364×10^[illegible]	1.337×10^[illegible]	7.459×10^{4}
32	4641	2.405×10^[illegible]	1.347×10^[illegible]	7.735×10^{4}
46	4499	2.479×10^[illegible]	1.365×10^[illegible]	6.749×10^{4}

Scheme 2. Anticorrosion mechanism of PU/MWNTs coating

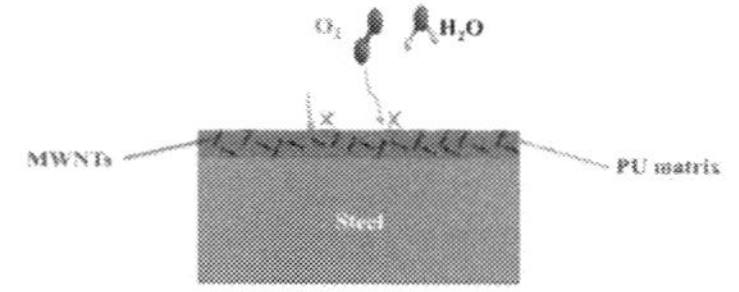

CONCLUDING REMARKS

(1) Significantly positive shift of nearly 1 V in *Eocp* and greatly reduced corrosion current with a high protection efficiency of 97.70% have been obtained in the PU/MWNTs composite coating;

(2) The composite film shows good chemical stability in the corrosive environment

(3) Anticorrosion properties originate from PU matrix via a solid barrier and well-dispersed MWNTs via increasing the diffusion path of corrosive ions

REFERENCES

1. D. A. Jones, Principles and Prevention of Corrosion, 1996.
2. D. E. Tallman, G. Spinks, A. Dominis and G. G. Wallace, J. Solid State Electrochem., 2002, 6, 73.
3. A. Baldissera and C. Ferreira, Prog. Org. Coat., 2012, 75, 241.
4. H. Wei, S.Y. Wei, Z. Guo, JMA. 2013, 1, 10805.

AKNOWLEDGEMETS

NSF CMMI-1030755

NORTHROP GRUMMAN

388750 Preparation of in Situ Anti-Sticking Supported Catalysts and Its Applications in Gas-Phase Polymerization

Abstract:

Gas-phase polymerization becomes dominant because of its economical and environmental benefits. However, the sticking characterization of the resulting polymers is a fatal problem that hinders the industrialization of gas-phase polymerization process for rubbers, such as polyisoprene and ethylene-propylene rubber.

An exploratory research was carried out on in-situ preparation of polyisoprene and ethylene-propylene rubber particles by gas-phase polymerization. A new method of preventing agglomeration of rubber particles was proposed. The selected catalyst system and nano-SiO_2 were both supported on micron-size silica to prepare the anti-sticking supported catalysts. During polymerization, nano-SiO_2stayed on the surface of the particles and formed a barrier layer, which should be effective for preventing agglomeration of the polymeric particles.

For the in-situ preparation of polyisoprene rubber particles, $Nd(O\textit{i}Pr)_3$/MAO/*t*BuCl was selected as the catalyst system. Morphology of products' particles, polyisoprene microstructure, molecular weight(MW) and molecular weight distribution(MWD), and polymerization activity were measured. It was found that with the increase of nano-SiO_2, the particle agglomeration is suppressed while polymerization activity decreases, and MW and MWD increase; the content of cis-1,4-unit tends to decrease. The microstructure of polyisoprene does not change significantly.

For the in-situ preparation of ethylene-propylene rubber particles by gas-phase polymerization, rac-$[Et(Ind)_2]ZrCl_2$/MAO was selected as the catalyst system. Morphology of ethylene-propylene rubber particles, microstructure, MW and MWD, and polymerization activity were characterized. Similarly, using nano-SiO_2 could effectively prevent the agglomeration of ethylene-propylene rubber particles. Polymerization activity decreased, while microstructure slightly changed when nano-$SiO_2$was introduced.

In summary, non-sticky polyisoprene rubber particles and ethylene-propylene rubber particles were in-situ prepared successfully in gas-phase polymerization by this novel method. A path to solve the common key problem of rubber particles' agglomeration was proposed.

"*Schizophyllan*" a Novel Polymer for EOR in High Temperature, High Salinity Carbonate Reservoirs: Adsorption characteristics over carbonate minerals

Mohammad Shoaib

Department of Chemical Engineering
University of Waterloo
Canada

Ahmed Abdala

Department of Chemical Engineering
The Petroleum Institute
Abu Dhabi, UAE

Ali Al Sumaiti

Department of Petroleum Engineering
The Petroleum Institute
Abu Dhabi, UAE

Polymer EOR has been proposed for some harsh (T~120°C, Salinity>200g/l) Middle East carbonate reservoirs. After screening a large number of polymers, a biopolymer *Schizophyllan* has been selected. This polymer is exhibiting excellent, intact viscosifying properties unaffected by long term thermal aging and high salinity up to a temperature of 135°C. Now, the next and one of the most important selection criterions is its adsorption properties over carbonate reservoir minerals. The adsorption level of this polymer over carbonate minerals will provide another step towards finalizing the polymer for field application. Adsorption at different natural (Calcite, Dolomite, Kaolin and Silica) and reservoir minerals has been studied. The effect of salinity, temperature, and initial concentration and background ions on the adsorption level is discussed.

386992 Designing Hydrophilic, Reactive Polymers for CO2 Capture and Acid Gas Transport

Abstract:

We have developed novel polymer materials based on the photopolymerization of imidazole-styrene monomers. These materials retain the inherent reactivity/basicity of the imidazole site as a means to transport CO2 from humid flue gas as bicarbonate anion. However, neat poly(imidazole-styrene) materials are glassy and hydrophobic, which limits their ability to provide facilitated transport of CO2 or other acid gases and absorb/retain water. To circumvent this issue, we are making the materials more hydrophilic and rubbery through partial neutralization with aqueous acids. This study focuses on the relationships between pKa, acid type and swelling behavior of the polymers.

Studying Stress Relaxation at Polymer Interfaces using FTIR-ATR Spectroscopy

Onyekachi Oparaji and Daniel Hallinan Jr.

Evolution of the structure of heterogeneous battery electrodes contributes significantly to fading of battery capacity. Both conventional electrodes and polymer-electrolyte-based electrodes rely on polymer binders to maintain intimate mixing of the components. However, large volume changes inherent in the charge and discharge of a battery can lead to delamination of the components from the polymer binder, rendering those components inaccessible. Therefore, our research focuses on studying the dissipation and storage of local stress gradients at polymer interfaces. We use a model styrene – ethylene oxide block copolymer and a selective solvent to preferentially swell one block, thereby imposing local stress gradients in the other block. Time-resolved Fourier Transform Infrared - Attenuated Total Reflectance (FTIR-ATR) spectroscopy is then used to measure stress relaxation in the unswollen block. In addition, structural characterization of morphological properties using small angle X-ray scattering will be reported. Understanding the storage and dissipation of stresses in polymer electrolyte will provide insight into battery fade and enable the design of longer-lasting polymer electrolyte binders.

390669 Atmospheric Plasma-Assisted Chemical Vapor Deposition

Abstract:

Recent work has sought to adapt plasma-assisted CVD and functionalization to ambient conditions with atmospheric non-thermal plasma discharge systems, for the in situ treatment of surfaces that are not amenable to processing under typical vacuum chamber CVD conditions. We present work on the design of low-cost systems—power supplies, gas and precursor feeds, and real-time analysis hardware—for the deposition of ultrathin, chemically-functionalized polymer films on heat-labile surfaces at atmospheric pressure.

The phase composition and lamellar morphology of linear and branched polyethylene: Morphology interpretation of TD-NMR and SAXS studies over broad range of crystallinity

R. Pokorný, J. Chmelař, P. Schneider, J. Kosek

Department of Chemical Engineering, Institute of Chemical Technology Prague, Technická 5, 166 28 Praha 6, Czech Republic, Tel.: 224 35 32 96, E-mail: Juraj.Kosek@vscht.cz

The semi-crystalline structure of polyethylene (PE) strongly affects its physical properties and also the transport of small-molecule penetrants in PE. The diffusion of penetrants in PE is important not only in PE manufacturing and down-stream processing, but also affects its application properties, such as permeability of PE membranes. As a semi-crystalline polymer, PE is composed of thin lamellar crystals with characteristic thickness approx. 10 nm. These lamellae are arranged into stacks and separated by an amorphous phase. The lamellae grow from a nucleation center, branch and twist, until they impinge on other lamellae growing from adjacent nucleation centers. Thus, the space filling spherulitic structure is formed.

Although the morphology of PE is extensively studied, most studies are focusing only on the crystallinity. Thus, they often non-critically adopt the classical three-component model of PE, which considers a crystalline and an amorphous phase separated by an interfacial component. However, this can lead to very high interfacial component content, with unrealistic values above 40 wt. % reported for PEs with lower crystallinities. To address this issue, we used low field ^{1}H NMR and small-angle X-ray scattering (SAXS) to obtain the phase contents and characteristic lamellar dimensions of PE samples with a wide crystallinity range, ranging from VLLDPE (35 wt. %) to HDPE (82 wt. %). Evaluating the ^{1}H NMR data measured for the HDPE samples, we observed only one amorphous component – a rigid amorphous phase, which consists of chains constrained between the neighboring lamellae in the lamellar stacks. However, a second amorphous component with significantly higher mobility was observed in the PE samples with crystallinity lower than 60%. We believe that this mobile amorphous phase, referred to as "free amorphous", consists of chains outside of the lamellar stacks. To support our hypothesis, we used SAXS to estimate the thicknesses of amorphous and crystalline domains in the stacks. These data were then used to calculate the fractions of the free amorphous chains outside of the lamellar stacks, obtaining excellent agreement with our ^{1}H NMR results and with a large set of literature data. Finally, the obtained experimental data were also compared with model predictions based on reconstructed 3D replicas of PE spherulites.

383127 Continuous Growth of Vertically Aligned SBS Via in-Situ Crosslinking in Electrospray

Abstract:

Electrospray has been recently developed as a novel technique for continuous deposition of ordered block copolymer thin films as an analog to physical vapor deposition. The development of well phase-separated microstructures is achieved by balancing thermal equilibration, deposition rate and residual solvent content, which are all readily tuned by spray parameters. We have demonstrated previously that well-ordered common diblock copolymers such as PS-b-PEO, PS-b-PMMA and PS-b-P4VP can be deposited by electrospray with film morphology strongly dependent on spray parameters. Here we explore the preservation of non-equilibrated vertical orientation through in-situ crosslinking in a cylinder-forming SBS (Styrene-Butadiene-Styrene) triblock copolymer deposited by electrospray. We performed parametric studies of solvent composition, flow rate, collection distance, deposition temperature on determining the surface morphology. Perpendicular morphology can be accomplished on preferential substrates through the evaporation of solvent from deposited material in the 'wet' spray regime, which then equilibrate with substrate over time and reorganize into parallel orientation under thermal annealing. The addition of thermal initiator, however, allows in-situ crosslinking of the polymer matrix thereafter locking in the as-sprayed morphology during deposition. By balancing the rate of crosslinking and ordering with the right amount of initiator, we achieve quasi-epitaxial growth of vertically oriented domain of SBS over micron-thick films.

382330 Solution Copolymerization of Ethylene and Propylene Catalyzed By [O-NS]TiCl3

Abstract:

The single-site catalytic system based on the metallocene complexes of group IV metals have been extensively studied. Recently, there has been a growing interest in developing new nonmetallocene catalysts, and in expanding beyond the first half of the transition metal series and various ligand backbones, as well as allowing access to previously inaccessible macromolecules. Of the successful nonmetallocene catalysts developed, the metal complexes possessing phenoxy-imine ligands are one of the promising examples. Group IV metal complexes $[O^{-}NX^{R}]TiCl_3$($O^{-}N$=phenoxy-imine, X=O, S, Se, R=alkyl)proved to be an excellent catalyst for the olefin polymerization.

In the present work, the catalyst $[O^{-}NS]TiCl_3$ was used for ethylene and propylene copolymerization with the MAO as cocatalyst. The effects of copolymerization conditions such as polymerization temperature, cocatalyst concentration and feed ratio of ethylene/propylene (E/P) were investigated in toluene solution. The resulting copolymers were characterized by gel permeation chromatography(GPC), differential scanning calorimetry(DSC) and ^{13}carbon nuclear magnetic resonance(^{13}C-NMR) etc. The reactivity ratio and characteristics of the final copolymers were also discussed.

For this catalytic system, the activity of the given catalyst reaches 1.6×10^{6}g Polymer/(mol Ti h atm) in the ethylene polymerization process, and it will reduce by 50% at most with the propylene insertion under similar conditions. Whereas, the catalytic system can't catalyze the propylene polymerization. It has been found that polymerization behaviors were mainly dependent on [Al] to [Ti] ratios, temperatures and monomer feed ratios of E/P. Optimum [Al]:[Ti] ratio and polymerization temperature for the highest yield of copolymer have been obtained. In particular, a series of E-P copolymers with M_n=10,000 40,000 was synthesized. Most E-P copolymer samples have polydispersity values close to a Schulz-Flory distribution($M_w/M_n\approx2$). Thermal behavior for these copolymers was studied by means of the DSC technique. It has been observed that T_m increases with increasing the ethylene content due to the crystallization of EEE block segments. However, with increasing the propylene content, two melting peaks appeared at 124°C and 104°C. ^{13}C-NMR results showed that the copolymers are essentially composed of long ethylene sequences with some isolated PP diad units and less amount of isolated propylene units, as concluded from the low values of EPE triads for both samples and the higher values for PPE+EPP. The PPP triad is negligible for all samples. Also, The obtained monomer reactivity ratios(r_E=13.7-18.6, r_P=1.1-1.5) indicated a long ethylene sequences in the E-P copolymers.

370839 A General Mathemathical Programming Formulation for the Synthesis of Polimer Nanofibers

Abstract:

The optimal synthesis of nanofibers has been the subject of several research efforts in recent days. This is because nanofibers represent a high potential to be used in biomedical applications due to the high surface area respect to the volume and their microporous structure, for this reason nanofibers are currently the fastest growing segment of the nanotechnology market. This is because they have potential applications in medicine, including artificial organ components, tissue engineering, implant materials, drug delivery and wound dressing. Among all the current techniques available for the synthesis of nanofibers electrospinning is the most widely studied technique. Different parameters influence the electrospinning process to yield nanofibers with good quality and diameter, and several studies have focused in analyzing the effect of these variables on morphology and fiber's diameter which are key parameters for potential biomedical applications. However, it is important to note that the cost has not been previously considered as key parameter in the synthesis of nanofibers.

Therefore, this paper presents a general mathematical programming formulation to determine the optimal operating conditions to synthesize nanofibers through the electrospinning process at the minimum cost. Mathematical programming has the advantage that it allows to manipulate a lot of variables and constraints to determine the optimal solution of a given problem and particularly, disjunctive programming formulations allow to easily representing a complex combinatorial problem. Several relationships based on experimental data for different natural polymers to determine the nanofiber diameter and costs are proposed. Also, a general optimization approach is proposed to trade off the relationships between cost and nanofiber diameter. A case study including the specific relationships for three natural polymers and five operating conditions is presented. The proposed approach is general and it can be applied to different cases.

385081 Modeling Polymer Translocation through Biological and Solid-State Nanopores

Abstract:

We have investigated the ubiquitous phenomenon of electrophoretically-driven transport of single macromolecules through single biological and solid-state nanopores, by using a combination of Langevin Dynamics simulations and Fokker-Planck formalism. In our theoretical/modeling work, we have explicitly accounted for geometry and charge decoration of the nanopores and the chemical sequences of the macromolecules which are transported by the single-file motion. Specifically, we will present results for the following three related issues: (1) Translocation of a uniformly charged flexible polyelectrolyte, such as ss-DNA, through a narrow cylindrical pore bearing charge patterns under an externally applied electric field. We find unexpected nonmonotonic dependence of the kinetics of transport on the period of patterning. An explanation of this novel phenomenon is derived by performing statistical mechanics calculation, based on polymer statistics and the Fokker-Planck representation of polymer transport. The role of sequences on the polyelectrolyte is also addressed. In addition, we report the consequences of the stochastic resonance between an oscillatory driving field and the ratcheting mechanism of a translocating polymer along a patterned pore. (2) In the interest of composing suitable strategies to optimize translocation speed of DNA in the context of DNA-sequencing, we have modeled the diffusion and mobility of DNA through a nanopore under a ratcheting force from a polymerase. Our work offers a quantitative measure of the signal-to-noise ratio for the readability of base calling in the electrophoresis- based DNA sequencing. (3) Translocation of branched polyelectrolyte molecules through nanopores. We find that the transport kinetics of branched polymers depends crucially on the architecture of the polymer and the dimensions of the nanopore, suggesting a new analytical technique to characterize such polymers in aqueous media, as an alternative to gel permeation chromatography.

388180 Atomistic Descriptors of Material Properties

Abstract:

Modeling and simulation of materials can be performed across various length scales. This presentation focuses on modeling at the smallest scale using Density Functional Theory (DFT) and other ab initio methods to compute the properties of individual atoms in materials. Atomic population analysis techniques partition the total electron and spin density distributions of a material into individual atomic contributions. Integrating the electron density and spin density assigned to each atom gives the net atomic charges and atomic spin moments, respectively. Atomic spin moments quantify the magnetic structure of a material. The amount of electron density overlap between two nearby atoms can be used to compute effective bond orders that are important for understanding the nature of chemical bonds in materials. The effective volume of each atom (computed by the third order radial moment of the atomic electron density distribution) can be used to compute the C_6 dispersion coefficients (Tkatchenko and Scheffler, Physical Review Letters, 2009, Vol. 102, 073005). The net atomic charges and C_6 dispersion coefficients can be used to parameterize force-fields used in classical molecular dynamics and Monte Carlo simulations. This talk will survey recent developments in atomic population analysis methods, discuss the relative merits of various approaches, and highlight important applications.

381710 Mathematical Model for Estimation of Heat Insulation Properties of Polymer Foams

Abstract:

Heat insulation of buildings is becoming increasingly popular in Europe and across the world. Rigid polymer foams are one of the most often used materials for this purpose. Due to their chemical composition and inner structure, they provide excellent resistance to convection, conduction and also heat radiation. However, a continuous effort is being made to further improve this resistance and to find cheaper ways to produce polymer foams. This contribution is focused on mapping of all important phenomena driving the heat transport in polymer foams and mathematical modeling of the morphology that would yield foam with the best possible insulation properties.

Currently, the most promising results are expected from foams with very small cell size, e.g., lower than 1 μm. In these foams, larger amount of phase interfaces should lead to more scattering and thus to the decrease of radiative heat flux. Moreover, when the cells are so small, their size becomes comparable to the mean free path of gas molecules. Therefore, the mechanism of heat conduction changes from diffusive to Knudsen mode, which results in the reduced gas conductivity. However, we show that the cell size must be chosen very carefully, because for polymers with low absorption coefficient the foam equivalent conductivity can significantly increase with the reduction of the cell size.

We have developed a computer program allowing to theoretically analyze and predict heat insulating properties of polymer foams. It simulates the coupled conduction-radiation heat transfer in computer reconstructed three-dimensional foams. The heat conduction is governed by Fourier law, in which the gas conductivity depends on the cell size because of the so-called Knudsen effect. We use the P_1-approximation to describe the heat radiation and derive the appropriate boundary conditions, which consider partial photon reflection on phase interfaces. Moreover, the reflectivity of phase interfaces accounts for wave interference phenomena in the thin polymer walls. Overall, our model doesn't use any empirical correlations, the only input parameters are the foam microstructure and physical properties of pure phases.

By using this model, we can find optimal porosity, cell size, wall thickness and content of polymer in the struts. Additionally, we show that for certain types of foams, the equivalent conductivity depends on the total foam thickness and that the reduction of cell size could lead also to worse insulating properties. And finally, we propose ways, how to avoid these negative effects.

371781 A Re-Examination of Strongly Confined DNA in Nanoslits

Abstract:

DNA mapping and barcoding are important complimentary technologies to traditional high-throughput sequencing for assessing structural variation in large genomes. Linearization of the wormlike polymer, DNA, is a key step in these techniques and is often accomplished through confinement of the polymer in a nanoscale device. However, fundamental properties such as the root-mean-square end-to-end distance for highly confined polymers in even the most basic slit geometry are still not completely understood. This is largely due to the large separation of length scales between the monomer size, confinement length scale and contour length that must be spanned in order to properly account for the entropy of the polymer. To address this, we have directly computed polymer statistics for very long wormlike chains confined in slits using an off-lattice implementation of the pruned-enriched Rosenbluth method (PERM). Unlike the Metropolis algorithm, PERM is a chain-growth method that efficiently generates statistics for confined chains for over four orders of magnitude in contour length. From our calculations, we find that a strongly confined wormlike chain can be thought of as a nearly two-dimensional wormlike chain composed of deflection segments. This description is in contrast to a prior theoretical description of this regime, which attempted to map the behavior directly to a two-dimensional self-avoiding walk. Using our computational results, we show that the latter approach incorrectly treats the excluded volume, which results from three dimensional interactions between deflection segments.

373371 An Interatomic Potential for the Interfacial Characterization of Titanium and Polyethylene Based on the Modified Embedded-Atom Method

Abstract:

The modified embedded-atom method (MEAM), which is a reactive semi-empirical interatomic potential and was developed originally for the atomistic simulations of metals and metal alloys as well as metal hydrides, oxides, nitrides, and carbides, was recently parameterized for saturated hydrocarbons and related polymers, such as polyethylene and polypropylene, by the same research group [1]. This interatomic potential successfully predicts the energetics and geometries of linear and cyclic alkanes as well as alkyl radicals and reproduces the pressure-volume-temperature (PVT) relationships for these materials with great accuracy. The key advantages of the MEAM potential are 1) single formalism for metals, hydrocarbons, polymers, etc. that enables the simulations of complex material systems involving defects, free surfaces, and interfaces, 2) reactive simulation involving bond breaking and bond formation suitable to study damage and failure as well as chemical reactions in the atomistic and molecular scale, 3) computational efficiency due its simple formalism. The MEAM potential is widely used for the atomistic simulations of metals and metal alloys and its extension to hydrocarbons has enabled great many potential applications for polymer/metal material systems. In this work, we studied the interfacial energetics of polyethylene in contact with elemental titanium (Ti) surface. Ultra-high molecular weight polyethylene (UHMWPE) and Ti are two materials used in many biomedical applications, such as prosthetic joint materials (knee replacement). To enable the interfacial simulation of single- and multi-chain PE in contact with a Ti surface, we parameterized the MEAM potential for the Ti/C/H system of elements starting with our recently published set of carbon, hydrogen, and CH diatomic parameters [1]. For Ti, we used the parameters published by Kim and Lee [2] for the Ti-C binary system. We utilized the first-principles interaction energy data for the Ti-C, and Ti-H diatomics as the fitting database. Once parameterized, we used the MEAM potential to perform NVT dynamics simulations to calculate the energetics of PE/Ti interfacial interactions as well as to determine the conformations of single- and multi-chain PE on the Ti surface. The details of potential parameterization and molecular dynamics simulations as well as key findings will be presented.

[1] S. Nouranian, M. A. Tschopp, S. R. Gwaltney, M. I. Baskes, and M. F. Horstemeyer, "An Interatomic Potential for Saturated Hydrocarbons Based on the Modified Embedded-Atom Method," Physical Chemistry Chemical Physics, vol. 16, pp. 6233-6249, 2014.

[2] Y.-M. Kim and B.-J. Lee, "Modified embedded-atom method interatomic potentials for the Ti–C and Ti–N binary systems," Acta Materialia, vol. 56, pp. 3481-3489, 2008.

Comparison of Monte Carlo and quasi Monte Carlo technique in structure and relaxing dynamics of polymer in dilute solution

AIChE Annual Meeting, Atlanta, GA, November 16-21, 2014
For submission to "Atomistic and Molecular Modeling and Simulation of Polymers"

Rajib Mukherjee and Urmila Diwekar
Vishwamitra Research Institute, Center for Uncertain Systems:
Tools for Optimization and Management (CUSTOM), 368,
56th Street, Clarendon Hills, IL-60514

Abstract:

The performance of various polymer products for nanotechnology, biomaterials, paints and coatings is largely governed by the characteristics of their surfaces and interfaces. Polymer surface are often processed from polymers in solutions. The functional properties of the polymer are largely dependent on the morphology of the polymers in solution. Thus the structure and dynamics of polymer in solvent solution is an important area of research. The structural properties of polymer in solution are especially important in case of surface science where the phase-separated morphology in the micro/nano scale dictates the properties of the product. To understand polymer properties from the molecular scale it is essential to model polymers in micro/nano scale. Modeling polymers in solution is an efficient way to determine the morphology and thus the properties of the products. It saves time as well as helps to design novel materials with desired properties.

Polymers in solution systems are generally modeled with bead spring model and Monte Carlo or importance sampling Monte Carlo simulations is used to find the optimal configuration where the energy of the system is minimized. Often in these simulations, random numbers are used in the Monte Carlo steps. Normally random numbers try to form clusters and do not cover the entire dimension of the system. Thus the minimum energy structures obtained from these simulations are not optimal configuration of the system. In the present work a coarse-grained lattice based importance sampling Monte Carlo (MC) simulation is used to investigate the behavior of a polymer and its interactions with solvents in the formation of a spinodally decomposed phase-separated morphology. For stochastic sampling of polymer, generally uniformly distributed random numbers are used for the choice of polymer lattice site and polymer movement. In the present work pseudo random numbers as well as quasi-random numbers generated from Hammersely sequence sampling (HSS) are used in the simulation steps for stochastic selection of polymers and its movements. The optimal configuration of the polymers obtained from pseudo random number and quasi-random number are compared. The result shows that simulation with HSS attains a lower energy state after initial quench. At the late stage of spinodal decomposition, the structure factor decrease-showing Ostwald ripening.

376403 Investigating the Effective Parameters in the Coil-Stretch Transition of High Molecular Weight Polystyrene Under Uniaxial Extensional Flow: A Hi-Fidelity Brownian Dynamics Approach

Abstract:

Excluded volume and hydrodynamic interactions play a central role in macromolecular dynamics under equilibrium and non-equilibrium settings. The high computational cost of incorporating the influence of hydrodynamic interaction in meso-scale simulation of polymer dynamics has motivated much research on development of high fidelity and cost efficient techniques. Among them, the Chebyshev polynomial based techniques and the Krylov subspace methods are most promising. To this end, in this study we have developed a series of semi-implicit predictor-corrector Brownian dynamics algorithms for bead-spring chain micromechanical model of polymers that utilizes either the Chebyshev or the Krylov framework. The algorithm with the highest efficiency and fidelity, namely, the Krylov subspace method, is used to simulate dilute solutions of high molecular weight polystyrene in uniaxial extensional flow [L. Li, R. G. Larson, and T. Sridhar, J. Rheol. 44, 291 (2000)]. Finally, the influence of different parameters, namely appropriate inclusion of excluded volume (EV) and hydrodynamic interactions (HI), the level of fine-graining, the type of EV potential, and the type of conservative spring force law on the observed extensional hardening of polystyrene dilute solutions over a broad molecular weight range will be discussed.

381701 Thermal Transport in Cross-Linked Natural Rubber Subjected to Uniaxial Elongation

Abstract:

It is well established that when elastomers are subjected to deformation, the orientation of polymer chain segments between cross links can be perturbed far from their equilibrium conformation. In previous work we have used a novel optical technique called Forced Rayleigh Scattering (FRS) to measure deformation-induced anisotropy in thermal conductivity. We have found that there exists an apparently universal relationship between the thermal conductivity and stress tensors in both linear and cross-linked polymers. More recently, have developed a technique based on Infrared Thermography (IRT) and found consistency with our earlier FRS results. We have also used IRT to investigate the dependence of heat capacity on deformation in elastomers. This phenomenon is related to the longstanding question of whether there is an internal energy contribution to rubber elasticity, and more generally in polymeric liquids. By tracking the temperature evolution of stretched samples heated by a laser beam, we are able to measure heat capacity changes with respect to the equilibrium value as a function of elongation. We find that heat capacity increases with stretch ratio in lightly cross-linked cis 1,4- polyisoprene specimens subjected to uniaxial extension. We evaluate our results with a simple thermodynamic analysis of the internal energy contribution to tension based on classical rubber elasticity results. Additionally, we find that the deviation from the equilibrium value of heat capacity is consistent with an independent set of experiments comparing anisotropy in thermal diffusivity and conductivity employing FRS and IRT techniques.

373449 Ab-Initio Crystallization of Alkanes: Structure and Kinetics of Nuclei Formation

Abstract:

Molecular dynamic simulations of the crystallization of eicosane have been carried out as an example of a low molecular weight n-alkane system and for a high molecular weight alkane as an analog of a semi-crystalline polyethylene. An alternative method for determining crystal domains and growing nuclei has been developed based on the eicosane system behavior. In the former case, density and crystal lattice parameters conformed well to previously reported experimental values. Moreover, crystallization and nuclei formation were monitored using a procedure developed herein. Analysis showed that metastable nuclei could be obtained at 250 K lasting appreciably long lifetimes of tens to hundreds of nanoseconds, which then exhibited an abrupt irreversible increase in size. The present study emphasizes the long quasiequilibrium of these crystallizing systems even at moderate supercooling.

Modeling Oxygen Permeability in Biodegradable Polymer Films for Food Packaging

1. TG | TOPICAL CONFERENCE: INNOVATIONS OF GREEN PROCESS ENGINEERING FOR SUSTAINABLE ENERGY AND ENVIRONMENT:

    (a) Conversion of Solid Wastes to Energy and/or Product This session solicits presentations on innovative processing technologies for conversion of solid wastes into reusable energy and/or products to reduce environmental footprints. The solid wastes of large volume and energy content are of interest, such as municipal solid wastes, agricultural wastes, forestry wastes, and industrial wastes (spent cars, tires). Hazardous and toxic wastes are not covered by this technical session.

        i. Chair Frank Petrocelli Principal Research Associate: Air Products and Chemicals, Inc 7201 Hamilton Boulevard Tonnage Gases, Equipment and Energy - Technology Allentown, PA 18195 Email: petrocfp@airproducts.com
        ii. Co-Chair Sulin Chen Washington State University LJ Smith Hall Department of Biological Systems Engineering Pullman, WA 99163 Email: chens@wsu.edu – Will not be published
        iii. Co-Chair Gibson Asuquo Project manager: Office of the Industrial Technology Program Department of Energy Golden, CO 80401 Email: gibson.asuquo@go.doe.gov – Will not be published

2. 15 | FOOD, PHARMACEUTICAL & BIOENGINEERING DIVISION: Advances in Food and Bioprocess Engineering

    (a) This session will highlight recent advances in bioprocess and food engineering. Papers addressing issues and novel technologies related to food and bioprocessing from both academics and industry will be considered for presentation in this session.

        i. Chair Ying Jin Postdoc: Ohio State University 140w 19th Ave, 125A Chemical engineering Columbus, OH 43210 Email: jyzr2011@gmail.com Alternate Email: jyzr2011@gmail.com – Will not be published

ii. Co-Chair Yinming Du Research Scientist: Givaudan Flavors Corporation OH Email: yinmingdu@gmail.com – Will not be published

3. TG | TOPICAL CONFERENCE: INNOVATIONS OF GREEN PROCESS ENGINEERING FOR SUSTAINABLE ENERGY AND ENVIRONMENT: Green Product Design

   (a) This session is targeted for discussions of the latest advancements in design and synthesis of green consumer products with decreased toxicity, and more sustainable usage and manufacturability. The ingredients of interest to this session are greener solvents, polymer building blocks, surfactants from renewable sources, bio-catalysis, fragrances, and bulk chemicals. The consumer products that reduce consumption of resources (water, energy) and minimize environmental emissions are also of interest, such as washers, driers, and refrigerators. Specific industry sectors of interest to this session are pharmaceuticals, CPG (cleaning, cosmetics, and food), textile, appliances and their raw material suppliers.

      i. Chair Kaustav Ghosh Lead Research Engineer: Whirlpool Corporation 750 Monte Road MD 5155 Benton Harbor, MI 49022 Email: kaustav_ghosh@whirlpool.com – Will not be published
      ii. Co-Chair Wei LIU Chief Engineer: Pacific Northwest National Laboratory 902 Battelle Boulevard Energy and Environment Richland, WA 99352 Email: wei.liu@pnl.gov – Will not be published
      iii. Co-Chair Z. Conrad Zhang Professor: Dalian National Laboratory for Clean Energy 457 Zhongshan Road DNL602 Bioenergy Dalian, 116023 China Email: zczhang@yahoo.com – Will not be published

4. O8 | MATERIALS ENGINEERING AND SCIENCES DIVISION:

   (a) Naturally-Derived and Sustainable Biomaterials Cosponsored By: Environmental Division (09) This session invites papers on developing materials derived from natural and sustainable sources as facilitators and promoters of healing and regeneration as well as for the delivery of prophylactic and therapeutic treatments. This session will focus on showing structure/property relationships of naturally-derived or naturally-inspired biomaterials and research involved in surmounting obstacles associated with the processing and broader use of sustainable biomaterials. Some examples of naturally-derived materials are as follows: bladder submucosa, small intestinal submucosa, collagen, cellulose, chitosan, and alginate. Research papers presenting in vitro and in vivo biocompatibility assessments and showing potential for use in regenerative medicine or drug delivery are encouraged to participate in this session.

i. Chair Zhiqiang Cao Assistant Professor: Wayne State University Department of Chemical Engineering and Materials Science Detroit, MI 48202 Email: zcao@wayne.edu – Will not be published

ii. Co-Chair Bret Ulery U Chicago Email: bulery@uchicago.edu – Will not be published

388054 Lithium Dendrite Growth through Polymer Electrolyte Membranes

Abstract:

High modulus block copolymer electrolytes have the ability to lengthen the cycle lifetime of rechargeable batteries by slowing lithium dendrite growth. However, batteries made with these polymer electrolyte membranes still eventually fail via dendritic short circuit. Hard X-ray microtomography was used to image dendrite growth *in situ* in symmetric lithium/polymer electrolyte/lithium samples. Each symmetric cell contained many dendrites. The volume of the dendrites was measured as a function of the charge passed through the electrolyte in order to determine the rate of dendrite growth in these systems. The dendrites appear to form on crystalline contaminants at the lithium metal electrode/electrolyte interface and then slowly form in the electrode before protruding into the electrolyte. Further experiments were performed to determine the effect of temperature and depth of discharge on dendrite morphology and growth rate. The dendrites have a "foam-like" structure where the interior of the foam cells is lithium metal and the exterior consists of elements from the lithium, polymer electrolyte, and salt. The size of the "cells" in the dendrite is correlated to the amount of lithium ions moved per cycle. These experimental results are compared with theoretical models for the growth of dendrites in high modulus polymer electrolyte systems.

386946 Polar Monomer – Salt Complexes: Ionic Liquid-like Polymers through Simple Blending

Abstract:

This study examines the photopolymerization and gas separation behaviors of imidazole monomers in the presence of bulky salts. We have observed improved monomer conversion compared to photopolymerization of these monomers in the absence of the salt. Additionally, poly(alkylimidazole)-styrene monomers were mixed with the aforementioned bulky salts to synthesize membranes and evaluated for room-temperature CO_2/N_2, and CO_2/CH_4 separation performance. The membranes were formed via UV-polymerization. Polymers were analyzed via IR spectroscopy to determine the degrees of polymerization, DSC to determine glass transition temperatures and scanning electron microscopy to observe surface features as well as polymer thickness. We will discuss the dependence of salt concentration on the magnitude of the improvements and potential mechanisms for this behavior.

365891 Understanding Organic Semiconductor Polymorphism Using High Speed in-Situ Optical and X-Ray Diffraction Methods

Abstract:

Organic electronics have been considered a leading candidate to make transparent and flexible electronics at a low cost. We have previously shown that the solution shearing method is a process that improves electrical performance for a range of OSCs, and the method is compatible with roll to roll industrial processing. This method can also tune the polymorph formation in OSCs, enabling high performance transistors without changing the OSC chemical structure. However, it is difficult to study the morphological and polymorph formations that enable high OTFT performance in situ. Not only does the thin film crystallize at a fast time scale, the evaporation front, where the crystal grows from the solution, is very small. The entire evaporation front can be less than 200 microns. Thus, the solution evolves into a crystallized thin film within seconds, and within an area less than 0.2 mm wide.

We use an X-ray 'microbeam' at the Cornell High Energy Synchrotron Source, with a beam width of < 20 microns, in conjunction with a high speed detector to resolve and follow crystallization from solution of the OSC during solution shearing. We have collected up to 100 frames per second X-ray images, and are able to create grazing incidence x-ray diffraction movies to easily see how crystallization occurs in the solution shearing system in real time. We also use an optical microscope trained at the evaporation front, which we can use to collect optical videos of the evaporation front at up to 10,000 frames per second. Being able to simultaneously study kinetic crystallization using both optical and X-ray movies helps us understand how different processing conditions result in various polymorphs. We study the model OSC 6,13-bis(triisopropyl)-silylethynyl pentacene (TIPS-pentacene) and show that confinement of the growing thin film plays a key role in forming metastable polymorphs, and that the film formation proceeds downwards from the air-solution interface. We generate metastable crystal polymorphs through other solution processing conditions as well. This is the first time such a fast rate of data collection has been utilized for grazing incidence X-ray diffraction.

Analysis of Charge Carrier Transport in Organic Photovoltaic Active Layers

Xu Han and Dimitrios Maroudas
Department of Chemical Engineering, University of Massachusetts, Amherst

Active layers consisting of organic polymers have generated significant interest in photovoltaic device fabrication. Organic photovoltaic (OPV) devices are much more cost-effective to fabricate than conventional inorganic photovoltaic (IPV) devices because of the easier large-scale manufacturing and the lower cost of OPV polymers. However, the efficiency of OPV devices is limited by the low charge carrier mobilities in the polymeric layers, as well as by carrier trapping in structural defects present in the layers. Experimentally, the time-of-flight (TOF) method is used to characterize excess charge carrier transport by producing transient photocurrents.

We present a systematic analysis of charge carrier transport in OPV devices based on deterministic, phenomenological charge carrier transport models. The models are used to simulate TOF experiments and predict photocurrent evolution in devices with active layers composed of P3HT thin films and nanoparticle assemblies, P3HT nanoparticle assemblies with dispersed PMMA nanoparticles, PBTDV films, and P3HT/PCBM polymer blends forming bulk heterojunctions. The models account for transient drift and diffusional transport of electrons and/or holes together with charge trapping and detrapping kinetics in the active layer, coupled self-consistently with Poisson's equation for the electric field in the layer. Dynamics of charge pair dissociation and recombination processes also are accounted for. Our models are validated by comparisons of their predictions with experimental measurements of transient photocurrents. For all the polymer layers examined, fitting the modeling predictions to the experimental data of photocurrent evolution is used to examine the propensity of the active-layer material to generate charge and form defects, determine the value of zero-field electron and hole mobilities and their Pool-Frenkel field dependence, analyze the kinetics of charge trapping and detrapping processes, and investigate charge pair dissociation as well as bimolecular and geminate recombination. We have studied effects of materials morphology and nanostructure by fitting transient photocurrent data in active layers consisting of both drop-cast thin films and nanoparticle assemblies and comparing the fitting outcomes. Moreover, in films consisting of nanoparticle assemblies, we have analyzed the effect on charge carrier transport of the presence of excess surfactant molecules, insulating (PMMA) nanoparticles, and nanoparticle sizes. The model predictions provide valuable input toward synthesis of organic active layers that lead to improved OPV device performance.

373099 Optimal Operating Policies for Producing Styrene and Methyl Methacrylate Copolymers Via RAFT Polymerization

Abstract:

Controlled radical polymerization (CRP) techniques offer the possibility of creating new materials with pre-specified properties and controlled molecular structures in mild operating conditions. These advantageous characteristics have given rise to an increasing interest in these synthetic methods, which is reflected in the numerous patents applications, journal articles and commercial products already been produced.[1, 2]

The polymerization control is achieved through the establishment of an equilibrium between active and inactive chains which is strongly shifted toward the dormant species. The active radical concentration reduction leads to a decrease in the total effect of termination, which results in a uniform growth of the polymer chains. Moreover, the lifetime of each chain is increased from a fraction of a second to hours, allowing experimental and industrial practitioners to manipulate the polymer structure through changes in operating conditions, such as reactants feeding policies.[3, 4] In this way it is possible to obtain copolymers with controlled molecular weights, molecular weight distribution (MWD), composition and chain architecture.[5]

Three CRP approaches have been more extensively studied: atom transfer radical polymerization (ATRP), nitroxide mediated polymerization (NMP), and reversible addition-fragmentation chain transfer (RAFT). The two former techniques rely upon adding an agent that reversibly terminates the radicals. Instead, RAFT polymerization is controlled through the addition of a chain transfer agent (CTA) that transfers the active radical to a large number of chains. RAFT polymerization is one of the most versatile approaches due to the large range of applicable monomers. Unlike ATRP, it does not require the use of transition metals. However, RAFT polymerization products have some potential for odor and discoloration if the dithioester group that acts as CTA leaches from the polymer chain.[4, 6]

Through a copolymerization process different comonomers can be combined into a single material in many diverse structures. By changing chain length, composition or sequence length distribution, unique products can be obtained. The resulting phase behavior influences the physical properties and ultimate applications.[7] Polystyrene-block-poly(methyl methacrylate), or pSt-*b*-pMMA, is of special interest for nanoscale patterning thanks to its ability to produce diverse morphological patterns and the ease of removal of pMMA domains by UV exposure.[8] To this end, block chain length, chain length ratios, polydispersity index, and block composition greatly impact the final nanoscale morphology.[9] On the other hand, gradient copolymers of the same comonomers (pSt-g-MMA) are hoped to be a superior class of blend stabilizers than their block counterpart due to the formation of broad interphase regions.[10]

Simulation of RAFT systems is particularly difficult due to the distinct features of its kinetic mechanism. The polymerization proceeds through the formation of a two-arm intermediate adduct which makes necessary the computing of a bivariate MWD even for homopolymerization reactions.[11] In previous works we presented a mathematical model able to predict the molecular structure of styrene and methyl methacrylate copolymerization via RAFT.[12] Average properties, such as molecular weights and global composition, were modeled using the well-known method of double index moments. In addition, the full bivariate MWD of the copolymer was obtained by means of 2-D probability generating functions (pgf). This technique proved to be capable of dealing with the complex kinetic mechanisms and obtain the MWD without simplifying assumptions or any a priori knowledge of the distribution shape.

In this work, the developed model is used to seek optimal operating policies to produce styrene and methyl methacrylate RAFT copolymers with controlled structure and pre-specified properties. The difference between the obtained properties and the desired ones is minimized by changing design variables (such as reactor type or lateral feeds position) and operating policies. For semibatch reactors, operating conditions such as initial charge and feeding policy were optimized. On the other hand, tubular reactor optimization space included the flowrate of the main feed, and the position, flowrate and composition of the lateral feeds. The optimization problem was formulated and solved in gPROMS (Process Systems Enterprise, Ltd.). Optimal operating policies were obtained for producing copolymers with controlled polydispersity and pre-determined copolymer structure: either total molecular weight and chain length ratios for block copolymers, or pre-specified compositional gradients for gradient copolymers. Modeling studies like the one presented in this work are of great help to find operating conditions that allow manufacturing unique structures designed beforehand. In this way, experimental trial and error procedures can be minimized, resulting in resource savings and facilitation of the scaling up of CRP techniques to an industrial level.

[1] W. Jakubowski, N. V. Tsarevsky, P. McCarthy, K. Matyjaszewski, *Materials Matters* **2010**, *5 (1)*, 16.

[2] K. Matyjaszewski, J. Spanswick, *Materials Today* **2005**, *8*, 26.

[3] M. Zhang, W. H. Ray, *J. Appl. Polym. Sci.* **2002**, *86*, 1630.

[4] W. A. Braunecker, K. Matyjaszewski, *Prog. Polym. Sci.* **2007**, *32*, 93.

[5] K. Matyjaszewski, K. A. Davis, *Statistical, Gradient, Block and Graft Copolymers by Controlled/Living Radical Polymerizations*, Springer-Verlag, Berlin, Germany, 1st edition, **2002**.

[6] S. Grajales, *Controlled Radical Polymerization Guide*, Sigma-Aldrich Co. LLC. Materials Science, **2012**.

[7] F. S. Bates, *Science* **1991**, *251*, 898.

[8] G. S. W. Craig, C. J. Thode, M. Serdar Onses, P. F. Nealey, *Materials Matters* **2011**, *6 (3)*, 82.

[9] W. Jakubowski, A. Juhari, A. Best, K. Koynov, T. Pakula, K. Matyjaszewski, *Polymer* **2008**, *49*, 1567.

[10] U. Beginn, *Colloid. Polym. Sci.* **2008**, *286*, 1465.

[11] C. Barner-Kowollik, *Handbook of RAFT Polymerization*, Wiley-VCH, Weinheim, Germany, 1st edition, **2008**.

[12] C. Fortunatti, C. Sarmoria, A. Brandolin, M. Asteasuain, *Computer Aided Process Engineering* **2014**, June 15-18 Budapest, Hungary. Accepted 6 pages paper.

392226 Crosslinking of a Positive Tone Polynorbornene low-k Dielectric

Abstract:

Research in microelectronics packaging has recently been driven by the reduction in size of the die the package contains. Die-die, interconnect-interconnect, and substrate-die distances are rapidly decreasing in size. As the electronic system shrinks, the need for photo-definable, permanent low-k dielectric materials becomes more important to electrically and mechanically isolate increasingly denser electrical pathways.

There are two photochemistries that make films photodefinable. Negative tone materials decrease in solubility in a developer when exposed to UV radiation. Positive tone materials increase in solubility when exposed to UV radiation. Positive tone photochemistry has a distinct advantage over negative tone because of a reduced sensitivity to particulates on a photo mask (1). Additionally, positive tone chemistries allow an aqueous develop instead of the organic solvents used in most negative tone systems that are an environmental concern.

Polynorbornene (PNB) has shown promise as a permanent dielectric material because of a relatively low dielectric constant, 2.2 (2). Synthesizing a PNB copolymer with substituted pendent groups allows cross-linking and photo-definability, although it raises the dielectric constant by introducing polarizable groups to the polymer. In this case, diazonaphthoquinone (DNQ), a well-studied positive tone photoactive compound, is used to shift the solubility of a film containing the PNB copolymer.

Previously, our group studied the effects of cure temperature on a negative tone formulation of PNB (3). A photo acid generator was used to create crosslinking via an epoxy ring-opening reaction. To quantify the cross-linking, the elastic moduli of thin films were determined. Over the cure temperatures investigated, a maximum elastic modulus of 2.8 GPa was observed at 160 °C. The modulus then showed a steady decline to 2.4 GPa at 240 °C. This degradation in modulus was assumed to be a result of the acidity of the film (3).

In this work, a positive tone material is crosslinked, making it permanent. With a neutral film, crosslinking likely occurred with DNQ acting as the cross-linker, giving a modulus of 4.0 GPa. When developed, DNQ-loaded films showed a decline in modulus to 2.6 GPa. This is likely due to the reactions of DNQ with aqueous base (4). An epoxy cross-linker added to the developed positive tone film helped cross-linking by inhibiting base uptake and allowing DNQ to act as a cross-linker. The positive tone material with epoxy cross-linker showed a minimum dielectric constant of 3.65 at the highest cure temperature, 220°C. With the correct processing conditions and additives, the modulus and dielectric constant of the PNB film can be controlled.

References

1. B. K. Mueller, E. Elce, A. M. Grillo, and P. A. Kohl, "Positive-tone, aqueous-developable, polynorbornene dielectric: Lithographic, and dissolution properties," *J. Appl. Polym. Sci.*, vol. 127, no. 6, pp. 4653–4661, (2013).

2. G. Maier and D.- Garching, "Low dielectric constant polymers for microelectronics," vol. 26, (2001).

3. M. Raeis-Zadeh, N. D. Melendez, Y.-C. Chen, and P. A. Kohl, "Aqueous-Develop, Photosensitive Polynorbornene Dielectric: Optimization of Mechanical and Electrical Properties," *J. Electron. Mater.*, vol. 40, no. 10, pp. 2126–2138, (2011).

4. M. Koshiba, M. Murata, M. Matsui, and Y. Harita, "Thermally Induced and Base Catalyzed Reactions of Naphthoquinone Diazides," Adv. Resist Technol. Process. V, vol. 920, pp. 364–371, (1988).

389651 Polymer-Based Manufacturing of Microscale Fuel Cells

Abstract:

The development of microscale fuel cells has been motivated by the increasing demand for high power density, portable electronic devices (*i.e.*, smartphones, tablet computers) with a fast recharge [1]. In comparison to rechargeable batteries, microscale fuel cells demonstrate superior energy densities and the potential to provide longer off-grid run times [2]. Additionally, fuel cells enable nearly instantaneous recharging by refilling or replacing a liquid-based fuel cartridge. Here we will demonstrate polymer-based manufacturing of a lightweight, microscale fuel cell for portable electronic applications.

At the macroscale, fuel cells are traditionally assembled manually in layer-by-layer fashion followed by clamping the layers between metal plates. Scaling to larger systems involves the stacking of an increasing number of layers, requiring increasingly heavy clamping constructions to secure the stack without leaks. We demonstrate an alternative approach to fuel cell manufacturing which utilizes polymeric layers to eliminate the heavy clamping structures needed in traditional fuel cell stacks [3]. This approach enables much smaller, lighter structures and thinner stacks, which is attractive in portable applications.

[1] E. Kjeang, N. Djilali, D. Sinton, Journal of Power Sources, 186 (2009) 353-369.

[2] C. Potera, Environmental Health Perspectives, 115 (2007) A38-A41.

[3] A.S. Hollinger, P.J.A. Kenis, Journal of Power Sources, 240 (2013) 486-493.

374452 Mechanical Properties of Porous Polymer Separator for Lithium-Ion Batteries

Abstract:

Aging and degradation of Li-ion batteries over many charge and use cycles are caused by both chemical and mechanical processes. The studies of the latter are usually limited to the consideration of the processes in electrodes, particularly the expansion of the electrodes due to lithium intercalation. However, expansion of the electrodes also causes the compression of the separator between them. Recent studies showed that the compression causes the separator pore shrinkage and worsening of the ion transport through it, which has significant impact on the battery performance [1–3]. Therefore, it is necessary to know the mechanical properties of the separators relevant to the battery operation conditions. So far, the reported mechanical properties are limited to those derived from the tensile test [4,5], while tensile tests do not represent the loads in a battery.

Here we present the results of compressive tests of commercial porous polypropylene (PP) separator Celgard 3501 at different strain rates. The experiments were carried out for dry polymers and immersed in fluid (DMC – typical organic solvent for electrolyte). We focused on two characteristic mechanical properties of the separator: effective Young's modulus determined in the region of small deformation, and yield stress. We found that both of these properties differ for the dry material and immersed in fluid. We also found strong dependence of both on the strain rate, particularly we observed the noticeable stiffening of the wet sample at high strain rates. This effect cannot be explained by viscoelastic properties of the polymer, and related to poroelastic behavior – fluid drainage from the pores.

We developed a quantitative model taking into account both non-linear viscoelastic and poroelastic properties of the separator. Our model is capable of determining the material parameters relevant to the compression of the separator during battery operation. We performed simulations using Dynaflow finite element code [6] and calculated the effective Young's modulus corresponding to the experimental conditions. Our model shows that poroelastic effects have to be taken into account when interpreting the experimental results on fluid-immersed compression testing of the separators at the strain rates above 10^{-3} s^{-1}. The results of our work will be further used in the development of predictive model for studying the coupled mechanics and electrochemistry of lithium-ion batteries. Such model can serve as a tool for optimization the materials and geometric parameters of lithium-ion cells to find a path towards extending their performance and lifecycle.

References

[1] C. T. Love, Thermomechanical analysis and durability of commercial micro-porous polymer Li-ion battery separators, Journal of Power Sources 196 (5) (2011) 2905–2912.
[2] C. Peabody, C. B. Arnold, The role of mechanically induced separator creep in lithium-ion battery capacity fade, Journal of Power Sources 196 (19) (2011) 8147–8153.
[3] Y. Pan, Z. Zhong, Modeling the Ion Transport Restriction in Mechanically Strained Separator Membranes, Journal of the Electrochemical Society 161 (4) (2014) A583–A586.
[4] A. Sheidaei, X. Xiao, X. Huang, J. Hitt, Mechanical behavior of a battery separator in electrolyte solutions, Journal of Power Sources 196 (20) (2011) 8728–8734.
[5] I. Avdeev, M. Martinsen, A. Francis, Rate- and Temperature-Dependent Material Behavior of a Multilayer Polymer Battery Separator, Journal of Materials Engineering and Performance 23 (1) (2014) 315–325.
[6] J. H. Prevost, DYNAFLOW: A Nonlinear Transient Finite Element Analysis Program. Department of Civil and Environmental Engineering, Princeton University, Princeton, NJ. http://www.princeton.edu/~dynaflow/ (last update 2014) (1981).

386650 Fabrication of Size-Controllable Highly Ordered P3HT Nanostructures and Analysis of Their Assembly

Abstract:

In previous studies, solution processing methods such as ultrasound irradiation or addition of poor solvents have been used to induce nano-scale structure formation in poly(3-hexylthiophene) (P3HT) thin-films. Concomitantly, P3HT charge carrier mobility has been reported to be enhanced, which indicates that the features formed during these processes are likely to be ordered P3HT chains that are stacked via pi-pi interactions, resulting in improved interchain charge carrier transport. Here, we present a comprehensive study of the thin-film structure-charge carrier transport property relationships within P3HT thin-films by precisely controlling development of highly ordered P3HT nanostructures. When utilized separately, ultrasound irradiation and poor solvent addition are limited in their ability to assemble P3HT nanostructures having a range of dimensions. However, in a synergistic approach, the combination of both methodologies, i.e., ultrasound irradiation coupled with poor solvent addition, P3HT nanostructures with controlled length, width and quantity have been fabricated. Field-effect hole mobility of P3HT thin-films composed of nanostructures prepared here has exhibited a clear trend related to the differentiated nanostructure dimensions and quantities. P3HT hole mobility with material having highly ordered nanostructural features has reached impressive maximum average value of $0.103 cm^2/V\text{-}s$ and corresponding mobility values have been correlated with AFM morphology and x-ray crystallography data. Further, systematic investigation based on AFM morphology, grazing incidence x-ray diffraction, and UV-Vis spectroscopy data has enabled us to successfully apply a conventional 2-step crystallization model including nucleation and growth steps to the self-assembly process of P3HT nanostructures.

384899 Long-Range Order in Solutions and Films of Poly(3-hexylthiophene): Role of Self-Assembly and Capillary Flow in Processing

Abstract:

Poly(3-hexylthiophene) (P3HT) has been studied as a model conjugated polymer to understand how processing of solutions can impact film morphology and charge carrier mobility for potentially inexpensive, large area, flexible and lightweight organic electronic devices such as organic light-emitting diodes (OLED), Organic Photovoltaics (OPV) and Organic Field-Effect Transistors (OFET). In this study, we find that self-assembly that occurs with aging time in P3HT solutions can facilitate the formation of ordered, birefringent fluids through capillary flow. Drawing these liquid crystalline fluids into thin films shows that order achieved in the solution can be maintained after solvent evaporation. Semiconducting organic materials with long range order and few grain boundaries are highly desirable for achieving fast charge transport over macroscopic length scales. Polarized optical microscopy (POM), UV-Vis Spectroscopy, Raman spectroscopy and dynamic light scattering are used to probe solution-state ordering while POM, atomic force microscopy, and OFET device fabrication and testing are used to probe the solid state ordering and the resulting charge carrier mobility.

A Blend Approach to P3HT Based Field Effect Transistor Performance Enhancement via Inclusion of 2,5-bis(3-dodecylthiophen-2-yl)thieno[3,2-b]thiophene

Ping-Hsun Chu[1], Lei Zhang[2], Jung Ok Park[3], Mohan Srinivasarao[3], Alejandro L. Briseño[2] and Elsa Reichmanis[1,3,4]
ereichmanis@chbe.gatech.edu

[1]School of Chemical and Biomolecular Engineering
Georgia Institute of Technology, Atlanta, GA
[2]Department of Polymer Science
University of Massachusetts, Amherst, Amherst, MA
[3]School of Materials Science and Engineering
Georgia Institute of Technology, Atlanta, GA
[4]School of Chemistry and Biochemistry
Georgia Institute of Technology, Atlanta, GA

Organic field effect transistors (OFET) are attracting significant attention due to their facile processability, chemical tunability and low cost for producing large area electronics. Both small molecules and polymers have been widely studied for this purpose. Small molecules usually provide superior electronic properties, but suffer limitations associated with crystal grain boundaries and high vacuum deposition steps. Polymers, however, demonstrate excellent film uniformity, but experience problems related to chain entanglements and broad polydispersity which inhibit crystallization. In this work, we are engaged in studying semiconducting small molecule-polymer blend thin-films and their applications in field-effect transistors. In contrast to traditional devices using a single component, the blend thin-films can take advantage of the pros of polymers and small molecules while simultaneously diminishing the impact of their cons. According to experimental results, electrical characteristics of OFET devices can be improved by using polymer-molecular semiconductor blend thin-films as the active layer. We demonstrated that incorporation of the small molecule, 2,5-bis(3-dodecylthiophen-2-yl)thieno[3,2-b]thiophene (BTTT-1), into poly(3-hexylthiophene) (P3HT) polymer thin-films can enhance the charge carrier mobility and lead to approximately a 20-fold decrease in threshold voltage. The performance enhancement was dependent on the small molecule volume ratio, and was maximized at a critical oligomer concentration of about 50%. This thin-film blend combines the high charge transport properties of the crystalline small molecule semiconductor with the facile solution processability of the polymer component. In this investigation, BTTT-1 was selected because of its planar symmetrical fused aromatic rigid frame, i.e., thieno[3,2-b]thiophene, which promotes the formation of highly ordered crystalline domains and relative to a single thiophene ring structure in the P3HT backbone, thienothiophene is more stable to oxidation. Moreover, the alkyl side chains may facilitate interdigitation and formation of a compact lamellar structure. By taking advantage of these features, the crystalline small molecule might enhance P3HT molecular ordering. Alternatively, the polymer matrix might serve as a bridge between the solution processed small molecule crystalline domains to decrease the impact of crystal grain boundaries. Correlations between molecular ordering and the charge transport characteristics of resultant thin-films with respect to the polymer:small molecule volume ratio will also be presented and discussed.

385544 Effect of the Reaction Conditions on the Morphology of Polyglycerol Produced from the Polymerization Reaction of Glycerol Using Sulfuric Acid As Catalyst

Abstract:

The effect of the reaction conditions in the morphology of polyglycerol was studied. Polymerization of glycerol was performed using sulfuric acid as catalyst and the factors studied were temperature and catalyst percentage. Previous studies on the synthesis of polyglycerol from glycerol, has been focused in the development of catalysts to achieve high conversions and high molecular weights. Production of high molecular weight polyglycerol using sulfuric acid as catalyst has been accomplished, and it has been suggested that the use of this catalyst promote branched and cyclic polymer structures [1]. To the authors' knowledge, there have not been studies focused in the morphology of polyglycerol synthesized from glycerol. Polyglycerol is a glycerol valued-added product, a promising monomer in the production of biodegradable polymers. Polyglycerol is form by an ether backbone and abundant pendant hydroxyl groups. These pendant hydroxyl groups make polyglycerol a building block for diverse polymeric complexes [2]; for instance, polyglycerol is used as starting material for drug delivery [3,4] temperature-responsive co-polymers [5], and dyes adsorption from aqueous media [6], among many other applications [7-10]. The morphology and functionality of polyglycerol, such as degree of branching, molecular weight distribution, and hydroxyl number, plays an important role in the polymer application. The study of the effect of reaction conditions on the polyglycerol morphology was performed by a factorial design experiment. The response variables were hydroxyl number, molecular weight distribution, and degree of branching. Hydroxyl number was calculated according to ASTM D4274 standard. The weight distribution was found using a Mattrix-Assisted Laser Desorption/ Ionization-Time-Of-Flight (MALDI-TOF), and the degree of branching was found using Nuclear Magnetic Resonance (NMR). Additional Characterization was performed using Thermogravimetric Analysis (TGA) and Differential Scanning Calorimetry (DSC). The results suggest that it is possible to tune polyglycerol morphology changing reaction conditions. Temperature of reaction affects the degree of branching and the hydroxyl number of polyglycerol; whereas the molecular weight distribution is not affected. Variations of the glass transition temperature in the studied polyglycerol suggest changes in polyglycerol structures.

References

1. Salehpour, S. and M.A. Dubé, Towards the Sustainable Production of Higher-Molecular-Weight Polyglycerol. Macromolecular Chemistry and Physics, 2011. 212(12): p. 1284-1293.

2. Wilms D., S.-E.S., Hyperbranched Polyglycerols: From the Controlled Synthesis of Biocompatible Polyether Polyols to Multipurpose Applications. Accounts of chemical research, 2010. 43(1): p. 129-141.

3. Steinhilber, D., et al., Surfactant free preparation of biodegradable dendritic polyglycerol nanogels by inverse nanoprecipitation for encapsulation and release of pharmaceutical biomacromolecules. J Control Release, 2012.

4. Calderon, M., et al., Development of efficient acid cleavable multifunctional prodrugs derived from dendritic polyglycerol with a poly(ethylene glycol) shell. J Control Release, 2011. 151(3): p. 295-301.

5. Taylor, D.K., et al., Temperature-Responsive Biocompatible Copolymers Incorporating Hyperbranched Polyglycerols for Adjustable Functionality. J Funct Biomater, 2011. 2(3): p. 173-194.

6. Chen, Z., et al., Multicarboxylic hyperbranched polyglycerol modified SBA-15 for the adsorption of cationic dyes and copper ions from aqueous media. Applied Surface Science, 2012. 258(13): p. 5291-5298.

7. Li, M., et al., Methotrexate-conjugated and hyperbranched polyglycerol-grafted Fe(3)O(4) magnetic nanoparticles for targeted anticancer effects. Eur J Pharm Sci, 2013. 48(1-2): p. 111-20.

8. Ahn, C.H., J.-D. Jeon, and S.-Y. Kwak, Photoelectrochemical effects of hyperbranched polyglycerol in gel electrolytes on the performance of dye-sensitized solar cells. Journal of Industrial and Engineering Chemistry, 2012. 18(6): p. 2184-2190.

9. Mamiński, M.Ł., et al., Fast-curing polyurethane adhesives derived from environmentally friendly hyperbranched polyglycerols – The effect of macromonomer structure. Biomass and Bioenergy, 2011. 35(10): p. 4461-4468.

10. Mamiński, M.ł., M. Czarzasta, and P. Parzuchowski, Wood adhesives derived from hyperbranched polyglycerol cross-linked with hexamethoxymethyl melamines. International Journal of Adhesion and Adhesives, 2011. 31(7): p. 704-707.

389297 Time Delay of Abnormal Potential Change in Situ Polymerization of Polyaniline on Graphene Oxide Nanosheet

Abstract:

Graphite oxide was prepared from natural flaked graphite according to a modified Hummers method. The resulting oxide was successfully exfoliated into Graphene oxide sheet by ultrasonication and then cut into the smaller. In situ polymerization of polyaniline after the oxidant, $(NH_4)_2S_2O_8$(APS), was added into the solution and stirred, the potential change of the mixture was extremely slow compared to removing nano Graphene oxide. So the time delay was abnormal. The resulting GO/Polyaniline composite was characterized by transmission electron microscopy (TEM), Atomic force microscopy (AFM), Fourier transform infrared spectroscopy (FTIR), Raman spectra, Thermo Gravimetric Analysis (TGA) and fluorescence spectra. According to the phenomenon, we proposed an explanation in the work. The procedure consisted of two steps of polymerization. In the first polymerization step, aniline absorbed on the graphene surface was polymerized and graphene oxide was the carrier of electron transfer. So that the potential of the solution was almost stable. In the second, in the solution polymerization was started. At the same time, the potential begun to change.

Polyurethane/Polysiloxane hybrid nanocomposite coating for aircraft appliocations

Abstract

Chromate conversion coating, which uses a multilayer coating consisting of a chromate pretreatment layer, an epoxy primer, and a polyurethane topcoat, is an effective coating for aircraft. However, Chromium oxo anions (Cr^{+6}) are confirmed human carcinogens, which has forced the aircraft industry to modify or replace the chromate conversion coating on its equipment. High solid content polyurethane/polysiloxane hybrid coating is considered a good candidate. But, the corrosion performance of the high solid content polyurethane/polysiloxane hybrid coating is lower than the chromate conversion coating, so further improvements or modifications are needed.

This study focused on the modification of the polyurethane/polysiloxane hybrid coating through the introduction of nanofillers and proposed a nanocomposite coating system. Platy nanofiller, such as thermally reduced graphene and chemically modified graphene oxide, and tubular nanofiller, such as oxidized carbon nanofibers and chemically modified carbon nanotubes, were considered. The nanofillers were dispersed in solvents using sonification.

Contact Information

Maher Alrashed

PhD Candidate

Polymer Engineering Department

The University of Akron, Akron OH

Cellphone: 216 – 835 – 3361

Email: mma62@zips.uakron.edu

383993 Confined Crystallization in Biocompatible Polymer Blend Thin Films

Abstract:

Interest in polymer blend thin film systems is rapidly growing for applications in biotechnology, electronics, and consumer goods. Blend system morphology depends on a number of factors, including the polymer-polymer interaction parameter and the composition of the blend. These factors dictate whether phase separation will occur via nucleation and growth or spinodal decomposition before the morphology coarsens. Polymer crystallinity, the ability to crosslink one of the polymer components, and substrate effects in thin films add complexity to the phase separation process but can lead to distinct and useful results. For example, the ability to tune the mechanical and optical properties of polymer blends by controlling the degree of crosslinking and the extent of crystallinity is beneficial. This project aims to study the interplay between the crosslinking of poly(dimethylsiloxane) (PDMS) and the crystallization of poly(caprolactone) (PCL). At room temperature, which is below the melting temperature of PCL, both processes occur simultaneously. By casting films above the melting temperature of PCL, we can disallow the competition between crosslinking and crystallization. By this method, we create a system wherein upon lowering the temperature the crystallization of PCL is confined by the crosslinked PDMS domains. Thus, we have a system for studying the interplay between crosslinking and crystallization and for examining the effect of 3-dimensional confinement on polymer crystallization morphology and kinetics.

380083 Transport Properties of Sulfonated Poly(ether ether ketone) Membranes with Counter-Ion Substitution

Abstract:

Transport properties of sulfonated poly(ether ether ketone) (SPEEK) were analyzed as a function of counter-ion substitution (Ba^{+2}, Ca^{+2}, Mg^{+2}, Mn^{+2}, Ni^{+2}, Cu^{+2}, Zn^{+2}, K^{+1}) for direct methanol fuel cell (DMFC) applications. The polymeric membranes were characterized using several techniques: elemental analysis (EA), thermogravimetric analysis (TGA), water swelling, and Fourier-transform infrared spectroscopy (FT-IR). The techniques were used to identify the changes in the chemical and thermal properties of the membranes. Proton conductivity and methanol permeability were measured to understand the influence of the sulfonation level and counter ion substitution on the transport mechanism of protons and methanol. Significant differences in their thermal, physical and transport properties were observed when SPEEK was exchanged with counter-ions. The most promising membrane studied was SPEEK exchanged with Ca^{+2}; this membrane presented a normalized selectivity (proton conductivity over methanol permeability divided by the respective values of Nafion®) of 36.7. In addition, the physical and transport properties of SPEEK suggest a different transport mechanism than other sulfonated polymers created by the synergism of the ether-sulfonic groups and morphology.

380309 Tin Fluorophosphate Microfiber Manufacutred By Melt State Centrifugal Forcespinning

Abstract:

Phosphate glasses are polymer-like materials that possess many unique properties such as optical transparency, good chemical durability, and high thermal expansion. These properties make phosphate glasses excellent candidates for many applications, ranging from optical components to inorganic-metal sealing materials to packaging. Phosphate glass in fiber form has interesting properties that are not present in the bulk material such as a high surface area to volume ratio and a high degree of anisotropy. However, the production of continuous inorganic phosphate glass fibers remains challenging due to the high melting transition of phosphate glass, typically over 1000 ^{o}C. Here we present the first demonstration of melt state forcespinning of phosphate glass microfibers from a specially synthesized tin fluorophosphate glass with a low glass transition temperature (T_g). These fibers exhibit excellent thermal stability after being thermally treated, and the T_g of the thermally treated fibers increases by 100 ^{o}C compared to the bulk material. The compositional and structural evolution of the fibers during thermal treatment and their influence on T_g will be discussed.

377192 A General Route to Prepare Metal-Organic-Framework Aerogel

Abstract:

A general route to prepare metal-organic-framework aerogel

Zhang LIU[1], Hao Chen[2,3], Wei HAN[2], King Lun YEUNG[1,2,*]

[1]Division of Environment, [2]Department of Chemical and Biomolecular Engineering, [3]Nano Science and Technology Program,

The Hong Kong University of Science and Technology, Clear Water Bay, Kowloon, Hong Kong, PR China

** Corresponding author: kekyeung@ust.hk*

Division: 08 Materials Engineering and Sciences Division

Subdivision: 08A02 Polymer Networks and Gels

Format: Poster

Metal-organic-framework (MOF) is a type of crystal materials that can produce porous structure and a high surface area. These remarkable properties make MOF become ideal candidates for fuel-gas storage media and high-capacity adsorbents. However, the merit of their properties is closely related to their pore structures and characteristics. Development of strategies to introduce hierarchical pores into MOFs will be beneficial for increasing MOFs capacity and improving mass transfer and diffusion ability. The current study aims to seek a general route to prepare hierarchical porous MOF aerogel structures. Several common MOFs such as Al, Fe -MOF (MIL-53), Cr-MOF (MIL-101), Co-MOF (PCN-9), and Zr-MOF (UiO-66), were selected as target MOFs and successfully synthesized via an as-developed two-step route: A propylene-oxide assisted sol-gel approach was employed to form metal-based alcogels and secondly a new synthesis approach was used to convert the metal alcogels to corresponding MOF aerogels. The XRD and SEM technology evidence that this route creates MOF aerogel with inter-connective crystal MOF particles, which allows the micro-pores, meso-pores, and macro-pores to exist simultaneously in MOF structures. The hierarchical porous structures thus provide MOFs improved mass transfer and adsorption capacity. This synthetic route will also find application in preparation of other MOFs aerogels.

384762 Patterning Thin Polymer Films By Photodirecting the Marangoni Effect Using Photosensitizers

Abstract:

The Marangoni effect causes liquid flow in response to gradients in surface energy. In this study, we direct this effect at a photosensitizer doped into poly(isobutyl methacrylate) (PiBMA) thin films by programming spatial patterns with near UV-visible light (NUV-vis) exposure to oxidize the sensitizer while the film in a glassy state. Upon heating the film to the liquid state, thermal decomposition of the sensitizer oxidizes the host polymer chains by grafting reactions and creates covalent bonds between the sensitizer and the PiBMA chain. As a result, a local surface energy gradient of the film is generated so that the polymer flows from low surface energy regions (unexposed to light) to high surface energy regions (exposed to light) due to the Marangoni effect. This flow creates smooth, three dimensional topographical features reflective of the light exposure pattern. Proper selection of the photosensitizing compound in an otherwise transparent polymer extends the use of this method to more readily available and safer visible light sources. In this work, such a photosensitizer-polymer system was successfully demonstrated by activating the sensitizer with sunlight.

385873 Bio Based Active Barrier Material's and Package Development

Abstract:

Polyethylene terephthalate (PET) is a semi crystalline thermoplastic polymer, which has attractive mechanical, chemical, thermal and barrier properties. It has a wide range of application including for fibers and packaging materials. The food and packing industry are interested in ways to reduce the permeability of various gases like oxygen, carbon-dioxide and water vapor, in PET to improve the barrier properties. There are different ways through which the permeability of gasses can be reduced including the use of nano material, fillers, increasing crystallinity, using reactive additives. The focus on this project is to select and incorporate a naturally derived antioxidant or oxygen scavenger into the PET through blending or reactions. Vitamin E, oleic acid, linoleic acid, linolenic acid are being used in this study. Thus far efforts have focused on characterizing the oxygen scavenging capacity of the compounds using a pressure cell. Linolenic acid has the highest capacity followed by linoleic acid. The scavengers were incorporated into polymer through reactive extrusion. Blends of oleic acid and linoleic acid in 0.5% by weight were prepared. The PET/scavengers were blown into bottles and their side walls used to measure the oxygen transmission rate (OTR). The addition of linoleic acid resulted in the largest decrease in oxygen transmission rate.

Abstract

Many of the excellent properties of highly crosslinked polymeric materials are due to their networks structures. In this study, network structures of three epoxy systems, Epon828-PACM, Epon836-PACM, and Epon1001F-PACM, were investigated via equilibrium swelling theory. Each of these systems were separately cured in presence of an inert solvent, THF, ranging from 0 to 92% by volume fraction of solvent. Experimental results showed that the conventional swelling theory is valid for specimens prepared in moderate dilute environments, i.e. up to approximately 60% solvent by vol. fraction. However, in extremely dilute environments, i.e. above 60% solvent, the computed molecular weight between crosslinks, or M_c, based on Bray-Merrill model, increase exponentially. This drastic increase in M_c values for the samples prepared in extremely dilute environments was investigated by the T_g measurement of the polymer phase (supercritically dried specimens). It was found that the measured M_c values could not predict the corresponding T_g of the systems according to the Fox and Loshaek equation, since the resulting systems produced in extremely dilute environments are highly porous after supercritical extraction/drying. Because of this, a modifying factor, based on the probability of finding elastic chains in a porous network, was incorporated in the Bray-Merrill model. The adjusted M_c values of each thermoset and the corresponding T_g values match via the prominent Fox and Loshaek equation. These modified M_c values indicate that, the polymer networks that form in presence of an inert solvent have relatively uniform molecular weight between crosslinks, irrespective of the during-cure solvent content.

Polyurethanes (PUs) are used for a variety of applications ranging from elastomers to adhesives. These materials are commonly synthesized by step-growth polymerization of multifunctional alcohols with polyisocyanates leading to a nanophase separated morphology, which strongly influences the physical properties. Additionally, the reactivity of the isocyanate group with other functional groups such as amines and water permits the use of a large range of starting materials leading to highly tunable thermal and physical properties. Isocyanates have recently come under increased regulatory pressure, leading to a search for alternative chemistries which can form PU or PU-like materials.

The major reaction investigated for possible PU replacements consists of the reaction between five-membered cyclic carbonates and multifunctional amines. These reactions form polyhydroxyurethanes (PHUs) which are analogous to PU except for an additional primary or secondary hydroxyl group adjacent to the traditional urethane linkage. We investigated two different types of PU replacement materials utilizing this reaction: PU/PHU hybrids and pure thermoplastic PHU materials. Although the PU/PHU hybrid initially uses isocyanate monomers, the prepolymer material, which is utilized in the final reaction, is purely cyclic carbonate functionalized and does not contain any isocyanate moieties. Multiple soft-segment backbone structures and molecular weights as well as a selection of hard segments and chain extenders were used in both sets of reactions. The resulting structure-property relationships in the PU/PHU hybrid and the pure PHU were investigated using a variety of experimental techniques, including attenuated total reflectance fourier-transform infrared spectroscopy (ATR-FTIR), dynamic mechanical analysis (DMA), differential scanning calorimetry (DSC), T-peel adhesive testing, and tensile mechanical property testing. The PU/PHU hybrid elastomer exhibited thermal and mechanical properties equal to, if not exceeding those of a traditional PU analog; specifically, these materials showed two glass transitions, tensile moduli around 25 MPa, and elongation at break values greater than 500%. The PHU/PU hybrid materials also exhibited peel strength values comparable to traditional PU on a variety of substrates. The pure PHU materials showed a range of properties that were highly dependent on the hard segment content of the final material, as well as the soft-segment polarity and structure. Structure-property relationships comparing PHU to conventional PU materials will be presented.

383237 Polymerization of Castor Oil in Supercritical Carbon Dioxide

Abstract:

Biobased polymers are desired to replace traditional polymers made from petrochemicals in order to be more environmentally friendly and decrease dependence on foreign oil. Vegetable oils are a suitable renewable biobased starting material due to their availability and low cost. In this work, castor oil is polymerized in supercritical CO_2. The effect of temperature, pressure, reaction time and initiator amount on the molecular weight is studied. The thermal properties of the polymers are compared.

390266 Polymorphism of Precision Halogen Substituted Polyethylenes Studied By Thermal, Microcopy and Spectroscopic Techniques

Abstract:

Polyethylenes with halogen precisely substitution along the methylene backbone are unique models to study the effect of chain-defects on crystallization of polymers. Packing of the Chlorine units in the crystallites of polyethylene with Chlorine substitution on every 15th backbone carbon was found to undergo a transition within one degree of undercooling. On crystallization with high undercooling the chains pack in an all-trans planar conformation (Form I). The crystals formed at higher temperatures pack in a non-planar herringbone-like structure (Form II). The FTIR spectra analysis in a broad range of crystallization temperatures is extended here to a series of polyethylenes with Cl substitution every 9th, 19th and 21st carbons. The unique polymorphic transition found before, is also found in all members of the series. The polymorphic transition temperatures and temperature range of coexistence of both forms are analyzed as a function of Chlorine contents in the series.

374514 Polymerized Ionic Liquid Block Copolymers As Solid-State Polymer Electrolytes for Lithium-Ion Batteries

Abstract:

Replacing liquid-based electrolytes with solid-state polymer electrolytes (SPEs) can alleviate safety and stability concerns, while offering other attractive properties, such as thin-film forming ability, flexibility, and transparency. Ideal SPEs expertly engineering within a solid-state battery will require high lithium ion conductivities and high mechanical strength to result in high overall storage capacity/energy density and high stability and cyclability in a lithium-ion battery. Block copolymers are a well-explored material platform for this application because it combines the properties of high ionic conductivity and high strength in a nanostructured solid-state film due to covalent connection of two dissimilar polymer chains in sequential chain architecture. Herein, we present new block copolymer chemistry, where the conductive block is a polymerized ion liquid (PIL), i.e., PIL block copolymer. PILs possess unique physiochemical properties, such as high solid-state ionic conductivity, high chemical, thermal, and electrochemical stability, and widely tunable physical properties. PIL block copolymers are a new type of SPE that combine the advantages of block copolymers (nanostructured morphology) and PILs (unique physiochemical properties). In this work, the PIL diblock copolymer poly(MMA-*b*-MUBIm-TFSI) with an ionic PIL component (1-[(2-methacryloyloxy)undecyl]-3-butylimidazolium bis(trifluoromethane)sulfonimide) (MUBIm-TFSI) and a non-ionic component (MMA) was synthesized *via* reverse addition fragmentation chain transfer (RAFT) polymerization technique followed by post-functionalization and ion exchange. SPEs consisting of this PIL block copolymer and lithium salt/IL were fabricated *via* solution casting at various lithium salt/IL compositions ranging from 0 to 0.5 mol ratio of (salt/IL)/polymer. Interestingly, the PIL diblock copolymer exhibited no microphase separation over the entire salt/IL composition range explored evidenced by no scattering peak in small-angle X-ray scattering (SAXS) and only one broad glass transition temperature (T_g) in differential scanning calorimetry (DSC). The T_g decreased from 39 to 10 °C with increasing salt/IL composition. The temperature-dependent ionic conductivity increased by three orders of magnitude with increasing salt/IL composition, with a half an order of magnitude increase from 0.3 to 0.4 mol ratio and over one and a half orders of magnitude increase from 0.4 to 0.5 mol ratio, which was surprising for this small T_g depression between each ratio. These results suggest that ionic conductivity can be significantly improved without microphase separation and that salt-polymer compatibility may play a significant role on conductivity.

Alkaline Fuel Cell Performance with a Polymerized Ionic Liquid as the Anion Exchange Membrane

Rishon R. Benjamin[1], Jacob R. Nykaza[1], Yossef A. Elabd[2]

[1] Drexel University, Department of Chemical & Biological Engineering, Philadelphia, PA 19104, U.S.A.

[2] Texas A & M University, Artie McFerrin Department of Chemical Engineering, College Station, TX 77843, U.S.A.

Proton Exchange Membrane Fuel Cell (PEMFC) versus Alkaline Fuel Cell (AFC)

PEMFC	AFC
Converts chemical fuel directly to electrical energy	Converts chemical fuel directly to electrical energy
Requires robust polymers with high proton conductivity	Requires robust polymers with high hydroxide conductivity
Requires extensive noble metal catalysts (Pt)	Can be used with inexpensive non-noble metal catalysts (e.g., Ni)

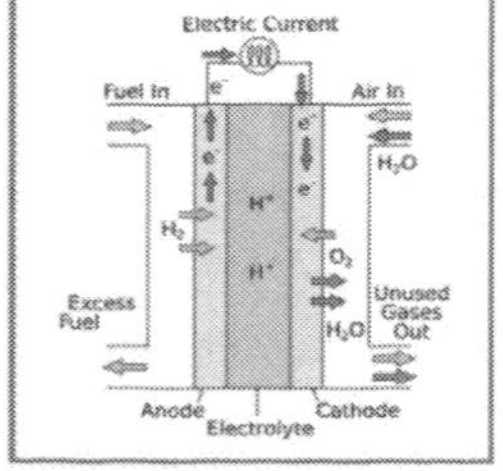

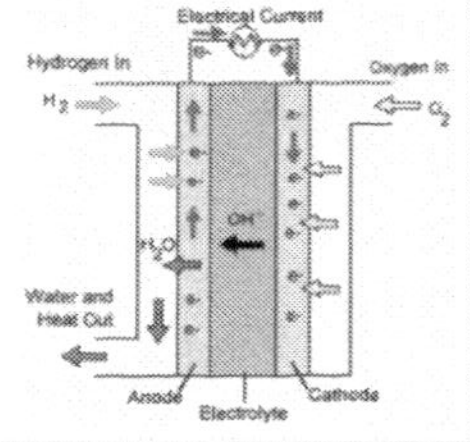

Polymerized Ionic Liquid (PIL) Block Copolymers

Polymerized ionic liquids (PILs), a new class of materials, are of interest for energy conversion and energy storage devices including fuel cells, batteries, super capacitors, and solar cells.

PIL block copolymers as anion exchange membranes (AEMs) in alkaline fuel cells are of particular interest due to the following unique properties:

- High conductivity and chemical stability
- Low volatility and non-flammability
- High mechanical and thermal stability
- Tunable nanostructured morphology and domain size
- Variable macroscopic properties due to nanostructured morphology

Objectives

- Synthesize PIL block copolymer AEMs
- Fabricate Membrane Electrode Assemblies (MEAs) with new membranes
- Measure fuel cell performance of new MEAs as a function of :
  - *Electrode composition*
  - ***Back pressure***
  - *Fuel purity*
  - *Membrane thickness*
  - ***PIL content in block copolymer***
  - *Heat press technique*

MEA Fabrication and Fuel Cell Conditions

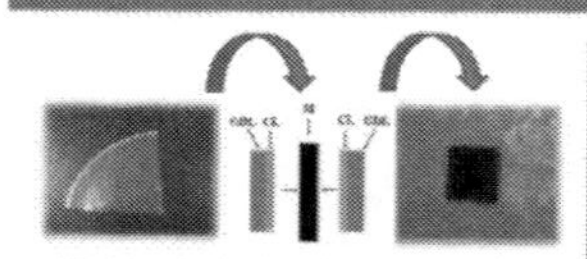

GDL: Microporous carbon-fiber treated with PTFE
CL: 30wt% polymer solution, 13wt% Pt in a 3:1 IPA:H_2O solution sonicated for 5 mins.
M: Solution cast using acetonitrile as the solvent. Annealed at 120 °C for 48 h and ion-exchanged as needed.

Unless otherwise mentioned, fuel cell tests were conducted with:

Anode/Cathode Temp: 60 °C
Cell Temp: 60 °C
Anode/Cathode back pressure: 25 psig
Anode Fuel Flow Rate: 0.42 L/min
Cathode Fuel Flow Rate: 1.00 L/min

The cell was allowed to equilibrate for 4 hours before collection.
The cathode and anode fuels were fully humidified prior to injection into the fuel cell.

Polymer Synthesis: Reversible Addition Fragmentation chain Transfer (RAFT) Polymerization

Synthesis of IL Monomer

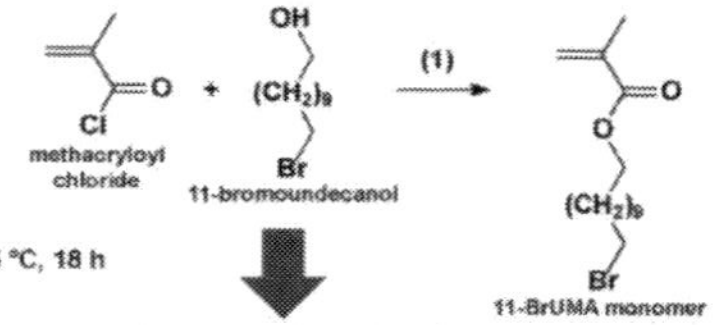

(1) triethylamine, DCM, 25 °C, 18 h

Synthesis of Neutral PIL Block Copolymer

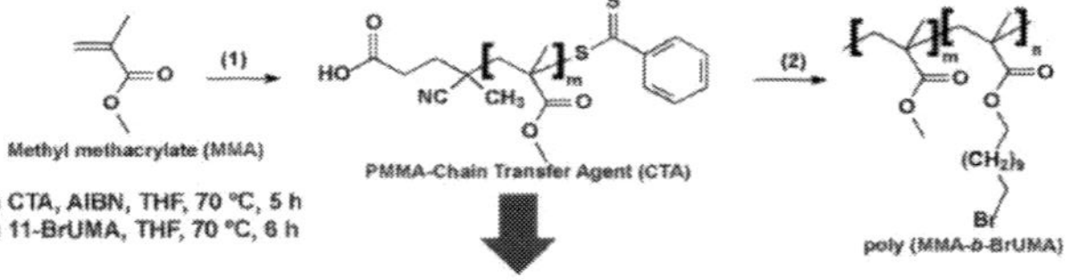

(1) CTA, AIBN, THF, 70 °C, 5 h
(2) 11-BrUMA, THF, 70 °C, 6 h

Functionalizing PIL Block Copolymer

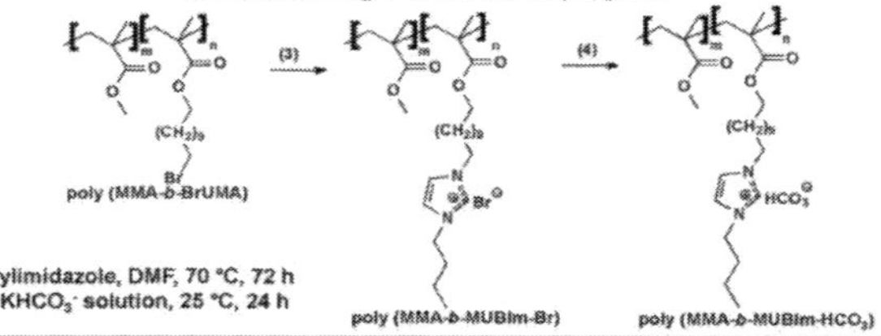

(3) 1-butylimidazole, DMF, 70 °C, 72 h
(4) 0.1M $KHCO_3$ solution, 25 °C, 24 h

Alkaline Fuel Cell Performance

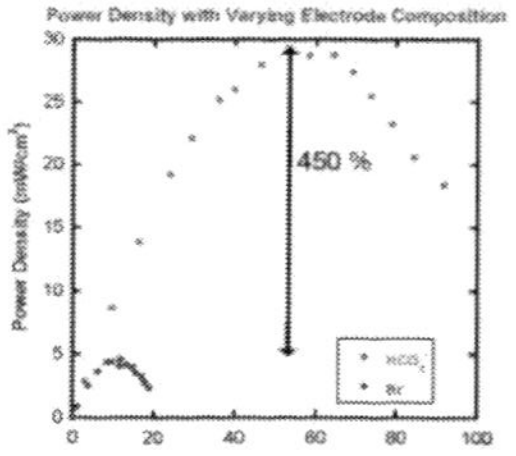

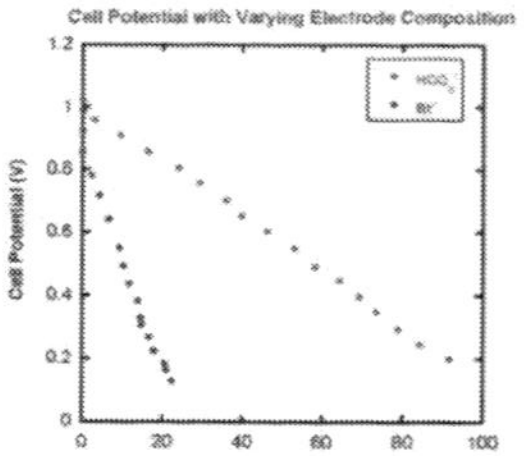

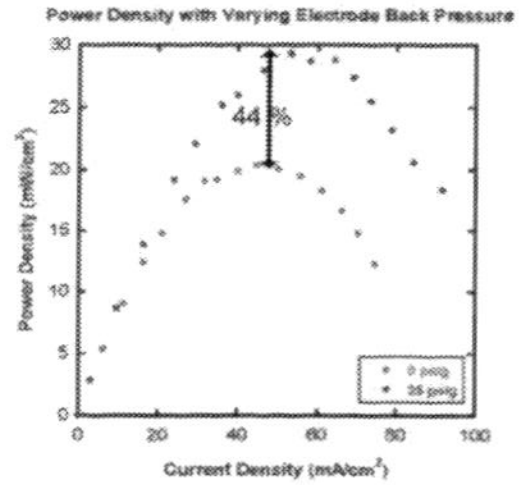

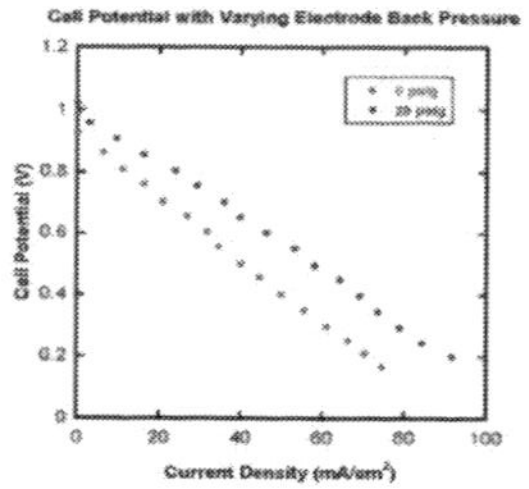

Alkaline Fuel Cell Performance

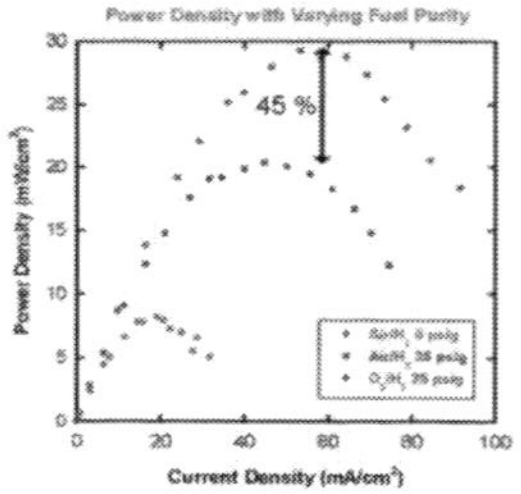

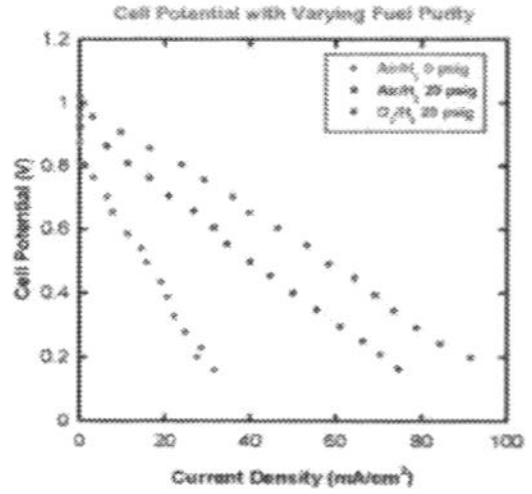

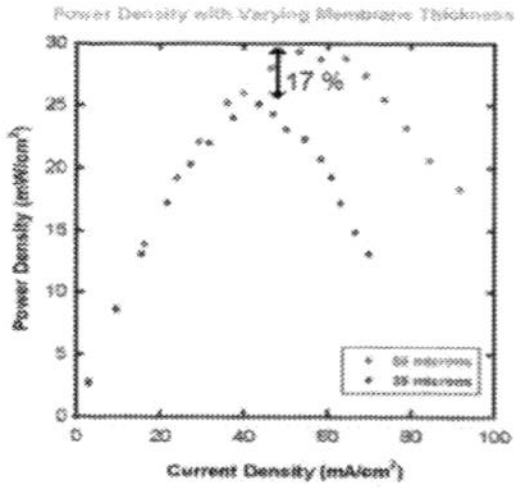

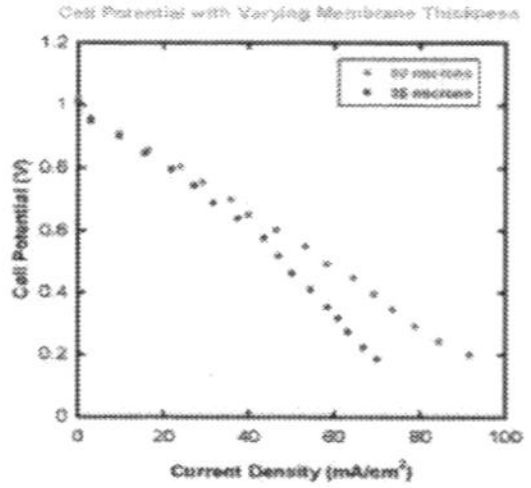

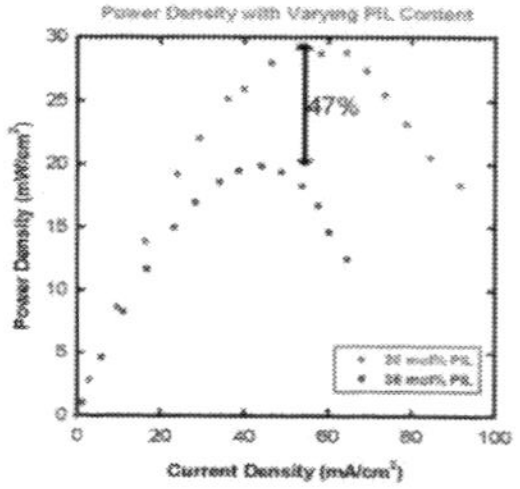

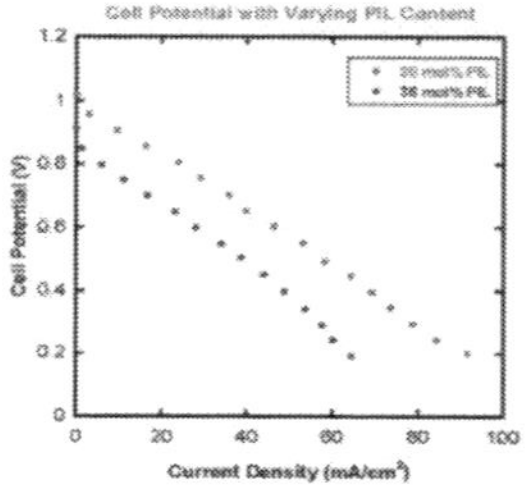

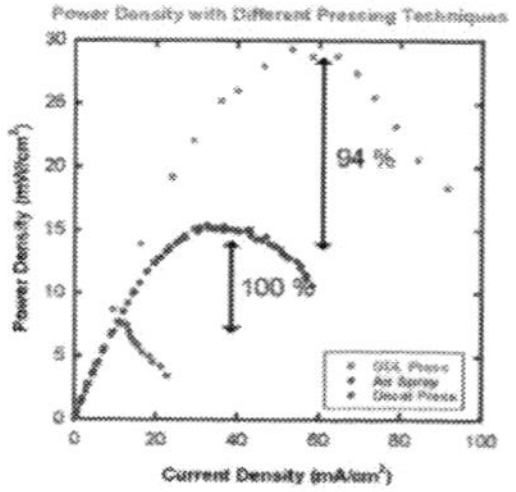

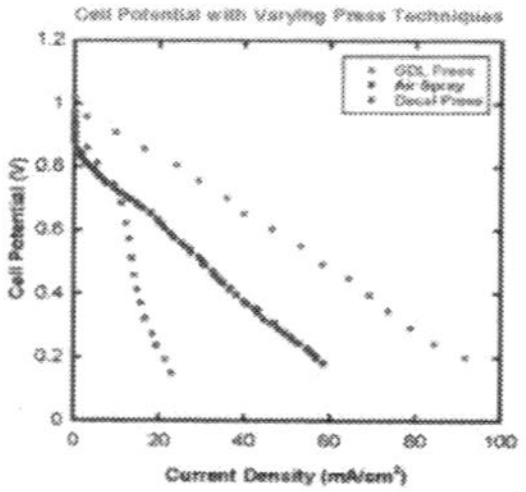

Conclusions

- ✓ Bicarbonate versus bromide form of polymer in electrodes improved AFC performance by 450%
- ✓ **Increasing back pressure from 0 to 25 psig increased AFC performance by 45%**
- ✓ Pure oxygen at the cathode resulted in a 45% increase in performance over air
- ✓ Increasing AEM thickness from 35 to 50 microns increased AFC performance by 17%
- ✓ Decreasing PIL content from 35 to 20 mol% increased AFC performance by 47%
- ✓ The GDL press MEA fabrication technique resulted in the highest AFC performance

Funding

AIChE
Drexel
Nanotechnology Institute
(NTI 1001-037)
(W911NF-07-1-0452)
AIChE

380025 Synthesis and Characterization of Poly(styrene-isobutylene-methyl vinyl ether) for Direct Methanol Fuel Cell Applications

Abstract:

Novel block copolymers based on poly(styrene) (PS), poly(isobutylene) (PIB) and poly(methyl vinyl ether) (PMVE) were synthesized using cationic polymerization. They were synthesized employing two different concentrations of PMVE. The resulting block copolymers were characterized using: Gel Permeation Chromatography (GPC), Fourier Transforms Infrared Spectroscopy (FT-IR), Thermogravimetric Analysis (TGA), and Differential Scanning Calorimetry (DSC). The block copolymers were then sulfonated and solvent casted into ionic polymer membranes (ionomers) for direct methanol fuel cell (DMFC) applications. The ionomers were characterized using several techniques: Elemental Analysis (EA), Thermogravimetric Analysis (TGA), water absorption, and Fourier-transform Infrared Spectroscopy (FT-IR). The techniques were used to identify the changes in the chemical and thermal properties of the membranes, as well as to evaluate changes in the resulting morphologies. Proton conductivity and methanol permeability were measured to understand how the different concentrations of PMVE influenced the transport mechanism of protons and methanol through the membrane for DMFC applications. Significant differences in their thermal, physical and transport properties were observed.

383599 Materials Design of Conjugated Block Copolymers for Photovoltaics

Abstract:

Weak intermolecular interactions and disorder at junctions of different organic materials limit the performance and stability of organic interfaces and hence the applicability of organic semiconductors to electronic devices. We have demonstrated control of donor-acceptor heterojunctions through microphase-separated conjugated block copolymers. When utilized as the active layer of photovoltaic cells, block copolymer-based devices demonstrate efficient photoconversion well beyond devices composed of homopolymer blends. The 3% block copolymer device efficiencies are achieved without the use of a fullerene acceptor. Resonant soft X-ray scattering and grazing-incidence X-ray diffraction results reveal that the efficient performance of block copolymer solar cells is due to self-assembly into mesoscale lamellar morphologies with primarily face-on crystallite orientations. We can build on these initial results with the combination of Density Functional Theory, Molecular Dynamics simulations and polymer theory to design donor-acceptor block copolymers with control of charge transfer processes. For example, interfaces in conjugated block copolymers are governed by chain flexibility and the interaction parameter. As such, we can present strategies to design block copolymers with suppression of bimolecular recombination through the molecular composition and microstructure.

386352 Development of Novel Alternating Ionic Polyimides for Gas Separation Membranes

Abstract:

Recently, several promising classes of promising polymer materials have emerged as advanced gas separation membranes. Here, we focus on combining two of these materials platforms in the form of novel polymers with alternating ionic and polyimide groups. These materials present opportunities for unique nano- and microstructures that can impart high fractional free volume and enhanced permeation of gases such as CO_2. Given the number of simple precursors available for study, a large number of unique materials is possible. This presentation will discuss our synthetic techniques, polymer physical properties, gas permeation and molecular modeling.

388443 Synthesis of Well-Defined Poly (Styrene-block-Methyl methacrylate) Copolymers By Classical Anionic and Controlled Radical Polymerization (CRP) Methods

Abstract:

Block copolymers constitute a fascinating group of polymeric materials belonging to the "soft matter" family. These materials are formed by blocks of polymeric chains from different monomers. In most cases, these blocks are thermodynamically incompatible, but the covalent bonds between them impose specific arrangements, generating particular morphologies both in bulk and in solution.

In the particular case of block copolymer films over specific substrates, much work has focused on manipulating the film thickness, the surface energies of each block, and the interfacial interactions of each block with the substrate to generate useful, perpendicularly (vertically) oriented morphologies with high degrees of long-range lateral order. (Widin, Kim et al. 2013)

In the present work, two different synthetic methods were used to produce well-defined diblock copolymers based on styrene and methylmethacrylate (PS-*b*-PMMA): classical anionic polymerization (high vacuum techniques), and controlled radical polymerization (CRP). The aim of this work is not only to obtain well-defined PS-*b*-PMMA but also to prove that the copolymers obtained by CRP exhibits the same behavior as those obtained by the classical anionic polymerization (Soeriyadi, Boyer et al. 2011; Gao, Zhang et al. 2011).

Anionic and controlled radical polymerization (CRP) techniques have allowed the routine preparation of block copolymers with two or three different monomer blocks. However, anionic polymerization must be performed under a rigorous set of polymerization conditions. For CRP, after each block addition, the copolymer requires purification to eliminate unreacted monomers prior to the next chain-extension step with another monomer.

The copolymers obtained were chemically characterized by nuclear magnetic resonance (^{1}H-NMR), Fourier transform infrared spectroscopy (FTIR), size-exclusion chromatography (SEC), and the morphological characterization was determinate by transmission electron microscopy (TEM) and scanning electron microscopy (SEM). From this analysis, the main features of the block copolymers (such as their chemical composition or morphologies and scattering size domains developed on thermal treatment) are discussed, consequently studied how the composition and the morphology of the copolymers relate with the final properties. Different morphologies can result when they are solution cast from different solvents.

Gao, J., Z. Zhang, et al. (2011). "Copper(0)-mediated living radical copolymerization of styrene and methyl methacrylate at ambient temperature." Macromolecules **44**(9): 3227-3232.

Soeriyadi, A. H., C. Boyer, et al. (2011). "High-order multiblock copolymers via iterative Cu(0)-mediated radical polymerizations (SET-LRP): Toward biological precision." Journal of the American Chemical Society **133**(29): 11128-11131.

Widin, J. M., M. Kim, et al. (2013). "Bulk and thin film morphological behavior of broad dispersity poly(styrene-b-methyl methacrylate) diblock copolymers." Macromolecules **46**(11): 4472-4480.

387431 X-Ray Microtomography of Lithium Dendrite Growth through Polystyrene-Block-Poly(ethylene oxide) Copolymer Electrolyte Membranes

Abstract:

High modulus block copolymer electrolytes have the ability to lengthen the cycle lifetime of rechargeable batteries by slowing lithium dendrite growth. However, batteries made with these polymer electrolyte membranes still eventually fail via dendritic short circuit. Hard X-ray microtomography was used to image dendrite growth *in situ* in symmetric lithium/polymer electrolyte/lithium samples. Each symmetric cell contained many dendrites. The volume of the dendrites was measured as a function of the charge passed through the electrolyte in order to determine the rate of dendrite growth in these systems. The dendrites appear to form on crystalline contaminants at the lithium metal electrode/electrolyte interface and then slowly grow in the electrode before protruding into the electrolyte. Further experiments were performed to determine the effect of temperature and depth of discharge on dendrite morphology and growth rate. The dendrites have a "foam-like" structure where the interior of the foam cells is lithium metal and the exterior consists of elements from the lithium, polymer electrolyte, and salt. The size of the "cells" in the dendrite is correlated to the amount of lithium ions moved per cycle. These experimental results are compared with theoretical models for the growth of dendrites in high modulus polymer electrolyte systems.

Synthesis and characterization of a family of Photosensitive alternating copolymers

Ishan Fursule and Bradley Berron

Department of Chemical and Materials Engineering, University of Kentucky.

Responsive polymers, membranes and coatings offer an opportunity to actively control the configuration of a molecule. Polymers chains with stimuli-responsive azobenzene side chains are widely being studied in the domain of photo and thermal sensitive polymers [1-4]. In solution phase they can be introduced to control the intrinsic viscosity, pH, and electrical conductivity [5]. Other promising applications that have been proposed such as molecular switches or photomechanical systems, liquid crystal display, optical data storage [6]. Here, we study the incorporation of these responsive groups into the polymer backbone, and study the role of polymer design on the temporal response of the material. In particular, we control the ratio of actuatable azobenzene material to non-actuatable alkyl chains. Both species are incorporated through an alternating copolymer structure, and the ratio of actuatable groups is adjusted through the length of the alkyl co-monomer. The change in spatial configuration (cis and trans) of azobenzene derived monomer can be triggered by irradiating it with UV (trans to cis) and restored back by either thermal relaxation (cis to trans). The role of crosslinking is studied by incorporating tri-functional alkyl monomers. The chain length of tri-functional alkyl monomer will define the degree to which its properties are changed on irradiating it with UV.

References

1] Almeria Natansohn and Paul Rochon (2002). Photoinduced Motions in Azo-Containing Polymers. *ACS, C. R.*, 102, 4139-4175.

2] Matsuoka, S. S. (2014). Photoresponsive Block Copolymer: Synthesis, Characterization, and Surface Activity Control. *Langmuir*, 30, 3957-3966.

3] Osamu Tsutsumi, T. S. (1997). Photochemical Phase Transition Behavior of Nematic Liquid Crystals with Azobenzene Moieties as Both Mesogens and Photosensitive Chromophores. *The Journal of Physical Chemistry*, 101, 1332-1337.

4] Ribagorda, E. M. (2012). Control over molecular motion using the cis–trans photoisomerization of the azo group. *Beilstein Journal of Organic Chemistry.*, 8, 1071-1090.

5] Yapei Wang, M. Z. (2009). Block copolymer aggregates with photo-responsive switches: Towards a controllable supramolecular container. *Polymer*, 50, 4821-4828.

6] Liangjing Fang, H. Z. (2013). Synthesis of Reactive Azobenzene Main-Chain Liquid Crystalline Polymers via Michael Addition Polymerization and Photomechanical Effects of Their Supramolecular Hydrogen-Bonded Fibers. *Macromolecules*, 46, 7650-7660.

380405 Quantitative Study on Branching Behavior in RAFT Copolymerization of a Vinyl/Divinyl System

Abstract:

Outstanding chemical and mechanical properties make highly branched polymers (HBP) one of the most intensively studied objects of the past few decades. Synthesis of HBPs via controlled/living radical copolymerization of vinyl and divinyl systems are of interest because of their facile synthetic routes. Although theoretical studies on the branching formation mechanism have been carried out, these theories require experimental evidences for validation. Herein, a series of HBPs were synthesized by reversible addition-fragmentation chain transfer (RAFT) solution copolymerization. Methyl methacrylate (MMA) and a disulfide-based dimethacrylate (DSDMA) cross-linker were used for the synthesis at 70 °C in toluene. 2-cyano-2-propyl dodecyl trithiocarbonate was chosen as a RAFT chain transfer agent and 2, 2'-azobis(2-methylpropionitrile) as an initiator. The selective cleavage of disulfide bonds of DSDMA at the cross-linkages enabled us to determine the weight fraction of branched polymers with different primary chains. Branching density (BD) and cyclization density (CD) were also determined accordingly by varying the initial monomer concentration between 15-45% and using [MMA]/[DSDMA] ratios of 100:(0.5, 1, 1.5). It was found that a higher monomer concentration enhances branching rather than cyclization, while increasing the divinyl content both increases BD and CD. Furthermore, cyclization is preferred for a longer primary chain length. We verified the random branching characteristics of RAFT copolymerization in vinyl/divinyl systems. This work has developed a feasible method to study the branching behavior and provided insight into the mechanism of RAFT copolymerization at the pre-gel stage.

STUDY OF BLOCK COPOLYMER MICROPHASE SEPARATION DURING FILM DRYING USING IN-SITU SAXS

Alicia R. Pape, John A. Pople, Donald G. Baird, and Stephen M. Martin

Department of Chemical Engineering
Virginia Tech
Blacksburg, VA 24061
Phone: (540) 231-6631; Fax: (540) 231-5022
Email: palic11@vt.edu

Solution casting is a commonly used method for the production of block copolymer membranes and thin films for various applications. While the properties of these films are determined in part by the microstructure, the kinetics of microstructure formation during the drying process are not well understood. We have studied the effect of the rate of microphase separation in solution-cast block copolymer films during solvent removal. The effects of solvent removal rate on microphase separation kinetics in thick (~400μm) films were studied using small-angle X-ray scattering on several block copolymer solutions. Ordering was studied in-situ for poly(methyl methacrylate-*b*-butyl acrylate), poly(styrene-*b*-butadiene), and poly(styrene-*b*-isoprene) solutions in toluene. A sweep gas flowed beneath the sample stage in the drying chamber to control solvent vapor concentration in the chamber and thus the drying rate in the film. This experimental setup has allowed us to directly study the effect of solvent removal rate on ordering in the samples, rather than relying on temperature to control the rate of solvent removal. Using the bulk solvent concentration in the films, we were able to estimate the partitioning of solvent between microdomains during the drying process. Partitioning results were used to determine the volume fraction of ordered microstructures in the films with respect to concentration and time. The ordering rate in the samples that were studied was found to increase with solvent removal rate independent of temperature.

ELECTRICAL CHARACTERIZATION OF SULFONATED POLY(STYRENE-ISOBUTYLENE-STYRENE) TRIBLOCK COPOLYMER THIN FILMS

Martha Rozo Medina[a] and Agnes M. Padovani[a,b]
[a] Chemical Engineering Department, University of Puerto Rico-Mayaguez
[b] General Engineering Department, University of Puerto Rico-Mayaguez

Polymers are a fundamental part of electronic systems and are widely used due to their low cost, ease of processing, and their wide range of mechanical and electrical properties. They have traditionally been used as passive components such as insulators and coatings, but more recent applications also include sensors, batteries, and low-temperature fuel cells. A thorough understanding of the electrical properties is necessary in order to improve their performance and to determine the best design conditions in any given application. The electrical properties of polymeric materials can be characterized through dielectric constant and loss tangent measurements. The first one represents the ability of the material to store potential energy, while the second one represents the losses related to heat dissipation factors. This work focuses on the electrical characterization of sulfonated poly (styrene-isobutylene-styrene) (SIBS) triblock copolymer thin films. More specifically, the dielectric properties of SIBS were evaluated at low frequencies and as a function of critical parameters such as the sulfonation percent and polymer concentration. The sulfonation percent was varied in the range of 0% to 91%, whereas the polymer concentration was studied in the range of 5% to 9%. For the electrical characterization, metal-insulator-metal (MIM) parallel-plate capacitors were fabricated using a combination of physical vapor metal deposition and spin-coating for the polymer films. The electrical measurements were performed using a Network Analyzer that can measure the complex impedance over frequencies in the range of 100 kHz to 1 MHz. The results show that the dielectric constant and loss tangent increase with sulfonation percent and decrease with increasing frequency. This work provides insights into the correlation of SIBS electrical properties and its chemical structure and morphology.

390366 Pattern Density Multiplication of Block Copolymer Directed Self-Assembly Using a Chemoepitaxial Guiding Underlayer with Topography

Abstract:

Pattern density multiplication in the directed self-assembly (DSA) of block copolymers (BCPs) is often achieved through a chemically patterned guiding underlayer, also known as a chemoepitaxial guiding underlayer. These underlayers are typically considered flat underlayers composed of a pinning stripe region (i.e. a region that is highly preferential to one block of the BCP) and a neutral stripe region (i.e. a region that is non-preferential or weakly preferential to the opposite block of the BCP as compared to the pinning stripe). However, in reality it is likely that some level of topography will be introduced when producing these chemoepitaxial underlayers, possibly by improper backfilling or etching, which then has the ability to affect the resulting microphase separation and alignment processes. This topography may also cause deformations in any resulting lamellae, which may interfere with subsequent pattern transfer etching steps. Due to the complexity of creating test patterns to study such effects and the difficulty in performing adequate metrology on such test samples, detailed mesoscale molecular dynamics simulations have been used in this work to probe the effect of topography on chemoepitaxial DSA processes. The current work presented in this paper systematically explores the effect that topography in these chemoepitaxial underlayers, such as a raised or lowered pinning stripe, has on the alignment of the BCP film. The effect of even small levels of topography is found to be potentially profound. For example, it is found that small amounts of topography which raise the pinning stripe can greatly alter the ideal composition of the neutral stripe in these systems, shifting the composition to a more neutral value. This topography can also cause increased defects buried within the film due to the compression of lamellae within the trenches of the topography. Likewise, processes that result in depressed pinning stripes also alter the process windows and behavior of such processes. A complete discussion of these effects will be provided.

377970 Decreasing Polymer Flexibility Improves Wetting and Dispersion of Polymer Grafted Particles in a Chemically Identical Polymer Matrix

Abstract:

We present a molecular dynamics simulation study of nanocomposites containing homopolymer grafted particles in a homopolymer matrix, where the graft and matrix chemistries are identical, to elucidate the effect of polymer flexibility on the wetting of the grafted layer by the matrix and the nanocomposite morphology. Decreasing flexibility of the graft and matrix causes increased wetting of the grafted layer by the matrix. This increased wetting of the grafted layer with decreasing flexibility is more significantly driven by decreasing the graft flexibility than by decreasing the matrix flexibility. This is due to a large increase in mixing entropy of the graft and matrix upon wetting rather than the reduction in conformational entropy loss of matrix upon wetting. Due to this improved wetting with decreasing flexibility of the graft and matrix, we observed increased particle dispersion in the polymer matrix.

Sorption in polyolefins: equilibria, diffusion and morphology

J. Chmelar, A. Zubov, R. Pokorny, J. Kosek

Institute of Chemical Technology Prague, Prague, Czech Republic

Polyolefins are commercially important due to their massive production and variety of applications. The two most common polyolefins, polyethylene and polypropylene, account for almost 50% of the annual worldwide plastics production. Sorption equilibria and transport of penetrants in polymers significantly influence polymerization processes and subsequent product treatment such as degassing. Sorption also enhances unwanted agglomeration of polymer particles in the reactors, which can lead to the shut-down of the whole production line.

This contribution presents two experimental techniques for the measurement of sorption and diffusion in polyolefin powder: a gravimetric apparatus and a pressure-decay apparatus. Both the sorption of pure compounds and co-sorption are addressed. The measured equilibrium data are correlated with PC-SAFT simulations, while several models are used for diffusion, namely (i) simple Ficks law, (ii) multigrain or bi-disperse model, and (iii) complex model of transport in reconstructed 3D porous particles.

In the case of polyolefins, the sorption processes depend on their morphology. The characterization of semi-crystalline polyolefin morphology is thus essential for the analysis of sorption and diffusion in polyolefins, as it provides input data for mathematical models and as the comparison of experimentally measured sorption data with the morphology enables to study the structure-property relationships. This contribution thus presents several advanced experimental techniques for the characterization of polyolefin morphology, namely: (i) X-ray micro-tomography (μCT), (ii) time-domain NMR (TD-NMR) and (iii) confocal Raman microscopy.

Figure. Simulation of propylene degassing from a porous polypropylene particle. The simulation considers a real 3D morphology based on micro-tomography data.

Design and Synthesis of Fluorinated Gradient Copolymer for Photo-Reversibly Switchable Hydrophobic Surface

Yin-Ning Zhou, Zheng-Hong Luo
Department of Chemical Engineering, Shanghai Jiao Tong University, Shanghai 200240, China

Introduction

Atom transfer radical polymerization (ATRP) technique provides an effective platform for the realization of the precision polymerization with gradient composition.[1,2]

However, in a ***batch process***, composition drift is very common due to the difference in the reactivity ratio of the monomers. To control copolymer composition, a ***semi-batch operation*** is commonly used. The copolymers with uniform composition or linear gradient composition can be successfully designed and controlled through optimizing monomer feeding to a semi-batch reactor.[3-5]

Light-responsive smart surfaces with tunable wettability have received increasing attention recently, due to their facile operation, limited environmental impact and potential applications.[6,7]

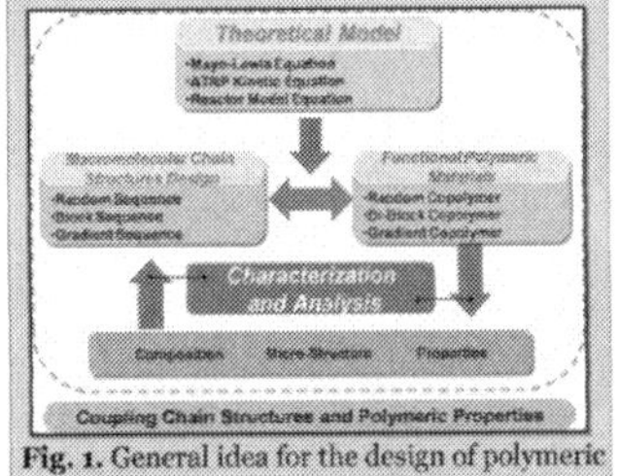

Fig. 1. General idea for the design of polymeric materials

Methods

ATRP Kinetic Model

Elementary reactions of ATRP

Type	Scheme
Initiation	$P_0X + C \xleftrightarrow{k_{a0}} P_0\cdot + CX$ $P_0\cdot + M_i \xrightarrow{k_{I,i}} P_{1,i}\cdot$
Propagation	$P_{r,i}\cdot + M_j \xrightarrow{k_{p,ij}} P_{r+1,j}\cdot$
ATRP equilibrium	$P_{r,i}X + C \xleftrightarrow[k_{da}]{k_{a}} P_{r,i}\cdot + CX$
Transfer	$P_{r,i}\cdot + M_j \xrightarrow{k_{tr,ij}} P_r + M\cdot$
Termination	$P_{r,i}\cdot + P_{s,j}\cdot \xrightarrow{k_{td,ij}} P_r + P_s$ $P_{r,i}\cdot + P_{s,j}\cdot \xrightarrow{k_{tc,ij}} P_{r+s}$

Kinetic Equations for Each Type of Chain Species

Type of chains	Mass balance equations
Propagating radical	$\frac{d[P_{r,i}\cdot]}{dt} = \sum_j k_{p,ji}[P_{r-1,j}\cdot][M_i] - \sum_j k_{p,ij}[P_{r,i}\cdot][M_j] + k_a[P_{r,i}X][C] - k_{da}[P_{r,i}\cdot][CX] - \sum_j k_{tr,ij}[P_{r,i}\cdot][M_j] - \sum_j \sum_s (k_{tc,ij} + k_{td,ij})[P_{r,i}\cdot][P_{s,j}\cdot]$
Dormant	$\frac{d[P_{r,i}X]}{dt} = k_{da}[P_{r,i}\cdot][CX] - k_a[P_{r,i}X][C]$
Dead	$\frac{d[P_r]}{dt} = \sum_j k_{tr,ij}[P_{r,i}\cdot][M_j] + \sum_i \sum_j \sum_{s=0}^{r} k_{tc,ij}[P_{s,i}\cdot][P_{r-s,j}\cdot] + \sum_i \sum_j \sum_{s=0}^{\infty} k_{td,ij}[P_{r,i}\cdot][P_{s,j}\cdot]$

Definition of moments

Propagating radical	Dormant	Dead
$\mu_i^n = \sum_{r=1}^{\infty} r^n[P_{r,i}\cdot]$	$\lambda_i^n = \sum_{r=1}^{\infty} r^n[P_{r,i}X]$	$\tau^n = \sum_{r=1}^{\infty} r^n[P_r]$

Semi-Batch Reactor Model

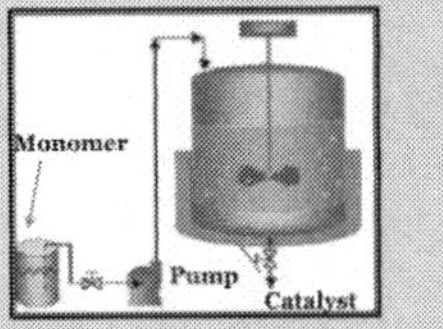

Fig. 2. Well-mixed isothermal tank reactor

Description	Equation
Evolution of reaction volume (V)	$\frac{dV}{dt} = V_f - \sum_{i=1} M_{wi} R_p V\left(\frac{1}{\rho_p} - \frac{1}{\rho_i}\right)$
Evolution of density in the reactor (ρ)	$\frac{d\rho}{dt} = \frac{V_f \rho_f}{V} - \frac{\rho}{V}\frac{dV}{dt}$
Mass balance equation for the *i* th species	$\frac{dC_i}{dt} = \frac{1}{V}\left(V_f C_{if} - C_i \frac{dV}{dt}\right) + R_i$

Introduction of Light-Responsive Units

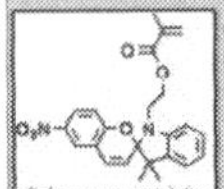
Spiropyrans containing methacrylate (SPMA)

- ☑ Broad spectrum range
- ☑ Near-quantitative in both directions
- ✗ Photo-degradation
- ✗ Thermal relaxation

The two Drawbacks can be reduced by incorporating SP into polymer

365 nm for ring opening [Hydrophilicity]

↕

520 nm for ring closing [Hydrophobicity]

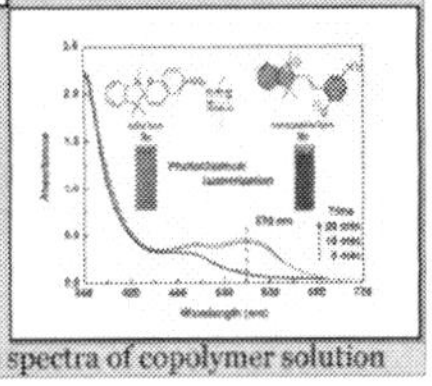

Fig. 3. UV-vis spectra of copolymer solution

Results

Fluorinated Copolymer With Gradient Composition

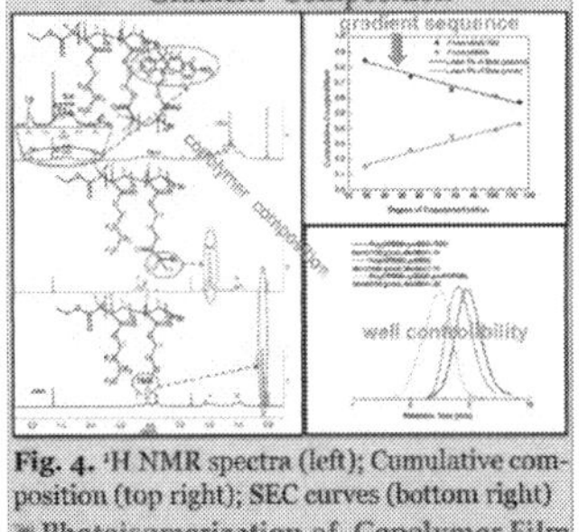

Fig. 4. ¹H NMR spectra (left); Cumulative composition (top right); SEC curves (bottom right)

Photoisomerization of Copolymer Film

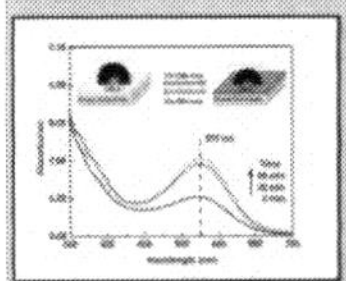

- Isomerization is slowed down by limited free volume and restricted conformational freedom.
- Water contact angle changes reversibly between 103.6° and 94° (△WCA=9.6°).

Fig. 5. UV–vis spectra of copolymer film

Controllable Wettability of Surface

Fig. 6. Photographs of water contact angle: blank (A1, B1); visual light irradiation 30 min (A2, B2); and UV irradiation 30 min (A3, B3)

- Flat substrate has limited wetting controllability, which is governed by surface chemical composition;
- Thanks to the functional film and surface roughening, the wetting controllability reaches 28.1°.

Surface Composition

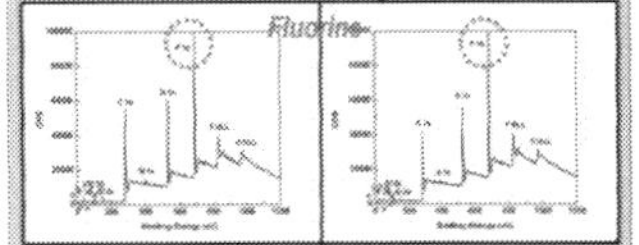

Substrate	C 1s	N 1s	O 1s	F 1s	W_F (%) (in bulk)
Flat Silicon	64.73	2.18	15.00	18.08	11.55
Etched Silicon	64.14	1.99	15.66	18.21	

- Similar compositions, but fluorinated segments enrich on surfaces

Surface Morphology

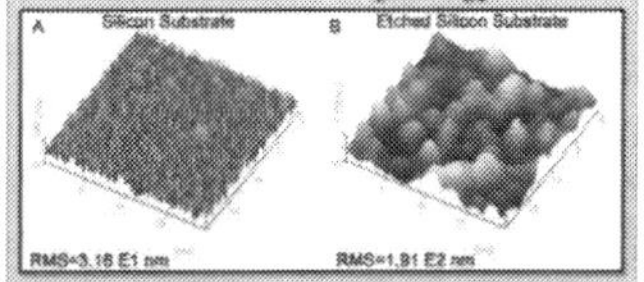

Fig. 6. AFM images of as-fabricated surfaces

- The root-mean-square roughness (RMS) is 191 nm for silicon surface with pyramids;
- In this situation, the water drop infiltrate the groove due to the photoinduced isomerization of SP, and thus leading to a higher area of solid–liquid interface.

Surface Morphology

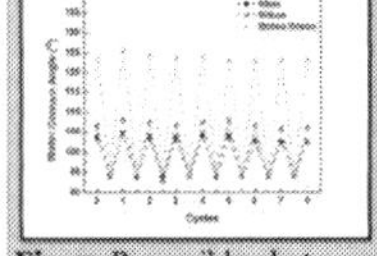

- The switchable wettability induced by alternately UV and visible light irradiations is reversible and stable.

Fig. 7. Reversible photoresponsive wettability

Conclusion

- ATRP is a robust "controlled" radical system for designing the copolymer with desired structures;
- Semi-batch operation is facile for controlling the copolymer composition;
- Thanks to the surface engineering technique, the roughened surface decorated with functional film possesses a fairly large variation range of Δ WCA (28.1°) for controlling surface wettability.

Acknowledgements

The authors thank the National Natural Science Foundation of China (No. 21276213) and the State-Key Laboratory of Chemical Engineering of Tsinghua University (No. SKL-ChE-13A05) for supporting this work.

References

1. Matyjaszewski K, Ziegler MJ, Arehart SV, Greszta D, Pakula T. Gradient copolymers by atom transfer radical copolymerization. J Phys Org Chem. 2000; 13:775-786.
2. Matyjaszewski K, Tsarevsky NV. Macromolecular engineering by atom transfer radical polymerization. J Am Chem Soc. 2014; 136:6513-6533.
3. Wang R, Luo Y, Li B, Sun X, Zhu S. Design and control of copolymer composition distribution in living radical polymerization using semibatch feeding policies: A model simulation. Macromol Theory Simul. 2006; 15: 356-368.
4. Wang R, Luo Y, Li B, Zhu S. Control of gradient copolymer composition in ATRP using semibatch feeding policy. AIChE J. 2007; 53: 174-186.
5. Zhou Y-N, Luo Z-H, Chen J-H. Theoretical modeling coupled with experimental study on the preparation and characterization comparison of fluorinated copolymers: Effect of chain structure on copolymer properties. AIChE J. 2013; 59:3019-3033.
6. Zhou Y-N, Li Jin-Jin, Zhang Qing, Luo Z-H. Light-responsive smart surface with controllable wettability and excellent stability. Langmuir 2014; 30:12236-12242.
7. Zhou Y-N, Li Jin-Jin, Zhang Qing, Luo Z-H. A Novel Fluorinated Polymeric Product for Photoreversibly Switchable Hydrophobic Surface. AIChE J. 2014; DOI: 10.1002/aic.14602.

370810 Synthesis of Silver Nanoparticles in Poly(acrylic acid) Networks and Their Catalytic Activity

Abstract:

Silver nanoparticles with narrow size distribution were prepared in poly(acrylic acid) networks with various amount of sodium borohydride ($NaBH_4$). Effect of sodium borohydride amount on the formation of silver nanoparticles was studied in detail. The size, size distribution, and stability of silver nanoparticles were tunable by the molar ratio of $NaBH_4/AgNO_3$ when less than 10, while its effect was weakened when it exceeded 10. This phenomenon could be characterized by UV spectrum and TEM. The catalytic activity of silver nanoparticles with various size and size distribution on the reaction of 4-nitrophenol's reduction was discussed, especially at the appearance of alpha-cyclodextrin.

Melt spinning of thermotropic liquid crystalline pitch: Study of process variables

Victor Bermudez and Amod A. Ogale, Chemical Engineering Department and Center for Advanced Engineering Fibers and Films (CAEFF), Clemson University, Clemson, SC 29634

Thermotropic liquid crystalline (mesophase) pitch exhibits planar orientation of the mesogens, which makes them suitable as a precursor for producing carbon fibers with significant graphitic content. The precursor fibers can be produced by melt-spinning, a rapid commercial process that enables reduction of processing costs. One of the main challenges in melt spinning of mesophase pitches is the understanding of the relationships between the process conditions and the microstructure of the resulting pitch fibers, which in turn controls properties of the final product.

Among various process conditions, the draw down ratio (DDR, defined as the ratio between the winding speed and the extrusion speed) is of particular importance because it has a direct effect on reducing the as-spun fiber diameter, which is critical for improving fiber strength. In the current study, the combined effect of capillary diameter of the multi-hole spinneret and the draw down ratio is being investigated to obtain precursor fibers of a given diameter. For flexible chain polymers, larger DDR is known to improve modulus and tenacity of fibers. However, there is only limited evidence for mesophase pitch, which suggests an opposite trend. Therefore, current studies are directed toward investigating the effect of DDR on tensile and transport properties of resulting carbon fibers derived from synthetic and petroleum-derived pitches.

372588 Orientational Order of Rheologically Modified Main Chain Liquid Crystalline Polymers Processed By Multilayer Coextrusion

Abstract:

Orientational Order of Rheologically Modified Main Chain Liquid Crystalline Polymers Processed by Multilayer Coextrusion

Zhenpeng Li[1], Zheng Zhou[2], Shannon Armstrong[2], Eric Baer[2], Donald R. Paul[1], Christopher J. Ellison[1]

[1] McKetta Department of Chemical Engineering, University of Texas at Austin, Austin, TX, 78712, USA

[2] Department of Macromolecular Science and Engineering, Case Western Reserve University, Cleveland, OH, 44016, USA

Abstract

Liquid crystalline polymers (LCPs) have been considered for potential applications as barrier films and high performance membranes due to their extraordinary barrier properties and chemical resistance. However, LCPs usually possess extremely low viscosities, which presents a challenge for implementing them into multilayer coextrusion due to the mismatch of their viscosities with other film components. In this study, a commercial grade diepoxy reagent was used to modify the rheological properties of a thermotropic main chain LCP. The effects of diepoxy concentration and reaction time on the resulting melt viscosity and liquid crystal structures of the LCP were investigated. High quality multilayer films were fabricated by coextruding the rheologically modified LCP with polypropylene-graft-maleic anhydride. Additionally, the orientational order of the LCP in the multilayer films was studied by polarized infrared spectroscopy and x-ray diffraction. It was found that the LCP chains tend to orient perpendicular to the extrusion direction instead of aligning along the shear direction, which was ascribed to a 'log-rolling' effect during processing.

Control of thermal degradation of poly(lactic acid) using functional polysilsesquioxane microspheres as chain extenders

Ting Han, Zhong Xin, Yaoqi Shi, Shicheng Zhao, Xin Meng*

State Key Laboratory of Chemical Engineering, School of Chemical Engineering, East China University of Science and Technology, Shanghai, People's Republic of China

*Corresponding author: xzh@ecust.edu.cn

Abstract. *In this paper, the functional polysilsesquioxane microspheres functionalized by poly(epoxypropoxy)silsesquioxane (PESQ) and poly(epoxypropoxy/amino)silsesquioxane (PEASQ) were grafted onto commercial PLA via functional group reactions through melt processing, and directly used as a novel chain extenders to improve the thermal degradation of PLA. During the whole reaction time 35 min, the torques of P-2PESQ and P-1PEASQ was 60% higher than that of Processed PLA. The addition of PESQ or PEASQ into PLA increased the onset temperature for thermal degradation.*

Poly(lactic acid) (PLA), a linear and aliphatic thermoplastic polyester that can either be semi-crystalline or amorphous depending on the stereopurity of the polymer backbone, can be synthesized from 100% renewable resources such as fermentation products of corn and sugar beets. However, the processing instability of PLA hinders its applications in practice by causing the cleavage of polymer chains and a decrease in molecular weight[1-4], eventually resulting in a deterioration of the rheological properties. Moreover, the low melt viscosity of PLA can also destroy its performance in the blow molding and foaming process[5].

In this paper, the functional polysilsesquioxane microspheres functionalized by poly(epoxypropoxy)silsesquioxane (PESQ) and poly(epoxypropoxy/amino)silsesquioxane (PEASQ) were grafted onto commercial PLA via functional group reactions through melt processing, and directly used as a novel chain extenders to improve the thermal degradation of PLA.

Results showed that the shear viscosity of PESQ modified PLA (PLA-PESQ) (i.e., 4500 Pa·s) was approx five times larger than that of neat PLA (i.e., 750 Pa·s), but was 33% smaller than that of PEASQ modified PLA (PLA-PEASQ). The molecular weight of both PLA-PESQ and PLA-PEASQ were about twice more than that of neat PLA, and the Melt Flow Rate (MFR) reduced to more than half of neat PLA. The enhanced performance was mainly attributed to the change of molecular structure caused by chain extenders.

Torque Evolutions. The properties of PLA-PESQ and PLA-PEASQ were further studied. During the whole reaction time 35 min, the torques of P-2PESQ and P-1PEASQ was 60 □ higher than that of Processed PLA.

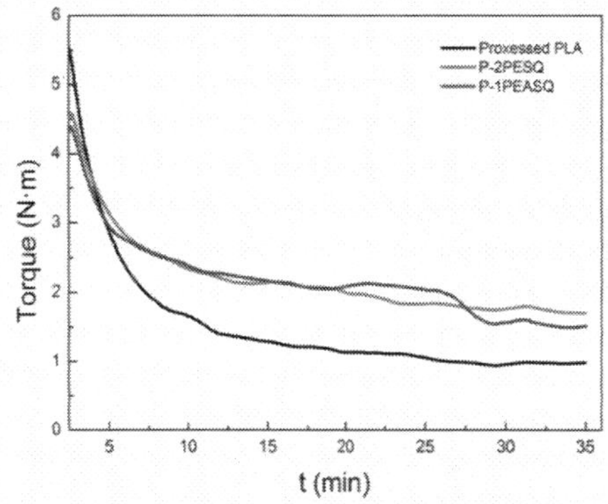

Fig. 1. Torque evolutions for Processed PLA and modified PLA reaction courses.

Thermal Gravimetric. Thermal gravimetric analysis revealed that the addition of PESQ or PEASQ into PLA increased the onset temperature for thermal degradation. PEASQ was found to be a more efficient chain extender under processing conditions, which the thermal decomposition temperatures, at 1%, 5% and 10% weight loss are 324 , 334 , 342 , respectively, increased 8 , 16 and 13 , compared to processed PLA. The PEASQ can be used as a potential chain extender for PLA.

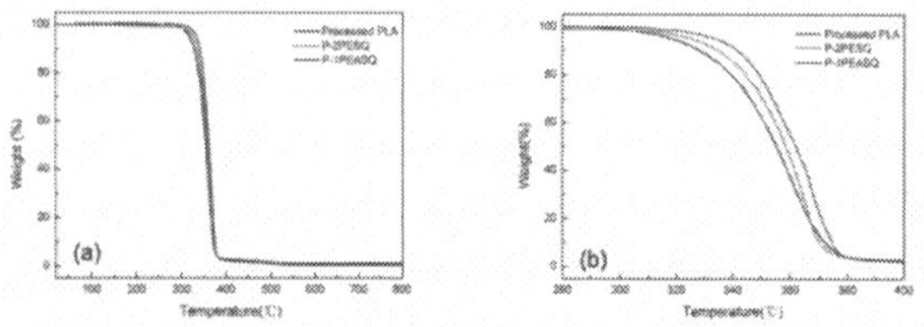

Fig. 2 TGA curves(a) and part(b) of Processed PLA and modified PLA

MFR. As shown in Table 1, the MFR of PLA processed for 35 min (Processed PLA) was 21.7 g·10min^{-1}, about twice than neat PLA. It meant that the heavy thermal degradation was happened to PLA during 35 min processed. As to the modified PLA, MFR were almost not changed, even a small amount of reduced, which showed fine performance in control thermal degradation. By comparison, PEASQ was a more effective additive than PESQ, due to its less dosage and better performance.

Mechanical Properties. Furthermore, the impact strength of PLA-PESQ and PLA-PEASQ (Table 2) increased to twice of that of neat PLA, indicating an dramaticlly enhanced toughness. The impact strength of P-2PESQ was increased by 52 J·m^{-1}, a little better than that of PLA-PEASQ.

Table 1 The MFR of Processed PLA and Modified PLA

Samples	MFR(g·10min^{-1})
Processed PLA	21.7
P-2PESQ	11.4
P-1PEASQ	10.9

All the samples were processed for 35 min at 200 □.

Table 2 The mechanical properties of Neat PLA and optimal modified PLA

Samples	Impact strength J·m^{-1}	Flexural modulus MPa	Tensile strength MPa
Neat PLA	25	2697	44
P-2PESQ	52	2701	40
P-1PEASQ	46	2677	39

Acknowledgments

This work is financially supported by the National Natural Science Funds of China (21306047, U1162110), Program of Shanghai Subject Chief Scientist (XD1401500), and Fundamental Research Funds for the central universities of China (WA1214051, WA1314051).

References

1. Lunt J. *J.* Polymer degradation and stability, 1998, 59(1): 145-152.
2. Tsuji H, Fukui I. *J.* Polymer, 2003, 44(10): 2891-2896.
3. Vert M, Schwarch G, Coudane J. *J.* Journal of Macromolecular Science, Part A: Pure and Applied Chemistry, 1995, 32(4): 787-796
4. Thomson R C, Yaszemski M J, Powers J M, et al. *J.* Journal of Biomaterials Science, Polymer Edition, 1996, 7(1): 23-38.
5. Di Y, Iannace S, Di Maio E, *J.* Macromolecular materials and engineering, 2005, 290(11): 1083-1090.

380552 Single Molecule Characterization of Dual-Colored DNA Comb Polymers

Abstract:

Single Molecule Characterization of Dual-Colored DNA Comb Polymers

Danielle J. Mai, Amanda B. Marciel, Charles M. Schroeder

We report the synthesis and single molecule characterization of two-color, comb-shaped branched biopolymers. In this work, we utilize a hybrid enzymatic-synthetic approach to graft "short" DNA branches to "long" DNA backbones, thereby producing macromolecular DNA comb polymers (**Figure 1**). The branches and backbones are synthesized via polymerase chain reaction with chemically modified deoxyribonucleotides (dNTPs) and primers: "short" branches consist of internal Cy5 labels and a terminal azide group, and "long" backbones contain internal dibenzylcyclooctyne (DBCO) groups and a terminal biotin tag. In this way, we utilize strain-promoted, Cu(I)-free [2+3] cycloaddition "click" chemistry for facile grafting of azide-terminated branches at DBCO sites along backbones. Copper-free click reactions are bio-orthogonal and nearly quantitative when carried out under mild conditions. Moreover, resulting comb polymers can be labeled with a DNA stain (e.g., SYTOX Green®) for dual-color imaging. To this end, we use single molecule fluorescence microscopy to directly observe comb polymers by tethering probe molecules to a surface with biotin-NeutrAvidin linkages, extending them with pressure-driven flow, and imaging them with fluorescence microscopy (**Figure 2**). Our imaging approach allows for characterization of these materials at the single molecule level (*e.g.*, quantification of polymer contour length and branch distributions for varying synthetic conditions). We are also extending this approach to single polymer rheology experiments, with specific aims to understand the effects of branching on relaxation timescales (surface tethered or free solution), steady-state extension in various flow fields, and dynamics in semi-dilute or concentrated solutions. In this way, our work will contribute to the overall understanding of topologically complex polymer melts and solutions.

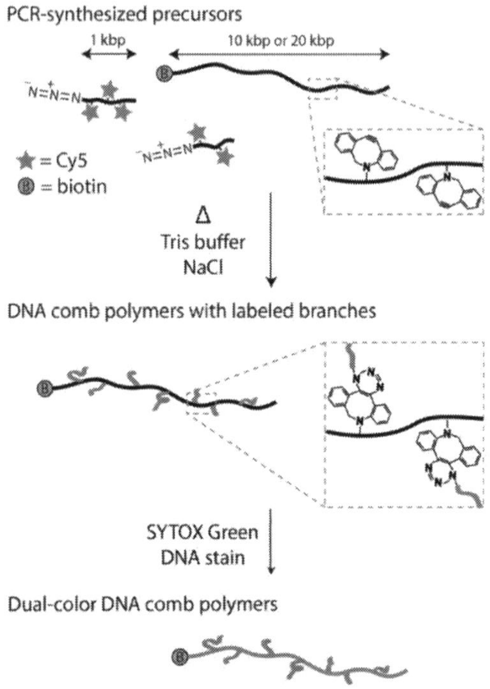

Figure 1. Synthesis of biotin-tagged, dual-color DNA comb polymer molecules.

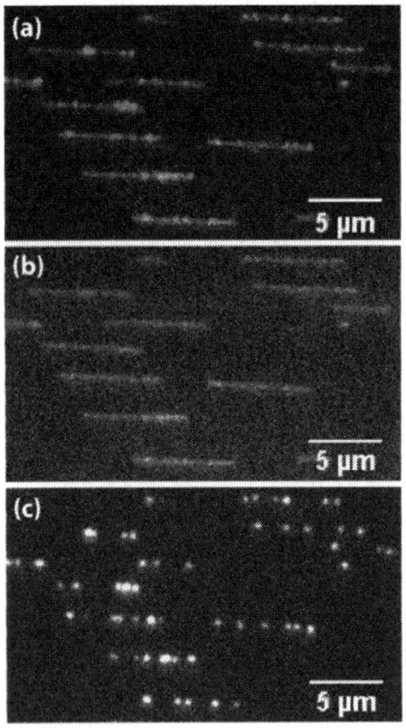

Figure 2. Single molecule images of tethered dual-color DNA comb extension under shear flow at a surface. (a) Co-localized composite image; (b) SYTOX Green-stained backbones and branches; (c) Cy5-labeled branches.

378622 Interaction of Comb-Type Copolymers with Long Chain Paraffin in Oil Observed By Epifluorescence Microscopy

Abstract:

Interactions between copolymer additives (maleic alkylamide and alpha-octadecene copolymers: MAC, AMAC, and NMAC) with different functional groups with long chain paraffin crystals in waxy model oils have been systematically investigated by epifluorescence optical microscopy. Ultraviolet (UV) and blue light (495 nm) were applied in this experiment. Both long chain paraffin crystals and the assembled polymers were found glowing blue with UV illumination, while the polymers emitted stronger yellow-green fluorescence light than crystals in blue light field. MAC, which has long paraffinic lateral chains, was found to be capable of changing the shape of wax crystals from granule-like into sharp needle-like, implying a co-crystallization mechanism in the wax forming process. Copolymers with aromatic groups on their backbones (AMAC and NMAC) were observed to behave in different ways from MAC. NMAC, which contains naphthalene groups and glowed deep yellow fluorescence light in blue light field, seems to be merely absorbed on the surface of wax crystal. AMAC, which has benzene groups, appeared to self-crystallize separately from wax and its aggregations were found inclined to adhere the wax crystals and even functioned as the center of gathering waxes. Epifluorescence microscopy method provides a new perspective to acquire an insight into the distribution of polymer additives, the assembly mechanism, and their effects on modifying the morphology of long chain paraffin crystals in waxy oil.

386385 Reversible Patterning and Actuation of Hydrogels By Electrically Assisted Ionoprinting

Abstract:

The ability to pattern, structure, re-shape and actuate hydrogels is important for biomimetics, soft robotics, cell scaffolding and biomaterials. We introduce an 'ionoprinting' technique with the capability to topographically structure and actuate hydrated gels in two and three dimensions by locally patterning ions via their directed injection and complexation, assisted by electric fields. Applying an oxidative bias to a patterned metal anode injects divalent ions into the charged polymer networks which form robust ionic crosslinks. The magnitude and duration of the applied current controls the degree of ionic binding to the polymer backbone. The bound charges change the local mechanical properties of the gel to induce relief patterns and in some cases evoke localized stress, causing rapid folding. These ionoprinted patterns are stable for months, yet the process is fully reversible by immersing the gel in a chelator solution. The ionically patterned hydrogels exhibit programmable temporal and spatial shape transitions, and serve as a basis for a new class of soft actuators that can gently manipulate objects both in air and in liquid solutions.

385740 Fundamentals and Applications of Hydrogel Actuation By Electric Fields Towards Soft Robotic Components

Abstract:

We present a novel class of "walking" gel prototypes of soft robotic devices comprised of cationic and anionic gel legs, which bend in response to the redistribution of mobile ions between the gel network and solution in an external field. Stimuli responsive hydrogels could be key components in the next generation of soft matter actuating and sensing devices. Furthermore, they could serve as "smart" biocompatible materials for mimicking dynamic structures found in nature since they transduce chemical energy into mechanical motion without the use of external mechanical input. We present our work manipulating the internal hydrogel stress on both the molecular and macro level utilizing electrical fields. Electrical fields provide excellent control over induced osmotic pressure and electrostatic interactions between the fixed charges within gel networks for preprogrammed motion. The magnitude of bending is a function of the induced osmotic pressure difference which is governed by the interaction between the mobile ions and fixed charge groups of the hydrogel. The sign of the fixed charges on the polyelectrolyte network determines the direction of bending under an applied external field. Therefore, we combined two oppositely charged gels to promote the motion of the legs in two directions. We characterized the electro-actuated response of the hydrogels as a function of charge density and external salt concentration. The unidirectional gel walker motion on flat elastomer substrates exemplifies a simple way to move and manipulate soft matter devices and robots in liquid media. Our current work aims to control the bending of thermo responsive gels by embedding inside them chain-like structures of latex colloidal particles assembled directionally using electric fields. When an external AC field is applied across a colloidal suspension, the particles align into chains due to the acquired dipoles. These chains can serve analogously to an endoskeleton within a gel matrix. As the gel shrinks in response to heat, the organized particle chains determine the final structure as opposed to the isotropic shrinking that would be observed in the absence of chain alignment. This approach will result in complex gel/particle composites whose shape change can be programmed internally using dielectrophoretic structuring. Thus, we are developing simple soft matter actuator devices and robotic components utilizing conventional polymers and external stimuli, which may serve as active components where conventional, stiff materials are inadequate.

Modeling the Mechanical Properties of Biopolymers for Automotive Applications

K. Mahdi[1], A. Elkamel[2], L. Simon[2], E. Tsai, V. Vinayagamoorthy[2], and I. Bagshaw[2]
[1]Chemical Engineering Department, Kuwait University, Kuwait
[2]Chemical Engineering Department, University of Waterloo, ONT, Canada

Abstract—The automotive industry is constantly looking for alternative solutions to reduce manufacturing cost and use renewable materials. Implementing agro-fibres as polymer fillers in thermoplastic matrix will satisfy the automotive criteria without sacrificing the mechanical properties currently set by the conventional fillers such as glass fibre, talc, or mica. This paper proposes the use of wheat straw as filler in polypropylene for automotive industry and investigates models for determining compositions of the materials to correspond to mechanical properties Wheat straw is an agriculture crop that is low in density, high in strength, and cheap to produce. These attributes out-compete the properties of petroleum-based fillers; however, the agro-fibre has shortcomings to creating a biocomposite. The wheat straw is highly water absorbing, deformation effects occur when the material is subjected to intense heat, and the supply of the crop is seasonal.

Varying weight percentages of wheat straw and polypropylene to create the biocomposites through an extrusion process performs data collection. The end products are molded into proper shapes for mechanical testing. Different modeling approached that include polynomial regression, artificial neural networks and support vector machines are investigated to prepare predictive models for the biocomposite properties. A comparison between the purely empirical methods shows that support vector machines produced the best model, followed by artificial neural networks, and then polynomial regression.

Keywords—biocomposites, mechanical properties, empirical modeling, neural networks, support vector machine, extrusion, automotive industry.

I. INTRODUCTION

THE modern automobile contains over 100kg of conventional plastics in mouldings, sound barriers and other components for such low stress applications. In many of these applications, the plastic is implemented in the form of a composite material, wherein traditional polymers are reinforced or otherwise filled with another material to change the mechanical properties or make the product more cost effective to the application. Nevertheless, the primary material cost associated with these products is petroleum. With the present volatility of the petroleum market and the world's energy needs only expected to rise amid the development of many large third-world economies, great value can be found through the more efficient use of crude oil as a resource. Figure 1 presents the volatility of crude oil price up to year 2008.

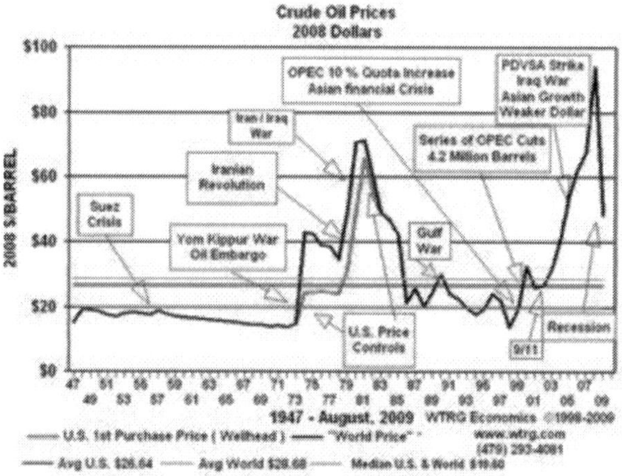

Fig. 1 Crude Oil Prices in 2008 (ISM, 2010)

Beyond efficiency, a developing environmentally minded group of 'green-thinkers' are also entering the market place, considering total carbon footprint among the other elements of automobile selection. To cater to this rising demographic, many automotive companies are beginning to feature the overall impact of their automobiles within the closed-loop context. Below, Figure 2 highlights the non-metal exterior components of the suzuki swift, which is an element of their in-depth literature of on the utility of this economy-class car.

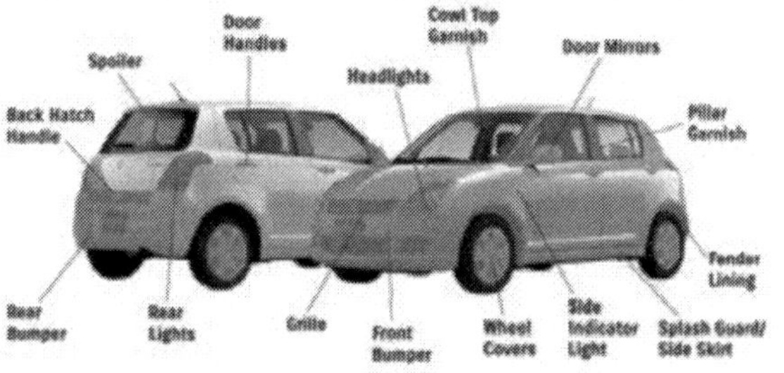

Fig. 2 Plastic Components of Suzuki Vehicle (Suzuki, 2009)

While the consumer would enjoy the added benefit of a lower cost product, which is superior in some regards (ie. weight), the automaker definitely receives additional credit for substituting in a green technology, from the standpoint of environmentally minded consumers. Thus, emerging biocomposite technologies give the automaker yet another method of product differentiation, separating themselves from their competitors by emphasizing the 'green' nature of their

K. M. Author is with the Chemical Engineering Department, Kuwait University, Kuwait (phone: 965.2498.5618;e-mail: khaled.mahdi@gmail.com).

A. E. Author is with The Chemical Engineering department, University of Waterloo, ON, Canada (e-mail: aelkamel@uwaterloo.ca).

L. S. Author is with The Chemical Engineering department, University of Waterloo, ON, Canada (e-mail: lsimon@uwaterloo.ca).

E. T., V. V., and I. B. authors were with The Chemical Engineering department, University of Waterloo, ON, Canada.

product. This can be considered an added driving force for the advancement of biocomposite technology, as automakers seek to increase market share and loyalty of their customers as environmental issues facing the world come to the forefront of people's consciences.

II. WHEAT STRAW POLYPROPYLENE BIOCOMPOSITES

In most cases, elements comprised of polymers and resins within an automobile are made of 2 key components. Specifically, an expensive (typically petroleum based) plastic resin makes up the matrix in which either cheap filler or a more expensive reinforcement material are suspended. The selection of the suspension material depends on the application of the product, with talc or calcium carbonate making up most of the mainstream fillers for low impact situations, and carbon fiber or fiberglass making up the majority of the high strength applications.

While the applications of low weight high strength composite materials is indeed an important aspect of the automotive industry, it is typically reserved for the chassis of high end sports cars and are thus not as widely applicable to the industry in general. In this case, the price of the automobile is more directed at maintaining the exclusive nature of owning one of these vehicles, and thus development within this field would be more towards performance and less towards value. Hence, the focus on cost reduction of automotive components has been shifted towards the discovery and understanding of a cheaper and lighter filler product, consistent with wheat straw in the application of bio-filled composite materials.

Wheat straw can be classified as more of a filler-type of suspension. Since the improvement in the mechanical properties of wheat-straw filled polypropylene is marginal at best, it would only result in a higher performing composite material in a class of component that you would fine talc/calcium carbonate filled components. Though wheat straw is in fact fibrous in nature, the fibres are fragile and breakable and should not be considered as a strengthening agent. This role should then be reserved for natural fibres such as hemp.

III. BENEFITS AND SHORTCOMINGS OF WHEAT STRAW BIOCOMPOSITES

As in most material substitutions, there are both benefits and drawbacks to the use of wheat straw as a filler replacement to a more traditional material such as talc or calcium carbonate. Since wheat straw will be used as a filler substitute for talc or calcium carbonate as a filler material, this section is set up as a comparison to the substitution effect of switching to wheat straw with the tradeoffs outlined below.

A. Benefits

A.1. Strength

Composites filled with wheat straw tend to be stronger in comparison to traditional composite materials. While the margin of strength increase may not be high enough to be suitable for high strength demand applications (carbon fiber etc.), it is promising for the potential of these materials, since their applications are not restricted in this class of automotive component through a relative strength basis. Thus, the versatility of biocomposites as potential replacements for existing components is high and favourable for those investing in the development of this technology.

A.2. Density

As mentioned previously, automobiles typically contain 100kg of composite material. The low density nature of wheat straw allows for the development of lighter components with the aggregate total expected to be in excess of 10kg, once the maximum rate of substitution has been achieved. While this improvement can be considered marginal at best from the standpoint of overall fuel economy of the vehicle, the fact remains that the substitution effect of replacing traditional composite materials with wheat straw biofilled composites results in a net gain/benefit to the consumer as well as the manufacturer.

A.3. Petroleum Substitution Rate

Due to the more favorable mechanical properties which wheat-filled composite materials offer, a higher filler to resin ratio may be achievable in many cases. This is very beneficial as it would drive down the cost of the product further while decreasing the variability in the price of the product, due to a reduced dependence upon the volatile crude oil markets. The knowledge of this benefit could have been used as a leverage point for the development team to gain funding from both private and public sectors, and likely has been given the primary objectives outlined in the Ontario BioCar initiative. (BioAge Group, 2009)

A.4. Cheaper

Since wheat straw is collected from a locally produced crop, the logistical cost of obtaining this material is lower. Since wheat straw is a side product to the high value crop of wheat, it is also comparatively cheaper than a mineral which is produced specifically (calcium carbonate, talc). Traditional applications of wheat straw in the agricultural context primarily include use as a bedding or absorbent material, and are priced low given this context. The reduced transportation and material costs of wheat straw, coupled with a higher rate of petroleum substitution could potentially make wheat straw-filled composite materials a cheaper alternative.

A.5. Greener

As a renewable resource, the use of wheat straw as a filler for composite materials reduces the environmental impact of manufacturing the component. Rather than using harmful mining techniques which unnecessarily releases toxins into the ground water and other mediums, while requiring high petroleum consumption to harvest the materials, wheat straw is harvested as a part of the routine of producing wheat for the food supply. Thus, the net effect of shifting material consumption from talc/calcium carbonate to wheat straw as the filler material of choice is positive in terms of environmental impact. One present application of biocomposite utilization within the Ford Flex (discussed at length within this work) is noted as reducing the production of carbon dioxide by roughly 1.5 tons per year. (BioAge Group, 2009)

B. Potential Short-Comings

B.1. Water Absorption

By nature, wheat straw is an excellent absorber of excess moisture. It is also, in fact, generally used as an absorbent bedding material when raising livestock. This however is not a desirable property for application in bio-composites, as the product can in some instances absorb large quantities of water through exposed straw particles at the surface of the component.

Though consistent stretches of 10 to 20 days at 95% relative humidity are not common in the context of Canadian weather patterns, the fact remains that absorption of the composite is a very real possibility. With a water absorption rate of up to 25% achievable, careful consideration of the likelyhood of this occurring and the direct effect it results in must be considered. Coupled with the fact that seasonal and abrupt variation in temperature also exists, the absorbed water could potentially pose a cracking or other degradation risk in the Canadian climate. In more tropical climates, full absorption and desorption of moisture from season to season could be a very real possibility, and thus may affect the suitability of this product in various regions.

B.2. Deformation Effects

The primary cause for concern with respect to the water absorption factor associated with bio-filled composites is the material expansion, which it causes. Specifically, in many applications, the components are fitted within the assembly with little additional room or tolerance. Thus, any expansion of the product due to water absorption could compromise the integrity of the component, or otherwise result in warping of the component, which could be unsuitable from both aesthetic and utility standpoints.

B.3. Inconsistencies of Supply

As with every crop, seasonal variation of the local crop quality can be different from year to year. The Figure shows below points out the regions of agricultural availability.

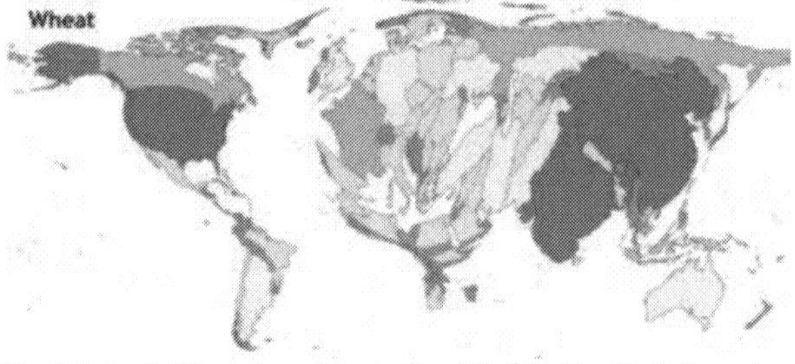

Fig. 3 Potential Depletion of Agricultural Soil Bodies (deCarbonnel, 2009)

B.4. Odor

The depletion of agricultural bodies causes an odor to be released. Rather than having the "new plastic" smell, the use of biofiller will create a smell similar like coffee. This may be an unattractive feature in a vehicle but it does not affect the performance of the material.

IV. EXPERIMENTAL

A. Raw Materials

Wheat straw is harvested and prepared by manually cutting the fibres into 5cm long pieces and then grinded in a steel rotary grinder. A 5 mm mesh sieves out unprocessed straws to be re-fed back into the grinder. The product is a powder, which is dried in an oven.

The thermoplastic is a high viscosity polypropylene impact polymer donated by A. Schulman Inc. The material is received in pellet size. It is denoted as virgin polypropylene because it is newly synthesized in the purest form.

Maleic anhydride polypropylene (MAPP) is combined with the biocomposite as an additive.

B. Compounding Materials

To create the wheat straw-polypropylene composite, the raw materials undergo an extrusion process. Extrusion is a method in which materials are heated and mixed to produce long strands of homogenized product by forcing through a cross section. The Haake Minilab Microcompounder, also known as the Minilab, carries out this process. Wheat straw powder and polypropylene pellets are fed into a compounder via a hopper and enter a heated barrel. Inside the machine, two co-rotating screws drive the substances in a cycling channel. The materials mix as they are melted until a uniform product is formed. The blended biocomposite is flushed out an exit hole at the end of the process as thin round bars, which can be pelletized for molding.

C. Testing

The strands of the mixed product from the Minilab are cut into pellets and compression molded into plates. The compression involves heating two metal plates and sandwiching the melted pellets into the mould figure. Over the span of several minutes, force is applied and released and re-applied again. This ensured that no air bubbles would be trapped inside the material during the process. The plates are formed and cooled. Using the MiniMat 2000 miniature material tester to determine flexural properties performs following ASTM D790 specifications as guidelines, tests.

Three mechanical properties are tested for the biocomposite: flexural strength, flexural modulus, and the yield strength. Flexural strength is the highest point of stress the material can resist under a load before breaking. Flexural modulus is defined by the tendency of the material to bend. Yield strength is the stress the material can withhold until it is deformed elastically and permanently.

V. MODELING METHODOLOGY AND RESULTS

As stated previously the objective of this paper is to develop a versatile model capable of determining mechanical properties of a biopolymer given product composition or vice versa. The specific mechanical properties that are being sought after are the Flexural Modulus, Yield Strength and Flexural Strength. Unfortunately, there are no theoretical frameworks for Yield Strength and Flexural Strength thus empirical modeling for these two mechanical properties will be done with purely empirical methods such as Polynomial Regression, Artificial Neural Networks and Support Vector Machines. However Flexural Modulus has a theoretical framework (Shackelford, 2005). Therefore, if a formula is developed it will have real worth if it buttresses with framework.

A. Polynomial Regression

Polynomial regression is the standard empirical modeling tool used in mixture experiments - specifically, linear regression using Schiff's canonical equations (Montgomery, 2005). The term linear regression may be slightly misleading as linear only refers to the methodology in which the parameters of the model are found. It has nothing to do with its ability to model non-linear phenomena. Polynomial regression is different from all the other modeling techniques discussed in this report, as it is the only modeling technique to assume a structure a priori. The structure assumed in mixtures problems is those of Scheff's canonical equations (Montgomery, 2005). Scheff's canonical equations have been used throughout industry and have a thorough run through a variety of mixture problems (Montgomery, 2005). Listed below are Scheff's canonical equations.

Equation 1: Linear

$$E(y) = \sum_{i=1}^{p} \beta_i x_i$$

Equation 2: Quadratic

$$E(y) = \sum_{i=1}^{p} \beta_i x_i + \sum_{i<j}^{p}\sum \beta_{ij} x_i x_j$$

Equation 3: Full Cubic

$$E(y) = \sum_{i=1}^{p} \beta_i x_i + \sum_{i<j}^{p}\sum \beta_{ij} x_i x_j + \sum_{i<j}^{p}\sum \delta_{ij} x_i x_j (x_i - x_j) + \sum_{i<j<k} \beta_{ijk} x_i x_j x_k$$

Scheff's polynomials can be very simple or complex. For example the linear form only captures individual effects but as the complexity increases; that is as the higher order terms come into play, more complexity can be captured. It can be expected that the non-linearity of the process will be captured in these higher ordered terms.

There are important reasons for the use of this type of modeling. While appealing to the authority of its longstanding use may seem facile, it should be recognized that its extensive use is indicative of some important characteristics. The first and foremost is its relatively simple ease of implementation. Calculating model parameters once the data scheme has been set out is quick, and this coupled with the assurance that parameters have been globally optimized. Another consequence of using Schiff's canonical forms is that individual effects and interactions between components can be easily mapped out. Thus, generalizations and transferability of results can happen more readily.

One disadvantage of modeling with this polynomial regression is that it is susceptible to outliers as it includes data without discretion. Another considerable disadvantage to standard polynomial regression is that it even with its higher order terms it cannot guarantee good fit. This can be a blessing or curse depending on the final outcome. Limiting its ability to fit a curve prevents overfitting, but if the data is such that the fit is inadequate, the model cannot aid us in regards to the basics of product design.

The polynomial regression structure that was selected was the quadratic form. This is shown in the equation below:

Equation 4: Quadratic Polynomial Regression

$$y = \beta_1 x_1 + \beta_2 x_2 + \beta_3 x_3 + \beta_4 x_1 x_2 + \beta_5 x_1 x_3 + \beta_6 x_2 x_3$$

Where x_1, x_2 and x_3 represent the process variables; that is the percentage of polypropylene, wheat straw, and MAPP.

The quadratic form was selected as it was seen as middle path between the simplicity of a linear form and the complexity of a cubic form. Essentially, what was hoped was that the quadratic form would capture enough of the non-linearity with its higher order terms without over-fitting.

The following is the procedure used in determining the parameters of the quadratic form of Scheff's canonical equations specifically for flexural modulus.

Equation 5: Quadratic Form of Scheff's Canonical Equations

$$X = [x_1\ x_2\ x_3\ x_1x_2\ x_1x_3\ x_2x_3]$$

$$y = [2390.17\ 2638.06\ 1902.52\ \ldots.\ 2501\ 2234.37\ 1962.14\ 1753.56]'$$

$$\beta = (X'X)^{-1} X\, y$$

$$\beta = 1.0e+005 * [\,1.6597\quad 1.1675\quad 0.0528\quad 0.6547\quad 0.0296\quad 0.0214]$$

Therefore the model using the quadratic form of Scheff's canonical equations is:

$$F.M = 1.0e+005\ (\,1.6598x_1 + 1.1675x_2 + 0.0528x_3 + 0.6547x_1x_2 + 0.0296x_1x_3 + 0.0214x_2x_3)$$

Likewise the regressions found for Flexural Strength and Yield Strength are:

$$F.S = 1.0e+003\ (\,4.4534x_1 + 3.1304x_2 + 0.1417x_3 + 1.7559x_1x_2 + 0.0795x_1x_3 + 0.0575x_2x_3)$$

$$Y.S = 1.0e+003\ (\,6.9221x_1 + 4.8676x_2 + 0.2203x_3 + 2.7299x_1x_2 + 0.1235x_1x_3 + 0.0894x_2x_3)$$

B. Artificial Neural Network

In regards to mixture experiments, Artificial Neural Networks have had limited applicability. However, their use in regression problems has been relatively much more popular. This paper will help build a case for their use in our particular problem. First, as the name implies, Artificial Neural Networks are simulated neural constructs, or in other words a simulated oversimplification of a brain. Thus one would expect that it would do well at what a brain does well at such as classification and poorly at what a typical computer does well (rapid calculation). This is typically the case, so for our purposes we are aiming for the neural network to produce a decent regression.

As stated previously, an artificial neural network is a simulated neural construct or a crude simulation of a brain, and thus it would be expected that the fundamental unit of an artificial neural network is a neuron. The neuron of an artificial neural network approximates roughly how a neuron in the brain would operate. In an artificial neural network, a neuron aggregates all of its weighted inputs and then applies a transfer function to the sum to form an output value. Typically, and in the case of our modeling the transfer function is a log-sigmoid transfer function. The first steps in neural networks then proceeded to use these neurons in groups. Feed forward architecture in neural networks connotes how the artificial neural network is implemented. Immediately, one can gather that in a feed forward network, the information travels from input to output; that is in the forward direction. It should also be noted that every neuron in the perceptron is connected to every neuron in the following layer. A perceptron works by setting different weightings to the input values, which connect to the neurons. If the sum of the weighted inputs is greater than a pre-determined bias of the neuron, a signal is sent, in this case as the final output.

This type of single layer perceptron was found to be lacking as it could not solve a variety of problems, most famously its inability to model a XOR function. Fortunately, it was found that with the addition of a hidden layer between input and output layer artificial neural networks could overcome the short comings of a simple single layer perceptron. This multilayer perceptron also gained the ability to model any function as it was proven to be universal approximator.

The methodology that is typically used in determining the weights of the connections in a multilayer perceptron is called the backpropagation algorithm.

The backpropagation algorithm works as follows:

1. Calculate output of neural network
2. Calculate error between training data and output.
3. Calculate what output should be each neuron, and the factor by which the output should be corrected by.
4. Adjust weights to lower error on individual neurons.
5. Assign responsibility of error to previous layer, with emphasis on stronger weighted inputs
6. Repeat from step 3 on neuron in previous layer, using the responsibility as found in present layer for error.

A multilayer perceptron coupled with the backpropagation algorithm is capable of solving various regression problems. However, there are fair number of problems and traps one should avoid.

One of the main advantages to an artificial neural network as a multilayer perceptron is its ability to approximate any function. The problem however lies in the fact that an artificial neural network is so good at being a function approximator that it causes overfitting. That is by excessively fitting the data it confuses noise and signal thereby decreasing its predicative strength. Another disadvantage and probably the most salient one is that once an adequate model is built it can be said to be only relevant to the data it was derived from. One cannot generalize a network nor seek an understanding by looking at the weightings and the connections found in a neural network. Finally, like a linear regression it is susceptible to outliers as it has no means by itself to throw out data which diverges wildly from the rest of the set.

A standard artificial neural network can also be written as the following in equation form (Samarasinghe, 2006):

Equation 6: ANN Equation

$$y(x, \omega, H) = g(\sum_{i=1}^{H} \omega_i f(\sum_{j=1}^{N_{in}} \omega_{ij} x_j - \theta_j) - \omega)$$

Where H is the number of hidden nodes, $f(x)$ is some sigmoidal function, $g(x)$ in our case a linear function.

It can be seen from stating an artificial neural network in this form that it appears to emulate a linear form $y = \omega x + \theta$. Therefore, this lends credence to the idea that the extraneous variables are the tools which will hold the non-linearity of the phenomena seeking to be modeled. Thus it is typically thought that the non-linearity is captured within the number of hidden nodes as well as the weights- though weights not as much as the hidden nodes. Too much non-linearity will cause overfitting thus in this report a common heuristic will be used to circumvent this problem. The standard methodology is to separate the data given into two sets: a training set and a validation set. The training set is used to train the network where as the validation set is used to check the model. If the network overtrains towards the training set of data its predictive strength would reduce and thus produce larger error for the validation set. The standard heuristic is to separate the data giving two thirds of the data to the training set and one third of the data to validation set (Samarasinghe, 2006). However, in this report the data is divided into 60% training data, 20% validation data and 20% test data when training artificial neural networks.

The weights as signified ω in equation are found through a variety of methods. However, in this report, the values for the weights are simply initiated through random number generator between two points. The focus then on the training primarily relies on the number of hidden layer neurons there are.

Figure 4 is the first artificial neural network built and its mechanical property it aims to capture is Flexural Modulus. The artificial neural networks of Yield Strength and Flexural Strength were found independently using the same methodology as this section however this section will mainly describe the process of generating a neural network for Flexural Modulus as the results found in one set mechanical properties was analogous to other mechanical properties. Therefore, for the sake of brevity this section will only consider Flexural Modulus. As can be seen the network in Figure 4 comprises of one hidden layer. Figure 5 shows how well this artificial neural network fits each of the data sets in terms of mean square error.

One major point that can elude most users who simply rely on their validation set is that they may accidentally overtrain their validation set at the cost of their training data. Focusing solely on minimizing the error on the validation set does this. Therefore in order to test this pitfall we used the methodology of graphing validation error against training error. From the previous modeling attempts it was seen that there was an adequate lower bound and upper bound to the number of hidden neurons that could possibly be used. Each number of neurons was considered between 10 hidden neurons and 20 neurons with vary weights for each number of neurons. The errors were averaged for each number of neurons and Figure 6 was produced.

The optimal number of neurons would produce a point closes to the origin. The number of hidden layer neurons that were found to be optimal was 11 neurons. Figure 7 shows the performance of 11 hidden layer neuron performances as well as the regression fit. The overall R^2 was found to 0.77 for the artificial neural network for the property of Flexural Modulus. Similarly, the optimal number nodes were found and an optimal artificial neural network was built for the properties of Flexural Strength and Yield Strength.

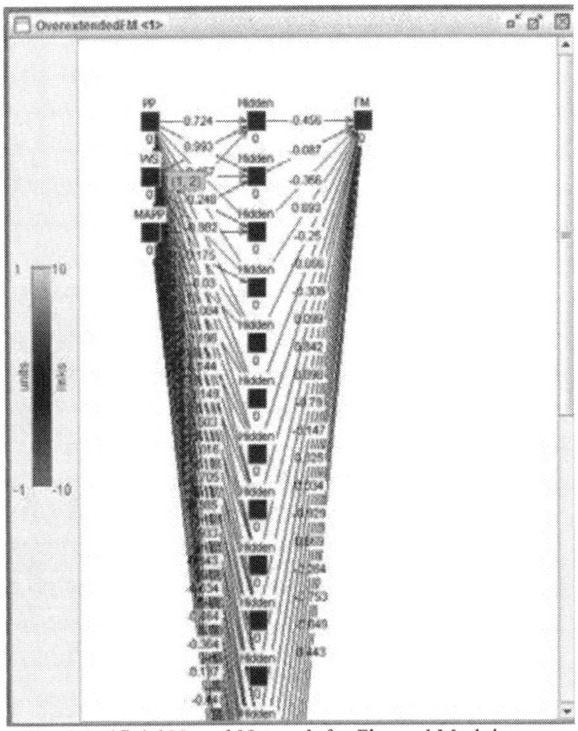

Fig. 4 Artificial Neural Network for Flexural Modulus

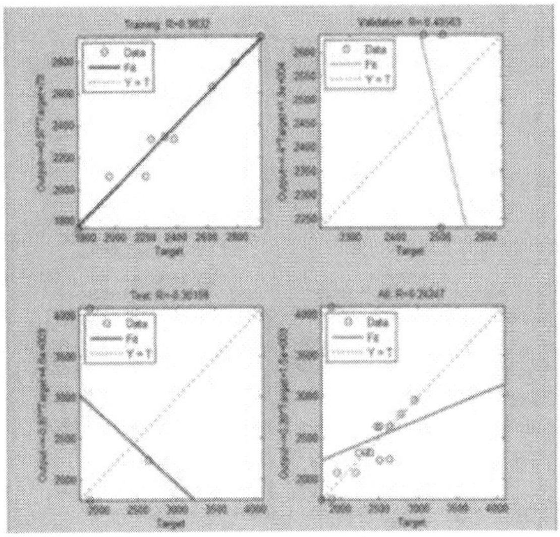

Fig. 2 Performance of Simple ANN

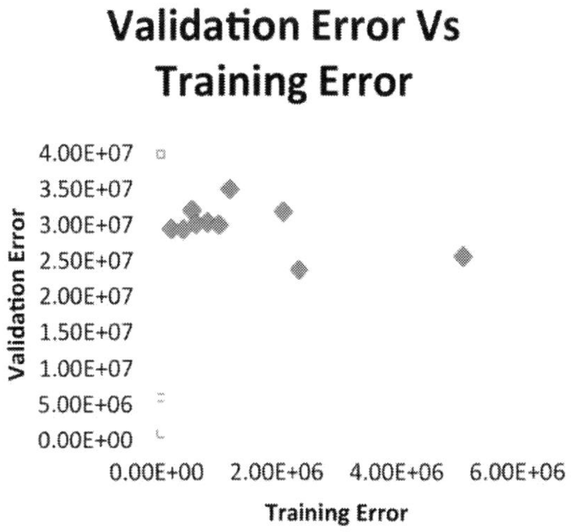

Fig. 6 Validation Error vs Training Error

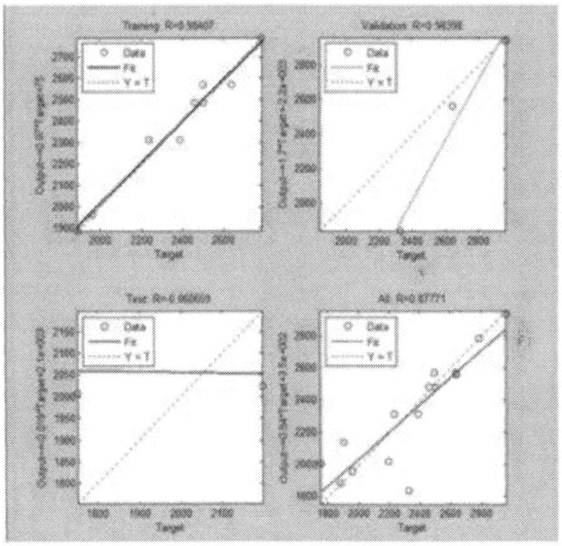

Fig. 7 Regression Plot for Optimal ANN

C. Support Vector Machine

Support Vector Machines (SVMs) are similar to ANNs as both methodologies aim at classifying data but SVMs are derived out from mathematical theory, whereas ANNs are empirical understanding of the brain. From a geometrical perspective, SVMs approach is to create a maximum margin between support planes to separate classes of data. If the plane completely bounds the class of data on one of its side, it "supports" that class of data sort to speak. To maximize the support planes margin towards the favored class of data, the planes are pushed to the brink of the class of data until they come in contact with small cluster of data points (support vectors); this is demonstrated in Figure 8. The closest points (support vectors) to the support planes are considered relevant and intuitively, the others are not part of the classification.

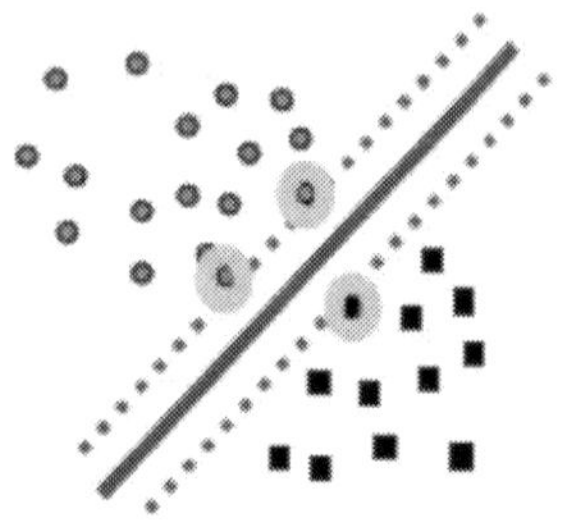

Fig. 8 Plane that maximizes margin (Bennett & Campbell, 2000)

The kernel trick is implemented to account for non-linear systems. This is performed by converting the non-linear data into a higher dimensional space where the maximum margin hyper plane could be used. Therefore the non-linear data is classified with a hyper plane.

Analogous to the classification solution, SMVs Regression only considers points by a preselected error margin from the predicted value and is able to produce a regression model with non-linear systems. A disadvantage of Support Vector Machines is the difficulty to make conceptual understanding, to know what is fundamentally occurring or the types of relationships that exist between the different components in a mixture experiment. This is attributed by the methodology of SMVs: the function of vectors is input variables and not a direct function.

The computing resources needed for SMVs do not depend on the dimensionality of the input space and so they are much less computationally expensive than ANNs. Support Vector Machines has better ability to handle over fitting compared to ANNs; hence SMVs are implemented to overcome this ANNs weakness.

The robustness of SVMs is accredited to its most powerful boon: preselecting error margin to choose data clusters (support vectors) and disregarding outliers. Thus SMVs can also be a tool to detect outliers; however this robust regression methodology is less interpretable to less robust alternatives.

Once again, for the sake brevity we will only illustrate the Flexural Modulus mechanical property when developing a support vector machine as the other to svms were built with the svm procedure and the modeling strength was just as positive as the gain found when modeling Flexural Modulus.

The two main factors that can be changed and that have an effect in support vector machine regression as supported by Tanagra is the kernel used by the support vector machine as well as the penalty cost. The kernel is analogous to the transfer function in an artificial neural network and the penalty cost refers to cost associated with attaining data outside the epsilon bandwidth of the regression line. The characteristics of SVM and Analysis of Variance for the training data are shown in Tables I and II, respectively. A radial basis function is used as the kernel. Through trial and error an appropriate penalty cost is found at 2.

TABLE I
Characteristics for Radial Basis kernel Support Vector Machine

Characteristic	Value
# support vectors	10

TABLE II
Analysis of Variance for Radial Basis kernel linear Support Vector Machine

Sum of Squares	
Total	1140526.0201
Error	107829.4584
Pseudo-R2	0.9055

The R2 value for the validation set was found to be 0.7535. The R2 value for the entire data set was found to be 0.8587, which is significantly higher to the value found for the artificial neural network.

D. Comparison

As stated in previous sections, the mechanical properties corresponded in the same manner to each of the modeling techniques. Figures 9 -11 show the estimated values from the model versus the exact value. A model with a perfect fit would produce a 45 degree line of data points. Essentially, these graphs encapsulate what has already been stated: polynomial regression is poor, artificial neural networks are much better though require time and investment to build, and support vector machines are better than both with minimal trial and error estimation of parameters.

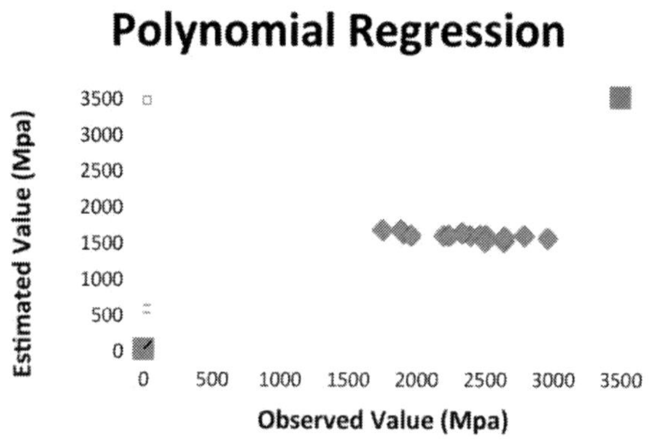

Fig. 9 Normalized Plot for Polynomial Regression

VI. CONCLUSIONS AND RECOMMENDATIONS

In this paper a versatile model capable of determining mechanical properties of a biopolymer given product composition or vice versa was thought. It was found that polynomial regression though used widely in industry, for the purposes of developing such a model specific to where the data was determined from was much poorer than artificial neural networks. Artificial neural networks were far superior to polynomial regression, however in terms of ease of development and overall fit support vector machines superseded artificial neural networks.

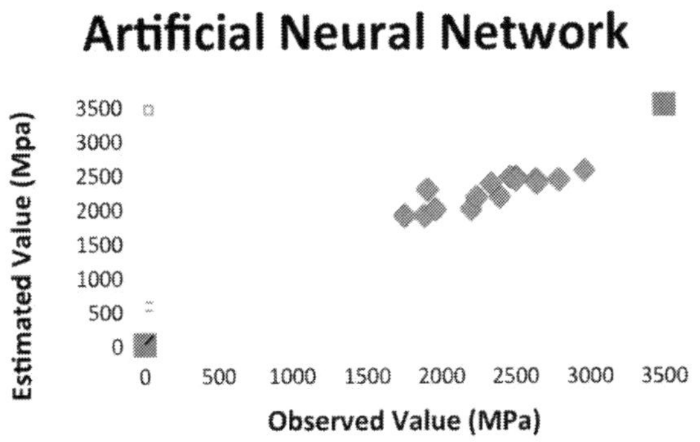

Fig. 10 Normalized Plot for ANN

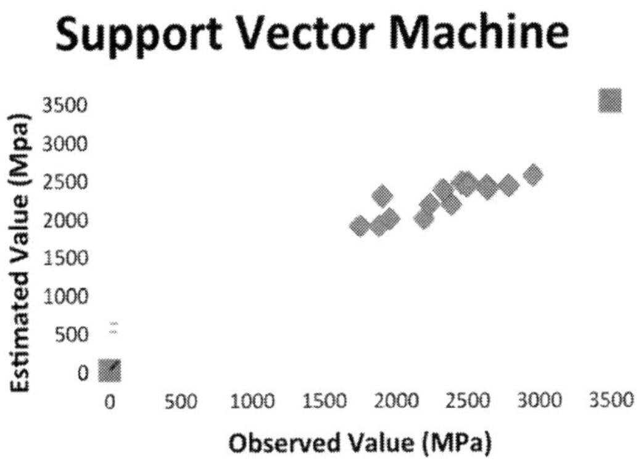

Fig. 9 Normalized Plot for Support Vector Machine

The most obvious recommendation would be to collect more data points. This would lead to better polynomial regressions, artificial neural networks and support vector machine models. Perhaps with more data the polynomial regression may become the superior method as its ease of implementation may out trump any marginal level of 'fit' strength an artificial neural network or support vector machine has.

REFERENCES

Bennett, K., & Campbell, C. (2000). Support vector machines: hype or hallelujah? *ACM SIGKDD Explorations Newsletter* .

deCarbonnel, E. (2009, Febuary 9). *Catastrophic Fall in 2009 Global Food Production*. Retrieved April 2, 2010, from The Market Oracle: http://www.marketoracle.co.uk/images/2009/Feb/Countries_by_agricultural_output%5B1%5D-747806.gif

Group, B. (2009, November 11). *Ford and Ontario Bio-Car Initiative Develop Wheat Straw-Reinforced Plastic; First Application in 2010 Flex*. Retrieved April 14, 2010, from Green Car Congress: http://www.greencarcongress.com/2009/11/biocar-20091111.html

Group, B. (2009, November 8). *Study: Bio-Based Plastics Could Ultimately Replace Up to 90% of Total Global Consumption of Plastics in 2007*. Retrieved April 12, 2010, from Green Car Congress: http://www.greencarcongress.com/2009/11/study-biobased-plastics-could-ultimately-replace-up-to-90-of-total-global-consumption-of-plastics-in.html#more

Guelph, U. o. (2010). *About Bioproducts at Guelph*. Retrieved April 21, 2010, from Bioproducts at Guelph: http://www.bioproductsatguelph.ca/biocar/about

ISM, O. (2010). *Price of Oil: Cause and Effect*. Retrieved 14 2010, April, from Oil ISM: http://www.oilism.com/oil/wp-content/uploads/2007/12/oilprice19701 1.gif

Montgomery, D. (2005). *Design and Analysis of Experiments*. Hoboken: John Wiley & Sons, Inc.

Samarasinghe, S. (2006). *Neural Networks for Applied Sciences and Engineering*. New York: Auerbach Publications, Taylor and Francis Group.

Shackelford, J. F. (2005). *Introduction to Materials Science for Engineers*. New Jersey: Pearson Prentice Hall.

Shakeri, A. a. (2010). Water Absorption and Thickness Swelling Behavior of Polypropylene Reinforced with Hybrid Recycled Newspaper and Glass Fiber. *Applied Composition Material* , 183-193.

Pan, Ming-Zhu, Ding-Gou Zhou, Mosto Pousmina, and S.Y. Zang. "Effect os Wheat Straw Fiber Content and Characteristics, and Coupling Agent Concentration on the Mechanical Properties of Wheat Straw Biber-Polypropylene Composites." *Journal of Applied Polymer Science*. 10 (2009): 1000-1007.

Zena Sin-Nga Ng. "Bulk Orientation of Agricultural Filler-Polypropylene Composites". Waterloo, Ontario, Canada. 2008.

Matweb Material Property Data. "Flexural Strength Testing of Plastics." Automation Creations Inc. 2010. http://www.matweb.com/reference/flexuralstrength.aspx

377552 Adhesion of Polymer/Laponite Nanocomposite Hydrogels As Determined By Jkr Instrument

Abstract:

Biocompatible nanocomposite hydrogels (NC gels) consisting of poly(acrylic acid) (PAA), poly(methacrylic acid), or poly(acrylic acid-co-N-isopropylacrylamide) and nano-sized clay (Laponite) were successfully synthesized by in-situ free radical polymerization of monomers in aqueous solutions of Laponite. The obtained NC gels were uniform and transparent. Their viscosity, storage modulus G', and loss modulus G" increased significantly upon increasing the content of Laponite and the dose of monomers. They showed tunable adhesion by changing the dose of Laponite and monomers, as well as pH as determined by 180° peel strength measurement. The maximal adhesion was shown when reaching a balance between cohesion and fluidity. A home-made JKR instrument was employed to study the surface adhesion behavior of the NC gels. The combination of peel strength, rheology, and Johnson-Kendall-Roberts (JKR) measurements offers the opportunity of insight into the mechanism of adhesion of hydrogels. The NC gels with tunable adhesion should be ideal candidates for dental adhesive, wound dressing, and tissue engineering.

367380 Blood-Cell Inspired Drug Delivery Systems for Improved Delivery of Polymeric Carriers

Abstract:

Polymeric particle drug delivery systems are widely explored to improve the biological outcome of encapsulated drugs for therapeutic effects. Unfortunately, nearly all synthetic materials, polymeric particles included, suffer from limited abilities *in vivo*. Poor vascular circulation, limited targeting and the inability to negotiate many biological barriers have prevented the overwhelming majority of polymeric particle drug delivery systems from entering the clinic. To further complicate this, the above requisites must be performed simultaneously while also limiting toxic effects to the patient. While it is theoretically possible to design such multifunctional carriers, as Mother Nature has provided us with many examples of successful carriers in the form of circulatory cells, an engineered example capable of these same functions has yet to be demonstrated. Here, to improve the delivery abilities of synthetic carriers we report blood-cell inspired drug delivery systems that are inspired by the natural delivery abilities of the main individual cellular components in blood: (i) erythrocytes (red blood cells), (ii) leukocytes (including monocytes) and (iii) thrombocytes (platelets). Two main strategies are employed to take advantage of blood-cell delivery mechanisms to improve the *in vivo* functions of synthetic carriers. The first strategy, known as "cellular hitchhiking", involves attaching synthetic materials (polymeric particles) to the surface of circulatory cells to transfer properties from cell to particle and subsequently alter the *in vivo* fate of the synthetic particles. Specific *in vivo* examples of cellular hitchhiking to be discussed include: (i) adsorption of nanoparticles to the surface of red blood cells for both prolonged circulation and lung targeting, and (ii) antibody-mediated attachment of flat disc shaped "cellular backpacks", fabricated via layer-by-layer assembly, to the surface of monocytes for specific targeting to inflamed tissues. The second strategy, the design of synthetic cells, involves incorporating key biophysical and biochemical design parameters of natural platelets into completely synthetic platelet-like nanoparticles. Synthetic platelets, fabricated via layer-by-layer assembly, mimic the shape, flexibility and complex biochemical interactions of real platelets. The design, characterization and *in vivo* hemostasis (prevention of blood loss in a tail transection model) abilities of synthetic platelets will be discussed. The use of circulatory blood cells, either: (i) directly, as in the case with cellular hitchhiking, or (ii) indirectly, as in the case with platelet inspired synthetic cells, offers a new design paradigm for nanomedicine.

388638 Optimization of Liposome-Encapsulated siRNA for Viral Gene Silencing

Abstract:

Inadequate nucleic acid delivery efficiency, caused by *in vivo* barriers such as nonspecific tissue interactions and extracellular and intracellular degradation, is the greatest obstacle to effective gene therapy. In particular, an important gene therapy tool used for specific silencing of gene expression, siRNA, is especially sensitive to extracellular degradation. Using a liposome formation method previously developed for plasmid DNA encapsulation, in which nucleic acids are first condensed into nanoparticles using a cationic polymer and then encapsulated in a lipid bilayer, siRNA was encapsulated for protective delivery in stealth liposomes. It has been found that the size and charge of siRNA nanoparticles can be controlled by changing the siRNA:polymer ratio, and plasmid DNA encapsulation can be controlled by changing lipid film content. Therefore, siRNA nanoparticles made with varying siRNA:polymer ratios were encapsulated in liposomes of varying lipid film content to optimize both siRNA yield and fraction of siRNA loaded liposomes. siRNA designed to silence the expression of the E7 gene, a human papillomavirus gene associated with tumorigenesis in HeLa cells, was delivered to HeLa cells via stealth liposome-encapsulated siRNA. In order to understand the effect of the material properties of liposome-encapsulated siRNA on transfection efficiency, liposomes made with various siRNA:polymer ratios and load fractions were delivered to cells, and mRNA silencing and cancer cell death was measured. Understanding how delivery vehicle assembly affects both vehicle properties and siRNA transfection efficiency is essential for design of effective *in vivo* gene delivery schemes.

389656 Functional Poly(α-hydroxyl acid)-Based Biomaterials for Drug and Gene Delivery

Abstract:

Biomedical applications of conventional Poly(α-hydroxyl acid)s (PAHAs) have been considerably restricted by their lack of functional groups. Therefore, recently our group has prepared a variety of PAHAs with pendent functionalities and further converted them into novel polymeric scaffolds for therapeutic delivery. Several types of well-defined polymer-drug conjugates (PDCs) with PAHA-based backbones and high drug loadings have been prepared. Among them, the PDCs with drug as monovalent pendent moieties can exhibit sustained drug release behavior, as well as enhanced therapeutic efficacy. Tertiary amine-functionalized cationic PAHAs have also been synthesized. With remarkable degradability, low cytotoxicity and the ability to adsorb genes via electrostatic interactions, these cationic PAHAs can serve as novel synthetic vectors to enable highly efficient delivery of plasmid DNA and siRNA. Well-defined PAHA-based nanoparticles and nanocapsules have been obtained by UV-induced thiol-ene cross-linking of allyl-functionalized PAHA-based precursor polymers in transparent miniemulsions. Specifically, converted from allyl/amine-functionalized PAHAs, cationic PAHA nanocapsules can not only encapsulate drug in their inner cavities but also adsorb genes on their cationic shells. These nanocapsules can effectively evade multi-drug resistance and co-deliver drug and gene into cancer cells.

374186 Lactobionic Acid Functionalized Chitosan-Graft- Beta Cyclodextrin-p53 Nanoparticles As Efficient Carriers for Targeted Gene Delivery

Abstract:

The p53 is a well-known tumor suppressor protein that induces apoptosis upon cellular stress (e.g. DNA damages) or regulates cell-cycle progression via activation of transcriptional factors such as p21. In normal cells, the p53 activities are regulated by expression of negative regulators, viz., murine double minute 2 (MDM2) and MDM4 proteins. However, previous studies demonstrated that p53 is the most frequently mutated or deleted transcriptional factor in human tumors. Therefore, restoration of the activity of p53 in cancer cells by delivering fresh and healthy p53 plasmid DNA can reactivate the cellular apoptosis pathway and increase cell sensitivity to chemotherapeutic agents.

This study aims to prepare a chitosan-based nano-carrier for the delivery of p53-plasmid DNA to liver cancer cells in order to reactivate the p53-dependent apoptosis pathway as well as to increase sensitivity of the cells to anti-cancer drugs. To this end, chitosan-graft-ß-cyclodextrin at four different ß-cyclodextrin/chitosan molar ratios were successfully prepared and characterized using Fourier transform infrared spectroscopy (FTIR), H^1NMR, elemental analysis, and X-ray diffraction. The cytoxicity of the prepared compounds was examined in three different cell lines (MCF7, C6, and HepG2), where chitosan and PEI (25 kDa) were used as negative controls. Next, adamantine- poly-ethylene-glycol 5000-disulfide bond- lactobionic acid (Ada-PEG5000-SS-LA, (a targeting ligand for HepG2 cells)) was successfully synthesized and used for functionalizing chitosan-graft-ß cyclodextrin nanoparticles encapsulating p53. Ada-PEG5000-LA functionalized nanoparticles were synthesized as control to test the effect of the S-S group on gene transfection efficiency. The nanoparticles were characterized using TEM and zeta-potential measurements. The ability of the functionalized nanoparticle to encapsulate and protect p53-pDNA was tested through gel retardation and PicoGreen assays. The in vitro transfection efficiency was carried out in HepG2 and Hela cell lines using pRL-CMV as reporter gene. The luciferase gene expression was quantified using commercial kits.

The results demonstrated that functionalized chitosan-graft-β cyclodextrin nanoparticles had no cytoxicity and exhibited higher transfection efficiency on HepG2 cells compared to chitosan.

377521 Synthesis and Characterization of Thermosensitive Biomaterials Based on Chitosan

Abstract:

Herein, we reported a novel biocompatibility and thermosensitive amphiphilic polymer based on chitosan, which was synthesized by grafting carboxyl-terminated poly (N-vinylcaprolactam) (PNVCL-COOH) onto N-phthaloylchitosan (PHCS) via a conjugation reaction between hydroxyl and carboxyl. As a biopolysaccharide extracted from crustaceans and one of the most abundant biopolymers in nature, chitosan is currently receiving a great deal of interest for medical and pharmaceutical applications due to its interesting intrinsic properties, such as no-toxic, biocompatible, biodegradable, and cost-effect. Poly(N-vinylcaprolactam) can response to the variations in temperature. It exhibits a phase separation around the lower critical solution temperature (LCST), which calls its perspectives for applications in biochemistry and medicine. In this work, the PNVCL chains were grafted onto the backbone of hydrophobic chitosan derivatives (PHCS) in order to obtain biocompatibility and thermosensitive amphiphilic polymers. FT-IR and 1H NMR techniques were utilized to confirm the molecular structure of PHCS-g-PNVCL. The LCST was analyzed by turbidity. This biocompatible polymer should be promising stimuli-responsive materials for controlled drug release.

375144 Combination Glaucoma Therapy By Vitamin E Nanobarrier Loaded Silicone-Hydrogel Contact Lenses

Abstract:

Combination Glaucoma Therapy by Vitamin E Nanobarrier Loaded Silicone-Hydrogel Contact Lenses

Glaucoma affects about 60.5 million people worldwide and causes blindness in about 8.2 million. Glaucoma cannot be cured but it can be managed by medications that reduce the intraocular pressure (IOP). Glaucoma drugs are exclusively delivered through eye drops which have a low patient compliance, particularly when more than one IOP reducing drug is required, which happens for about 50% of patients in the US. In addition to poor compliance, drug delivery via eye drops has low bioavailability and potential for side effects. To overcome all the limitations of eye drops we propose to use extended wear contact lenses for glaucoma therapy. This talk will discuss our latest efforts in managing glaucoma through simultaneous delivery of two different glaucoma drugs from contact lenses. We will show that drug delivery by contact lenses can increase bioavailability, improve drug stability and compliance, and reduce side effects. To our knowledge, this is the first study on extended delivery of multiple ophthalmic drugs from the same delivery device.

Extended wear contact lenses are loaded individually with timolol and dorzolamide to determine the partition coefficient and release profiles. The drugs are then loaded simultaneously such that the mass of drug loaded for each drug matches the therapeutic dose. The release profiles are also measured for simultaneous drug loading. Commercial contact lenses cannot provide extended delivery of drugs. To extend the release duration we loaded the lenses with vitamin E diffusion barriers, which increases the release time due to the increase in path length of the drug molecules. The therapeutic effect and safety of the lenses was also evaluated in beagle dogs.

In vitro release studies show that the release duration of the two drugs from lenses without modification is less than 3 hr. With diffusion barriers, the release duration increases to about 72 hr. The drug release profiles are diffusion-controlled in both cases, with vitamin E incorporation reducing the effective diffusivity due to the barrier-effect, which increases the tortuosity. The relative improvements in drug delivery rates are similar for both drugs, further suggesting that the mechanism for the increase in drug release durations in geometric rather than based on direct interaction of the drugs with vitamin E. Preliminary evidence suggests that the vitamin E forms thin barriers with high aspect ratio of about 50 at the interface between the silicone and the hydrogel phases in the contact lenses. Vitamin E incorporation does not have a significant impact on any critical contact lens property, except ion permeability, but even the reduced ion permeability is adequate for *in vivo* applications. There are some additional benefits of vitamin E incorporation such as improved ultraviolet blocking, reduced modulus and reduced water loss rates, both of which could improve comfort. The *in vivo* studies showed that the vitamin E loaded contact lenses are safe as there was no adverse event in the entire study with 10 dogs lasting about 10 days. *In vivo* studies show that the drug eluding lenses can significantly reduce the IOP comparable to b.i.d. delivery of eye drops, even though the lenses were loaded with only 1/6 of the drug loaded in eye drops. Lenses without vitamin E have to be replaced daily, while vitamin E loaded lenses can be worn continuously for 3 days with constant IOP reduction. The IOP reduction with all contact lens studies was more stable compared to the eye-drops and the magnitude was about twice that for the eye drop studies. The studies with vitamin E loaded lenses showed that the IOP reduction effect was maintained for more than a week after the contact lenses were removed. This was attributed to creation of depots in the eyes during the lens-wear duration with subsequent release from the depots towards the target duration.

Based on our studies, we conclude that vitamin E modified lenses are promising for extended multidrug delivery, which may greatly improve therapeutic effect and patient compliance. With this technique, extended delivery of other multidrug system or even multiple comfort enhancers is also possible.

379023 On the Dissolution Stability of Ionically Crosslinked Chitosan Nanoparticles

Abstract:

Ionically crosslinked chitosan nanoparticles are widely explored as drug delivery vehicles. They are biocompatible, mucoadhesive, facilitate transport across epithelial membranes, and can be formed through simple mixing of chitosan solutions with multivalent anions, such as tripolyphosphate (TPP) and pyrophosphate (PPi). To ensure that these particles release their payloads at target rates, it is essential to control their dissolution stability. Yet, literature on this subject remains confounded by opposing reports (with some indicating such particles to remain stable at physiological pH and ionic strength, while others reporting them to dissolve). To address this, we used isothermal titration calorimetry (ITC), dynamic light scattering (DLS) and UV-Vis spectroscopy to investigate the effect of chitosan degree of deacetylation (DD) on chitosan/TPP nanoparticle stability. This revealed the stability of chitosan/TPP particles to be very sensitive to chitosan DD (which is often not reported, and may underlie some of the conflicting findings). Particles formed using high DD-values remained intact, while those formed using lower DD-values dissolved at physiological ionic strengths (where the dissolution likely reflected the weaker chitosan/TPP binding). These experiments also showed that cumulant analysis of DLS data (which is often used for chitosan/TPP nanoparticle characterization without supporting techniques) can lead to misleading conclusions on their stability, and might be another source of conflicting results. Finally, to advance the understanding of chitosan nanoparticle dissolution, we investigated the reversibility of their ionic crosslinking (using chitosan/PPi nanoparticles as the model system). This indicated the crosslinking to be irreversible, where the particles remained partly intact even when the PPi concentration was reduced below that required for their formation. This irreversibility was attributed to the cooperativity of chitosan/PPi binding and modeled using the Bragg-Williams theory.

375787 Co-Delivery of Salinomycin and Doxorubicin to Target Both Breast Cancer and Cancer Stem Cells Via Crosslinked Multilamellar Liposomal Vesicles (cMLVs)

Abstract:

Breast cancer is the most frequently diagnosed cancer in women of all female cancer. Chemotherapy has been widely used in the treatment of various stages of breast cancer. The major challenge on breast cancer treatment is the failure of chemotherapy, which is likely due to the presence of the cancer stem cells (CSCs). There are evidences shown that "CSC-enriched side populations" which are a subpopulation of cancer cells, however, they have ability to self- renew and to differentiate into the heterogeneous lineages of cancer cells that comprise the tumor. Currently available chemotherapies can target fast growing tumor bulk, but CSCs could remain un-attacked, causing a cancer resistance, recurrence and distal metastasis later on due to stronger capability in DNA damage repair. Salinomycin is a polyether antibiotic isolated from *Streptomyces Albus*, has recently been shown to possess anti-tumor properties by selectively inhibiting cancer stem cells population. Studies have shown that salinomycin selectively targets and decreases CSCs in breast cancer cells as well as others. Further studies have demonstrated the use of salinomcin in combinatorial therapy with conventional chemotherapeutics. However, due to hydrophobicity of salinomycin, effective in vivo delivery remains a huge challenge. In Gupta et al. study, free salinomycin was administered by intraperitoneal injection in vivo with the aid of ethanol due to its poor aqueous solubility. To overcome this issue, there is a need to develop a nanoparticle delivery system for hydrophobic drug such as salinomycin in anti-cancer therapy. Besides, nanoparticle delivery system has shown significantly higher efficacy in delivering therapeutics compared to free drugs by prolonging drug circulation time, stabilizing release of drugs, reducing systemic toxicity, increasing drug penetration by EPR effect and allowing higher accumulation of drugs at the target sites. In present study, to further confirm the CSCs targeting effect of salinomycin in combinatorial therapy with doxorubicin, a novel drug delivery system, crosslinked multilamellar liposomal vesicles (cMLVs) are utilized to encapsulate salinomycin (hydrophobic) and doxorubicin (hydrophilic),enabling these drugs to be delivered simultaneously to the site of interest in a single carrier for investigating the in-depth characterization of dual- inhibitory effects on both CSCs and breast cancer cells by in vitro and in vivo breast cancer models.

Gupta PB, Onder TT, Jiang G, Tao K, Kuperwasser C, Weinberg RA, et al. Identification of selective inhibitors of cancer stem cells by high-throughputscreening. Cell 2009;138:645e59.

Joo K, Xiao L, Liu S, Liu Y, Lee C, Conti P, et al. Crosslinked multilamellar liposomes for controlled delivery of anticancer drugs. Biomaterials. 2013;34:3098-109.

Liu Y, Fang J, Kim Y, et al. Codelivery of Doxorubicin and Paclitaxel by cross-linked multilamellar liposome enables synergistic antitumor activity. Mol Pharm. 2014;11(5):1651-61.

378202 The Chemistry of the Terminal Surface Groups of PAMAM Dendrimers Determine the Microstructure of the Grafted Peg Layer

Abstract:

Poly(amido amine) (PAMAM) dendrimers are promising nanocarriers in a wide range of biomedical applications including gene and drug delivery and as imaging agents. PAMAM dendrimers are characterized by high size uniformity, low polydispersity and multifunctional surface groups that are readily modifiable. Poly (ethylene glycol) (PEG) is a versatile ligand and widely used to modify polymeric nanocarriers. PEGylation of PAMAM dendrimer nanocarriers (DNC) can increase the aqueous solubility of the nanocarriers when hydrophobic therapeutics are conjugated, it helps protect the conjugated payload, improves biocompatibility of the DNCs, can function as a linker to better present targeting ligands, and helps modulate extra- and intra-cellular transport. The microstructure of the PEG layer grafted onto PAMAM dendrimers will impact how the DNCs interact with the physiological environment, and therefore, the therapeutic efficacy of the drug molecules. Understanding the microstructure of PEGylated PAMAM dendrimers is, thus, of great relevance in the context of drug / gene delivery applications.

In this work, molecular dynamic (MD) simulations were employed to investigate the effect of the chemistry of the terminal surface groups of PAMAM on the microstructure of the PEG graft layer. Generation 3 (or 2.5) PAMAM dendrimers with amine (G3NH2), hydroxyl (G3OH) and carboxyl (G2.5COOH) surface groups were selected for this study to represent cationic ($G3NH3^+$), neutral (G3OH) and anionic ($G2.5COO^-$) dendrimers – state of the dendrimers under physiological conditions. All dendrimers were PEGylated with 8 units of PEG, with 1000Da (PEG1000). We observed that $G3NH3^+$ and $G2.5COO^-$ assume a significantly more "open" structure compared to G3OH. The chemistry of the terminal surface group was seen to greatly impact the structure of the grafted PEG layer. The PEG layer grafted onto G3NH3+ is fairly collapsed due to strong interaction with the ether groups. Conversely the COO^- groups in the anionic dendrimer tend to help expand the PEG graft layer. These qualitative results are related to quantitative properties of the microstructure such as radial distribution functions. These results are relevant as they allow us to guide the design of DNCs and their conjugates for drug delivery applications.

379064 Gene Silencing Ability of Multivalent siRNA-PAMAM Dendrimer Conjugates

Abstract:

RNA interference (RNAi) promotes silencing of gene expression in a post-transcriptional manner. RNAi is a potential therapeutic in the treatment of several medically relevant pulmonary diseases, such as lung cancer, cystic fibrosis, asthma, chronic obstructive pulmonary disease, respiratory syncytial virus, and severe acute respiratory syndrome viral infection, and there are several ongoing clinical trials related to lung and siRNA. However, progress in this area has been hindered largely by the lack of efficient carriers capable of overcoming the lung architecture, extra and intracellular barriers present in the lung tissue to the efficient delivery of siRNA to the cell cytosol, and also due to formulation challenges.

In this work we demonstrate the gene silencing ability of multivalent siRNA-poly(amidoamine) (PAMAM) dendrimer conjugates. siRNA was conjugated onto generation four, amine-terminated PAMAM dendrimers (G4NH2) through a reducible disulfide bond for facilitated cytosolic release. The conjugates had a hydrodynamic diameter of 10 nm, and high siRNA loading efficiency. Gene silencing in airway epithelial cells showed a marked dose response as a function of the conjugate concentration. An efficiency of 55% was achieved at a G4NH2 molar concentrations several thousand-fold lower than its IC50, thus suggesting the potential of the strategy. We also show that the gene silencing ability of siRNA is not affected by exposing the biomolecule to hydrofluoroalkanes propellants used in pressurized metered-dose inhalers, thus paving the way to the translational aspects of this work related to the oral inhalation delivery of siRNA. This work is relevant as it represents a novel platform for the delivery of therapeutic siRNA to the lungs, which has potential in the treatment of a large number of medically relevant diseases.

Key words: pulmonary siRNA delivery, alveolar epithelium, siRNA-dendrimer conjugates, poly(amidoamine) (PAMAM) dendrimers, pressurized-metered dose inhalers (pMDIs); gene silencing

Facile Synthesis and Biomedical Applications of Novel Dendrimers

Jianbin Tang Mingzhou Ye, Yuxin Han, and Youqing Shen*
Zhejiang University, Hangzhou, 310027, China.

Biography

Presenter: *Dr. Jianbin Tang*

2001-2006 Ph.D student at Zhejiang University; 2006-2008 Postdoctoral research associate at University of Wyoming; 2008-Present Associate Professor at Zhejiang University

Dendrimers characterized with spherical hyperbranched structures of monodisperse molecular weight have received substantial research interest due to their wide applications, especially its potential biomedical application as carriers for drug delivery[1], gene therapy[2], MRI contrast agent[3], and photodynamic therapy[4]. A variety of dendrimers such as polyamidoamine, polylysine, polyester, polyglycerol, and triazine dendrimers were synthesized and investigated for biomedical application[5]. However, the difficulty of synthesis and poor biocompatibility hampered their translation into clinical application.

In this presentation, biodegradable and biocompatible polyester dendrimers (PEDs) and polyglycerol dendrimers (PGDs) with β-cyclodextrin (β-CD) as the core were facilely synthesized for the first time. Since β-CD has 21 functional groups, we substantially reduced the synthesis steps of the dendrimers. Moreover, the large and rigid structure of β-CD made us able to isolate and purify the dendrimers by simply precipitation, avoiding of the complex purifications such as column chromatography or reverse phase column chromatography. In addition, the biocompatibility of the dendrimers was enhanced by modifying the surface of dendrimers with zwitterionic amino acid, which provided non-fouling properties against protein adhesion. These dendrimers were investigated as nanocarriers of drug and imaging probes and showed great advantages over other carriers.

References

1. Kesharwani, P.; Jain, K.; Jain, N. K., Dendrimer as nanocarrier for drug delivery. *Prog. Polym. Sci.***2014,** *39* (2), 268-307.

2. Tian, W.-d.; Ma, Y.-q., Theoretical and computational studies of dendrimers as delivery vectors. *Chem. Soc. Rev.* **2013,** *42* (2), 705-727.

3. Tang, J. B.; Sheng, Y. Q.; Hu, H. J.; Shen, Y. Q., Macromolecular MRI contrast agents: Structures, properties and applications. *Prog. Polym. Sci.* **2013,** *38* (3-4), 462-502.

4. Jang, W. D.; Nakagishi, Y.; Nishiyama, N.; Kawauchi, S.; Morimoto, Y.; Kikuchi, M.; Kataoka, K., Polyion complex micelles for photodynamic therapy: Incorporation of dendritic photosensitizer excitable at long wavelength relevant to improved tissue-penetrating property. *J. Control.Release* **2006,** *113* (1), 73-79;

5. Medina, S. H.; El-Sayed, M. E. H., Dendrimers as Carriers for Delivery of Chemotherapeutic Agents. *Chem. Rev.* **2009,** *109* (7), 3141-3157.

382349 Controlled and Targeted Drug Delivery to Tumor Cells Via Mucoadhesive Chitosan and Chlorotoxin

Abstract:

The aim of this research is to develop a localized drug delivery system consisting of non-ionic surfactant vesicles (niosomes) with chlorotoxin (CTX) embedded in a chitosan hydrogel with enhanced targeting ability for the treatment of cancerous tumors. We found the specific accumulation of chitosan on the surface of ovarian epithelial carcinoma cells known to have high expression of mucin antigen MUC1. This behavior was not observed on the normal ovarian epithelial cells. This finding indicates the capability of chitosan to target tumor cells expressing MUC1. Besides, the incorporation of CTX (a 36-amino acid peptide capable of specific binding to tumor cells of neuroectodermal origin but not to normal cells) along with niosomes into the chitosan hydrogel has been used as the second targeting strategy to further improve the specific delivery of drugs to tumor cells such as glioma.

Attenuated Total Reflectance-Fourier Transform Infra-Red (ATR-FTIR) spectroscopy was used to investigate the possible molecular interaction between the chitosan hydrogel and various cell lines. Cell based Enzyme Linked Immunosorbent Assay (Cell-ELISA) was used for quantitative analysis of the expression level of MUC1 in different cell lines in order to understand the mechanism involved in interactions between chitosan and specific tumor cells. To examine the effect of CTX on the release rates, *in vitro*release studies were performed. Transmission Electron Microscopy (TEM) was used to assess the morphology of CTX and nisomes as well as chitosan hydrogel embedded with niosomes and CTX.

Preliminary results of Cell-ELISA revealed approximately 1.5 fold higher expression level of MUC1 in ovarian epithelial carcinoma cell line (OV2008) when compared to normal ovarian epithelial cell line (MCC3) implicating higher affinity of chitosan for OV2008 than MCC3 cell lines. This finding is consistent with ATR-FTIR results as well as the previously observed higher chitosan accumulation on the surface of OV2008 compared to MCC3, confirming the mucoadhesive property of chitosan and indicating its specificity in targeting MUC1 overexpressing tumor cells. *In vitro*release studies indicate that embedding CTX along with the niosomes does not disturb the controlled release from the chitosan network. In fact, attachment of CTX to the surface of niosomes shown by TEM imaging improves the stability of niosomes resulting in extended release rates.

Based on these findings, it is anticipated that the drug delivery system described here would improve the uptake of drugs into epithelial and neuroectodermal origin tumor cells due to the enhanced tumor targeting and its controlled and sustained drug release.

383665 Systemic and Local Biodistribution of Poly(amidoamine) Dendrimer and Highly Pegylated Poly(amidoamine) Dendrimer Via Pulmonary Delivery

Abstract:

Many pulmonary diseases including lung infections, cystic fibrosis, chronic obstructive pulmonary disease and various lung cancers are stubborn and even life-threatened, leaving patients in severe physical and spiritual pain. Many drug-loaded nanocarriers (NCs) against the diseases via intravenous or oral administration result in high systemic concentration but a relatively low amount of drug locally. In contrast, lung is a potentially ideal target for delivering the drugs and NCs directly due to the capability to avoid first-pass metabolism, enable rapid therapeutic action, high local drug concentration and minimization of systemic adsorption allowing for decreased side effects. Thus, elucidating the systemic and local distribution of NCs via pulmonary delivery is fundamental and required research before real application.

In this work, the systemic and local pulmonary distribution of poly(amidoamine) dendrimer nanocarriers (DNCs) was investigated in detail. 3rd generation DNCs and highly PEGylated DNCs were prepared and labeled by fluorescent probe Cy3. The DNCs were delivered into the lungs via both pharyngeal aspiration (PA) and tail intravenous injection (IV). Pharmacokinetics result showed that the plasma concentration of the DNCs via IV in systemic circulation was much higher than that via PA route within 6.5 hours, and PEGylation conferred DNCs with significantly higher plasma concentration and longer blood circulation residence in both administration routes. Ex-vivo tissue imaging and DNCs extraction from tissues showed liver, kidneys, and spleen were major accumulating site of the DNCs in IV route, while the DNCs given via PA were mainly found in lungs and lymph nodes. In addition, cell tagging in conjunction with flow cytometry technique indicated the DNCs were taken up by all four tagged pulmonary cell populations: myeloid, epithelial, endothelial and ciliated cells. Furthermore, PEGyaltion help the DNCs escape from myeloid cell trapping to some extent. Therefore, the findings that DNCs can accumulate in lungs and lymph nodes may imply pulmonary delivery is a potentially ideal polymeric nanocarrier administration route for treating lung and lymphatic diseases.

Key words: Biodistribution; Pulmonary delivery; PAMAM dendrimer; Cell tagging; Lymph nodes

384939 A Versatile Peptide – Polysaccharide Platform for Nucleic Acids Delivery

Abstract:

Gene therapy has drawn significant interest in the past two decades since it provides promising strategy to treat both genetic disorders and acquired diseases. However, the transfer of gene therapy to clinical applications is troubled with many difficulties, since many current systems are of toxicity, low transfection efficiency and low biodegradability. To address these challenges, we developed a polysaccharide-peptide hybrid system as a safe and efficient vector for plasmid DNA and microRNA delivery, and investigated the structure-function-cytotoxicity relationship of this polysaccharide-peptide hybrid system. Our results indicate that this polysaccharide-peptide hybrid system is a very promising platform for safe and efficient delivery of a broad range of nucleic acids.

Silica/biopolymer Nanocomposites for Enhanced Oral Drug Delivery

Xin Fan; Young Suk Choi; Allan E. David

Department of Chemical Engineering, Auburn University

Oral delivery is the most desirable route for drug administration because of its convenience and improved patient compliance. Delivery of drugs, particularly macromolecular drugs, through the oral route, however, faces several problems. Poor stability, solubility and permeability of these drugs often limit their bioavailability to sub-therapeutic concentrations. Nanocomposites have attracted increasing attention due to their ability to protect drugs from degradation in the gastrointestinal tract and to improve penetration across the mucus barriers. Utilization of the mucoadhesive properties of biopolymers or the "stealth" properties of poly (ethylene glycol) (PEG) has been shown to further improve oral drug delivery. It is possible that a balance between the stealth and mucoadhesive properties of the nanocomposites would enhance oral drug bioavailability. These nanocomposties can be produced by water in oil microemulsion method, which is isotropic, thermodynamically stable and suitable to assist the sol–gel procedure to get silica based nanocomposites. Surface properties, particle size, and composition on the mucoadhesive properties of nanocomposite materials were studied to elucidate the nature of the interaction between nanocomposite surfaces and biological membranes, and to use the understanding gained to optimize the oral delivery of drugs.

Key words: Oral drug delivery; Nanocomposite; Mucoadhesion

Self- Assembled Lipid Grafted Aminoglycoside Polymeric Micelles for Simultaneous Co-Delivery of Chemotherapeutic Drugs and DNA: Synthesis and Characterization In Vitro

Bhavani Miryala[a], Matthew D. Christensen[a] and Kaushal Rege[a]

Chemical Engineering, Arizona State University, Tempe, AZ 85287.

Abstract

A library of lipid grafted aminoglycoside polymers was successfully synthesized, and their application for co-delivery of chemotherapeutic drugs and DNA simultaneously was investigated. The lipid-polymer conjugates were characterized using ^{1}H NMR and FTIR spectroscopy. The lipopolymers could self -assemble into micelles with positive charges. The size and zeta potential of the micelles were in the range of 20-170 nm and 0.2 to 50 mV. In vitro transgene expression evaluation in multiple cancer cells indicated that the self-assembled micelles could serve as potential gene delivery vectors. The effect of lipid sub substitution on the transfection efficiencies of the polymers is studied in multiple cancer cells. Hydrophobic drug entrapment (Doxorubicin and Nile red) and co-delivery with the gene was also studied. The self-assembled lipopolymeric micelles could entrap hydrophobic drugs efficiently, while maintaining high gene transfection efficiency.

380387 Hydrotalcite Film Fabrication By Secondary Crystal Growth and Analysis of Crystal Growth Kinetics

Abstract:

Hydrotalcite films have attracted considerable attention due to their unique properties and industrial application such as catalysts, anti-coating materials, electrodes, sensors, and membranes. Hydrotalcite films are usually prepared by two conventional methods: physical deposition or in-situ synthesis. In the physical deposition method, film is formed by deposition of pre-synthesized seeds using solvent evaporation, spin coating, layer-by-layer assembly or Langmuir-Blodgett method and the properties of film can be controlled in the hydrotalcite synthesis process. On the other hand, an in-situ synthesis can be applied by directly synthesizing hydrotalcite film on the substrate and this method enhances the mechanical stability of the film. However, these conventional methods have difficulties due to the weak interaction between particles and limited synthetic conditions. In this study, to overcome these drawbacks, we fabricated hydrotalcite film by means of secondary crystal growth for the first time. First, hydrotalcite seeds were synthesized and deposited on the alumina substrate, and then hydrotalcite film was fabricated by hydrothermal treatment in an autoclave for the growth of hydrotalcite. In the secondary crystal growth, the morphology change of hydrotalcite film was observed by scanning electron microscopy (SEM) and the structural change of hydrotalcite film was confirmed by X-ray diffraction (XRD). For the kinetic analysis, particle size and size distribution were measured from SEM images with time and different kinetic models were fitted to the experimental data to find the best model for describing secondary crystal growth process.

Synthesis of mesoporous materials by ion exchange resin

Santamaría. E, Maestro.A, Gutiérrez, J.M* and González. C.
Chemical Engineering Department, Faculty of Chemistry, University of Barcelona, Martí i Franqués 1-11, Barcelona 08028, Spain
**e-mail: josemaria.gutierrez@ub.edu*

Structured mesoporous silica was successfully synthesized using ion exchange resin as proton source to avoid the use of HCl during the synthesis and preventing the ions being freed in the media. The material was obtained from sodium silicate solution, which acts as a silica source and poly(ethylene oxide)-poly(propylene oxide)- poly(ethylene oxide) ($EO_{19}PO_{39}EO_{19}$ denoted as P84) was used as structure-directing agent. The influence of two composition variables was studied: the ratio of silica used and the ratio of surfactant. The obtained materials were characterized through, SEM, TEM, SAXS and nitrogen adsorption-desorption isotherms. The use of sodium silicate solution as silica source instead of TEOS or TMOS and the possibility of obtaining a material through an ion exchange resin is important from and application point of view due to relative cheap raw materials and equipments, and the proved technology of resin ionic exchange.

In order to approach the industrial viability, mesoporous materials were synthesized in an ion exchange resin column simulating the industrial process. The same resin was used for all the experiments after washing with NaOH solution, regeneration with HCl solution, and finally washing with water

Height of exchange zone in the column was determined for different concentrations of silicate and different flow rates following the pH of outlet solution. No precipitation of silica was observed in the column.

The effect of the surfactant concentration and the sodium silicate solution were studied. In figure 1, an example of mesoporous silica as seen by TEM and SEM is shown. It can be observed that the material presents meso-structure at the nanometric scale and is formed by aggregates at the micrometric scale.

Figure 1. TEM image (left) and SEM image (right) of the obtained mesoporous materials.

Keywords: Silica, ion exchange resin, mesostructured material.

Financial support from CYCYT CTQ2011-29336-C03-02 is gratefully acknowledged

383363 Thermodynamic Measurements of Cation Exchange in Chalcogenide Nanocrystals

Abstract:

A variety of methods have been developed to synthesize nanomaterials of various sizes and composition. Often these methods rely on a hot injection synthesis scheme, but these are generally developed on a case-by-case basis and not appropriate for all compositions of matter. Cation exchange of nanocrystals is a useful method for the modification of well-defined template materials into a number of derivatives with controlled composition and crystallinity that are otherwise unavailable via traditional synthetic methods. While a general thermodynamic understanding of the processes that occur in cation exchange is established, very little quantitative data of the thermodynamics of cation exchange has been published.

Recently, we have applied isothermal titration calorimetry (ITC) to study the thermodynamics of CdSe cation exchange. These ITC experiments involve the incremental titration of a cation source into a cell containing CdSe nanocrystals and measurement of the heat evolved. With incremental titration, the enthalpy and the equilibrium constant for a reaction are simultaneously measured. We have applied this method to measure how nanocrystal size, cation source (Ag, Cu), and temperature impact the thermodynamics of the reaction. We will summarize the similarities and differences between the thermodynamics of cation exchange in nanoscale and bulk materials. The fundamental thermodynamic data measured in this work will enable the development of more informative synthesis-structure relationships in cation-exchange.

384781 Understanding the Fundamentals of the "Hercynite Cycle" and It Operational Behavior Under Pseudo-Isothermal Water Splitting Conditions

Abstract:

economically, efficiently and cleanly. Two-step, metal oxide based STWS cycles generate H_2 by a metal oxide undergoing sequential high temperature reduction and water re-oxidation. The "hercynite cycle", which is based on iron reduction in a spinel structure, relies on intimate contact between $CoFe_2O_4$ and Al_2O_3 phases though the exact mechanism was not well understood. In this work we study both the mechanism of the "hercynite cycle" and its operation under pseudo-isothermal conditions. Through a combination of computational and experimental studies using periodic boundary condition density functional theory, high temperature XRD and EDS we have determined that the hercynite cycle operates via an O-vacancy mechanism rather than the displacement reaction mechanism as previously suggested. When operating under PITWS conditions we were able to produce H_2 in excess of 200 μmol/g, which enables the Hercynite cycle to maintain the high production rates of isothermal water splitting but with higher production rates.

Innovative application of plasma spray process: Manufacture of economic and scalable ceramic microfiltration membranes

Girish Ramakrishnan[a], Gopal Dwivedi[b], Qiyuan Wu[a], Jiajie Cen[a], Yue Zhao[a] Sanjay Sampath[b], Alexander Orlov[a]

[a] Department of Material Science and Engineering, Stony Brook University, NY
[b] Center of Thermal Spray Research, Stony Brook University, NY

We investigate the potential use of thermal sprayed Alumina membranes for water filtration. Membranes were produced by plasma and flame spray processes using both (α and γ alumina) sources. To further optimize the membranes for microfiltration applications we varied the spray parameters including spray distance, plasma power, thickness etc. Chemical and morphological characterizations of membrane properties were also carried out alongside to understand the effect of changing these process parameters on the membrane performance. In order to evaluate the filtration efficiency of the prepared materials flux, permeability and rejection rate were measured. Water flux for membranes prepared by plasma spray was on average comparable to that of the commercially available inorganic membranes with the advantage of being produced by highly scalable method of manufacturing at a fraction of the cost.

388004 Agglomeration Study of Aluminum Oxide Microparticles By Emulsion Polymerization

Abstract:

Agglomeration processes may be classified according to the principal mechanism by which the particles formingthe required product are joined. The more stable bonding forces are obtained by sintering, but these joins may also be formed by other processes, if binders harden or crystallize.

Some applications of alumina particles agglomeration with particle size control are adsorption of Cd, Cu and Asto reduce the environmental impact in some countries; porosity increased with application in filters, electrical sensors, catalysts and supplies and using abrasives for mechanical polishing in microelectronics parts and generalpolishing industry.

Currently most existing agglomeration methods focus on the use of antisolvents, sintering and fluidized beds,and other techniques that require expensive equipment at high pressures and temperatures.

This research has an important technological interest as there are few vendors offering agglomerated aluminumoxide. The proposed technique is emulsion polymerization which aims to minimize the above problems.

Polymerization is carried out in the presence of a monomer (epoxy) plus two catalysts (2-methyl imidazole and quaternary ammonium salt). A solution of surfactant is used to control the polymerization and particle size.

The research involves the study of the behavior of pure compounds by Differential Scanning Calorimetry (DSC); determining the polymerizing temperature by DSC at different concentrations (5%, 7.5%, 10%) with the twocatalysts employed. In addition the polymerized microdrops with both catalysts are shown and agglomerated particles of aluminum oxide and the methodology used to obtain them.

Resorcinol formaldehyde xerogel derived carbon nanoparticles as anode material for lithium ion

M. Kakunuri [a], S. Vennamalla [a] and Chandra S. Sharma [a*]

[a] Department of Chemical Engineering, Indian Institute of Technology, Hyderabad, Telangana., India.

Preparation of resorcinol formaldehyde organic gel

- The first stage includes preparation of sol mixture followed by gelation and curing.
- In second stage these wet gels will be dried either supercritically (in presence of carbon dioxide) or subcritically (in presence of atmosphere or nitrogen) to get aerogel or xerogel, respectively.
- Polycondensation of resorcinol and formaldehyde results in different type of gels with varying pore size and morphology depends on reactant concentrations, pH of solution, curing and drying.

1

Experimental Details

Materials used: Resorcinol (R), Formaldehyde (F), DI water and K_2CO_3 as catalyst (C)

Molar ratios 1R: 2F: 27.03W: 0.04C

Hydrogel prepared using sol-gel technique

Inverse emulsion polymerization

Water Phase : RF sol (0.5 ml)
Oil phase: Cyclohexane (25 ml)
Amphiphilic surface active agent: Span 80 (12.5 ml)

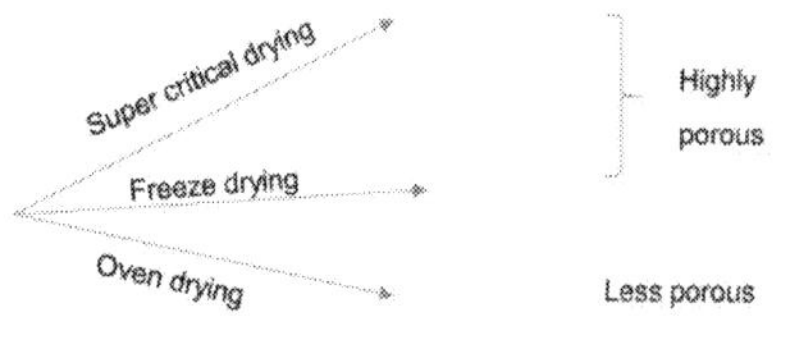

Substrate:

Stainless steel (SS) foil (polycrystalline SS 316, 100 μm thick) , MTI corp., USA.

2

Synthesis of RF Xerogel nanoparticles

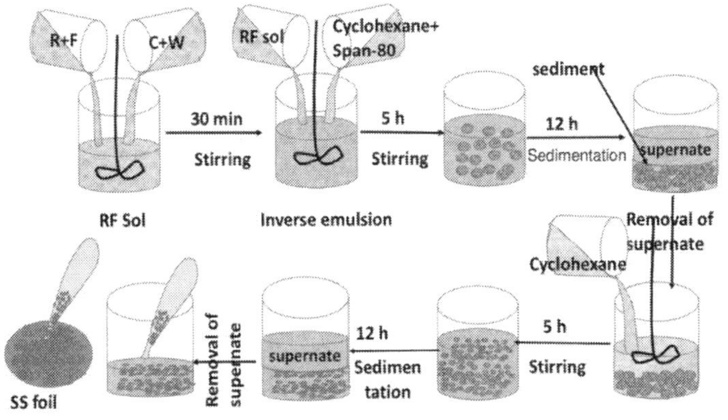

Molar ratio
1R: 2F: 27.03W: 0.04C

Sub-Critical Drying:
60 °C for 12 h (Oven)

Pyrolysis:
900 °C (In inert atmosphere)

Fig: Schematic showing the various steps of synthesis of RF xerogel nanoparticles using repetitive inverse emulsion polymerization

3

Morphology of RF xerogel derived carbon nanoparticles

Fig: Low and high magnification FESEM images of carbon xerogel nanoparticles network (a-b) sample collected from sediment after 12 h sedimentation of first cycle; (c-d) sample collected after 12 h sedimentation of second cycle.

4

Structural characterization of RF xerogel derived carbon nanoparticles

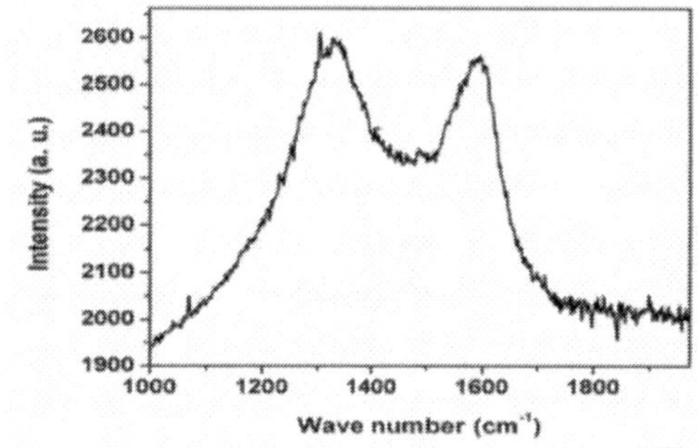

Fig: Raman spectrum

(ID/IG) =1.01,

In-plane crystallite thickness = 4.23 nm

Mean stack height : 10 A°

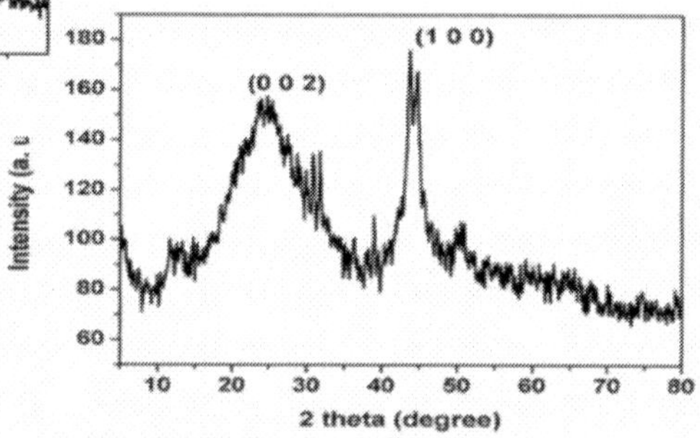

Fig: XRD spectrum

5

Electrochemical Performance of carbon xerogel nanoparticles

Counter electrode: Li foil, Electrolyte: $LiPF_6$

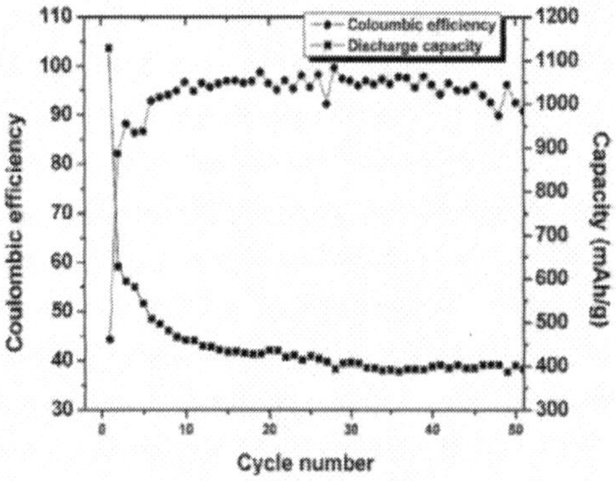

Fig: Cycling performance of carbon xerogel nanoparticles

Current density: 37.2 mA/g (0.1 C)

Galvanostatic charge discharge

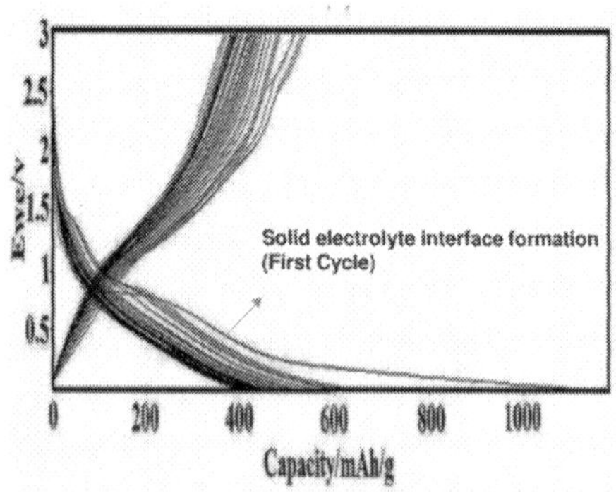

Fig: Charge-discharge profiles of RF derived carbon xerogel nanoparticles at 0.1 C rate in the voltage range of 0.01-3.0 V.

6

Summary

- We have successfully prepared carbon xerogel nanoparticles by inverse emulsification with the help of repetitive sedimentation
- XRD and RAMAN confirmed that the carbon nanoparticles are hard carbon (non-graphitic) particles.
- Reversible lithium intercalation capacity was stabilized about 400 mAh/g (> graphite intercalation capacity) after 15 cycles.

7

389075 Synthesis and Microstructural Analysis of Sr and La Doped Synroc Powders Made By Solution Combustion Synthesis

Abstract:

Perovskite ($CaTiO_3$) together with hollandite, $BaAl_2Ti_6O_{16}$, and zirconolite, $CaZrTi_2O_7$ are main phases in SYNROC (synthetic Rock). Most of the elements in HLW can be immobilized into solid solutions in these phases. This synthesis process is simpler, the synthesized synroc can immobilize waste loading up to 30wt% Sr and La with satisfied physical and chemical properties (density and porosity). The proper selection of the type and composition of the fuels are important to obtain the crystalline solid solution of Synroc. When glycine or mixtures of organic acid with stoichiometric composition were used, the solid solution of Synroc was entirely produced. The microstructure analysis by XRD and SEM/EDS shows that the major phases are perovskite, hollandite and zirconolite which well agrees with the design. It proves that combustion synthesis offer a suitable Sr and La-waste into synroc which is favorable for geological disposal.

ANALYSIS OF CEMENTICIOUS MATERIALS REINFORCED WITH FIBERS

R. A Tolosa[1, 2, 6], N.P Arias[1, 5], O.Giraldo[3, 5], C.A. Cardona[4, 6, *]

[1]Departamento de Ingeniería Eléctrica, Electrónica y Computación.
[2]Escuela de Arquitectura y Urbanismo.
[3]Departamento de Física y Química.
[4]Departamento de Ingeniería Química
[5]Laboratorio de Materiales Nanoestructurados y Funcionales.
[6]Instituto de Biotecnología y Agroindustria.
Universidad Nacional de Colombia Sede Manizales.
*Corresponding autor: ccardonaal@unal.edu.co

Nowadays, with the aim to improve the performance of the cementicious materials, several strategies are used such as add fibers to the initial mixture and so modified the mechanical properties. Fibers such as steel fiber, aramid, nanocarbon, AR-glass, and polymeric fibers are used to control the cracking. Similarly, other classes of fibers, such as biopolymeric or lignocellulosic fibers, have increased in interest in recent decades for use to reinforce concrete due to its good mechanical properties, low density, low cost, and recyclability, but efforts are required to improve the adherence of the fibers with the matrix.

When is considering the reason why the study of natural fibers takes validity, is worth noting they become valuable when it is talking about to sustainability, renewable raw materials, biodegradability, and recyclability. In the past was questioned their use due the characteristics variability, which depend on the type of plant, growth conditions, weather, and postharvest handling. Nonetheless, according to the results reported by ACI for their implementation as reinforcements in various matrixes, the composites showed relevant parameters such as tensile strength, compression, fatigue, fracture, increased modulus of elasticity, where these are noted for their contribution to the mechanical improvement. Also, the reinforcement increase the capacity against external actions (Control cracking, impact resistance, abrasion resistance, thermal expansion and contraction). For their use it has been established microstructural aspects, like the fiber orientation according to the loads direction, contributing to achieve better rigidity and strength of the composite material. Nonetheless, the lack of uniformity remains as a palpable problem; characteristics like chemical composition, crystallinity, surface properties, diameter, cross section shape, length, strength, and rigidity, determine different behaviors, affecting directly the composite reproducibility and the production of this material to industrial level, where it is seek for high levels of standardization.

Also. given the bioorganic nature of this lignocellulosic polymer, it requires a modification of its surface to improve its interfacial properties, thereby turning it into a new functional biopolymer material. In this way, in the present work fibers from Colombian wastes from *Angustifolia kunth Bamboo*, banana and plantain are modified with manganese oxide with the aim to improve the interfacial adherence with the cementicious matrix. Different physicochemical techniques such as atomic absorption, x ray diffraction, scanning electron microscopy, were used to characterize the raw and modified fibers. Also flexural strength was used in the case of *Angustifolia Kunth* bamboo to characterize the mechanical properties of the composite material. The results showed the presence of manganese oxide nanoparticles in covered fibers and it was confirmed by the SEM results (Figure 1). The crystallinity index of the fibers exhibited variations related to the chemical treatment as it is shown in the diffraction pattern for angustifolia kunth bamboo (Figure 2). The mechanical assays at 24 h of the composite materials indicated that the addition of fiber improves the ductility of the composite (Figure 3), with high values of the total energy absorption and the energy ductility index (Figure 4).

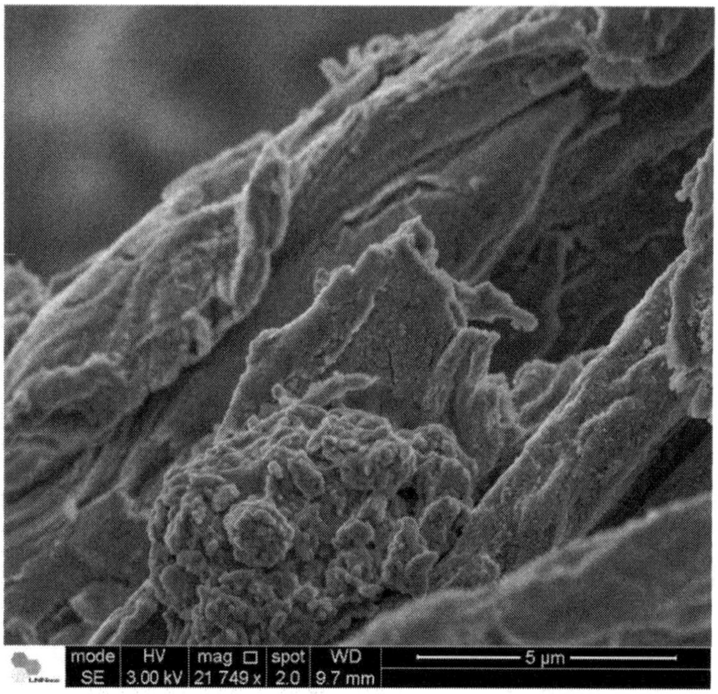

Figure 1. SEM microscopy

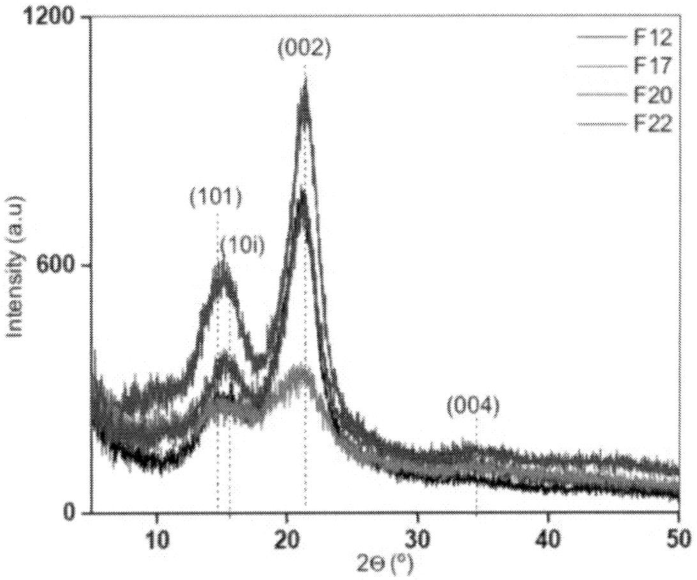

Figure 2. XRD pattern of angustifolia kunth modified with manganese oxide at different pH solution

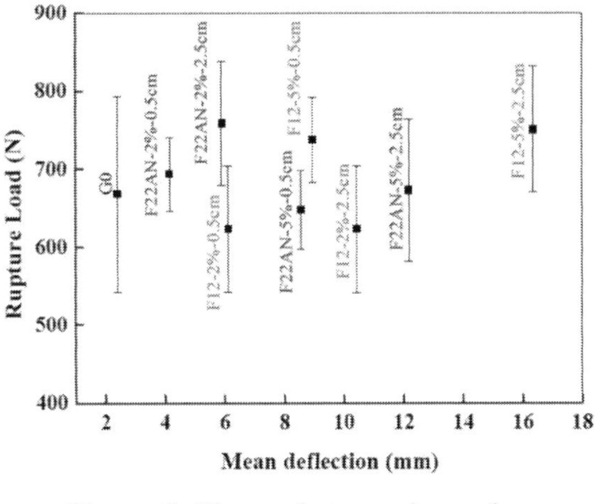

Figure 3. Flexural strength results

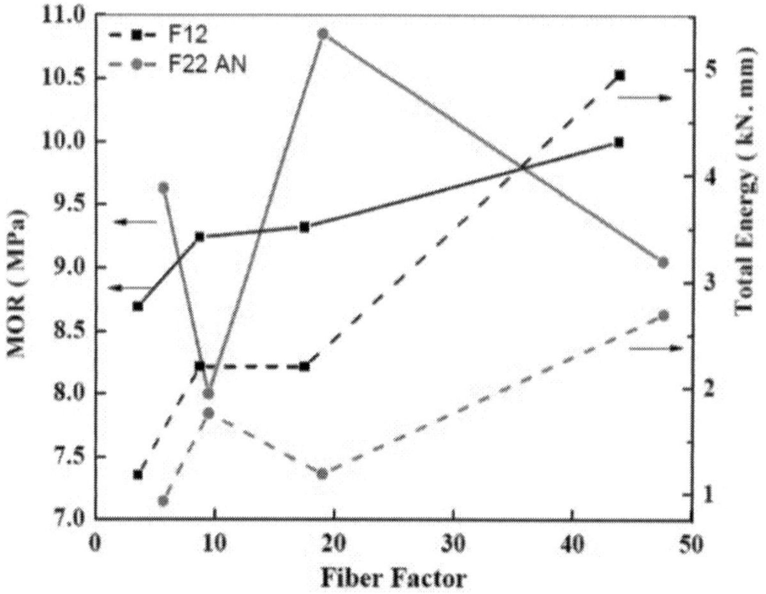

Figure 4. Module of rupture and Total energy results

390465 A Novel Computer-Aided Molecular Design Approach to Design New Non-Intuitive Chemical Admixtures for Cement

Abstract:

Cementitious materials are among the broad categories that make up the ceramic industry. Portland cement concrete is the literal foundation of our modern society used in the construction of most buildings, foundations, bridge decks, and pavements. However, the desired performance of those structures is not always attained. In order to control and alter the properties of concrete, an array of chemical admixtures are added to improve performance.

The task of searching for new admixture molecules for concrete can be a daunting, costly, and lengthy process. Most of the time, the process involves incrementally permuting an existing molecular scaffold that is known to impart a desired property. This incremental approach involves intuition and expert knowledge, more so than scientific formalisms, and limits the likelihood of discovering next-generation highly active substances. Alternatively, a computer-aided molecular design (CAMD) strategy that uses an inverse quantitative structure property relationship (I-QSPR) is proposed for the design of new chemical admixtures for concrete. The CAMD technique is a powerful tool that applies information about how well certain chemicals perform in a given application, encodes their molecular structure using what is referred to as "molecular descriptors," relates the descriptors to its performance, then uses that information to generate new chemicals with optimal predicted properties of interest. In this work, the CAMD algorithm was used to identify two different types of chemical admixtures for cement with optimal properties, (1) shrinkage-reducing admixtures (SRAs) and (2) water-reducing admixtures (WRAs).

SRAs reduce the autogenous and the drying shrinkage of cementitious material. SRAs are nonionic surfactants that are effective at reducing the surface tension of the pore solution of concrete. The I-QSPR approach was employed to identify non-intuitive compounds that were predicted by the QSPR model to be effective at reducing the surface tension of water. Five compounds were then tested for their effect at reducing autogenous and drying shrinkage of cement paste and compared to the effect of two commercially available SRAs. Results for both types of shrinkage indicate that the designed compounds perform similar to commercial admixtures, yet have different chemical functionalities.

WRAs are added to cementitious material in order to increase flowability. However, a negative side-effect of an increased flowability of the material is an increase of setting time, the time required for cementitious material to stiffen. Accordingly, the I-QSPR algorithm was employed to identify structures with high flowability and controlled set time. As a result, over 670 structures with predicted optimal flowability and setting time were identified. Six structures were selected and tested for their effect on the flowability and the setting time of cement paste. Experimental results show the power of the CAMD strategy at identifying novel and non-intuitive structures capable of increasing flowability, but at the time, have controlled setting time of the cement paste.

FIRST-PRINSIPLES STUDY OF ELECTRONIC AND OPTICAL PROPERTIES IN A NEW SEMICONDUCTING OXYTELLURIDE Ba_2OTe

A first-principles study has been performed to investigate the electronic and optical properties of a new semiconducting material, Ba_2OTe. In order to compare and get the right band-gap, we employed both the traditional PBE type GGA functional and modified Becke-Johnson (mBJ) potential with the full potential linearized augmented plane wave (FP-LAPW) method implemented in the WIEN2K code within the framework of density functional theory (DFT). Our band structure calculations show a direct band gap with 1.9 eV (PBE) and 2.6 eV (mBJ) at Γ point, respectively, which is formed by Ba-d states in the conduction band and Te-p states in the valence band. The dielectric tensors are derived within the random phase approximation (RPA). The results of the dielectric function, absorption coefficient, and reflectivity are all along both x and z directions. The optical band gap calculated (2.6 eV) agree well with the experimental value (~3 eV).

371617 Optofluidic Encapsulation of Photon Upconversion Systems Towards Solar Energy Harvesting

Abstract:

Photon upconversion (UC), the emission of light at shorter wavelength than the excitation, has the potential for overcoming the thermodynamic limits of sunlight-powered devices and processes. However, conventional UC systems such as those using two-photon absorption and second-harmonic generation require high excitation intensities and as a result may be undesirable. An attractive route to lowering the incident power density for UC lies in harnessing energy transfer through triplet-triplet annihilation (TTA).

The efficiencies of TTA-UC are determined by effective diffusion of the triplet excitons to have spatial overlap between the donor and acceptor wave functions for energy transfer. To maximize energy migration, molecular diffusivity within an inert medium is of paramount importance, especially in solid-state matrices for practical operation. Although many rubbery polymers with a low glass-transition temperature have been investigated as a potential matrix, their restricted molecular mobility resulted in much lower UC efficiency than liquid solutions.

Here, we have developed an alternative UC system composed of uniform capsules fabricated via a microfluidic emulsion approach using a photocurable resin. The capsules comprise a core-shell structure consisting of a fluidic active core which allows for highly efficient molecular interactions required for TTA-UC thus preventing the large decrease in the efficiency observed in solids, and an elastomeric shell for facile device integration with sufficient mechanical strength and air stability. We have investigated photochemical properties related to diffusive energy-transfer-driven photoluminescence in a bi-molecular UC system to utilize sub-bandgap photons of the solar spectrum.

380347 ZnO Nanostructure Optimization to Enhance High Efficiency Inverted Organic Solar Cells

Abstract:

Much scientific effort has been directed toward developing organic solar cells (OSCs) because of their significant potential to lower manufacturing costs of large area, lightweight, and highly flexible photovoltaics. Inverted OSCs have gained considerable attention as they offer increased device stability and processing advantages. ZnO serves as an effective electron extraction layer in inverted OSCs while efficiently blocking holes. Two ways to fabricate ZnO films includes using a sol-gel method or by depositing pre-synthesized nanoparticles (NPs) and the morphology and thickness of these films greatly impacts light absorption and charge collection. In this work, we used transfer matrix optical modeling and tailored film processing techniques to optimize these ZnO films for high efficiency OSCs. We compared the performance of commercially available ZnO NPs with in-house synthesized ZnO NPs and sol-gel derived ZnO films. Atomic force microscopy measurements show that controlled static or dynamic baking conditions for the sol-gel ZnO films can yield either unstructured films or nanostructured films with "nanoridges", respectively. The optical properties of these films were measured with ellipsometry and used in the optical modeling to analyze and rationally design the devices. We used a blend of [6,6]-phenyl C71-butyric acid methyl ester (PC71BM) and the low-bandgap semiconducting polymer thieno[3,4-b]thiophene/benzodithiophene (PTB7) to make the bulk heterojunction active layer and MoO3 as the top hole collecting layer with a Ag anode. When using ITO-coated glass as the cathodic substrate, we showed an optimized PCE as high as 8.4% using statically baked, sol-gel derived ZnO. The sol-gel derived ZnO was not compatible with flexible PET substrates due to the high baking temperature (150°C) causing the PET to warp. When using flexible ITO-coated PET substrates and in-house synthesized ZnO NPs, the optimized device performance reached 6.9% PCE, which is exceptional performance for flexible OSCs. The optical and electrical properties of these films will be discussed and tied to device performance measurements. These results show the importance of interfacial layer design to device performance and provide a simple way to optimize ZnO films for high performance inverted OSCs.

381948 Reducing Electron Recombination in Dye Sensitized Solar Cells: A Comparative Study

Abstract:

Establishing a compact barrier layer on the interfaces of photoanode is an effective approach to reduce the electron recombination in dye sensitized solar cells (DSSCs). However, electron recombination on transparent conductive oxide (TCO) and on TiO_2 has not been compared to identify the dominant recombination process. In this work, by depositing HfO_2 as barrier layers via atomic layer deposition (ALD) on tin-doped indium oxide (ITO) and TiO_2, the recombination processes on the two interfaces are compared. 2 cycles of ALD deposition on ITO and TiO_2 lead to their respective highest efficiency of 7.10% and 7.25%, corresponding to an enhancement of 10% and 10.4%, respectively. With more than 2 ALD cycles, the barrier layers showed blocking effect regardless of the location of deposition, resulting in decrease of efficiencies. Fill factor (FF) was seen to change with shunt resistance (R_{sh}) and series resistance (R_{se}), which in turn depend on the location of barrier layers. Depositing barrier layer on ITO did not impact R_{sh} and R_{se}, leaving FF unchanged. Depositing barrier layer on TiO_2 leads to decrease of R_{sh} and increase of R_{se}, which ultimately reduced FF.

390018 UV Treatment on Organic Solar Cell for Improving Performance

Abstract:

We reported UV irradiation treatment step during the fabrication of organic solar cells can enhance the equipment performance. Previously confirmed, the original fullerenes, such as C60 and C70, and its derivatives, such as PCBM, can undergo photochemical conversion when exposed to visible light, especially ultraviolet radiation. Fullerene monomers can be transformed into oligomers two dimers or higher oligomers. Thermal annealing has been proved to be one of the processing steps of P3HT: PCBM solar cell effectively. But under thermal annealing, PCBM molecules have been shown to diffuse within the blend to form micron-sized fullerene crystals, resulting in severe degradation of device performance. Before the device thermal annealing, ultraviolet ray treatment, found that short light treatment time have improve the device's performance approximately. UV irradiation treatment time has a best value on the performance of the device which could enhance the effect of the best.

391801 Formation of Three-Dimensional Semiconductor Metalattices Via High-Pressure Chemical Vapor Deposition in Nano-Templates

Abstract:

Three-dimensional semiconductor metalattices can be fabricated by high pressure chemical vapor deposition within nano-templates such as those provided by nano-opals. In the deposition process, a high-pressure (~30 MPa) germane/helium mixture flows through the nano-opal template at elevated temperature. Germane decomposes at the surface of silica spheres to deposit a germanium film. Because the pore structure within the template is characterized by passages whose cross-sectional areas vary substantially in the longitudinal direction, the throats or mesopores in the entry region of the template can get blocked when the time scale for reaction is small compared to that for transport of the precursor. This can lead to limited deposition in the larger pores by cutting off the precursor supply when diffusion is the dominant mechanism for precursor transport within the pores, thereby ruling out complete filling of the nano-template. We model the opal structure as an ideal face-centered-cubic packing of silica spheres, and perform numerical simulations of germane transport and decomposition within the model structure. Our results show that complete infiltration of nano-opals can be achieved at high pressures by establishing flow of the precursor within the opal to supply precursor to the larger pores by convection.

SPINEL DECORATED ALIGNED CARBON NANOTUBE ARRAYS AS SUPERCAPACITOR ELECTRODES

ABSTRACT

Accompanying the need for green energy generation is a need for energy storage systems since most green energy sources like solar and wind are intermittent. In recent times, much research focus has been placed on providing more durable and higher capacity energy storage devices like supercapacitors which has led to the design of better electrode materials for these supercapacitors. However, these materials, despite displaying improved capacitance, have still not been deemed fit for profitable production of capacitors on an industrial scale. Graphitic carbon nanomaterials have been tipped as superior materials for making supercapacitors because of their unique electrical properties and pore structure. We report the use of aligned carbon nanotubes as supercapacitor electrodes and highlight their superiority, as a result of a better ordered pore structure, to entangled or randomly oriented carbon nanotubes for electrode applications. Also, the enhanced capacitance of these electrodes after electrochemical decoration with transition metal oxides, specifically spinels, is studied and reported.

385495 Electrosprayed MnO2 Electrodes for Supercapacitors

Abstract:

The electric double-layer capacitors, also referred to as supercapacitors, represent a group of emerging energy storage technology with the properties filling the gap between rechargeable batteries and classical capacitors. The charge separation in the electric double-layer on the interface of porous electrodes soaked with electrolyte results in the fast energy storage/release (specific power > 10^3 W/kg) and extended capacitance of up to hundreds of F/g of electrode material. The specific capacitance of supercapacitors can be further extended by the pseudo-capacitive contribution of faradaic processes taking place on the electrode-electrolyte interface or in the surface layer of the electrode[1]. Resulting properties of such systems are beneficially employed in wide range application, e.g., recuperation of breaking energy and delivery of starting energy for electric vehicles.

In our contribution we prepared porous MnO_2 electrodes for supercapacitors using electrospraying of aqueous solution of Mn-based precursor. Manganese was selected for its low costs, environmentally friendly nature and pseudo-capacitive behavior. The electrospraying method enables the direct and homogeneous deposition manganese oxide particles of uniform size in the range of 10-100 nm. Within the study, the electrospraying procedure was optimized with respect to the composition of precursor solution and its flow-rate, operating voltage of electrospray, inter-electrode distance and amount of deposited matter. The post-treatment of the electrode by potentiodynamic cycling in three-electrode arrangement was used to achieve the required MnO_2 layer of porous structure. Subsequently, the electrodes were characterized in a laboratory supercapacitors cell using aqueous solution of Na_2SO_4 as electrolyte. Both symmetric and asymmetric configurations of supercapacitor were tested, the latter one using carbon-based counter electrode. The specific capacitance, resistance and cycle stability of the prepared super-capacitors were measured.

Conway, B.E., V. Birss, and J. Wojtowicz, The role and utilization of pseudocapacitance for energy storage by supercapacitors. Journal of Power Sources 1997, 66 (1–2), p. 1-14.

380914 Fabrication of Graphene Quantum Dots (GQDs)

Abstract:

The unique structure of monolayered graphene was composed of a one-atom thick two-dimensional crystal of sp^2 carbon atoms arranged in a honeycomb lattice had been attributed to its extraordinary electronic properties such as high intrinsic mobility, and excellent thermal and electrical conductivity. Those characteristics of graphene had led to extensive applications in the fields of electronic devices, including photovoltaic cells, supercapacitors and flexible touch screens. In addition to the superlative electronic properties, recent studies have demonstrated the graphene's photoluminescent (PL) characteristics, which could expand its application area in optical related fields. The PL property of graphene was derived by controlling the zero-band gap of graphene. Since the bandgap could be tuned by the size, shape, and fraction of the sp2 domains in the sp3 matrix, a variety of graphene derivatives were explored as PL graphene moieties.

We demonstrated one step way to fabricate two different sizes of graphene quantum dots (GQDs) through modified Hummers oxidation from graphene oxide, which had many advantages of simple, low cost, in larger scale with a high yield comparing to the previously reported methods. Several instruments were employed to characterize the composition and morphology of resultants. Bright blue luminescent from GQDs were obtained with a high yield of 34.8%. Moreover, the size of GQDs has affected the light wavelength and proposed mechanism will be discussed in detail during the meeting. The discovery of the new PL from GQDs may expand the application of graphene-based materials to other fields such as optoelectronics and biological labeling.

Self-assembly Synthesis Co@N-C sphere/N-graphene Nanocomposites Electrocatalyst for Oxygen Reduction Reaction

Dongdong Li, Liang-xin Ding, Haihui Wang*

School of Chemistry & Chemical Engineering, South China University of Technology,

Guangzhou 510640, China

* hhwang@scut.edu.cn

Catalyst for oxygen reduction reaction (ORR) is one of the key technologies in developing alkaline-based fuel cells and metal-air batteries. At present, developing a simple and cost-effective protocol for the production of non-precious-metal electrocatalysts for ORR is still a challenge. Herein, we reported a facile self-assembly method synthesis of Co@N-C sphere/N-graphene (Co@N-C/N-G) nanocomposites electrocatalyst. The synthesis involves the pyrolysis of graphene oxide, $CoCl_2$ and dicyandiamide at 900 °C for 2 h, and subsequent acide treatment of the pyrolysis product in order to dissolve the Co without protecting of carbon shell. The catalyst with the Co sphere encapsulated in N-C shell grown on N-G sheets was prepared, the N-C shell was used to protect the Co, the Co@N-C spheres on the graphene were used to prevent the stacking of graphene, and the N-graphene were used to enhance the electric conductivity of Co@N-C. In alkaline media, the nanocomposite catalyst exhibits higher catalytic activity than single materials containing only Co@N-C or N-G, and closing catalytic activity with the commercial Pt/C. In addition, the catalyst shows more superior stability and better suitability as a cathode catalyst for direct methanol than Pt/C for the ORR. The high catalytic activity and stability arises from the synergetic coupling effect between the Co@N-C sphere nanoparticles and graphene nanosheet. Furthermore, this material was synthesized through a simple procedure, which gives a great promise for large scale production.

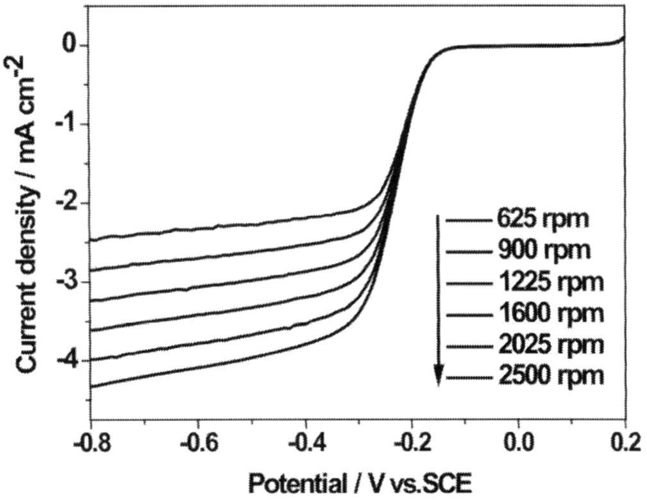

Fig.1 LSVs of the Co@N-C/N-G catalyst in an O_2-saturated 0.1 M KOH solution at various rotational rates at a scan rate of 5 mV/s.

References

[1] Zhang, S.; Zhang, H.; Liu, Q.; Chen, S., Fe–N doped carbon nanotube/graphene composite: facile synthesis and superior electrocatalytic activity. *J. Mater. Chem. A* **2013,** *1* (10), 3302-3308.

[2] Zhu, C.; Dong, S., Energetic Carbon-based Hybrids: Green and Facile Synthesis from Soy Milk and Extraordinary Electrocatalytic towards ORR. *Nanoscale* **2014,** 6, 2964-2970.

[3] Hu, Y.; Jensen, J. O.; Zhang, W.; Cleemann, L. N.; Xing, W.; Bjerrum, N. J.; Li, Q., Hollow Spheres of Iron Carbide Nanoparticles Encased in Graphitic Layers as Oxygen Reduction Catalysts. *Angew. Chem. Int. Ed.* **2014,** *53* (14), 3675-3679.

[4] Choi, C. H.; Park, S. H.; Woo, S. I., Binary and ternary doping of nitrogen, boron, and phosphorus into carbon for enhancing electrochemical oxygen reduction activity. *ACS Nano* **2012,** *6* (8), 7084-7091.

387791 Performance and Stability of PEM Fuel Cell Catalyst Synthesized Using ALD

Abstract:

Several platinum nanoparticle catalysts supported on XC72 were created using various chemistries during atomic layer deposition (ALD). The resulting platinum loading, particle size and particle dispersion were dependent on the chemistry of the deposition. The catalytic performance of each kind of particle was assessed in a PEM fuel cell. ALD catalysts were found to be comparable to commercial Pt/C electrocatalysts. Thermal stability tests determined platinum nanoparticle growth was the slowest for platinum particles made using hydrogen as a second precursor.

376456 Silicon@Carbon Nanomaterials Application in Anode for Li-Ion Batteries

Abstract:

a core–shell structured Silicon@Carbon nanocomposite was prepared by using a simply Stöber method assisted approach. Stöber method was firstly introduced to synchronously synthesize uniform Monodisperse SiO_2@Resorcinol–Formaldehyde Resin and tetraethoxysilane (TEOS), resorcinol and formaldehyde were used starting materials. And then an easier magnesiothermic reduction technique was selected to promote the development of Silicon@Carbon nanocomposite. Moreover, we can obtain Yolk-shell/Hollow-shell Silicon@Carbon nanostructures from the SiO_2@ Resorcinol–Formaldehyde Resin by modified the preparation procedure. As an anode material in Li-ion batteries, these Silicon@Carbon core-shell nanomaterials show highly stable and high capacity retention rate, which suggest that the novel hybrid materials have alluring prospect for electrochemical energy storage applications.

388352 Exploring Epitaxial Relationships Between Growth Orientations of InAs Nanowires and Au Surfaces

Abstract:

The control in growing large vertical arrays of semiconducting nanowires with specific orientations is imperative to build nanowire based nanoscale devices for commercial logic gate applications. There has been recent evidence in the literature that the metal particle influences the final structure and orientation of the as-grown nanowires through an epitaxial relationship. In our experiments on InAs nanowire (NW) growth on shaped Au nanoparticles, we have observed preliminary evidence of preferential growth of <110> nanowires on shaped Au nanoparticles compared to that on spherical nanoparticles. Since shaped Au nanoparticles have more Au (111) and Au (100) surfaces than the spherical Au nanoparticles, our target is to explore whether an epitaxial relationship can be established between the orientation of the InAs NW growth direction and the surface facets of the gold nanoparticle on which the NW grows.

In this work, we have established the relative binding strengths of ZB and WZ H-terminated InAs NW fragments with particular facets of an Au nanoparticle using density functional theory (DFT) calculations. In particular, we have first established the binding strengths of (i) an H-terminated [0001], [10-10] and [11-20] oriented wurtzite (WZ) InAs NW fragments with Au(111) and Au(100) surfaces and, (ii) an H-terminated <111> , <100> and <110> oriented zinc blende (ZB) InAs NW fragments with Au(111) and Au(100) surfaces. The results of the DFT calculations show that the <111> oriented nanowire fragments for both the wurtzite and zinc blende structures are more stable than the <100> and <110> oriented nanowire fragments and particularly, within the size range of the study (< 5 nm), the wurtzite phase is more stable than the zinc blende phase. Again, the relative stability of the <111> nanowire fragment on the Au(111) surface and that of the <100> and <110> nanowire fragment on the Au(100) surface is demonstrated by calculating the binding energies, excess energies and electronic charge redistribution of various nanowire fragments on the different Au surfaces. The results of the DFT calculations provide a very exciting prospect of growing certain nanowire orientations by selectively tailoring the metal catalyst surface to have specific surface facets. This would ultimately allow for catalyst selection and design.

Ultra-high Density Nanoparticles with Perfect Ordering and Controllable Dimensions

Cheng Xu*, Luping Li, Yang Zhao, and Kirk J. Ziegler

Department of Chemical Engineering, University of Florida, 1006 Center Drive, Gainesville, FL 32611, US

*Presenting Author (Ph: 352-392-3122; Email: chengotopia@ufl.edu)

Abstract:

Fabrication of nanoparticles with perfect ordering on large-area substrates has gained interest for various applications in nanotechnology. These nanoparticles can be used for photoelectric devices, nanoplasmonics, data storage, and biological sensors as well as catalysts to control nanowire growth. Available techniques include combination of block copolymer template and either E-beam or focused ion beam lithography. However, it is difficult to control the thickness of the block copolymer template and subsequently use the template at high temperatures. In this study, we report a strategy of using anodized alumina template (AAO) to generate close-packed hexagonal patterns with tunable size and thickness and then applying the AAO template as a mask to perform pattern transfer from AAO to a foreign substrate. First, to demonstrate the transfer process, ultra-thin AAO templates were fabricated on aluminum using 2-step anodization. The dimensions of AAO templates could be controlled by the type of electrolyte, concentration of electrolyte, anodization voltage, temperature, and stirring rate. The close-packed hexagonal pattern provides maximum pore density that can reach10^{10}-10^{12} cm^{-2}. With the protection of a polystyrene thin film, AAO was transferred to a foreign substrate after dissolving away aluminum. Then the polystyrene was removed using ozone plasma, and gold nanoparticles were fabricated using e-beam evaporation or sputtering followed by annealing. As shown in Figure 1, the deposited gold nanoparticles have identical density and dimension as AAO template. These gold nanoparticles were used as catalysts for vapor-liquid-solid (VLS) ZnO nanowire growth and also for fabrication of nanoplasmonic devices. This technique provides a new route of fabricating large-scale, high-density nanoparticles with tunable sizes

Figure 1. Au nanoparticles with perfect ordering using AAO template as a mask.

Keywords: nanoparticles, vapor-liquid-solid nanowire growth, ZnO nanowires, anodized alumina, nanoplasmonics

393994 Synthesis and radio frequency oxygen-plasma treatment of graphene for electrodes of electrochemical capacitors

Abstract:

Synthesis and radio frequency oxygen-plasma treatment of graphene for electrodes of electrochemical capacitors

Chuen-Chang Lin[1,*] and Sheng-Yen Fan[1]

[1]*Department of Chemical & Materials Engineering, National Yunlin University of Science and Technology, 123 University Road Sec. 3, Douliu, Yunlin 64002, Taiwan*

[*]E-mail: linchuen@yuntech.edu.tw

1. Introduction

The capacitive properties of graphene films depended on the number of graphene layers and the graphene films with three layers possessed higher specific capacitance than those of thicker graphene films [1]. Mono-layered graphene was synthesized with a very low precursor supply, but a higher flow resulted in few layer graphene [2]. Furthermore, the large enhancement in capacitance may be attributed to the possible pseudocapacitance because of the availability of oxygen-containing functional groups [3]. In order to add oxygenated functionalities to the surface of grapheme, it positioned ~30 cm downstream from the plasma source for minimizing surface damage was modified by microwave radical generator oxygen–plasma surface treatment at different times [4]. Thus, in this research, graphene was grown on the Nickel foam using thermal chemical vapor deposition (CVD) at different volume flow rates of Ar, then RF (radio frequency) oxygen-plasma was used to modify the surface properties of graphene, such as functional groups at different power levels/times, and finally maximum capacitance at better conditions was sought.

2. Experiment

The pretreated Nickel foam substrate was heated at 1000 °C in H_2 (100 sccm) and Ar (250 sccm) for 10 min to reduce the surface oxide layer. Next, graphene was grown on the annealed Nickel foam using CVD with a gas mixture of CH_4 (6 sccm), H_2 (100 sccm), as well as different volume flow rates (250, 400, and 550 sccm) of Ar for 10 min at 1000 °C, and then cooling to ambient temperatures at a rate of 10 °C /min in Ar with the same volume flow rate as graphene grown. Finally, the graphene electrode was modified by RF plasma with 30 sccm of oxygen at different power levels (50, 75, and 100 W) and times (3, 10, and 30 min). The cyclic voltammetry was undertaken with a 6 M aqueous electrolyte (KOH) for the prepared graphene electrodes.

3. Results and Discussion

There is some contribution to the specific capacitance from foam Ni, it seems unimportant (See Fig. 1). The graphene grown with 400 sccm of Ar possessed the highest specific capacitance (See Fig. 1). The reason behind this behavior may be explained as follows. The ratio of I_{2D}/I_G (0.68) for the graphene grown with 400 sccm of Ar is about equal to the ratio of I_{2D}/I_G (0.704) for the graphene grown with 550 sccm of Ar, then they are almost the same layer numbers, which are fewer than that for the graphene grown with 250 sccm of Ar due to lower ratio of I_{2D}/I_G (0.442). The fewer the layer number, the lighter the weight, which leads to increasing the specific capacitance for the graphene grown with 400 sccm of Ar. The graphene grown with 400 sccm of Ar possessed more wrinkles than others. The wrinkles in the graphene enlarge the specific surface area, which also leads to increasing the specific capacitance for the graphene grown with 400 sccm of Ar. Furthermore, higher power leads to higher specific capacitance and longer time also leads to higher specific capacitance (See Fig. 2). The reason behind this behavior may be that the higher the power and time, the greater the percentage of the carbonyl functional group (C O) (See Fig. 3) at which fast faradic reactions take place and give rise to pseudo-capacitance, leading to higher specific capacitance.

Figure 1. The effects of the graphene grown with different volume flow rates of Ar and foam Ni on specific capacitance.

Figure 2. The effects of power and time on the specific capacitance of the graphene grown with 400 sccm of Ar.

Figure 3. The effects of power and time on the carbonyl functional group of the graphene grown with 400 sccm of Ar.

References

1) W. Chen, Z. Fan, G. Zeng, and Z. Lai, Journal of Power Sources, 225 (2013) 251-256.

2) S. Kumar, N. McEvoy, T. Lutz, G. P. Keeley, V. Nicolosi, C. P. Murray, W. J. Blau, and G. S. Duesberg, Chem. Commun., 46 (2010) 1422-1424.

3) A. K. Mishra and S. Ramaprabhu, The Journal of Physical Chemistry C, 115 (2011) 14006-14013.

4) N. McEvoy, H. Nolan, N. A. Kumar, T. Hallam, and G. S. Duesberg, Carbon, 54 (2013) 283-290.

364917 Enhancing Adsorption of Heavy Metal Ions By L-Tryptophan Functionalized Graphene Oxide

Abstract:

Heavy metal ions caused water pollution has currently attracted much attention due to their toxic effect to human health and other organisms in the environment. Adsorption technique has considered as the most adopted removing metal ion method because its cost effective, simple operation and high efficiency. Graphene oxide has been widely used as adsorbent in wastewater treatment due to super large surface area and flat structure. However, low removal efficiency is exhibited under low pH values in existing adsorbents. Introducing amine and carboxyl groups which show strong affinity for heavy metal ions, onto graphene oxide to strengthen adsorption, L-tryptophan can be a good candidate. Firstly, L-tryptophan functionalized graphene oxide was prepared and characterized by Fourier Transform Infrared Spectroscopy, X-ray diffraction, Raman Spectroscopy and transmission electron microscopy. L-tryptophan was chemically grafted on graphene oxide by the nucleophilic substitution reaction of amine groups with epoxy groups. Then, Pb(II) and Cu(II) were treated as target metal ions and L-tryptophan functionalized graphene oxide was adopted as adsorbent. The effects of some factors, including pH, contact time, ion concentration and adsorbent dosage, on adsorption of Pb(II) and Cu(II) were investigated. L-tryptophan functionalized graphene oxide sample, which has more ion interaction sites, presented high removal efficiency over 90% at pH 4 and 5 for Pb(II) and Cu(II). The adsorbent also demonstrates extremely fast Pb- and Cu-removal within 40min. Lastly, the adsorption kinetics of Pb(II) and Cu(II) onto L-tryptophan functionalized graphene oxide were examined and could be well fitted to pseudo-second-order model. The results suggest that the L-tryptophan functionalized graphene oxide was suitable for removal of Pb(II) and Cu(II) with low pH.

386701 Electroless Deposition of Silver on Graphene Nanosheets

Abstract:

Silver-coated graphene nanosheets (GN) will be prepared by using electroless deposition of solid silver onto the GNs. Silver is selected because it has higher conductivity and less expensive than other metals frequently used for coating such as gold. Electroless deposition is a low cost and efficient way of coating silver . Successive centrifugation and sonication is used to clean and separate the silver-coated GNs. The GNs will be viewed under SEM to determine if a uniform coating has been obtained. Preliminary Experiments have been performed with carbon nanotubes. The silver-coated GNs should show significant improvements in electrical and thermal conductivity over uncoated GNs and these properties will be tested. Silver coated GNs will be dispersed in polymers to enhance the electrical, thermal, and mechanical properties.

389905 Preparation and Properties of Sulfonic Reduction of Graphene-PVA Films

Abstract:

The fundamental properties of graphene are making it an attractive material for a wide variety of applications. Reduction of graphene (rGO) was composited after graphene oxide (GO) partially reduced by sodium borohydride , then reacted with the sulfanilic acid, prepared sulfonic rGO. The effects of ultrasonication pre-treatment of graphene oxide (GO) on the properties of sulfonic rGO/poly(vinyl alcohol) (PVA) composites were studied. Results show that prepared large graphene oxide layer by the improved hummers method, we can see through the transmission electron microscope (TEM) to get the sulfonic rGO diameter about 15-30 um, PVA was dissolved under heating in water, obtained an PVA aqueous solution. In the case of stirring the sulfonic rGO aqueous solution was slowly added in PVA aqueous solution. The mixed solution was coated on glass or metal sheet peeled after air drying, obtained sulfonic rGO-PVA film. We observed the feature of the sulfonic rGO-PVA film surface and cross section by scanning electron microscope (SEM). The film has good mechanical properties,this research have a profound impact for mass production of graphene.

389411 Hierarchical Porous Reduced Graphene Oxide/Polyaniline Composite By a Facile Method

Abstract:

Hydrogel of Na_2CO_3, $CaCl_2$ and graphene oxide was synthesized by the hydrothermal method using glucose as the reducing agent. The reducing process was accompanied by generation of $CaCO_3$ nanoparticles. Hierarchical porous reduced graphene oxide/Polyaniline Composite was obtained by dilute polymerization in the presence of the hydrogel. The structure and composition of the nanocomposite has been characterized by transmission electron microscopy (TEM), Atomic force microscopy (AFM), Fourier transform infrared spectroscopy (FTIR), Raman spectra, Thermo Gravimetric Analysis (TGA) and fluorescence spectra. The composite was used as an electrode material for supercapacitors exhibited a synergistic effect of carbonaceous nanomaterials and conducting polymers.

Biopolymer – Thermally Reduced Graphene Nanocomposites: Structural Characterization and Properties

Vikas Mittal, Gisha Luckachan and Ali Chaudhry
Chemical Engineering Department,
The Petroleum Institute, Abu Dhabi, UAE
vmittal@pi.ac.ae

27

PLA/PBAT/PA Graphene Composites

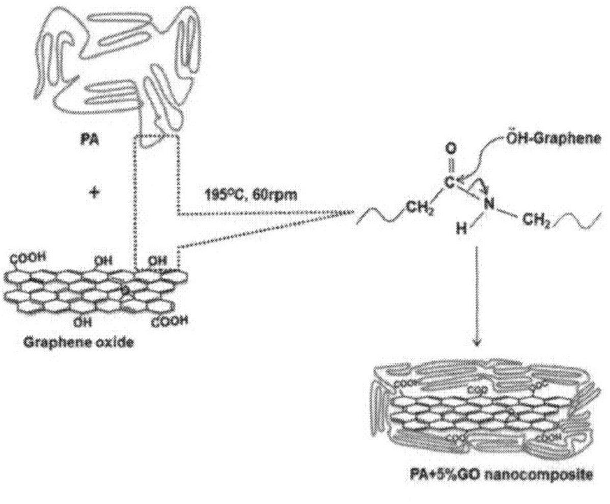

Representation of interaction of the PA chains with graphene surface

28

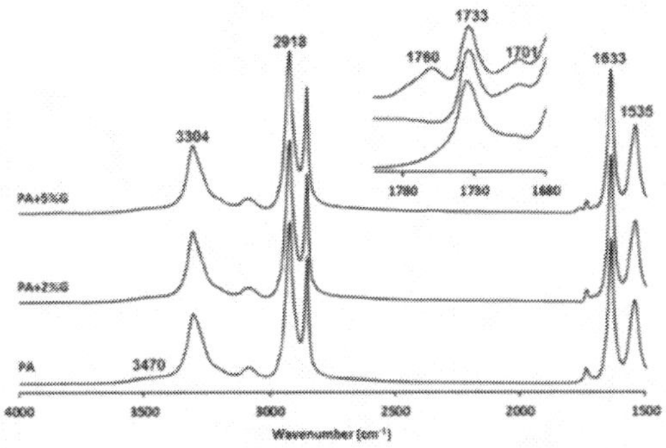

Amide bands of PA and PA+G nanocomposites

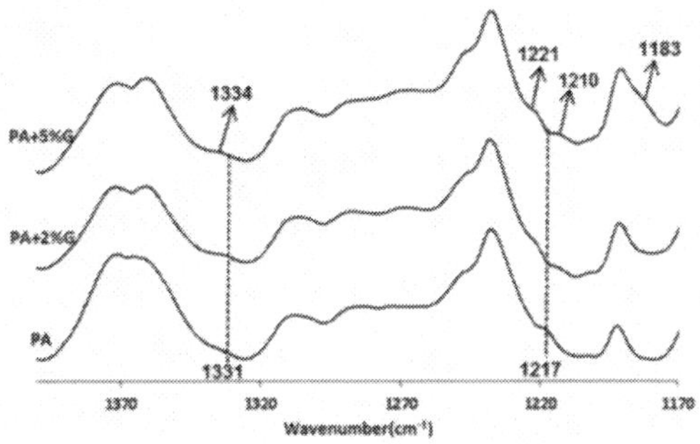

IR spectra of PA and PA+GO nanocomposites in the region of 1400 cm^{-1} to 1170 cm^{-1}

29

NMR Characterization

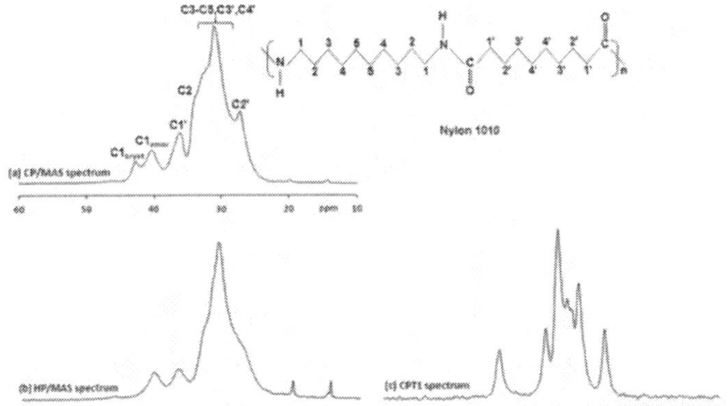

Solid state ^{13}C NMR spectra of polyamide; (a) CP/MAS spectrum, (b) HP/MAS spectrum and (c) CPT1 spectrum

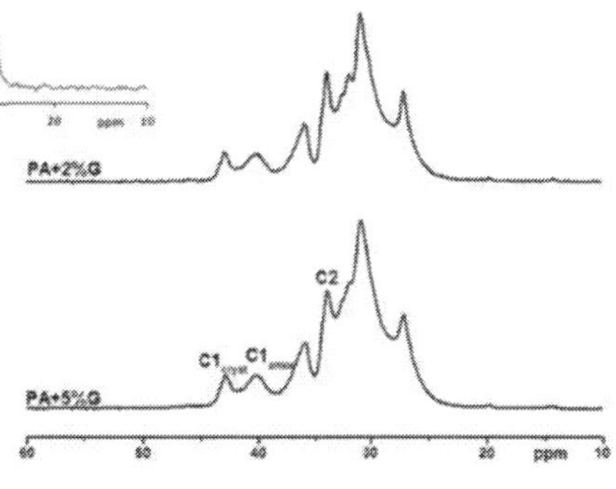

Solid state ^{13}C CP/MAS spectrum of PA nanocomposites

Morphology

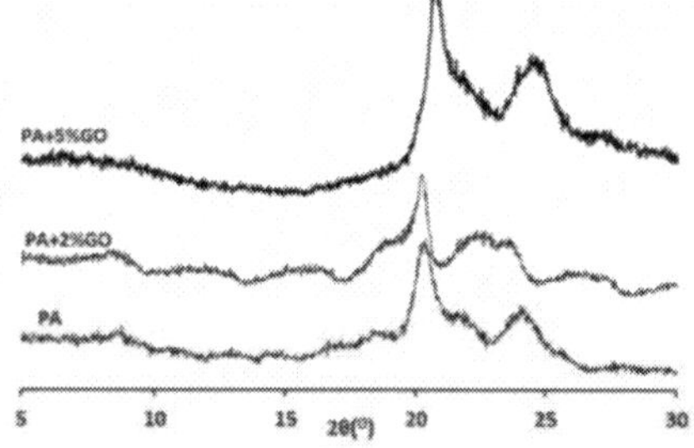

X-ray diffraction patterns of PA and PA+G nanocomposites

TEM micrographs of nanocomposites containing 5 wt% of graphene oxide; (a), (b) Ecoflex; (c),(d) PA and (e),(f) PLA

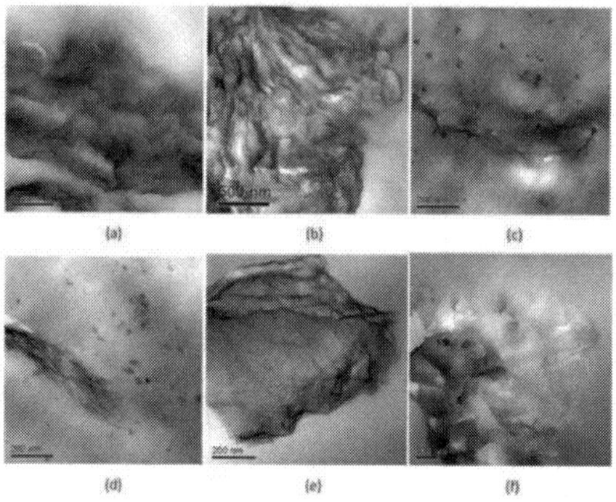

Rheological Properties

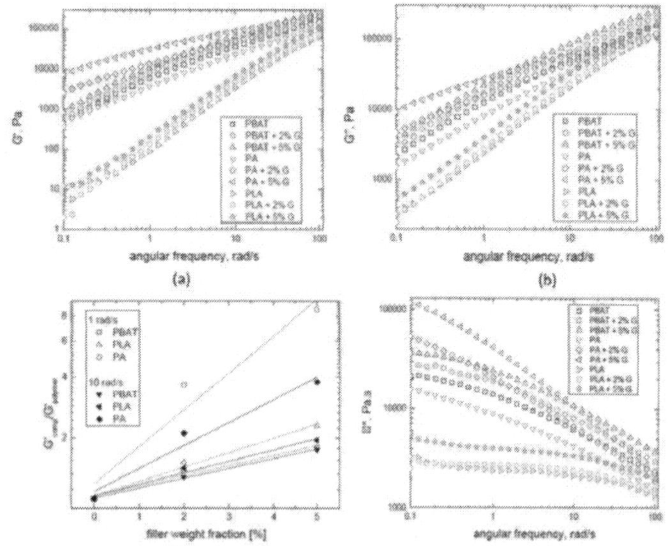

(a) Storage modulus (G') and (b) loss modulus (G'') of the pure polymers and nanocomposites as a function of angular frequency; (c) relative storage modulus as a function of filler weight fraction at 1 rad/s and 10 rad/s; (d) complex viscosity of the pure polymers and nanocomposites as a function of angular frequency

32

Mechanical Properties

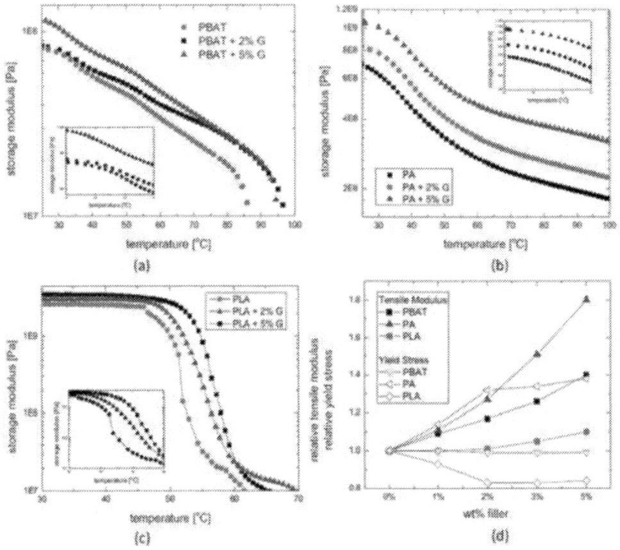

Storage modulus of pure polymers and nanocomposites analyzed from DMA (a) Ecoflex, (b) PA and (c) PLA; (d) relative tensile modulus and yield stress of the composites as a function of filler weight fraction

33

Mechanical Properties

Composite	Tensile modulus[a], MPa	Yield stress[b], MPa	Yield strain[c], %	Stress at break[d], MPa	Elongation[e], mm
Ecoflex	78	9.1	13.1	20.9	88.7
Ecoflex+2%GO	91	9.1	11.3	17.5	83.8
Ecoflex+5%GO	110	9.0	9.6	15.0	57.7
PA	990	32.5	15.3	45.1	58.0
PA+2%GO	1258	42.9	15.2	46.0	25.9
PA+5%GO	1780	44.7	14.3	50.5	23.2
PLA	3306	70.3	3.0	63.7	1.6
PLA+2%GO	3335	58.6	2.4	55.6	1.3
PLA+5%GO	3614	59.2	2.4	57.0	1.3

[a,c,e]Relative probable error 5%.
[b]Relative probable error 2%.
[d]Relative probable error 15%.

34

Thermal Properties

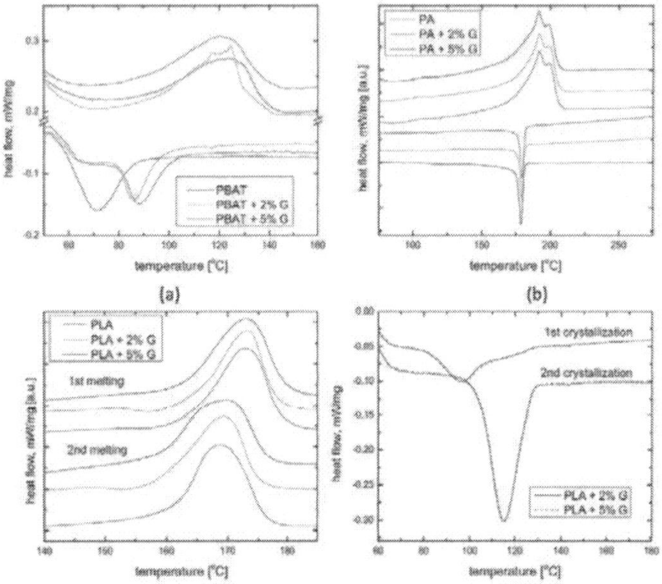

DSC melting and crystallization transitions of pure polymer and nanocomposites; (a) Ecoflex and (b) PA. Figure (c) and (d) represent respectively the melting and crystallization transitions of pure PLA and nanocomposites for first and second runs

Thermal Properties

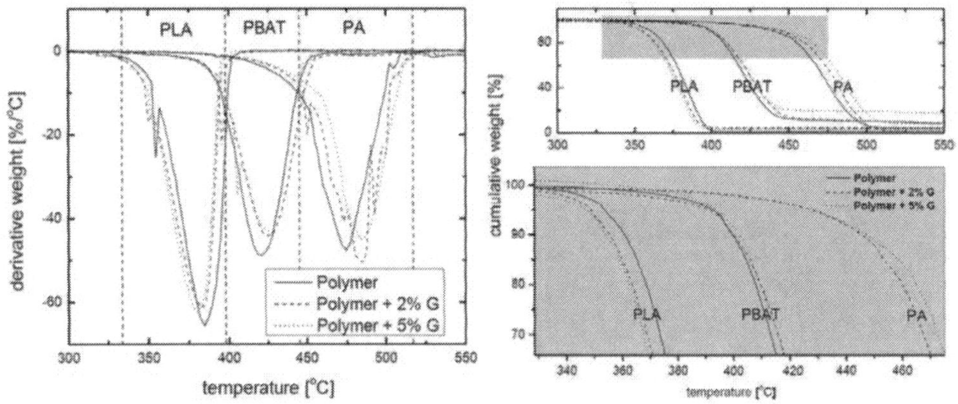

Derivative and cumulative TGA thermograms for pure polymers and nanocomposites

Thank You

37

375399 Porous Polyaniline Nanofiber/Graphene Mutilayer Electrodes for Energy Storage

Abstract:

Polyaniline (PANI), one of the most extensively studied conducting polymers, is a good candidate for electrode materials for energy storage application due to its high theoretical capacity (294 mAh/g), low cost and ease of synthesis. However, the performance of PANI electrode was not satisfactory so far. For example, its cycling stability is quite low due to gradual oxidation of PANI during cycling. In order to enhance its electrochemical performances, the development of hybrid electrode materials is urgent.

Graphene have been attracted great attention for energy storage applications due to its excellent properties such as high conductivity and high surface area. Most commonly, graphene is produced from graphene oxide by chemical reduction using toxic reducing agents. In addition, chemically reduced graphene oxide has limited processability because of strong pi-pi interactions between graphene layers. This limited processability hampers its wide application for synthesizing hybrid materials.

Here, we demonstrate porous polyaniline nanofiber (PANI-NF)/graphene multilayer electrodes using layer-by-layer (LbL) assembly for energy storage systems. In order to circumvent the limited processability of reduce graphene oxide, graphene oxide was directly used to build up PANI-NF/GO LbL films followed by electrochemical reduction. PANI-NF/electrochemically reduced graphene oxide (ERGO) LbL electrodes shows exceptionally high capacity, specific energy, and specific power. The PANI-NF/ERGO LbL electrodes also possess remarkably enhanced cycling stability as compared to PANI-NFs alone. The results suggest that our porous PANI-NF/ERGO LbL electrodes are promising electrodes for high performance energy storage systems.

389802 Fabrication of Large Area Transparent Conductive Graphene/Agnws Films Via Liquid Phase Deposition (LPD) Method

Abstract:

Graphene, a novel 2-D material, has exhibited many fantastic physical and chemical properties; AgNWs has excellent conductive properties, so both of them give rise to the great performance of transparent conductive grapheme films. we introduce a new way, which is that the GO/AgNWs solution drops on the liquid surface, dispersed evenly, slowly lower level until GO/AgNWs deposited on the substrate, then reduced by hydrazine hydrate , hydroiodic acid ,or high temperature .The films via liquid phase deposition were exhibited high conductive and good transparent. We can also easily prepared high quality grapheme films which has different sizes by the method. It is extremely attractive to industrial production, can greatly accelerate the process of industrialization of grapheme as well.

373819 Stabilization of Oxygen-Deficient Structure for Conducting $Li_4Ti_5O_{12-d}$ By Molybdenum Doping in a Reducing Atmosphere

Abstract:

$Li_4Ti_5O_{12}$(LTO) is recognized as being one of the most promising anode materials for high power Li ion batteries; however, its insulating nature is a major drawback. In recent years, a simple thermal treatment carried out in a reducing atmosphere has been shown to generate oxygen vacancies (Vo) for increasing the electronic conductivity of this material. Such structural defects, however, lead to re-oxidization over time, causing serious deterioration in anode performance. Herein, we report a unique approach to increasing the electronic conductivity with simultaneous improvement in structural stability. Doping of LTO with Mo in a reducing atmosphere resulted in extra charges at Ti sites caused by charge compensation by the homogeneously distributed Mo^{6+} ions, being delocalized over the entire lattice, with fewer oxygen vacancies (Vo) generated. Using this simple method, a marked increase in electronic conductivity was achieved, in addition to an extremely high rate capability, with no performance deterioration over time.

373831 Anomalous Decrease in Structural Disorder Due to Charge Redistribution in Cr-Doped Li4Ti5O12 Negative-Electrode Materials for High-Rate Li-Ion Batteries

Abstract:

Since one of the main drawbacks of Li4Ti5O12 as a negative-electrode material is its low electronic conductivity, several researchers have attempted to improve the conductivity by narrowing the band gap through transition-metal doping. Herein, we report another, more significant effect of doping in addition to the band gap narrowing, namely, an anomalous decrease in the structural disorder in Li4Ti5O12 upon Cr3+ion doping. Although it is generally recognized that doping with heterogeneous elements increases the structural disorder, the Cr3+ion doping in Li4Ti5O12 demonstrated an unexpected structural phenomenon. From the results of various structural analyses using a synchrotron beam, such anomalous structural changes were revealed to originate from charge redistribution at nearby Ti4+ions. Finally, the capacity was markedly enhanced, especially at high C-rates (125 mAh/g for 10Ccharge/10Cdischarge, 145 mAh/g for 1Ccharge/50Cdischarge) because of both the band gap narrowing and the increased ionic diffusivity due to the decreased structural disorder, but was decreased instead for too-high doping levels.

374286 Fabrication of Silica-Coated Paraffin Microparticles As Phase Change Materials Using Membrane Emulsification

Abstract:

Energy shortage is one of global problem. Therefore energy-storage system is more important. In this study, microencapsulated phase change materials(MEPCMs) were prepared by membrane emulsification technique using the Shirasu porous glass (SPG) membrane. MEPCMs were obtained 5~6 μm-scaled paraffin particles by adding TEOS.Particle size analyzer (PSA), scanning electron microscopy (SEM), Fourier transform infrared spectroscopy (FT-IR) and differential scanning calorimetry (DSC) were used for characterization of paraffin. In summary, MEPCMs which havehigh latent heat capacity can be used for energy storage system.

Micro-textured BORON NITRIDE NanoPLATELET MODIFIED POLYETHYLENE Films

Ozgun Ozdemir and Amod A. Ogale*

Department of Chemical Engineering, and Center for Advanced Engineering Fibers and Films (CAEFF), Clemson University, Clemson, SC 29634

Abstract

Linear low density polyethylene (LLDPE) micro-textured films filled with boron nitride nanoplatelets (BNNP) were produced by continuous melt extrusion. Nanoparticles displayed a significant extent of dispersion inside the matrix. The addition of BNNP led to more than 10-fold increase of the in-plane thermal conductivity (TC) of the nanocomposite (7.7 W/m.K *vs* 0.3 W/m.K for pure LLDPE), and 1.3-fold increase of through thickness TC. Micro-textured films were fabricated by a micro-patterned die to increase convective heat transfer rate as a consequence of increased surface area. Nanoplatelets were aligned parallel to the film machine direction. Film stiffness and tensile strength are comparable to the base LLDPE.

Processing

Twin screw extruder

SEM micrograph of lateral surface of microtextured BNN-PE film

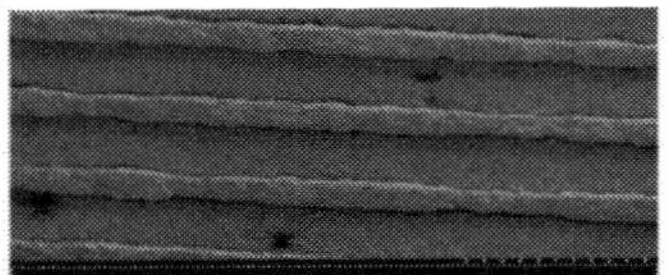

Cross-section of the micro-textured die

Thermal Conductivity

Sample	Inplane Thermal Conductivity (W/m.K)	Through thickness Thermal Conductivity (W/m.K)
Textured BNN-LLDPE	11.5±5.2	1.7±0.1
Non-textured BNN-LLDPE	7.7±1.1	0.8±0.1
Base LLDPE	isotropic	0.3±0.1

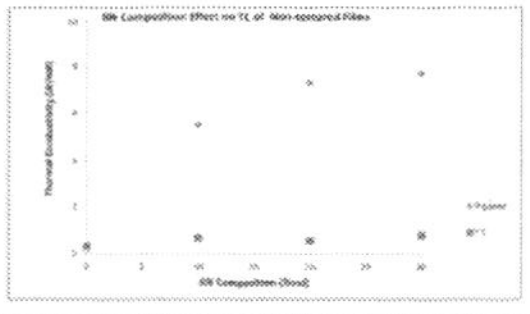

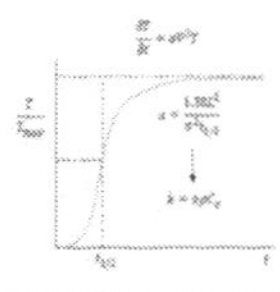

Introduction

- There is a dramatic increase in heat generated per unit area as electronic devices become miniaturized
- LLDPE has been attractive due to its light weight, ease of processability, excellent flexibility and low cost; however, it is thermally insulating
- One approach to increase the thermal conductivity of a polymer is to introduce high-thermal-conductivity fillers. Another approach for accelerating heat dissipation is to increase the surface area of the heat exchanger/material
- Boron Nitride nanoplatelet (BNN) have a high thermal conductivity and electrical resistivity, low thermal expansion coefficient, and are nontoxic, nonreactive, thermally and chemically stable.
- Thus, BN-based composite materials are of value for thermal management in microelectronics packaging.

Objectives

- Continuous melt extrusion of micro-textured LLDPE nano-composite films
- Characterization of the transport properties

Microstructure

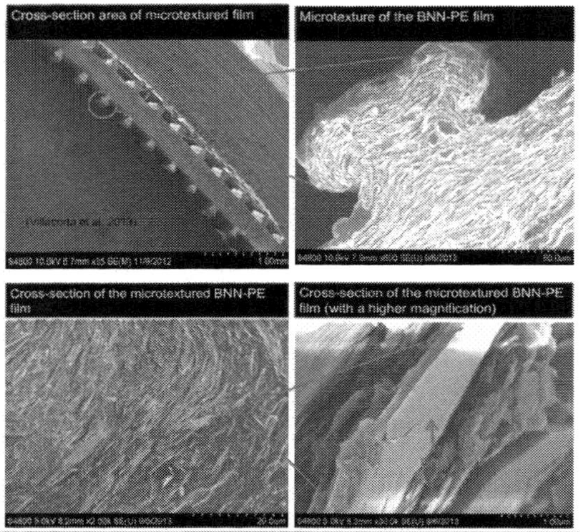

Cross-section area of microtextured film

Microtexture of the BNN-PE film

Cross-section of the microtextured BNN-PE film

Cross-section of the microtextured BNN-PE film (with a higher magnification)

Experimental

- The compounding of BNN into LLDPE matrix and melt-extrusion of composite films was conducted in a DSM micro compounder through a micro-machined trapezoidal textured die (courtesy Hoowaki, LLC) attached to DSM micro-scale twin screw extruder
- The microstructure was studied using S4800 scanning electron microscope
- Tensile properties of the films were measured on ATS 900 tensile testing machine according to ASTM D882
- 1 cm x 1 cm square samples were obtained for through-thickness thermal conductivity and 25 mm diameter disks for in-plane measurements. Laser flash analysis were conducted on samples by using Netzch LFA 447

Mechanical properties

Sample	Tensile Modulus (GPa)	Tensile Strength (MPa)	Strain to Failure (%)
Textured-BN	0.7 ± 0.2	14.7 ± 1.2	2.4 ± 0.6
Non-textured BN	1.3 ± 0. 2	23.1 ± 4.4	2.6 ± 0.2
Base PE	0.4 ± 0.0	23.4 ± 0.4	687.0 ± 21.6

- BNN addition resulted in an increase of modulus (compared to that of base LLDPE)
- Tensile strength of the nanocomposite films was comparable to those of LLDPE, but the nanocomposites displayed significantly lower ductility
- Textures led to 35% decrease of tensile strength
- BNN aligned parallel to the film direction due to shear and extensional stresses encountered during extrusion and film drawing
- In-plane TC is approximately 8 times higher than through-thickness TC, which is consistent with the in-plane orientation of BNN
- Although nanocomposite films lose ductility, their stiffness and tensile strength are comparable to the base LLDPE

Future work

- ❑ Experimental comparison of heat transfer rate between non-textured and micro-textured films
- ❑ Investigation the hybrid effect of BNN and GNP on thermal and electrical conductivity of microtextured LLDPE composite films microtextured LLDPE composite films

References

1. Villacorta B, Hulseman S, Cannon A, Hulseman S, and Ogale A. , "Continuously Extruded Micro-Textured Polypropylene Films", Polym Eng Sci 54: 2147-2154, 2013.
2. Chan E., Leung S, Khan M, Naguib H, Dawson F, Adinkrah V and Hayward L. "Fabrication and characterization of ceramic-filled thermoplastics composites with enhanced multifunctional properties"Journal of Thermoplastic Composite Material, 2012.
3. Zhang X, Shen L, Wu H, Guo S. " Enhanced thermally conductivity and mechanical properties of polyethylene/boron nitride composites through multistage stretching extrusion" Composite Science and Technology, 2013.

CLEMSON UNIVERSITY

CAEFF CENTER FOR ADVANCED ENGINEERING FIBERS AND FILMS

Self-humidifying Nafion-zeolite Composite Membrane

Viola SIM[1], Wei HAN[1], Ho Yee POON[1], Yakub FAM[1], King Lun YEUNG[1,2,*]

[1]*Department of Chemical and Biomolecular Engineering,* [2]*Division of Environment, The Hong Kong University of Science and Technology, Clear Water Bay, Kowloon, Hong Kong, P.R. China.*

[*]*Corresponding author: kekyeung@ust.hk*

Session: 08 MATERIALS ENGINEERING AND SCIENCES DIVISION

Topic: 08004 Poster Session: Materials Engineering & Sciences (08F Composite Materials)

Presentation type: Poster

Abstract

Perfluorosulfonic acid (PFSA) membrane possesses high proton conductivity and long term stability under full hydration condition. A lack of water decreases its proton conductivity sharply, and excessive water results in electrode flooding, therefore lowering cell performance. Traditional PEMFCs utilize external humidifying devices to keep PFSA membrane humid, which increases cell cost and lowers overall energy efficiency.

Our lab previously reported a self-humidifying membrane by confining Nafion into zeolite-coated micron-sized pores. The zeolite coating served as a water-retention layer to aid in the self-humidification property of the cell. The self-humidifying PEMFC was assembled by using porous stainless steel electrodes with Pt/C catalyst layers and self-humidfying membrane.

Here various zeolites have been prepared and tested in our PEMFCs, including MFIs like Sil-1 and HZSM-5, Faujisites like NaX, NaY and others like NaA. These various zeolites with different water uptake abilities are chosen to observe the effect of zeolite type on the performance. The proton conductivities of different PEMFCs are tested with different techniques including electrochemical impedance spectroscopy and ohmic polarization areas of I-V curves, and compared to the results of PEMFCs with commercial Nafion 117 and Nafion 112 membranes. The role of confinement is also investigated in the improvement of the performance of self-humidifying PEMFC.

379610 Mechanical Properties of Polyhydroxyamide/Carbon Nanotube Composite Fibers: Experiment-Computation Collaborative Approach

Abstract:

Polymer composite fibers including carbon nanotubes (CNTs) have attracted great attention from industry and academia as a new class of composite fiber materials because of its excellent mechanical and thermal properties. CNTs have also remarkable electrical conductivity, in contrast to the commercial polymeric fibers exhibiting electrical insulation behavior. The structure and property of polymer/CNTs composite fibers are highly dependent on the amount, alignment, and dispersion of CNTs, and interfacial interaction between polymer and CNTs, which are affected not only by type, geometry and functional group of CNTs, but also by processing conditions such as spinning method and post-treatment. In particular, the interaction between polymer and CNTs is considered as a critical factor to be controlled and investigated, because it strongly influences the mechanical properties of final composite fibers. Although many experimental research results provide valuable and unique information about the structure-property relationship, it still needs to understand an atomistic-level behavior for interfacial interaction between polymer and CNTs. In order to delve the influences and reinforcing mechanism of CNTs on the mechanical properties of polymer composite fibers, in this study, polyhydroxyamide (PHA), a possible precursor which could be converted to polybenzoxazole (PBO), and multi-walled carbon nanotubes (MWCNTs) composite fibers are prepared by wet-spinning and their structural features, thermal properties and mechanical performances are then investigated as a function of various MWCNTs contents. Also, molecular dynamics simulations were performed to simulate and understand the influences and reinforcing mechanism of the CNTs on the mechanical properties of the composite fibers. In experimental results, SEM images show that the MWCNTs are well dispersed in the PHA matrix without forming aggregates. As a result, the mechanical properties such as initial modulus and tensile strength of the composite fibers were substantially enhanced compared with those of neat PHA fiber. For example, initial modulus, elongation at break, and tensile strength of the composite fibers were improved more than 19.6, 15.6, and 70.1%, respectively, by incorporating up to 2.0 wt% MWCNTs. In addition, an elastic loop test showed that the compressive strength of the composite fibers were noticeably higher than that of neat PHA fiber. The improved overall performances of the composite fibers were believed to be due to the good dispersion and specific interaction between the PHA and the MWCNTs.

379851 Synthesis and Characterization of Polystyrene/ZIF-8 Core-Shell Particles By Solvothermal Method

Abstract:

Metal-organic frameworks (MOFs) have attracted much attention because of their tunable properties and various applications. MOFs are porous material consisting of metal ions (Zn, Cu, Co, etc.) or metal clusters interconnected by organic molecules (ligands), showing organic-inorganic hybrid structures. Unlike other materials, MOFs are noted to show a good framework flexibility and shrinkage/expansion due to connection with guest molecules. Zeolitic imidazolate framework-8 (ZIF-8) is a sub class of MOFs, having an open channel structure composed of Zn ion and 2-methylimidazole. In this research, we investigated solvothermal synthesis of ZIF-8 particles, polystyrene/ZIF-8 core-shell particles, and hollow ZIF-8 particles. The crystal sizes of ZIF-8 increased with solvothermal temperature from 50°C to 70°C. The polystyrene/ZIF-8 core-shell particles were synthesized by mixing polystyrene core particles and ZIF-8 precursor solution, followed by solvothermal treatment for 10 min at various temperatures (50°C, 60°C, 70°C). The ZIF-8 shell thickness increased with the temperature and the number of coating steps. To fabricate hollow ZIF-8 particles, polystyrene cores of the polystyrene/ZIF-8 core-shell particles were removed by washing them in N,N'-dimethylformamide (DMF). As synthesized ZIF-8 particles, polystyrene/ZIF-8 core-shell particle and hollow ZIF-8 particles were characterized by XRD, FE-SEM, TEM, BET, TGA, and particles size analysis. These ZIF-8 particles in core-shell and hollow forms can offer unique opportunities for gas separation, adsorption, gas storage, sensor and catalyst applications.

379909 Synthesis of Molybdenum Oxide Particles By Ultrasonic Spray Pyrolysis and Their Partial Oxidation Activity

Abstract:

Molybdenum oxides exist mainly in two forms: MoO_2 and MoO_3. MoO_3 has been used as a semiconductor, a field-emitter, an electrochromic material, photochromic material, and a gas sensor. MoO_2 exhibits metal-like electronic conductivity because of the existence of delocalized electrons in its valence band. Both MoO_3 and MoO_2 have been used as a catalyst in hydrocarbon reforming processes. In particular, MoO_2 has attracted a lot of attention due to coke resistance and sulphur tolerance during the partial oxidation of hydrocarbons. Although commercial MoO_2 has a very small BET surface area (< 10 m^2/g), some researchers reported solvothermal synthesis of MoO_2 nanoparticles with a high BET surface area of about 50 m^2/g. However, the yield of MoO_2 nanoparticles by solvothermal synthesis was less than 1 g/batch. In this study, we developed a large scale production process for molybdenum oxides by ultrasonic spray pyrolysis. The particle size and phase structure of the product molybdenum oxides were affected by precursor concentration, pyrolysis temperature, carrier gas flow rate, etc. As-synthesized MoO_2 particles were analyzed by BET, XRD, EDX, TGA, and FT-IR. Partial oxidation reaction was studied to test the catalytic performance of the as-synthesized MoO_2 using n-dodecane as a fuel which is a model compound of bio-diesel. The effect of O_2/C ratio, reaction temperature, and weight-hourly space velocity (WHSV) on the conversion and H_2 production were systematically investigated. H_2 and CO yields, and carbon conversion over as-synthesized MoO_2 were calculated and compared with those over commercial one. The as-synthesized MoO_2 particles showed promising reforming performance without any coke formation.

385138 Influence of "Free" Ionic Liquid Concentration on the Photopolymerization Kinetics of Ionic Liquid Monomers

Abstract:

In recent years, ionic liquids (ILs) have been explored as materials in a variety of applications. Although the focus of much of this research has been the use of bulk ionic liquids in electrochemical, separations, and solvent applications, polymers have also been synthesized from ionic liquids containing polymerizable functionalities. Polymerized ionic liquids (Poly(ILs)) have been explored for use in fields as diverse as catalysis, energy generation, and biological engineering. However, despite the potential applications of these materials, little research has been completed on the polymerization kinetics of poly(ILs). Previous research has demonstrated that ionic liquid solvents increase the radical polymerization rate of a variety of vinyl monomers. In our study, a series of styrene functionalized branched, cyclic, and *n*-alkyl imidazolium bis(trifluoromethylsulfonyl)imide (Tf_2N) ionic liquids were photopolymerized in the presence of the non-polymerizable ionic liquid [C_2mim][Tf_2N]. Mixtures of varying polymerizable ionic liquid:non-polymerizable ionic liquid molar ratio were polymerized and the conversion of the monomers measured using real-time Fourier transform infrared spectroscopy (FTIR). These data were used to determine the effects of non-polymerizable ionic liquid solvents on both the monomer conversion and reaction rate of the IL monomers.

387423 Size and Zeta Potential Behavior of Gold Nanoparticles – Polyelectrolytes Complexation

Abstract:

A complexation of nanoparticles with polyelectrolytes is a key factor in the process such stabilization and dispersion of nanoparticles in aquatic environment. The charged nanoparticles significantly affect the colloidal properties of a polyelectrolyte in water. The important physicochemical properties such as the polyelectrolyte size and zeta potential, nanoparticle size, nanoparticles zeta potential changes due to nanoparticles-polyelectrolyte complexation. Dynamic light scattering (DLS) and zeta potential techniques were used for investigation of aggregate size and zeta potential of citrate-coated gold nanoparticles 50 nm (GNP-50) in complexation with humic acid (HA), poly(styrene maleic acid sulfonate) (PSMA), polyethylenimine (PEI) and polyacrylic acid sodium salt (PAA) separately. The aggregate size and zeta potential of gold nanoparticles and each polyelectrolyte were measured from the acidic (pH=2) to basic (pH=14) range and compared to complex of mixture of them. For the complex of GNP-PEI and GNP-HA , the size and zeta potential of the complex showed a new trend or synergistic response marked by a unique shift in the isoelectric point of the polyelectrolyte when adsorbed onto the nanoparticle. For the poly(styrene-maleic anhydride sulfonate) material no changes in size and zeta potential were seen. This behavior is explained by comparison to literature-based on Monte Carlo simulations.

370889 Synthesis of Ag2O Nano-Particles in Spherical Polyelectrolyte Brushes

Abstract:

Ag_2O nano-particles were synthesized by chemical precipitation using spherical polyelectrolyte brushes (SPB) as nano-reactors. SPB was consisting of polystyrene core on which poly(acrylic acid) were attached. SPB had been used not only as a template but also provided great extent of stability to Ag_2O nano-particles. Average diameter of Ag_2O nano particles were around 10 nm which were confirmed by TEM images. This stable dispersion of Ag_2O-SPB composite is envisioned to use in application of photo-catalysis of waste water because Ag_2O is visible light photo-catalyst with narrow band gap of 1.3eV.

389736 Non-PFC Plasma Chemistries for Patterning Low-k Materials

Abstract:

Low-k materials, such as fluorine-doped and carbon-doped silicon dioxide, exhibit reduced dielectric constants necessary to curtail parasitic capacitance and avoid crosstalk in devices, keeping pace with the trend of increasing device densities. SF_6 and perfluorocarbons (PFC) gases, which are primarily used in plasma etch of interlayer dielectric materials, generally have high global warming potentials (GWP), making their increased usage undesirable. This work focuses on evaluating etch chemistries from a thermodynamic standpoint for back end of line (BEOL) applications in patterning of proposed low-k carbon-doped silica compounds with varying carbon content.

Group and bond additivity methods were used to estimate the Gibbs free energies of formation for these carbon-doped compounds. PFC and non-PFC etchants with H_2/NH_3 were assessed through the use of volatility diagrams comparing partial pressures of volatile etch products as a function of etchant partial pressure at 300K. Minimization of Gibbs free energy was employed to calculate the equilibrium distribution of species in the etch system across a range of temperatures. NF_3 and CF_3I were identified as potentially viable for etching carbon-doped silica. NF_3, a non-PFC gas with an atmospheric presence of 1ppt, GWP of 16,800, and much greater abatement efficacy, is the most effective etchant in pure form, producing volatile etch product pressures between six and eight times that produced by CF_4. CF_3I, despite being an iodofluorocarbon gas, exhibited a GWP of unity and displays reduced damage to doped carbon, making it preferable in etching carbon-doped silica. Pure CF_3I shows reduced product pressure with increasing carbon content; however, addition of H_2 and NH_3 improves its performance for the three most highly doped silica compounds. Given the higher cost associated with using NF_3 and CF_3I, the introduction of an additive (H_2 or NH_3) was assessed. Addition of H_2 and NH_3 generally showed an increase in partial pressures predicted by the volatility diagrams, with H_2 producing a much more significant increase than the pure etchant or the addition of NH_3. Preliminary experimental results comparing etch rates for moderately to highly carbon-doped silica samples with 20 sccm CF_4 and hydrogen addition generally agree with theorized predictions. Varying the feed composition between 0%, 20%, and 50% H_2, etch rates of 38, 45 and 49 nm/min were recorded for lightly doped silica, and 49, 73 and 128 nm/min were measured for heavily carbon-doped silica.

Investigation of Composition and Processing Parameters on Mechanical Properties of Magadiite/SBR Composites

Yating Mao[1], Shigeng Li[1], Hans-Conrad zur Loye[2], and Harry J. Ploehn[1]

[1]Department of Chemical Engineering and [2]Department of Chemistry & Biochemistry, University of South Carolina, Columbia, SC 29208

This work investigates mechanical reinforcement and energy dissipation in styrene-butadiene rubber (SBR) composites containing sulfur-functional, organosilane-modified magadiite (MGD), a layered silicate ($Na_2Si_{14}O_{29}\cdot 9H_2O$). Using MGD gives us plate-like active filler particles with surface chemistry similar to precipitated silica. In previous research, we demonstrated that MGD/SBR elastomers have better mechanical properties than silica/SBR composites with the same filler weight loading. MGD has significantly higher specific surface area than silica, and both the SI-69 coupling agent and SBR can access the MGD interlayer space, resulting in superior reinforcement. This work aims for a deeper understanding of reinforcement mechanisms in MGD/SBR composites by investigating the influence of composition and processing parameters on the elastomer structure and properties. The main composition parameter is the type of alkyl ammonium ion used to prop open the MGD interlayer space. We treat MGD with cetyl trimethyl ammonium (CTA), dodecyl pyridinium (DP) and hexadecyl ammonium (HDA) cations to produce "organo-MGD" materials with varying initial interlayer spacing. Using these materials, we explore the hypothesis that larger organo-MGD spacing will permit greater access of SBR into the MGD interlayer gaps. We investigate a second composition parameter, the phr loading of SI-69, to assess the importance of chemical cross-linking between MGD and SBR. The main processing parameter is the batch mixing time, to see if longer batch mixing facilitates MGD expansion and ingress of SI-69 and SBR pre-polymer into the interlayers. We prepare cross-linked MGD/SBR samples with varying organo-MGD materials, varying levels of SI-69 addition, and varying mixing time to study the influence of those factors on mechanical reinforcement. All samples are characterized by XRD after mixing, milling, and curing steps to help us understand at what stage SBR enters the MGD interlayer. DMA and tensile testing quantify the mechanical properties. By exploring the roles of composition and processing factors on the mechanical properties of MGD/SBR composites, this work will help us better understand the impact of non-spherical filler particle shape on reinforcement mechanisms and energy dissipation in platelet-filled elastomers.

387400 Nonclassical Pathways of Zeolite Growth from Amorphous Precursors to Crystalline Products

Abstract:

Crystallization commonly occurs through classical pathways that involve the addition of atoms or molecules to a crystal surface that grows by layer nucleation and spreading. Zeolites can be categorized in a growing list of materials (mostly natural minerals) that form through nonclassical routes [1] involving amorphous precursors that undergo structural disorder-to-order transformation(s). Here we present a case study of zeolite L (LTL framework type) crystallization wherein we track the formation and evolution of amorphous precursors. We show that the amorphous material undergoes a series of structural transformations, forming worm-like particles (WLPs) that progressively decrease in population over the course of zeolite nucleation and growth. The evolution of pre-nucleation amorphous precursors was evaluated on the basis of their temporal change(s) in morphology, particle size, and degree of aggregation (i.e. fractal dimension) to explore the pathway(s) of amorphous-to-crystalline phase transformation. The putative mechanism of growth involves the initial aggregation of colloidal silica particles, which subsequently form WLPs. The latter appear to provide nutrients (i.e. silica and alumina) to growing crystals, potentially serving as interfaces for heterogeneous nucleation and the growth of LTL crystals via solution-mediated processes. We have coupled these studies with a systematic investigation of zeolite growth modifiers (ZGMs), which are site-specific molecules that bind to growing crystal surfaces and alter their size and morphology [2]. Here we will discuss how ZGMs affect the formation of WLPs and their structural transformation to LTL crystals. In addition, we will discuss the broader relevance of nonclassical growth pathways for other zeolite framework types. Collectively, these mechanistic studies provide an improved fundamental understanding of zeolite crystallization, which is paramount to the rational design of microporous materials with tailored physicochemical properties for improved performance in a variety of commercial applications.

References:

[1] Lupulescu, A.I. and Rimer, J.D., In Situ Imaging of Silicalite-1 Surface Growth Reveals the Mechanism of Crystallization, *Science* 2014 (In Press)

[2] Lupulescu, A.I., Kumar, M., Rimer, J.D., A Facile Strategy to Design Zeolite L Crystals with Tunable Morphology and Surface Architecture, *J. Amer. Chem. Soc.* 135 (2013) 6608-6617

Solution–Processable Exfoliated Zeolite Nanosheets Purified by Density Gradient Centrifugation

Kumar Varoon Agrawal, **Lorraine F. Francis, Michael Tsapatsis**

Department of Chemical Engineering and Materials Science, University of Minnesota, 151 Amundson Hall, 421 Washington Avenue SE, Minneapolis, MN 55455, USA.
E-mail: varoo001@umn.edu

Sub-100 nm thick zeolite films are desirable for achieving high molecular throughput membranes. These thin films can be fabricated by assembly of 1.5 unit cell thick, high-aspect-ratio zeolite nanosheets. We recently reported synthesis of structurally and morphologically intact MFI-nanosheets [1] by exfoliation of multilamellar MFI [2] using melt blending with polystyrene [3]. Exfoliated nanosheets were extracted out of the polymer matrix by solution processing. Nanosheets were purified from polystyrene and unexfoliated particles by non-linear density gradient centrifugation (DGC) [4]. Purified nanosheets were characterized by transmission electron microscopy (TEM), and electron diffraction (ED) to confirm that purified nanosheets were highly crystalline.

Sub-100 nm thick zeolite films were fabricated on porous supports by a filtration of nanosheets suspension. Film thickness was simply controlled by adjusting the amount of nanosheets used in the filter coating. Scanning electron microscopy (SEM) images of the nanosheet film show that MFI-nanosheets cover the entire support while orienting themselves along their *b*-axis. Interparticle gaps in the nanosheets films were reduced by hydrothermal treatment of the nanosheet films, which led to *b*-oriented MFI membranes.

References:

[1] K. Varoon, X. Zhang, B. Elyassi et al., *Science*, **2011**, *334*, 72-75.
[2] M. Choi, K. Na, J. Kim, Y. Sakamoto, O. Terasaki, R. Ryoo, *Nature*, **2009**, *461*, 246-249.
[3] Maheshwari S, Jordan E, Kumar S, et al., *Journal of the Americal Chemical Society*. **2008**, *130*, 1507-1516.
[4] Agrawal KV, Topuz B, Jiang Z, et al., *AIChE Journal*, **2013**;*59*, 3458-3467.

381210 Facile Synthesis of Organosilica Nanoparticles Having Hollow Interiors and Porous Shells in Buffer Solution Using Colloidal Silica As Templates

Abstract:

Hollow nanoporous silica nanoparticles possess hollow interiors with nanoporous shells. They have become of particular interest for numerous potential applications including in catalysis, bio-sensing, drug delivery and anti-reflection (AR) coatings. Hollow structures can provide large cavities in nanometer-scale that can be accessed through nanoporous shells. Instead of silica, bridged silsesquioxanes (hereafter denoted organosilica), prepared by hydrolysis and polycondensation of precursors with the general formula $(R'O)_3Si–R–Si(OR')_3$(R = organic bridge; R' = alkoxyl groups), have several advantages as shell components, including facile functionalization and potential to modify mechanical and physical properties of the particles. A general method for preparing hollow organosilica nanoparticles requires the use of soft-templates such as block copolymer micelles.

Herein, we report a facile method for preparing monodisperse, hollow organosilica nanoparticles using silica nanospheres as hard templates [1]. A shell having tiny particles of organosilica was successfully formed on hard templates by stirring a biphasic mixture of bridged alkoxysilane precursors and an aqueous dispersion of hard templates, yielding silica-organosilica core-shell nanoparticles. The internal silica cores were selectively removed by treating the core-shell nanoparticles in a basic solution at elevated pH owing to the difference in hydrolytic stability between silica and organosilica, yielding the nanoporous organosilica nanoparticles with hollow interiors. Our newly developed method can be performed in a one-pot fashion and is applicable to bridged silanes with different organic functional group.

[1] N. Koike, T. Ikuno, T. Okubo, A. Shimojima, Chem. Commun. 2013, 49, 4998–5000.

Effects of Nanopore Size on the Interactions of Lipid Bilayers with Mesoporous Silica Particles

Daniel M. Schlipf, Stephen E. Rankin, Barbara L. Knutson
Department of Chemical and Materials Engineering
University of Kentucky Lexington, KY 40506
Corresponding Author: bknutson@engr.uky.edu

Lipid bilayers immobilized on a variety of synthetic surfaces have applications in the field of sensing, separations and small molecule discovery. When immobilized on a porous platform, the bilayer provides a mechanism to screen small molecules and encapsulate proteins within the pore volume. On average, bilayers are only a few nano-meters thick, making their interaction with nano-meter diameter pores of great importance to their intended use. Mesoporous silica provides a unique immobilization platform due to its tunable properties such as pore sizes, morphologies and surface functionality. This research addresses a current challenge in the study of the interactions of lipid bilayers with nano-meter scale pores, the limited approaches to link the tunable pore properties of mesoporous silica with the visualization techniques to characterize the lipid bilayers. Within our laboratory, hydrothermal aging of porous particles during synthesis has proven to be a very effective technique to tailor pore diameters on the 3-15 nano-meter scale, which also spans the scale of bilayer thickness. Additionally, our spherical particle synthesis results in large (7.5 μm) particle diameters, which are amenable to visual confirmation and bilayer localization using confocal scanning laser microscopy.

This investigation examines the effect of pore size on the surface and interior properties of supported and tethered lipid bilayers on mesoporous silica particles. Particles are synthesized using hydrothermal aging which provides nano-meter control over pore diameter between 3 and 15 nm. The lipid dipalmitoylphosphatidylcholine (DPPC) was chosen for its biological relevancy and well understood bilayer properties, which when formed are approximately 5 nm thick. Lipid silane 13-(chlorodimethylsilylmethyl)heptacosane is used as covalent tether of the membrane to the inorganic support, and the properties of supported and tethered bilayers are compared. Nanoporous particles are synthesized with particle diameters on the micron scale for visual characterization and bilayer localization is confirmed using confocal scanning laser microscopy. The diffusivity of lipids within the particle is determined at the surface and in the interior of the particle using Fluorescence Recovery After Photobleaching (FRAP). Size based nano-pore inclusion and exclusion on bilayer diffusivity both on the surface of and within porous particles is investigated.T Membrane fluidity is also evaluate as a function of surface density of the tethered lipid silane.

386158 Ordered Mesoporous Materials By Template-Free Nanoparticle Assembly

Abstract:

Assembly of binary silica nanoparticle superlattices (BSNS) of large (A) and small (B) size is demonstrated as a facile and template-free route to three-dimensionally ordered mesoporous (3DOm) silicas, the pore topology of which derives from the interconnected interstices of the resulting ordered nanoparticulate structures. Even without adoption of common stabilization, for example charge- or steric-stabilization (e.g., polymeric passivation layers) or solvent index matching for minimizing van der Waals interactions, we find that the assembly of binary silica nanoparticles from aqueous solutions of basic amino acids obeys predicted hard-sphere phase behavior. Specifically, the size ratio governs symmetry of the system among AB, AB_2, and AB_{13} stoichiometric assemblies that are isostructural with NaCl, AlB_2, and $NaZn_{13}$, with deviations from hard sphere predictions manifested by interstitial solids in the case of the smallest particle size ratios considered. The facile nanoparticle assembly and its concomitant structural diversity translate to tunability of the mesopore topology. High yield binary assembly is found to be robust to both sub- and super-stoichiometric precursor mixtures and to systematic changes (i.e., reduction) in nanoparticle size, the latter providing a facile handle for self-templated tunability of mesopore size.

The enhanced hydrothermal stability of the BSNS materials, owing to their thicker pore walls, will be described. In an extension aimed at realizing bottom-up strategies for stable, BSNS-supported catalytic materials, this talk will also describe how mechanistic insight into BSNS assembly can be exploited to expand compositional diversity of the structures specifically by co-assembly of binary metal and siliceous nanoparticles. Finally, use of the BSNS structures and their metal-embedded counterparts as sacrificial templates for carbon replication will also be described as a means for deriving bi-modal mesoporous carbon materials with and without template-transferred metal functionality. The overarching goal of this talk, therefore, will be to establish a new paradigm for bottom-up assembly of adsorbents and supported catalysts endowed with large pore volume, controllable 3D-ordered mesopore topology, and tunable uniformly distributed function.

Detailed step-by-step reactions towards the synthesis of a MOF

David C. Cantu*[a], B. Peter McGrail[b], Vassiliki-Alexandra Glezakou[a]
[a]Physical Sciences Division, [b]Energy and Environment Division
Pacific Northwest National Laboratory, Richland WA 99352

david.cantu@pnnl.gov

Based on density functional theory calculations and molecular simulation, a detailed mechanism on the formation of the secondary building unit of MIL-101 is presented. MIL-101, a chromium terephthalate metal-organic framework, is one of the most porous materials known, with high gas storage capacities. A thorough and precise knowledge of MOF synthesis is currently beyond our understanding due to the complexity of the electronic states of the metal centers, metal-anion-solvent interactions and the high number of competing simultaneous events that occur during MOF formation. The secondary building unit (SBU) in MIL-101 consists of three octahedrally coordinated chromium centers linked by a central oxygen, and the adjoining terephthalate linkers. Formed SBUs are key for nucleation, the rate-limiting step in the synthesis of most MOFs. We propose a series of reactions that lead to the formation of the MIL-101 SBU. Initial reactions that form the metal center have the highest energy barrier, and are followed by multiple competing linker addition reactions. Based on transition state theory for individual reactions and the Arrhenius equation, a simple kinetic model for overall SBU formation was built, showing SBU formation rates similar to known nucleation rates.

389754 Synthesis of Hierarchical Sn-MFI As Lewis Acid Catalysts for Isomerization of Cellulosic Sugars

Abstract:

Zeolites are crystalline microporous materials with distinct pore dimensions, connectivity and framework compositions, which are widely used in petrochemical and biomass processing. However, the catalytic performance of zeolites is frequently hindered by the slow molecular diffusion in the microporous networks (0.5 nm to 2 nm), in particular when bulky molecules are reacted. A proven strategy to overcome the diffusion limitation is to create secondary mesopores into zeolites, leading to the formation of hierarchical porous structures. The secondary mesopores could enhance the diffusion of molecules to the active sites and desorption of the formed products.

In this work, hierarchical stannosilicate molecular sieves with ordered mesoporosity and MFI topology (three dimensionally ordered mesoporous imprinted (3DOm-i) Sn-MFI) have been successfully synthesized within the confined space of three dimensionally ordered mesoporous (3DOm) carbon via a seeded growth approach. The obtained 3DOm-i Sn-MFI consists of uniform 30 nm spherical opaline structures having ordered mesopores with a pore size from 4 nm to 11 nm. The 3DOm-i Sn-MFI exhibits higher catalytic activity for the isomerization of cellulosic sugars than the conventional 300 nm Sn-MFI and 60 μm Sn-MFI catalysts. It is revealed that there is no diffusion limitation for the isomerization of a triose sugar, conversion of dihydroxyacetone (DHA) into methyl lactate (ML). 3DOm-i Sn-MFI offers significant improvements for the isomerizations of C_5 and C_6 sugars, such as xylose and glucose, by greatly promoting molecular transport. The reaction rate of xylose on 3DOm-i Sn-MFI is at least 20 times higher than that on conventional bulky sized Sn-MFI. The reaction rate for glucose is also enhanced over 3DOm-i Sn-MFI, but to a lesser extent as compared to the reaction of xylose. This is possibly because it is difficult for glucose to diffuse into the 10 membered-ring pore of MFI, and the reaction might only occur on the external surface of the Sn-MFI catalysts. Furthermore, the combination of seeded growth with confined synthesis provides us with synthesizing hierarchical Sn-MFI using commercially available carbon materials such as carbon black and activated carbon, indicating that the synthesis methodology is versatile and reliable for tailoring the structure of hierarchical zeolites.

364039 Invited Talk: Gas Sorption and Permeation in Thin Films of High Free Volume Polymers

Abstract:

Separation of gases by polymer membranes has become a well established industrial practice. To achieve high productivity, or high fluxes of gas permeation, the active portion of these membranes must be very thin, of the order of 100 nanometers. In most cases, the preferred polymers are in the glassy state where properties depend on prior history. A great deal of research over the past decade has shown that thin films (< 1,000 nm) show much more rapid physical aging than observed in bulk, and generally thin films can have different permeation properties than thick films (10's of microns). There has been much interest in identifying new polymers for membrane applications that provide a more advantageous combination of permeability and selectivity than those currently used commercially. One direction of this research has been to develop high free volume polymers that have exceptionally high gas permeabilities and very high glass transitions. We have examined the gas permeation behavior of a number of these high free volume polymers in both thick and thin films forms and have compared their physical aging rates. The materials examined include polyimides, poly[1-(trimethylsilyl)-1-propyne], PTMSP; PIM-1 (PIM = "polymers of intrinsic microporosity); amorphous Teflons; and a polybenzoxazole made by thermal rearrangement (TR). The thin films showed much lower permeabilities than thick films especially for PTMSP and PIM-1. For a wide range of polymers, including those mentioned above, the physical aging rate was found to increase strongly with the glass transition temperature of the polymer relative to the temperature of observation.

Diffusion and Migration in Ionic Membranes

Michael A. Hickner, Associate Professor
Department of Materials Science and Engineering
Department of Chemical Engineering
The Pennsylvania State University
University Park, PA 16802 USA

Membranes serve important functions as ion and water diffusion barriers in water purification, energy storage and conversion systems, and food processing applications. We have explored the factors that control the ionic resistance and permselectivity of ion exchange membranes for both sodium chloride and ammonium bicarbonate processes. Our data has quantified the resistance-permselectivity tradeoff for a wide range of membranes and demonstrated that both the water uptake and fixed charge density are important parameters in determining overall membrane performance. Interestingly, when the resistance of an anion or cation exchange membrane is measured between two compartments of different concentration, like in reverse electrodialysis (RED), the resistance is much greater than what is reported by the manufacturers. We will report the underlying phenomena governing ion exchange membrane behavior and show how this fundamental knowledge can be used to understand RED stack resistance.

379910 Ionic Conductivity and Gas Permeability of Polymerized Ionic Liquid Block Copolymer Membranes for Energy Applications

Abstract:

Polymer membranes for many energy applications, such as solar-to-hydrogen fuel production, require ionic conductivity while acting as gas diffusion barriers. We have synthesized a diblock copolymer consisting of poly(styrene-*block*-(4-(2-methacrylamidoethyl)-imidazolium trifluoroacetate) by treating poly(styrene-*block*-histamine methacrylamide) (PS-*b*-PHMA) with trifluoroacetic acid. The PS block serves as the structural support while the imidazolium derivative is an ion-conducting polymerized ionic liquid (PIL). Small angle X-ray scattering and transmission electron microscopy demonstrate that the block copolymer self-assembles into well-ordered nanostructures, with lamellae and hexagonally packed cylindrical morphologies. The ionic conductivities of the PS-*b*-PHMA materials were as high as 2×10^{-4} S/cm while an order of magnitude increase in conductivity was observed upon conversion to PS-*b*-PIL. The ionic conductivity of the PS-*b*-PIL increased by a factor of ~4 up to 1.2×10^{-3} S/cm as the PIL domain size increased from 20 to 40 nm. These insights allow for the rational design of high performance ion conducting membranes with even greater conductivities via precise morphological control.

Oxygen diffusion measurements were also performed on these materials as well as Nafion, the canonical cation exchange membrane, using a transient electrochemical reduction method. We provide evidence for dual-mode diffusion whereby distinct diffusion coefficients are observed for oxygen moving through the hydrophilic channels and fluorocarbon matrix. These diffusive modes can be tuned by thermal annealing which changes both the percent crystallinity and crystallite size (determined by WAXS and DSC). Thus, the oxygen permeation can be decreased by more than a factor of 2 in a facile manner. Our results provide key insight into the balance of ionic conductivity and gas diffusion in membranes for energy applications.

Water Sorption and Transport in Nafion Thin Films: Understanding the Effect of Confinement

Eric M. Davis, Nichole K. Nadermann, Edwin P. Chan, Christopher M. Stafford, and Kirt A. Page

Materials Science and Engineering Division, National Institute of Standards and Technology, Gaithersburg, MD 20899

Perfluorinated ionomers, specifically Nafion, are the most widely used ion exchange membranes for fuel cell applications. For these applications, Nafion is typically incorporated into the catalyst ink layer as a binder and proton conductor, thus water and ion transport are strongly affected by the extent of thin film confinement on the nanometer size scale. Therefore, developing a deeper understanding of the fundamental water transport mechanisms in Nafion thin films is critical. In this study, the diffusion of water vapor into Nafion thin films was measured using time-resolved *in-situ* polarization modulation infrared reflection adsorption spectroscopy (PM-IRRAS). Unlike other gravimetric techniques used to measured water sorption in thin films (e.g., quartz crystal microbalance), PM-IRRAS provides molecular-level resolution of the Nafion and water in real-time during the water sorption experiment. Specifically, the diffusion of both H_2O and D_2O vapor in dry and hydrated Nafion films was captured through a series of integral water sorption experiments. The effective diffusion coefficient of water in the Nafion thin films was many orders a magnitude less than those reported in bulk films and was dependent on the initial hydration level; however, little difference was observed between the diffusion coefficient for H_2O and D_2O. Furthermore, water-induced swelling (i.e., polymer relaxation) of the Nafion thin film was observed, where this relaxation process was determined to be orders of magnitude slower than those calculated for bulk films.

Additionally, liquid H_2O diffusion in fully hydrated membranes (both nano-scale and bulk) was captured using poroelastic relaxation indentation (PRI). Unlike PM-IRRAS, PRI measures in-plane diffusion, which is especially useful for thin films as it measures in-plane transport on the micrometer length scale. While this indentation technique has been proven to be a useful tool in measuring mechanical and transport properties of hydrated, homogeneous soft materials (e.g., hydrogels), this is the first demonstration of the technique for studying water transport in a heterogeneous, phase-segregated polymer. The effective diffusion coefficient of liquid water in bulk Nafion was similar to those reported in the literature, validating the use of PRI to investigate water diffusion in heterogeneous soft materials. Furthermore, the effective water diffusion coefficient in the Nafion thin films fell between those reported for bulk Nafion and values calculated from PM-IRRAS experiments. These results suggest that while in-plane diffusion in Nafion thin films takes place over larger length scales, confinement in the through-plane direction significantly affects water diffusion in the in-plane direction. The results of this study indicate that PRI can potentially serve as a powerful technique in studying water transport in other phase-segregated materials such as block copolymers and heterogeneous gels.

375741 Kinetic Water Sorption in Poly(ethylene furanoate) Compared to Poly(ethylene terephthalate)

Abstract:

Poly(ethylene furanoate) (PEF), which is a bio-based, performance-enhanced furanic analog to poly(ethylene terephthalate) (PET), is being investigated as a possible replacement for PET in the beverage packaging industry. The current work investigates concentration dependence in diffusion coefficients for water at 35°C in both amorphous PEF and PET over the entire water activity range. PEF exhibits increased chain rigidity compared to PET as evidenced by reduced diffusion coefficients over the entire water activity range. Transport is governed by simple Fickian diffusion up to ~0.6 activity, after which long-term non-Fickian relaxations induce morphological changes in both polymers. Various diffusion models are used to describe the kinetic sorption curves, including the simple Fickian model, the Berens-Hopfenberg model, and the model proposed by Long and Richman. Additionally, activation energies of water diffusion calculated from measurements at 15, 25, 35, and 45°C and a sorption interval of 0.1 – 0.2 activity will be presented.

Triptycene-containing Polyimide Membranes: Synthesis and Gas Transport Properties

Jennifer Wiegand (jwiegan2@nd.edu), Patrick Buckley (pbuckley@nd.edu), Ruilan Guo* (rguo@nd.edu)

Department of Chemical and Biomolecular Engineering, University of Notre Dame Notre Dame, IN 46556-5637

This paper reports fundamental studies on the synthesis and characterization of a series of new triptycene-containing polyimides that have great potential for energy efficient membrane gas separation applications. These newly designed and synthesized robust membranes exhibit high gas permeability as well as high selectivity as compared with several commercial polymeric membranes such as Matrimid® polyimide and polysulfones. We demonstrate that fractional free volume, which is the key structural factor determining a glassy polymer membrane's transport properties, can be feasibly tuned in its size and size distribution via incorporation of select side substituent groups neighboring the triptycene units, or by changing the linking geometry of triptycene units along the polymer backbone structure. Depending on the variations of the macromolecular structure, these polyimide membranes show correspondingly varied and tunable physical and transport properties, which provides a new dimension in molecular design for structural manipulations at the molecular level to allow facile construction of tailored free volume architecture, thus promoting fast transport and selective separations. Molecular design, synthesis and characterization of these new triptycene-containing polyimides will be presented. Fundamental transport properties including pure gas permeation data (with H_2, N_2, O_2, CO_2, and CH_4) and sorption data will be discussed for these polyimide membranes to explore the fundamental relationship between microscopic structures with macroscopic transport properties for these triptycene-containing polyimides.

References:

1. Sanders, D.F.; Smith, Z.P.; Guo, R.; Robeson, L.M.; McGrath, J.E.; Paul, D.R.; Freeman, B.D., Energy efficient polymeric gas separation membranes for a sustainable future: A review, *Polymer*, **2013**, *54(18)*, 4729–4761
2. Wiegand, J.R.; Smith, Z.P.; Liu, Q.; Freeman, B.D.; Patterson, C.; Guo, R., Synthesis and characterization of triptycene containing polyimides with tunable fractional free volume for gas separation membranes, *submitted*, **2014**.

380606 Diffusion of Small Molecules in Polymer Glasses

Abstract:

The Free Volume Model (FVM) proposed by Cohen and Turnbull over 50 years ago had provided a simple picture that relates the structural properties of amorphous glasses to the diffusing behaviors of small molecules in them. Albeit the semi-phenomenological nature of the model, FVM has met with great success in matching experimental results and providing guidance in designing new materials for membrane separations. However, accumulating experimental data on diffusion of gases through polymer membranes have increasingly suggested inconsistencies of the model. In attempts to correct these deviations within the framework of FVM, effects of activation energy, cavity connectivity and accessible free volume had been introduced. These modifications are largely phenomenological due to the lack of a systematic understanding of the diffusion process in glassy polymers. In this study, using molecular dynamics simulations we aim to provide a coherent picture of small molecules diffusing in polymer glasses. Our results suggest that there are three distinct diffusion regimes depending on the relative size of the diffusant and "cavities". The dominant mechanism that controls diffusion in each regime is identified respectively. Based on the identified mechanisms, deviations from the FVM can now be systematically understood and predicted. Our results provide a basis on which further development of more comprehensive diffusion models will be constructed, a work that is undergoing.

371827 Diffusion and Mobility of ssDNA Confined in a Nanopore Under a Ratcheting Force from an Enzyme

Abstract:

The idea of DNA sequencing based on single-file translocation of the DNA through nanopores under the action of electric fields has received much attention over the past two decades due to the societal need for high speed and high-throughput sequencing. However, interrogating individual bases as they traverse through the pore with the required signal to noise ratio has been a major problem, as the typical speeds are too high. Recently, the use of phi29 polymerase in conjunction with a protein pore (alpha-hemolysin) in the presence of an electric field has been shown to slow down the polymer translocation speed enabling reasonably successful base-calling. We have evaluated the signal-to-noise ratio of such a construct by performing Langevin Dynamics simulations on models of this construct. By monitoring the contributions of the conformational fluctuations of the polymer, we have computed the diffusional behavior of monomers of the chain under varying speeds of the polymerase activity and externally imposed voltage gradients. The fractal dimensions of polymer dynamics at different stages of translocation are also obtained. Our simulations show that even if the translocation speed is slowed considerably by using the polymerase-nanopore construct, the conformational fluctuations of ssDNA inside the pore are always present at high levels resulting in high levels of noise in the detection signal. We report a strategy to reduce the noise by specifically binding the wiggling tail of the polymer outside the pore with specific proteins in order to suppress the conformational fluctuations inside the pore.

390415 Diffusion of Entangled Rod-Coil Block Copolymers

Abstract:

Polymer science is exploring advanced materials which combine functional domains such as proteins and semiconducting polymers with traditional flexible polymers onto the same molecule. While thermodynamic assemblies of different geometries introduce many interesting new phenomena such as entropic packing and liquid crystalline interactions, dynamic effects are also important to understand for optimal design of material mechanics, processing kinetics, and because of the new physics that directly arises from the motion of multiple domains of dissimilar geometries.

In this talk, we describe a reptation theory for entangled rod-coil block copolymers as a model for this wider class of functional polymeric materials. The large geometrical differences between rigid rods and Gaussian coils cause significant nonlinearities in dynamical behavior as these two motifs are combined on the same molecule. In particular, our theory hypothesizes that the motion of rod-coils is slowed relative to rod and coil homopolymers because of a mismatch between the curvature of the rod and coil entanglement tubes. In the small rod limit where the rod is a perturbation on coil motion, the randomly varying curvature of the coil's tube presents entropic barriers to the reptation of the rod, modifying the unhindered motion of the coil along its tube into an activated reptation process. In the large rod limit where the coil is a perturbation on rod motion, the long rod cannot rotate around the surrounding entanglements so motion is only possible when the coil moves into a straightened entanglement tube in an arm retraction process. These mechanisms were verified by tracer diffusion measurements using molecular dynamics simulation and forced Rayleigh scattering experiments. Additional simulations using a coarse-grained slip-spring model provide direct evidence for the activated reptation mechanism.

364286 Atomic Layer Deposited Highly Dispersed Platinum Nanoparticles Supported on Multiwalled Carbon Nanotubes for Liquid Phase Reactions

Abstract:

Highly dispersed platinum nanoparticles were deposited on gram quantities of non-functionalized multiwalled carbon nanotubes (MWCNTs) by atomic layer deposition (ALD) in a fluidized bed reactor at 300 °C. (Methylcyclopentadienyl) trimethylplatinum and oxygen were used as precursors. The results of TEM analysis showed that ~1.3 nm Pt nanoparticles were highly dispersed on non-functionalized MWCNTs. The porous structures of MWCNTs did not change with the deposition of Pt nanoparticles. For comparison, the commercial 3 wt.% Pt/C catalyst was also characterized. The ALD-prepared Pt/MWCNT was used for the hydrogenation of xylose to xylitol and hydrogen transfer of ketones. The ALD-prepared Pt/MWCNT showed the best catalytic performance with 100 % conversion of xylose and 99.3 % selectivity to xylitol, compared to commercially available Pt/C, Ru/C, and Raney Ni catalysts. The heterogeneous hydrogen-transfer reactions of acetophenone indicated that an acetophenone conversion of 100% and a 1-phenylethanol selectivity of 99.0% could be obtained with a ketone to Pt mass ratio of 24,690 and a ketone to KOH mass ratio of 22 at 150 °C for 5 h. The stability of ALD produced Pt/MWCNT catalyst was much higher than that of the commercial Pt/C, due to the presence of surface defects on the MWCNTs and the strong metal–support interaction for the ALD-prepared Pt/MWCNT catalyst.

Temporal and Spatial Reactors for Atomic Layer Deposition on Particles

J. Ruud van Ommen, Fabio Grillo, David Valdesueiro,
Dirkjan Kooijman, Mojgan Talebi, Aris Goulas

Delft University of Technology, ChemE, Julianalaan 136, 2628 BL Delft, the Netherlands
j.r.vanommen@tudelft.nl

Whereas most research on atomic layer deposition (ALD) is aimed at depositing ultrathin films on wafers and other flat substrates, ALD can also be used to provide the surface of particles with either an ultrathin continuous coating or a decoration of nanoclusters [1,2,3]. When carried out in a so-called fluidized bed, ALD is an attractive way of producing nanostructured particles, for example for catalytic applications, with excellent scale-up potential. In such a system, the particles are suspended in an upward gas flow. Nanoparticles – contrary to what is typically observed for micron-sized particles – are fluidized as very dilute agglomerates with distinctive fluidization characteristics. We will discuss the challenges related to fluidization and coating of nanoparticles.

Like most conventional ALD reactors, fluidized-bed ALD reactors are typically operated in a temporal mode: the pulses of the different gaseous reactants are delivered subsequently in time. An alternative is to separate the administering of the reactants in space [4]. We recently developed a spatial ALD reactor for particles. We will discuss spatial ALD of Pt on TiO_2 nanoparticles that are pneumatically transported through the reactor with nitrogen at atmospheric pressure. These nanoparticles are blown through a tube as very porous agglomerates, ensuring good accessibility to the precursor/co-reactant ($MeCpPtMe_3$ and O_2), which are injected along the tube. Optimizing the operating conditions, we obtained an even spread of 1-2 nm nanoclusters of Pt over the TiO_2 nanoparticles (see Fig. 1). This material showed to be a very active photocatalyst. Reaction tests demonstrated an increase of the photocatalytic activity with a factor 5, compared to native TiO_2 nanoparticles. With the current setup, we are able to produce nanostructured particles via ALD at a production rate of about 1 g/min. However, the experience with pneumatic transport from other fields opens up promising scale-up prospects for continuous particle ALD.

References

[1] D. M. King, X. Liang, A. W. Weimer, Powder Technol. 221 (2012) 13-25.

[2] D. Longrie, D. Deduytsche, C. Detavernier, J. Vac. Sci. Tech. A 32 (2014) 010802.

[3] A. Goulas, J.R. van Ommen, KONA Powder and Particle J. 31 (2014) 234-246.

[4] W.M.M. Kessels, M. Putkonen, MRS Bull. 36 (2011) 907-913.

Figure 1. TEM image of the Pt nanoclusters, deposited on TiO_2 nanoparticles by ALD in a spatial reactor.

377799 Metallic Co ALD Catalysts for Fuels from Syngas

Abstract:

Direct deposition of a metallic Co by ALD has been demonstrated using low-cost cobaltocene precursor and H_2; this catalyst preparation method eliminates reduction of oxide and improves catalytic productivity. Eliminating high-temperature treatment of catalyst avoids sintering and incorporation of metal into the support (aluminate). Two methods of depositing metallic Co are discussed: on $Al_2O_3$and on W ALD. Both surfaces have exhibited deposition of metallic Co; however, the deposition reactions are vastly different.

Application of the Co catalyst deposited by metallic ALD has multiple advantages. First, cobaltocene is a less-expensive reactant than other proposed chemistries for metallic Co ALD. Second, metallic Co ALD will be less likely to sinter than incipient wetness catalysts because reduction temperatures are avoided. Third, the deposition of metallic Co compared to Co_3O_4 ALD provides higher productivity of Co catalyst because the metal does not form a mixed oxide with the support.

Chemical vapour impregnation- a versatile route to high redox activity nanoparticle catalysts

Michael M Forde PhD.

Graham J. Hutchings, Chris K. Kiely, Qian He, Nikolaos Dimitratos, Ren Su and Lokesh Kesavan

AIChE Meeting (Atlanta) November 2014

Talk Outline

- Overview of our catalyst preparation techniques
- Characterisation of CVI materials- examples
- Catalysis with CVI materials
  - Toluene oxidation
  - Levulinic acid hydrogenation
  - H_2 evolution by photocatalysts

Sol-immobilisation (SI)

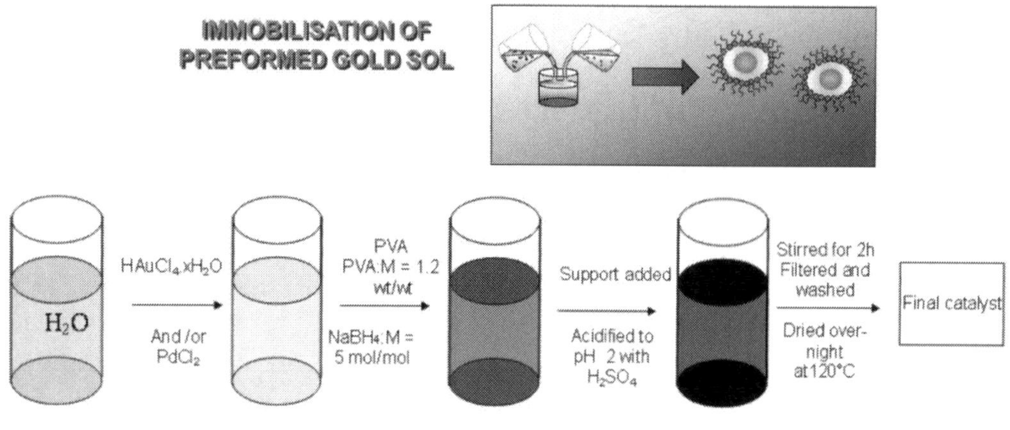

- Final catalyst not heat treated
- Can prepare bi and tri metallics in this way
- Order of metal precursor addition and reduction affects morphology

Core-shell structure by SI

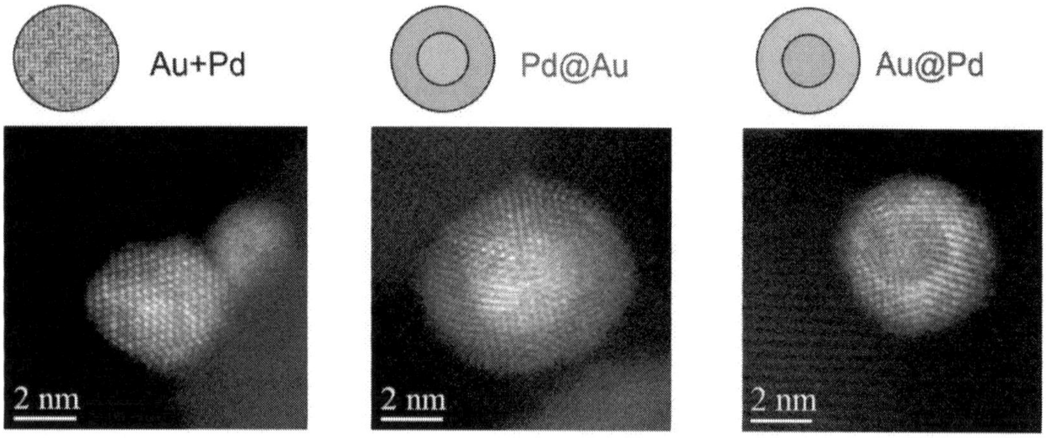

Varying precursor addition order and reduction can be used to change morphology

Go to method for making small nanoparticles

Issues

- ➢ Surface ligands present
- ➢ Spectator species present (chloride etc)
- ➢ Materials not heat stable
- ➢ Lengthy synthesis
- ➢ Reproducibility and general application

Issues common to many techniques!

Chemical Vapour Impregnation (CVI)

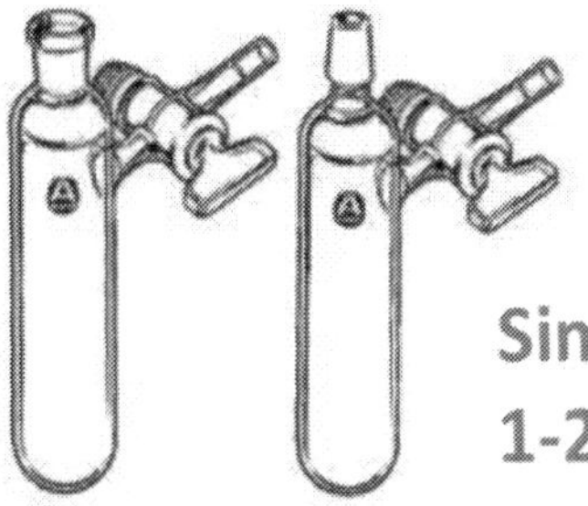

Simple, scalable, reproducible technique
1-2h for deposition

- ✓ Physically mix precursor and organometallic precursor
- ✓ Evacuate tube, heat under constant vacuum
- ✓ Remove vacuum, cool and remove material
- ✓ Heat treat to produce fine dispersion of nanoparticles on
- ✓ support

Features of CVI

- ✓ Can deposit wide range of metals on almost any support
  - ✓ We've done Fe, Cu, Pt, Pd, Rh, Ru, Ag, Zn, Ga
- ✓ Very fine metal dispersion
  - ✓ Size depends on nature of metal, treatment conditions, loading
- ✓ Can be used to prepare bi and trimetallics
- ✓ Can be used to modify one material by selective deposition of a second metal.. Core shell structures
- ✓ Contamination free support
- ✓ Clean surface to NP's

Pd and Pt - examples

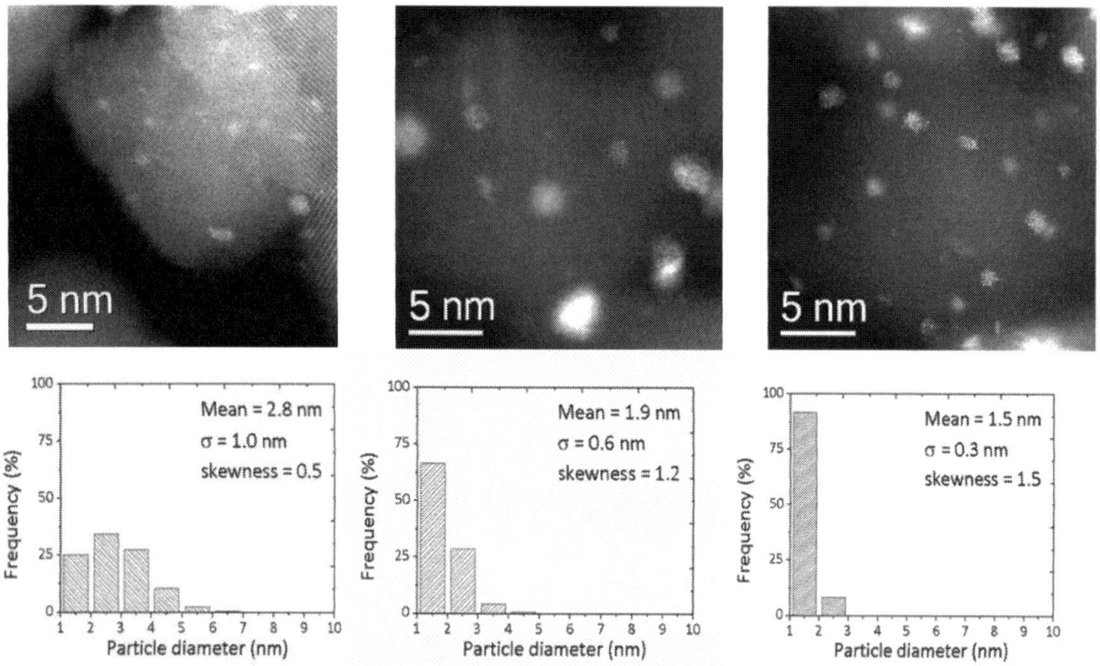

2.5%Pd/TiO_2 1.25%Pd-1.25%Pt/TiO_2 2.5%Pt/TiO_2

Pd and Pt examples

Pd

Pt

Pd-Pt

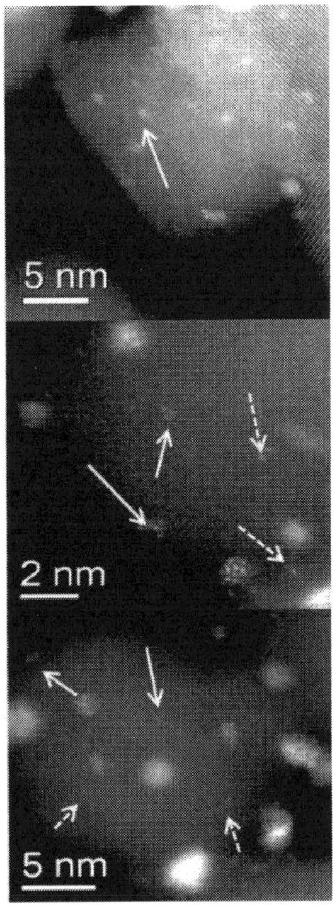

Sub-nm clusters

Single atoms

Still observed after 400°C treatment

Characterisation CVI - examples

Pd only **Pt only** **Pd-Pt**

BF

HAADF

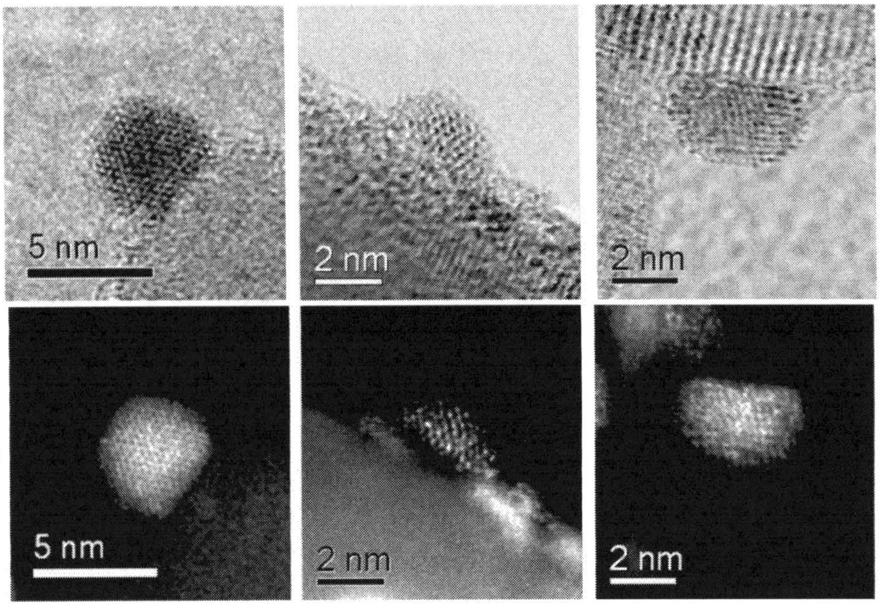

Clean nanoparticles in all cases

Mixed Pd-Pt - example

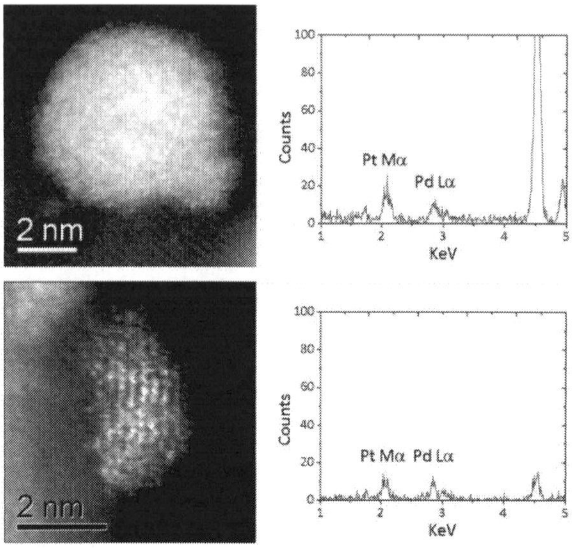

$1.25\%Pd\text{-}1.25\%Pt/TiO_2$

One step deposition

Homogeneous alloy expected

Most particles found to be **homogeneous alloys**

Occasional Pt and Pd only particle observed

Modifying Au catalysts

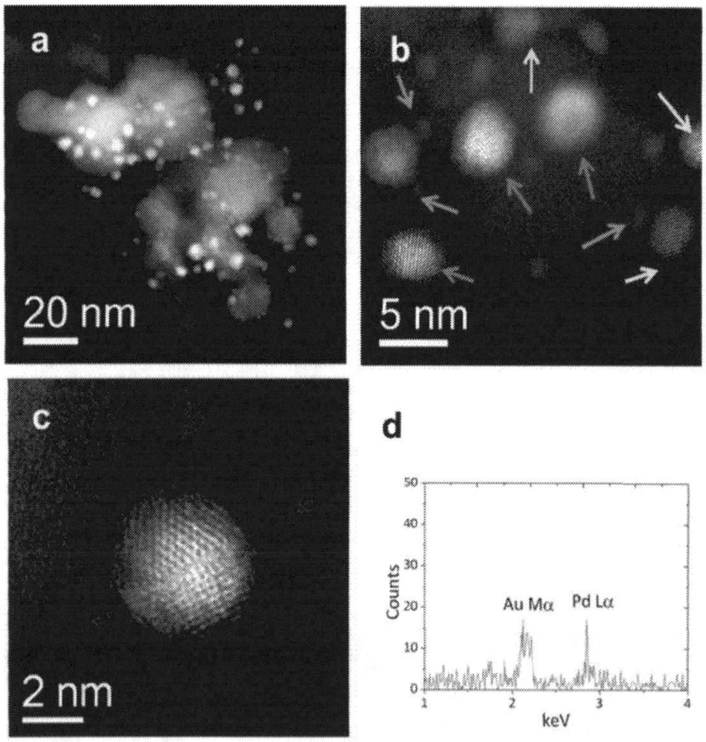

Core Shell Pd on Au

Au only NP

Pd only NP

Modifying Au catalysts

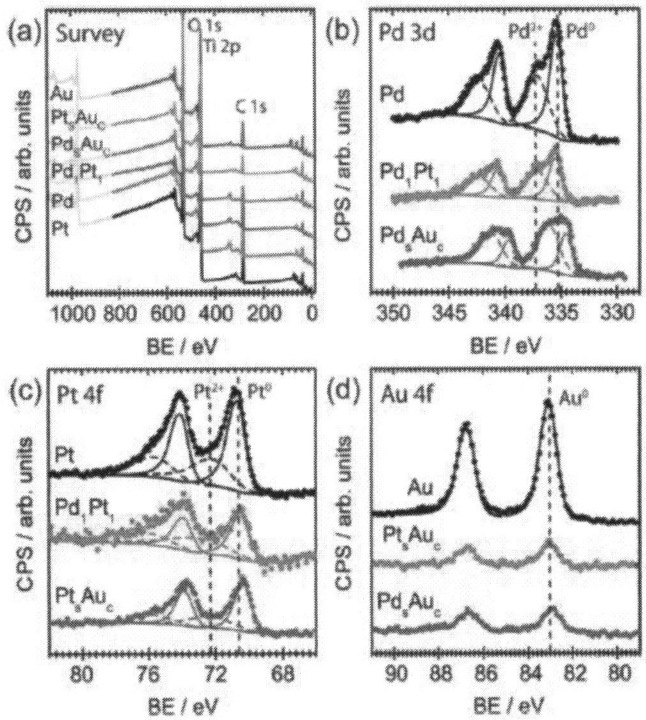

In hybrid Au peak is reduced severely… Evidence of core shell structure

Pd can be found in +2/0 oxidation state (age)

Pt always found in 0 oxidation state

Characterisation- Fe/SiO_2 by CVI

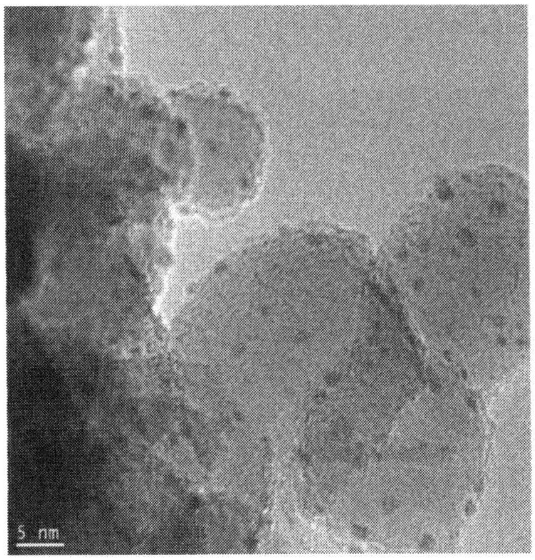

Uniform distribution of FexOy
Particles ~ 3nm

Size affected by metal loading

If heat treatment is done under H_2
All surface Fe disappears

TOLUENE OXIDATION

ACS Nano, 2014, 8, 957-969
Angew. Chem. Int. Ed., 2012, 51, 5981–5985
Phys. Chem. Chem. Phys., 2013,15, 10636-10644
Science, 2011, 331, 195

Catalysis- Toluene Oxidation

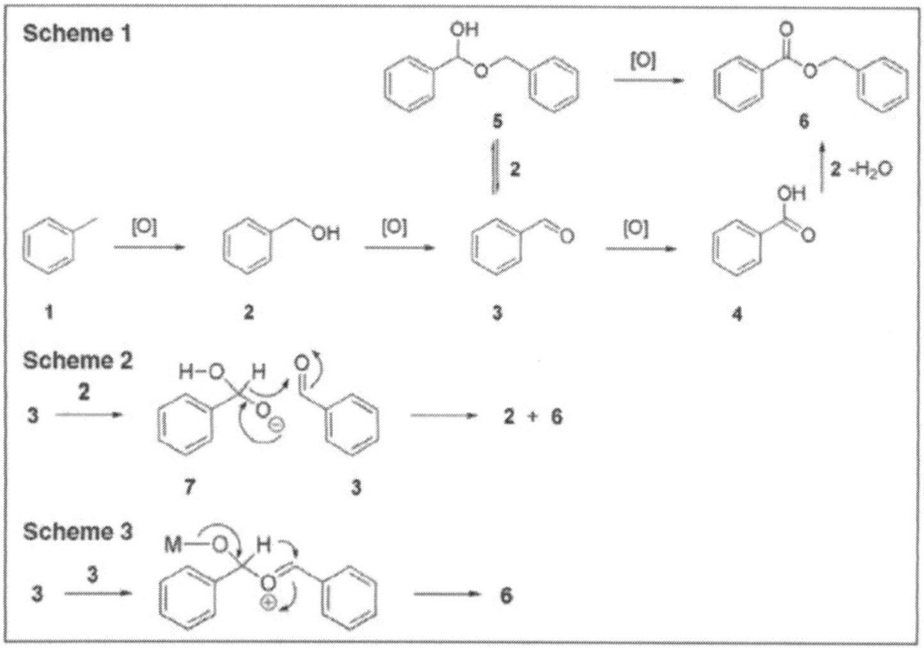

Toluene oxidation- 1wt% (AuPd)/C by SI

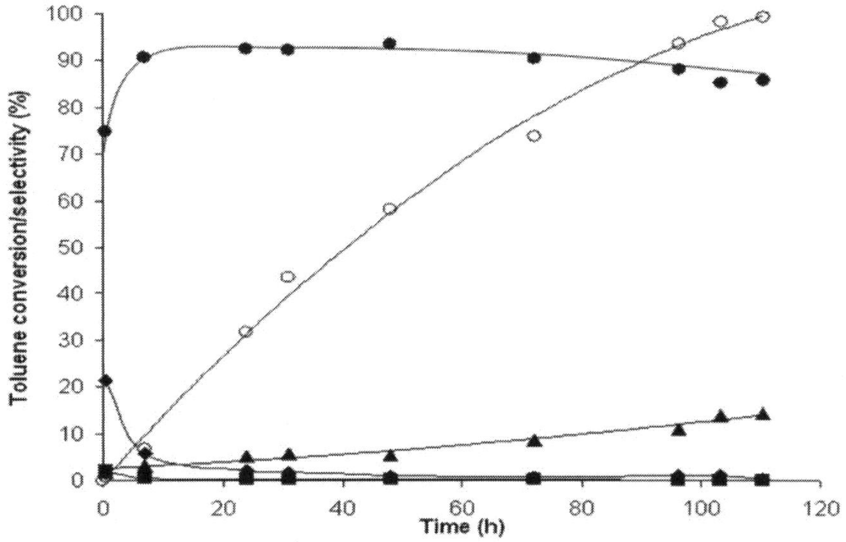

Key result- High conversion to benzylbenzoate
Solvent free, base free, 160°C, 10bar O_2 (continuous)

Toluene oxidation- 1.25%Pd-1.25%Pt/TiO_2 CVI

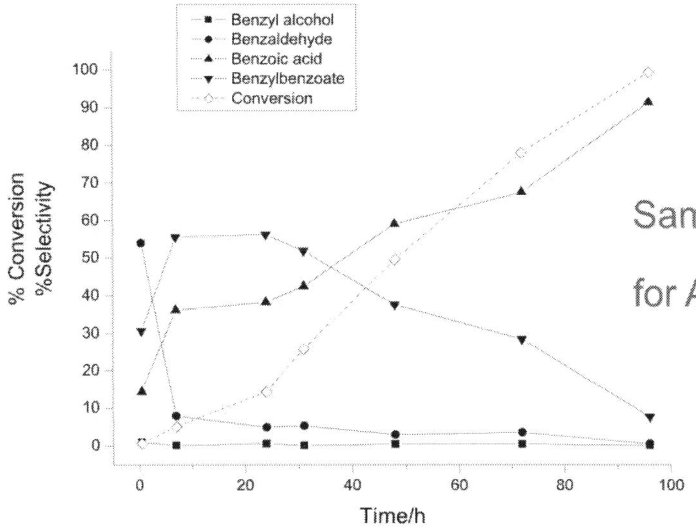

Same conditions as

for AuPd/C (SI) catalyst

Excellent activity
Different product distribution
Pd-Pt/TiO_2 or Pd-Pt/C by SI does not achieve high conversion

Toluene oxidation with TBHP and Pd-Pt/TiO_2

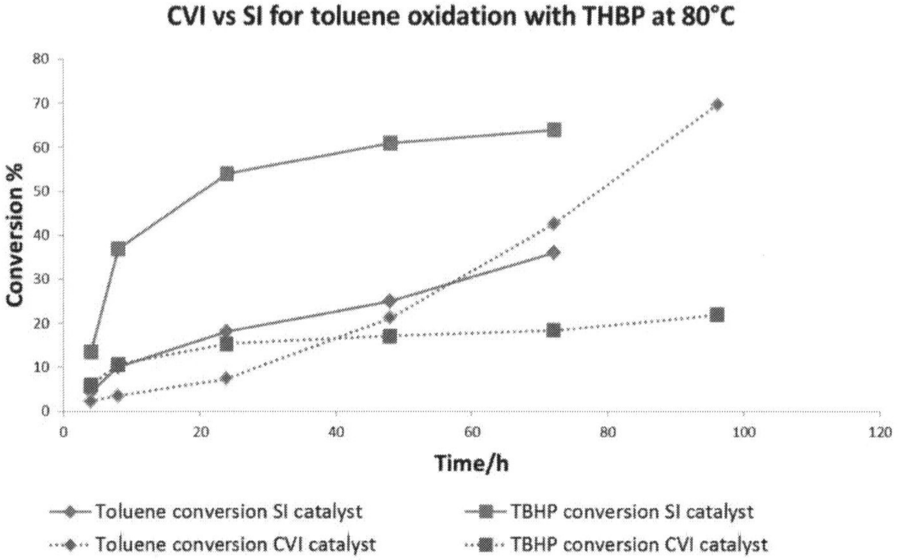

CVI catalyst is activating O_2...under N_2 toluene conversion drops, TBHP conversion increases

Higher conversion achieved with CVI catalyst

Levulinic Acid Hydrogenation

ACS Nano, 2014, 8, 957-969

Hydrogenation Catalysis- Levulinic Acid

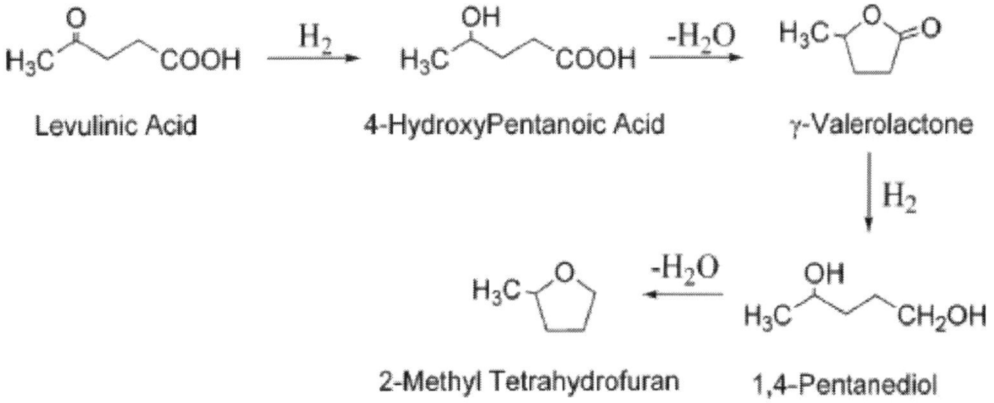

High rate, milder conditions, Lower excess of H_2

Levulinic Acid Hydrogenation

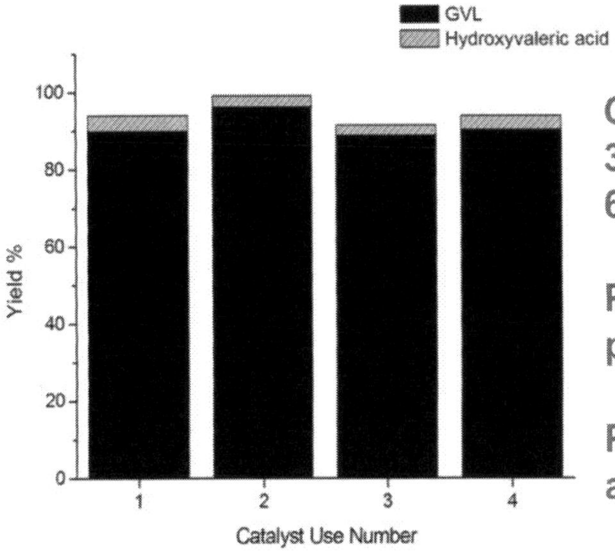

Catalyst even works at 30°C !
60% yield GVL in 20h

Pd-Pt has good performance at 150°C

Pd only performs well above 180°C

Pt/TiO_2 catalyst, 100 °C, H_2:LA= 2:1, 4 h

Photocatalysis

Dalton Trans., 2014,43, 14976-14982

H_2 from ethanol-water

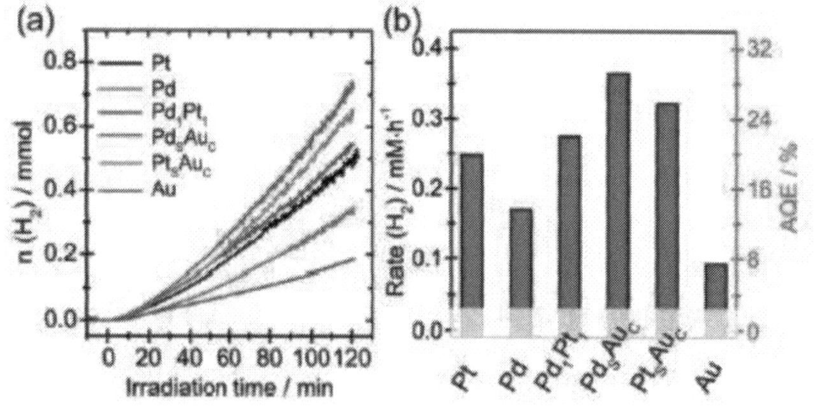

Beneficial effect of core shell structure observed
UV led used as light source
No dark reaction observed

Summary

- Simple, scalable techniques
- Well defined nanoparticles
- Flexibility to make exotic nano-architectures
- Clean, stable nanoparticles
- Wide activity for oxidation, hydrogenation, photo-catalysis

388051 Preparation of Carbon Supported Pt Nanoparticles By Atomic Layer Deposition

Abstract:

Hydrogen fuel cells offer an efficient use of a more energy dense fuel compared to combustion engines using gasoline. However, adaptation of fuel cells to power vehicles depends on several factors, one of which is economic synthesis of each component. Atomic layer deposition (ALD) using a fluidized bed reactor is an economic and scalable method for manufacturing the noble metal nanoparticle catalyst. This work furthers understanding of reported mechanisms of platinum ALD by investigating reactions on a common carbon support, XC72-R. By-products of reactions during ALD using trimethyl(methylcyclopentadienyl)platinum(IV) ($MeCpPtMe_3$) and either oxygen or hydrogen as reactants on un-modified carbon and carbon functionalized using nitric acid are inspected using *in situ* mass spectrometry. This understanding will help fine-tune parameters during a commercial catalyst synthesis process. Platinum weight loading, average particle size and particle dispersion depend on the chemistry of the pre-processing and precursor reactions. Oxygen becomes integrated into the carbon during functionalization and binds platinum tighter than just carbon; therefore particles aggregate less during deposition on functionalized carbon. Combustion of the ligand of $MeCpPtMe_3$ attached to carbon, as well as substrate oxidation during the oxygen dose, contribute to more particle aggregation versus when hydrogen is utilized for ligand removal. Mechanisms for each ALD condition are proposed.

Particle Size and Support Effect of Platinum catalyst in a Proton Exchange Membrane Fuel Cell

Shibely R. Saha[1] and Dongmei Li[1]*
[1]*University of Wyoming, 1000 E. University Ave. Laramie, WY 82071*
*dli1@uwyo.edu

Platinum (Pt) loading in proton exchange membrane fuel cells (PEMFC) is reduced and utilization increased by studying particle size and support effect on Pt catalysis. Generally, smaller Pt particle size in the PEMFC electrode catalyst layer provides increased reaction sites, thus maximizing the utilization and minimizing the loading. Another important factor that contributes to efficient catalyst use is its distribution upon the support, as well as how supported catalysts are dispersed in the gas diffusion layer (GDL). Catalysts that are well-distributed on its support and well-dispersed on the electrode provide desired contact for the reactants and improved mass transport of the reactants to the active catalytic sites [1].

To investigate the impact of support on Pt catalyst, transition metal carbides (TMCs), such as Molybdenum Carbide (Mo_2C), were selected as a support for Pt, due to their similarity in electronic and geometric structure to Pt [2] . Although previous research has used TMCs, including Mo_2C, as Pt supports, low TMC surface area has limited the catalytic performance[3]. To address this challenge, nanosize Pt has been discretely deposited upon the surface of a Mo_2C hollow nanotube support (confirmed by HRTEM) via island growth mechanisms of Atomic Layer Deposition (ALD). Nanosize Pt particles supported onto hollow nanotube Mo_2C (also referred as co-catalyst) with sub-monolayer coverage are presumed to further increase the reactivity of the catalyst, possibly resulting from both increased active Pt sites and enhanced interaction between the Pt particle and support. The reactivity of the co-catalyst towards Hydrogen Evolution Reaction (HER) and Hydrogen Oxidation Reaction (HOR) has been evaluated and compared to that of commercial 20% Pt supported on carbon black powder by electrochemical characterization using a three-electrode cell. This electrochemical characterization showed significantly higher HOR and HER activity from the co-catalyst with 2.4% Pt loading (confirmed by Atomic Absorption Spectroscopy (AAS)), compared to 20% Pt/C, demonstrating its potential applicability as PEMFC anode.

To evaluate co-catalyst performance in a device setting, PEMFC membrane electrode assembly (MEA) was fabricated using the co-catalyst as PEMFC anode. Device performance demonstrated a significant increase in maximum power per mg of Pt, compared with commercial 20% Pt supported on carbon black powder. As such, both electrochemistry cell and device testing results revealed that the sub-monolayer Pt coverage (a.k.a. discrete islands) on nanoscale Mo_2C tubes generated a synergetic effect that requires further systematic theoretical and experimental study to understand.

To investigate how particle size and distribution of the co-catalyst on electrode affects the device performance of PEMFC MEA, ultrasonic AccuMist (Sono-Tek) spray technique was applied, where the catalyst can be atomized, minimizing co-catalyst agglomeration, often observed in conventional brush painting approach. Scanning Electron Microscopy (SEM) confirmed that smaller particle size was generated on the electrode with ultrasonic AccuMist spray approach, with 37% increase in maximum power per mg of Pt in the electrode compared to that by conventional brush painting technique.

References

1. Lister, S; McLeanG., *Journal of Power sources*, p.61-76, 130(2004)
2. Levy, *R.B., Science*, p. 547-549, 181(1973)
3. Esposito, D.V.; Hunt, S.T.; Kimmel, Y.C.; and Chen, J.G.; *J. Am. Chem.Soc.,* p. 3025−3033, 134(2012)

390688 Carboxyl Multi-Walled Carbon Nanotubes Stabilized Palladium Nanocatalysts for Fuel Cell Application

Abstract:

Carboxyl Multi-walled carbon nanotubes (MWNTs-COOH) supported palladium (Pd) nanoparticles (NPs) catalysts (Pd/MWNTs-COOH) were prepared by a one-pot solution-based method, i.e., direct decomposing palladium acetylacetonate ($Pd(acac)_2$) in a refluxing xylene solution in the presence of MWNTs-COOH without any additives. Pd/MWNTs-COOH with different Pd loadings were prepared by altering the amount of $Pd(acac)_2$. Different annealing temperature effect was also investigated on Pd/MWNTs-COOH in a reducing atmosphere. X-ray diffraction (XRD) results indicate a successful deposition of Pd NPs on MWNTs-COOH. Transmission electron microscope (TEM) micrographs revealed that Pd NPs with increasing size were uniformly grown onto MWNTs surface with the increase of Pd loadings, and varying Pd particle sizes were also observed implying varying steps of Pd NPs as diffusing, contacting and coercing on tube wall surface with the increase of annealing temperature. The loading of Pd was calculated from the thermal gravimetric analysis (TGA), the variation of –COOH was investigated by Raman spectroscopy and X-ray photoelectron spectroscopy (XPS). The electrocatalytic activities of the Pd/MWNTs-COOH catalysts were evaluated for methanol oxidation and ethanol oxidation in KOH solutions. For methanol oxidation, bare Pd NPs and Pd/MWNTs (MWNTs without -COOH) was also synthesized and compared, parameters including the concentration of methanol and KOH were investigated, showing that the best catalyst performance was achieved by the balance between the adsorption of hydroxyl and methanol. A pronounced influence of temperature on methanol oxidation has been manifested, which indicates the positive role of increasing temperature. For ethanol oxidation, Cyclic voltammetry (CV), chronoamperometry (CA), electrochemical impedance spectroscopy (EIS) and Tafel were introduced to characterize the electro-performances, Pd/MWNTs-COOH with initial precursor ratio of 2:1 gave the best performance, 100 °C annealed Pd/MWNTs-COOH also gave best performance compared with other temperature annealed ones

384084 Surface Grafting and Reactive Modification of Azlactone-Based Polymer Brushes

Abstract:

Reactive modification of polymer thin films provides a useful route to confer new properties to the underlying material, with the range and strength of interactions across the interface dictated by the display of functional groups along the surface. Layers of end-tethered polymer chains, or polymer brushes, were created by grafting chains of poly(4,4-dimethyl-2-vinylazlactone) (pVDMA) onto silicon substrates and in situ functionalization studied through the use of ellipsometry and neutron reflectivity measurements as a function of parameters that set brush structure. The results indicate that the grafting density of chains and size of the functionalizing agent play major roles in governing the extent of functionalization, and with all but the smallest functionalizing agents, reactive modification is incomplete. In addition to developing an average view of the in situ functionalization, the sensitivity of neutron scattering methods to isotopic substitution (of D for H) provides insight into the location of functional groups installed within the pVDMA brush by reactive modification. In total, these studies provide perspectives on the relationship between extent of reaction, display of functional motifs, and properties of the interfacial layer.

390181 pH and Thermo-Responsive Behavior of Amino (Meth)Acrylate Polymer Brushes on Silicon Substrates By in-Situ Ellipsometry and Atomic Force Microscopy

Abstract:

Stimuli responsive polymers have potential use in a wide range of applications, from nanometer-scale drug delivery to flow control in microfluidic devices. In this study, surface-confined atom transfer radical polymerization (ATRP) and single electron transfer living radical polymerization (SET-LRP) were used to synthesize a series of surface-grafted poly(amino (meth)acrylate) brushes from silicon substrates that were pre-modified with silane self-assembled monolayers (SAMs). Poly(2-(dimethylamino)ethyl methacrylate) (PDMAEMA), poly(2-(dimethylamino)ethyl acrylate) (PDMAEA), poly(2-(diethylamino)ethyl methacrylate) (PDEAEMA), and poly(2-(tert-butylamino)ethyl methacrylate) (PTBAEMA) were synthesized and characterized using in-situ techniques. In-situ spectroscopic ellipsometry studies were performed using a temperature-controlled liquid flow cell to collect dynamic and equilibrium data under various pH/ionic strength conditions. Multilayer models were generated based on a Gaussian general-oscillator (GenOsc) dispersion model that accounted for the polymer, initiator, and silane SAM layers. In-situ atomic force microscopy (AFM) measurements were also collected under different pH and temperature conditions using a liquid cell. AFM was used to identify changes in the polymer brush structure, including phase separation/aggregation behavior changes. In addition to measuring thickness changes, response rates, and stimuli conditions for these surface-tethered pH and temperature responsive poly(amino (meth)acrylates) brushes, for the first time application of the GenOsc model to evaluate the optical constants of surface-tethered polymer brushes is presented.

Closed Batch iCVD of Ultra-Thin and Conformal Polymer Films

*Nan Chen, Christy D. Petruczok, Karen K. Gleason**

Initiated chemical vapor deposition (iCVD) has been widely used to produce various functional and conformal thin films with solvent-free processing. This process has historically been operated as a continuous flow (CF) process. Here, we demonstrate a modified fabrication process based on iCVD for producing ultra-thin and conformal polymer films. This so-called "closed batch" (CB) iCVD process provides better control on the thickness of ultra-thin polymer films while using significantly less reactant than the conventional CF iCVD. In the current work, we synthesized four different polymers, poly(N-isopropylacrylamide), poly(trivinyl-trimethyl-cyclotrisiloxane), poly(1H,1H,2H,2H-perfluorodecyl acrylate), and poly(ε-caprolactone) (PCL) by both CB and CF iCVD. The resulting polymers from CB process are functionally identical to those from CF process, and CB films exhibited the desirable conformality characteristic of the conventional CF iCVD. In CB iCVD process, ultra-thin (< 30 nm) films can be controllably and reproducibly deposited, and there is no prior optimization process required to achieve uniform films, which can largely reduce the equipment cost in commercial applications. We also found the CB iCVD films are extremely smooth, exhibiting RMS roughness values between 0.4 and 0.7 nm. Use of the CB process improves reaction yield by factors of 10 to 200 for the four different film chemistries and decreases material cost per 100 nm of film by one to two orders of magnitude. Also in this work, we firstly synthesized PCL, a biodegradable material, using all-dry iCVD process. The CB process makes PCL economically applicable in iCVD as the high cost of the PCL monomer, 2-methylene-1,3-dioxepane (MDO).

382381 Effect of Oxygen on Admicellar Polymerization of Styrene on Silica Surfaces

Abstract:

Previous studies on the formation of ultrathin polymer films on surfaces utilizing admicellar polymerization, the surface analog of emulsion polymerization, relied on unusually high ratios of initiator to monomer for free radical polymerization. The effect of oxygen on the polymer formed through admicellar polymerization on silica particles was studied using cetyltrimethylammonium bromide (CTAB) as the adsorbed surfactant bilayer template, styrene as the monomer, and azobisisobutyronitrile as the free radical initiator. The solutions were deoxygenated by purging the headspace with nitrogen prior to the initiation of the admicellar polymerization of styrene. The formed polymer thin film was extracted from the silica surface with refluxing tetrahydrofuran and characterized using Fourier transform infrared (FTIR) spectroscopy, thermogravimetric analysis (TGA), and gel permeation chromatography (GPC). Samples were compared to control experiments performed without deoxygenation. The results demonstrated that the purging of oxygen significantly decreased termination resulting in the formation of higher molecular weight polymer. Surface characterization of the formed polymer films on the silica was accomplished with atomic force microscopy (AFM).

378938 Humidity-Dependent Compression-Induced Glass Transition of the Air-Water Interfacial Monolayers of Poly(D,L-lactic acid-ran-glycolic acid) (PLGA)

Abstract:

Constant rate compression isotherms of the air-water interfacial monolayers of poly(D,L-lactic acid-*ran*-glycolic acid) (PLGA) show a distinct feature of having an exponential increase in surface pressure in the high surface polymer concentration regime. We have previously demonstrated that this abrupt upturn in surface pressure is linked to the glass transition of the monolayer, but the detailed mechanism of this process had not been fully understood. In order to obtain molecular-level understanding of this behavior, we performed extensive characterizations of the surface mechanical, structural and rheological properties of PLGA monolayers at the air-water interface, using combined experimental techniques including the Langmuir film balance, X-ray reflectivity and double-wall-ring interfacial rheometry methods. We observed that the mechanical and structural responses of the PLGA monolayer are significantly dependent upon the rate of monolayer compression; the glass transition was induced in the monolayer only at fast compression rates. Surprisingly, we found that this deformation rate dependence is also dependent upon the humidity of the environment. With water acting as a plasticizer for the PLGA material, the diffusion of water molecules through the PLGA monolayer appears to be the key factor determining the glass transformation property and thus the mechanical response of the PLGA film against lateral compression. Based on our combined results, we hypothesize the following mechanism for the compression-induced glass transformation of the PLGA monolayer; (1) initially, a humidified/non-glassy monolayer is formed in the full surface-coverage region (where the surface pressure shows a plateau) during compression; (2) further compression leads to the collapse of the PLGA chains and the formation of new surfaces on the air side of the monolayer, and this newly formed top layer of the PLGA film is transiently glassy in character because the water evaporation rate in the top surface region is momentarily faster than the humidification rate (due to the initial roughness of the newly formed surface); (3) after some period of time, the top layer itself becomes humidified through diffusion of water from the subphase, and thus it becomes non-glassy, leading to the relaxation of the applied compressive stress.

Abstract title

iCVD Synthesis and Deposition of Polyglycidol as a Polymer Electrolyte in Dye Sensitized Solar Cells

Authors

Chia-Yun Hsieh, Kenneth K.S. Lau*

Keywords

Polyglycidol, Polymer Electrolyte, Initiated Chemical Vapor Deposition, Dye Sensitized Solar Cell.

Section

Polymer Thin Films and Interfaces (08A04)

Abstract

Polyglycidol (PGL) is a derivative of polyethylene oxide (PEO) containing repeat units of –COC– in the backbone along with –CH_2OH side groups. It has the potential of being a good ion conductor given the presence of a PEO-like ether backbone together with pendant hydroxyl groups. Unlike PEO which is well-known to crystallize easily and hinder ion transport, PGL avoids crystallization due to the steric interference of the side group. In addition, during polymerization of PGL from the ring opening polymerization of glycidol monomer, it is possible for the PGL to undergo branching and crosslinking as a result of reactions through the OH groups. This also further suppresses any crystallization of PGL and allows it to be potentially a good polymer electrolyte for applications in electrochemical devices like the dye sensitized solar cell.

In this study, we report for the first time the synthesis of polyglycidol via initiated chemical vapor deposition (iCVD) that does not rely on the use of liquid solvents. iCVD relies on gas-to-surface reactions that simultaneously leads to polymerization and polymer growth as a thin film. Based on FTIR and XPS analysis, iCVD PGL is well defined with the expected stoichiometry. XRD confirms PGL to be amorphous with no crystalline structure. For application in DSSCs, there needs to be effective integration of the polymer electrolyte with the mesoporous TiO_2 anode, which has been a challenge with conventional methods using polymer solutions due to surface tension and wettability issues. With iCVD, we demonstrate good pore filling by carefully selecting reaction-limited synthesis conditions that provide faster mass transport of reactive species. Through TGA and SEM analysis, up to 92% of the total pore volume within the TiO_2 mesoporous layer is occupied by PGL. We expect PGL to be a good candidate as a polymer electrolyte DSSCs.

Photo-Definable, Low-k Polynorbornene Dielectric for Electronics Packaging

B.K. Mueller, J.M. Schwartz, A.E. Sutlief, P.A. Kohl

School of Chemical and Biomolecular Engineering
Georgia Institute of Technology
311 Ferst Dr., Atlanta, GA 30332

Permanent dielectric materials are necessary in microelectronic devices to separate electronic components. Low dielectric constant (low-k) materials are especially important because they electrically isolate on-chip, chip-to-chip, chip-to-package, or on-package interconnections. Polymers are widely used for these purposes, as they offer a low stress, easily processed alternative to inorganic materials. Additionally, permanent low-k polymeric materials can often be formulated to be directly patternable by photolithography.

Currently, most positive tone, permanent dielectrics utilize the diazonaphthoquinone-based photochemistry, which provides relatively low sensitivity ($D_{100} > 300$ mJ/cm^2) and low contrast (< 2.5). In order to pattern very thick films or high aspect ratio structures, materials with greatly improved lithographic properties are needed. This can be accomplished with a well-known photochemistry known as chemical amplification. In this case, a photo-generated acid catalyzed a deprotection reaction that leads to the solubility switch of the film. The low loading of the photoactive compound provides low absorbing, high sensitivity, and high contrast films. However, cross-linking of these positive tone films after patterning is also desirable, and the presence of the strong acid makes cross-linking difficult.

To this end, our lab has developed a functional chemistry for chemically amplified photo-patterning and subsequent cross-linking. Of particular interest is the low dielectric constant of the resulting polymeric film, ca. 2.2 to 2.8 depending on composition, without added porogens. This was achieved on a polynorbornene backbone with pendent tert-butyl ester and fluoroalcohol moieties. Films of this polymer are cast with small loadings of a photoacid generator (PAG). These films are originally insoluble in aqueous base. Exposure of the PAG to ultraviolet light produces a strong acid that catalyzes the deprotection of the tert-butyl ester. This deprotection results in a carboxylic acid, increasing the solubility of the film in aqueous base. A film with 3 parts per hundred PAG was measured to have a D_{100} of 8.09 mJ/cm^2, and contrast of $\geq$ 14.2. These values are 10x improved over those of existing positive tone dielectrics.

Since these polymer films are permanent structures in microelectronic devices, they are thermally cured to impart good thermal, mechanical, and electrical properties. After photolithography, the unexposed regions of the dielectric are thermally cured for 2 hours at 250°C. During this step, the PAG thermally forms the acid catalyst and deprotects the tert-butyl ester. The resulting carboxylic acid undergoes a Fischer esterification with the alcohol functionality in the solid state to form a covalent cross-linked network. This reaction raises the glass transition temperature of the polymer and reduces the number of polar hydroxyl and carboxyl groups. The esterification is also acid catalyzed but is much slower than the deprotection reaction so that the exposed regions can be deprotected without being cross-linked. This is the first chemically amplified, cross-linkable dielectric.

This polynorbornene dielectric differs from other positive tone polymeric materials, because the curing reaction occurs between two pendent functional groups. As such, it is desired to understand how each of these groups influences the mechanical and electrical properties.

It was found that the monomer ratio of the tert-butyl ester and the fluoroalcohol had a significant effect on physical properties. Increasing the fluoroalcohol content decreased the dielectric constant, likely because of the high carbon-fluorine bond content. However, increasing the fluoroalcohol decreased modulus and hardness. This is likely due to the fact that the carboxylic acid will readily cross-link with another carboxylic acid to form an anhydride, confirmed by infrared spectroscopy. A polymer with 60 mol% fluoroalcohol was measured to have a dielectric constant of 2.78, a modulus of 2.95 GPa, and a hardness of 243 MPa. These values are similar to or exceed those of existing positive tone dielectrics.

In order to further improve electrical and mechanical properties, a third monomer was incorporated into the polynorbornene backbone. This third monomer has an alkyl pendent group which does not participate in photolithography or cross-linking. This monomer addition does not come without challenges, namely in solubility. The alkyl monomer lacks an acidic proton and lowers solubility in aqueous base. The physical properties of this dielectric will be presented. Further, the challenges associated with this chemistry will be addressed in an effort to formulate a robust dielectric with improved lithographic, mechanical, and electrical properties.

References
1. Tummalla, R. R.; Swaminathan, M. Introduction to System-on-Package; McGraw-Hill: New York, 2008.
2. Reichmanis, E.; Houlihan, F. M.; Nalamasu, O.; Neenan, T. X. Chem. Mater. 1991, 3, 394
3. Y. So, E. J. Stark, Y. Li, S. Kisting, A. Achen, K. Baranek, D. Scheck, J. Hetzner, J. J. Folkenroth, M. Töpper, and T. Baumgartner, *IEEE Trans. Adv. Packag.* 2006, 29, 4, 741–750.
4. X. Z. Jin and H. Ishii, "Novel positive-type photosensitive polyimide with low dielectric constant," *J. Appl. Polym. Sci.* 2005, 98, 15–21.
5. B. K. Mueller and P. A. Kohl, "Chemically amplified, positive tone, cross-linkable thick-film polymer," *J. Appl. Polym. Sci.* 2013, 130, 2, 759–765.

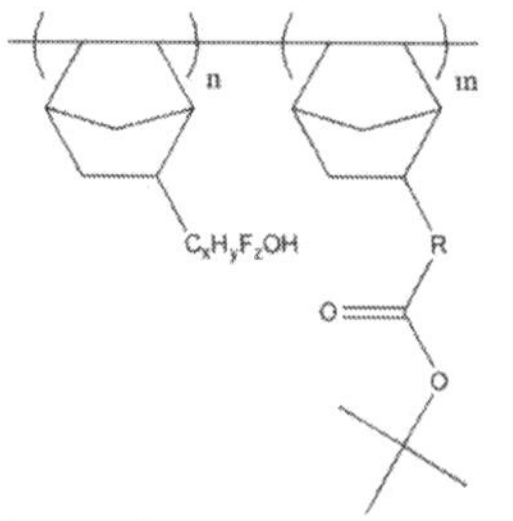

Figure 1: Chemical structure of polynorbornene terpolymer with fluoroalcohol, t-butyl ester, and alkyl pendent groups

381079 Imprinting of Metal Receptors into Layer-By-Layer Polyelectrolyte Films for Antifouling Applications

Abstract:

Imprinting of Cu^{2+} selective layer-by-layer (LbL) films have been investigated and their antifouling performance has been evaluated. Adsorption of polycation having histidne grafted poly (allylamine) (PAH) and methyl ester containing polyanions yields polymer films with histidine ligands that bind metal ions. Ester groups within the polyanion are favorable for covalent cross linking with the histidine grafted polycation. Two methods of LbL film formation have been demonstrated alternate deposition poly anion with high density methyl esters and poly cation with histidine ligand (non-imprinted LbL_A), PAH-Histidine is coordinated with Cu^{2+} ions in solution and then assembled into multilayer films with poly anion containing high density of methyl esters (imprinted LbL_B). In the end, both the LbL films are cross linked at mild temperature to yield the covalent bridging of the layers. Comparative study of the non-imprinted LbL_A films and imprinted LbL_B films for the Cu^{2+} ions binding capacity, leaching rate and stability of the films have been investigated and demonstrated that the imprinted films are better than the non-imprinted films; thereby the processing method of LbL assembly template is vital for imprinting of metals.

387257 Biocompatibility and Wear Properties of Polymer Coatings Created with Atmospheric Pressure Plasma

Abstract:

Atmospheric pressure plasma was used to alter the surface of high molecular weight polyethylene (HMWPE) by grafting various biocompatible polymers, to provide a more hydrophilic surface to improve wear resistance for use in applications, such as prosthetics and orthopedic implants.

For this study, the organic coatings included biocompatible polymers such as poly(2-hydroxyethylmethacrylate), polyethylenimine, and polyethylene glycol. Atmospheric pressure plasma was used, along with an in-house constructed spray delivery system, to graft the molecules to HMWPE substrates. The surface of the substrates was maintained at a temperature below 80°C during the coating process. Coatings were characterized by Fourier transform infrared spectroscopy (FTIR), contact angle analysis, and adhesion testing. A significant decrease in contact angle was noted for various coatings produced with this method, indicating increased hydrophilicity. Plasma processing conditions, specifically the pretreatment of the substrate and the input power, affected the adhesion and uniformity of the polymerized layer. Trends varied for the polymers used in this study, indicating the need for optimization of plasma parameters for the desired polymer coating.

The biocompatibility of the surface was characterized using various techniques described in literature. The wear properties of the coatings were determined by weight loss under conditions of a random motion pin-on-plate apparatus with the coated polyethylene plaques immersed in serum albumin solution. The results indicate that the biocompatible polymer coatings on HMWPE surfaces may improve the performance and longevity of these surfaces when used in biomedical applications such as orthopedic implants and implanted medical devices.

388063 Enhanced Fog Water Collection from Wire Mesh Structures

Abstract:

Water scarcity is a growing problem in the world today and fog represents an untapped source of water in arid, coastal regions. Plants and animals in these environments have evolved to use chemical and textural features on their surfaces to harvest this necessary resource. In this work, we examine the role of surface wettability and substrate geometry on the overall fog collection efficiency of woven and vertical wire mesh networks. An aerosolized water stream was directed at these meshes in a controlled environment to measure their fog collection efficiency. We utilized a blend of poly(ethyl methacrylate) and a low surface energy molecule (fluorodecyl POSS) to tune the surface wettability of the fog collecting meshes, so that they exhibited low contact angle hysteresis with appreciable surface adhesion, which led to enhanced fog collection efficiency. Low contact angle hysteresis improves drainage of captured water off of the mesh structures and high adhesion decreases the propensity of the captured droplets to be re-entrained in the effluent fog stream. We also extend a previously developed framework for predicting the collection efficiency of mesh based fog collectors. The results of preliminary pilot-scale field experiments using these design elements will also be described.

356991 Effects of Synthetic Polymers on Mechanical Strength and Water Absorption of Hardened Gypsum Pieces Made from Calcium Sulfate Hemihydrates

Abstract:

α- or β-Calcium sulfate hemihydrate (α-H or β-H) is commonly used as gypsum-modeling material for ceramics, which require high mechanical strength and water absorption. Polymers may affect the gypsum performance and ratio of water to hemihydrate (W/H). In this experiment, the effect of W/H, the addition and structure of some synthetic polymers on the properties of hardened gypsum piece were investigated. It was found that α-H and β-H are hardly workable in common W/H region; their performances are strictly relevant to W/H. At a W/H as high as 0.7, the more hydrophilic polymers provided an increase in the mechanical strength and water absorption; four sulphur polycarboxylates synthesized were found to maintain or enhance water absorption of β-H, but only the one without carboxyl group led to an increase of the mechanical strength. The results suggest that keeping sulphur and PVA residues in the molecular design will be necessary for polymer structure applied in gypsum modeling material.

Catacarb Carry over incident causing gigantic production loss and Fire Incident

Muhammad Hashim, Fatima Fertilizer Company Ltd. Sadiqabad Pakistan

Abstract

From 31st August to 15th September 2012 Fatima Fertilizer Ammonia Plant faced massive Catacarb foaming in Towers subsequently solution carry over to Methanator 4 times during this period causing gigantic Production Loss and downstream Expander System Fire Incident. This paper describes how the problem was analyzed, the sequence of events, which tests were performed after consulting with Vendor and how the problem was eventually solved.

Introduction

Fatima Fertilizer Ammonia Plant is designed by CF Braun & Co. Alhambra California built in 1967. This plant owned by Exxon and Kemira in 1985. Fatima took over this plant from Netherlands to Pakistan in 2007 that started successful production in 15th March 2010 with design capacity 1500 MTPD. This paper encourages you to answer following questions;

1. Importance of activated carbon filtration how badly it can hit the Catacarb solution if not in service during normal operation.
2. Importance of Color index measurement besides of other tests in Catacarb Analysis.
3. Methantor trip Logic Modification in order to avoid Methantor run away reaction
4. Expander trip logic modification to avoid oil console over pressurization / Fire Incident

Incident Description

Ammonia Plant was in restart after 10 days planned shutdown, while restarting the ammonia plant from 29th August 2012, various problems were encountered including foaming in Catacarb causing excessive carryover of solution repeatedly to Methantor, Methantor run away reaction and Cold box Expander process gas breakthrough from HP drain to oil console causing huge fire incident. Issues were resolved accordingly and Production resumed on 15th Sept. 2012.

Sequence of Event:

31st August at 1730 hrs; Catacarb solution carried over with process gas to downstream exchangers and Methantor R-104 due to foaming in CO2 absorber. R-104 was immediately isolated, depressurized and drained.
1st September at 1430 hrs; Catacarb solution carryover was again observed due to foaming causing high level of CO2 absorber downstream knock out vessel D-133.
2nd September at 1030 hrs; Due to poor performance of Methantor feed effluent exchanger E-120A/B. Flushing of E-120A/B was carried out with turbine condensate.

3rd September at 1540 hrs; Even with Excessive shock dozing of antifoam in catacarb foaming could not be controlled and Methantor bed temperature ran away to maximum value on first bed TI-025B 516 °C due to high CO_2 slip from absorber. Methantor trip security I-6 was manually actuated due to high temperatures of Methantor.
4th September at 0915 hrs; Cold box expander located at downstream of Methantor; Huge Fire was occurred at C-103 oil console due to oil splashing on hot steam lines caused by over-pressurization of oil console due to process gas at -140°C entrance into sour oil HP drain line because C-103 was not isolated timely from field during emergency trip of Methantor because of catacarb carry over again Consequently Expander seal, some instrument cables and instruments around C-103 were damaged which were replaced.
12th September at 1850 hrs; Catacarb foaming still not came in control even taking all control measures air injection, reducing circulation rates, increasing anti foam doze etc. Vendor E&A was also consulted but in vain so it was decided to replace 40% of the total catacarb charge.
15th September at 2330 hrs; Ammonia production was resumed under low load operation with controlled foaming and bearable flooding.

Root Cause analysis of Catacarb Foaming and Expander FIRE

1. **Solution foaming tendency:**

   Fatima CATACARB system was equipped with a carbon filter to treat around 30 m3/hr of solution continuously. The best defense against foaming was extensive carbon and mechanical filtration. Fatima had kept the activated carbon filter idle since start up based on their Lab Results. Without carbon filtration, the solution foaming tendency had been rising; Organic contaminants gradually accumulated in the solution, for example: derivatives of heavier hydrocarbons, amine degradation products, glycols from antifoam, alcohols from shift reactors etc. They reduced solution surface tension and increase foaming tendency.

2. **Passivation**

   Fatima did passivate the system during startup by circulating the solution for 41 hours. Due to lack of accurate available inhibitor information from lab, the effectiveness of passivation was unknown. Fortunately, foaming from high iron and suspended solids alerted the operation to increase air injection and later performed another passivation. Consequently, the system regained corrosion protection with iron dropped to 60-70 ppm. Otherwise, if corrosion proceeds, it not only creates heavy foam but also may damage the semi-lean tower.

3. **Local Corrosion:**

   Review of operating procedures and solution chemistry before shutdown did not show any abnormality however local corrosion might be resulted from bed churning in semi-lean regenerator during regeneration due to excessive liquid level. Since no solution was returned to the storage, excessive liquid level in the semi-lean regenerator submerged the reboiler vapor return nozzles and caused water hammering on packed beds and packing. Through violent scratching and abrasion against carbon steel vessel walls and among carbon steel rings, passivation films breached and minor corrosion starts. Lab results clearly indicated high iron and suspended solids. However, the analytical method for available inhibitor was interfered by dark solution color and did not detect any difference.

4. Suspended solids:

Once corrosion happens, Fe+2 will fast buildup in the solution. Through oxidation, Fe+2 is converted to Fe+3 and precipitated from the solution as iron oxides.

4.1 Laboratory sample results were misguided the process team when sample were sent to nearby industries then Suspended solids were reported 1560 ppm against the recommended value <100 ppm while Fatima Lab was reporting 119 ppm during foaming incident due to difference in analytical method.

4.2 Color index method was not being in used since commissioning as it was not mentioned in Vendor analysis guidelines and during foaming incident when color index analysis were carried out it reached to 12800 Hz

5. Foaming tendency by Shake Test:

Foam shake test help operators to develop close understanding of frothiness of solution but this technique was not being used. Only foaming tendency was being checked by sparger test. Shake test usually offers advance warning against pending problems.

Laboratory Results:

Test	Catacarb Results Comparison (Engro / Fatima)								
Date: 5th Sep		Catacarb Lean T-101			Catacarb Semi-Lean T-102			Engro Semi-Lean Tower	
	Date	by Fatima	By Engro		by Fatima	By Engro		Engro Results	
Time (Hrs)		0500	Value	unit	0500	Value	UOM	05-09-2012	05-09-2012
K_2CO_3	%	19.5			14.3				
$KHCO_3$	%	8.4			18.0				
Equivalent K_2CO_3	%	25.3	24.27	cc	26.7	26.0	%	26.8	26.2
V^{+5}	%	0.39	0.29	%	0.30	0.29	%	0.4	
V^{+4}	%	0.01	0.09	%	0.02	0.08	%	0.04	
Iron Total	ppm	~68	64	ppm	~91	62.01	ppm	53.2	
Amine	%	0.83	0.98	%	0.90	1.05	%		1.1
Chlorides	ppm		22	ppm		24.0	ppm		
Formate	%		1.39	%		1.49	%	2.5	2.5
Foaming(Height)	ml	Nil	4	cm	Nil	4.0	cm		2.5
Foaming(Break Time)	Sec	Nil	9.1	sec	Nil	9.6	sec		5.0
pH								10.1	10.0
Color	Hazen	22820	1.192	Abs	23360	1.313	Abs	0.365	0.366
S.Solids	ppm	119	1240	ppm	166	1560.0	ppm	42.0	

6. Methantor Trip Logic:

Due to Catacarb carry over during flooding from CO2 Absorber CO/CO2 slippages increased temperature and Methantor trip logic was built in only on high temperatures but no tripping upon high level of upstream KO drum as per latest design practices.

7. **Expander Trip Logic:**

Upon Methantor tripping there was no built in trip logic of Expander C-103 tripping. Once Methantor tripped seal gas to Expander was cut, It was operator actions to trip C-103 from DCS and close expander inlet outlet valves from field in order to avoid inadvertent mixing of oil and cold process gas. As its mixing could cause freezing and cold process gas ingress into oil console system could cause console over pressurization and Fire Incident.

Root Cause:

1. Different methods for analyzing Total Vanadium and Suspended solids misguided the whole TEAM when samples compared with the nearby industries then came to know the BIG difference between both results.
2. Putting Carbon filtration system out of service since commissioning thinking no need of it till the time laboratory results are normal whereas mechanical filtration system remained in service.
3. No built in Methantor Trip Logic upon high level of Absorber KO drum caused Runaway reaction
4. No built in Expander Trip Logic upon Methantor trip caused oil freezing in Expander and delay in operators action to isolate C-103 caused FIRE incident.

Corrective Actions:

1. Installation of one more carbon filter and Put both carbon filters into continuous service with an additional upstream mechanical filter of 5-10 microns retention for suspended solids and a downstream filter of 1-2 microns retention for carbon fines. Normal Results of Fe and SS were achieved in late October 2012.
2. Replacement of activated carbons at least once every 6 months.
3. Analyzing available inhibitor according to E&A's newly revised redox titration method in future.
4. Performing shake test on semi-lean sample to get foam time in each shift to get advance warning against foaming issue
5. Monitor iron and keep it below the suggested 100 ppm against possible corrosion by monitoring total Vanadium contents and injecting air as per requirement.
6. Monitor suspended solids and keep it below the recommended 100 ppm against possible corrosion through stringent monitoring and control on Filtration Mechanism
7. Drain solution back to storage if plant outage is more than a week.
8. Regenerate solution after shutdown by keeping reboiler in operation for 2-4 hours so that bicarbonate conversion drops below 20%.
9. Clean the open sump from time to time and provide a SS liner to avoid in-leaks of contaminants in next Turnaround.
10. Modification of Methantor trip logic upon high level of Absorber downstream KO drum high level.
11. Expander C-103 logic modification to automatically trip C-103 on

11.1 DP across seal gas and HP seal oil (PDI-080) < 0.2kg/cm2

11.2 DP across seal gas and expander casing (PDI-081) drops below 0.1kg/cm2.

12. Installation of degasifying tank at sour oil HP drain line to avoid Oil console over pressurization and consequent fire due to higher seal gas venting.

Conclusion

Although the foaming and flooding instance was very costly and inconvenient we had to suffer production loss from 31st August to 12th October partially or completely but it served as a good reminder that CATACARB needs adequate cares to sustain safe operation. Fortunately, there was no physical damage to equipment.

Cold box Operational Problems causing significant Production Loss

Muhammad Hashim, Fatima Fertilizer Company Ltd. Sadiqabad Pakistan

Abstract

From 23rd March to April 2014 Fatima Fertilizer Cold box operation became mystery for us, We faced cold box frequent transition from Good mode to Bad mode operation with unknown reasons causing significant Production Loss. Investigation on the basis of Why Tree Analysis was carried out and Root cause was identified. This paper describes how the problem was analyzed, the sequence of events, which tests were performed to rule out one by one all issues and how the key factor was eventually solved.

Introduction

The feed gas, containing 3% nitrogen, is cooled to a temperature around -140° C in the first heat exchanger section by heat exchange with the two effluent streams. It then passes through the turbo expander to provide the net refrigeration. The feed is subsequently cooled to around -175°C in second section of the heat exchanger and flows then into the rectifier column where the liquid portion (mainly nitrogen) drops into the bottom and the Vapor flows upward. Liquid from the bottom is flashed at reduced pressure via the Joule-Thompson valve; the resulting drop in temperature provides the driving force in the condenser. Reflux from this condenser washes methane and argon out of the up flowing vapors. As a consequence, the rectifier produces a hydrogen-nitrogen mixture in the ratio three to one, with only little residual argon and methane. This stream and the waste gas stream from the condenser flow upward through both exchangers and cool the feed.

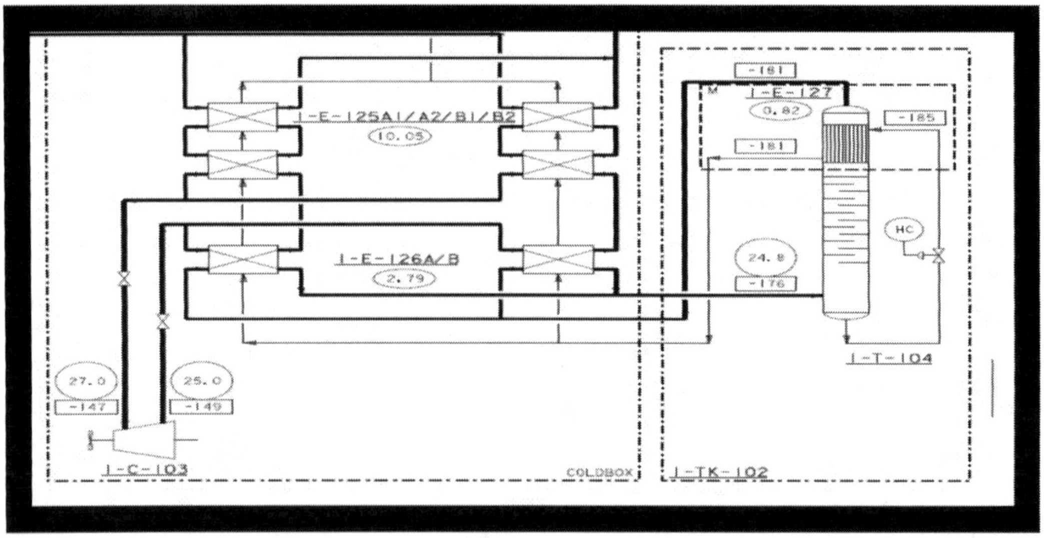

Problem Description

Cold box feed gas inlet path frequent frosting in E-125A2/B2 and bad mode operation in E-126 A/B used to observe each after 1~2 months due to un-known reason resulting in gradual decrease in performance of middle exchangers. Ultimately heat transfer duty of middle exchangers E-125A2/B2 drop below a certain value consequently level in tower (T-104) start depleting rapidly. Due to drop in level it becomes difficult to maintain H/N ratio for Ammonia Synthesis leading to back end shutdown followed by cold box warm up. Production to Production loss is 0.17 Billion after every 1~2 months period.

Sequence of Event

On 23rd March 2013 Delta T across E-125A2/B2 exchangers dropped badly and duty shifted on upper exchangers. Consequently delta P across expander increased and level in tower start depleting sharply, H/N ratio changed from 3:1 to 2.5:1 and ammonia production stopped. Once again Bad Mode situation happened after Turnaround on 27th May 2013 this time temperature profile across both cores varied surprisingly shown in graphs.

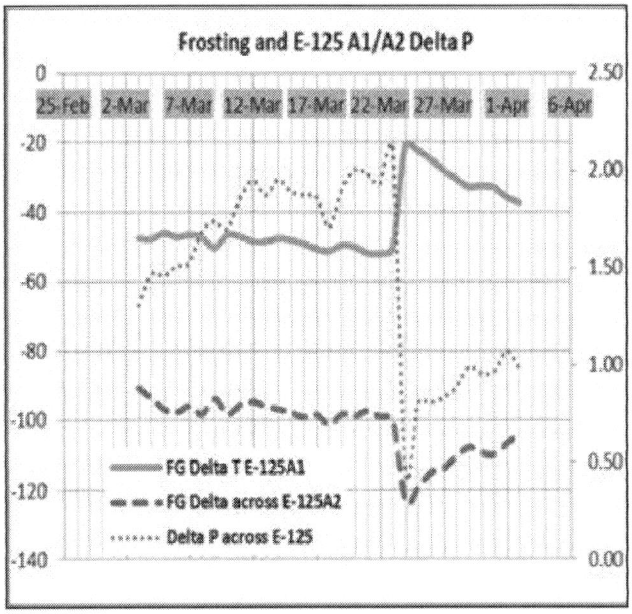

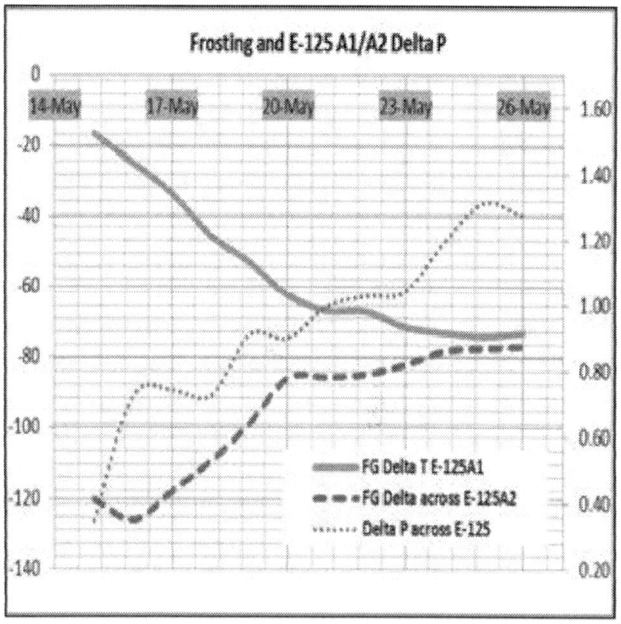

Possible Causes

Possible causes for Exchangers bad mode operations were think through to pin point for addressing the actual root causes. Following causes are noted;

1. Moisture slip from upstream driers section causing frosting in Cold box
2. Ammonia slip from upstream driers section causing frosting in Cold box
3. Carbon dioxide slip from upstream driers section causing frosting in Cold box

4. Driers desiccant dust from filters FIL-104A/B outlet
5. C-103 Expander oil film in Cold box exchangers E-126A/B
6. Waste gas Mal distribution from E-127 to E-126 A/B
7. Mechanical Failure of Cold Box Exchangers

Why Tree Analysis

Based upon above possible causes Why Tree Analysis was carried out and root cause was find out.

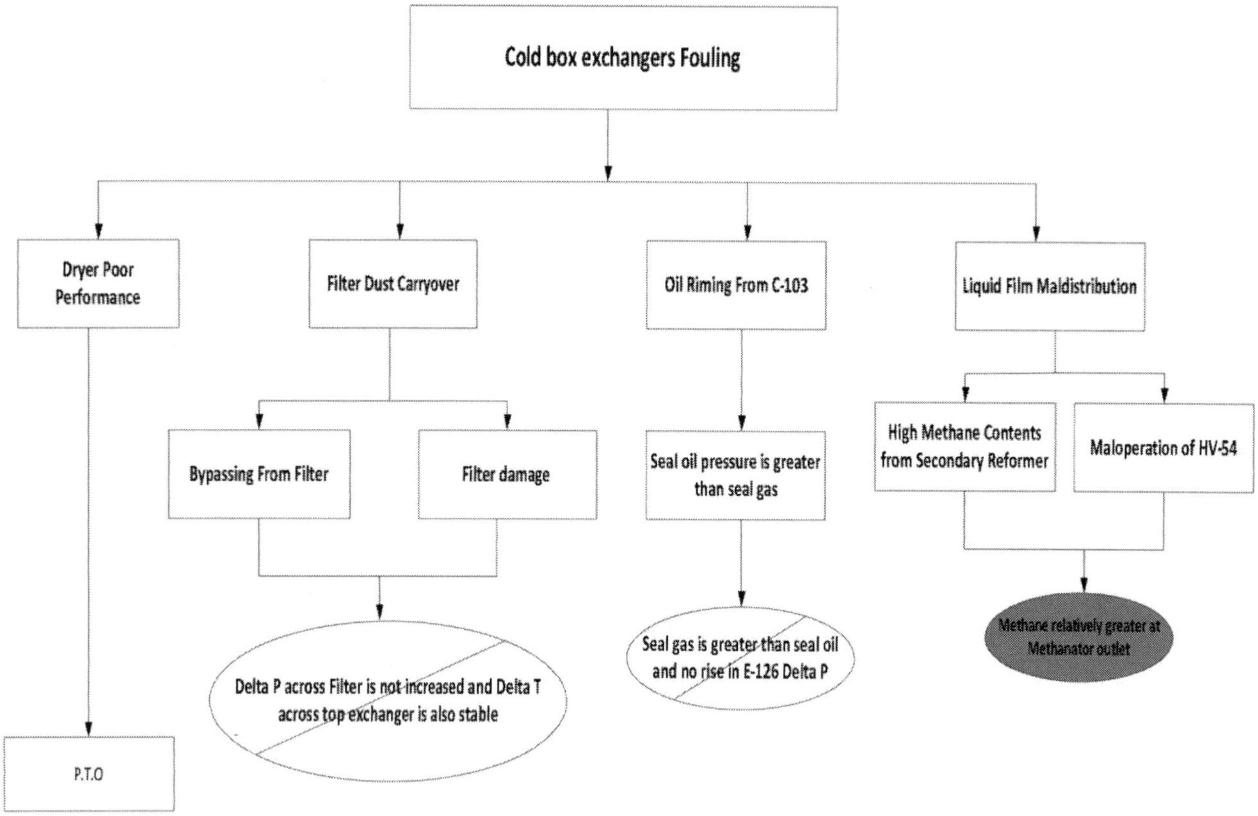

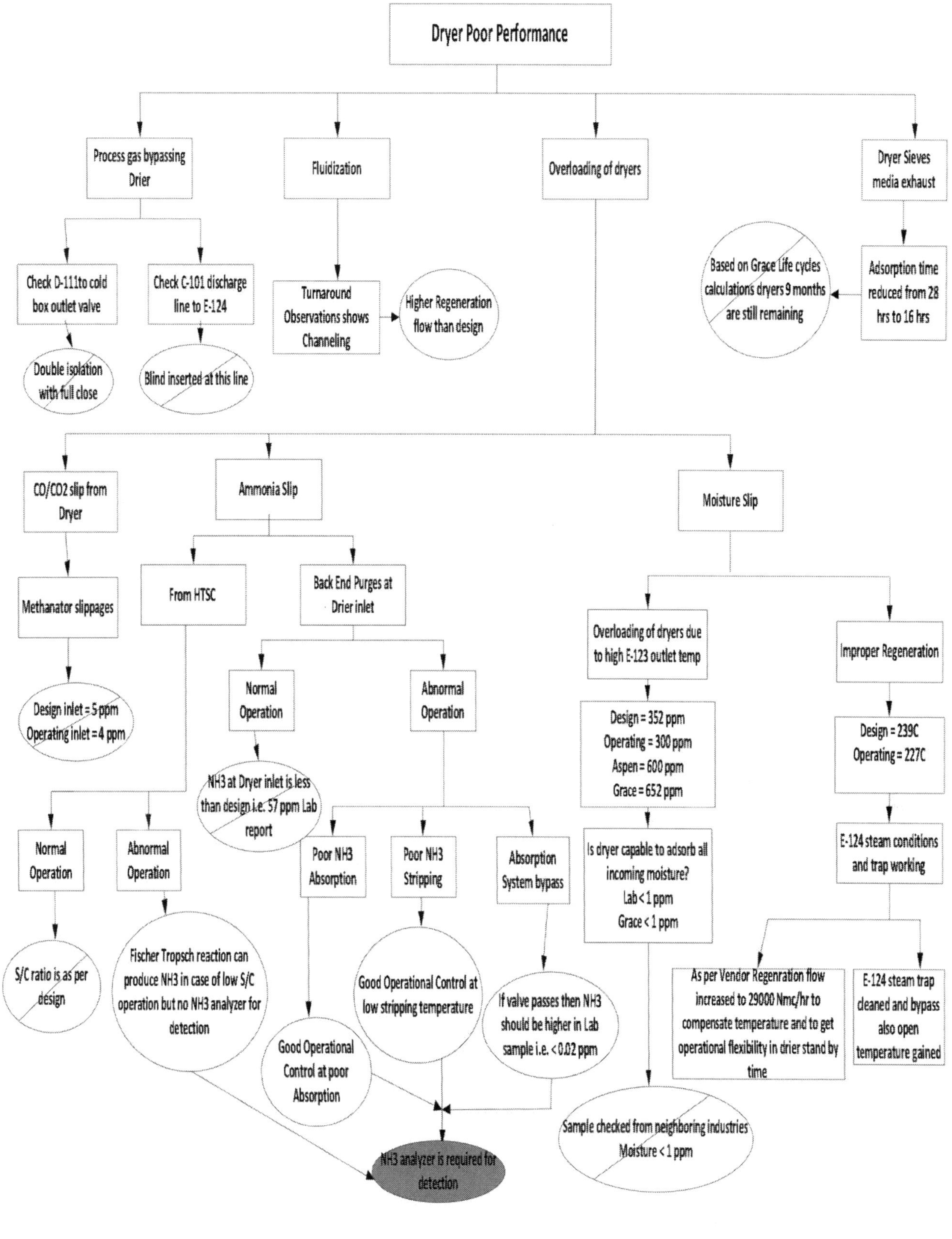

Dryer Poor Performance
Process gas bypassing Drier
Fluidization
Overloading of dryers
Dryer Sieves media exhaust
Check D-111to cold box outlet valve
Check C-101 discharge line to E-124
Turnaround Observations shows Channeling
Higher Regeneration flow than design
Based on Grace Life cycles calculations dryers 9 months are still remaining
Adsorption time reduced from 28 hrs to 16 hrs
Double isolation with full close
Blind inserted at this line
CO/CO2 slip from Dryer
Ammonia Slip
Moisture Slip
Methanator slippages
From HTSC
Back End Purges at Drier inlet
Overloading of dryers due to high E-123 outlet temp
Improper Regeneration
Design inlet = 5 ppm
Operating inlet = 4 ppm
Normal Operation
Abnormal Operation
Design = 352 ppm
Operating = 300 ppm
Aspen = 600 ppm
Grace = 652 ppm
Design = 239C
Operating = 227C
NH3 at Dryer inlet is less than design i.e. 57 ppm Lab report
Normal Operation
Abnormal Operation
Poor NH3 Absorption
Poor NH3 Stripping
Absorption System bypass
Is dryer capable to adsorb all incoming moisture?
Lab < 1 ppm
Grace < 1 ppm
E-124 steam conditions and trap working
S/C ratio is as per design
Fischer Tropsch reaction can produce NH3 in case of low S/C operation but no NH3 analyzer for detection
Good Operational Control at low stripping temperature
If valve passes then NH3 should be higher in Lab sample i.e. < 0.02 ppm
As per Vendor Regenration flow increased to 29000 Nmc/hr to compensate temperature and to get operational flexibility in drier stand by time
E-124 steam trap cleaned and bypass also open temperature gained
Good Operational Control at poor Absorption
Sample checked from neighboring industries
Moisture < 1 ppm
NH3 analyzer is required for detection

Recommendations related to Operational Control

Based upon above RCA following recommendations were extracted that derived the whole plan.

1. Heat of vaporization of methane is 125Cal/gr @ -80C vs. 37 Cal/gr @ -80C for N2.
 so Keep Secondary Reformer Gas to Air ratio in such a way that Methane slip remain within design limit. Lower methane slip is beneficial for cold box.
2. Operate Cold box with waste gas pressure as low as possible to achieve maximum vaporization from Cold box Exchangers.
3. Drier regeneration gas flow should be kept as per design 20000 NMc/hr rather to operate at 29000 NMc/hr during cooling cycle to avoid channeling in Drier.
4. Drier End of Run sampling frequency regime for NH_3 analysis and cross verification on Moisture Analyzer.
5. ΔP across Driers outlet Filters monitoring to avoid dust fusion in cold box
6. Operating seal gas pressure greater than seal oil.

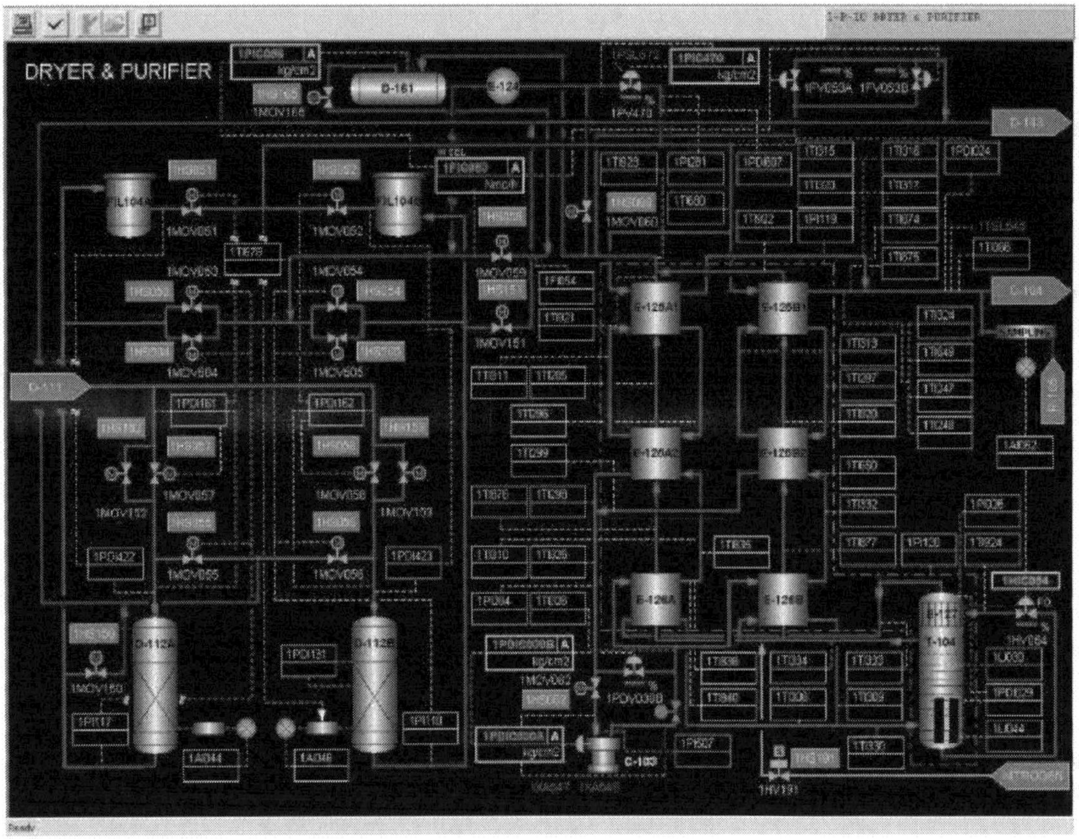

Recommendations related to Engineering Control

Following suggestions are proposed for Drier and Adsorbent Vendors for sustainable Cold box plant operations.

1. Moisture, CO_2 and NH_3 analyzers should be installed at outlet of drier for good operational control at cold box inlet instead of Lab sampling regime.
2. Drier depressurization should always be kept in downward flow to avoid bed lifting.
3. Drier regeneration heating path should be design with upward flow and cooling path with downward flow so that during downward flow flow rate can be increased and time of regeneration may reduce.
4. For Drier pressurization and Depressurization control valves should be proposed instead of gate valves for controlled pressurization and depressurization otherwise channeling may occur.
5. Drier upstream moisture separator should be design with dual density demister pads to avoid chances moisture slip from drier.
6. Drier inlet must contain distribution nozzle for gas equal distribution on bed.
7. Drier adsorbents should be designed in such a way that stand by drier timing should be ~ 10 hours for creating margin in operational flexibility.

All designers to close the above Engineering Controls points in design phase for better services to customer.

Conclusion

Although cold box frosting and bad mode operational instance was very costly and inconvenient we had to suffer production loss from 23rd March to April 2014 partially or completely but it served as a good reminder that Drier needs adequate cares to sustain safe operation of Cold box. Fortunately, there was no physical damage to Cold box equipment.

Urea Manufacture by Novel Techniques Utilizing Alternative Raw Material Sources

SUST-E Urea Process

Taibah Jaffery
Supervisor: Dr. Arshad Hussain

Fatima
Fertilizer Company Limited

Sponsor

* This work has been conducted entirely by the National University of Sciences and Technology, Pakistan. Fatima Group does not own a share in this technology.

2

Fatima Group

Fatima Fertilizer Company Limited

- One of the largest business groups in Pakistan with diversified businesses in fertilizers, textiles, and sugar industries.
- Primary business includes manufacture of compound fertilizers- annual capacity 2.1 MTonnes
- Fatima Group has experience in undertaking large-scale green-field projects in Pakistan
- FFL was conceived at a cost of M USD 750 in December 2003

Fatima Group

Fertilizers

Pakarab Fertilizers

Fatima Fertilizer

Sugar

Fatima Sugar Mills

Trading

Reliance Commodities

Textiles

Reliance Weaving

Fazal Cloth Mills

Energy

Fatima Energy

3

Fatima Fertilizers Company

Vision, Mission & Core Values

Fatima Fertilizer Company Limited

Vision

To be a world class manufacturer of fertilizers and ancillary products, with a focus on safety, quality and contribution to national economic growth and development. We will care for the environment and the communities we work in while continuing to create shareholders' value.

Mission

- To be the preferred fertilizer company for farmers, business associates and suppliers through quality and service.
- To provide employees an exciting, enabling and supportive environment to excel in, be innovative, entrepreneurial in an ethical and safe working place based on meritocracy and equal opportunity.
- To be a responsible corporate citizen with a concern for the environment and the communities we deal with.

Core Values

Integrity: Our actions are driven by honesty, ethics, fairness and transparency
Innovation: We encourage creativity and recognize new ideas
Teamwork: We work collectively towards a common goal
Safety, Health and Environment: We care for people and communities around us
Customer Focus: We believe in listening to our customer
Excellence: We strive to excel in everything we do
Valuing People: We value our people as our greatest resource

4

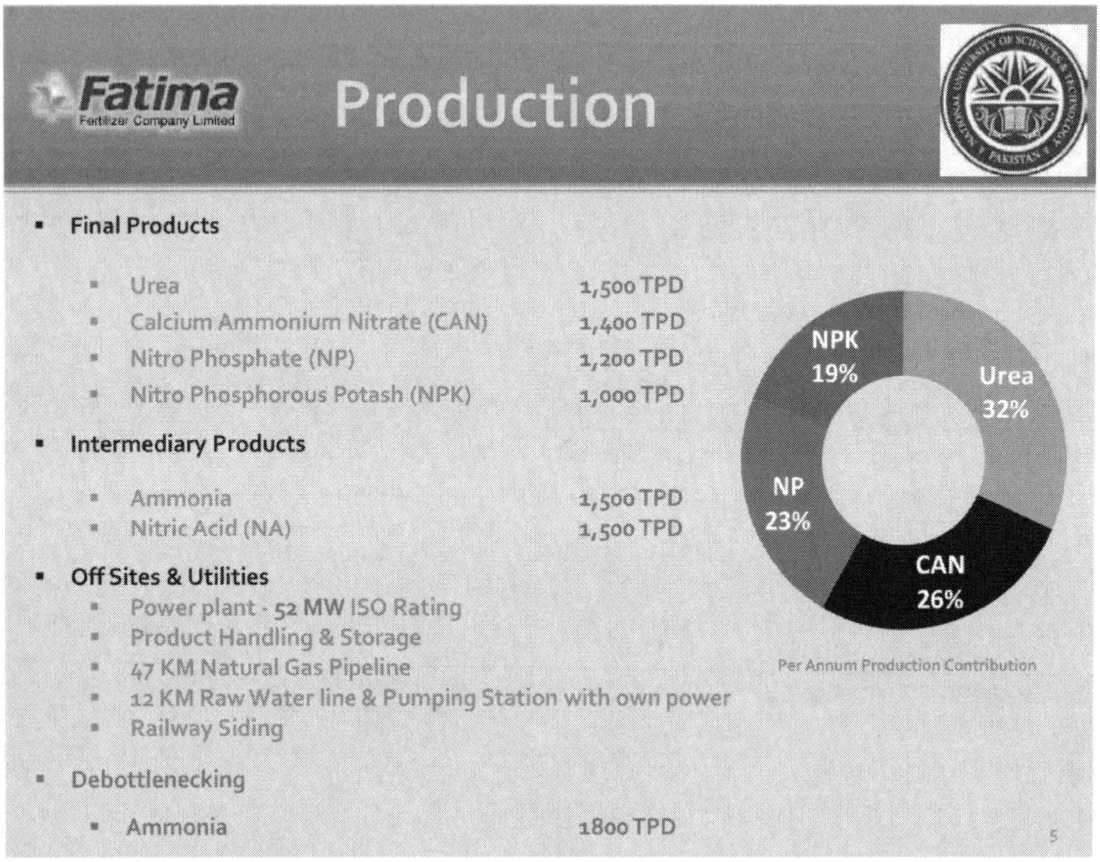
Fatima
Fertilizer Company Limited
Production
Final Products
Urea 1,500 TPD
Calcium Ammonium Nitrate (CAN) 1,400 TPD
Nitro Phosphate (NP) 1,200 TPD
Nitro Phosphorous Potash (NPK) 1,000 TPD
Intermediary Products
Ammonia 1,500 TPD
Nitric Acid (NA) 1,500 TPD
Off Sites & Utilities
Power plant - 52 MW ISO Rating
Product Handling & Storage
47 KM Natural Gas Pipeline
12 KM Raw Water line & Pumping Station with own power
Railway Siding
Debottlenecking
Ammonia 1800 TPD
NPK 19%
Urea 32%
NP 23%
CAN 26%
Per Annum Production Contribution
5

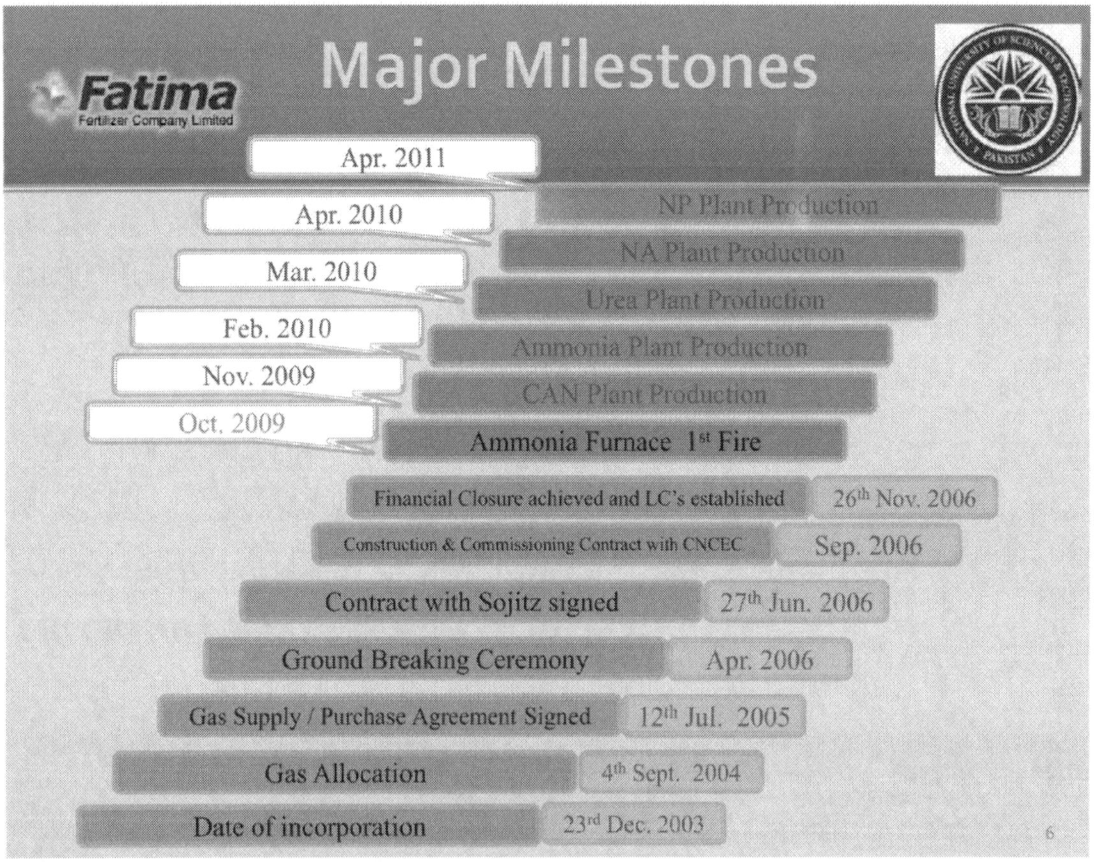
Fatima
Fertilizer Company Limited
Major Milestones
Apr. 2011
NP Plant Production
Apr. 2010
NA Plant Production
Mar. 2010
Urea Plant Production
Feb. 2010
Ammonia Plant Production
Nov. 2009
CAN Plant Production
Oct. 2009
Ammonia Furnace 1st Fire
Financial Closure achieved and LC's established
26th Nov. 2006
Construction & Commissioning Contract with CNCEC
Sep. 2006
Contract with Sojitz signed
27th Jun. 2006
Ground Breaking Ceremony
Apr. 2006
Gas Supply / Purchase Agreement Signed
12th Jul. 2005
Gas Allocation
4th Sept. 2004
Date of incorporation
23rd Dec. 2003
6

Fatima Fertilizer Company Limited

FFL Site Safety

PSM

Lagging Indicators

Target 0.27

Total Recordable Injury Rate (Year to date)

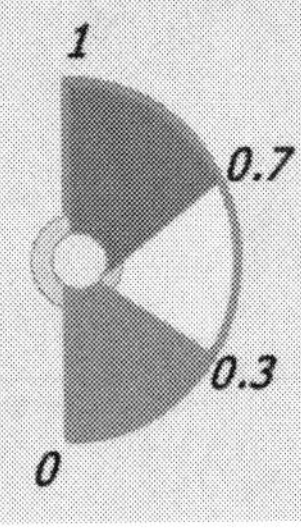

FFL 0.00
Cont. 0.087
Comb. TRIR = 0.061

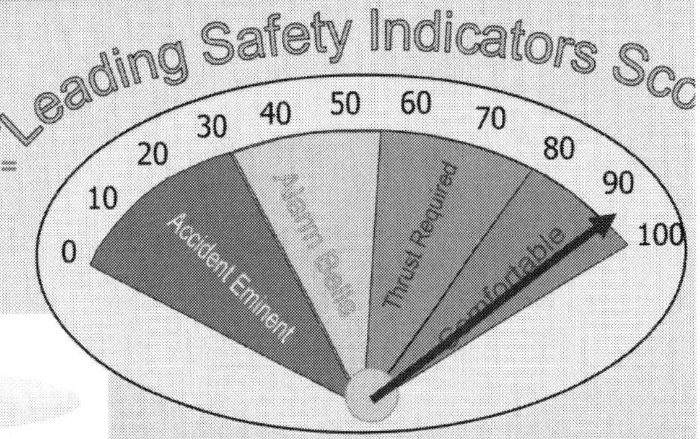

Current PSM
Level at 2013 start

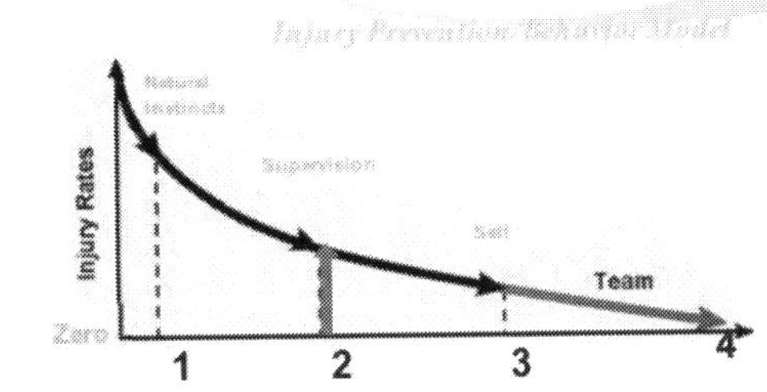

- SAFE WORKING HRS FFL = 6.80 Million
- SAFE WORKING HRS CONT. = 13.89 Million
- SAFE WORKING HRS COMBINE = 20.7 Million

Current 93
Target 80

7

Fatima Fertilizer Company Limited

Future Projects

Debottlenecking

- Ammonia DBN
- Phase 1 completion in Dec 2015 - 1500-->1650 TPD
- Phase 2 completion in 2017 - 1650-->1800 TPD

Midwest Fertilizers Company Project

- Ammonia 1500 TPD
- NA 1500 TPD
- UAN1500 TPD

8

Objective

- To make the Urea manufacturing process inherently independent of non-renewable feed sources (Natural Gas, Coal, etc)
- Sustainable and environment friendly production with minimum net energy requirement
- Minimizing carbon emissions in the atmosphere - Goal of regulating authorities: achieving 90% recovery of CO_2 with maximum 35% increase in electricity cost [1]
- Carbon Sequestration

9

Challenges

- Clean Hydrogen Source
- Clean Energy Source
- Flue Gas Separation into Carbon dioxide and Nitrogen

10

Proposed Solution

- SUST-E Urea Manufacturing Process
- Al-steam reaction for combined production of Hydrogen and Energy
- Membrane Separation of Flue Gases from Thermal Power Plant, Cement Industry, etc.
- Replace Front-end (Steam reforming)
- Renewable energy e.g. CSP for additional energy requirements

$$2Al + 3H_2O \rightarrow Al_2O_3 + 3H_2 + Heat$$

$$\text{Flue Gas} \rightarrow CO_2 + N_2$$

$$N_2 + 3H_2 \rightarrow 2NH_3 + Heat$$

$$2NH_3 + CO_2 \rightarrow NH_2COONH_4 \leftrightarrow CO(NH_2)_2 + H_2O$$

Scope

- Providing a workable design for the new synthesis route
- Investigation of technical feasibility of 7000 TPD Urea and 4000 TPD Ammonia production
- Investigation of economic feasibility of 7000 TPD and 4000 TPD production
- Providing direction for further research
- Tools – MATLAB, ASPEN PLUS

Process Development

Process Research & Innovation	Pilot Studies	Technology Transfer and Manufacturing	Process Intensification & Microprocess Engineering	Product Design

The Sust-E Urea Process

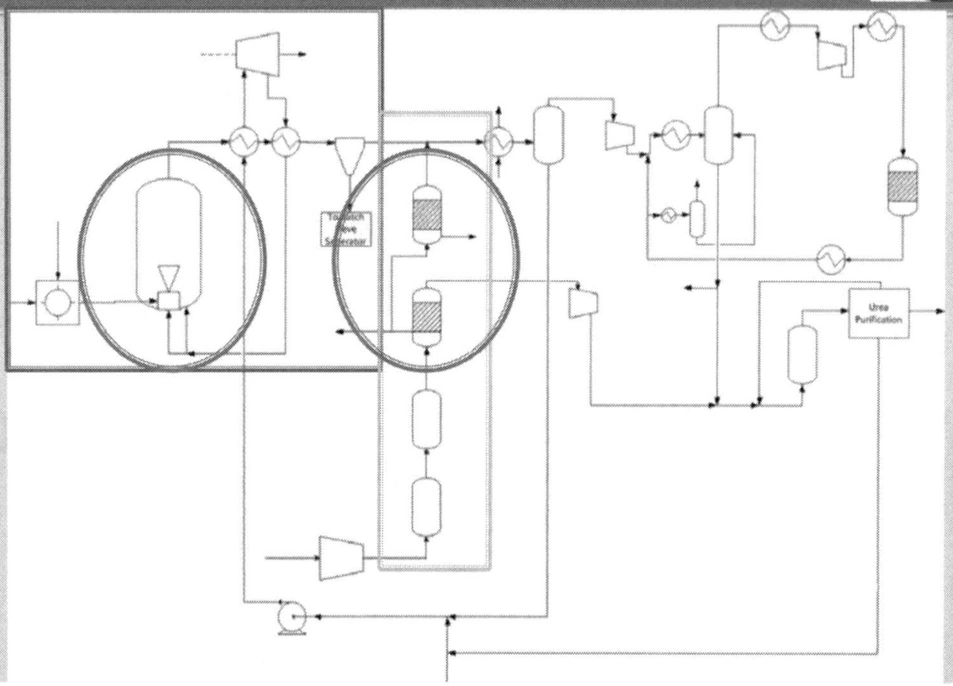

13

Raw Materials Chosen

- Aluminum ingots (grinding tool based on Watanabe's work) [2]
- Flue Gas from coal based plants
- Water
- Energy source – Al/H_2O reaction

14

Aluminium- Steam Reactor

- $2Al + 3H_2O \rightarrow Al_2O_3 + 3H_2$ above 480 °C [3-6]
- Reactants are **fed** at only 600 °F (316 °C) and steam is fed at 150 psig via nozzle
- The Reactor is a combustor (in plug flow)
- 15442 tubes of 10" dia and height of 4ft, arranged in 16 floors, 34 rows of 29 tubes each
- Temperature of 2900 K achieved(> Al melting temperature) [4-5]
- The residence time is 60 ms and conversion is 95%.[4-5]
- Heat of reaction 1100-1250 MW, Heat Loss = 405-680 MW
- ΔH > 540 MW

15

Membrane Separation

- Coal based flue gas
- Two stage, co-current process using PVAm membrane [7-8]
- Nitrogen purification using Carbon Molecular Sieve Membrane (Celo550) [9]
- Multi-component model used to for designing using MATLAB
- Membrane Area (m^2): PVAm Stage1= 2594122, Stage2= 627118
- Celo550 Area ~ 27,922,908.05 ft^2
- 94.6 % pure CO_2 & 94.5 % pure N_2 recovered after membrane separation.
- Celo 550 counter-current membrane separator gives 99.99% pure nitrogen.

16

PFD for Membrane Separation

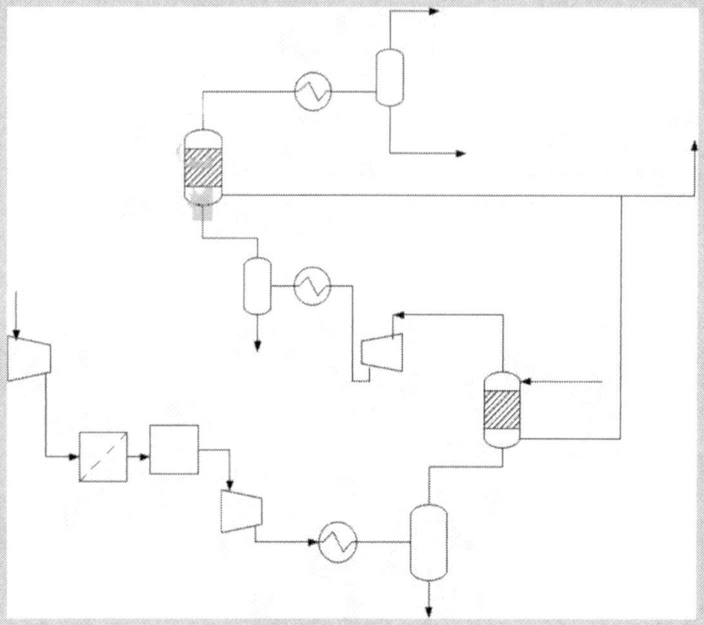

17

Process Integration

- Process Integration for reduction of energy and capital consumption.
- Reaction heat used to heat up feed as well as produce excess electricity
- Water vapor from exhaust steam turbine condensate used as sweep gas

18

Consumption and Production

Urea (99.8%)	7000 TPD
Ammonia (including inventory)	4000
Aluminium	6670
High Purity Alumina	11973
H_2O	4265
H_2	710
N_2	3290
CO_2	5119
Excess N_2	30008
Flue gas	39852 (1206331.27 Nm^3/hr)

19

Energy Efficiency

- Released by exothermic reaction: 1100-1200 MW
- Electricity production: 315 MW including 129 MW required for membrane separation
- Requirement for Al-Steam Feed: 260-280 MW
- Total major plant electricity requirements (grinding, vacuum systems, etc): 174.3 MW
- Surplus: 140.7 MW

*Turbine and generator efficiencies have already been accounted for

20

Cost Feasibility

- Total Plant Investment: $ 138 billion (at current cost of Alumina) – will be reduced by:
  - Increase in supply of alumina refractory after establishment of Sust-E Urea Technology
  - More efficient reactor design with better heat recovery and lower surface area
- Highly expensive bi-product (High Purity Alumina)
- Payback time: 2 yrs
- ROR = $ 69 billion/yr (can be achieved at $32/kg of alumina (34% of current cost) and $0.20/kg of Urea)

21

Sustainability

- Inherently independent of non-renewable resources - capable of running on solar power
- 0.73kg CO_2/kg Urea consumed, hence CO_2 emissions reduced.
- Feed Aluminum can be produced using renewable energy sources
- Eliminates requirement for catalysts (and hence emissions related to their production)

22

Conclusion

- The process is technically capable of achieving process requirements (purity, temperature, pressure) for ammonia and urea manufacture
- Economically feasible
- Energy Efficient

23

Areas for Further Research

- Pilot Studies
- Optimization of reactor design for reduced heat loss and surface area (and hence cost)
- Cost Models for Aluminum and High Purity Alumina nano-particles after establishment of Sust-E Urea plants
- Scale-up of electrolysis cells using non-consumable cathodes for Aluminum production
- Higher efficiency membranes

24

References

- [1] Favre E, "Carbon dioxide recovery from post-combustion processes: Can gas permeation membranes compete with absorption?"
- [2] Watanabe M, Ximeng J, Ryuichi S, "Method for generating hydrogen utilizing activated aluminum fine particles." US patent. US 2006/0034756 A1; 2006
- [3] Petrovic J, Thomas G, "Reaction of Aluminum with Water to Produce Hydrogen – A Study of Issues Related to the Use of Aluminum for On-Board Vehicular Hydrogen Storage", U. S. Department of Energy
- [4] Foote, J. P, Thompson, B. R, Lineberry, J. T, "Combustion of Aluminum with Steam for Underwater Propulsion"
- [5] Franzoni F, et al., Combined hydrogen production and power generation from aluminum combustion with water: Analysis of the concept, International Journal of Hydrogen Energy (2010),
- [6] Golovitchev, V. I, Al/H2O Combustion, Report No. 2, Combustion Engine Research Centre, Chalmers University of Technology
- [7] Hussain, A., A single stage membrane process for CO2 capture from flue gas by a facilitated transport membrane, Journal of Separation Science & Technology 47 (2012) 1857-1865.
- [8] Hussain, A., Hägg, M.-B., A feasibility study of CO2 capture from flue gas by a facilitated transport membrane, Journal of Membrane Science 359 (2010) 140-148.
- [9] Campo M. C, Magalhães F. D, Mendes A, Separation of nitrogen from air by carbon molecular sieve membranes

25

Questions?

26

Crystalline strontium hafnate, $SrHfO_3$ (SHO), is an ideal candidate to study as a suitable high-*k* gate dielectric on Ge. SHO ($a \sim 4.069$ Å) has a reasonable lattice match to the Ge (001) surface ($a/\sqrt{2} \sim 3.992$ Å), yielding a ~1.9% compressive strain in the epitaxial film. SHO shows a high permittivity (k~20) with appropriate band alignment (~1 eV offset) to Ge. In addition, the crystalline nature of the SHO film is expected to drastically reduce the interface trap density at the oxide-Ge interface. We will report our recent results on the growth, characterization, and electrical performance, of epitaxial SHO films and heterostructures for next-generation high-*k* dielectrics on Ge.

In our recent publication, we reported on the direct growth of crystalline strontium titanate, $SrTiO_3$ (STO), on Ge via atomic layer deposition (*Adv. Mater. Interfaces* doi: 10.1002/admi.201400081). Electrical measurements of a 15-nm thick undoped STO film show a large dielectric constant (k~90), but high leakage current (~10 A/cm^2 at +1 eV). In the present work, the unfavorable conduction band offset (and high leakage current) of STO on Ge is circumvented by growing the Hf-based perovskite, SHO. For the growth of SHO, we employ strontium bis(triisopropylcyclopentadienyl), hafnium formamidinate precursors and water. After thermal deoxidation, the Ge substrate is transferred *in vacuo* to the deposition chamber where a thin film of $SrHfO_3$ (2-4 nm) is deposited by ALD at 225 °C. Following post-deposition annealing at 700 °C, the perovskite film becomes crystalline with epitaxial registry to the underlying Ge (001) substrate. *In situ* x-ray photoelectron spectroscopy confirms stoichiometric to Sr-rich films with no GeO_x formation or carbon impurities.

Ex situ x-ray diffraction confirms the perovskite structure and orientation of the SHO film. Thicker SHO films (above 2 nm) appear to show a relaxed lattice constant, indicating relaxation of the epitaxial film above the critical thickness. The electrical performance of several SHO films and heterostructures will be presented. In general, the leakage current is reduced by several orders of magnitude for the SHO films versus STO on Ge. The current work demonstrates the promise for crystalline oxides grown by ALD on Ge for advanced semiconductor devices, including high-mobility Ge-based transistors.

High mobility amorphous oxide semiconductors for transparent thin film transistor (TFT) applications: Metallization selection and TFT device performance

Sunghwan Lee[1,2], *Karen K. Gleason*[1] *and David C. Paine*[2]

[1] *Department of Chemical Engineering, Massachusetts Institute of Technology Cambridge, MA 02139, USA*
[2] *School of Engineering, Brown University, Providence, RI 02912, USA*

Amorphous oxide semiconductor (AOS)[1,2] materials based on In_2O_3 are candidates for thin film transistor (TFT) channel applications that are now being implemented in high performance active-matrix liquid crystal display and active-matrix organic light emitting diodes technologies. This class of materials has a high field effect mobility (μ_{FE}~10-25 cm^2/Vsec)[3-5] in both the amorphous and crystalline states and is an alternative to present-day a-Si TFTs (<1 cm^2/Vsec) technology. Additional advantages of AOS materials include low temperature processing (RT–200 °C), isotropic wet etch characteristics, and high optical transparency (>85 % in the visible regime) all of which make this material suitable for large area, flexible, and transparent devices on inexpensive polymer substrates.

We have investigated the electrical and structural stability of amorphous In–Zn–O (a-IZO) TFTs. Unlike the pure amorphous In_2O_3 which is unstable and crystallizes to the bixbyite phase at <150 °C, the addition of 10 wt% ZnO to In_2O_3 (a-IZO) stabilizes the amorphous phase to temperatures up to 500 °C. In previous work, we reported[6] high performance a-IZO n–MOSFETs which operate in depletion-mode (V_{Th}~ –3.21 V) and have saturation field effect mobility of (μ_{sat}) of ~18 cm^2/Vsec and on/off ratio of >10^6. These demonstration devices were fabricated on thermally oxidized Si that used the Si as a bottom-gate, thermal SiO_2 as the gate dielectric, semiconducting a-IZO as the channel and conducting a-IZO as the source/drain metallization. We have also deposited TFTs with sputtered dielectric layers to demonstrate an all-room temperature TFT fabrication process.

One challenge is in the selection of metallization materials that have a thermally and electrically stable interface with a-IZO since post-speosition annealing is often used to improve device performance characteristics such as μ_{sat} and sub-threshold slope. These post-deposition process temperatures are well below the crystallization temperature of In_2O_3-based AOS materials. However, we have reported that the stability of the amorphous structure is dependent on the metallization material selected.

In this presentation, we present a study of the microstructural and electrical stability of a-IZO channel materials metallized with Mo, Ti, and demonstrate a simple thermodynamics-based approach to the selection of appropriate metallization materials. In situ TEM, along with XRD and AFM studies, are presented that show that the crystallization temperature of a-IZO metallized with Ti is sharply reduced (T<200 °C) while Mo and conducting a-IZO preserve the stability of the amorphous structure. Further, these studies show that low temperature crystallization of a-IZO metallized with Ti is followed by the formation of an interfacial Ti-oxide layer (TiO_2) and the growth of oxide needles laterally past the Ti contact into the bare back-gated channel region. We also compare the effect of two classes of metallization materials (reactive vs. non-reactive) on the performance of amorphous IZO TFTs which show significantly different device characteristics: reactive Ti-metallized IZO TFTs showed larger device threshold voltages and weaker drain current saturation than those of stable Mo/IZO devices. These differences are attributed to an increase in channel carrier density due to the formation of TiO_2 in reactive Ti/IZO TFTs. These findings of the effect of stable/unstable metallization on the a-IZO TFT device performance may be significant to the selection of metallization materials for the stability of the channel carrier density and threshold voltages in amorphous In-(Ga)-Zn-O TFT applications.

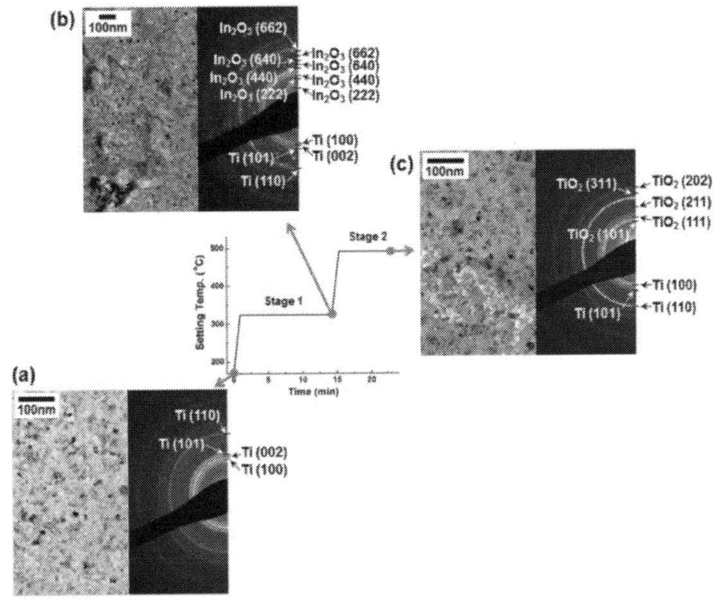

Figure. 1 *In-situ* heating TEM analysis of bilayers of Ti/IZO shows the morphology of reaction and reaction phase of (a) as-deposited state, (b) stage 1 annealed at ~320 °C (nominal) and (c) stage 2 annealed at ~470 °C (nominal).

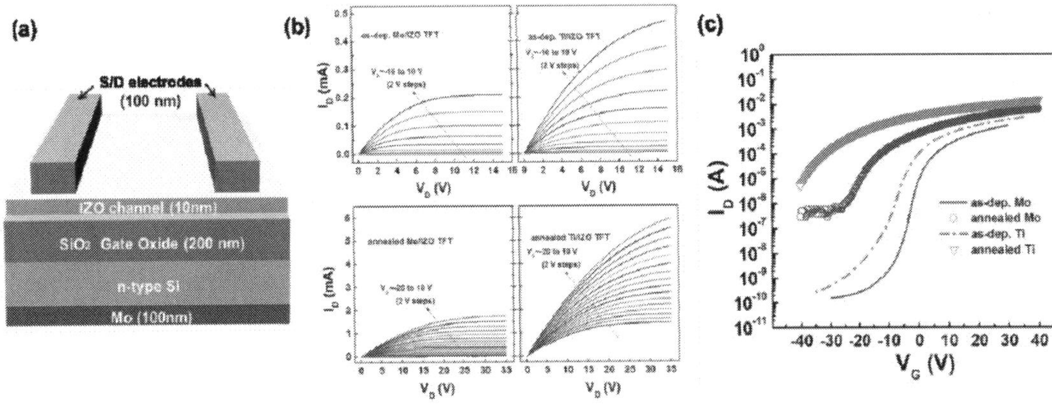

Figure. 2 (a) Schematic of bottom-gated amorphous IZO TFT structure with the aspect ratio of W/L=500 μm/50 μm., (b) Output characteristics of as-deposited (a) Mo/IZO, (b) Ti/IZO and annealed (b) Mo/IZO, (d) Ti/IZO TFT devices, and (c) Transfer characteristic plots of I_D vs V_G from Mo/IZO and Ti/IZO TFTs before and after annealing

Table. 1 Electrical resistivity (ρ) and work function (Φ_m) of potential IZO metallization materials and associated standard Gibbs free energy of reaction (ΔG°_{rxn}) for the formation of oxide via reaction with In_2O_3 ($\Delta G_F^{\circ} = -744.9$ kJ/mol) at 200 °C.

Metal	Mg	Hf	Al	Zr	Ti	Mn	Cr	Ga	Zn	In	W	Mo	Sn	Ni	Cu	Ag
ρ (μΩ cm)	4.39	33.1	2.82	42.1	42.0	144.0	12.5	27.0	5.9	8.37	5.28	5.34	11.5	6.93	1.67	1.58
Φ_m (eV)	3.61	3.65	4.20	4.00	4.20	4.08	4.44	4.35	4.11	4.08	4.55	4.33	4.43	5.24	4.65	4.29
Oxide	MgO	HfO_2	Al_2O_3	ZrO_2	TiO_2	MnO	Cr_2O_3	Ga_2O_3	ZnO	In_2O_3	WO_2	MoO_2 MoO_3	SnO_2	NiO	CuO Cu_2O	Ag_2O
ΔG°_{rxn} (kJ/mol)	-876.1	-807.6	-752.6	-738.4	-509.7	-274.5	-235.3	-170.1	-133.7	0	23.1	24.5 152.1	55.6	190.1	434.4 376.7	776.3

1 K. Nomura, H. Ohta, A. Takagi, T. Kamiya, M. Hirano, and H. Hosono, Nature **432,** 488 (2004).

2 S. Lee, K. Park, and D. C. Paine, Journal of Materials Research **27,** 2299 (2012).

3 K. Nomura, T. Kamiya, H. Ohta, M. Hirano, and H. Hosono, Applied Physics Letters **93,** 192107 (2008).

4 S. Lee, H. Park, and D. C. Paine, Thin Solid Films **520,** 3769 (2012).

5 P. Barquinha, G. Goncalves, L. Pereira, R. Martins, and E. Fortunato, Thin Solid Films **515,** 8450 (2007).

6 S. Lee, B. Bierig, and D. C. Paine, Thin Solid Films **520,** 3764 (2012).

366470 Design Rules for Stabilizing Polar Metal Oxide Surfaces: Adsorption of O2 on Zn-Terminated Polar ZnO(0001)

Abstract:

For oxide semiconductors with appreciable ionic character, under-coordination of the surface atoms leads to thermodynamic instability that is typically restored by reconstruction, faceting, or extensive surface defect creation. Developing design rules for stabilizing polar metal oxide surfaces that avoid these phenomena could offer novel protocols for applications such as improved nanostructure growth and design of photocatalytic heterostructures. The present work describes calculations by density functional theory for Zn-terminated polar ZnO(0001) which demonstrate that stabilization via chemisorbed O_2 together with vacancy formation is energetically as favorable as stabilization by vacancies alone. The stabilization mechanism including adsorption is so effective that it promotes O_2 adsorption to an extent that is not possible on non-polar ZnO. Experimental evidence for such stabilization behavior is presented based on measurements of O_2 adsorption on polar ZnO(0001) via the optical modulation technique of photoreflectance. The measured isotherms yield a sizable adsorption enthalpy of adsorption near 1.8eV, confirming a strong interaction with the polar surface.

387865 Electron/Hole Injection Drives the Ultrafast Phase Transition of VO2

Abstract:

Interplay among the microscopic degrees of freedom in transition-metal oxides can generate macroscopic quantum phenomena that provide functionality in electronic and photonic devices. Here, we report density functional calculations and molecular dynamics simulations of VO_2, which undergoes a semiconductor-to-metal phase transition accompanied by a monoclinic-to-rutile structural change at 68 °C. We find that the lattice vibration at the critical temperature generates a metallic state in the monoclinic structure, which may explain the observed metallic intermediate phase in experiments. Moreover, we find that the electron/hole doping strongly couples with the lattice vibration causing collapse of one particular phonon mode and stimulating the structural phase transition. Molecular dynamics simulations show a temperature-dependence of the required carrier density for the phonon collapse, that is, at higher temperature, fewer free carriers are required. We show that the abrupt change of the vibration results from the weakening of the V-V bonds induced by the hole doping. The tunable phase transition is valuable for various application of VO_2, such as memristors and molecular sensors.

Support was provided by the DTRA Grant HDTRA1-10-1-0047, NSF Grant DMR-1207241, and McMinn Endowment at Vanderbilt University. DFT calculations were performed at the DoD AFRL.

Epitaxial growth of Fe_3O_4 thin film on SiC (0001)

Negar Hamedani Golshan, Katherine S. Ziemer
Northeastern University Chemical Engineering Department, Boston MA, USA

High quality epitaxial growth of Fe_3O_4 film on semiconductors will lead to next generation of spintronic devices such as spin-FETs, spin filters, and magnetic tunnel junctions (MTJs). Integration of Fe_3O_4 has been attracting interest for these applications not only because of its high Curie temperature of 860K and 100% polarization at room temperature, but also its predicted half metallic properties at room temperature and a negative spin polarization. However, these properties have been very difficult to realize in thin film form, and device design strategies need high-quality thin films of Fe_3O_4. The most common reason reported in literature for failure of the films to achieve theoretical performance is that current growth techniques produce films with antiphase boundaries (APB). These APBs have a strong antiferromagnetic coupling that negatively impact the magnetic and transport properties of epitaxial Fe_3O_4 films. The most significant impacts are negative magnetoresistance (MR) in a range between -0.7 and -1 %, and unsaturated magnetization even at high fields of 8T [1]. Therefore, searching for any progress to reduce APBs is of essential importance for the applications of Fe_3O_4 films.

Epitaxial Fe_3O_4 films have been grown mostly on MgO either as a substrate or template layer because of the very small lattice mismatch (0.3 %) between Fe_3O_4 and MgO. However, the performance of today's spintronic devices would be enhanced by integrating Fe_3O_4 on wide band gap semiconductors which can function in high-temperature, high-power, and high-frequency environments. In addition long spin relaxation time in wide bandgap semiconductors is beneficial where spin transport at large length scales is required. Consequently the combination of half-metallic Fe_3O_4 and wide bandgap SiC could be one of the most promising magnetic semiconductor hybrid systems for development of future spintronic devices.

In this study, through molecular beam epitaxy (MBE), the use of a magnesium oxide (MgO) template layer and the interface formation mechanisms have been investigated for effective heteroepitaxy of high-quality magnetite (Fe_3O_4) on 6H-SiC. 6H-SiC(0001) substrates are cleaned in an *ex-situ* hydrogen furnace. The resulting 6H-SiC surfaces are smooth with a uniformly stepped surface and a $\sqrt{3}\times\sqrt{3}$ R30° surface reconstruction with a silicate adlayer, verified by reflection high energy electron diffraction (RHEED) and x-ray photoelectron spectroscopy (XPS). High quality, single crystalline MgO(111) is obtained with a smooth surface (RMS<0.5 nm) and a stepped morphology conformal to the underlying 6*H*-SiC morphology, but is inherently twinned due to the ionic nature of a (111) oriented rock salt structure. The smooth, conformal 2-D growth of MgO requires the presence of atomic oxygen in a Mg-adsorption controlled mechanism, and grows in tension with a 3.3% lattice mismatch. This presentation will show the relationship between processing conditions and FexOy chemistry and structure and discuss strategies for enabling the reduction of APBs in Fe_3O_4 on SiC.

1. D Gilks, L Lari, J Naughton, O Cespedes, Z Cai, A Gerber, S M Thompson, K Ziemer and V K Lazarov, *Origin of anomalous magnetite properties in crystallographic matched heterostructures: Fe_3O_4(111)/$MgAl_2O_4$(111)*. J. Phys: Condensed Matter 2013. **25**.

389711 Tungsten Trioxide/Zinc Tungstate Bilayers: Electrochromic Behaviors, Energy Storage and Electron Transfer

Abstract:

Materials with integrated functions of energy storage and electrochromism are highly desirable from the point of view of sustainable development.[1, 2] Herein, pair-sequentially spin-coated tungsten trioxide (WO_3) and zinc tungstate ($ZnWO_4$) bilayer films onto indium tin oxide (ITO) coated glass slides have been prepared via sol-gel methods followed by annealing. The bilayers of $ZnWO_4/WO_3$ (the bilayer film consisting of an inner layer of $ZnWO_4$ and an outer layer of WO_3 on the ITO glass) and $WO_3/ZnWO_4$ (the bilayer film consisting of an inner layer of WO_3 and an outer layer of $ZnWO_4$ on the ITO glass) exhibit integrated functions of electrochromic and energy storage behaviors as indicated by the *in situ* spectroelectrochemistry and cyclic voltammetry (CV) results. Accordingly, blue color was observed for the bilayer films at -1 V in 0.5 M H_2SO_4 solution. An areal capacitance of 140 and 230 $\mu F/cm^2$ was obtained for the $ZnWO_4/WO_3$, and $WO_3/ZnWO_4$ film, respectively, at a scan rate of 0.05 V/s in the CV measurements. The CV results also unveiled the electron transfer behavior between the semiconductor films in the oxidation process, suggesting a sequence-dependent electrochemical response in the bilayer films. Meanwhile, methylene blue (MB) was used as an indicator to study the electron transfer phenomenon during the reduction process at negative potentials of -0.4 and -0.8 V, in 0.5 M Na_2SO_4. The results indicated that the electrons transfer across the bilayers was enhanced at more negative potentials.

References

1. H. Wei, X. Yan, S. Wu, Z. Luo, S. Wei and Z. Guo, *The Journal of Physical Chemistry C*, 2012, **116**, 25052-25064.

2. H. Wei, J. Zhu, S. Wu, S. Wei and Z. Guo, *Polymer*, 2013, **54**, 1820-1831.

Nanoporous Silica Thin Film Coatings for Optical Devices

Silica sol-gels are an affordable and customizable approach for creating uniform oxide thin films on substrates. One can tailor the optical and mechanical properties of the silica film by altering the catalyst and precursor, and by tuning the relative concentrations of solvent to water. Additionally, by introducing dopants to the initial solution, optically active and high refractive index materials can be created; by including surfactants, it is possible to form nanopores in the oxide film. This flexibility has enabled silica sol-gel films to find applications in a wide range of fields related to integrated photonics– such as high refractive index sol-gels in integrated waveguides and rare earth doped sol-gels as microlasers.

One example of utilizing the porous thin film structure is in biodetection and catalysis characterization. The nano-structured surface retains extremely large surface area, which translates to high collection efficiency. If used in the design of an optical biosensor or chemical sensor, the signal could be significantly enhanced. Additionally, assuming the pores are filled with air, the effective refractive index is lower than silica, creating a low index material. Therefore, this material could be used as a cladding layer on an integrated device; however, none of these applications have been pursued to date.

The reason for avoiding porous materials in integrated optical devices is somewhat obvious. In all integrated optical devices, surface scattering leads to optical loss. Theoretically, the pores should behave as surface scattering sites, thereby degrading the device performance. However, recent work has shown that small (sub-wavelength) defects have minimal impact on the device performance. Therefore, assuming the pores are on the nm-scale, they should have minimal impact on the device behavior. This discovery paves the way for combining porous materials with optical devices. However, to achieve this goal, it is necessary to both develop a nanoporous material and to make it compatible with optical devices.

One method of creating a porous silica sol-gel film is to intercalate a surfactant (head-tail carbon chains) into the precursor. As the sol-gel ages, the surfactant self-assembles into a micelle structure. Upon sintering, the micelles are removed, leaving a porous film. Therefore, the pore diameter is related to the initial surfactant chain length and concentration. However, time and temperature during aging are also critical parameters. If either parameter is incorrect, the micelles will not form or they will agglomerate. In the former case, the film will not have pores; in the latter case, the pores will be micro-scale and non-spherical. Previous work focused on characterizing this thermodynamically-governed process on planar surfaces. However, integrated optical devices are non-planar; as such, it is expected that the pore structure will change on the device surface.

In the present work, we synthesize nanoporous coatings for silica whispering gallery mode optical microresonators. These types of optical microresonators are fabricated in circular geometries, and are able to confine light for long periods of time. They have been used in many applications, including biodetection and telecommunications. As such, they are an ideal platform for these initial development efforts. In order to thoroughly explore the ability to make nanoporous films for these devices we use both spherical and toroidal resonators in place of the conventional planar substrate.

Briefly, we use an acid catalyzed hydrolysis-condensation reaction to synthesize the silica sol-gel. In our method, tetraethyl orthosilicate (TEOS) is the precursor, and hydrochloric acid (HCl) is the catalyst. To form the pores, the surfactant dodecyltrimethylammonium bromide (DTAB) is added to the solution. After aging, the sol-gel solution is deposited on the wafer or device surface, and the sample is annealed in a tube furnace.

It is important to note that the radius of curvature of the toroidal device surface is significantly smaller than the sphere. Therefore, the synthesis and annealing conditions required to maintain pores on the toroid should be different than those for the sphere or even a flat wafer. To explore this dependence, the impact of several different parameters on the final pore size and size distribution are studied. The aging time and temperature are varied to optimize micelle formation and the subsequent pore formation. Additionally, the surfactant concentration is varied to quantitatively determine its effect on pore size. Finally, the effect of the water concentration on the hydrolysis-condensation sol-gel reaction is studied. As water concentration increases, more hydroxide groups are introduced to the solution, which should significantly increase matrix networking and alter the sol-gel structure. The final films are quantitatively characterized both on the control wafers and on the spherical and toroidal optical microresonators using high resolution scanning electron microscopy and analyzed through an image analysis software.

387386 Polyaniline – Metal Oxides Nanocomposites with Unique Physicochemical Properties

Abstract:

The unique physicochemical properties of polyaniline (PANI)/metal oxides polymer nanocomposites (PNCs) were systematically studied with the PNCs containing different kinds of metal oxides. PANI PNCs with non magnetic nanoparticles (NPs)-barium titanate ($BaTiO_3$) and magnetic NPs-manganese ferrite ($MnFe_2O_4$) were synthesized by a surface initiated polymerization method. Giant magnetoresistance (GMR) was observed in the PNCs, and the value was found to depending on the species and the loading of NPs. With the magnetic field of 9E5 Oe, the GMR can reach to -34% for PNCs with 30.0 wt% $MnFe_2O_4$ NPs and 30% for PNCs with 1.0 wt% $BaTiO_3$ NPs. And the positive and negative GMR was analyzed theoretically from the wave functional shrinkage model and forward interference model, respectively. Compared with the negative real permittivity of PNCs with $BaTiO_3$ NPs, all the PNCs with $MnFe_2O_4$ NPs showed high positive dielectric permittivity and high dielectric loss value, which is due to the interfacial polarization. The electron transportation behavior of the PNCs was studied and for both PNCs, the temperature dependent resistivity result reveals a 3-d variable range hopping (VRH) electron transport mechanism. By varying the loading level of NPs, the crystalline structure, thermal stability and morphology of the PNCs were also systematically studied. And the peak shift observed in the FT-IR spectra of both PNCs indicated a strong interaction between the formed PANI and the metal oxides NPs.

AICHE FALL MEETING 2014

ABSTRACT

Light emitting and organic light emitting diodes are a fascinating area of research and have attracted a large amount of interest because of their use in high efficiency lighting[1-3]. In the present work, we mainly focus on three main technical objectives relevant to the engineering of new materials for hybrid light organic-inorganic Quantum Dot LEDs. First, we emphasize on the synthesis of novel colloidal quantum dots (CdSe and ZnSe QDs) to use with functionalized conducting polymers in the creation of hybrid OLEDs. This objective will explore the critical role of the polymer/quantum dot interface and its ultimate role in device efficiency. Our second aim is focused onto the exploration of spin coating and inkjet deposition to lay down the quantum dot and polymer layers in discrete layers to produce light and finally we have explored the possible economical ways of integration and characterization of these novel materials into an active lighting device structure.

REFERENCES

[1] J. Kido, M. Kimura, K. Nagai, "Multilayer white light-emitting organic electroluminescent device," *Science,* vol 267, pp. 1332-1334, 1995.

[2] W. Ki, J. Li, "A semiconductor bulk material that emits direct white light", *Journal of the American Chemical Society,* vol *130*, pp. 8114, 2008.

[3] H. J. Bolink, H. Brine, E. Coronado, M. Sessolo, "Ionically Assisted Charge Injection in Hybrid Organic-Inorganic Light-Emitting Diodes", *ACS Applied Materials & Interfaces,* vol. 2, pp. 2694-2698, 2010.

384739 Bi-Continuous, Ultra-Large-Pore Carbons By Template-Replica Co-Assembly

Abstract:

Among classes of mesoporous materials bearing high surface area, three-dimensionally ordered pore topology, and large pore volumes, ordered mesoporous carbons (OMCs) have attracted specific interest due to their low cost, thermal and physicochemical stability and electrical conductivity, properties that make them attractive for applications as diverse as electrodes in energy conversion and storage devices, supercapacitors, catalysts, and adsorbents. So-called 'soft-templating' or 'direct' strategies that exploit the self-assembly phase behavior of copolymer surfactants as templates and polymerizable carbon precursors by evaporation induced self-assembly (EISA) can be used to realize ordered mesoporous carbons. While such processes enjoy scalability, owing to their reliance on 'one-pot' solution phase thermodynamics, they do face several critical drawbacks including the cost of template sacrifice, pore shrinkage, limits on carbon source variability and thus on the tunability of intrinsic carbon properties, and difficulties in realizing stable, ultra-large pore topologies for applications requiring access by bulk molecules or ease of solution/solvent infiltration.

While the alternative 'hard'-templating approach provides promise for addressing these drawbacks, it is generally discounted due to its multi-step nature and possible reliance on surfactant-templated fabrication of ordered mesoporous silica precursors as template materials. Inspired by the simplicity of 'one-pot' soft-templating strategies, this talk will highlight progress toward realizing a facile 'one-pot' hard-templating method capable of deriving bi-continuous, three-dimensionally ordered porous carbon thin films and powders. Specifically, we will show how pre-formed size tunable silica nanoparticles (ca. 20-50 nm) can be co-assembled from aqueous solutions with a carbon source into powders and thin films bearing a three-dimensionally ordered, bi-continuous mesopore topology.

We will describe the pseudo-phase behavior governing this co-assembly, the order-disorder transition and particle-particle spacing of which is modulated by adsorption of the carbon source to the assembling silica nanoparticle template and its concomitant modulation of the particle-particle interactions. Mechanistic insight into the formation of these materials will employ comprehensive analysis spanning HPLC adsorption studies, nitrogen and argon physisorption, XPS, cyclic voltammetry, and high-resolution electron microscopy. We exploit the high surface area, apparently tunable sp^2-hybridized carbon content, and facile and low-cost 'one-pot' synthesis of these ordered mesoporous materials for applications spanning low-cost electrodes in both supercapacitor and dye-sensitized solar cell applications, in both cases showing promising performance relative to laboratory and literature standards.

On the Rotational Intergrowth of Hierarchical FAU/EMT

Authors: M. Khaleel[1], M. Tsapatsis[1]

[1]*Department of Chemical Engineering and Materials Science, University of Minnesota, Minneapolis, Minnesota, USA.*

Keywords: hierarchical, nanosheets, rotational intergrowth, formation mechanism

Abstract

Hierarchical zeolites is a class of materials that combine zeolite micropores with meso and/or macropores.[1,2] These materials exhibit higher reaction rates,[3,4] improved selectivity,[5,6] resistance to deactivation,[6-8] and novel adsorption behavior[9] compared to microporous-only zeolites. Several approaches were reported for the synthesis of hierarchical zeolites at lab scale, [10] but have not been scaled up for industrial applications because of safety and cost constraints. We believe that repetitive branching by rotational intergrowths[4,11,12] holds promise for industrial implementation due to its simplicity (one-step synthesis) and lower cost (simple structure directing agents or additives) compared to templating approaches.

In a recent publication, Inayat et al. reported the synthesis of hierarchical arrangement of faujasite nanosheets as a type X zeolite (Si/Al<1.5).[11] The house-of-card nanosheet structure combines faujasite micropores (0.74 nm), with mesopores (ca. 7 nm) contained within the sheets (sheet thickness of □ 100-200 nm) and macropores (ca. 200 nm) resulting from the intersection of the sheets. However, a plausible scenario for how branching takes place was not presented up to now. Since understanding the branching mechanism may enable one to control the nanosheet thickness, branching frequency and other structural characteristics, we attempt to provide such understanding through a structural study using Transmission Electron Microscopy imaging and diffraction and comparison to simulations.

Our results show that T-H-Faujasite: (1) at the particle level has a cuboctahedron morphology consistent with cubic FAU; (2) each cuboctahedron particle consists of a skeletal arrangement of nanosheets interpenetrating at 70.5° or 109.5°; (3) each nanosheet is thin along $<111>_{cubic}$, and is predominantly composed of FAU, with EMT present in small isolated domains that do not give rise to XRD peaks. We propose that the presence of the hardly detectable small EMT domains is what directs the formation of anisotropic nanosheets and creates the growth instability leading to branching. This is a type of dendritic growth that is similar, yet distinct, to that observed in many crystals including snowflakes. It demonstrates, along with earlier work on MFI/MEL,[4] that there is a direct link between zeolite polytypes and the repetitive branching mechanism leading to hierarchical structures.

References

[1] A. Corma, *Chem. Rev.* 1997, *97*, 2373-2419.

[2] M. Davis, *Nature* 2002, *417*, 813-821.

[3] S. van Donk, A. Broersma, O. Gijzeman, J. van Bokhoven, J. Bitter, K. de Jong, *J. Catal.* 2001, *204*, 272-280.

[4] X. Zhang, D. Liu, D. Xu, S. Asahina, K. Cychosz, K. Agrawal, Y. Wahedi, A. Bhan, S. Hashimi, O. Terasaki, M. Thommes, M. Tsapatsis, *Science* 2012, *336*, 1684-1687.

[5] C. Christensen, K. Johannsen, I. Schmidt, C. Christensen, *J. Am. Chem. Soc.* 2003, *125*, 13370-13371.

[6] a) J. Garcia-Martinez, M. Johnson, J. Valla, K. Li, J. Ying, *Catal.: Sci. Technol.* 2012, *2*, 987-994; b) J. Garcia-Martinez, K. Li, G. Krishnaiah, *Chem. Commun.* 2012, *48*, 11841-11843.

[7] R. Srivastava, M. Choi, R. Ryoo, *Chem. Commun.* 2006, 4489-4491.

[8] M. Choi, K. Na, J. Kim, Y. Sakamoto, O. Terasaki, R. Ryoo, *Nature* 2009, *461*, 246-250.

[9] D. Xu, G. Swindlehurst, H. Wu, D. Olson, X. Zhang, M. Tsapatsis, *Adv. Funct. Mater.* 2014, *24*, 201–208.

[10] a) R. Chal, C. Gerardin, M. Bulut, S. van Donk, *ChemCatChem* 2011, *3*, 67-81; b) Y. Tao, H. Kanoh, L. Abrams, K. Kaneko, *Chem. Rev.* 2006, *106*, 896-910; c) J. Perez-Ramirez, C. Christensen, K. Egeblad, C. Christensen, J. Groen, *Chem. Soc. Rev.* 2008, *37*, 2530-2542; d) J. Čejka, S. Mintova, *Catal. Rev.* 2007, *49*, 457-509; e) W. Schmidt, *ChemCatChem* 2009, *1*, 53-67; f) S. Lopez-Orozco, A. Inayat, A. Schwab, T. Selvam, W. Schwieger, *Adv. Funct. Mater.* 2011, *23*, 2602-2615.

[11] A. Inayat, I. Knoke, E. Spiecker, W. Schwieger, *Angew. Chem.* 2012, *124*, 1998-2002; *Angew. Chem. Int. Ed.* 2012, *51*, 1962-1965.

[12] W. Chaikittisilp, Y. Suzuki, R. Mukti, T. Suzuki, K. Sugita, K. Itabashi, A. Shimojima, T. Okubo, *Angew. Chem.* 2013, *125*, 3439-3443; *Angew. Chem. Int. Ed.* 2013, *52*, 3355-3359.

363840 Hollow Spherical Rare-Earth-Doped Yttrium Oxysulfate: A Novel Structure for Upconversion

Abstract:

Rare-earth (RE) ions exhibit unique luminescent properties, including the ability of converting the near infrared long-wavelength excitation radiation into shorter visible wavelengths by photon upconversion (UC) due to the incompletely filled 4f electron shell. Compared to organic fluorophores and semiconducting nanocrystals, rare-earth-based nanomaterials display higher photochemical stability, sharp emission bandwidths, large anti-Stokes shifts and Stokes (up to 500 nm) that separate discrete emission peaks from the ultraviolet/infrared excitation. In recent years, rare-earth-doped UC materials have been increasingly applied as a novel type of luminescent optical labels in biological assays and medical imaging and have become promising alternatives to organic fluorophores and quantum dots, etc. Nonetheless, these rare-earth-doped systems were mainly based on the $NaREF_4$, $LiYF_4$, REF_3, etc. crystal structure. There are few reports on rare-earth-based $RE_2O_2SO_4$ structure as an alternative for the UC application.

In this work, a facile hydrothermal route followed by annealing was applied for the preparation of monoclinic yttrium oxysulfate ($Y_2O_2SO_4$) hollow spheres with other rare earth ions (Yb^{3+} and Er^{3+}) doped. The formation of hollow spheres may involve the Ostwald ripening. The hybrid materials were firstly demonstrated for upconversion application. The compatibility of host crystal structure enables the easy doping of a second rare-earth metal element without significantly changing the host lattice. The photoluminescent properties were affected by the ratio and concentration of doped rare earth metal ions due to the featured energy transfer and symmetry of crystal field. The type of luminescent center and the crystallinity of samples were also proved to have remarkable influences on the optical properties of as-prepared products.

389070 Structural Modification of Metal Organic Frameworks for Steam Stable CO2 Adsorbents

Abstract:

The heightened global concern over increasing CO_2 concentrations in the atmosphere have led to intensive research on CO_2 mitigation strategies, specifically on the ways to capture CO_2 from post-combustion flue gas. Porous adsorbent materials may represent promising pathway to capture the greenhouse gas in a simple and cost-efficient way.[1] Among adsorbents, metal organic frameworks (MOFs) featuring high surface areas and chemical tunability have demonstrated competitive CO_2 adsorption capacities under the CO_2 capture conditions simulating the flue gas.[2] However, most of the research employing MOFs as a CO_2 adsorbent have been carried out under dry conditions, which is far from the realistic conditions. To use MOFs for practical CO_2 capture applications such as for the flue gas, the material stability of MOFs under humid conditions should be evaluated.

In this work, structural changes and CO_2 capture performance of MOFs was studied under humid CO_2 capture conditions using accelerated steam treatment tests. Traces of structural degradation based on XRD patterns and porosity measurements as well as dramatic loss of its CO_2 capture capacity was observed when the bare Mg/DOBDC was exposed to the humidity. To improve materials stability, pore surface modification with ethylene diamine groups was performed using post-synthesis functionalization procedures used in the previous work.[3] Materials stability of the Mg/DOBDC under humid conditions was assessed before and after functionalization and analyzed using characterization techniques including XRD, surface area and pore size analyzer, TGA, etc. Amine functionalization was hypothesized to shield open metal sites of the framework from water molecules therefore protecting material from degradation. It was found that while some degradation occurred still, CO_2 capacities were almost completely regenerated after steam treatment.

(1) Choi, S.; Drese, J. H.; Jones, C. W. *ChemSusChem* **2009**, *2*, 796.

(2) Sumida, K.; Rogow, D. L.; Mason, J. A.; McDonald, T. M.; Bloch, E. D.; Herm, Z. R.; Bae, T. H.; Long, J. R. *Chemical Reviews* **2012**, *112*, 724.

(3) Choi, S.; Watanabe, T.; Bae, T.-H.; Sholl, D. S.; Jones, C. W. *The Journal of Physical Chemistry Letters* **2012**, *3*, 1136.

384550 Formation of Porous Materials Using a Lattice Model

Abstract:

Zeolites are of great importance in modern technology with applications in catalysis, separations, biosensing and microelectronics. Despite the importance of zeolites, our understanding of how they form, and of the thermodynamics that govern their crystallization remain rather poor. Developing such understanding would assist in the fabrication of new zeolites with tailor-made structures for advanced applications. We aim to address these issues with our molecular modeling work.

We focus our attention on all silica materials, where we have adopted an atomic representation of the silicic acid molecule by representing it as a unit cell on the body-centered cubic lattice. The silicon atom occupies the center of the unit cell and the hydroxyl groups are located at the corners. The condensation reaction is modeled as the sharing of two hydroxyl groups from two different tetrahedral on a single lattice site. This model has proven useful in understanding a variety of processes such as silica polmerization at the iso-electric point, formation of surfactant-templated mesoporous materials, and the thermodynamics of ordered crystalline ground states, particularly zeolite-analogs. Recently, we have extended this model to study silica polymerization under various silica concentrations and pH values, finding gel states and nanoparticle states with surprising trends in nanoparticle size.

We have also applied this model to simulate the formation of crystalline zeolite-analogs using Parallel Tempering Monte Carlo simulations of our lattice model. We have observed a variety of crystalline ground states, including layered and fully connected three dimensional structures, under various conditions in the presence of a model of a structure directing agent (SDA). Our aim is to elucidate the role of SDA in governing the zeolite-analog topology. We will report on our progress to date in this research program.

385676 In-Situ Gisaxs Investigation of Mesostructure Formation during Low-Temperature Aging in Mesoporous Titania Thin Films

Abstract:

The this study, we investigate the mesostructure formation mechanism during aging in mesoporous titania thin films with oriented hexagonal close packed (HCP) cylindrical nanopores. Aging of titania films in a refrigerator under high humidity (approx. 94%) is an essential part of the formation of well-ordered mesostructured in titania films using titanium ethoxide as the precursor and Pluronic P123 as the pore template (e.g. Koganti et al., *Nano Letters*2006, 11, 2567), but the kinetics of mesostructure development and orientation have not been investigated. Preliminary investigations in our group showed that significant changes in the ordered mesostructure can continue for at least 20 minutes after film deposition, so this is the relevant time scale for further investigation. To provide orthogonal alignment of the pores, surface modification of substrates with crosslinked P123 has been used to provide a surface that is chemically neutral towards the PEO and PPO blocks of P123.

Here, a GISAXS study has been performed in which in-situ measurements are done immediately after coating a film inside of a controlled-environment chamber, to understand the rate of mesostructure development and orientation during aging. The GISAXS pattern shows rods consistent with an orthogonally aligned mesophase on both sides of the beam stopper, indicating that the cylindrical nanopores are aligned orthogonal to the modified substrate. The intensity of these peaks has been used to study the time dependent mesostructure development during aging. We find that mesostructure development begins after approximately 10-12 minutes of aging in a high-humidity environment, and that after this time lapse, the mesostructure starts developing rapidly. The orthogonal hexagonal mesostructure is the first ordered phase to form by a disorder-order transition. The experiments are performed not only for films on plain glass (orientation parallel to the film) but also for films on modified glass with orthogonally aligned cylindrical pores in order to the effect of pore orientation during aging. In-situ GiSAXS has also been done during aging at room temperature to study the effect of aging temperature over mesostructure development and orientation. From the temperature study, we find direct evidence that aging at 4 °C is indeed necessary for ordered mesostructure development and that aging at this temperature helps to provide orthogonal orientation of the cylindrical nanopores.

Microcontact Printing of Polyelectrolytes on PEG for Micropatterning Nanoparticles, DNA, Proteins and Cells

Zhibin Wang

11/17/2014

Department of Chemical and Biomedical Engineering

FAMU-FSU College of Engineering

Florida State University

1

Microcontact printing of polyelectrolytes for micropatterning

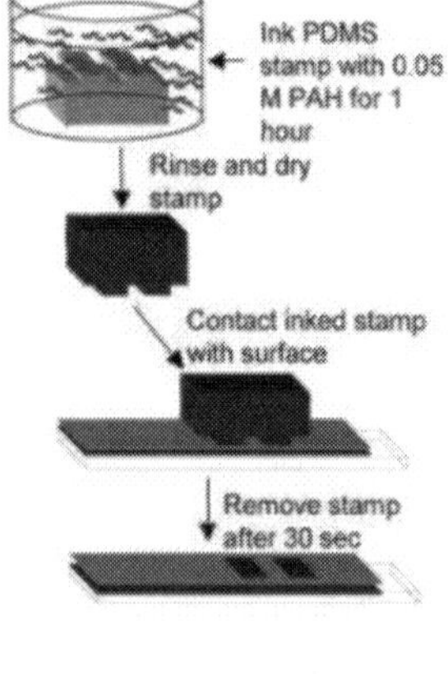

P. T. Hammond et al. *Langmuir* 2003

AFM image of PAH stripes on PEM surface

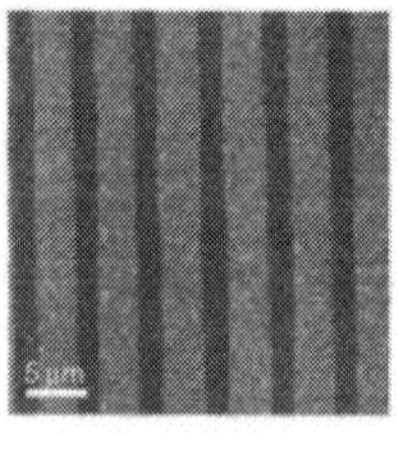

Fluorescence image of protein microarray

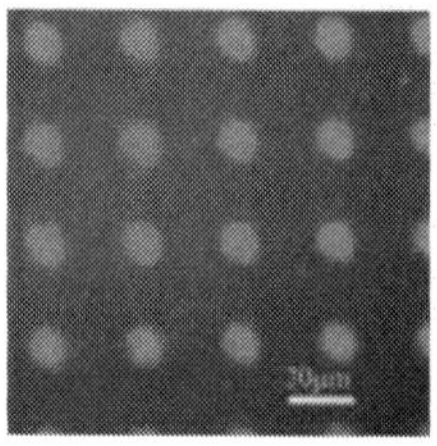

P. T. Hammond et al. *Biomacro.* 2004

AFM image of DNA nano-array

Cha et al. *ACS Nano.* 2009

PAH: Poly(allylamine hydrochloride)
PEM: Poly electrolyte multilayer

Polyethylene glycol grafted surface for micropatterning

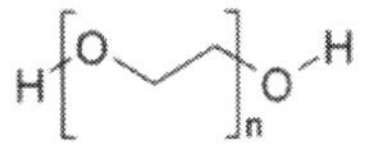

Polyethylene glycol (PEG)

- Hydrophilicity
- Flexibility
- Nontoxicity
- Nonimmunogenecity

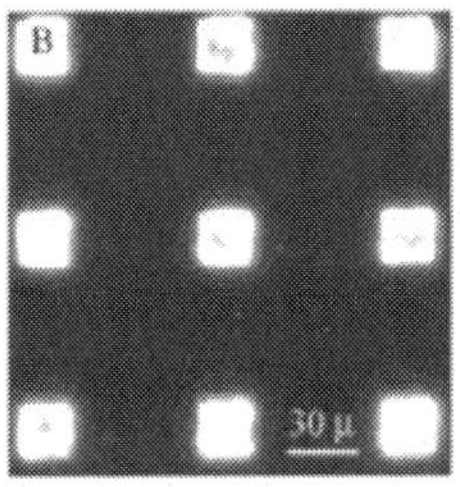

Zhang et al. *Langmuir* 2002

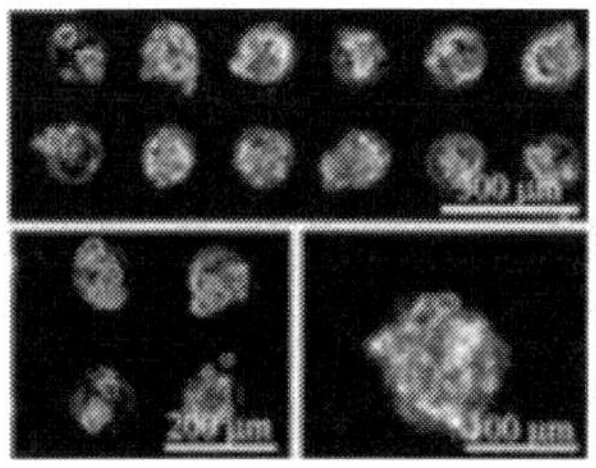

Yoshimoto et al. Lab Chip 2009

Steric repulsion (against van der Walls and hydrophobic interaction)

- **Osmotic (solvation of PEG chains)**
- **Elastic (conformational entropy of PEG chains)**

Fabrication of PEG grafted surface for micropatterning

Self-assembly monolayer of short chain PEG

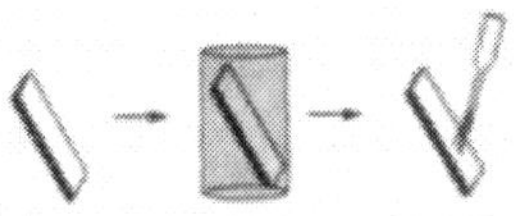

Freshly prepared gold slide — Organic thiol oligo (ethylene oxide) solution overight — Rinse and dry

M. Whitesides *et al. J. Am. Chem. Soc.* 1993

Photolithography

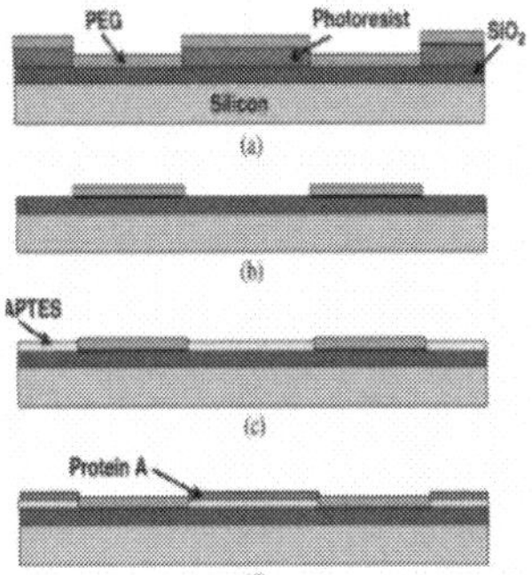

Zohar et al. *Nanotech.* 2005

Soft lithography

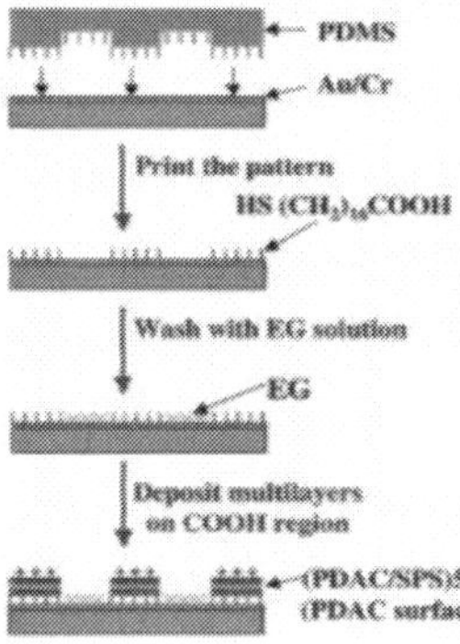

P. T. Hammond et al. *Adv. Mater.* 2002

Drawbacks: Expensive facilities; Harsh condition; Multiple processing steps

Microcontact printing of polyelectrolyte on PEG

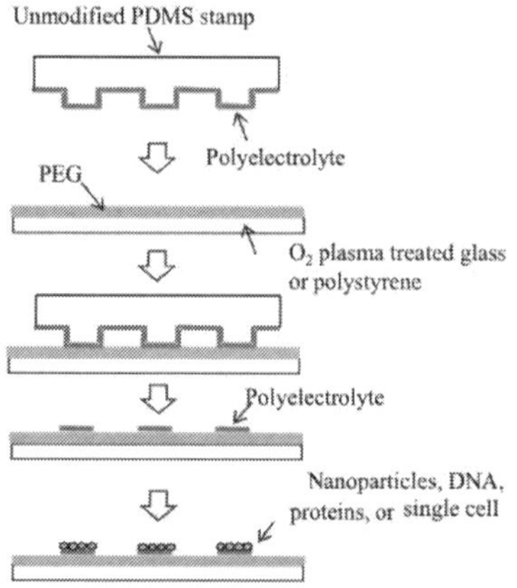

Schematic of microcontact printing of polyelectrolyte on PEG

Characterization of PEG grafted surface

Short chain of PEG silane

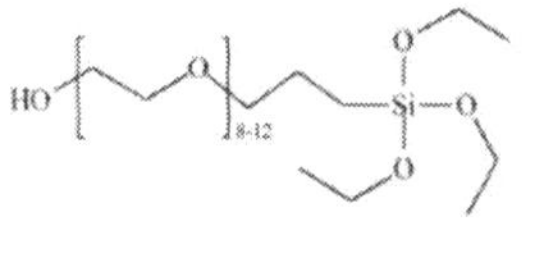

Contact angle measurement of PEG grafted surface

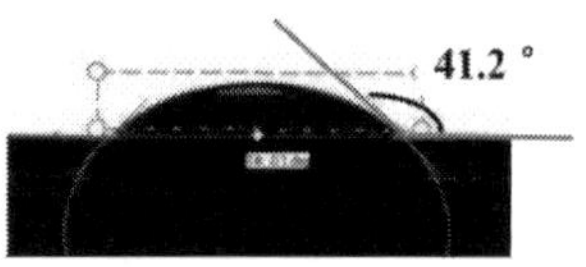

AFM image of PEG grafted surface

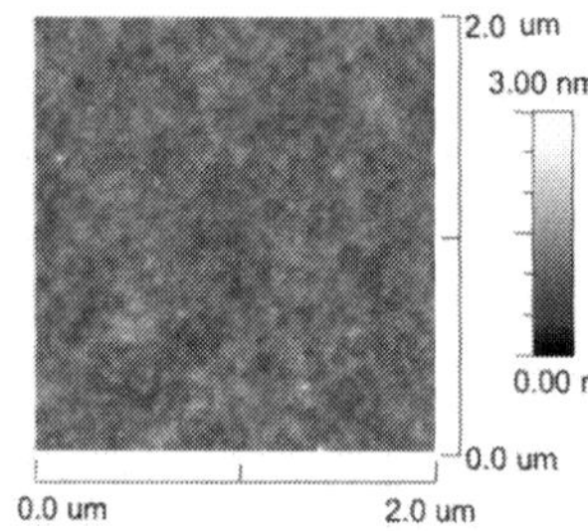

Microcontact printing of PAH on PEG

Fluorescence and AFM images of poly(allylamine chloride) (PAH) dots on PEG 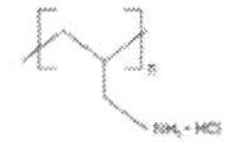

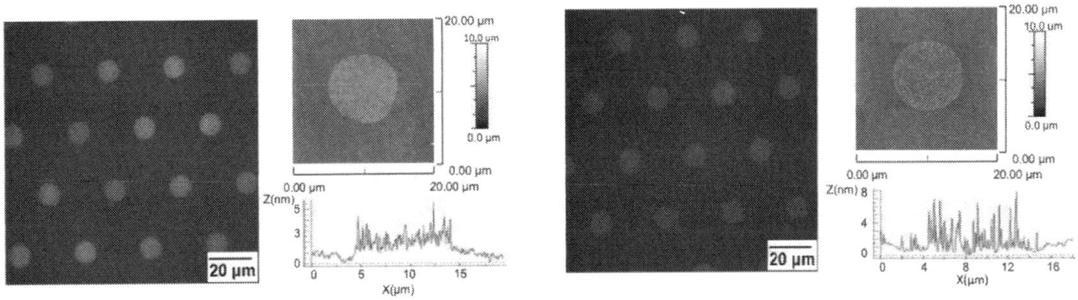

Before (left) and after (right) 1 day incubation in PBS

Z Wang et al. *Soft Matter* 2012

Microcontact printing of polyelectrolyte on PEG

Branched poly(ethylene imine)-RITC

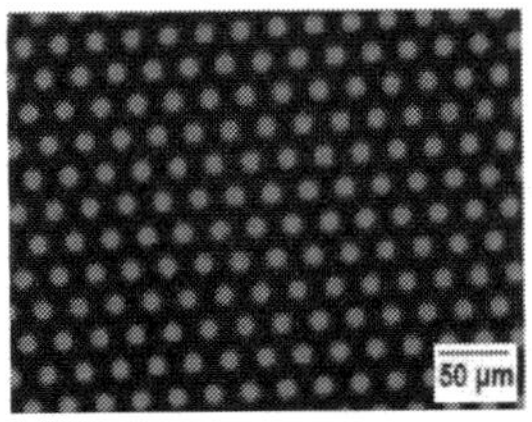

Linear poly(ethylene imine)-RITC

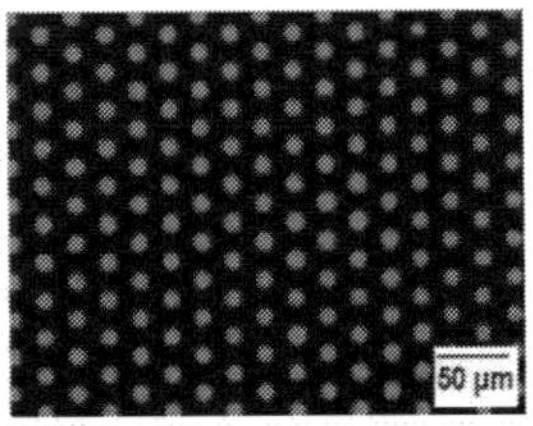

Chitosan (red) and poly(lysine) (green)

Stretched DNA bundles (YOYO-1)

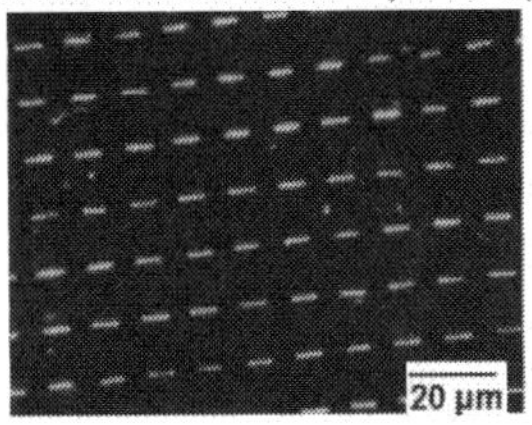

Bilayer of Poly(sodium 4-styrene sulfonate)/PAH-RITC

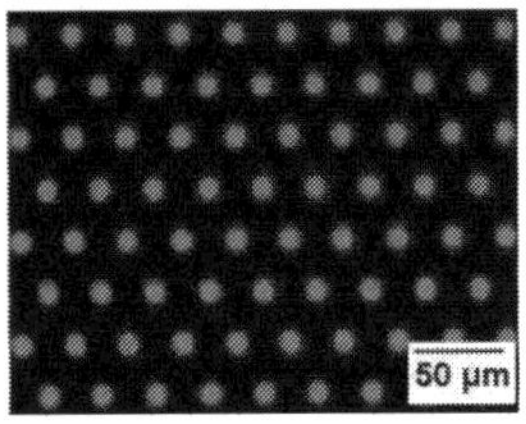

Printed polyelectrolyte on PEG for micropatterning nanoparticles and biomaterials

Quantum dots (red)

µCP of PDAC Poly(diallydimethylammonium chloride)

DNA molecule (YOYO-1)

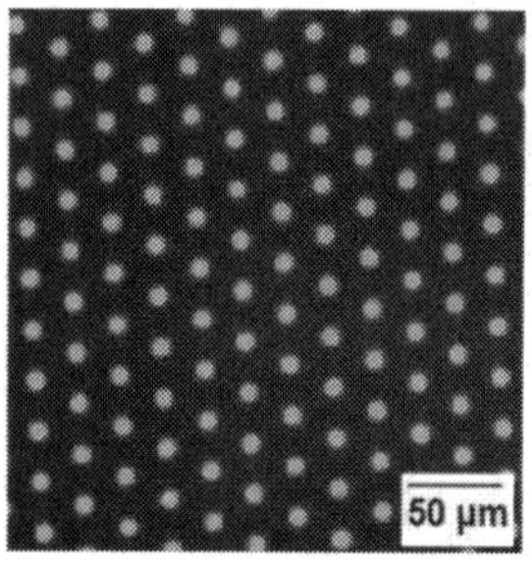

µCP of PAH

Neutravidin (red) and biotin (green)

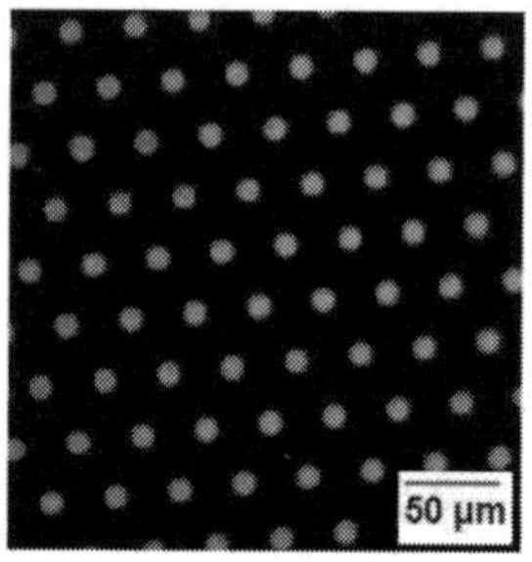

µCP of PAH and modified by glutaraldehyde

Printed polyelectrolyte on PEG for micropatterning live cells

K562 cell patterned by µCP of PAH on PEG grafted glass

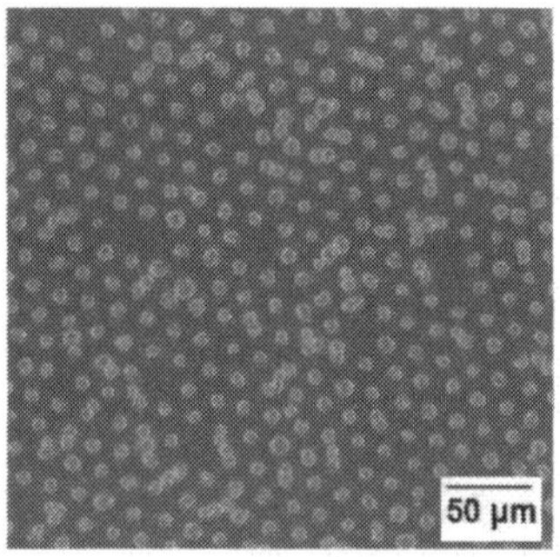

K562 cell patterned by µCP of PAH on PEG grafted polystyrene

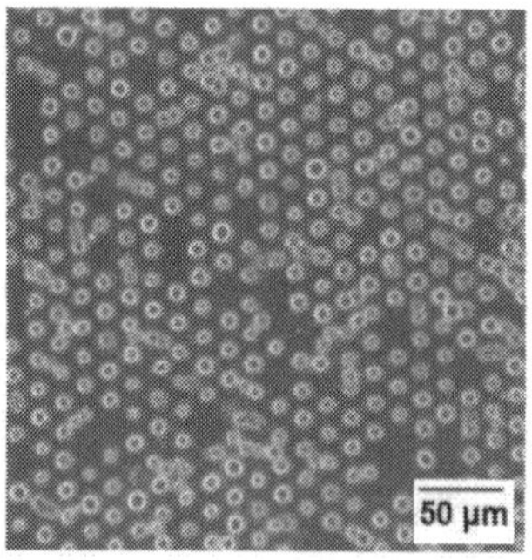

Patterned cells viability (24 h)

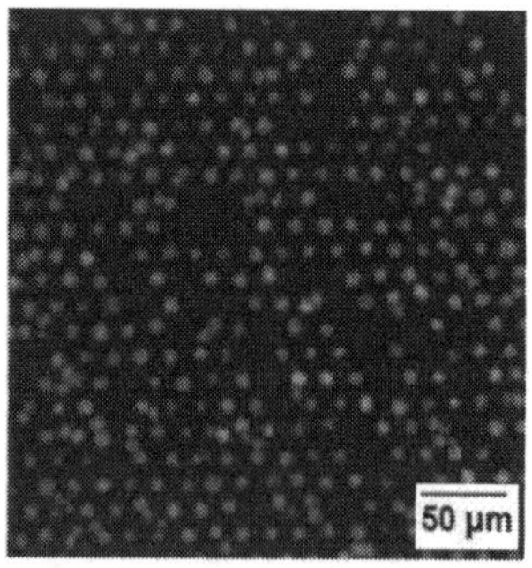

Green: Acridine orange
Red: Propidium iodide

Summary

- A novel μCP technique have been developed based on printing polyelectrolytes directly on a PEG-silane SAM using unmodified PDMS stamps.
- Monolayer-like structures of PAH, PLL, chitosan, linear PEI, branched PEI and PDAC, PAH/PSS bilayer, and stretched DNA molecular bundles have been printed.
- The printed structures are stable under aqueous conditions and allow for micropatterning of nanoparticles, DNA, protein, and live cells.
- This method is potentially useful in fields of tissue engineering and biomedical devices.

Acknowledgement

Dr. Guan's group:

Dr. Jingjiao Guan (Chemical & Biomedical Engineering, FSU)

Dr. Peipei Zhang (University of Maryland)

Dr. Brett Kirland (FDLE)

Dr. Yingru Liu

Dr. Eric Lochner (Physics, FSU)

Dr. Ken Knappenberger (Chemistry, FSU)

Florida HEALTH

The Florida Department of Health

Thank you

PEG-based Tumor Millibeads for Three-dimensional Cancer Cell Culture

Shantanu Pradhan, Jacob M Clary, Chloe S Chaudhury, Elizabeth A Lipke

Abstract:

Models that mimic the pathophysiology of human cancers are critical to advancing cancer research and drug discovery. In this research, we have developed a novel cancer model (called, 'tumor millibead model') through the use of poly(ethylene glycol)(PEG)-based hydrogel biomaterial for three-dimensional (3D) cancer cell culture, which can be potentially used in the drug discovery and development process. Whereas most current cancer research relies on two-dimensional (2D) cultured cells or animal models, the use of biomaterials to create tissue-engineered cancer models provides advantages in replicating the tumor microenvironment. Animal and 2D cell culture models have some inherent limitations in their ability to mimic the native human cancer pathophysiology. 2D models lack the complexity of the 3D extracellular microenvironment, an important player in the tumorigenic progression. Animal models are physiologically different from the microenvironmental conditions of the human body and, hence, frequently fail to provide accurate information about the future clinical efficacy of the tested drugs. Therefore, there is a clear need for the development of cancer models which can provide physiologically and clinically relevant information about drug action and efficacy. In order to address this problem, the emerging field of cancer tissue engineering is focused on incorporating the principles of cancer biology in tissue engineering to create 3D *in vitro* engineered cancer tissue through the use of hydrogel biomaterials, facilitating the investigation of tumorigenic phenomena and testing efficacy of new anti-cancer drugs.

In order to create the tumor millibead model reported in the study, we have developed a novel method for the fabrication of millimeter-scale hydrogel spheres (called 'hydrogel millibeads') using a surface-tension based, liquid-liquid dual phase system. This technique allows the creation of highly uniform, PEG-based hydrogel spheroids, which are significantly larger in size compared to PEG-based hydrogel microspheres obtained by previous fabrication methods. In addition, the ability to culture cancer cells within the hydrogel millibeads provides a unique cancer model which can mimic the millimeter-sized tumors in the human body.

Briefly, poly(ethylene glycol diacrylate) (PEGDA) precursor solution of a fixed volume (2, 5 or 10 μl) was pipetted on the surface of an oil phase solution in a Petri dish, where the precursor droplet floated at the air-oil interface due to surface tension differences between air, oil and aqueous phase. The droplets were crosslinked in presence of light of wavelength 365-700 nm for 10 seconds to form hydrogel millibeads. They were subsequently harvested and washed with DMEM to remove all oil. The hydrogel millibeads obtained were larger in size, more uniform in shape with better ease of size control and consistency between fabrication batches. The swelling behavior and diffusion characteristics of hydrogel millibeads were correlated to their initial fabrication size. Human breast cancer MCF7 cells were encapsulated within PEGDA millibeads to generate 'tumor millibeads', which were maintained in long-term 3D culture. Fluorescence imaging and SEM imaging revealed the spherical morphology and uniform distribution of cells in tumor millibeads. Live/Dead staining revealed the presence of a necrotic core of cells surrounded by healthy viable cell layers at the periphery of the millibeads, which is a prominent feature observed in many native and *in vivo* tumors.

Overall, in this study, we have demonstrated a novel technique for the fabrication of hydrogel millibeads which can support long-term 3D cancer cell culture. Important characteristics of the native tumor microenvironment, including tumor heterogeneity, necrotic core formation and 3D cancer cell morphology, are closely simulated within the tumor millibeads. In addition, use of the PEG-based hydrogel system allows for the biochemical and mechanical modifications of the model engineered tissue to match the properties of native cancer tissue in the human body. Biochemical modifications include the use of PEG-peptide or PEG-protein conjugates that allow cellular adhesion, degradation and remodeling of the engineered tissue. Mechanical modifications include tuning of the matrix stiffness to match the native stiffness of cancer tissue by controlling relative PEG concentration, crosslinking density and chain length. This engineered tumor model could be potentially used to obtain physiologically and clinically relevant information from future drug studies. The millibead fabrication technique can be scaled up to create engineered 3D cancer tissue models for high-throughput drug-testing applications. Hydrogel millibeads can also be potentially used to encapsulate suitable biomolecules or therapeutics for controlled delivery at the site of interest. Hence, PEG-based hydrogel biomaterials can significantly contribute to the creation of useful cancer models, and thereby advance cancer research, and drug discovery and development.

398269 Injectable Thermo-Sensitive Hydrogel As an Adjuvant: In Situ Modulation of Dendritic Cells for Cancer Vaccine

Abstract:

Attempts to develop cell-based cancer vaccines have shown limited efficacy due to the failure of transplanted dendritic cells (DCs) to survive long enough to reach the lymph nodes. The development of materials that can modulate DCs *in situ* to enhance antigen uptake and presentation has emerged as a novel method toward developing high quality cancer vaccines. Here, we propose a two-step hybrid strategy to produce a more robust cell-based cancer vaccine *in situ*. First, a significant number of DCs are recruited to an injectable thermo-sensitive mPEG-PLGA hydrogel through sustained release of chemo-attractants, like granulocyte-macrophage colony-stimulating factor (GM-CSF). Then, these resident DCs can be loaded with cancer antigens through the use of viral or non-viral vectors. We demonstrate that GM-CSF releasing mPEG-PLGA hydrogels successfully recruit and house DCs and macrophages, allowing subsequent introduction of antigens by vectors to activate the resident cells, thus initiating antigen presentation and boosting immune response. Moreover, this two-step hybrid strategy generates a high level of specific anti-tumor immunity demonstrated in both prophylactic and therapeutic models of melanoma. This injectable thermo-sensitive hydrogel shows great promise as an adjuvant for cancer vaccines, potentially providing a new approach for cell therapies through *in situ* modulation of cells.

378592 Award Submission: Water Content and Solute Partitioning in Hybrid Silicone-Hydrogels

Abstract:

With roughly six times the amount of corneal oxygen supply compared to HEMA-based soft contact lenses, silicone-hydrogel (SiHy) soft contact lenses (SCLs) provide comfort for extended wear, while maintaining ocular health [1-3]. Despite their commercial importance, the structure and morphology of SiHy SCLs is not well understood. It is, however, widely accepted that silicone-based SCLs should contain co-continuous hydrophilic and hydrophobic polymer phases [3]. The hybrid structure of SiHys allows for sufficient salt transport through the water phase [4] localized in hydrophilic domains, as well as, sufficient oxygen permeability through the silicone domains [2,3]. Additionally, partitioning of care-solution and wetting agents in SCLs for improved on-eye lens performance [5, 6] is dictated largely by the water content [7] and, thus, the SiHy microstructure. To our knowledge, no studies attempt prediction of water uptake or solute partitioning in silicone-based hydrogels.

We investigate equilibrium water and solute uptake in SiHys both experimentally and theoretically. Specifically, we focus on quantitatively predicting how initial SiHy formulation design affects equilibrium water content and partitioning of aqueous and oleophilic solutes of similar size (~100-300 Da). The SiHy materials studied were synthesized with free radical polymerization and consist of varying hydrophilic fraction (HEMA, MAA) and silicone fraction (Tris, mPDMS) with a block-copolymer macromer to ensure a co-continuous structure. Equilibrium water content was measured gravimetrically. UV/vis spectrophotometry and fluorescence confocal laser scanning microscopy were employed to measure partition coefficients for aqueous and oleophilic solutes respectively. Water content is predicted assuming that water is confined to the hydrophilic phase in the gel. Likewise, aqueous solutes have a high affinity for water, and dominantly partition in the hydrophilic phase as well. Conversely, oleophilic solutes predominantly partition into the silicone phase due to their low water solubility. To account for aqueous-solute exclusion and oloephilic-solute-adsorption in the silicone phase, extended Enhancement Factor Partitioning Theory [7] is adopted. Comparison of theory for water content and solute partitioning to the experimental data is excellent.

[1] Tighe B. Silicone hydrogels: structure properties and behavior. In: *Silicone Hydrogels: Continuous Wear Contact Lenses*; Sweeney DF, Ed. Butterworth-Heinemann: Oxford, 2004; pp.1-27

[2] Stapleton F, et al. Silicone hydrogel contact lenses and the ocular surface. *The Ocular surface*. 2006; 4; 24.

[3] Nicolson PC, Vogt J. Soft contact lens polymers: an evolution. *Biomaterials*. 2001; 22; 3273.

[4] Guan L, et al. Permeability and partition coefficient of aqueous sodium chloride in soft contact lenses. Journal of Applied Polymer Science. 2011; 122; 1457.

[5] Jones L, et al. The TFOS international workshop on contact lens discomfort: report of the contact lens materials, design, and care subcommittee. Invest Ophthalmol Vis Sci 2013; 54:TFOS37-70.

[6] Ketelson HA, Meadows DL, Stone RP. Dynamic wettability properties of a soft contact lens hydrogel. Colloid Surface B. 2005; 50; 1.

[7] Dursch TJ, et al. Water-soluble drug partitioning and adsorption in HEMA/MAA hydrogels. Biomaterials. 2014; 35; 620.

Superior Dispersion and Property Enhancements in Nanocrystalline Cellulose-Polyolefin Biocomposites Prepared via Solid-State Shear Pulverization

Krishnan Iyer[1], Gregory Schueneman[3] and John Torkelson[1,2]
[1]Department of Chemical and Biological Engineering, Evanston, IL 60208, USA
[2]Department of Materials Science and Engineering, Evanston, IL 60208, USA
[3]Forest Products Laboratory, US Forest Service, Madison, WI, USA
Contact information:
Krishnan Iyer – kai@u.northwestern.edu
Gregory Schueneman - gtschueneman@fs.fed.us
John Torkelson – j-torkelson@northwestern.edu

Cellulose, the main component of plant cell walls, constitutes the most abundant natural polymer resource available worldwide. It is the basic building block for textiles and for paper. Due to its remarkable strength with stiffness values exceeding 150 GPa, cellulose has garnered a lot of interest as a filler for polymers. Ever since the discovery of colloidal cellulose particles by Ranby in 1950's, the process of producing defect-free, rod-like crystalline nanocellulose has undergone considerable changes. Such cellulosic nanocrystals are known by different names such as nanowhiskers, nanocrystals, nanofibrils and nanocrystallites. Here, we refer to them as nanocrystalline cellulose (NCC). Such nanocrystals have large aspect ratios, being 3-5 nm in diameter and 100-300 nm in length. The unique physical properties of NCC have thus inspired several literature studies to explore its application as novel green nano-reinforcement in the production of polymer nanocomposites. Green nanofillers offer several sustainable benefits such as low cost, light weight, less energy for processing and reduced wear to instruments.

As a green filler, NCC has been incorporated into a variety of polymer matrices. Due to its hydrophilic nature, NCC is easily dispersed in water. Therefore, many studies have attempted to produce polymer-NCC composites with water-soluble polymers such as poly(vinyl alcohol). Favier et al. (Polym Adv Technol 1995;6:351–5) reported the first production of styrene-butadiene rubber (SBR)-NCC composites with water-based emulsion systems. In addition, NCC has been dispersed in several organic solvents, producing composites with enhanced mechanical properties. Compared to solution processing, relatively few studies have investigated the possibility of conventional melt-processing techniques such as twin-screw extrusion (TSE) for producing NCC-based nanocomposites. Achieving good dispersion of NCC in non-polar polyolefin matrices is a major challenge with TSE because of the hydroxyl groups on the filler surface and the often-agglomerated nature of NCC. Furthermore, NCC has poor thermal stability with degradation temperature as low as ~180 °C. This poses issues associated with filler degradation during melt processing for the production of composites.

Poor filler dispersion and degradation of NCC are detrimental to mechanical performance of polyolefin-NCC composites. Ljungberg et al. (Polymer 2006;47:6285–92) and Bahar et al. (J Appl Polym Sci 2012;125:2882–9) produced composites of PP with NCC using solution-processing techniques observing minor to no property enhancements. Durfesne and coworkers (Polymer 2009;50:4552–63) employed TSE to produce polyethylene-NCC composites with only modest increment in mechanical properties due to large filler agglomeration during processing. Surface treatment with organic acid chloride aliphatic chains of different size did not result in any further improvement in properties. Oksman and coworkers (Compos Sci Technol 2006;66:2776–84) employed a novel liquid feeding technique in which a NCC dispersion in water with plasticizer (poly(vinyl alcohol)) is pumped into molten poly(lactic acid) during extrusion. Such techniques still cause severe filler aggregation with several added challenges in removing moisture and in suppressing polymer degradation. Recently, Lin et al. (Macromolecules 2013;46:5570–83) used a wrapping technique in which NCCs were coated with low molecular weight polyethylene oxide to suppress NCC degradation during processing. However, filler agglomeration prevails due to incompatibility and inability of melt-mixing to provide sufficiently large shear forces to break up filler agglomerates. Therefore, there is a need for a solvent-free, continuous and industrially scalable processing technique for the production of such green polymer biocomposites.

Solid-state shear pulverization (SSSP) is a novel processing technique, which employs a modified twin-screw extruder where materials are processed under near-ambient temperature conditions. During SSSP, polymers are processed in their solid state, and large compressional and shear forces are imparted to the material. The repeated fragmentation and fusion of the material result in effective break-up of filler agglomerates. Near-ambient conditions in SSSP provide intimate mixing between polymer and nanofiller without regard to viscosity mismatch or thermodynamic limitations. Additionally, such conditions can prevent filler degradation of cellulosic materials, with effective particle break-up and dispersion, all of which are advantageous in the production of polyolefin-NCC biocomposites. In the past, these unique advantages of SSSP processing have been employed for the successful production of the following:

1. *Polymer blends:*
SSSP processing has been successfully used for the compatiblization of immiscible polymer blends with in-situ formation of block copolymers as a result of inter-radical coupling during the mechanochemical treatment.

2. *Nanocomposites:*
Fine dispersion and exfoliation of a variety of fillers ranging from platelet such as graphite and clay to agglomerated carbon nanotubes have been observed with dramatic mechanical property enhancements.

3. *Green composites:*
Recently, SSSP has been employed for the production of green polymer composites based on renewable waste materials such as rice husk ash with major enhancements in mechanical properties. Well-dispersed rice husk ash acts as a substituent for commercially available silica commonly employed as filler for plastics.

4. *Polar functionalization of polyolefins:*
Taking advantage of the unique radical chemistries associated with near-ambient temperature processing conditions in SSSP, polar functionalization of polyolefins (e.g., polypropylene) with ester or maleic anhydride moieties can be achieved with dramatically suppressed molecular weight reduction compared to melt processed systems.

Here we demonstrate the effective dispersion and major suppression of NCC degradation in green polyolefin-NCC biocomposites via SSSP with up to 10 wt% filler content.

Excellent dispersion and retention of pristine cellulose color are observed in both low-density polyethylene (LDPE) and PP matrices through optical microscopy and scanning electron microscopy. Synergistic property enhancements due to effective dispersion are demonstrated via a variety of characterization techniques such as differential scanning calorimetry (DSC) and tensile testing. Well-dispersed NCC acts as an excellent nucleating agent for PP crystallization with dramatically reduced crystallization half-times. In short, the superior filler dispersion and suppressed filler degradation give rise to synergistic enhancements in Young's modulus and excellent retention of elongation at break for 10 wt% NCC loadings in both PP and LDPE. Previously, poor outcomes were reported by Menez et al. when incorporating unmodified NCC in LDPE produced by TSE (Polymer 2009;50:4552–63). They observed only a modest 14 % increase in Young's modulus and a 85 % decrease in elongation at break relative to the neat polymer. In comparison, SSSP processed LDPE-NCC composite with 10 wt% NCC show up to 69 % increase in Young's modulus and no decrease in elongation at break relative to that of the neat polymer. To the best of our knowledge, the mechanical property enhancements shown by SSSP-processed polyolefin-NCC composites are superior to those reported in the research literature for similar composites made by melt processing. The degradation behavior of NCC, LDPE-NCC and PP-NCC systems is analyzed under nitrogen and air atmospheres using a thermogravimetric analyzer (TGA). Interestingly, LDPE-NCC composites show enhanced thermal stability relative to neat LDPE due to char formation from well-dispersed NCC. This study thus demonstrates the promise of SSSP processing in exploiting the full potential of this novel green nanofiller.

398273 Self-Assembled Protein-Inorganic Hybrid Supraparticles for Robust Protein Immobilization

Abstract:

Biomineralization is a process during which mineral clusters and biomacromoecules self-assemble into nanostructured composite materials. Proteins serve dual roles in these hybrid materials, first as templating agents to regulate structural and physical properties of the material, and second as ligands for biological communication. Inspired by this natural process, we developed a method to fabricate self-assembled and porous hybrid supraparticles with specific affinity recognition for robust protein immobilization. First, we constructed a recombinant fusion protein that contains naturally derived leucine zipper coiled coil motifs, which can enable non-covalent conjugation of proteins via a high affinity interaction between the zippers. In the fusion protein, two identical leucine zipper motifs were combined with a random-coil midblock. Second, the engineered protein directs formation of flower-like nanohybrids that are further self-organized into porous supraparticles. When calcium phosphate was precipitated in the presence of the fusion protein, it provided nucleation sites and directed the growth of calcium phosphate petals into flower-like nanohybrids. The van der Waals interactions between the flower-shaped nanostructures resulted in their colloidal assembly into interlocked chains, which were packed into hierarchically structured and porous supraparticles on the micrometer length scale. The protein-inorganic supraparticles are biocompatible, and provide large surface on which leucine zipper motifs, the high-affinity conjugation sites, are available. Globular proteins fused with the leucine zippers can be immobilized on the supraparticles via simple incubation in aqueous solution. In particular, immobilization of enzymes can enhance their stability and catalytic activity, thus providing benefits for many applications. Thus, our approach provides a versatile method to fabricate complex biomaterials by mimicking both specific biomolecular recognition and the natural biomineralization process for formation of hybrid nanostructures.

398274 Molecular Description of LCST Behavior of Elastin-like Peptides Poly(VPGVG) and Poly(VGPVG)

Abstract:

The elastin-like peptide (ELP) is temperature sensitive biopolymer which undergoes conformational transition at critical temperature. Elastin-like polypeptides (ELPs) with the repeat sequence of VPGVG are widely used as a model system for investigation of lower critical solution temperature (LCST) transition behavior. Recent experimental observations show that the order of amino acids in the ELP sequence also plays an important role since $(VGPVG)_n$ and $(VPGVG)_n$ peptides show the differences in thermo-transition behavior. In order to elucidate the factors driving the conformational transitions of ELPs we performed molecular dynamics simulations of $(VPGVG)_n$ and $(VGPVG)_n$ peptides with different length and at a range of temperatures. Comprehensive analyses of the hydrophobicity, shape, size, hydration and dynamics of $(VPGVG)_n$ and $(VGPVG)_n$ suggest that the transition behavior includes the thermal disruption of water network, loss of hydration, and increase in peptide hydrophobicity. Our simulations show that single poly(VGPVG) is more hydrophilic than poly(VPGVG) at high temperatures while similar hydrophobicity of two types of ELPs at low temperatures is observed. Simulations of the peptide-peptide association demonstrated that the aggregates formed by double poly(VGPVG) chains are more densely associated than poly(VPGVG) at high temperature. Even more compact association of double poly(VGPVG) was observed upon cooling the aggregates obtained from the double-peptide simulation. In the contrary, poly(VPGVG) aggregates dissociate upon cooling which is consistent with the reversibility exhibited in experiments. The results from our study provide an atomic-level description of the thermo-responsive conformational properties of elastin-like peptides and explain experimental observations. Support for this research was provided by the NSF's Research Triangle MRSEC (DMR-1121107).

398275 Identifying the Differentiation and Proliferation Stage of Single Hematopoietic Stem Cells Using Raman Microspectroscopy

Abstract:

The ability to monitor the differentiation and proliferation status of individual hematopoietic stem cells is critical to understanding the effects of various external factors, such as matrix properties and cytokines, on hematopoiesis. Raman microspectroscopy is a promising method for single cell identification and characterization. Unlike other invasive methods, Raman microscopy has the ability to detect changes in the composition of living cells without compromising the cells' morphology, proliferation, or pluripotency. However, identifying cells according to their Raman spectra is difficult due to a lack of spectral bands that are unique to a single cell line. Here, we report our ability to identify the differentiation stage of individual mouse hematopoietic stem cells and mature cells by multivariate analysis of their Raman spectra. Confocal Raman microscopy was used to obtain spectral fingerprints of single cells in various stages of hematopoiesis. A model of the cell type-specific variance in the Raman spectra was constructed using partial least-squares discriminant analysis (PLS-DA) and enabled accurately identifying cells of "unknown" differentiation stage, including closely-related long- and short-term hematopoietic stem cells with low prediction error (<10%). Our high identification accuracy cannot be attributed to spectral contributions from the polyacrylamide gel substrates, which varied in stiffness and thickness, and culture media. In addition, principal component analysis was used to identify the cell cycle stage of each single cell based on the principal component corresponding to DNA-related Raman peaks. Stem cells were found to be more proliferative than mature cells. Noteworthy, the Raman contributions that correlated to cell cycle phase did not significantly interfere with identification of cell differentiation stage. This method allows for non-invasive monitoring of stem cell differentiation over time with location specificity, and can be used to correlate external stimuli with stem cell fate decisions.

373320 Invited Talk: Pattern Formation in Block Copolymer and Grafted Nanoparticle Thin Films

Abstract:

This presentation focuses on simulations of pattern formation in block copolymer and grafted nanoparticle thin films. In the block copolymer simulations, we utilize a recently developed theoretically informed coarse graining model [1], which preserves chain connectivity, microphase separation and finite compressibility of the melt. In static block copolymer films, we find, in agreement with experiments, a remarkable cascade of morphological transitions from parallel to perpendicular orientations when the film height is varied [2]. We demonstrate that the emergence of a stable parallel configuration stems from entropic effects due to the conformational asymmetry and the confinement of the system. For the sheared films, we find that the thin films exhibit shear thinning behavior strongly correlated with the disentanglement and shear alignment of the constituent polymers, and show that shear can be employed to induce long-range ordering to the spontaneously self-assembled microdomains [3]. Finally, a "top down" coarse-graining methodlogy is utilized to investigate the self assembly of solvent-free grafted nanoparticles into thin free-standing films. We find that the nanoparticles self-assemble into a variety of morphologies ranging from dispersed particles, finite stripes, long strings, to percolating networks [4].

[1] D. Q. Pike, F. A. Detcheverry, M. Müller, and J. J. de Pablo, "Theoretically informed coarse grain simulations of polymeric systems," *J. Chem. Phys.*, **131**:084903 (2009).
[2] A. Nikoubashman, R. A. Register, and A. Z. Panagiotopoulos, "Sequential Domain Realignment Driven by Conformational Asymmetry in Block Copolymer Thin Films," *Macromolecules*, **47**:1193-8 (2014).
[3] A. Nikoubashman, R. L. Davis, P. M. Chaikin, B. T. Michal, R. A. Register, and A. Z. Panagiotopoulos, "Thin Films of Homopolymers and Cylinder-Forming Diblock Copolymers under Shear," submitted (2014).
[4] T. Lafitte, S. K. Kumar, and A. Z. Panagiotopoulos, "Self-assembly of polymer-grafted nanoparticles in thin films," *Soft Matter*, **10**:786-94 (2014).

377955 Effect of Polydispersity in Grafts and Matrix on the Morphology of Polymer Grafted Nanoparticle Filled Polymer Nanocomposites

Abstract:

Polymer nanocomposites are a class of materials that consist of a polymer matrix embedded with nanoscale fillers or additives that enhance the inherent properties of the matrix polymer. To engineer polymer nanocomposites for specific applications with target macroscopic properties (e.g. photovoltaics, photonics, automobile parts) it is important to have design rules that relate molecular features to equilibrium morphology of the composite. In this poster we will present our recent theory and simulation work on composites containing polymer grafted nanoparticles in a chemically identical polymer matrix, showing how polydispersity in graft and matrix polymers affects morphology of the polymer nanocomposite. Polydisperse grafted polymers stabilize the dispersed morphology of homopolymer grafted nanoparticles in a chemically identical homopolymer matrix, due to improved wetting of the polydisperse grafted layer by the monodisperse matrix. In the case of polydisperse matrix polymers, modeled as a mixture of short and long chains, the short matrix chains preferentially wet the monodisperse grafted layer (at high grafting density) more than the long matrix chains. Despite the preferential wetting by the short than long chains, matrix length bidispersity does not significantly change the overall wetting of the grafted layer, and as a result does not alter the tendency for dispersion/aggregation of the grafted particles. Furthermore, graft length bidispersity is better than matrix length bidispersity for stabilizing dispersions of polymer grafted nanoparticles in a chemically identical polymer matrix.

378463 Systematic and Simulation-Free Coarse Graining of Polymer Melts

Abstract:

Full atomistic simulations of many-chain systems such as polymer melts are not feasible at present due to their formidable computational requirements. Coarse-grained (CG) models have to be used instead, where each segment represents, for example, the center-of-mass of a group of atoms/monomers. Such models, however, reduce the chain conformational entropy, which plays an essential role in the thermodynamic behavior of polymeric systems. On the other hand, in most work on coarse graining, molecular simulations are used to obtain the structural and/or thermodynamic properties of both original and CG systems that need to be matched, which is not only computationally very expensive but also non-transferable. In this work we propose a systematic and simulation-free strategy for both structure- and relative-entropy-based coarse graining of homopolymer melts. We use integral-equation theories, instead of molecular simulations, for both original and CG systems, and examine in detail how the CG potential varies with N (the number of CG segments on each chain) and how well the CG models can reproduce the structural and thermodynamic properties of the original system. The structure-based approach reproduces the structural but not thermodynamic properties of the original system due to the information loss of coarse-graining, and the relative-entropy-based approach minimizes such information loss. Our systematic and simulation-free coarse-graining strategy is much faster than those using molecular simulations, and provides a quantitative basis for choosing small N-values that can still capture the chain conformational entropy, a characteristics of polymers.

A key objectives of the Materials Genome Initiative (MGI) is to create a coherent framework to perform coarse-grained (CG) simulations in a streamlined and efficient manner to reduce the time and cost of novel material design. In order to gain widespread application in the industrial sector it is imperative to know the limitations of coarse-grained potentials. In this study we apply methods of Uncertainty Quantification (UQ) to the CG potential of polystyrene at a resolution of 1 CG site per monomer. While various methods exist to parameterize CG potentials, we focus on the multiscale coarse-graining technique of Izvekov *et al.* [J. Chem. Phys. **120**, 10896 (2004)]. Baysian methods are used to assess the uncertainties in (i) the CG force field, and (ii) its associated predictions for the density. We draw conclusions with regards to formulating and optimizing the CG potentials in order to calculate the glass transition temperature. We conclude with an outlook at how these analyses can applied to other properties to better understand the boundaries of applicability of these CG potentials.

386983 Multiresolution Modeling of Polymers: Wavelet-Based Reconstruction

Abstract:

In contrast to multiscale methods, which often encompass multiple simulation techniques, multiresolution models use the same general modeling technique with representations of a system at varying length and time scales. We present work on a complete framework for modeling of semidilute polymer solutions, containing both coarse-graining and reconstruction steps. The coarse-graining approach is based on the wavelet-accelerated Monte Carlo (WAMC) method [1], which forms a hierarchy of resolutions to model polymers at length scales that cannot be reached via atomistic or even "standard" coarse-grained simulations. Using multiple stages of resolution, it is possible to simulate polymers of up to millions of repeat units in length. Although previously applied only to individual chain statistics [2], we show here how it can be extended to the study of polymer solutions. Bonded and non-bonded potentials between coarse-grained superatoms are computed. The non-bonded potential is computed using the same approach used previously in the study of single chains while the bonded potentials are computed using reverse Monte Carlo. A universal scaling function is obtained so that potentials do not need to be recomputed as the scale of the system is changed. To model polymer solutions, the intermolecular potential between the CG beads is assumed to be equal to the non-bonded potential, which is a reasonable approximation in case of semidilute systems. Using these potentials, various radial distribution functions are calculated from the coarse-grained representation at different resolutions, which compare favorably with the results from the atomistic simulations. We show that coarse-grained polymer solutions can reproduce results obtained from the simulations of the more detailed atomistic system to a reasonable degree of accuracy.

The reconstruction proceeds in much the same way: using probability distributions obtained from the coarse-graining procedure with respect to bond lengths, angles, and torsions, as well as the non-bonded potentials, we can reconstruct a more detailed version of the polymer that is consistent both with geometric constraints as well as energetic considerations. Using a quality estimator model for the reconstruction process, we can successfully "reverse map" entire atomistic configurations from coarse-grained descriptions. Applications for the reconstruction of atomistic models from united-atom models and detailed models from coarse-grained models will be shown, using both semidilute polymer solutions and self-assembled monolayers as test cases.

372533 Derivation of Coarse-Grained Lipid Potentials Using Multi-State Iterative Boltzmann Inversion

Abstract:

In this work, we present intermolecular potentials for coarse-grained (CG) models of the lipids present in the skin, specifically ceramides, cholesterol and free fatty acids. CG models were developed in order to make it feasible to simulate the timescales required to capture self-assembly behavior. The potentials were derived via the multi-state iterative Boltzmann inversion (IBI) method [1]. This approach is a modification of the original iterative Boltzmann inversion method [2] that involves iteratively adjusting a CG potential such that it simultaneously matches the structure of target atomistic systems at different thermodynamic state points and/or in different thermodynamic ensembles. This approach minimizes state dependence, producing robust, generally applicable CG potentials. Here, transferable potentials have been derived for CG models of a free fatty acid, two different ceramides, and cholesterol. We also show how the multi-state IBI method allows us to develop potentials that accurately reproduce structural properties beyond the radial distribution function used in the derivation of the potentials (e.g., tilt angle, nematic order). The properties of both pre- and self-assembled CG lipids mixed with water is also explored and the behavior compared to both atomistic and experimental results.

[1] T.C. Moore, C.R. Iacovella, and C. McCabe. "Derivation of coarse-grained potentials via multistate iterative Boltzmann inversion." (submitted)

[2] D. Reith, M. Putz, and F. Muller-Plathe, J. Comp. Chem. 24, 1624 (2003).

384981 Structure and Rheology of Polymer Solutions from Coarse-Grained Molecular Dynamics Simulations: Effects of Polymer Concentration, Solvent Quality and Geometric Confinement

Abstract:

Structure, dynamics and rheology of solutions of flexible linear polymers are investigated using coarse-grained (CG) molecular models and molecular dynamics (MD) simulations in presence of explicit solvent mediated interactions. MARTINI force field is employed to describe the polymer, solvent and the underlying physico-chemical interactions. The CG models are validated against atomistic ones by comparing the predictions of certain structure parameters such as persistence length, radius of gyration and radial distribution functions of the monomeric units. First, we will present results for the dynamics of a single polymer chain in shear flow. The effects of chain length and shear rate on the configuration statistics, e.g. tumbling frequency and orientation distribution of the end-to-end vector, will be presented and compared to experimental observations and predictions of stochastic dynamics simulations. Further, the effects of solvent-polymer interactions under good, theta and poor solvent conditions as well as geometric confinement in presence of favorable, neutral and unfavorable polymer-wall interactions on the configuration dynamics of a single polymer chain will be also discussed. Specifically, the role of solvent quality will be shown to have a pronounced effect on coil-stretch transition in shear flow. CGMD predictions for the relationship between the zero-shear viscosity and polymer concentration in dilute and semi-dilute regimes will be presented and compared to experiment results. Shear thinning behavior is observed in both dilute and semi-dilute solutions in non-equilibrium molecular dynamics simulations. Possible approaches using MD simulation data to parameterize phenomenological constitutive models such as the Carreau-Yasuda model will also be discussed. (Support from the National Science Foundation through grants CBET-1055219 and CDI 1049489 is gratefully acknowledged).

Molecular Simulation Based Analysis of Stress Dissipation and Polymer Strength Near a Surface

- Sunil Kumar, Sudip K. Pattanayek* and Sanat Mohanty!
Dept. of Chemical Engineering, Indian Institute of Technology, New Delhi, 110016 India

Understanding of behavior of polymers under external force in the vicinity of surfaces is important to the development of high performance polymer systems required for the development of Pressure Sensitive Adhesives (PSAs)[1, 2] and design of high performance nano-composites[3,4] for increased toughness etc. The development of the above applications require the detailed understanding of the mechanisms by which the nano-surface – polymer interactions help to dissipate the energy of the stress. A number of experimental[5-17] and simulation studies[19-33] have been reported to understand the structure and dynamics of polymers under the application of stresses, especially close to surfaces. Experimentally researchers have studied the effect of force in the perpendicular to the surface on polymer chain. They found that the dissipation of energy depends on the structure of polymers during pulling, interaction among polymer chains and between polymer chain and surface. Creton and his group[5] have studied the responses of polymer with different cohesive energy at constant applied velocity. They pointed out that as the surface forces decrease; the polymer sees an adhesive failure in that it is cleanly detached from the surface. However, as surface interactions become more favorable between the polymer and the surface, the polymer is stretched before forming fibrils and subsequently being cleanly displaced. The processes of fibrillation and of formation of cavities are important pathways to dissipation of energy and impact the properties of the polymer formulation.

We have studied pulling of a single polymer chain adsorbed randomly on a graphene surface using molecular dynamics simulation. Force is applied parallel and perpendicular to the surface in separate experiments. The choice of polymer (polyethylene) allows simulations of simple straight chains that have primarily van der Waals interactions with the surface. A Graphene surface is chosen because it results in a well-defined, regular surface structure. We note that graphite based nanoparticles have also been in use in adhesive formulations. The behavior of polyethylene on graphene surfaces is perhaps the simplest of systems that allows us to focus on the role of surface interactions with the polymer and polymer-polymer interactions (both of which are tunable through the simulation parameters) and their roles in debonding mechanisms and strength of the bond.

We implemented constant temperature molecular dynamics simulations[34, 35] in this study. The procedure adopted can be found in our recent publications[36]. Briefly, we put a randomly generated polymer chain containing N connected beads, where each bead corresponds to united atoms CH_2 in a simulation box (362Å × 362Å × 362Å). The boundaries are assumed to be repulsive. The N value chosen for all simulations in this study is 1000. In addition, a square graphene sheet (edge length = 100Å) consisting of a network of hexagonally orientated (defined by Al-Haik *et al.*[37]) carbon atoms (sp^2 hybridized) was kept at the center in the X-Y plane of the simulation box and the polymer chain placed above the surface. The graphene sheet is kept stationary in the simulation box. The coordinates of the beads in three dimensional is generated by using VMD[38] (Visual Molecular Dynamics) and used in the MD simulations box. We assumed that mass of beads of polymer and graphene sheet is m.

I. Results

A. Chain displacement due to application of external force parallel to surface (F_{xext}):

F_{xext} is applied on all the beads of the polymer chain along the x direction (parallel to the surface). Snapshots of the polymer at different time steps for external force values are shown in figure 1. At low values of F_{xext} (for example, F_{xext} = 0.025 kcal/mol.Å), the chain slides slowly along one face of the surface, comes to the edge and crawls to the other face of the plane. At a higher applied force the polymer evolves into a fibrillated structure, and then escapes from the surface. Figure 2 shows the variation of nonbonded polymer-surface energy (E_{PS}) with time at different values of the applied external force.

Chain displacement due to application of force perpendicular to surface (F_{zext}):

Continuing this analysis of the behavior of a polymer chain close to a surface, the impact of the chain in the presence of a force perpendicular to the plane of the surface is studied. An external force F_{zext} (perpendicular to the plane of the surface) is applied on all beads of an equilibrated polymer chain (equilibration time t=1500ps). Figure 3 shows snapshots of the polymer chain on the graphene surface at different times and with increasing value of the applied force. We have measured non bonded interaction potential energy between polymer- graphene surface (E_{PS}) as shown in figure 4. At a very low value of external force F_{zext} (for example, F_{zext} = 0.1 kcal/mol.Å), the polymer chain does not escape from the surface. At the higher values of the external force (F_{zext} = 1.0 kcal/mol.Å), the polymer chain takes much less time to escape from the surface completely.

II. Conclusions

The stiffness of the polymer, the relative energy of interaction between the polymer moiety and between the polymer and surface, all affect the energy dissipated during the process of stress transfer. While these ideas were presented and developed for macroscopic polymer systems by Creton *et al.* [9,10] , this study shows that the mechanisms of stress dissipation hold true for single polymers in the vicinity of a nano-surface. The above analysis also implies that the presence of nano-surfaces affects the nature of polymer response to stresses and thus will affect the polymer failure. Choice of the polymer, the nano-surface (and how it interacts with the polymer) affects the mode of failure for both PSAs and composites. Thus, this paper provides a composite designed with the primary parameters of composite design.

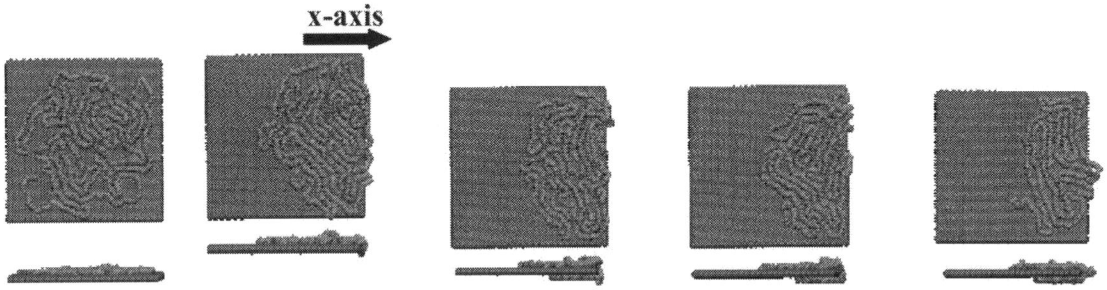

1500ps 1510ps 1520ps 1550ps 1600ps

(a) $F_{xext} = 0.025$

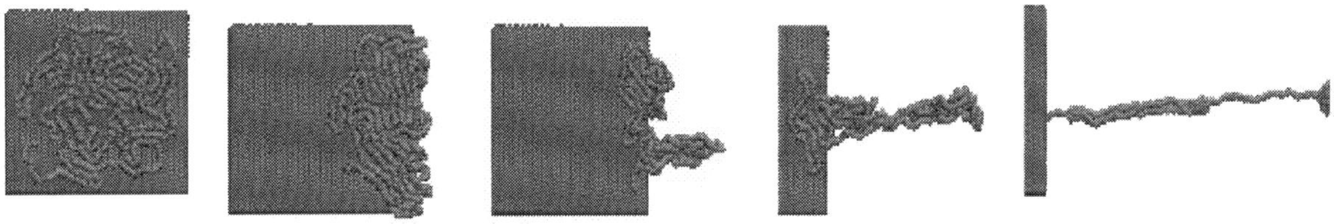

1500ps 1510ps 1520ps 1530ps 1550ps

(b) $F_{xext} = 0.05$

1500ps 1505ps 1510ps 1515ps 1530ps

(c) $F_{xext} = 0.10$

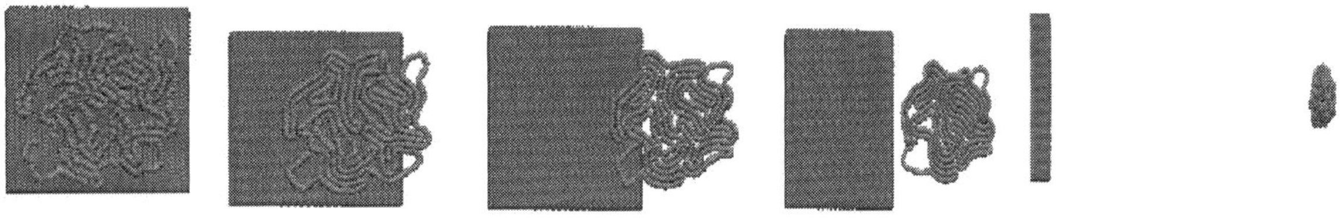

1500ps 1502ps 1505ps 1507ps 1525ps

(d) $F_{xext} = 1.00$

Figure 1: Snapshot of polymer chain pulling parallel to the surface (x-axis) at different values of external force, F_{xext}: (a) $F_{xext} = 0.025$, (b) $F_{xext} = 0.05$, (c) $F_{xext} = 0.1$ and (d) $F_{xext} = 1.0$ kcal/mol.Å and at fixed values of $\varepsilon_{PP} = 0.3968$, $\varepsilon_{PS} = 0.30$, $k_{\varphi} = 2.0$ kcal/mol and T=300K.

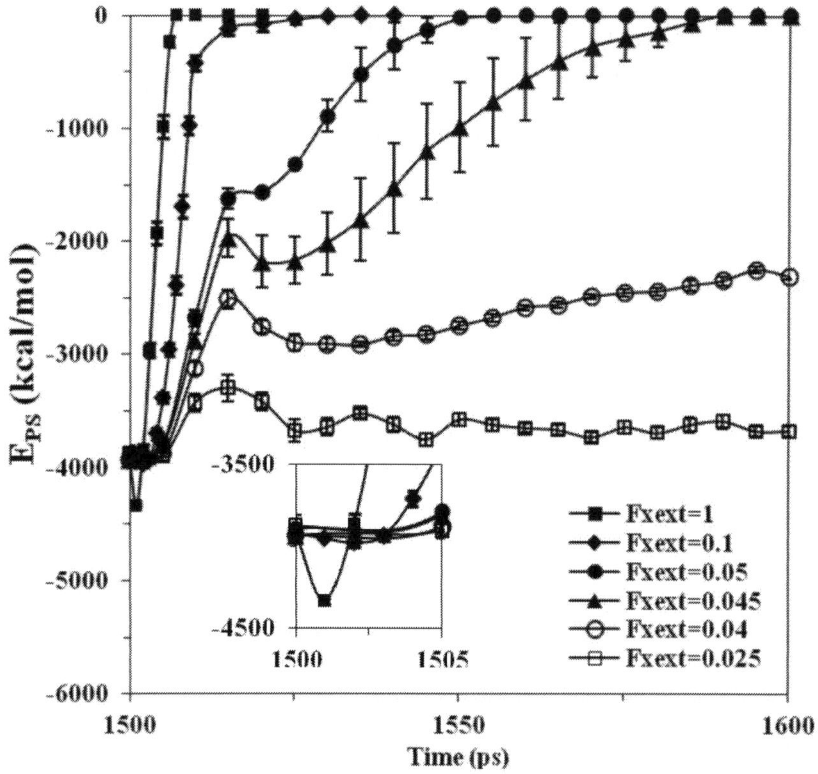

Figure 2: Non bonded energy profile between polymer beads and surface (E_{PS}) during polymer chain pulling parallel to the surface (x-axis) by various values of external forces F_{xext} (kcal/mol.Å).

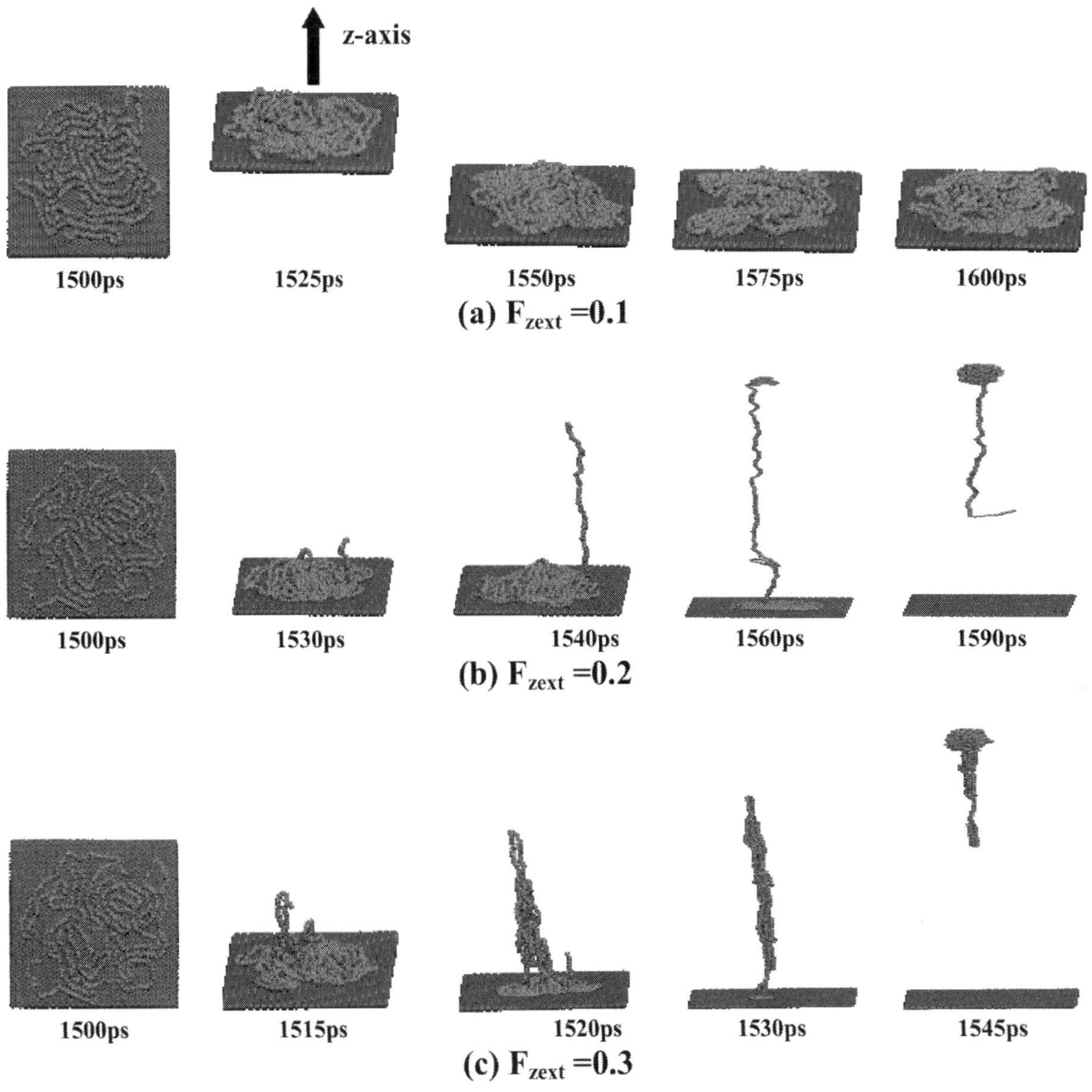
z-axis
1500ps
1525ps
1550ps
1575ps
1600ps
(a) F_{zext} =0.1
1500ps
1530ps
1540ps
1560ps
1590ps
(b) F_{zext} =0.2
1500ps
1515ps
1520ps
1530ps
1545ps
(c) F_{zext} =0.3

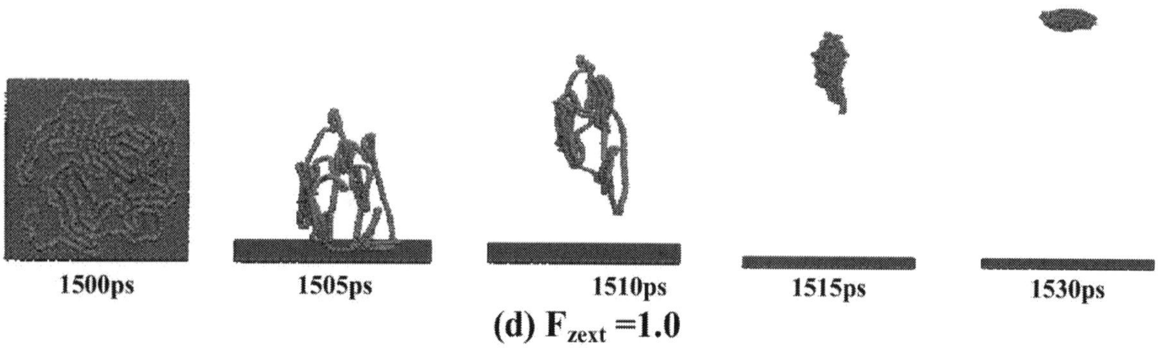

Figure 3: Snapshot of polymer chain pulling perpendicular to the surface (z-axis) by different values of external tensile force, F_{zext}: (a) $F_{zext} = 0.1$, (b) $F_{zext} = 0.2$, (c) $F_{zext} = 0.3$ and (d) $F_{zext} =$ 1.0 kcal/mol.Å at fixed $\varepsilon_{PP} = 0.3968$, $\varepsilon_{PS} = 0.30$, $K\varphi$=2.0 kcal/mol and T=300K.

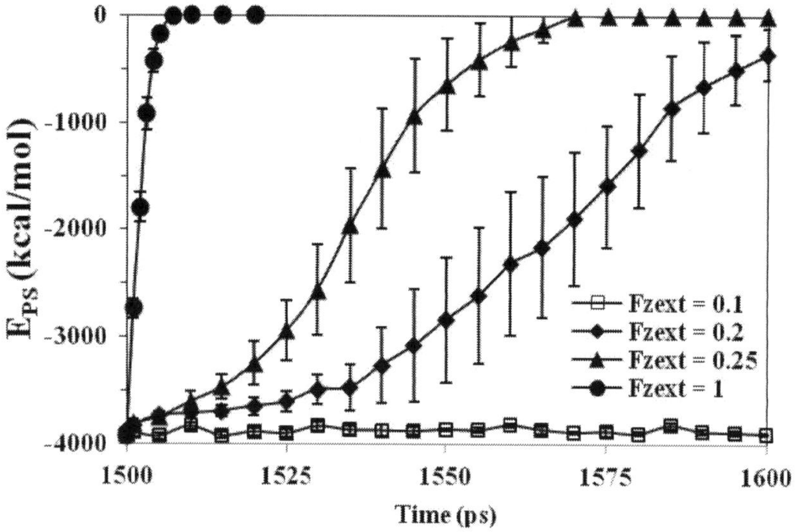

Figure 4: Non bonded energy profile between polymer beads and surface (E_{PS}) during polymer chain pulling perpendicular to the surface (z-axis) by various values of external forces F_{zext} (kcal/mol.Å).

References

[1]P. Tordjeman, E. Papon, and J-J. Villenave, J. Chem. Phys. **113**, 10712 (2000)

[2]A. Zosel, Colloid Polym. Sci. **263**, 541 (1985)

[3]S. D. Tobing and A.J. Klein, Appl. Polym. Sci. **79**, 2230 (2001).

[4]T. N. Krupenkin and G. H. Fredrickson, Macromolecules **32**, 5029 (1999)

[5]A. Lopez, E. D. -Contraires, E. Canetta, C. Creton, J. L. Keddie, and J. M. Asua, Langmuir, **27**, 3878 (2011)

[6]R. P. Wool and S. P. Bunker, J. Adhesion **83**, 907 (2007)

[7]R. Vendamme and W. Eevers, Macromolecules **46**, 3395(2013)

[8]S. Yamamoto, Y. Tsujii, and T. Fukuda, Macromolecules **33**, 5995 (2000)

[9]T. Hugel, M. Grosholz, H. C.-Schaumann, A. Pfau, H. Gaub, and M. Seitz, Macromolecules **34**,1039(2001)

[10]J. E. Bemis, B. B. Akhremitchev, and G. C. Walker, Langmuir **15**, 2799(1999)

[11]B. J. Haupt, T. J. Senden, and E. M. Sevick, Langmuir **18**, 2174 (2002)

[12]B. J. Haupt, J. Ennis, and E. M. Sevick, Langmuir **15**, 3886 (1999)

[13]C. Ortiz and G. Hadziioannou, Macromolecules **32**,780 (1999)

[14]F. Ritort, J. Phys.: Condens. Matter **18**, R531 (2006)

[15]D. Soto, G. Hill, A. Parness, N. Esparza, M. Cutkosky, and T. Kenny, App. Phys. Lett. **97**, 053701 (2010)

[16]F. Kuhner, M. Erdmann, L. Sonnenberg, A. Serr, J. Morfill, and H. E. Gaub, Langmuir, **22**, 11180 (2006)

[17]Y. Gao, G. Sirinakis, Y. Zhang, J. Am. Chem. Soc. **133**, 12749 (2011)

382695 Multiscale Simulations of Thermal Annealing of P3HT:PCBM Active Layers in Bulk Heterojunctions

Abstract:

Using large-scale coarse-grained molecular dynamics simulations, we have investigated the thermal annealing of poly(3-hexylthiophene) (P3HT) and phenyl-C61-butyric acid methyl ester (PCBM) blends in the presence of a substrate. The simulations are in agreement with neutron reflectivity (NR) and near edge X-ray fine structure (NEXAFS) experiments and reveal a vertical composition profile of the bulk heterojunction normal to the substrate with enrichment of PCBM near the substrate and air interfaces. We also demonstrate that the addition of short P3HT chains, as a third component of the blend, can be used to alter the morphology. The short chains are found to efficiently migrate to the P3HT/PCBM interface. To model the P3HT/PCBM blend interface, atomistic simulations of 3, 6, or 9 P3HT repeat units in contact with PCBM were performed. These simulations enable the sharpness of the donor/acceptor interface and orientation of thiophene rings at the interface to be examined as a function of P3HT size, and point to the sensitivity of chain orientation on the optimization of interfacial structure.

The development of non-precious transition metal catalysts is of great interest as next-generation technologies such as fuel cells and electrolyzers continue to grow in the alternative energy commercial market. Many commercial fuel cells, and in particular, acidic polymer electrolyte fuel cells, use platinum or other precious metals as the catalyst material. In addition, research that supports industries that have traditionally used costly, high-energy processes for chemical production is now considering alternative methods that would use less energy. For example, ammonia is currently produced using the Haber-Bosch process, where nitrogen gas is reduced to ammonia at high temperature and high pressure. For both energy conversion and chemical production applications, electrocatalysis under alkaline conditions avoids high energy conditions of elevated temperature and pressure, and alkaline electrocatalysis can be performed with non-precious metal catalysts. However, key challenges in the use of non-precious metal catalysts include material performance, as well as material stability and lifetime.

Here, we will present on a set of iron-nickel core-shell nanoparticles that have shown promise as electrocatalysts under alkaline conditions for both fuel oxidation and water electrolysis. The nanoparticle performance can be tuned for either fuel oxidation or oxygen evolution during water electrolysis through modification of the oxygen content and oxide morphology of the nanoparticles. Furthermore, we are now working on optimizing this metal combination for electrocatalytic production of ammonia from nitrogen. While the performance appears to be tailorable through optimization of the nanoparticle synthesis procedure, nanoparticle lifetime continues to be a challenge. Over a period of several months, the performance of the nanoparticles decreases, and an increase in the oxygen content and presence of oxides is observed. The nanoparticles appear to be polycrystalline with a nanocrystalline domain size of approximately 2-3 nm, and portions of the nanoparticles may be amorphous. The presence of nanocrystallinity and/or amorphous character suggests that the nanoparticles are thermodynamically metastable, and increased stability is likely necessary to retain catalytic performance over time. However, the crystalline/amorphous nature may also be the cause of the observed performance in electrocatalysis applications. An annealing study will be discussed as a possible method to stabilize the nanoparticle structure, as well as the effect of annealing on catalytic performance.

376598 Controlled Deposition of Iron Oxide Nanoparticles Using a Novel Gas-Expanded Liquid (GXL) Process to Generate Supported Fischer-Tropsch Catalysts

Abstract:

Fischer-Tropsch synthesis is a catalytic process used in industry to convert carbon monoxide and hydrogen to heavier hydrocarbons. Active metals like iron, cobalt, nickel and ruthenium are required as catalysts for this process. Supporting these metals on an oxidic support (silica/alumina) has many advantages including an influence in mass selectivity due to the pore structure of the support and stability of the support. Attrition or fracturing inside a fixed-bed or slurry reactor is also reduced. Nano-sized particles also offer several advantages due to their easy dispersability in organic solvents which makes homogeneous loading on supports easier. The usual methods of supported catalyst preparation (ex. incipient wetness impregnation, ion exchange and coprecipitation) rely on various solvent-based transport phenomenon including diffusion and capillary action to disperse the iron into the catalyst pores. Gas expanded liquids (GXLs) have various advantages over conventional solvents due to their pressure tunable transport properties. The diffusion of various molecules is usually enhanced in a GXL due to the lower density/viscosity of solvents at high pressure. We have developed a novel method to controllably deposit iron oxide nanoparticles on a support in the presence of a GXL. This method allows for a more uniform deposition of the metal particles into the pores of the support hence improving the performance of the catalyst in Fischer-Tropsch synthesis. In this study we have carried out preliminary experiments on the synthesis of these supported iron catalysts and tested the effect of various operating parameters on the performance of this catalyst under Fischer-Trosch synthesis conditions.

379001 Controlling Metal Nanoparticles in Fischer Tropsch Synthesis

Abstract:

Controlling Metal Nanoparticles in Fischer Tropsch Synthesis

Jorge Gascon, Freek Kapteijn

The growing concerns about oil depletion have spurred worldwide interest in finding alternative feedstocks for important petrochemical commodities and fuels. Fischer-Tropsch synthesis (FTS) is a surface polymerization reaction where syngas (a mixture of CO and H_2), derived from natural gas, coal, or biomass, is catalytically converted to hydrocarbons over Co, Ru, or Fe catalysts.[1,2] This technology has long been considered as an alternative process to produce long chain paraffins *via* the so-called gas-to-liquids (GTL) process. Owing to their high activity and selectivity to long chain hydrocarbons, Co-based catalysts are the preferred choice. However, earth abundant and cheap Fe displays superior properties over Co-based catalysts, especially for syngas derived from coal (CTL) or biomass (BTL). As compared to Co, Fe is more resistant to poisoning by contaminants that may be present in the syngas feed such as sulfur and ammonia. In addition, Fe-based catalysts are active in the water-gas-shift reaction under typical FTS conditions, enabling the *in-situ* re-adjustment of the H_2/CO molar ratio for the conversion of hydrogen lean syngas originating from coal or biomass.[1] Furthermore, the product slate of Fe-based catalysts is more directed to short chain (unsaturated) hydrocarbons and short chain oxygenates, both among the most important chemical building blocks.

The main challenge in the design of stable FTS catalysts lies in overcoming the high deactivation rate due to sintering, carbon deposition and phase changes. During this lecture, we will highlight different synthesis strategies able to deliver very stable, active and selective FTS supported catalysts.

References

(1) Abelló, S.; Montané, D. ChemSusChem 2011, 4, 1538.

(2) Sun, B.; Xu, K.; Nguyen, L.; Qiao, M.; Tao, F. ChemCatChem 2012, 4, 1498.

Coke-Resistant Core-Shell Catalysts for Dry Reforming of Methane

*Junshe Zhang, Fanxing Li**

Department of Chemical and Biomolecular Engineering
North Carolina State University, Raleigh, NC 27606
*fli5@ncsu.edu

Dry reforming of methane (DRM) represents a promising approach for carbon dioxide utilization and syngas generation. One of the key challenges that hinders the commercialization of DRM resides in catalyst performance. While supported noble metal catalysts are active and less susceptible to coke formation, they are highly expensive. Compared with noble metals, Ni-based catalysts are more commercially attractive. However, industrial application of Ni catalysts in dry reforming reactions is limited by its poor stability, caused by sintering of active metallic particles and carbon deposition on the metal surface.

Previous studies have indicated that coke formation is primarily catalyzed by step and edge sites. In addition, coke formation over small Ni particles is not energetically favored. To date, effective options to inhibit coke formation include poisoning coke-formation sites with sulfur, introducing alkali, alkaline earth or transition metal promoters, chemically or physically dispersing Ni in other metal oxides, confining Ni particles in mesoporous materials, and encapsulating catalyst particles in an inorganic shell. In the current study, we demonstrate that core-shell structured $NiO@SiO_2$ nanoparticles can be fabricated from NiO nanoparticles by the sol-gel method. The core-shell catalysts exhibit excellent catalytic performance toward dry reforming of methane with carbon dioxide at 850 °C; carbon deposition is negligible in 39 h on stream.

390429 Reactions Catalyzed By Metal Particles Supported on Functionalized Carbon Nanotubes in Biphasic Systems

Abstract:

Carbon nanotubes in biphasic systems have recently gained attention as Pickering emulsion stabilizers in the areas of underground oil recovery and heterogeneous catalyst supports due to their ability to undergo functionalization and tune the surface chemistry while maintaining the nanotube's unique electronic and physical properties [1-4]. While a great deal of literature exists for nanoparticle stabilized Pickering emulsions, very little is known about the fundamental behavior of nanotubes as Pickering emulsion stabilizers. In this contribution, Pd nanoparticles are supported on multiwalled carbon nanotubes with varying degrees of defects and hydrophilic functional groups. A combination of detailed emulsion characterization is coupled with phase specific chemical reactions to quantify both the location of functionalized carbon nanotubes at an oil/water interface, as well as the role of diffusion across the interface for enhanced reaction control.

A combination of acoustic spectroscopy, optical microscopy, SEM and TEM images of thin cross sections of the Pickering emulsions is used to quantify the influence of various surface functional groups on the interfacial area stabilized by carbon nanotube emulsions, as well as their orientation at the oil/water interface. This information is coupled with chemical reactions of olefins and small oxygenates with exclusive solubility in either the oil or water phase to measure the number of catalytic particles in each phase, as well as the rate of diffusion across the oil/water interface. Finally, this information is used to prepare novel hybrid emulsions containing blends of carbon nanotubes with various functional groups to create a gradient in catalytic particles along the oil/water interface enabling the enhanced control of chemical reactions in biphasic systems.

References:

[1] Crossley, S., Faria, J., Shen, M., & Resasco, D. E. (2010). Solid nanoparticles that catalyze biofuel upgrade reactions at the water/oil interface. *Science*, *327*(5961), 68-72.

[2] Shen, M., & Resasco, D. E. (2009). Emulsions Stabilized by Carbon Nanotube– Silica Nanohybrids. *Langmuir*, *25*(18), 10843-10851.

[3]Drexler, S., Faria, J., Ruiz, M. P., Harwell, J. H., & Resasco, D. E. (2012). Amphiphilic nanohybrid catalysts for reactions at the water/oil interface in subsurface reservoirs. *Energy & Fuels*, *26*(4), 2231-2241.

[4]Shi, D., Faria Albanese, J. A., Pham, T. N., & Resasco, D. E. (2014). Enhanced Activity and Selectivity of Fischer-Tropsch Synthesis Catalysts in Water/Oil Emulsions. *ACS Catalysis*.

388691 Solid State Chemistry Mass Production of Platinum Group Metal Catalysts with Tailored Particle Morphology

Abstract:

Catalytic property of platinum group metal (PGM) nanoparticles can be altered significantly by the nature of facets exposed. For example, Pt (100) plane excels in activity/selectivity compared to other planes in many reactions, including ring-opening hydrogenation of pyrroles, benzene hydrogenation, methanol oxidation (MOR), and electro-oxidation of ammonia (AOR). Pt-Ni (111) surface can exhibit exceptionally high activity towards oxygen reduction reaction (ORR) for polymer electrolyte membrane fuel cells (PEMFCs). The findings have stimulated the exploration of new methods for preparing PGM catalysts with tailored particle morphology, because the usage of PGMs and thus the cost can be largely decreased. However, to date, there have been no feasible methods for cost-effective and mass production of the shaped PGM catalysts.

We realize mass production of PGM catalysts with tailored particle morphology by developing a robust, green, and low-cost solid state chemistry approach. A variety of shaped PGM catalysts, including cubic Pt/SiO_2, cubic Pt/C, tetrahedral Pd/C, octahedral Pt-Ni/C, and cubic Pt-Cu/C, have been demonstrated for preparation. The experiments suggest that the formation of shaped PGM nanoparticles is resultant of employing both CO and H_2 gases, wherein H_2 aids transportation and reduction of the metal precursors on support and CO is responsible for the particle morphology formation. The prepared octahedral $Pt_{1.5}Ni/C$ catalyst exhibits high ORR activities of 3.99 mA/cm^2 Pt and 1.96 A/mg Pt at 0.90 V vs. RHE, which are about 20 and 10 times the values for commercial Pt/C specifically. The cubic Pt/C catalyst shows 1.44 mA/cm^2 at 0.6 V vs. RHE in AOR, which is five times that of 0.30 mA/cm^2 using commercial Pt/C.

OPTIMIZATION OF GREEN SYNTHESIS OF GOLD NANOPARTICLES USING *DELONIX REGIA* LEAF EXTRACT AND EVALUATION OF THEIR CATALYTIC ACTIVITY

*Preeti Dauthal, Mausumi Mukhopadhyay**

S.V. National Institute of Technology, Surat, Gujarat, India
Tel.: +91 261 2201645; Fax: +91 261 2227334
*Email: mmu@ched.svnit.ac.in; mausumi_mukhopadhyay@yahoo.com

Introduction

Noble metals such as of gold, palladium, platinum and silver with nano-size dimensions have received considerable attention due to their endless promises in the field of material science. Among these noble metal nanoparticles (NPs), particularly gold, nanoparticles (Au-NPs) plays a significant role in various catalytic reactions due to high surface area to volume ratio, energetic surface atoms and great surface energy. Biofabrication methods for the synthesis of these noble nanoparticles are not only a noble approach to synthesize benevolent nano-structures, but also a footstep to diminish the utilization of toxic chemical and generation of perilous by-products, which are hazardous to environment and public health. These methods are considered as benign, sustainable, cost-effective and ecologically sound. Biofabrication of nanoparticles using microorganisms is reported during recent studies (1, 2). However, use of the plant extracts as the potential resource for fabrication of NPs is gaining significant importance as unlike microbial fabrication these methods do not involve isolation and culture preparation techniques.

In a recent study, antioxidant potential of *Delonix regia* (*D. regia*) has been utilized for cost-effective and eco-benign fabrication of palladium nanoparticles (3). However, literature survey revealed that there is no report available for biofabrication of Au-NPs using *D. regia.* Thus, the present study is intended to fabricate Au-NPs by utilizing the metal reducing potential of *D. regia* plant extracts. Catalytic potential of fabricated nanoparticles is evaluated for reduction of anthropogenic pollutant 2-Nitroaniline (2-NA). Use of biofabricated nanoparticles as a catalyst in bioremediation reaction is a cost effective, sustainable and eco-friendly alternative (3-6).

Results and discussion

The effects of the concentration of *D. regia* leaf extracts, reaction temperature, reaction time and pH were evaluated for biofabrication of Au-NPs. Optimization of above reaction conditions was carried out based on surface plasmon resonance (SPR) using UV-visible spectroscopy analysis. Further biofabrication of Au-NPs were carried out using following optimized conditions: 30 °C reaction temperature, 4 mL of 1×10^{-3} M $HAuCl_4.3H_2O$, 0.5 mL of the leaf extract, pH 6 and 10 min reaction time.

Characterization of nanoparticles

Dynamic Light Scattering (DLS), ζ potential, Transmission Electron Microscopy (TEM), X-ray diffraction (XRD), Energy Disperse X-ray Spectroscopy (EDX) and Fourier Transform Infrared Spectroscopy (FTIR) techniques were adopted for evaluation of size, stability, morphology, crystal nature and purity of the biofabricated Au-NPs at optimized condition. UV-visible spectra represented SPR peak at 542 nm which suggested the reduction of gold ions (Au^{3+}) into zero-valent gold (Au^{0}). The results were rapid for the biofabrication of spherical Au-NPs of about 4-24 nm (by TEM) size within 10 min of reaction time. In the TEM micrograph, the edges of the nanoparticles were lighter compared to that of the centre, suggested capping and adherence of bio-organics of leaf extracts on the surface of Au-NPs (7). This observation was further supported by the FTIR, EDX and DLS analysis. Crystalline nature of nanoparticles was evident to a bright circular spot in the Selected Area Electron Diffraction (SAED) pattern and broad diffraction peaks in the XRD pattern (Figure 1).

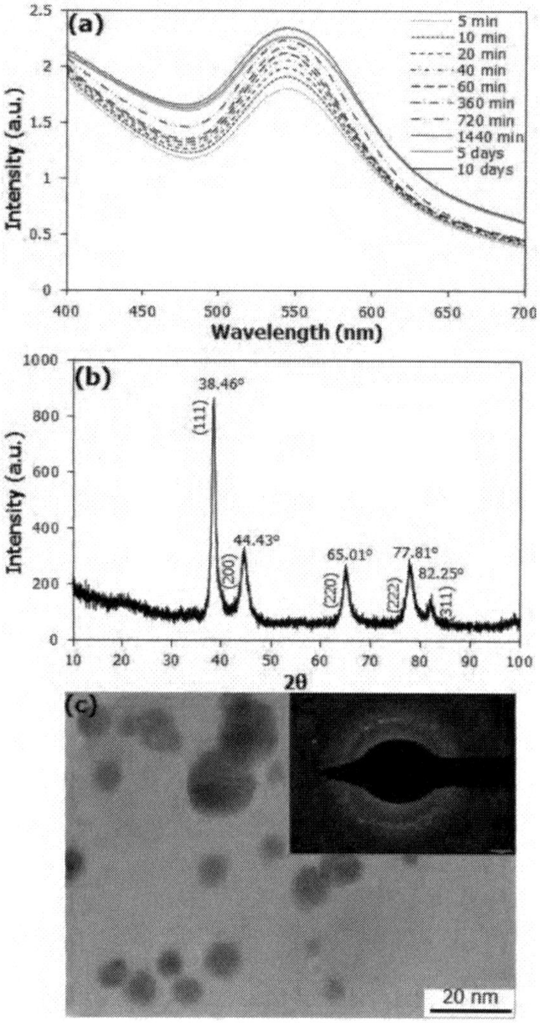

Figure 1. (a) UV-visible spectra (b) TEM micrograph (c) XRD spectra of biofabricated Au-NPs.

XRD pattern represented the five distinctive peaks present at 38.46° (111), 44.43° (200), 65.01° (220), 77.81° (311) and 82.25° (222) positions (JCPDS No. 04–0784), suggested the presence of crystalline gold at zero oxidation state. The lattice constant calculated from these patterns was 4.069 Å, confirmed the face centered cubic (fcc) geometry oriented along the (111) plane. The average particle size was calculated as 25.77 nm with the line width of the maximum intensity (111) diffraction peak by Debye Scherer equation. Thus, the XRD pattern proved strong evidence in favor of the TEM analysis for the presence of crystalline Au-NPs.

The appearance of optical absorption peak at 2.2 keV in EDX spectra further confirmed the presence of Au-NPs. High negative ζ potential (-15 mV) on the surface of biofabricated Au-NPs suggested long-term stability of colloidal Au-NPs. Particle size distribution of colloidal Au-NPs in DLS analysis was found in the range of 12-55 nm with an average particle size of about 39 nm with polydispersity index (PdI) 0.279, which appears to be higher as compare to TEM and XRD measurements. This was due to interference created by electric double layer on charged particles and overlapping NPs. Particles size observed in DLS is usually higher as the various forces of interaction in the solution like van der Waals forces of attraction is prominent (8). However, aggregation of NPs was avoided in the TEM imaging technique lead more precise observation (3).

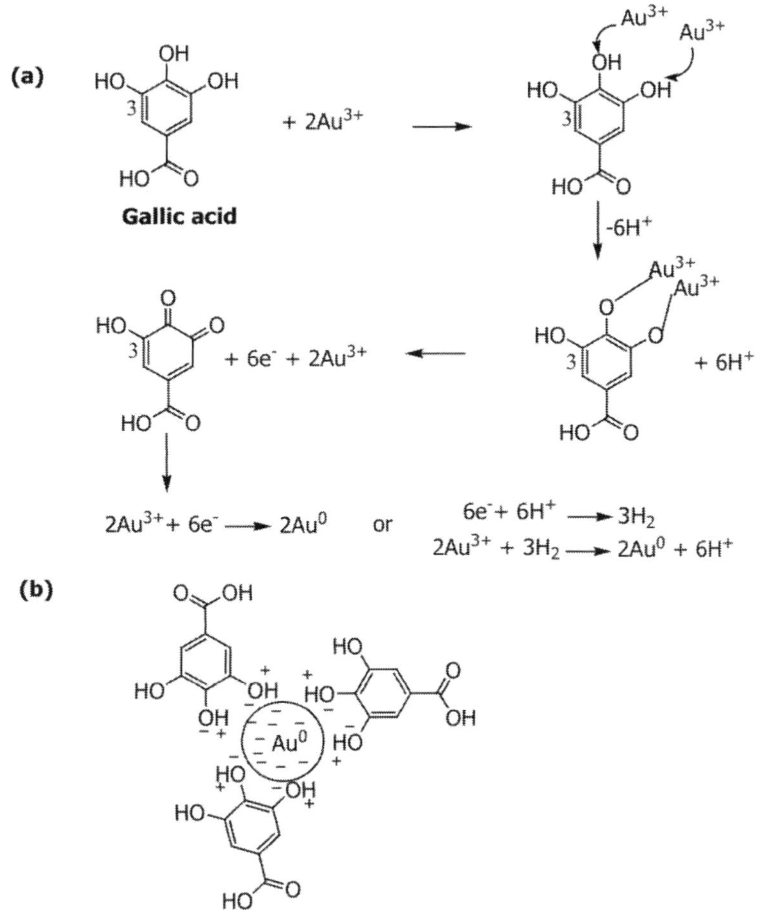

Figure 2. Probable reaction mechanism for (a) bio-reduction and (b) stabilization of Au-NPs using gallic acid.

FTIR analysis confirmed further the role of polar polyphenolic compounds of *D. regia* leaf extracts for biofabrication of Au-NPs and their chemical frameworks were also provided effective wrapping around the nanoparticles and robustness against agglomeration. Presence of water soluble polyphenolic compounds, including protocatechuic acid, gallic acid and hydroxybenzoic acid (phenolic acid) and catechins (flavonoids) are reported earlier in leaf extract of *D. regia* by HPLC analysis (3). Schematic representation of gallic acid mediated reduction of Au^{3+} ion and formation of respective quinone are shown in Figure 2.

Long-term stability (upto 10 days) of the biofabricated Au-NPs was also confirmed by UV-visible and ζ potential analysis. No significant change was observed in the SPR position and intensity, suggested long-term stability of nanoparticles. The ζ potential was found to be -14.3 mV even after 10 days of the incubation period.

Catalytic activity of nanoparticles

Catalytic activity of biofabricated Au-NPs was evaluated for the reduction of anthropogenic pollutant 2-NA. These nanoparticles were demonstrated to have a good catalytic activity for the reduction of 2-NA to 1, 2 Benzenediamine. Figure 3 represented reduction of 2-NA with $NaBH_4$ in aqueous environment at the surface of biofabricated Au-NPs as the catalyst.

The reduction of reaction followed pseudo first-order kinetics with respect to the substrates. It was also found that high concentration of Au-NPs and high temperature reduced the reaction time for catalytic reduction and increased the reduction rate. This observation suggested that catalytic reduction usually took place at the surface of catalyst (9). Catalytic reduction of 2-NA was found to increase by a factor of 1.6-1.9 for every 10 °C rise in temperature. The pseudo first-order rate constant was estimated to be 12.28×10^{-2} min^{-1} at room temperature (28 ± 2 °C) for this catalytic reduction reaction. The activation energy (E_a) for this catalytic reduction was calculated as 44.26 kJ mol^{-1}.

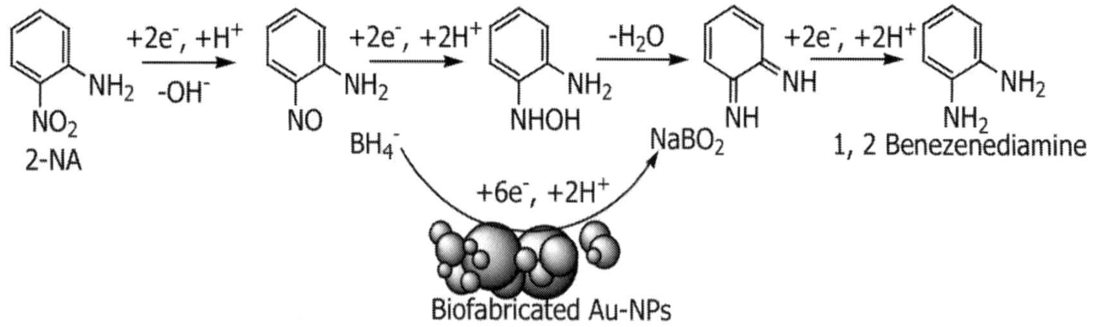

Figure 3. Schematic catalytic reduction of 2-NA to 1, 2 benzenediamine in the presence of $NaBH_4$ and biofabricated Au-NPs as the catalyst.

Au-NPs were recycled up to 10 cycles. It was observed that no significant change observed in the catalytic activity of the recycled catalysts up to 10 cycles. The reaction rate constant obtained at seventh cycle was found to be 10.57×10^{-2} min^{-1}. However, catalytic activity of biofabricated Au-NPs decreased slightly after 7 cycles of reduction reaction. This was due to the agglomeration of NPs (10, 11) or not fully recovered during recycling process. Thus, biofabricated Au-NPs proved as a potent recyclable catalyst for the industrial applications. These results suggested the fabrication of stable Au-NPs with the available biological resource *D. regia*, which offers an inexpensive, sustainable and eco-friendly alternative for fabrication of catalytically active Au-NPs.

References

1. Srivastava, Nishant and Mausumi Mukhopadhyay (2013), 'Biosynthesis and Structural Characterization of Selenium Nanoparticles Mediated by *Zooglea ramigera*," Powder Technology, 244, pp. 26-29.
2. Srivastava, Nishant and Mausumi Mukhopadhyay (2014), 'Biosynthesis of SnO_2 Nanoparticles Using Bacterium *Erwinia herbicola* and their Photocatalytic Activity for Degradation of Dyes," Industrial & Engineering Chemistry Research, 53, pp. 13971-13979.
3. Dauthal, Preeti and Mausumi Mukhopadhyay (2013), "Biosynthesis of Palladium Nanoparticles using *Delonix regia* Leaf Extract and its Catalytic Activity for Nitro-aromatics Hydrogenation," Industrial & Engineering Chemistry Research, 52, pp. 18131-18139.
4. Dauthal, Preeti and Mausumi Mukhopadhyay (2012), "*Prunus domestica* Fruit Extract-Mediated Synthesis of Gold Nanoparticles and Its Catalytic Activity for 4-Nitrophenol Reduction," Industrial & Engineering Chemistry Research, 51, pp. 13014-13020.
5. Sharma, Nilesh C., Shivendra V. Sahi, Sudip Nath, Jason G. Parsons, Jorge L. Gardea-Torresdey, Tarasankar Pal (2007), "Synthesis of Plant-Mediated Gold Nanoparticles and Catalytic Role of Biomatrix-Embedded Nanomaterials," Environmental Science & Technology, 41, pp. 5137-5142.
6. Dauthal, Preeti and Mausumi Mukhopadhyay (2014), 'Biofabrication, Characterization, and Possible Bio-reduction Mechanism of Platinum Nanoparticles Mediated by Agro-industrial Waste and their Catalytic Activity," Journal of Industrial and Engineering Chemistry, DOI: 10.1016/j.jiec.2014.07.009.
7. Suresh, Anil K., Dale A. Pelletier, Wei Wang, Michael L. Broich, Ji-Won Moon, Baohua Gu, David P. Allison, David C. Joy, Tommy J. Phelps and Mitchel J. Doktycz (2011), 'Biofabrication of Discrete Spherical Gold Nanoparticles using the Metal-Reducing Bacterium *Shewanella oneidensis*," Acta Biomaterialia, 7, pp. 2148-2152.
8. Dauthal, Preeti and Mausumi Mukhopadhyay (2013), 'In-vitro Free Radical Scavenging Activity of Biosynthesized Gold and Silver Nanoparticles using *Prunus armeniaca* (Apricot) Fruit Extract," Journal of Nanoparticle Research, 15, pp. 1366-1376.
9. Murugadoss, Arumugam and Arun Chattopadhyay (2008), "Surface Area Controlled Differential Catalytic Activities of One-Dimensional Chain-like Arrays of Gold Nanoparticles, " The Journal of Physical Chemistry C, 112, pp. 11265-11271.
10. Kuroda, Kyoko, Tamao Ishida and Masatake Haruta (2009), "Reduction of 4-Nitrophenol to 4-Aminophenol over Au Nanoparticles Deposited on PMMA," Journal of Molecular Catalysis A: Chemical, 298, pp. 7-11.
11. Koga, Hirotaka and Takuya Kitaoka (2011), "One-Step Synthesis of Gold Nanocatalysts on a Microstructured Paper Matrix for the Reduction of 4-Nitrophenol," Chemical Engineering Journal, 168, pp. 420-425.

Lead Chalcogenide Quantum Dot Arrays for Next Generation Photovoltaics

Rachelle Ihly, Sanjini Nanayakkara, Ashley Marshall, Jianbo Gao, Jianbing Zhang, Ryan Crisp, Matthew Beard, Joseph M. Luther

National Renewable Energy Laboratory, Golden, CO, USA

Photovoltaic devices with conductive films of colloidally synthesized PbSe quantum dots (QDs) possess external quantum efficiencies in the blue region of the solar spectrum greater than 100% due to multiple exciton generation (MEG) (where a high-energy photon can produce multiple electron-hole pairs). This greatly motivates continued research on this type of solar cell that has the potential to achieve >40% power conversion efficiency. The state-of-the-art (highest overall efficiency) optimized structure for lead chalcogenide QD solar cells uses a variety of interfacial layers that play an important role in the device functionality. The p-n heterojunction model is often used to describe the operation of QD solar cells despite the complex electronic structure of a disordered array of QDs acting as a macroscopic thin-film semiconductor (4). Advancements in device efficiency could follow better understanding of energetics along interfaces, throughout coupled films, and within individual nanostructures. Atomic force microscopy (AFM) techniques offer exceptional spatial resolution that can resolve such properties within devices and individual structures. Scanning kelvin probe microscopy (SKPM) is one such technique that can accomplish these goals. We have correlated the contact potential difference between a conductive AFM tip and the layers within an operating colloidal QD solar cell with device cross-section exposed. Utilizing scanning Kelvin probe microscopy (SKPM), we have correlated the contact potential difference between a conductive AFM tip and the layers within an operating QD solar cell cross-section to effectively map the potential throughout the device as a function of bias applied to the electrodes. This allows us to directly visualize regions of localized band bending and space charge and to help understand the operational mechanism behind such device architectures.

Periodic Nanostructured CdS/PbS Quantum Dot Solar Cells in Substrate Configuration

Kevin M. McPeak,[1] Ava Faridi,[1] Aurelio Rossinelli[1], Michael Heinrich[1], Sriharsha Jayanti[1], Daniele Braga[1], Christian D. van Engers[1], Yaroslav Romanyuk[2], and David J. Norris[1]

[1]Optical Materials Engineering Laboratory, Department of Mechanical and Process Engineering, ETH Zurich,

[2]Laboratory for Thin Films and Photovoltaics, Empa – Swiss Federal Laboratories for Materials Science and Technology, Switzerland

Quantum dot (QD) solar cells are promising candidates for low-cost, terawatt scale solar power generation, which is critical to meet the energy requirements of the 21st century. Unfortunately, current QD cells suffer from incomplete light absorption and charge transport issues which limit their overall photo-conversion efficiency. Advanced photon management techniques, such as periodic nanostructuring, in-couple and trap light in the absorber layer. Trapping light enhances absorption while reducing the thickness requirements of the absorbing material. Reducing the absorber thickness has the added benefit of mitigating bulk charge recombination and effecting production cost.

Herein we develop a CdS/PbS QD bilayer heterojunction solar cell in substrate configuration fabricated at temperatures below 100°C using chemical bath deposition for the CdS layer. The optimized planar cell achieved a photo-conversion efficiency of 3.7%. Template stripping was used to fabricate high quality, periodic nanostructured gold electrodes from reusable silicon templates. Periodic nanostructured cells, in substrate configuration, were fabricated on the template stripped gold electrodes with the same CdS and PbS volume as the planar cell. The nanostructured cells had a photo-conversion efficiency of 4.2%. Our approach to periodic nanostructured QD cells in substrate configuration can be adapted to different PbS QD heteropartners and is amenable to roll-to-roll processing on low-cost, flexible substrates.

Quantum dot solar cells offer much potential to provide transformative light-harvesting and conversion technology. However, limitations in the transfer and transport of electrons and holes in these nanostructured solar cells lead to low power conversion efficiencies. In particular, even though electron transfer from CdSe quantum dots to TiO_2 occurs on the order of 10^{10}-10^{11} s^{-1}, hole transfer to polysulfide electrolyte is much slower with rate constants approaching 10^8 s^{-1}. The back transfer of electrons to the oxidized polysulfide species is also a hindrance in obtaining higher photocurrents. Here we present an interfacial modification of CdSe with a thin Cu_xS layer to serve as a hole mediator between CdSe and polysulfide, which leads to higher power conversion efficiency. Transient absorption spectroscopy was used to demonstrate the hole transfer from CdSe to Cu_xS. Flash photolysis further shows on longer timescales the rate of electron-hole recombination is suppressed. These results confirm the transfer of holes from CdSe to Cu_xS and improved charge separation are responsible for the increased power conversion efficiency.

388101 CdTe Nanocrystals in Ink-Based Photovoltaics: A Study of Grain Growth and Device Architecture

Abstract:

We present on the use of cadmium telluride (CdTe) nanocrystal colloids as a solution-processable "ink" for large-grain CdTe absorber layers in solar cells. After layer-by-layer processing into thin film devices, the resulting grain structure and solar cell performance are found to depend on the initial nanocrystal size, shape, and crystal structure. Inks composed of predominantly wurtzite tetrapod-shaped nanocrystals exhibit better device performance compared to inks composed of irregular faceted nanocrystals or spherical zincblende nanocrystals despite the fact that the final sintered film has a zincblende crystal structure. Five different working device architectures were investigated. Surprisingly, we find the indium tin oxide/CdTe/zinc oxide structure leads to our best performing device architecture (with efficiency >11%) compared to others including two with a cadmium sulfide (CdS) *n*-type layer typically used in high efficiency sublimation-grown CdTe solar cells. Moreover, devices without CdS have improved response at short wavelengths.

Enhancement of Optoelectronic Quality and Solar Cell Efficiency from $Cu_2ZnSn(S_xSe_{1-x})_4$ Absorbers by Controlling Crystallization and Sintering with S, Se, SnS, and SnSe Volatile Species

B. Selin Tosun[a], *Hugh W. Hillhouse*[a].

[a]*Department of Chemical Engineering, University of Washington, Molecular Engineering & Sciences Building, 4000 15th Ave NE, Seattle, WA 98195.*

The photo conversion efficiencies (PCE) of earth abundant Cu, Zn, Sn, S, and Se, $CZT(S_xSe_{1-x})_4$ based solar cells rapidly developed to ~ 12.6 % due to effort and excitement generated as a result of their high potential in terawatt-scale energy conversion. Yet, the $CZT(S_xSe_{1-x})_4$ still does not come close to single-junction detailed balance, Shockley-Queisser limit (~ 31 %), mostly due the low open circuit voltages (< 0.5 Volts). Solution based depositions, such as colloidal nanocrystal or molecular based inks, provide potentially lower-cost and higher-throughput alternative routes to conventional vacuum based processes while resulting in high optoelectronic quality $CZT(S_xSe_{1-x})_4$ films. Herein, we focus on the molecular solution route using dimethylsulfoxide, a safer and environmentally friendly alternative to the current state-of-the-art hydrazine route. We show the effect of utilizing volatile species SnSe, SnS, and S under different process conditions on the optoelectronic quality of $CZT(S_xSe_{1-x})_4$ films. The absolute photoluminescence intensity, where the quasi-Fermi-level-splitting (QFLS) is shown to increase from 0.34 to 0.46, indicates the improvement in the optoelectronic quality of these absorber layers. This effect is also seen in the open circuit voltage (V_{OC}) in the completed devices. The V_{OC} is observed to increase from 0.400 V to above 0.480 V (all with a constant band gap of ~1.11 eV). The addition of volatile species to the sintering environment creates equilibria that removes the driving force for Sn loss. This minimizes the decomposition of $CZT(S_xSe_{1-x})_4$, and improves the shunt resistance; and hence the fill factor and conversion efficiency. It is also believed to reduce the potential fluctuations in the band gap due to impurity related donors and acceptors. PCE from solar cells completed with these absorber films reached 10.8 % without using an anti-reflection coating, which is would be approximately 11.6% with antireflective coating based on total area. In addition, in this study, we also observed an increase in the V_{OC} of the solar cells with the thickness of the $CZT(S_xSe_{1-x})_4$ absorber layer. However, forming large $CZT(S_xSe_{1-x})_4$ grains in films thicker than 1 μm with molecular ink route is challenging. In the molecular ink route, CZTS is deposited upon annealing and evaporation of the solvent on molybdenum coated soda-lime glass substrates. This results in films with nano-scale grains. This leads to bimodal distributions in the microstructure of the $CZT(S_xSe_{1-x})_4$ films thicker than 1 μm, where the top layer of the absorber layer results in large grains and the lower layer results in smaller grains after selenium incorporation. In this talk, we show that the utilization of the SnSe, SnS, and S with Se results in an increase in the grain size; grains larger than 1 μm (both height and width) are achieved. The change in the microstructure via the utilization of these volatile species is attributed to the formation of sulfide species with the alkali-metals (e.g. Na_2S) coming from the soda-lime glass which assists the grain growth as previously observed in CIGS based absorber materials. $CZT(S_xSe_{1-x})_4$ is a very similar system to CIGS solar cells (PCE ~20.4 %), which achieved the highest efficiency in the thin film solar cell market and became the major competitor of the current market domineer c-Si solar cell efficiencies (PCE ~25 %). The improved microstructure and V_{OC}, and therefore the PCE, with a safe and environmentally benign solvent is expected to lead further investigation on the $CZT(S_xSe_{1-x})_4$ device development and to lower the $/W value on the $CZT(S_xSe_{1-x})_4$ solar cells to more compatible price ranges with the high temperature processed (~3000 °C) c-Si solar cells.

Nanoparticle-based earth-abundant CZTSSe and CZTGeSSe solar cells: current progress and efficiency limitations

Charles J. Hages, Wei-Chang Yang, Nathaniel J. Carter, Eric A. Stach, Thomas Unold, Rakesh Agrawal

Recent progress in thin-film solar cell research has indicated the potential of $Cu_2ZnSn(S,Se)_4$ (CZTSSe) as an earth-abundant alternative to $Cu(In,Ga)Se_2$ (CIGSe) and CdTe based solar absorbers. One promising technology for absorber fabrication of efficient CZTSSe devices is nanoparticle ink-based processing, where the benefits of solution phase material synthesis and film formation along with reactive sintering techniques have been used to achieve 9.0% (9.8%) and 9.4% (10.1%) total-area (active-area) power-conversion efficiencies for CZTSSe and $Cu_2Zn(Sn,Ge)(S,Se)_4$ (CZTGeSSe), respectively.[1,2] While these results are among the highest efficiencies reported for this material system, a significant performance gap still exists between CZTSSe and CIGSSe/CdTe devices. In this work we present a unique approach to improved material quality through semiconductor alloying of CZTSSe and $Cu_2ZnGe(S,Se)_4$ (CZGeSSe), demonstrate novel electrical and material characterization of these materials to identity present limitations to improved device performance, and discuss a path forward to achieve state-of-the-art CZTSSe based solar cells from improved absorber processing.

One successful approach for improved material quality of CZTSSe-based solar cells has been through alloying of similar kesterite materials. Notably, partial cation substitution of Sn with Ge to form the CZTGeSSe absorber has demonstrated solar cells with improved electrical performance compared to traditional Ge-free CZTSSe devices.[1] In addition to the ability for band gap tuning/grading through control of the [Ge]/[Sn] ratio in the absorber – similar to [In]/[Ga] substitution in CIGSe absorbers – CZTGeSSe also benefits from a reduction in the concentration of multivalent Sn atoms in the crystal lattice, which can contribute to unfavorable defect formation. Presently, improved performance of Ge-alloyed CZTGeSSe absorbers compared to CZTSSe is chiefly the result of an improved minority carrier lifetime for this absorber. However, despite improved material quality and device performance, CZTGeSSe absorbers are limited by notable phase segregation/compositional fluctuations/alloy fluctuations in the bulk absorber relative to CZTSSe as well as surface elemental losses of Ge.[1,3]

To investigate the fundamental limitations to improved device performance for this material system, temperature-dependent optoelectronic characterization was conducted on CZTSSe and CZTGeSSe devices with varying degrees of Ge-incorporation. This analysis has identified limiting non-ideal diode behavior for the kesterite solar cells, which requires modified electrical characterization analysis techniques; traditional analysis techniques, commonly applied to CZTSSe devices, are found to be insufficient to describe device behavior as well as misrepresent the origin of device limitations.[3] Non-ideal behavior is considered in terms of temperature-dependent band edges, electrostatic and band gap potential fluctuations, low minority carrier lifetimes, and tunneling-enhanced recombination. Following modified analysis techniques, we find that CZTSSe-based devices (despite significant voltage limitations) are not characterized with significant reductions in the zero-Kelvin recombination potential ($E_{A,0}$), commonly associated with interface recombination limitations. However, significant voltage limitations are identified for CZTSSe and CZTGeSSe due to potential fluctuations in the bulk absorbers;[3] the origin of potential fluctuations is considered in terms of segregated phases, alloy compositional variations, non-stoichiometry, lattice strain, and compensating donors/accepters. In addition to current-voltage analysis, the presence of significant potential fluctuations is further verified from analysis of temperature/intensity-dependent photoluminescence and quantum efficiency measurements, in addition to material characterization using STEM-EDX compositional analysis.

As potential fluctuations are found to be the most significant limitation to improved device performance in CZTSSe-based devices, current research is aimed at enhancing the bulk material properties of the kesterite absorbers through improved absorber processing. On this front, analysis of phase stability, defect formation, and cation mobility is being applied to achieve systematic control of nucleation and growth during both nanoparticle synthesis and the reactive sintering mechanism. While the identified bulk material limitations in CZTSSe are expected to apply to absorbers fabricated from various processing technologies, we believe that solution phase synthesis techniques represent a promising approach for the formation of high-quality multinary absorbers with controllable defect formation and phase stability. Ultimately, improved performance for CZTSSe and CZTGeSSe-based solar cells is expected from optimized processing conditions to reduce cation disorder and secondary phase formation.

[1] – C. J. Hages, S. Levcenco, C. K. Miskin, J. H. Alsmeier, D. Abou-Ras, R. G. Wilks, M. Bar, T. Unold, R. Agrawal, "Improved performance of Ge-alloyed CZTGeSSe thin-film soalr cells through control of elemental losses", *Progress in Photovoltaics: Research and Applications*, (2013). DOI: 10.1002/pip.2442

[2] – C. K. Miskin, W.-C. Yang, C. J. Hages, N. J. Carter, C. S. Joglekar, E. A. Stach, R. Agrawal, "9.0% efficient $Cu_2ZnSn(S,Se)_4$ solar cells from selenized nanoparticle inks", *Progress in Photovoltics: Research and Applications*, (2014). DOI: 10.1002/pip.2472

[3] – C. J. Hages, N. J. Carter, R. Agrawal, T. Unold, "Generalized current-voltage analysis and efficiency limitations in non-ideal solar cells: case of CZTSSe and CZTGeSSe", *Journal of Applied Physics*. Submitted

392168 Thin Films from Nanocrystals Synthesized in the Gas Phase and Coated By Atomic Layer Deposition: A Case Study of Transparent Conducting ZnO

Abstract:

Thin films comprised of nanocrystals find applications in a wide variety of optoelectronic devices such as thin film transistors, solar cells and light-emitting diodes. Gas phase aerosol deposition of nanocrystal films is particularly attractive because deposition rates as high as 100 nm/s are possible even at low temperatures. The surfaces of the nanocrystals play a significant role in performance and functionality of these films. Understanding the origin of surface defects and developing strategies to mitigate them is crucial for successful applications. In this talk, we describe a new strategy for depositing films comprised of nanocrystals and for eliminating surface traps that degrade their performance. Specifically, the nanocrystals are synthesized in the gas-phase and deposited on suitable substrates through supersonic expansion and impaction. Following, the surfaces of the nanocrystals are coated by atomic layer deposition (ALD). This is a versatile approach since there are many material options for forming both the nanocrystal network and the ALD coating. In the particular example that will be discussed in this talk, ZnO nanocrystals are synthesized in a nonthermal plasma containing Ar, O_2 and diethylzinc. Using this approach, we demonstrate fast deposition of nanocrystalline films of ZnO, an earth-abundant, nontoxic, low cost material that can be used in thin film transistors or as a transparent conducting oxide (TCO). Without intentional doping or gating to fill traps, ZnO nanocrystal films have historically been too resistive because of low electron mobility and low electron concentration. We found that surface hydroxyl groups and adsorbed water molecules act as surface traps, and are responsible for the high electrical resistivity observed in ZnO nanocrystal films. We eliminate these surface traps by infilling the ZnO nanocrystal matrix with amorphous Al_2O_3 deposited using ALD. The ALD film serves two purposes. First, trimethylaluminum reacts with adsorbed OH and removes it from the surface. Second, the Al_2O_3 coating act as a barrier against water adsorption. After coating with Al_2O_3, the ZnO nanocrystal network becomes conductive and exhibits a relatively high Hall effect electron mobility, 0.1 $cm^2\ V^{-1}\ s^{-1}$ for as-deposited particle networks and as high as 3.0 $cm^2\ V^{-1}\ s^{-1}$ for annealed particle networks. We will discuss the effects of nanocrystal size and film porosity on electrical conductivity. Larger particles and less porous films result in higher conductivity.

377716 Orientation of Cylinder-Forming Poly(styrene)-Poly(n-hexylmethacrylate) Diblock Thin Films Pre- and Post-Shear

Abstract:

Block copolymer thin films provide an attractive method for nanopatterning at size scales inaccessible to conventional photolithographic techniques.[1] Block copolymers can microphase separate to form dense, periodic microdomains on the size scale of 10-100 nm.[2] To serve effectively as templates, however, the need to impart well-defined positional and orientational order to these microdomains is paramount. One method to achieve long-range orientational order in block copolymer thin films is through the use of shear, which has previously been shown to preferentially orient the microdomains of certain sphere-[3], cylinder-[4], and lamellae-forming[5] systems in the direction of applied shear. Numerous experimental factors influence the ease with which orientation is achieved as well as the ultimate quality of alignment observed. The present work aims to investigate several of these factors by studying the thin film morphology, in both the unaligned (thermally annealed) and shear-aligned states, of cylinder-forming poly(styrene)-poly(*n*-hexylmethacrylate) (PS-PHMA) copolymers which are of interest for nanolithographic applications such as the production of nanowire polarizing grids for deep ultraviolet light.[6] The tendency of PS-PHMA diblocks not to terrace (form "islands" or "holes") at reasonable annealing conditions, but instead to form mixed morphological patterns (mixtures of in-plane and out-of-plane cylinders or spheres) without discrete variation in film thickness makes it challenging to identify the precise monolayer film thickness at which maximum alignment quality is expected to occur. Furthermore, we investigate the effect of block copolymer composition on these thickness dependencies; in particular, we expect that moving the polymer more firmly into the cylinder-forming phase space might reduce the ability of the microdomains to restructure into spheres, and thus dampen the degree to which orientational switching between in-plane and out-of-plane cylinders is possible. Once the optimal integer layer film thicknesses are identified we subsequently examine both the limitations to the maximum achievable alignment quality as well as the ease with which these systems are aligned (by looking at alignment quality as a function of applied shear).

A series of cylinder-forming PS-PHMAs (PS minority block) with comparable molecular weights and narrow dispersities, but which systematically vary in composition to span the entire cylinder-forming phase space, were synthesized via living anionic polymerization. PS-PHMA thin films (~ 10-200 nm) with a thickness gradient were cast via flowcoating[7,8]; this allows for rapid investigation of thin film morphology, via atomic force microscopy, as a function of film thickness t. In unsheared films, the PS-PHMA diblock with the lowest styrene weight fraction (w = 0.21) showed morphologies ranging from complete coverage of dots (either spheres or out-of-plane cylinders) to complete coverage of lines (in-plane cylinders), or mixtures of the two, as t was varied. The fractional coverage of lines (F) was computed for each micrograph and examined vs. t. The data showed a strong dependence of F on t which included complete switching between $F = 1$ (all in-plane cylinders) at approximately integer multiples of the bulk interplanar spacing (d) to $F = 0$ (all out-of-plane cylinders or spheres) at incommensurate film thicknesses. The strength of this dependency decreased with increased t; at $t/d > 3$ only in-plane cylinders are observed. Performing a similar analysis on the three PS-PHMAs with higher w (w = 0.26, 0.30, 0.35) shows that as w increases the effects of thickness on in- vs. out-of-plane orientation gradually diminish such that for $w \geq 0.30$ the orientation is nearly thickness independent ($F \sim 1$ for all $t/d > 1$). This supports our hypothesis that by increasing w, thus moving away from the cylinder-sphere phase boundary, one can modulate the range of film thicknesses over which in-plane vs. out-of-plane cylinders are observed.

These PS-PHMA films were then sheared transverse to the thickness gradient, and the quality of their alignment quantified via an orientational order parameter $\psi_{2\alpha}$ ($\alpha = 1$ for analyzing all-line patterns, $\alpha = 3$ for mixed dot-line patterns). The PS-PHMA with the lowest w, which showed the strongest dependence of F on t/d, showed a similar trend in ψ_6 vs. t/d; the thicknesses which showed the highest fraction of in-plane cylinders pre-shear possessed the highest quality of alignment post-shear. The three higher-w polymers all showed relatively high alignment quality independent of film thickness ($\psi_6 > 0.9$ for all $t/d > 1$). For all polymers studied, generally the best alignment occurred near integer ratios of t/d. In these cases only in-plane cylinders are observed and the only lattice defects present are isolated dislocations. The areal density of these dislocations was computed, and ψ_2 plotted vs. this density. The slopes of the best-fit lines to these data reveal the influence of a single dislocation on the lattice order, while the y-intercepts indicate the maximum quality of alignment achievable in the absence of any lattice defects; the source of the differences between the y-intercept and unity are attributed to remaining fluctuations in the cylinders' trajectories. No significant differences in the slope or y-intercept were observed between monolayers ($t/d = 1$) and multilayers (t/d = 2, 3, 4) of the same polymer. In addition, while the lowest-w polymer showed both a lower y-intercept and a steeper slope, data for the other three PS-PHMAs possessed slopes and intercepts which were statistically indistinguishable (this common slope is in good agreement with that predicted from elastic continuum simulations[9]).

Finally, experiments to determine the ease with which alignment is achieved were performed by shearing monolayers of each PS-PHMA using a parallel plate rheometer, with a macroscopic (~ 0.2 mm) layer of viscous silicone oil between the stationary film and the top rotating rheometer plate. This allows for the application of a stress (σ) gradient from zero at the rotation axis to σ_{max} at the rheometer plate edge, thus enabling the rapid investigation of ψ_2 vs. σ. At low stresses no alignment is observed until a threshold value of σ is reached. At high stresses the alignment quality plateaus at $\psi_2 \approx 0.99$. While all four polymers showed similar alignment quality at high σ, the stress needed for alignment to first occur increases monotonically with w. In addition, the two lower-w PS-PHMAs possess very sharp transitions between the unaligned and aligned states while the two higher-w polymers show much more gradual transitions. These curves were compared to a phenomenological melting-recrystallization model for microdomain alignment.[10,11] The model contains two key parameters: a critical stress σ_c which relates the stress needed for alignment to begin to occur and a rate constant Γ which describes the rate of melting and recrystallization, and which governs the sharpness of the transition (in σ) between the unaligned and aligned states for a given shearing time. The data for all four polymers are well described by the model and the resulting best-fit parameters are used to quantitatively compare across the materials. As w increases, σ_c steadily increases, meaning that a higher stress is required to initiate alignment as the cylinder (PS) fraction increases. The rate constant Γ is more difficult to extract precisely from the data, but the two polymers with lower w show values of Γ smaller by 1-2 orders of magnitude than for the two polymers of higher w.

In conclusion we find that for unaligned films of PS-PHMA the in- vs. out-of-plane orientation of cylinders is highly sensitive to film thickness, but this sensitivity is substantially reduced with increased PS fraction. In general, thicknesses which possess the highest fractional coverage of in-plane cylinders pre-shear show the highest quality of alignment post-shear. For films of commensurate thicknesses (t/d = integer), all four PS-PHMAs show high alignment quality ($\psi_2 > 0.99$) of in-plane cylinders post-shear, although the lowest-w polymer shows consistently poorer alignment than the others. This alignment quality is limited by isolated dislocations and fluctuations in the cylinders' trajectories. Finally by examining alignment quality as a function of applied stress we observe that as PS fraction increases the stress and time needed for alignment to occur both increase and this behavior is well described by a melting-recrystallization model.

This work was supported by the National Science Foundation (MRSEC Program) through the Princeton Center for Complex Materials (DMR-0819860).

REFERENCES

1. Segalman, R. A. *Mater. Sci. Eng. R* **2005**, *48*, 191-226.
2. Hamley, I. W. *The Physics of Block Copolymers*; Oxford Univ. Press: Oxford, 1998.
3. Angelescu, D. E.; Waller, J. H.; Register, R. A.; Chaikin, P. M.; *Adv. Mater.* **2005**, *17*, 1878-1881.
4. Angelescu, D. E.; Waller, J. H.; Adamson, D. H.; Deshpande, P.; Cho, S. Y.; Register, R. A.; Chaikin, P. M. *Adv. Mater.* **2004**, *16*, 1736-1740.
5. Pujari, S.; Keaton, M. A.; Chaikin, P. M.; Register, R. A. *Soft Matter* **2012**, *8*, 5358-5363.
6. Papalia, J. M.; Adamson, D. H.; Chaikin, P. A.; Register, R. A. *J. Appl. Phys.* **2010**, *107*, 084305.
7. Stafford, C., M.; Roskov, K. E.; Epps III, T. H.; Fasolka, M. J. *Rev. Sci. Instrum.* **2006**, *77*, 023908.
8. Davis, R. L.; Jayaraman, S.; Chaikin, P. M.; Register, R. A. *Langmuir* **2014**, DOI: 10.1021/la501247x.
9. Marencic, A. P.; Chaikin, P. M.; Register, R. A. *Phys. Rev. E* **2012**, *86*, 021507.
10. Wu, M. W.; Register, R. A.; Chaikin, P. M. *Phys. Rev. E* **2006**, *74*, 040801.
11. Pelletier, V.; Adamson, D. H.; Register, R. A.; Chaikin, P. M. *Appl. Phys. Lett.* **2007**, *90*, 163105.

388959 Evolutionary Optimization of Directed Self-Assembly on Chemically Patterned Substrate

Abstract:

Directed self-assembly (DSA) of block copolymers on chemical patterns with density-multiplication is of considerable interest for sub-lithographic patterning. One of the central challenges for DSA is to optimize pattern design to achieve ideal, nearly defect-free patterns that meet industrial demands. We have addressed this problem by combining 3-D simulations of molecular assembling processes with on-the-fly optimization based on state-of-the-art evolutionary algorithms. We demonstrate the principle by optimizing the width and selectivity of chemically patterned substrates used to guide multiblock copolymers to assemble and to achieve both 3X and 4X density multiplication. We also explored the processing window by studying the effects of polymer composition, block fraction, and molecular architecture. The results compare favorably with experiments. The coupled simulation-evolution method provides a general and efficient framework for the design of complex device-oriented structures.

369532 Evaporation-Induced Self-Organized Surface Structures on Pre-Cast Polymer Thin Films

Abstract:

In this study, we fabricated well-ordered concentric rings with unconventional gradient on pre-cast PMMA films via evaporating a pure toluene droplet which was confined by a circular copper ring. Concentric rings were formed both inside and outside the copper ring due to the "stick-slip" motion of contact lines. Both the wavelength and the amplitude of the concentric rings inside and outside the copper ring decreases as the concentric rings approach to the copper ring. For the external rings, the wavelength and amplitude of the rings of larger diameters which are far away from the copper ring are larger than those of smaller diameters closer to the copper ring. This behavior is similar to the gradient rings fabricated by other methods.[1, 2] On the other side, the internal rings exhibit the opposite trend: the rings of smaller diameters have larger wavelength and amplitude. The concentric rings display asymmetric characteristics with the reference to the copper ring. The external rings have larger wavelength and amplitude than the corresponding internal rings approximately with the same distance to the copper ring. The result suggests that positive curvature (for the external rings) tends to induce concentric rings of larger wavelength and amplitude than negative curvature (for the internal rings). The characteristics of the concentric rings can also be controlled by the thickness of the pre-cast films. Concentric rings of larger wavelength and amplitude can be formed on a thicker film when other experimental conditions remain the same. In addition, a featureless zone was formed in the center of the copper ring due to the evaporative dewetting. The size of the featureless zone is proportional to the size of copper ring. The "ring-on-film" template has the potential to fabricate regular concentric rings with controllable characteristics in a simple way.

References

(1) J. Xu, J. F. Xia, S. W. Hong, Z. Q. Lin, F. Qiu and Y. L. Yang, Self-assembly of gradient concentric rings via solvent evaporation from a capillary bridge, *Physical Review Letters*, **2006**, *96*.

(2) Z. Q. Lin and S. Granick, Patterns formed by droplet evaporation from a restricted geometry, *Journal of the American Chemical Society*, **2005**, *127*, 2816-2817.

Multiple Phases of Block Copolymer Micelles formed via Electrospray-Enabled Interfacial Instability

Mathew Souva[1], Gauri Nabar[1], Jessica Winter[1,2], Barbara Wyslouzil[1,3]

The Ohio State University [1]William G. Lowrie Department of Chemical and Biomolecular Engineering, [2]Department of Biomedical Engineering, [3]Department of Chemistry and Biochemistry

With application in biological imaging, energy platforms, and computational systems, nanoparticles continue to be an important area of technology development. Both development and use of functional particles are areas for further nanotechnology research. Here, we describe a promising method to synthesize nanoparticles that combines electrospray and an interfacial instability process to form micelle structures from amphiphilic block copolymers. In this approach, block copolymers in an organic phase are introduced to an aqueous solution and form micron-scale emulsion droplets. Transient interfacial surface tension changes then result in micelle formation. The technique can be adapted to encapsulate smaller functional nanoparticles within micelles, including quantum dots (QDs) and superparamagnetic iron oxide nanoparticles (SPIONS), effectively co-localizing fluorescent and magnetic material in a ~30 nm diameter particle.

Previously, micelle formation via interfacial instability was performed in a batch process and resulted in spherical micelles comprised of poly(styrene-block-ethylene oxide) (PS-b-PEO). In this study, we use the electrospray method and control both the temperature and polymer loading in order to increase the rate of PS-b-PEO micelle production as well as access different polymer assembly phases. We used the electrospray synthesis approach with stable spray conditions that mimic batch settings, and produced spherical micelles at an increased number and rate.

By increasing polymer concentration in the organic phase of the electrospray and controlling the temperature of the emulsion, we were able to further increase the polymer throughput and the production rate of the spherical micelle formation process. This result alone is encouraging, and suggests optimal conditions for spherical micelle production in the electrospray process. Further variation of temperature and concentration yielded wormlike micelles, a finding consistent with previous research on amphiphilic polymer phase diagrams. This technology can be used to create empty and loaded micelles in either phase that have potential application in drug delivery, imaging, separations, and diagnostics.

392219 Polymer-based Air-gap MEMS Packaging Compatible with Epoxy-overmolding on Lead-frame Substrates

Abstract:

Packaging of microelectromechanical systems (MEMS) is essential for preserving device performance, providing physical protection and enabling electrical connections to devices. MEMS packaging sector is estimated to grow 2x faster than the integrated circuit (IC) packaging sector, because of ever-increasing penetration of MEMS technology into mobile and consumer applications in recent years [1]. Despite its growth, MEMS packaging is still one of the least explored components of MEMS technology [2]. Moreover, MEMS packaging can compromise up to 20-40% of the total material and assembly cost [3]. Besides, conventional MEMS packaging techniques involving ceramic, metal or plastic packages has challenges either of size limitations, high-temperature requirements, and packages being device-specific [4]. In addition, the surface of the MEMS devices should stay clean after packaging because the atmospheric contaminants and moisture can create dampening effects and oxidation, which might interfere with device performance and reliability. Therefore, there is a clear need for low-cost, clean, IC-compatible, high volume, wafer-level packaging solutions.

The goal of this study is to develop a MEMS packaging technique using polymer-based air-gaps as a cost-effective method. The intention is to stabilize and secure the movable MEMS components inside a protective envelope so that the MEMS chip can then be treated like an integrated circuit (IC) and enjoy the benefits of low-cost, high-volume IC packaging, such as lead-frame, or plastic overmolding packaging. The air-gap creation process uses thermal decomposition of patterned sacrificial polymer layer, poly(propylene carbonate) (PPC), which is encapsulated with an overcoat material, BCB (Cyclotene), through which PPC decomposition products permeate out, finally leaving an air cavity on top of MEMS device enclosed with overcoat. The process involves coating MEMS device with PPC, coating and patterning a thin pattern-transfer layer of BCB overcoat on top of PPC, reactive-ion etching by O_2 plasma for patterning PPC, coating and patterning overcoat layer of BCB, then simultaneous thermal decomposition of PPC and curing of BCB (at around 250°C for 2-4 hours in N_2 purged oven). PPC acts a temporary place-holder for defining air-gap region, whereas BCB provides mechanical stability to package. The application of the process on bare silicon substrates enabled 5-10 μm high air-gap creation in square or circular features up to 2 mm in size with no cracking or deformation in BCB overcoat. BCB is found to be chemically compatible with PPC providing uniform, conformal and adherent coating on top of PPC. Nanoindentation measurements revealed a mechanically stable and flexible BCB overcoat layer on top of the air-gap. The surface of silicon in the air-gap region was found to be clean after removal of BCB overcoat, with only a 5-10 nm thick high molecular weight hydrocarbon by X-ray photoelectron spectroscopic (XPS) measurements. The packaging process was also performed on actual working MEMS resonators of sizes 200 μm x 100-160 μm. Comparing the before-packaging and after-packaging resonant spectra measurements, we have observed a minimal decrease in resonant frequency (<0.1%), and little-to-no change in quality factor (<5.2%) in working MEMS resonators. Current focus is on implementing the developed packaging process into a lead frame packaging scheme with epoxy molding to achieve near-hermetic, actual MEMS devices. During epoxy molding, in situ decomposition of PPC and curing of BCB will occur simultaneously with curing of epoxy molding compound.

Overall, a new MEMS packaging technique has been studied in order to achieve low-cost, clean, IC-compatible MEMS packages. The technique is easy-to-implement, and readily scalable to wafer-level packaging. The packaging of actual MEMS devices with the use of the developed packaging technique along with conventional lead frame packaging with epoxy molding is underway.

References

1. MEMS Packaging Market and Technology Trends Report, Yole Développement, 2012.

2. M. Gaitan, MEMS Technology Roadmapping, Nano-Tec Workshop 3, 31 May 2012.

3. R.H. Grace, M. A. Maher, "MEMS: Think outside the chip... at the package level", Hearst Electronic Products 2010, http://www.electronicproducts.com/Packaging_and_Hardware/Prototyping_Tools_Equipment_Services/MEMS_Think_outside_the_chip_at_the_package_level.aspx

4. P. Monajemi et al., "Characterization of a Polymer-Based MEMS Packaging Technique", 11th International Symposium on Advanced Packaging Materials: Processes, Properties and Interface, 2006.

"Different Glass Transition Temperature Confinement Effect Caused by Residual Surfactant Capping the Free Surface in Emulsion Polymerized Polystyrene Films"

Abstract

When the confining dimension is decreased in polymers films or particles, a higher proportion of the material becomes exposed to an interface resulting in a material that possesses properties not representative of the bulk. In 1994, Keddie et al. first described the glass transition temperature (Tg)-confinement effect of thin polystyrene (PS) films supported on a silica substrate using ellipsometry by observing as much as a 29 K reduction in Tg from bulk PS for a 10-nm-thick film. Since then, numerous experiments and simulations have demonstrated that Tg reduction originates from the increased chain mobility at the free surface while any Tg enhancements observed originate from the decreased chain mobility due to secondary interactions between the polymer and its environment.

By eliminating the free surface, it is expected that the Tg-confinement effect would be suppressed. In a study by Forrest and coworkers (*Phys. Rev. Lett.* **2003**, 91, 235701), they observed that properly capped PS films with a thin Au layer exhibited no thickness dependence of Tg. In another study by Priestley and coworkers (*Macromolecules*, **2011**, 44, 4001-4006), PS particles made from surfactant-free emulsion polymerization suspended in aqueous solutions showed Tg reductions with decreasing diameter. However, capping the particles with a hard silica shell resulted in invariant Tg with particle size.

As an indispensible ingredient in conventional emulsion polymerization, surfactants act to lower interfacial energy between polymer particles and their environment by producing stable latexes. Due to the amphiphilic nature of surfactants, complete removal of the surfactant is a difficult and time-consuming process. In 2013, Zhang et al. (*J. Polym. Sci. Part B Polym. Phys.* **2013**, *51*, 574–586) acknowledged that, "the presence of surfactant molecule on the surface of polymer nanoparticles as a stabilizing agent in emulsion polymerization may significantly modify the observed confinement effects and needs to be removed prior to measurements, which could be a difficult task." To date, few studies have directly focused on the influence of surfactant on the Tg-confinement effect in nanoconfined films.

In this work, we systematically study the Tg-confinement effect of PS produced via conventional emulsion polymerization (E-PS) containing 3.2, 0.023, 0.0058, and 0.0031 wt% sodium dodecylsulfate (SDS) content. A portion of the initial latex was destabilized using a multivalent ion after polymerization and rigorously washed and purified in an attempt to completely remove all surfactant. Using a modified Epton Method, we quantified the amount of surfactant remaining in the washed and unwashed samples. It was discovered that each time the assay was performed and the polymer solution precipitated, the amount of SDS remaining in bulk E-PS was reduced. Using ellipsometry and fluorescence spectroscopy, we found that bulk E-PS samples containing 3.2 and 0.023 wt% SDS showed heavily suppressed Tg-confinement effect and that each subsequent assay and precipitation led to an E-PS sample that exhibited a Tg-confinement effect closer to that seen in PS films synthesized by anionic polymerization or conventional free radical polymerization. These results indicate that the presence of residual surfactants can have an enormous impact on the observed Tg of E-PS films.

Finally, we have used x-ray photoelectron spectroscopy (XPS) depth profiling to demonstrate that the remaining SDS present in the film preferentially resides at the free surface following spin coating and annealing and that there is little to no SDS present in the interior of the film. We conclude that the presence of the surfactant residing at the top surface of the film results in the formation of a capping layer that effectively constricts the mobility of chain segments at the free surface. It is interesting to note that this suppression of the Tg-confinement effect does not occur with post-polymerization addition of SDS to PS made by anionic or conventional free radical polymerization.

382748 Patterning Polystyrene Thin Films By Photodirecting Marangoni Convective Flow

Abstract:

The Marangoni effect causes liquids to flow in response to local surface energy gradients. We recently developed a method for photochemically directing this effect by generating prescribed surface energy spatial patterns in glassy polystyrene (PS) thin films. UV irradiation through a photo-mask selectively dehydrogenates the PS backbone locally, thus increasing surface energy in the UV exposed regions compared to the unexposed regions. As a result, the polymer flows from low to high surface energy regions according to the Marangoni effect upon subsequent thermal annealing above the glass transition temperature. This flow creates smooth, three dimensional topography reflective of the original light exposure pattern. A theoretical framework that gives a more thorough understanding of the physics of this process, its limits and ways to apply it efficiently for various target metrics, such as the feature periodicity and feature height will also be presented along with comparisons between simulative predictions and experimental observations.

Finally, while PS dehydrogenation can be used to produce a variety of topographical patterns, incorporating the near UV and visible light photosensitizer 9,10-dibromo-anthracene (DBA) into PS films enables significant improvements in feature aspect ratio compared to the absence of such a sensitizer. This suggests a judicious selection of the photosensitizing compound in an otherwise transparent polymer expands the use of this method with more readily available and safer visible light sources.

372915 Properties of Fluorescent Polymer Film with Three-Dimensionally Ordered Nanopores and Its Application in Explosive Detection

Abstract:

We demonstrate here how a simple dip-coating process of a mixture of polymer and fluorophore onto a glass slide generates a self-assembled fluorescent nanostructured film expressing regular nanopores. Morphology investigations through SEM, AFM and confocal fluorescence microscopy reveal that the fluorescent polymer films consist of a high-surface, three-dimensional nanoporous array of holes, enabling that the fluorescence of this material is rapidly and selectively quenched by nitroaromatic vapors. The morphology of the polymer film was controlled by variation of the dip-coating parameters and the the polymer : fluorophore ratio. This ratio also controls the fluorescence quenching efficiency of the sensing film. We demonstrate the possible molecular origins of this through structural XRD studies as well as investigations of the electronic structure (optical properties, band gap and conduction band determinations) of the polymer film. The findings here point toward a general method for the facile realization of well-defined three-dimensional high surface sensor materials with optimized electronic properties.

390356 Understanding Defect Annealing Kinetics in Self-Assembled Block Copolymers Using a Coarse Grained Block-Copolymer Model

Abstract:

Directed self-assembly (DSA) of block copolymers (BCPs) is a promising technique for producing nanoscale regular patterns. In particular, DSA of block copolymers is a leading candidate for extending current optical lithography capabilities below the 20 nm feature size scale for producing device features in semiconductor manufacturing. However, development of such techniques can be a time consuming and difficult proposition given the number of possible material and process variable combinations to explore, the limited knowledge of the fundamental behavior of such systems, and the difficulty in observing the assembled materials. Thus, the development of such DSA techniques could benefit greatly from predictive computer simulation methods that can faithfully represent the behavior of such materials and processes. A coarse grained molecular dynamics (MD) model combined with realistic potentials for polymer behavior can potentially provide more accurate simulations of the inherent polymer behavior, dynamics, and equilibrium states without a need to guess modes of molecular movement and without oversimplifying interatomic interactions, with the added benefit of the inherent dynamic nature of MD which allows easier analysis of dynamic properties, such as diffusivity or kinetics. These advantages make studying dynamic states important in such BCP-DSA processes particularly attractive. Defectivity is one of the primary hurdles for the implementation of BCP-DSA into semiconductor fabrication. Both equilibrium defectivity (as measured through free energy calculations) and dynamic defectivity must be considered when designing processes for BCP-DSA. A coarse grained MD model has been used in this work and allows for easy study of defect anneal rates as a function of χ, N, and various underlayer properties. A study of the most relevant parameters to anneal rates, as well as underlying causes for variations of kinetic rates in thin film conformations, will be presented.

389216 Effect of Block Properties and Solvent Quality on Self-Assembly of Polyelectrolyte Block Copolymers

Abstract:

Polyelectrolyte block copolymers, which combine structural features of polyelectrolyte, block copolymers and surfactants, may self-assemble in a variety of nanoaggregates in aqueous environment, such as micelles, vesicles, lamellar mesophases or micellar aggregates. The morphology and size of formed aggregates are determined by the characteristically complex equilibrium of noncovalent forces (electrostatic, steric, hydrogen bonding, Van der Waals, and hydrophobic interactions). The strength of repulsive Coulomb interactions between the polyelectrolyte segments can be efficiently tuned by variations in ionic strength or/and pH in the aqueous solution. In order to explore the self-assembly process of polyelectrolyte block copolymers, we developed Dissipative Particle Dynamics (DPD) model of polyelectrolyte block copolymers with incorporated electrostatic interactions to achieve a good balance between reasonable physical description and computational feasibility. We applied this coarse-grained model to explore the influence of block length, block architecture, and solvent quality on the properties of the assemblies formed in aqueous solutions. Our DPD model enables us to obtain the main characteristics of the micelles formed (the aggregation number, the corona and core sizes, and anisotropy) as a function of the block length and salt concentration. Based on a comprehensive set of data obtained we constructed a morphological diagram of polyelectrolyte block copolymers in aqueous solution. The coarse-grained modeling and simulation, which is demonstrated as a complimentary approach in addition to experimental and theoretical methods, can deliver insight into self-assembly processes of block copolymers and provide evaluation of the size of aggregates obtained along with their scaling relation representation. The simulation results suggest that this coarse-grained simulation scheme gives a route wherein one can effectively and efficiently capture the self-assembly behaviors of polyelectrolyte block copolymers.

382473 The Distribution of Nanoparticles in Inhomogeneous Polymer Melts

Abstract:

As polymer nanocomposite materials have found their way into an ever-growing number of applications, it has become clear that the dispersion state of the nanoparticles can play a key role in the resulting material properties. In some instances, it is ideal to have well-dispersed nanoparticles in a polymer matrix to facilitate high loading of the particles, while in other cases, self-assembled nanoparticle structures are preferred. One issue that continues to challenge the experimental design of polymer nanocomposites is the lack of efficient computational and theoretical techniques capable of predicting their equilibrium structure and thermodynamics. This problem is exacerbated in inhomogeneous systems, where nanoparticles may be distributed amongst multiple polymer phases. For many polymer formulations and alloys, polymer field theory represents a powerful framework for the study of inhomogeneous polymers. In this talk, I will describe our efforts to extend the field-theoretic simulations framework to include both bare and grafted nanoparticles, where polymer chains can be grafted to the surface of hard nanoparticles. Our method is compatible with both self-consistent field theory and fully fluctuating field theoretic simulations. Demonstrative calculations will be shown for the distribution of grafted nanoparticles in homopolymer thin films as well as binary homopolymer blends, where our results agree very well with recent experiments.

Deformation studies and viscoelastic properties of 1,4-polybutadiene chains via Rotational Isomeric State statistics

ABSTRACT

Rolling resistance in rubber tires leads to losses in vehicle fuel economy. Rolling resistance on the macroscale connects directly to deformation-induced changes in elastomer chain conformations on the microscale through changes in entropy. Chain configurations were generated using Flory's Rotational Isomeric State Approach, which weights particular torsional states along the backbone. As a prior result, we found that chain swelling with heating was attributed to increased sizes of 'taut' conformations with increasing temperature. Uniaxial deformation was applied to an ensemble of chains generated under unperturbed conditions, with extension applied along x axis and compression along y and z axes. Probability density distributions of the end-to-end vectors in x, y, and z directions were quantified to compute deformation forces. Gaussian model assumes each chain to behave as an ideal chain. Gaussian simulation and Gaussian model comparison was used to verify the simulation approach and agreement was found between them. We then extended the simulation approach towards applying deformation on cis and trans chains generated by our Rotational Isomeric State code. For extension and compression, cis chains showed a higher probability of occurrence than the trans chains at shorter chain sizes, and at medium and longer chain sizes, trans chains showed a higher probability of occurrence than the cis chains. At shorter chain sizes, simulation and Gaussian results predict similar probability densities. At medium chain sizes, simulation results predict higher probability densities than the Gaussian model while at longer chain sizes, the Gaussian model predicts higher probability densities than the simulation results. Deformation forces are linear in the size range of undeformed chains, showing a greater magnitude for cis than for trans and Gaussian chains. Probability weighted total deformation forces increase with increasing deformation. For a fixed extension, deformation forces decrease with increasing chain length and increase with increasing temperature. In both cases, forces computed for our simulation results are greater than the ones computed using the Gaussian model. Currently we are looking at changes in probability density distributions with temperature under deformation, changes in size and shape characteristics of chains under deformation. Quantifying the deformation forces allow us to compute viscoelastic properties of chains such as storage and loss moduli.

390299 Protracted Colored Noise Dynamics for Polymer Systems

Abstract:

Mean field potentials and Monte Carlo simulations have been used to study polymeric systems in a number of contexts and applications in recent years. While these methods provide great computational expedience, they include less accurate potentials and/or require intelligently selected moves that can affect the simulation outcomes. Conversely, molecular dynamics (MD) combined with realistic potentials can provide more accurate moves and states without a need for carefully calculated and selected moves, but such detail is gained in general at the expense of computational efficiency. In polymeric systems, use of MD techniques can be challenging because of hindered chain diffusion, which can make equilibration of polymeric systems challenging and slow. If the fluctuation distribution can be broadened intelligently in MD of polymers, then it may be possible for diffusion to be greatly increased and this leading to significantly faster equilibration while not sacrificing the basic ability to achieve proper thermodynamic sampling of equilibrium states in the system. To achieve this intelligent fluctuation perturbation, in this work a time and directionally correlated colored noise is used instead of the white noise that is used in Langevin dynamics. In the polymer simulations discussed in this work (which use a freely rotating mesoscale chain model), this correlated noise is applied along the polymer backbone contour over some finite time scale to encourage a reptation-like movement which drastically increases polymer diffusion. The parameter space for this technique has been explored and will be discussed, with a focus on understanding how the length of time correlation and strength of the force noise impacts the simulation behavior. The impact of using such techniques for accelerating MD simulations in polymer systems will be demonstrated in the context of block-copolymer system simulations. In particular, the use of these methods to explore the behavior of directed self-assembly of block copolymers as a method to extend high resolution lithography techniques will be highlighted.

Molecular simulation of solvent diffusion in polymer solution casting processes

Li Xi[§] and Bernhardt L. Trout[*]

[§]xili@mcmaster.ca, Chemical Engineering, McMaster University

[*]trout@mit.edu, Chemical Engineering, MIT

2014 AIChE Annual Meeting

Atlanta, GA, Nov. 18, 2014

McMaster University

http://xiresearch.weebly.com 1

Background: continuous manufacturing and solution casting

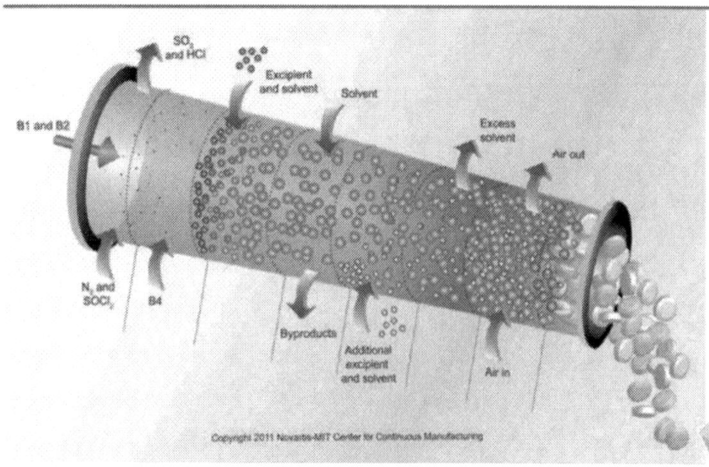

Mascia et al., *Angew. Chem. -Int. Edit.*, 2013

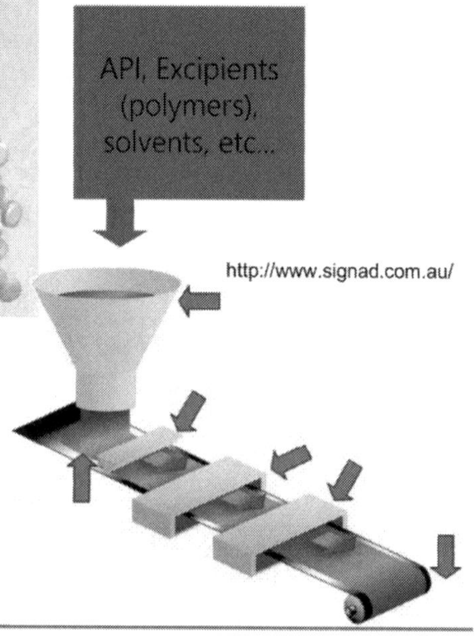

http://www.signad.com.au/

McMaster University

Background: diffusion in polymers

- Different mechanisms at the two concentration limits.
- Penetrant diffusivity drops by orders of magnitude in polymer matrices.

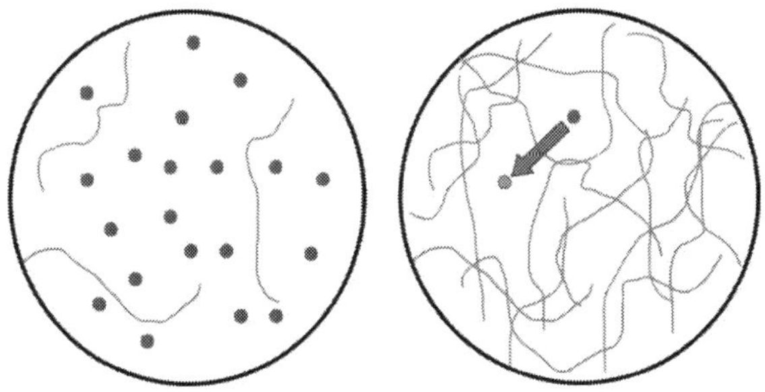

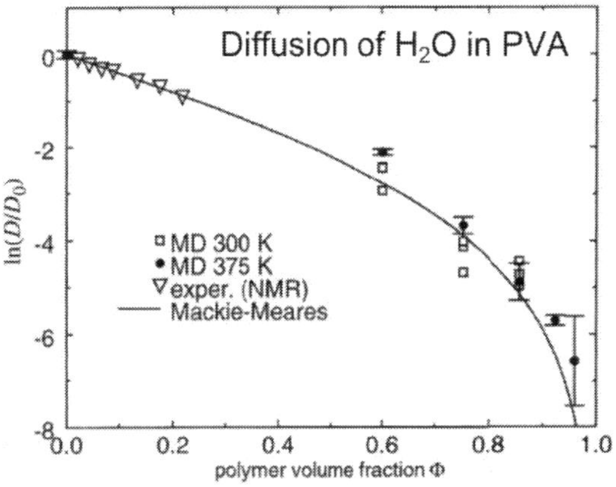

Müller-Plathe, *J. Membr. Sci.* 1998

Molecular model

- Simulation system: PVA + H_2O
- Molecular dynamics:
  - Full-atom representation.
  - Polymer-consistent force field (PCFF).
    - Maple et al. J. Comp. Chem. 1994
    - Hwang et al. J. Am. Chem. Soc. 1994
  - LAMMPS package
    - Plimpton, J. Comp. Phys., 1995

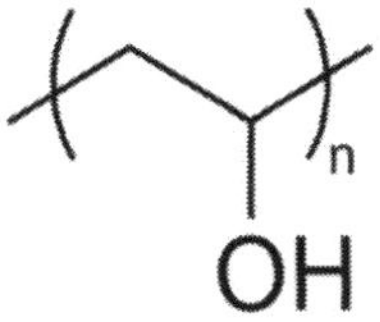

poly(vinyl alcohol)

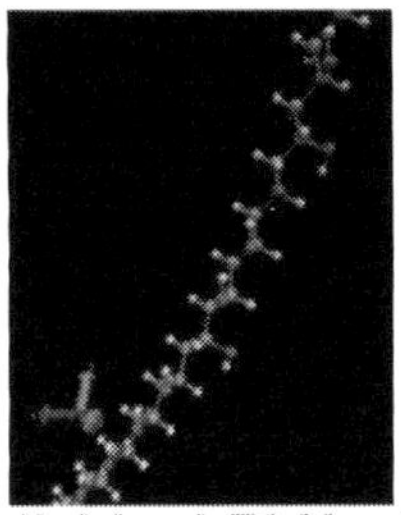

individual PVA chain

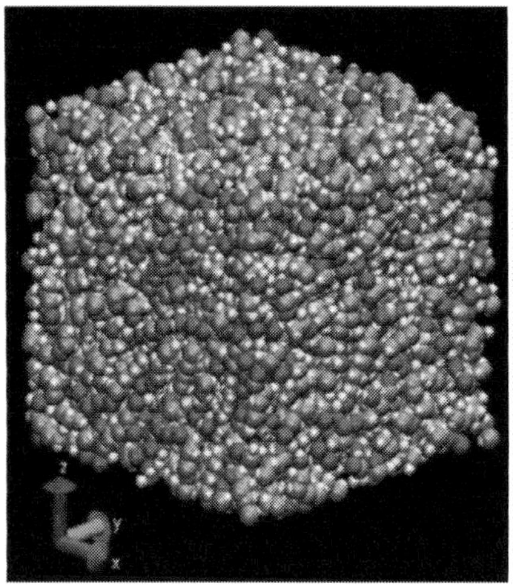

amorphous PVA cell

C_s->0 limit: trajectories of H_2O molecules

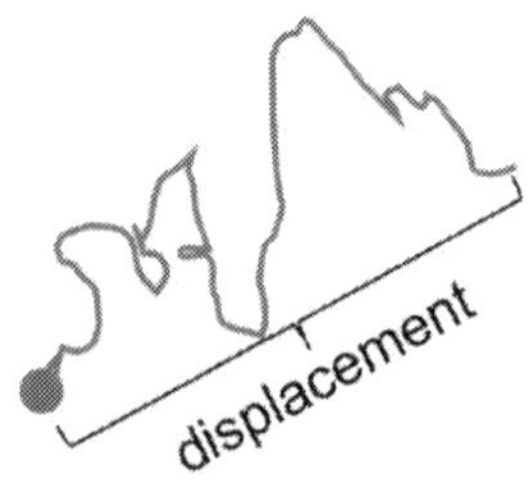

- Diffusion by "hopping" (C_s->0, below T_g)
- Separation of time scales:
  - τ_{wait} ~ O(10)ns.
  - τ_{hop} ~ O(1)ps.

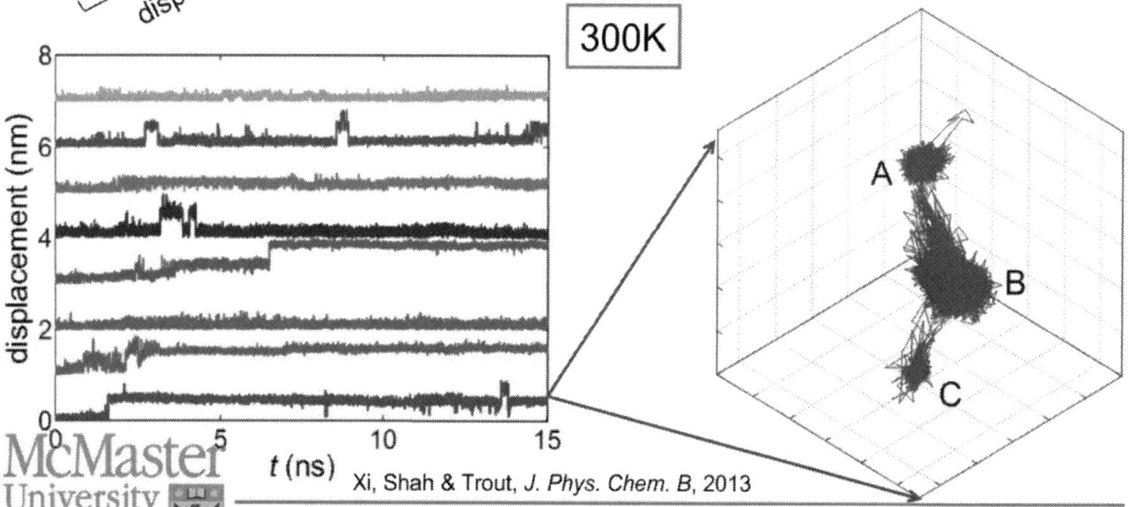

Xi, Shah & Trout, *J. Phys. Chem. B*, 2013

C_s->0 limit: detailed view of one typical hopping event

- Cavities change their configurations as the H_2O molecule makes the move.
- Highly cooperative motions.

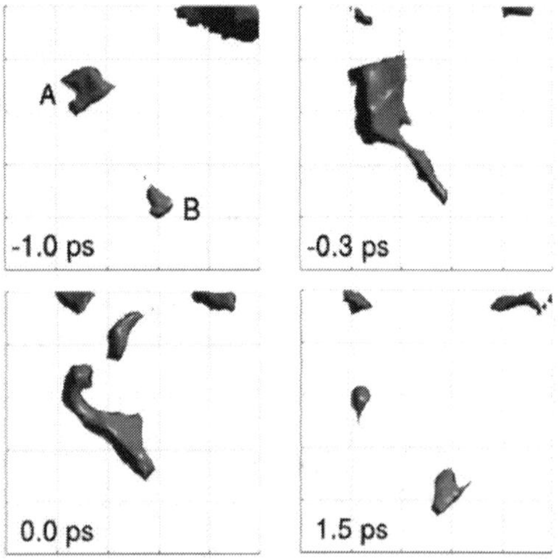

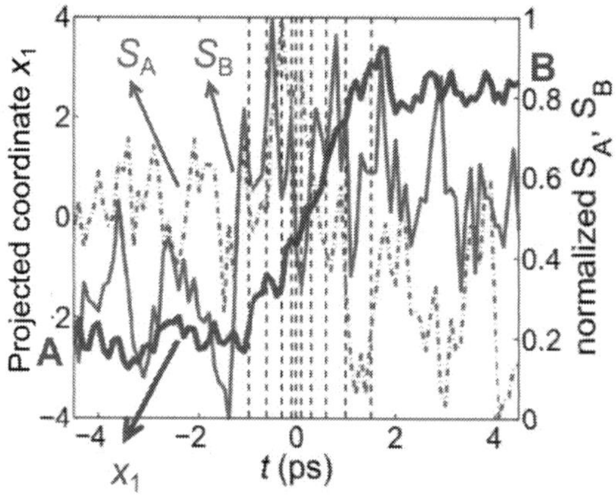

Statistical analysis of the transition path ensemble is needed.

Xi, Shah & Trout, *J. Phys. Chem. B*, 2013

Getting statistical samples of rare-event transitions

Rare events: overcoming a high free energy barrier

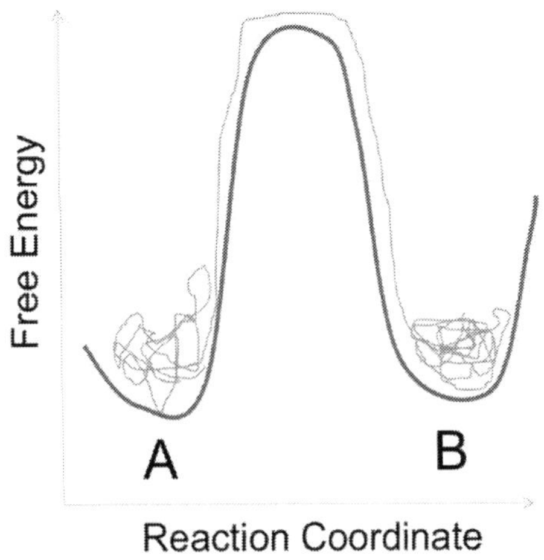

Aimless shooting

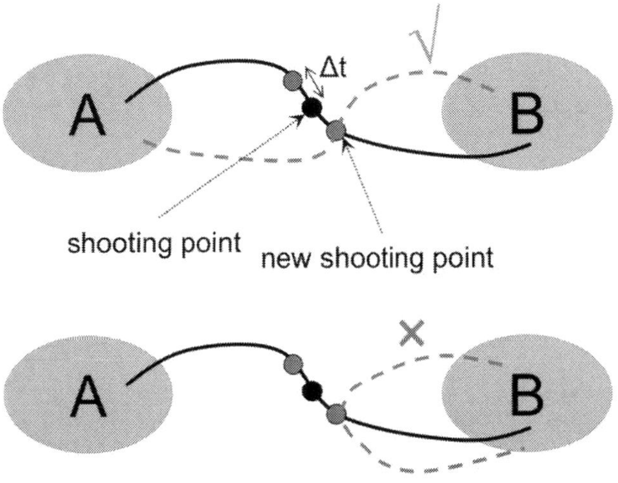

Peters & Trout *J. Chem. Phys.*, 2006
Peters, Beckham & Trout *J. Chem. Phys.*, 2007

C_s->0 limit: transition path ensemble for a hopping event

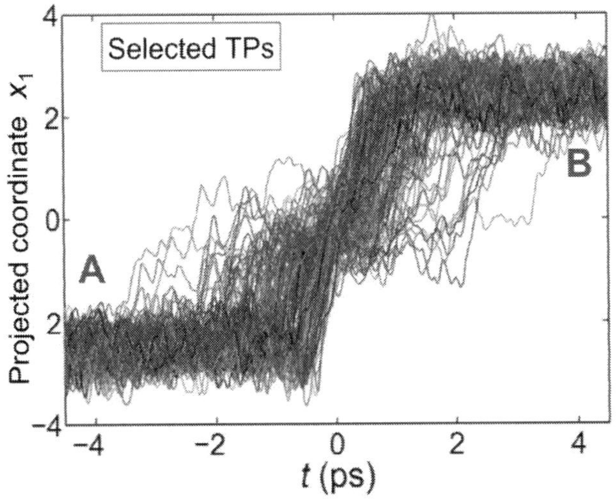

Summary of AS results

Result	N_{traj}
Forward in A *and* backward in B	421
Forward in B *and* backward in A	401
Forward in A *and* backward in A	315
Forward in B *and* backward in B	797
Accepted	822 (41.1%)
Inconclusive	66 (3.3%)
Total number of trajectories	2000

Xi, Shah & Trout, *J. Phys. Chem. B*, 2013

C_s->0 limit: statistical analysis with likelihood maximization

- Define a set of relevant collective variables (CVs): $q_1, \ldots, q_n$.
- Likelihood maximization tests different combinations of CVs, and quantifies their predictive power.

Reaction coordinate
(transition progress)

$$r(\boldsymbol{q}) = \alpha_0 + \sum_{k=1}^{m} \alpha_k q_k$$

H₂O Pos. → q_1 H₂O Ori. → q_2 H₂O-PVA interact. → q_3, q_4

rank	$\ln(L)$	q_1	q_2	q_3	q_4
1	-1139.80	x_1	v_2	$D(O_w, H_{p1:14})$	$D(H_{w2}, O_{p1:14})$
2	-1143.45	x_1	v_2	$D(O_w, H_{p1:14})$	$D(H_{w2}, O_{p1:392})$
3	-1144.11	w_1	$D(O_w, H_{p1:14})$	$D(H_{w1}, O_{p1:391})$	$D(H_{w2}, O_{p1:14})$
4	-1144.46	x_1	v_2	$D(O_w, H_{p1:14})$	$D(H_{w2}, O_{p1:887})$
5	-1144.47	x_1	v_2	w_3	$D(O_w, H_{p1:14})$
6	-1144.53	x_1	v_2	$D(O_w, H_{p1:14})$	$D(H_{w1}, O_{p1:391})$
7	-1144.76	x_1	v_2	$D(O_w, H_{p1:14})$	$D(H_{w1}, O_{p1:392})$
8	-1144.78	x_1	v_2	$D(O_w, H_{p1:14})$	$D(H_{w1}, O_{p1:14})$
9	-1144.79	x_1	v_2	$D(O_w, H_{p1:14})$	$D(H_{w2}, O_{p1:391})$
10	-1145.03	x_1	v_2	w_1	$D(O_w, H_{p1:14})$

Xi, Shah & Trout, *J. Phys. Chem. B*, 2013

C_s->0 limit: hopping mechanism

- Transition determined by the penetrant configuration and its electrostatic interactions with the polymer.
- Polymer configuration adjusts in a mostly passive manner to the penetrant movement.

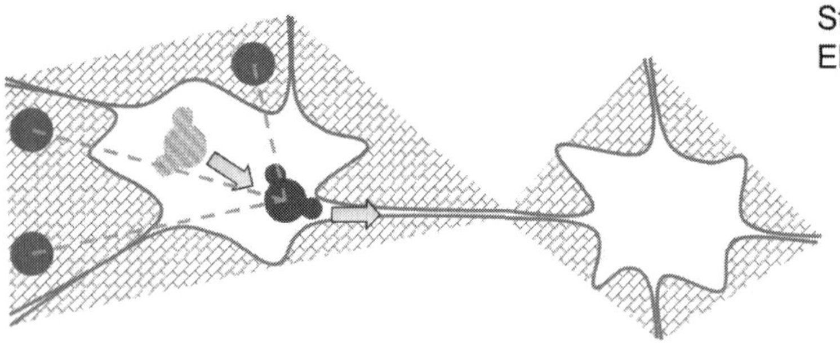

Steric effects
Electrostatic interactions

Finitely low C_s regime: H_2O trajectories

- Slow diffusion dominated by hopping for C_s up to 10%.

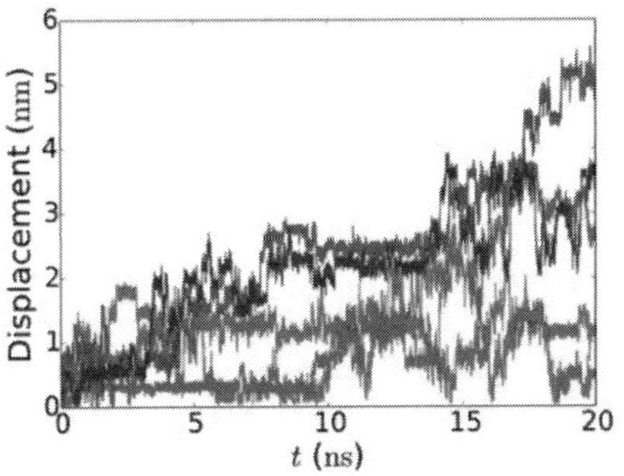

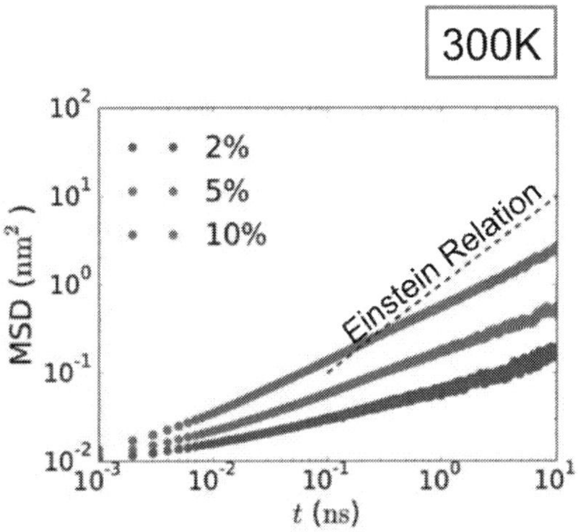

Finitely low C_s regime: clustering of H_2O molecules

- Clusters increase in size with higher C_s.
- A critical C_s for percolation?

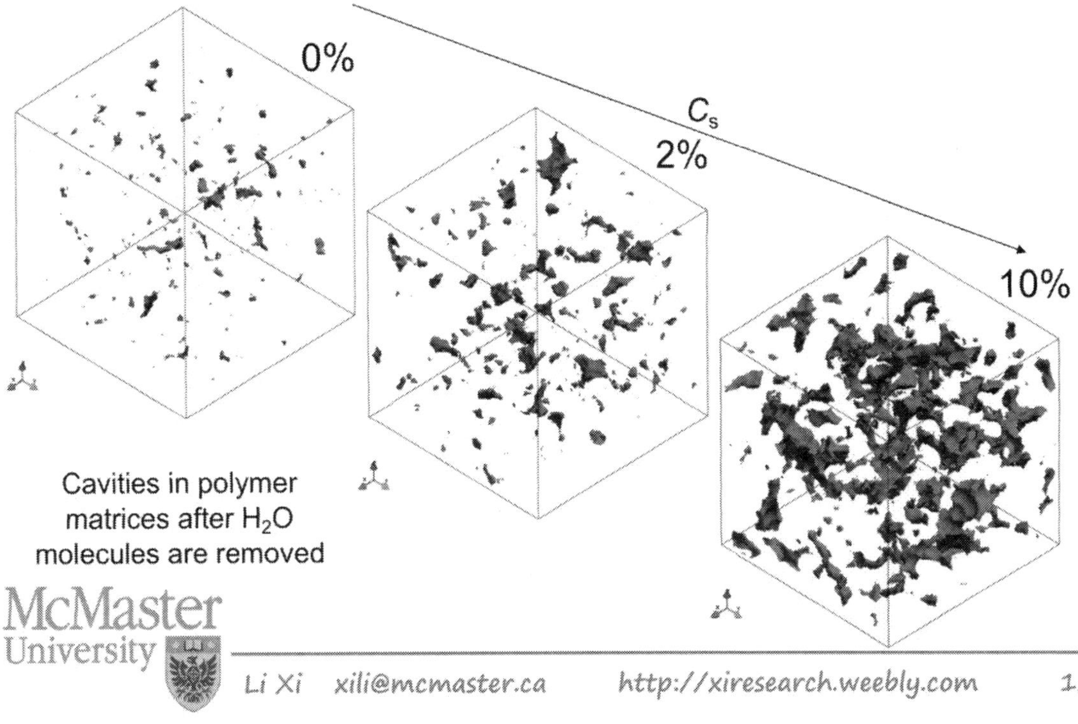

Cavities in polymer matrices after H_2O molecules are removed

Finitely low C_s regime: tunneling of H_2O trajectories

- Preferred H_2O pathways - or tunnels - connect the clusters.
- Clusters and tunnels form a diffusion network.

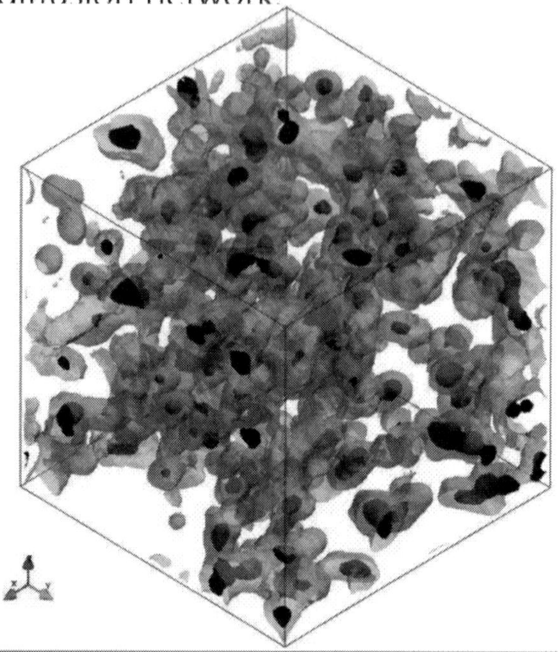

Isosurfaces of H_2O density over a period of time

Summary

- In glassy polymers, at low C_s, solvent diffusion is extremely slow, which may be modeled as random hops on a network.
- C_s ->0 limit: hopping of individual solvent molecules between cavities.
- Finitely low C_s regime: hopping along tunnels connecting solvent clusters.
- Eventually the solvent clusters will connect and percolate the domain.
- Aimless shooting + likelihood maximization: statistically identify the dominant factors in complex transitions.

Acknowledgment:

NOVARTIS

Nonequilibrium molecular dynamics (NEMD) simulations of an entangled $C_{700}H_{1402}$ linear polyethylene system containing 276 chains (corresponding to 193200 particles) were performed to investigate the chain dynamics over a wide range of Weissenberg numbers (*Wi*) under steady shearing flow. The simulations allow for a detailed examination of the effects of individual chain dynamics on many fluid properties, such as the rheological characteristic functions, chain configuration and orientation, topology, and multiple timescales, all as functions of *Wi*. A key finding is that the probability distribution function of the end-to-end vector, $|\boldsymbol{R}_{ete}|$, at high *Wi* is bimodal, as was the case for the unentangled ($C_{78}H_{158}$) and moderately entangled ($C_{400}H_{802}$) melts examined in prior simulation studies. The low $|\boldsymbol{R}_{ete}|$ peak is associated once again with the dynamical rotation/retraction cycles experienced by individual chains, and the high $|\boldsymbol{R}_{ete}|$ peak corresponds to the highly stretched and oriented macromolecules. To understand the underlying physics of the aforementioned phenomena, the relevant system timescales were determined. Specifically, the autocorrelation function of the unit end-to-end vector showed a periodic behavior at high *Wi* which was associated with the onset of "vorticity excursion" of individual chains within a background of highly stretched chains. The disengagement (τ_d) and rotational (τ_{rot}) relaxation times of the system were extracted by fitting a functional form of $A\exp(-t/\tau_d)\cos(2\pi t/\tau_{rot})$ to the autocorrelation function data. The number of entanglements and other topological features of the liquids were also computed as functions of *Wi*, and examined with regard to reptation and anisotropic diffusion theory. The key implications of these simulations with regard to the current state of rheological theory will also be discussed.

372815 Atomistic Simulations of Thermoresponsive Poly(N-isopropylacrylamide) Polymers

Abstract:

Poly(*N*-isopropylacrylamide) (PNIPAM) is a thermoresponsive polymer that undergoes a coil-to-globule transition in water around 32 °C. Because its structure is sensitive to external stimuli, PNIPAM-based systems (e.g., hydrogels, micelles, brushes) are of interest for applications like drug delivery, biosensors, and tissue engineering. The transition of the polymer from an extended coil to a collapsed globule conformation at the lower critical solution temperature (LCST) results from the interplay of polymer–polymer and polymer–water interactions. While it is suggested that the polymer dehydrates upon heating above the LCST, a full understanding of the structural changes in the polymer is still elusive. In this regard, we perform atomistic molecular dynamics simulations to study single oligomers of PNIPAM solvated in water at temperatures below and above the LCST. In addition, we treat a surfactant molecule that is a diblock of PNIPAM and an alkane chain. At higher temperatures, we observed a collapse of the PNIPAM structure in accordance with a decrease in polymer–water hydrogen bonds, an increase in the intramolecular polymer–polymer hydrogen bonds, and a decrease in the hydration number. These results are in agreement with experimental data. Additional insight is provided into the nature of the water around the polymer, such as the formation of a hydrogen bond network below the LCST. Lastly, we explore the formation of micelles from PNIPAM-based surfactants with a temperature-dependent hydrophilicity.

Sandia National Laboratories is a multi-program laboratory managed and operated by Sandia Corporation, a wholly owned subsidiary of Lockheed Martin Corporation, for the U.S. Department of Energy's National Nuclear Security Administration under contract DE-AC04-94AL85000.

371196 Self-Entanglement of a Single Polymer Chain Confined in a Cubic Box

Abstract:

We study the self-entanglement of a single linear polymer chain confined to a cubic box (L * L * L) using the bond-fluctuation lattice model and primitive path analysis. We probe chains with number of monomers between N = 30 and 750, and degree of confinement L/R_{g0} between 0.4 and 12, where R_{g0} is the radius of gyration of an unconfined polymer. We found that the conformational properties R_g/R_{g0} and L_p/R_{g0}, where L_p is the average primitive path length, collapsed onto a single master curve as a function of the degree of confinement. In the strongly confined regime, $L/R_{g0} << 1$, we found that $R_g/R_{g0} \sim (L/R_{g0})^{0.8}$, and $L_p/R_{g0} \sim (L/R_{g0})^{-2}$. We verify that simulation methodology used is quantitatively consistent with experimental data and the Colby-Rubinstein entanglement model for unconfined concentrated polymer solutions. The most significant difference between unconfined and confined systems is the variation of L_p with monomer density φ; $L_p \sim \varphi^{5/9}$, in the former, and $L_p \sim \varphi^{2/3}$, in the latter.

378529 Avidity-Controlled Hydrogels for Delivery of Induced Pluripotent Stem Cell-Derived Endothelial Cells and Growth Factors

Abstract:

We have developed protein-polyethylene glycol (PEG) hybrid hydrogels that form physical gels upon mixing to control the co-delivery of cells and growth factors. This hydrogel is a mixture of a linear, engineered protein that undergoes a sol-gel phase transition upon noncovalent complexation with an 8-arm PEG that has been decorated with either one or two repeats of a proline-rich peptide (P1 or P2, respectively) . Because of the dynamic nature of the molecular recognition between the two components, the hydrogels are reversibly shear-thinning and self-healing. The P2 variants exhibited higher storage moduli and slower release of encapsulated vascular endothelial growth factor (VEGF) compared to the P1 variants, demonstrating the ability to tune the hydrogel bulk properties through a biomimetic peptide-avidity strategy. Human induced pluripotent stem cell-derived endothelial cells (hiPSC-ECs) adopted a well-spread morphology within the three-dimensional hydrogels, which also provided significant protection from cell damage during ejection through a fine-gauge syringe needle. In a mouse hindlimb ischemia model of peripheral arterial disease, co-delivery of hiPSC-ECs and VEGF within the hydrogel was found to reduce inflammation and promote muscle tissue regeneration compared to a saline control.

Phenotypic Modulation of Smooth Muscle Cells on Biodegradable Elastomeric Substrates

Shanfeng Wang

Department of Materials Science and Engineering, The University of Tennessee, Knoxville, TN 37996

Here I present our recent findings on controlling primary rat smooth muscle cell (SMC) behavior using novel biodegradable photo-crosslinked polymer substrates with controllable chemistry, topology, and stiffness. We have synthesized photo-crosslinkable PCL triacrylates (PCLTAs) with six molecular weights of 2000-20000 g/mol and further investigated SMC adhesion, spreading, migration, proliferation, phenotypic modulation, gene expression of synthetic and contractile markers, and protein expression of calponin on PCLTA networks in both static and dynamic modes. SMCs cultured on stiffer semi-crystalline photo-crosslinked PCLTA substrates exhibited stronger stress fibers, larger expanding area, faster growing, and higher expression of contractile phenotype markers. In comparison, the cells cultured on the softer, amorphous substrates had a smaller projecting area and a lower growth rate. Surface grafting of PEG and preparing substrates with microgrooves, micro-pillars, micro-pores, nanowires, or nanopores have been achieved on PCLTA networks for regulating SMCs and other cell types.

387037 Fabrication of Highly Aligned Collagen Sponges from Self-Assembled, Fibrillar Collagen Gels

Abstract:

Background

Collagen is an attractive biomaterial for many tissue engineering applications because it is biocompatible, supports cellular infiltration and growth, and can be remodeled through natural enzymatic pathways. In addition, collagen has the potential to direct regeneration when aligned fibers provide contact guidance fields. The ability to control cellular alignment and orientation is a critical design factor for scaffolds of many tissue types including nerves and smooth muscle. Alignment of fibrillar collagen has most often been generated during self-assembly with external potential fields, or post self-assembly by external mechanical forces or by cellular traction. Fiber alignment can also be generated via electrospinning, although native collagen fiber architecture is not preserved. We have developed a simple method for fabrication of highly aligned collagen sponge-like scaffolds using freezing sublimation. Unlike other approaches where collagen is not driven to self-assemble but rather is frozen as a suspension, and aligned pores are generated via a temperature gradient, our approach generates alignment from self-assembled fibrillar gels. In this study, we have characterized the architecture and fabrication parameters of our novel aligned collagen fiber-like scaffolds and characterized their ability to direct and enhance the outgrowth of neural cell explants.

Methods

Scaffold Fabrication: Highly aligned collagen sponge-like scaffolds were fabricated from self-assembled, fibrillar collagen hydrogels. Briefly, type-I bovine collagen (Elastin Products Company, Owensville, MO) was reconstituted in 0.02N acetic acid at 3.0 mg/mL. Buffered hydrogel solutions were prepared using the following ratios: 20 μL HEPES:130 μL 0.15 N NaOH:100 μL 10x MEM: 56 μL M199: 10 μL L-glutamine: 10 μL penicillin/streptomycin: 677

μL 3.0 mg/mL collagen solution. Cylindrical silicone conduits of various diameter (2380 μm, 3175 μm, 4762 μm, and 6350 μm) and aspect ratio were filled with hydrogel solution and incubated at 37°C for 1 hour to allow self-assembly. Conduits were then moved into a -80°C freezer for 3 hours followed by lyophilization overnight.

Parameter Variation in Scaffold Fabrication: Variations to this scaffold fabrication process were made to evaluate the effect of freezing temperature and freezing rate. In the freezing temperature study, silicone conduits were filled and self-assembled as described previously, and placed in a -80°C or -20°C for three hours or snap frozen in liquid nitrogen. To examine the effect of a reduced freezing rate, scaffolds were frozen in a Cryo 1°C Freezing Container (Nalgene) at either -20°C or -80°C. To fabricate randomly aligned collagen sponge control scaffolds, hydrogels were prepared as described as above and placed in 15.6mm diameter wells in a 24 well plate, self-assembled at 37°C for 1 hour, frozen at -80°C, and lyophilized overnight.

Scanning Electron Microscopy: Non-hydrated scaffolds were prepared as described above and then sputter coated (SCD 004, Balzers Union Limited, Balzers, Liechtenstein) with gold/palladium and their surface topography imaged via SEM (Amray 1830I, Amray Inc. Bedford, MA). Hydrated scaffolds were prepared as described above and then incubated in phosphate buffered saline at 37°C for 1 hour. Scaffolds were then dehydrated in a series of aqueous acetone solutions (25, 50, 75, 95, and 100%) for 15 minutes each and left in 100% acetone overnight. Samples were critical point dried (CPD 020, Balzers Union Limited, Balzers, Liechtenstein), and sputter coated with gold/palladium and imaged via SEM.

Neural Explant Culture: Dorsal root ganglia (DRG) were isolated from pathogen-free chick eggs at embryonic day 8 (Charles River Laboratories, Cambridge, MA) and cultured in DMEM containing 10% fetal bovine serum (FBS), 1% L-glutamine, 1% penicillin/streptomycin supplemented with 100ng/mL nerve growth factor. DRGs were plated on aligned sponge-like collagen scaffolds as well as randomly aligned collagen sponge scaffolds and maintained in culture for 3 days.

*Immunocytochemistry*Cultures were fixed in 4% paraformaldehyde solution and then stained immunohistochemically for neurofilament heavy chain (NFM200) and observed with confocal fluorescent microscopy.

Results and Discussion

SEM imaging revealed that fabricated collagen sponges demonstrated highly aligned collagen fiber-like architecture with long range uniformity. Scaffolds appear as a cable-like bundle of aligned fiber-like structures. Further, cross-sections revealed a highly porous interior within each fiber-like structure. Fabricated scaffold diameter demonstrated a strong dependence on the diameter of the silicone conduits used for fabrication. Scaffolds fabricated in 2380 μm, 3175 μm, 4762 μm, and 6350 μm silicone conduits had average diameters of 515 μm, 680 μm, 1560 μm, and 2200 μm respectively, shrinking 20-35% the size of the initial cylindrical hydrogel. Upon hydrating the scaffolds, the diameter of scaffolds fabricated in 2380 μm, 3175 μm, 4762 μm, and 6350 μm silicone conduits swelled to 767 μm, 1094 μm, 2107 μm, and 2756 μm respectively. Additionally, the diameter of the individual fiber-like structures that comprise the scaffold also varied with silicone conduit diameter. Average fiber size for scaffolds fabricated in 2380μm, 3175μm, 4762μm, and 6350μm silicone conduits were 62 μm, 71 μm, 103 μm and 192 μm respectively. Scaffolds with similar fiber-like morphology were obtained at each of the three freezing temperatures examined (LN2, -20°C freezer, and -80°C freezer). Reducing the freezing rate to 1°C/min had no observable effect on fiber or scaffold diameter; however it did seem to smooth the edges of the fiber-like features within the scaffolds as compared to scaffolds which were placed directly in the freezer. SEM imaging of the hydrated scaffolds revealed that the fiber-like topography and the porous interior of the scaffolds was maintained following hydration. Aligned collagen sponge-like scaffolds scaffolds supported the growth of axons from chick embryo dorsal root ganglia, and the directional growth was enhanced over that within random collagen sponge scaffolds. This fabrication method to generate highly aligned collagen sponge-like scaffolds can be used to guide axon outgrowth or Schwann cell migration. Overall, this methodology offers rapid production of aligned collagen fibers that retain the strength and biocompatibility compared to current collagen-based scaffold techniques without the need for specialized equipment. Through specific variation of fabrication parameters, the size and topography of resulting scaffolds can be tuned to suit a variety of needs and applications.

Conclusion

Herein we present a novel method for the generation of highly aligned collagen fiber structures. Aligned collagen fiber scaffolds fabricated through this simple, rapid technique have the potential to serve as building blocks for regenerative medicine strategies across a variety of tissue types where tissue anisotropy is critical to functional outcome, such as engineering tendons and ligaments or nerve guidance conduits.

Acknowledgements

This work is supported by National Institutes of Health grant NIH-NINDS 1R01NS078385

Peripheral Nerve Extracellular Matrix Hydrogels for Treating Spinal Cord Injury

R. Chase Cornelison[1], Vivian Nguyen[2], & Christine E. Schmidt[2]
[1]McKetta Department of Chemical Engineering, University of Texas at Austin, Austin, TX
[2]J. Crayton Pruitt Family Department of Biomedical Engineering, University of Florida, Gainesville, FL

Introduction: Spinal cord injury (SCI) is extremely debilitating and is characterized by cavity formation within the tissue. Using experimental injury models, researchers have shown that grafts of peripheral nerve tissue exhibit vast potential to induce regeneration of spinal axons, in part because of the growth-supportive molecules of the extracellular matrix (ECM). Unfortunately, fresh tissue is often immunogenic upon implantation. Our lab previously developed an optimized method to remove immunogenic cellular components from tissue while preserving the native structure of the ECM [1], and this protocol is now used clinically to prepare ECM grafts for bridging human peripheral nerve defects. In applying this ECM scaffold to repair the injured spinal cord, it is desirable to create an injectable formulation that can applied into the cavity without the need for incising host tissue. To this end, we have created an injectable hydrogel derived from peripheral nerve ECM which exhibits thermal gelation under physiological conditions. Herein we detail the chemical, mechanical, and topographical characterization of this novel material as well as demonstrate in vitro biocompatibility.

Materials and Methods: Peripheral nerves from Sprague Dawley rats were decellularized according to the protocol developed in the Schmidt lab [1]. Acellular nerves were lyophilized, micronized, and digested using a pepsin solution at a pH of 2. After complete digestion, the enzyme was deactivated by neutralizing the pH to 7.4. Thermal gelation was then induced by incubation at 37 °C in a humidified environment. The protein profile of nerve ECM was identified using tandem liquid chromatography-mass spectrometry (LC-MS/MS). Additionally, collagen and glycosaminoglycan content was determined using Sircol and Blyscan assay kits (Biocolor Ltd; UK). An Instron 5542 (Norwood, MA) was used to examine the mechanical properties of this ECM hydrogel, and the topographical features were evaluated via scanning electron microscopy (SEM) after sample dehydration. To examine biocompatibility, primary rat Schwann cells were cultured on nerve ECM substrates in supplemented media, and cell survival was assessed using live/dead staining. Schwann cell metabolic activity in 3D hydrogels was also evaluated using AlamarBlue (Invitrogen).

Results: Protein identification reveals that our decellularization method preserves the relative tissue composition, e.g. the ratio of collagen I to laminin in fresh and acellular nerve is 1.6 and 1.4, respectively. In addition, acellular nerve tissue is shown to retain 83% and 61% of the native tissue's original collagen and glycosaminoglycan content, respectively. SEM images demonstrate that nerve ECM hydrogels are fibrous structures with fiber diameter on the same length scale as that found within native peripheral nerve, roughly 100-150 nm (Figure 1). Additionally, the compressive modulus for nerve ECM hydrogels was found to be 4.5 kPa. Using a live/dead assay, it was found that nerve ECM substrates are biocompatible, supporting 80% survival of primary rat Schwann cells compared to poly-D-lysine positive controls. Cells cultured on these matrices also exhibited an elongated morphology characteristic of Schwann cells whereas cells on dilute (1:10) matrices frequently did not.

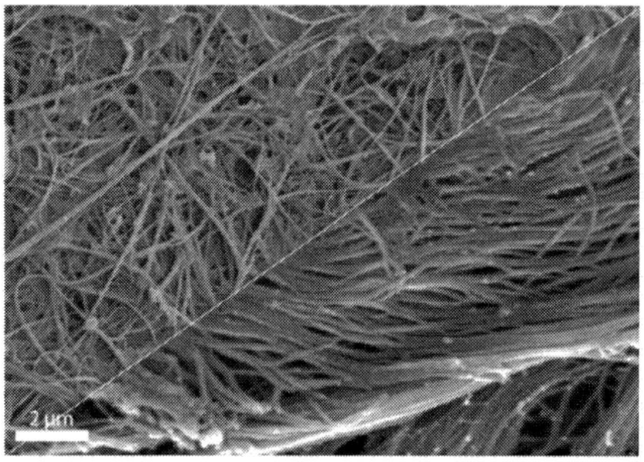

Figure 1: Scanning electron micrographs of peripheral nerve ECM hydrogel (top left) and fresh peripheral nerve (10 µm section, bottom right).

Conclusions: We have developed a protocol to process acellular peripheral nerves into an injectable, in situ gelling hydrogel scaffold. We have shown that this nerve ECM hydrogel approximates the chemical, mechanical, and topographical features of native neural tissue. The collagen and glycosaminoglycan content of peripheral nerve is comparable to that reported for similar tissues including brain and spinal cord [2], supporting the hypothesis that this ECM hydrogel will mimic the native spinal cord to promote tissue regeneration. In addition, the hydrogels exhibit a compressive modulus similar to that reported for native rat neural tissue [3] and optimal for neurite outgrowth [4]. We hypothesize this material will mimic the native tissue environment in vivo to promote tissue repair following SCI, thereby offering a clinically-relevant material to treat debilitating neural injury.

References:

[1] Hudson TW et al. Tissue Eng. 2004, 10 (1346-58).
[2] Medberry CJ et al. Biomaterials. 2013, 34(1033-40).
[3] Seidlits SK et al. Biomaterials. 2011, 31(3930-3940).
[4] Man AJ et al. Tissue Eng. 2011, 17(2931-2942).

Development of an A-Cellular Off the Shelf Vascular Graft.

Maxwell T Koobatian[1] Randall Smith[2] Sindhu Row[4] Stelios Andreadis [2,4]
Daniel Swartz[1,3]
[1]State University of New York at Buffalo Department of Physiology and Bio-Physics, Buffalo, New York, USA
[2]State University of New York at Buffalo Department of Biomedical Engineering, Buffalo, New York, USA
[3]Women and Children's Hospital of Buffalo Department of Pediatrics, Buffalo, New York, USA
[4] State University of New York at Buffalo Department of Chemical and Bio- Engineering, Buffalo, New York, USA

Introduction:

Currently, the options available for providing patients suffering from vascular diseases are limited. While autologous sources (such as saphenous veins) are the most common source to replace diseased blood vessels, not all patients have suitable vessels available. Furthermore, synthetic grafts/TEV's (Tissue Engineered Vessels) have a high rate of failure and often cannot be used to replace small diameter vessels (<5mm in diameter). Development of a small diameter A-Cellular Vascular Graft (ACVG) capable of "Off the Shelf" use that remains patent after implantation, would be a great advancement to the field of vascular research and biomedical engineering. Here we present such a vascular graft capable of implantation into a large animal ovine model. We demonstrate complete endothelialization after 1 month over the entire length of the ACVG and the tissue explant is capable of responding to receptor and non-receptor mediated vasoagonists. Furthermore, our ACVG is capable of rapid remodeling, and closely resembles native tissue upon explant at 3 months post implantation.

Materials and Methods:

Small-Intestinal Submucosa (SIS) tubes were acquired in tubular form from Cook-Biotech Inc. SIS tubes were approximately 4.75mm in diameter. Vascular Endothelial Growth Factor (VEGF) and Heparin were immobilized on the luminal surface and then implanted into the carotid artery of a large animal ovine model and explanted after 1 week, 1 month and 3 month time points.

Results and Discussion:

Heparin and VEGF was immobilized onto tubular SIS using EDC/NHS crosslinking chemistry. Amount of immobilized heparin and VEGF was then determined using a colorimetric assay and in-direct ELISA respectively. ACVGs were then implanted into the carotid artery of healthy one year old sheep. Flow probes, catheters, and sonometric crystals were implanted with the ACVG to acquire flow-rates, pressure and graft diameter throughout the experiment. We demonstrate that VEGF rapidly immobilizes circulating Endothelial Progenitor Cells (EPC) and monocytes responsible for the development of a functional endothelium. After one month implantation, the ACVG showed complete endothelialization of the lumen and migration of smooth muscle and macrophage cells from surrounding tissue throughout the medial layer. After 3 months, endothelialization was retained and cell migration of smooth muscle and macrophage cells was observed to be similar to native vasculature. Explanted tissues possessed mechanical properties, responded to receptor mediated vasoagonists, and grossly appeared to be similar to native tissue after explantation.

Conclusions:

The development of an "off the shelf" A-Cellular Vascular Graft has been successfully evaluated in a large animal model. The immobilization of heparin and VEGF enables our TEVs to rapidly endothelize and remodel after implantation. Future work includes the investigation and role of circulating monocytes and macrophages in the populating of cell free vascular grafts after implantation.

Acknowledgments

Special thanks for the assistance of Carmon Koenigsknecht, Randall Smith and Sindhu Row and the funding from SUNY Buffalo CTSA.

387406 Combined Biomolecule Delivery Improves BMP-Induced Osteogenesis

Abstract:

The use of bone morphogenic proteins (BMPs) shows promise in therapies for improving bone regeneration; however, high supraphysiological concentrations required for a desired osteoinductive effect, costs, and patient variability have prevented the full advantages of BMP-based therapeutics from being realized. Thus, one strategy is to deliver synergistic molecules with BMP to enhance efficacy and lower doses. In this work, a matrix metalloprotease (MMP)-sensitive hyaluronic acid (HA)-based hydrogel was used for the delivery of both stromal cell-derived factor-1 alpha (SDF-1α) and BMP-2 towards improving BMP-induced osteogenesis. SDF-1α has been shown to play an important role in stem cell trafficking and HA hydrogels are known to increase extracellular matrix production. Additionally, our lab has previously shown that HA is capable of inducing cell chemotaxis through the CD44 receptor.

Maleimide-modified HA (MaHA) was cross-linked using a difunctional MMP-sensitive peptide according to an addition reaction, allowing for protease-mediated hydrogel degradation and growth factor release. The compressive and rheological properties of MaHA hydrogels were characterized as a function of hydrogel polymer content (1-5 wt%) and functionalization (10-40%f), where mechanical properties and gelation time were strongly dependent on cross-linking density. To determine growth factor release, MaHA hydrogels loaded with either 100 ng/scaffold of SDF-1α or BMP-2 were degraded in 10, 2, or 1 U/ml type II collagenase (non-specific MMP degradation) and compared to degradation in buffer alone. In general, BMP-2 and SDF-1α release corresponded closely with hydrogel degradation profiles, where limited release occurred through diffusion in the absence of MMPs. In the presence of MMPs, hydrogel degradation and growth factor release was controlled as a function of both collagenase concentration and hydrogel formulation, where faster growth factor release occurred at higher collagenase concentrations and for hydrogels with lower cross-link densities. In vitro SDF-1α bioactivity was determined by quantifying cellular invasion or outgrowth from a hydrogel-encapsulated cell pellet into the surrounding hydrogel over time. After four days, significant invasion from the cell pellet was observed for hydrogels loaded with 250 ng of SDF-1α and limited invasion was seen without any SDF-1α loaded into the hydrogels. Lastly, a critical-sized cranial defect was used to determine the effect of growth factor delivery on bone formation in vivo. The animal group with combined SDF-1α and BMP-2 delivery showed significantly higher bone formation when compared to groups loaded with the same BMP-2 or SDF-1α concentration delivered separately or the empty defect group. Furthermore, histology indicated increased in vivo cellular invasion for hydrogels loaded with SDF-1α.

370243 Bioprinting with Chitosan-Gelatin Thermo-Sensitive Hydrogels

Abstract:

Conventional cell-seeding methods are inadequate in the development of *in vitro* tissues, because they involve random placement of cells and therefore lack the precision necessary for spatial control. Bioprinting is a novel technology, which can be used to pattern scaffolding biomaterials and cells with high precision to build 3D structures *in vitro*. However, the limitation of this approach is providing a conducive environment during and after the printing process. Some have used various biomaterials that change from solution phase to hydrogel phase with the initiation of chemical reactions by photonic energy. We evaluated the possibility of using thermo-sensitive chitosan-gelatin hydrogels formed at physiological conditions. Hydrogels were prepared with chitosan-gelatin along with Transglutaminase and DMEM powder in presence of 2-glycerol phosphate in sterile condition. We conducted a parametric study of printing speed and needle length to analyze and quantify cell viability. Cell viability can be controlled by varying the parameters such as printing speed and needle length. Cell viability of three different cell types was evaluated at day 2 and day 5. This study gave percentage of live, apoptotic, and necrotic cells to the process parameters. The increase in percentage of injured cells at day 5 indicated necessity to study the effects of cell density and also study about hydrogel materials. Material composition and architecture of the hydrogels can influence the fate and function of the cells. Material architecture studied using scanning electron microscope (SEM) shown that the hydrogels are porous.

Human umbilical vein endothelial cells (HUVEC-2) were cultured in Medium 200PRF supplemented with LSGS following vendor's protocols (BD Biosciences). The final concentrations of the components in the supplemented medium are: fetal bovine serum (FBS) 2% v/v, hydrocortisone 1 µg/mL, human epidermal growth factor 10 ng/ml, basic fibroblast growth factor (bFGF) 3 ng/mL, and heparin 10 µg/mL. Cells were maintained at 37°C, 5% CO_2/95% air, humidified cell culture incubator and fed with fresh medium every 36 hours. When confluent, cells were dissociated with trypsin and neutralized with trypsin neutralizer solution.

IMR-32 n*euroblasts* and HepG2 (HB-8065) cells (ATCC) were cultured with media containing EMEM following vendor's protocol. In brief, cells were maintained in EMEM containing 10% FBS at 37°C, 5% CO_2/95% air, humidified cell culture incubator and fresh media was supplemented every two days. When confluent, cells were detached from TCP using trypsin and neutralized with growth media. All viable cells were counted using trypan blue dye exclusion assay.

Cell viability was evaluated using live/dead cell viability assay kit (L34951, life technologies) according to manufactures protocol. Briefly, Percentages of late apoptotic and necrotic (low level green-stained with SYTOX green), and metabolically active/live (red-stained with C12-resazurin red), cells were assessed using live/dead cell viability assay kit. Samples were processed according to the manufacturer's recommendations. Briefly, cell-laden hydrogels were printed, cultured, and detached from TCP using trypsin. Cells were centrifuged and dispersed in 100 µL PBS. One µL of 50 µM C12-resazurin red stain and 1 µL of 1 µM SYTOX green stain were added to 100 µL of cell suspension. Cells were incubated at 37°C in an atmosphere of 5% CO_2 for 15 minutes. The stained cells were analyzed according to the vendor's protocol using FACSCalibur flow cytometer.

The results showed that decreased the residence time of the cells during bioprinting (by decreased needle length) increased the cell viability. To understand the effect of printing speed on cell viability, cell-laden hydrogels were printed varying volume ejected through the needle over time. The results showed that the cell viability is a strong function of printing speed. Decreased printing speed increased cell viability. However, at day 5, decreased cell viability was observed, probably due to very high cell density; therefore the cell viability will be evaluated at lower cell density. Cells seeded in hydrogels showed lower cell viability compared to the hydrogels printed at lower speeds; which indicated that this bioprinting technique did not cause injury to the cells. These results also suggested that there is a necessity to improve/optimize the hydrogel material combinations. Separately we also traced the cells with fluorescence and SEM pictures showed that the hydrogels are porous.

In summary, CG hydrogels provide significant potential for bioprinting. We demonstrated that this bioprinting technique can be used to control cell viability by varying process parameters such as printing speed and needle length. This approach had shown that with lower printing and smaller needle length we can have viable cells to form *in vitro* vascularization. Future studies will focus on testing stability of hydrogels by optimizing the combinations of hydrogel materials and evaluating cell functioning. Improving cell viability by varying cell density during bioprinting will be evaluated.

Biofabrication of 3D Collagen Scaffold Mimicking *In Vivo* Tissue Architectures

Veronica Rodriguez-Rivera
Ph.D Chemical Engineering
Chemical Engineering
Advisor: Dr. John W. Weidner and Dr. Michael J. Yost
University of South Carolina-Chemical Engineering Department
Medical University of South Carolina-Surgery Department
November 18, 2014

UNIVERSITY OF SOUTH CAROLINA

MUSC MEDICAL UNIVERSITY *of* SOUTH CAROLINA

Clinical Relevance

- ❑Next generation of biomaterials
  - ❑Organ donor shortage
    - ❑Liver disease
    - ❑Kidney Failure
    - ❑Coronary heart diseases
  - ❑Damage/Lost Tissue

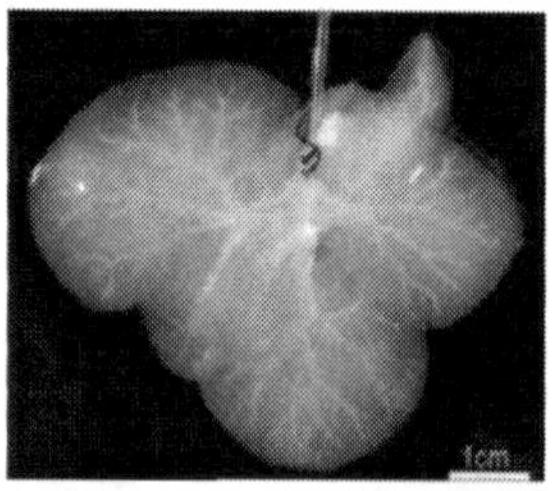

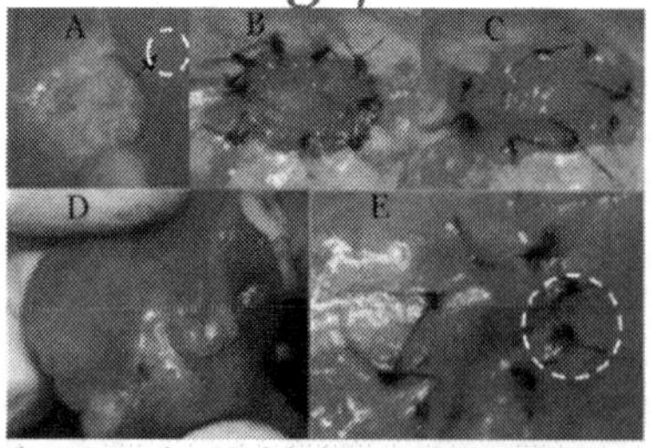

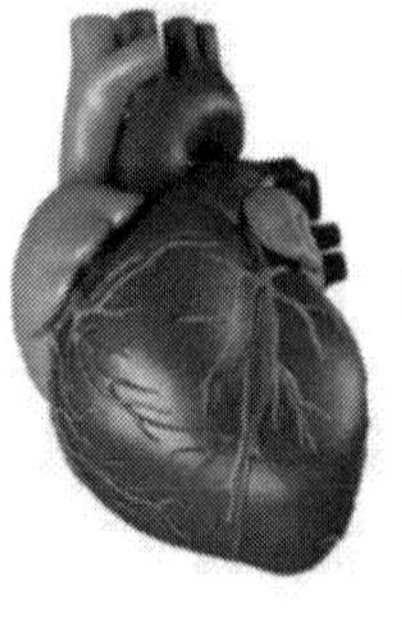

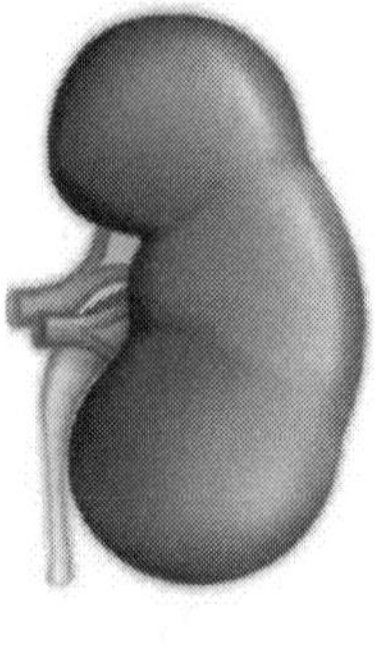

Fann, Stephen A. ,et al. A Model of Tissue-Engineering Ventral Hernia Repair. *Journal of Investigative Surgery*. 2006;19; 193-205; Baptista, P.M., et al., *The use of whole organ decellularization for the generation of a vascularized liver organoid*. Hepatology, 2011. **53**(2): p. 604-17.

Biofabrication Process

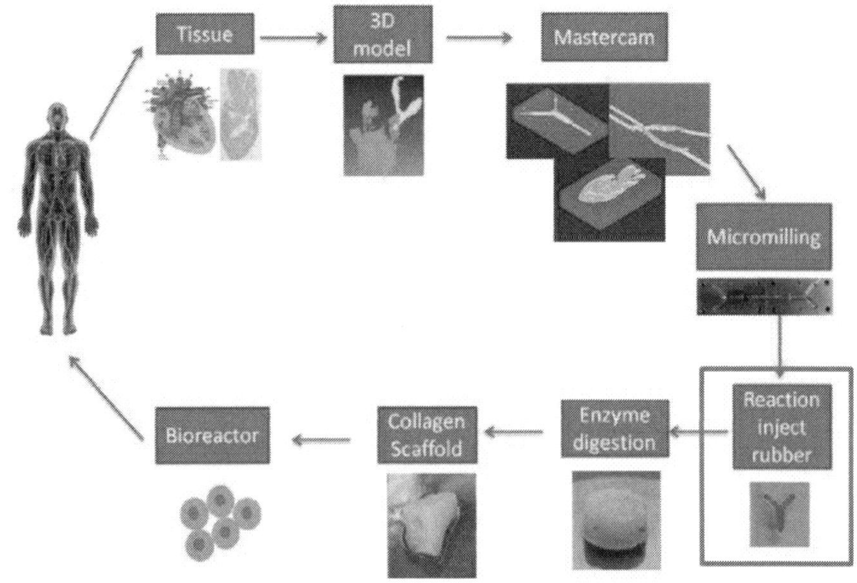

BSA Rubber

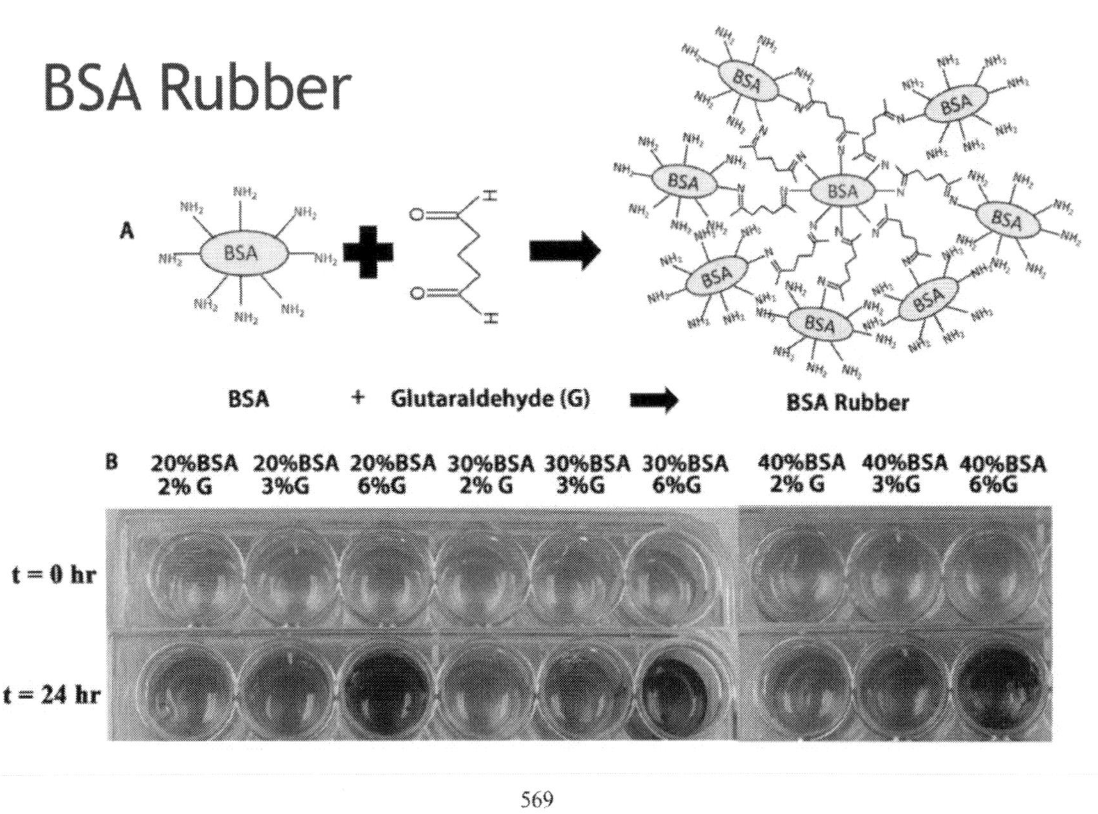

BSA Rubber Parameters

BSA
- 20%, 30%, and 40%

Glutaraldehyde
- 2%, 3%, and 6%

Solvent
- 1X PBS, 2X PBS, and DI

Enzyme
- 0.25 to 1% trypsin

Enzyme pH
- 6.5 to 7.8

Digestion temperature
- 30-37C

BSA:Glutaraldehyde ratio
- 1:1, 5:1, 4:1, and 10:1

Exposure time
- 15 – 72 hrs

Rate of enzymatic degradation

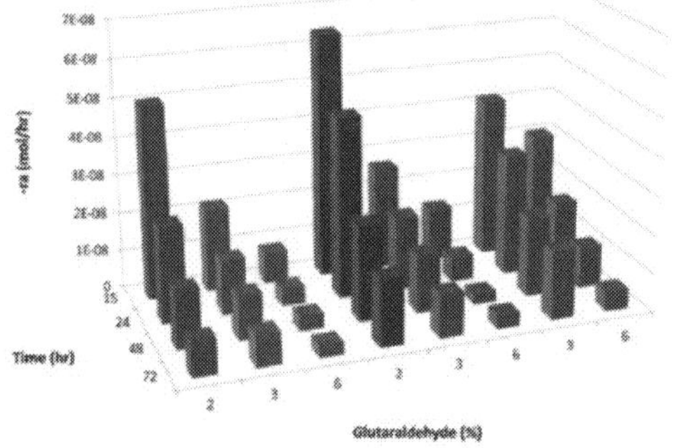

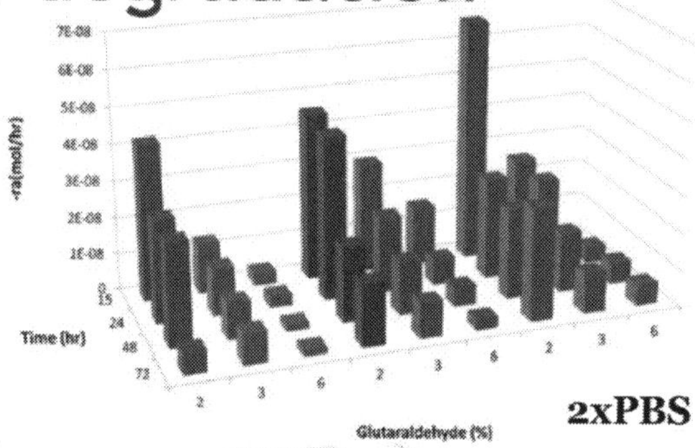

2xPBS

1xPBS

↑ **% Glutaradehyde** ↓ **-rA**

↑ **% BSA** ↑ **-rA**

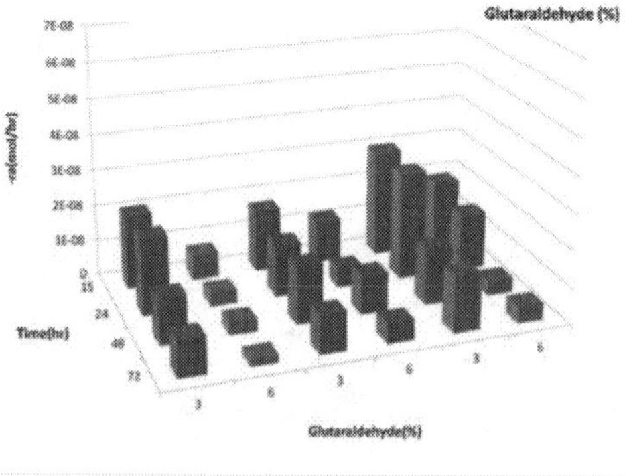

Blue- 20%BSA
Purple -30%BSA
Red-40% BSA

DI

Elastic Modulus

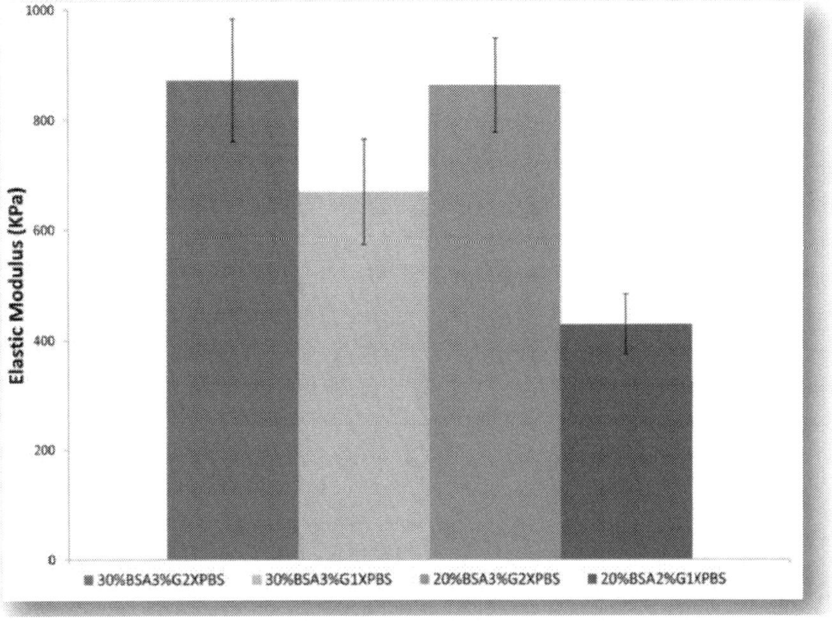

Biofabrication Process

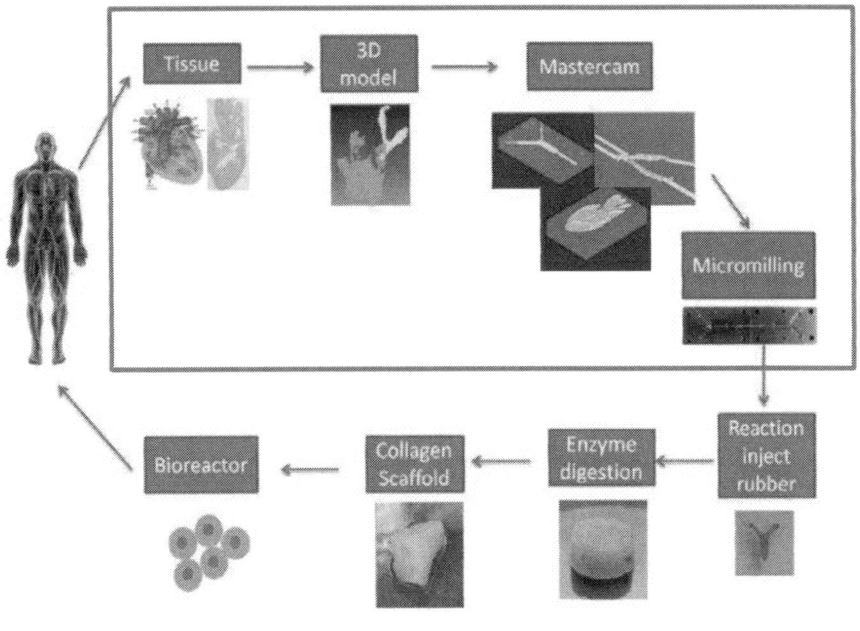

Renal CT Angiograms

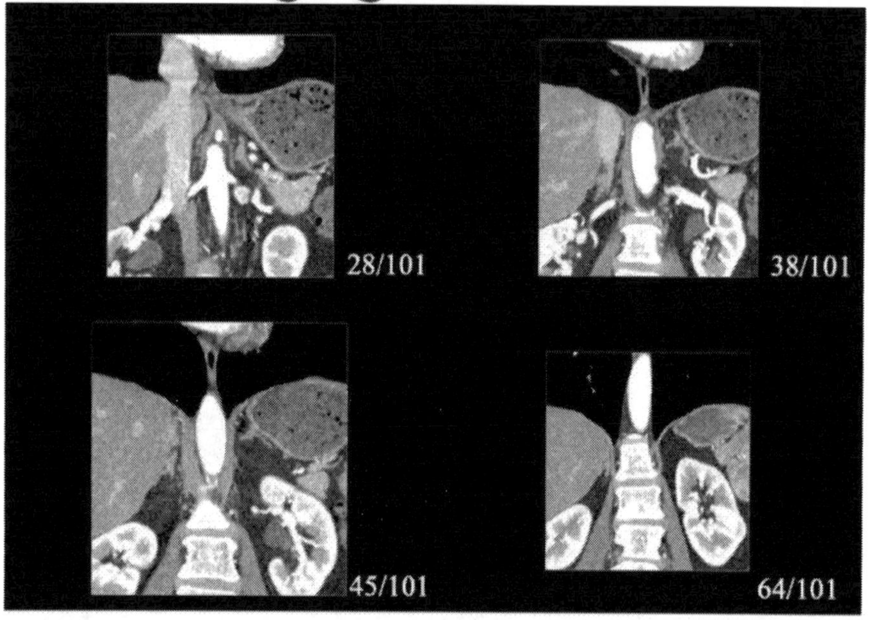

3D Reconstruction Movie

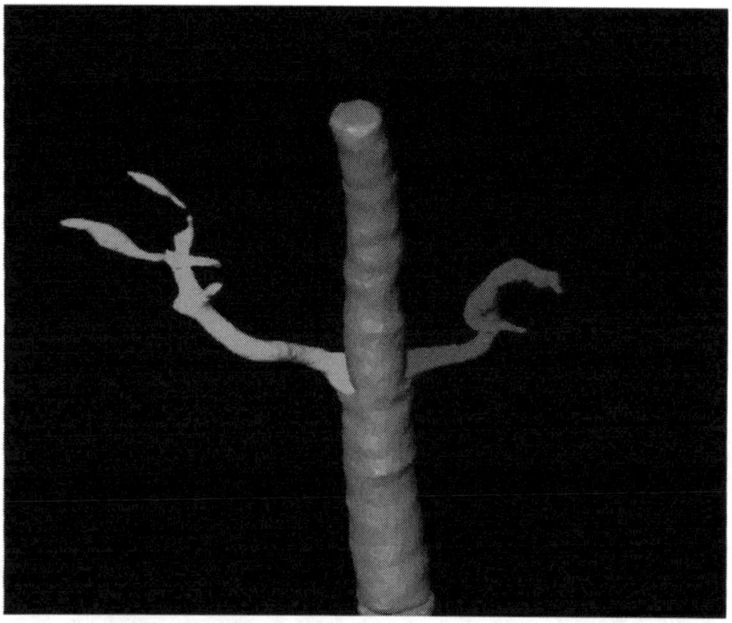

Renal Artery Model

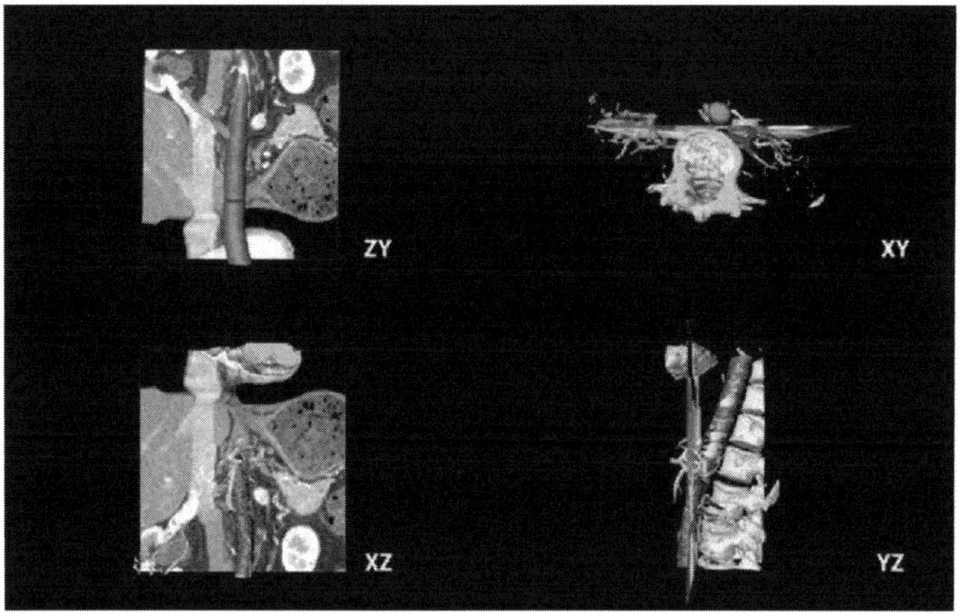

Renal Artery Model

Area	Diameter (mm)
Abdominal Aorta (Pink)	11.18-18.32
Left Renal Artery (Blue)	5.84
Right Renal Artery (Green)	5.95

3D Architectures in PLA Mold

Branched Architectures

Y Mold

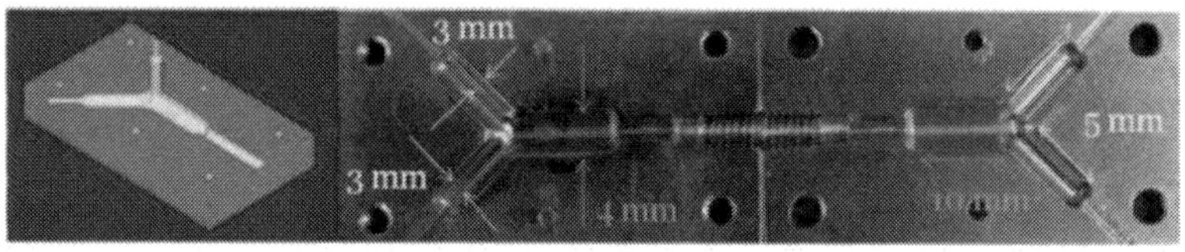

Stability Piece

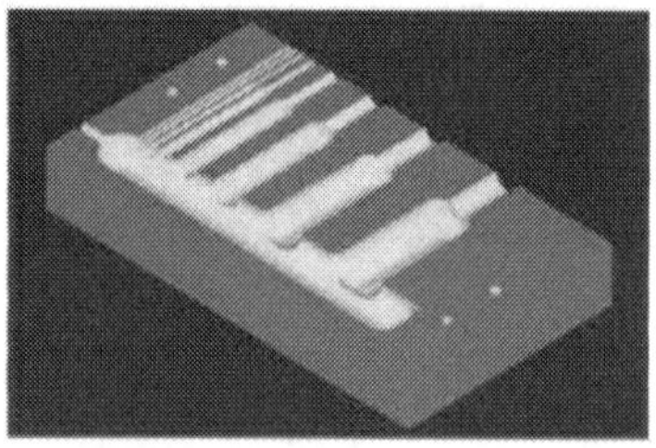

Channels of 0.3 -5 mm

Branched Architectures in Collagen Hydrogel

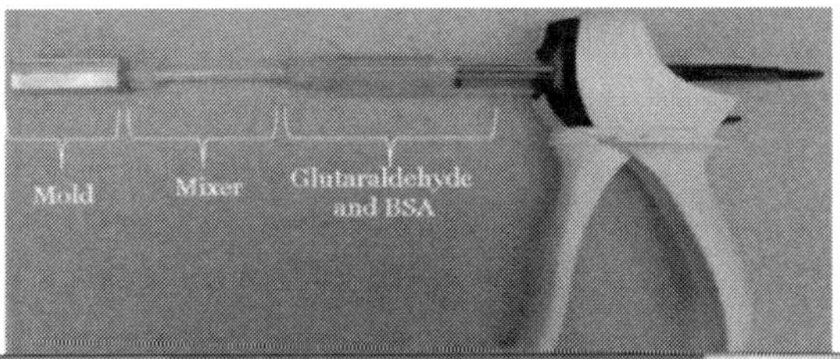

Future Work

- ❑3D Collagen Scaffolds
  - ❑BSA rubber on Renal Artery Molds
  - ❑Variety of tissue
  - ❑Elasticity of collagen construct
  - ❑Population of Cells

Acknowledgment

- Yost's Lab
  - Dr. Michael J. Yost
  - Heather Bainbridge
  - Matt Rhett, PhD
  - Jeffrey Garner
  - Sarah Grace Dennis
  - Dr. Bennett W. Calder
  - Dr. Stephen Fann
  - Alexander Wagner
  - Mary Morales
- Dr. John W. Weidner
- Dr. Gadala-Maria
  - Mohsen Nikkhoo
- Dr. Thomas Trusk
  - Advanced Tissue Biofabrication Center (ATBC)
- Funding
  - Alfred P. Sloan Foundation
  - South East Alliance for Graduate Education and the Professoriate (SEAGEP)
  - NIH-NIDCR IR01DE019355 (MJ Yost, PI)
  - NSF-EPSCoR(EPS-0903795)

384712 Role of Casting Solvent on Nanoparticle Dispersion in Polymer Nanocomposites

Abstract:

We investigate the influence of casting solvent on the final spatial distribution of nanoparticles (NPs) in polymer nanocomposites (PNCs). The nanocomposites formed from bare silica NPs and poly(2-vinylpyridine) (P2VP) were cast from two different solvents: methylethylketone (MEK) and pyridine, which are either theta or good solvents for both the NPs and the polymer. In MEK, we show that P2VP strongly adsorbs onto the silica surface, creating a temporally stable, bound polymer layer. The resulting “hairy” particles are sterically stabilized against agglomeration and thus good NPs dispersion is achieved at a low silica loading. With high silica content, due to the local bridging of silica NPs by P2VP chains, the composites demix into NP-rich and NP-lean phases. On the contrary, in pyridine, P2VP does not adsorb onto the silica surface. As a result, the phase behavior in such a case is governed by a subtle balance among electrostatic repulsion, Van der Waals attraction, polymer-induced depletion forces, and the kinetic slowdown of diffusion-limited NP aggregation. These initial, solvent-driven non-equilibrium agglomerated NP states are difficult to completely anneal away within a reasonable experimental time scale, especially for PNCs containing high molecular weight (MW) matrices. We further show that, in the case of using low MW matrices, the NP structure starts to approach its thermodynamic equilibrium state upon sufficient thermal annealing. These results emphasize the crucial role played by the casting solvent in the spatial dispersion of NPs in a polymer matrix.

382725 Improving the Dispersion of Nanoclays in PET/Clay Nanocomposites

Abstract:

Thermally stable surfactants were used to modify clay nanoparticles for use in preparation of poly(ethylene terephthalate)/clay nanocomposites. Nanoclays with smaller particle size distributions were prepared with a two-step centrifugation method that removed large particles from commercial montmorillonite (MMT). Scanning electron microscopy of aqueous dispersions of MMT and centrifuged clay (CMMT) illustrated that the average particle size of CMMT in water was much lower than that of MMT in water. Both CMMT and MMT were modified with two different thermally stable surfactants (imidazolium and phosphonium surfactants). PET/clay nanocomposites with surfactant modified MMT and CMMT were prepared by using a twin screw extruder. Fourier transform infrared spectroscopy and X-ray diffraction results revealed that the surfactants were associated with the clay surface and that the gallery spacing of the layered silicates was expanded. Thermal gravimetric analysis results indicated that clays modified with thermally stable surfactants degraded above 300 °C and could survive PET processing temperatures. Transmission electron microscopy revealed an improvement in the dispersion of centrifuged nanoclays (modified with both imidazolium and phosphonium surfactants) into the polymer matrix compared to non-centrifuged modified MMT with larger particle sizes. XRD indicated that particle size distributions have a significant effect on the dispersion of nanoclays modified with imidazolium surfactants. There was, however, a less important effect of centrifugation on the dispersion of nanoclays modified with phosphonium surfactants. DSC data also revealed that initial clay particle size distributions have a significant effect on the rate of crystallization of nanoclays modified with imidazolium surfactants.

385677 Nanoparticle Dispersion in Hard Porous Media As a Model of Composites

Abstract:

The transport of sub-micron particles in complex and highly confined porous media impacts processing of polymer nanocomposites. Specifically, attaining the optimal thermal, mechanical, electrical, and/or optical properties of polymer nanocomposites requires excellent control over the distribution of nanoparticles during processing in polymer solutions or melts. In the limit of strong confinement, in which the size of nanoparticles is comparable to typical length scales within the complex confining medium, the local mechanisms that influence nanoparticle transport remain poorly understood. Here, we use optical microscopy to probe the diffusive and transport properties of particles of size 200—400 nm in hard porous media that serve as models of fibers or other stiff components in a composite. We fabricate arrays of nanoposts that are arranged in a square lattice and employ both particle-tracking and differential dynamic microscopy to obtain information on the nanoparticle dynamics from microscopy movies. In both quiescent diffusion and in flow-driven transport through hard media, we generally find that the dynamics of the nanoparticles become increasingly slowed and stretched as the particles are more strongly confined within the post arrays. Under strong confinement, nanoparticles preferentially disperse along rather than normal to the direction of flow; this finding deviates from the dispersion behaviors observed for small solute molecules. These results have important implications for processing of nanocomposites, as the particle size will strongly affect the ability of nanomaterials to be uniformly dispersed in a composite matrix.

AIChE Annual Meeting 11/18/14

PRINCETON School of Engineering and Applied Science

Polymer-Grafted Nanoparticles as Patchy Colloids

Nathan A. Mahynski
Athanassios Z. Panagiotopoulos

Dept. of Chemical & Biological Engineering
Princeton University

---

1 AIChE Annual Meeting 11/18/14

"Nanoparticle Amphiphiles"

PRINCETON School of Engineering and Applied Science

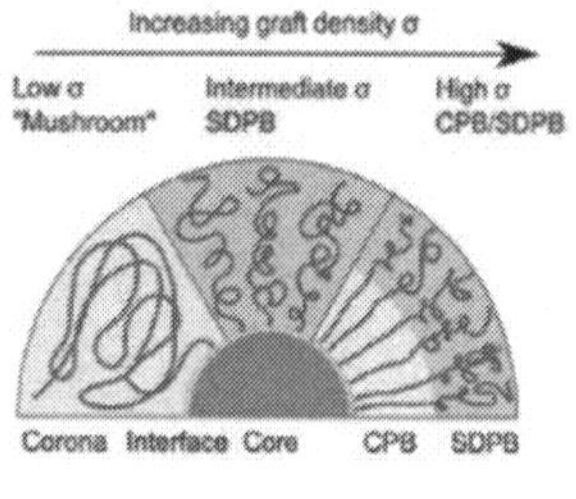

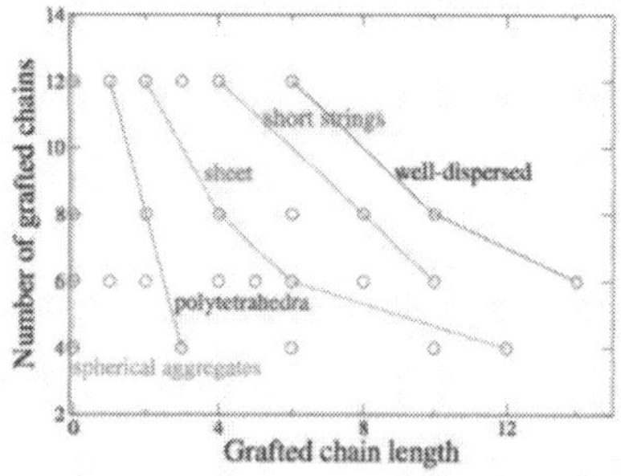

- Grafted Nanoparticles = inorganic core + organic polymer corona
- Dense shielding leads to good dispersion in organic matrices
- Sparse coronas lead to morphological changes usually attributed microphase separation

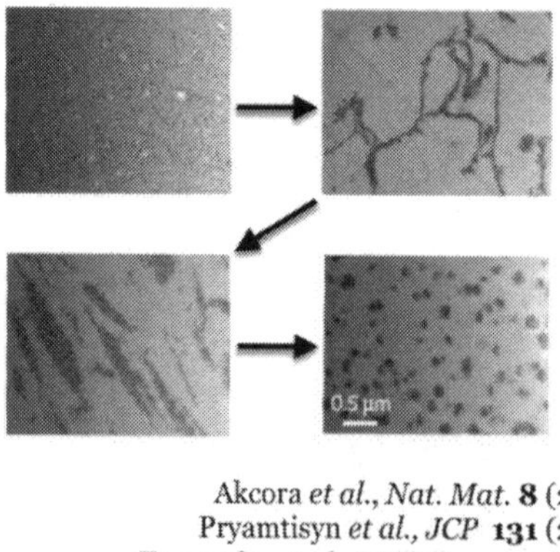

Akcora *et al.*, *Nat. Mat.* **8** (2009)
Pryamtisyn *et al.*, *JCP* **131** (2009)
Fernandes *et al.*, *MRS Comm.* **3** (2013)

11/18/14 AIChE Annual Meeting 2

vs. "Patchy Particles"

PRINCETON School of Engineering and Applied Science

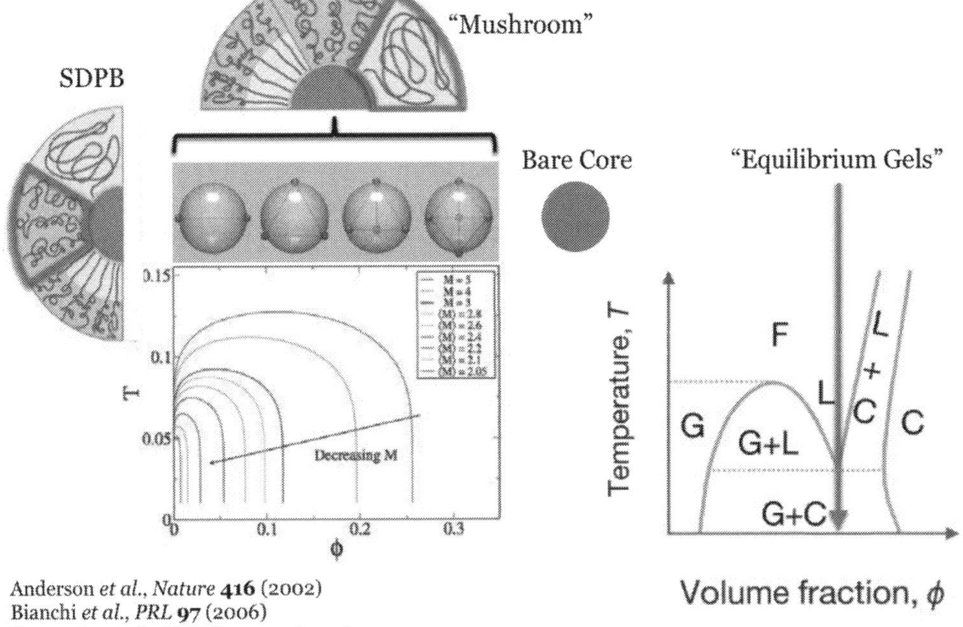

Anderson *et al.*, *Nature* **416** (2002)
Bianchi *et al.*, *PRL* **97** (2006)
Fernandes et *al.*, *MRS Comm.* **3** (2013)

11/18/14 AIChE Annual Meeting 3

Laponite Clay, "Empty Liquids"

PRINCETON School of Engineering and Applied Science

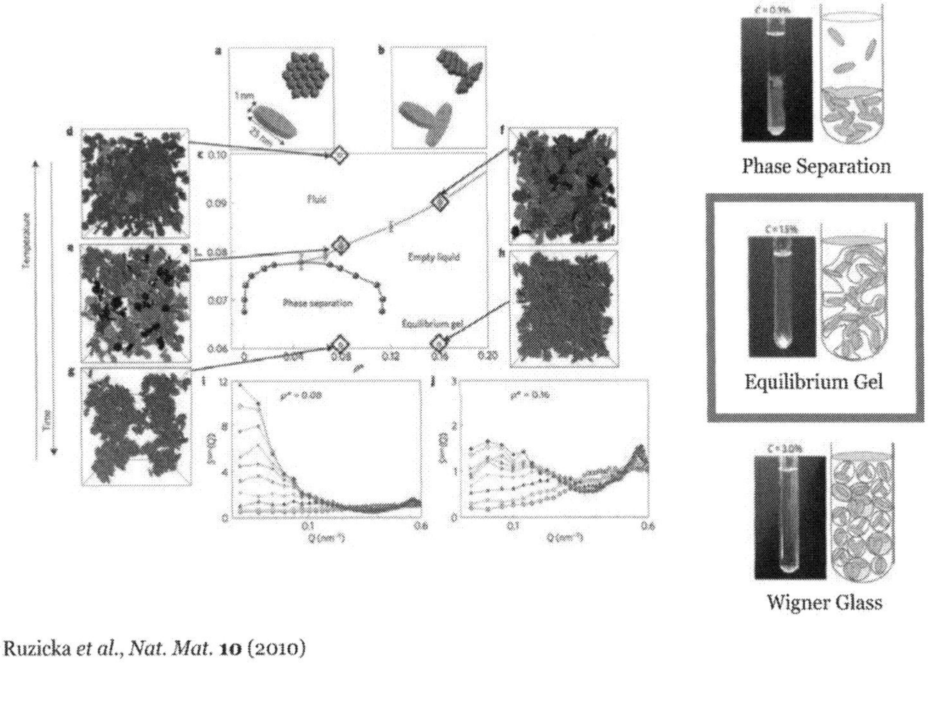

Ruzicka *et al.*, *Nat. Mat.* **10** (2010)

Which Paradigm is "Correct"?

PRINCETON School of Engineering and Applied Science

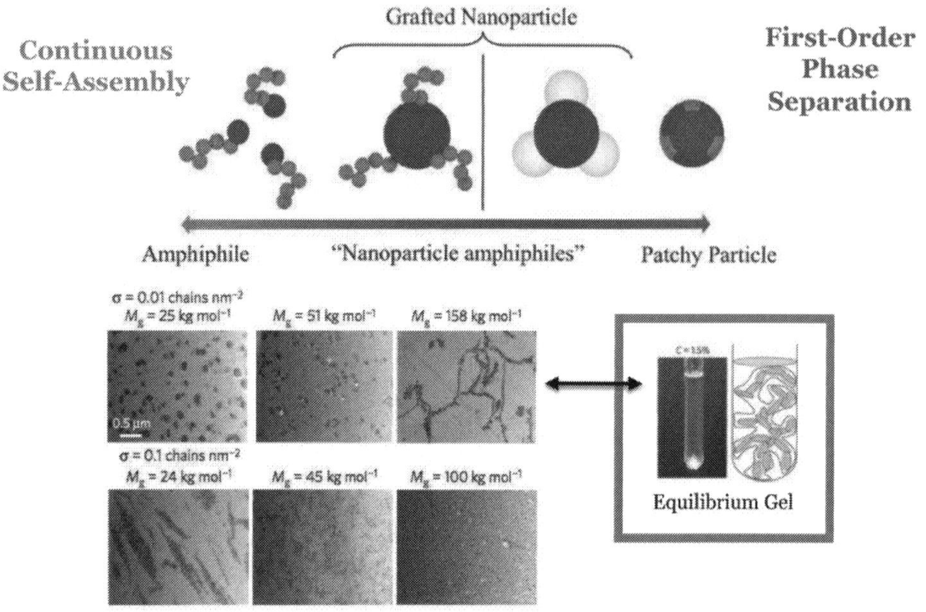

Akcora *et al., Nat. Mat.* **8** (2009)
Ruzicka *et al., Nat. Mat.* **10** (2010)

Simulation Technique

PRINCETON School of Engineering and Applied Science

- Lattice-based grand canonical Monte Carlo simulations
- Histogram-reweighting to identify binodals and mixed-field, finite-size to accurately identify critical points

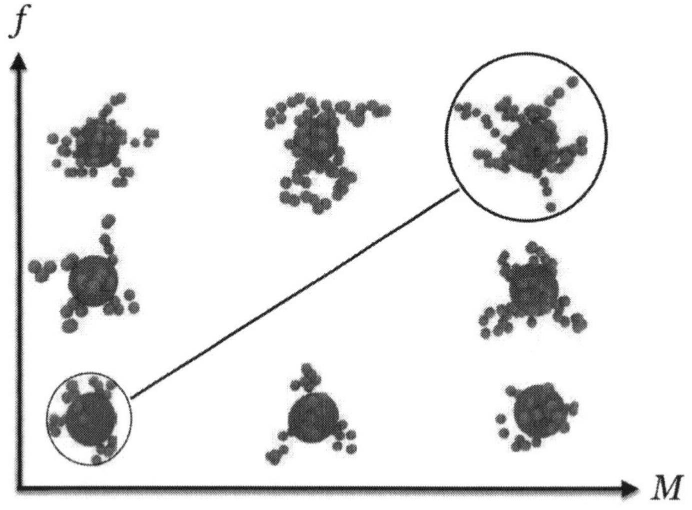

Phase and Morphology Diagram

PRINCETON School of Engineering and Applied Science

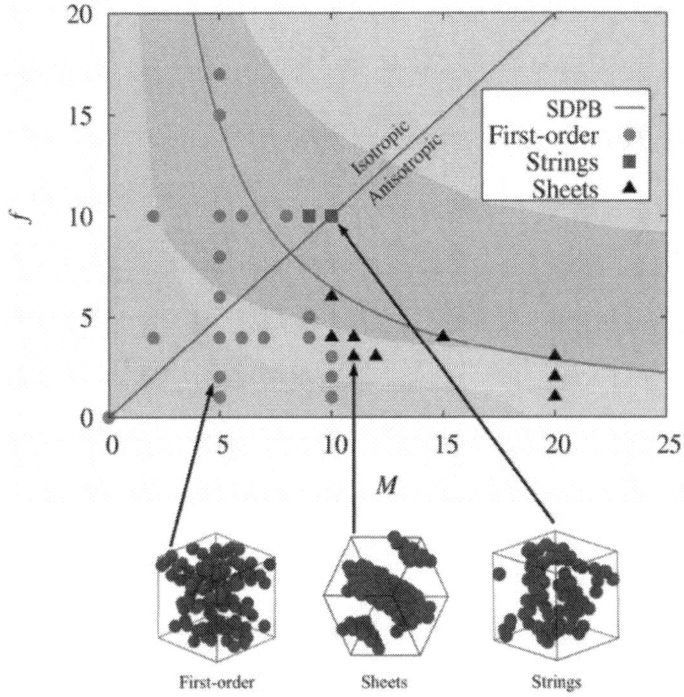

Isotropic Shielding

PRINCETON School of Engineering and Applied Science

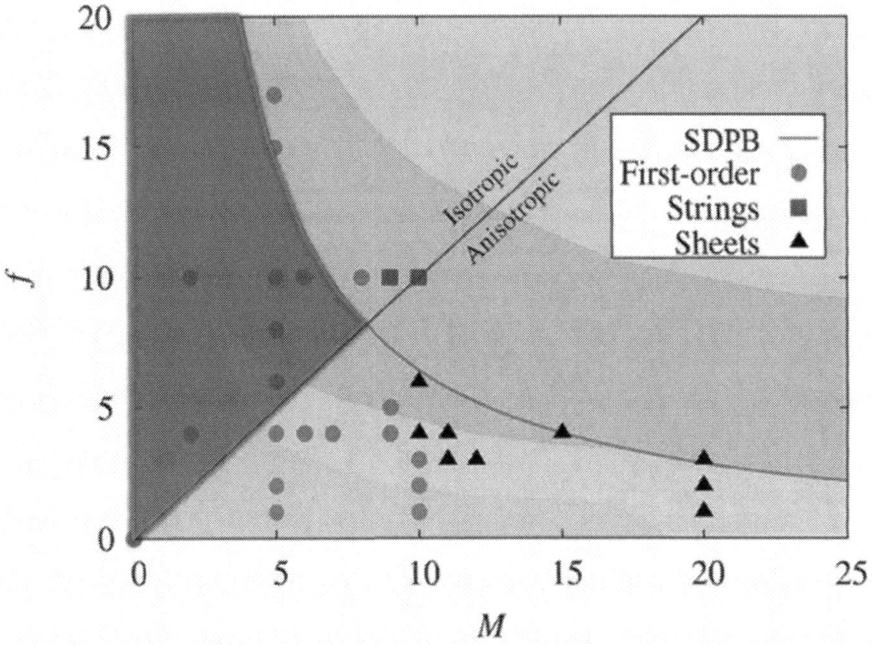

11/18/14

Isotropic Shielding

PRINCETON School of Engineering and Applied Science

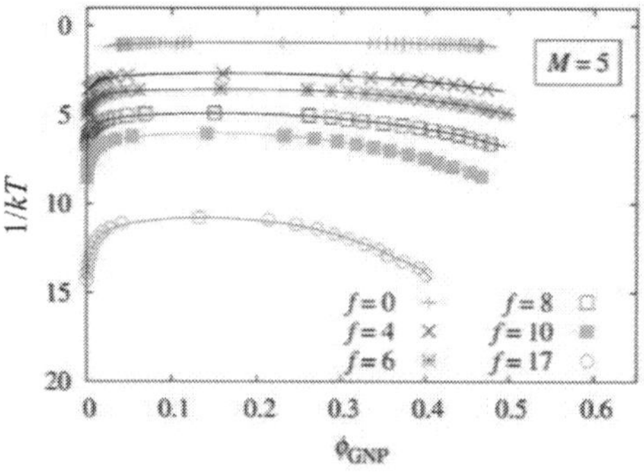

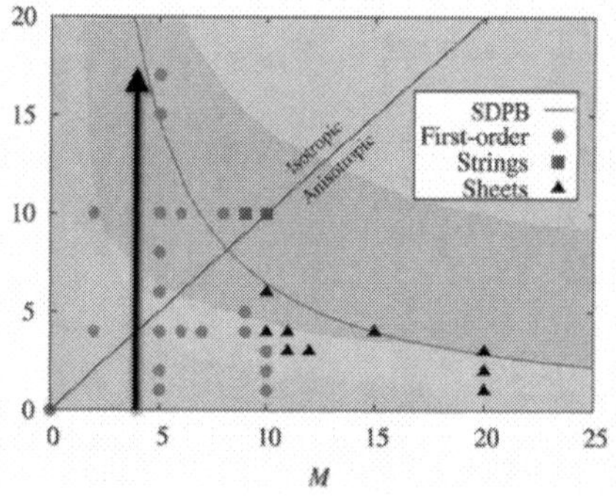

AIChE Annual Meeting 8

11/18/14

Anisotropic Shielding

PRINCETON School of Engineering and Applied Science

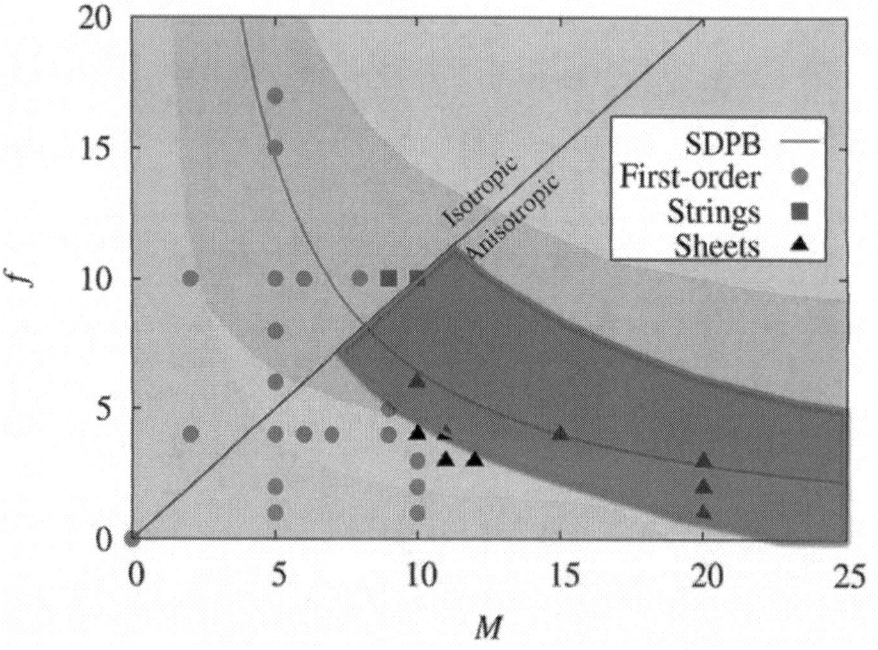

AIChE Annual Meeting 9

10 AIChE Annual Meeting 11/18/14

Anisotropic Shielding

PRINCETON School of Engineering and Applied Science

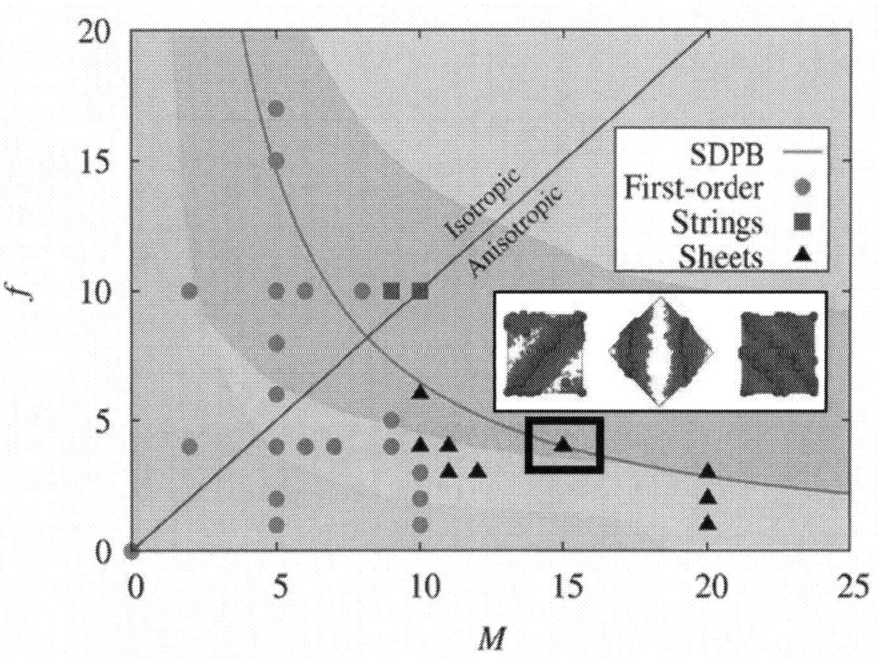

11 AIChE Annual Meeting 11/18/14

"Patchiness" in the Crossover?

PRINCETON School of Engineering and Applied Science

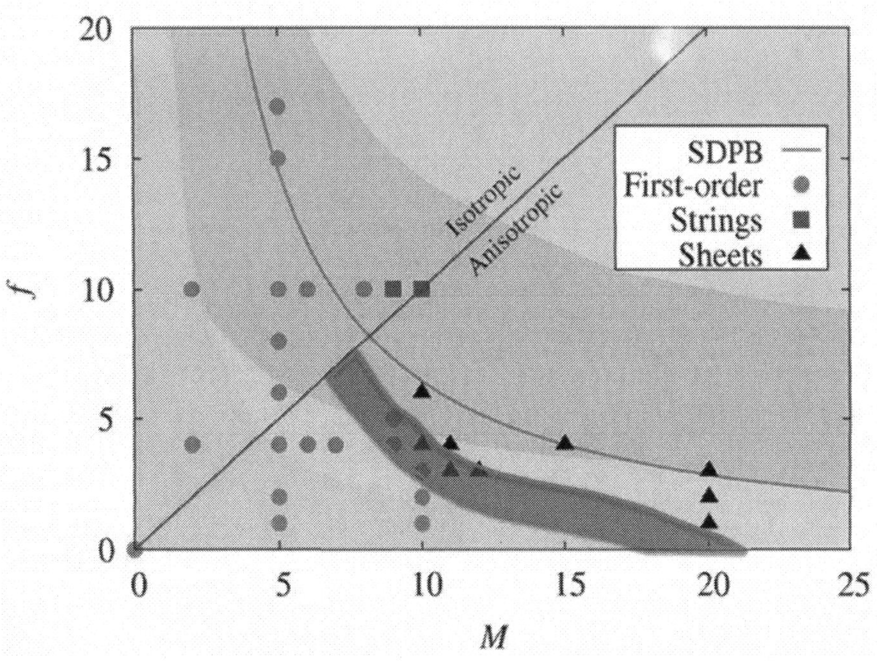

11/18/14 AIChE Annual Meeting 12

Patchy Behavior?

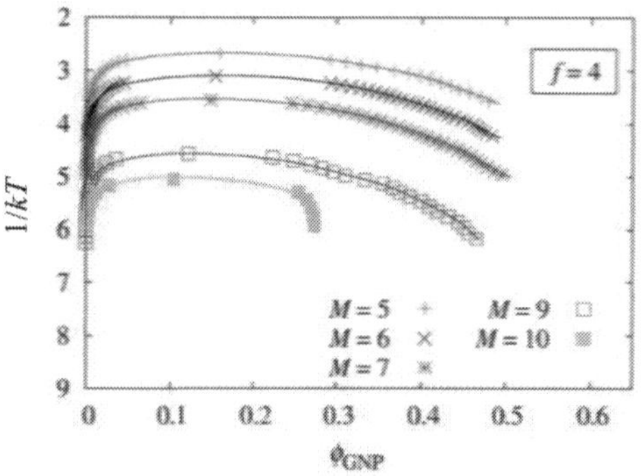

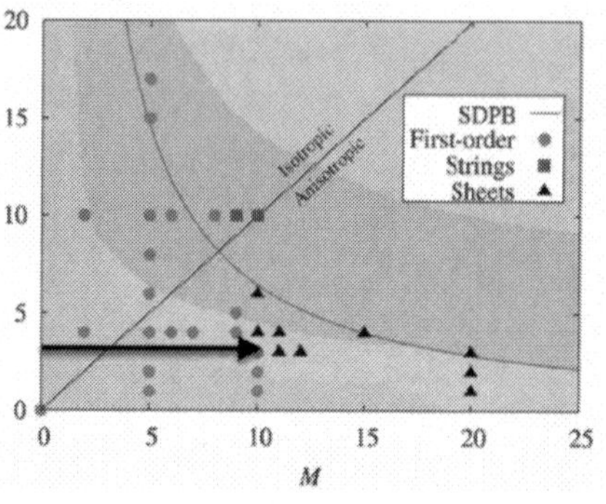

11/18/14 AIChE Annual Meeting 13

Not in the Thermodynamic Limit

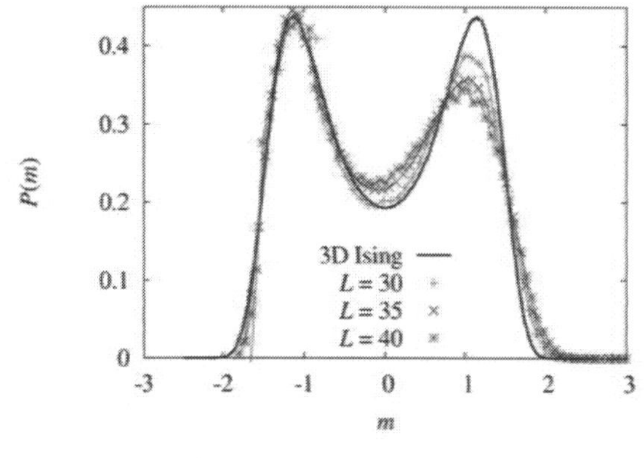

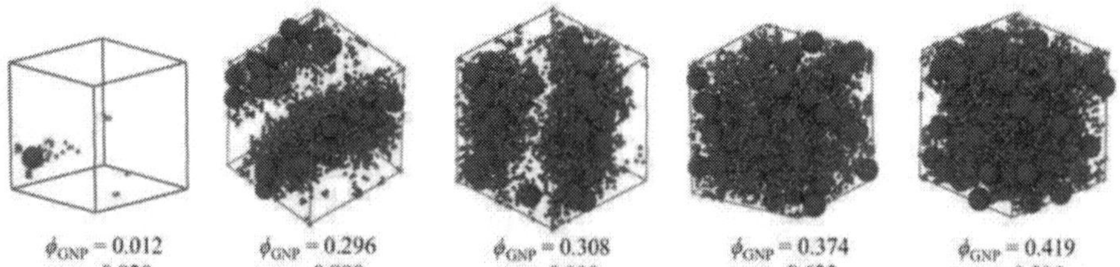

$\phi_{GNP} = 0.012$, $\mu = -0.920$ | $\phi_{GNP} = 0.296$, $\mu = -0.890$ | $\phi_{GNP} = 0.308$, $\mu = -0.800$ | $\phi_{GNP} = 0.374$, $\mu = -0.632$ | $\phi_{GNP} = 0.419$, $\mu = -0.395$

Not in the Thermodynamic Limit

PRINCETON School of Engineering and Applied Science

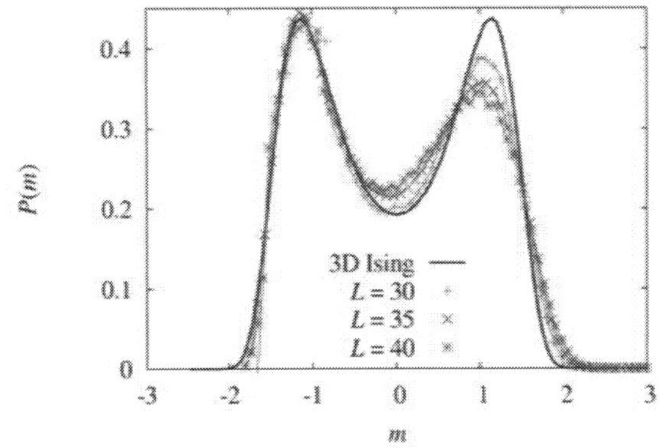

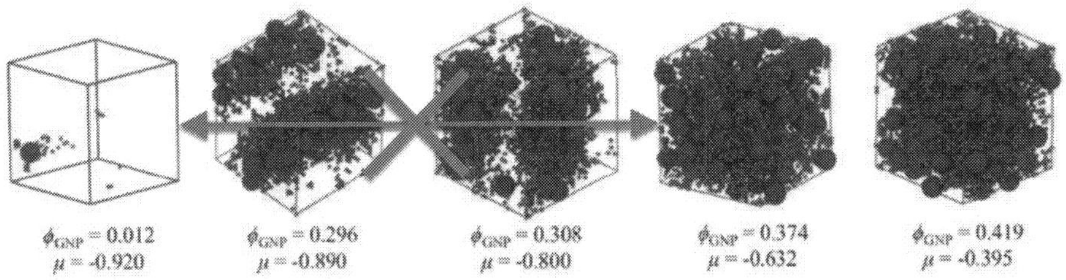

Approaching the "Crossover"

PRINCETON School of Engineering and Applied Science

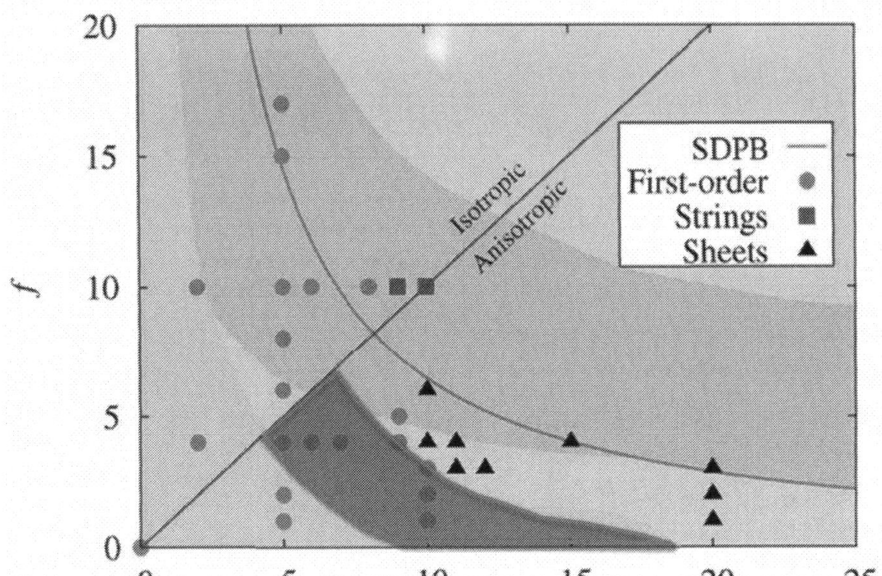

11/18/14

AIChE Annual Meeting

16

Competition Approaching the "Crossover"

PRINCETON School of Engineering and Applied Science

$f = 3$, $M = 10$, $L = 30$

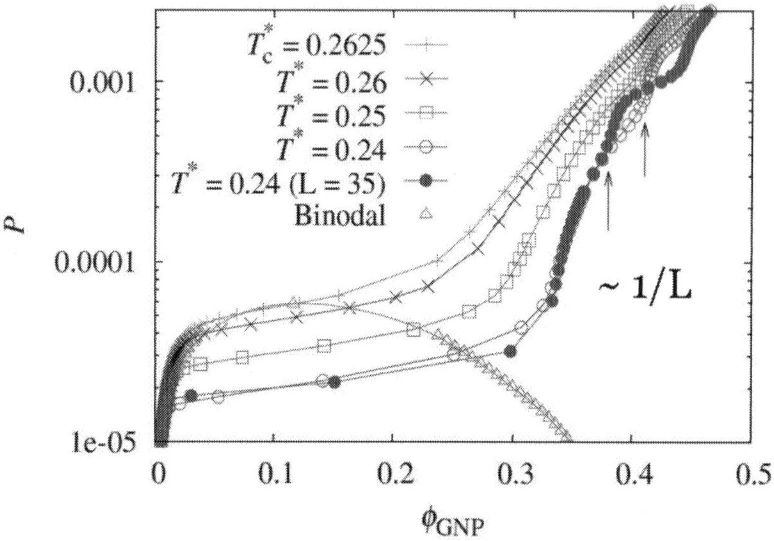

11/18/14

AIChE Annual Meeting

17

Competition Approaching the "Crossover"

PRINCETON School of Engineering and Applied Science

$f = 5$, $M = 9$, $L = 38$

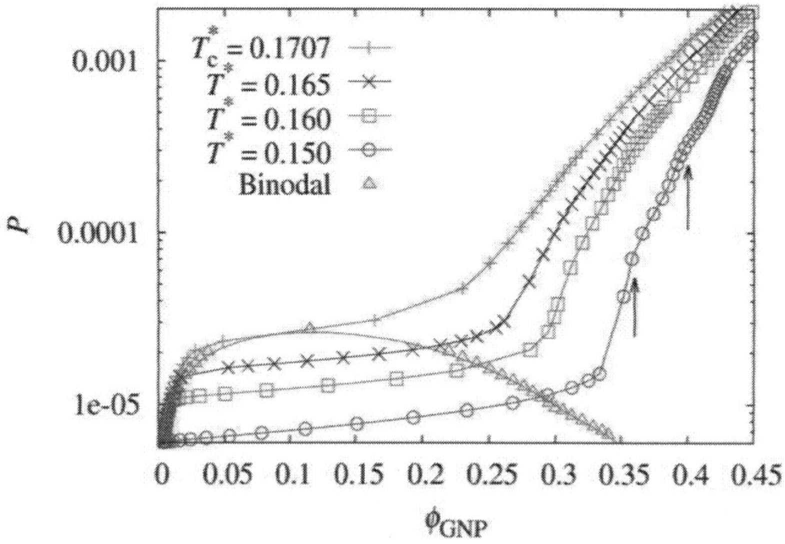

11/18/14 AIChE Annual Meeting 18

Fundamental Differences

PRINCETON School of Engineering and Applied Science

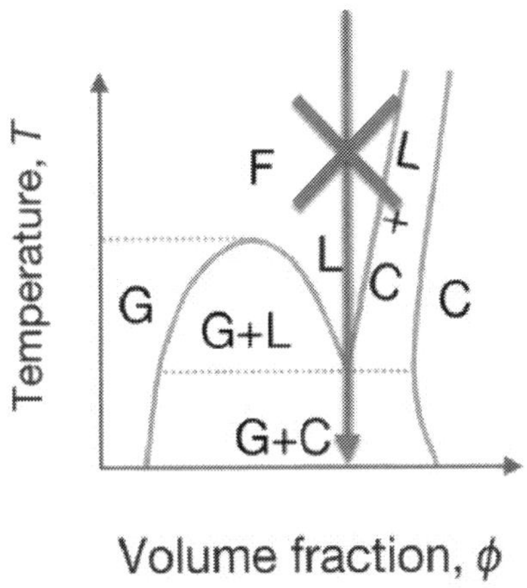

- No signs of a "collapsing binodal" in the thermodynamic limit as expected for patchy particles
- Fluid-fluid phase behavior seems to be bounded below, potentially be self-assembly (work is ongoing)

11/18/14 AIChE Annual Meeting 19

Conclusions

PRINCETON School of Engineering and Applied Science

- We explored (*f*, *M*) space for GNPs in terms of their the thermodynamic behavior
- "(ani-)Isotropic" shielding helps justify where phase separation and self-assembly each occur
- Near "patchy crossover" GNPs display both self-assembly and first-order phase separation behavior
- However, that competition does not result in thermodynamics reminiscent of "patchy particle" (*i.e.* equilibrium gels)

20 AIChE Annual Meeting 11/18/14

Acknowledgements

PRINCETON School of Engineering and Applied Science

Prof. A. Z. Panagiotopoulos

National Science Foundation

NSF

21 AIChE Annual Meeting 11/18/14

Questions?

PRINCETON School of Engineering and Applied Science

388696 Reinforced Magnetic Epoxy Nanocomposites with Conductive Polypyrrole Nanocoating on Nanomagnetite As a Coupling Agent

Abstract:

The new function of polypyrrole (PPy) to serve as a coupling agent has been demonstrated in preparing conductive epoxy resin nanocomposites with PPy coating on magnetite (f-Fe_3O_4) nanoparticles. Compared with pure epoxy suspension, the viscosity of epoxy nanosuspensions with 5.0 and 10.0 wt% f-Fe_3O_4 nanoparticles was almost the same. The glass transition temperature (T_g) of cured solid epoxy polymer nanocomposites (PNCs) with 30.0 wt% f-Fe_3O_4 nanoparticles was 22.3 °C higher than that of cured pure epoxy (88.2 °C) in the dynamic mechanical analysis (DMA).[1] At 50 wt% weight loss, the decomposition temperature of the cured epoxy PNCs with 30.0 wt% f-Fe_3O_4 nanoparticle was delayed from 421.5 (cured pure epoxy) to 483.4 °C. Compared with the tensile strength of 83.0 MPa for the cured pure epoxy,[2] the enhanced tensile strength of 91.7 MPa observed in the cured epoxy PNCs with 5.0 wt% f-Fe_3O_4 nanoparticles is attributed to the introduced interfacial interaction between the nanoparticles and the host epoxy matrix, through the PPy nanocoating, as observed in the fracture surface. The volume resistivity of the cured epoxy PNCs with 30.0 wt% f-Fe_3O_4 nanoparticles decreased almost 7 orders of magnitude compared with the cured pure epoxy (1.6×10^{13} μΩ cm). The cured epoxy PNCs exhibited good magnetic properties; the surface functionality and epoxy polymer had little effect on the magnetic moment of the Fe_3O_4 nanoparticles. The cured epoxy PNCs with f-Fe_3O_4nanoparticles showed positive permittivity, and the permittivity increased with increasing particle loading. The role of PPy nanocoating on the nanocomposite formation mechanism was investigated by using the FT-IR and TGA tests.

(1) Gu, H.; Guo, J.; He, Q.; Tadakamalla, S.; Zhang, X.; Yan, X.; Huang, Y.; Colorado, H. A.; Wei, S.; Guo, Z.: Flame-retardant epoxy resin nanocomposites reinforced with polyaniline-stabilized silica nanoparticles. *Industrial & Engineering Chemistry Research* **2013**, *52*, 7718-7728.

(2) Gu, H.; Tadakamalla, S.; Zhang, X.; Huang, Y.; Jiang, Y.; Colorado, H. A.; Luo, Z.; Wei, S.; Guo, Z.: Epoxy resin nanosuspensions and reinforced nanocomposites from polyaniline stabilized multi-walled carbon nanotubes. *Journal of Materials Chemistry C* **2013**, *1*, 729-743.

378723 Preparation and Characterization of SF-PLGA Film By TIPS Method

Abstract:

Besides excellent biocompatibility, a useful tissue engineering scaffold should provide suitable micropores and fibrous structure to mimic the natural extracellular matrix, inducing desired cellular activities and guiding tissue regeneration. Silk fibroin and PLGA have been widely used in tissue engineering because of excellent biocompatibility and mechanical property. In this work, a porous and fibrous silk fibroin (SF)-PLGA [poly(lactic-*co*-glycolic acid)] scaffold prepared by thermally induced phase separation (TIPS) is described.

Firstly, to study the effects of volume ratio of hexafluoro-2-propanol (HFIP)/water (v/v), concentration of (SF+PLGA)/solvent (w/v), weight ratio of SF/PLGA (w/w) and phase separation temperature (PST) on the pore size and porosity of SF/PLGA scaffolds, single-factor experiments were performed to determine suitable ranges of the four factors. Based on that, a 2^4 factorial experiment was used to analyze the effect of these parameters on the pore size and porosity. The results showed that the porous and fibrous scaffold with suitable pore size (25.53±2.18 μm) and high porosity (84.83±9.05%) could be made by the TIPS method when HFIP/water volume ratio is 2.25/1 (v/v), concentration of (SF+PLGA)/solvent (w/v) is 0.09 g/mL, weight ratio of SF/PLGA is 5/8 (w/w) and phase separation temperature is -50°C. SEM photographs showed the microfibrous and nanofibrous structures could be made successfully. Solvent residue analysis showed that the HFIP residue was 300 ppm, which was proved to be no cytotoxicity. FTIR and XRPD results showed that the structure of SF changed after TIPS process.

Secondly, based on the results of 2^4 factorial experiments, scaffolds with different weight ratios of SF/PLGA were studied to optimize materials with better properties for skin dressings. Mechanical properties showed that the range of tensile strength is 1035±306 to 75±21 kPa, the tensile strain range is 34.34±13.62~119.10±16.82%. The results of water vapor permeability, water absorption rate, and water loss rate showed that the optimal weight ratio of SF/PLGA is 8/8 (Level: 1), producing a material that can absorb wound skin exudate to prevent infection and maintain wound humidity. The water vapor transmission rate is 5591.88±56.70 g/m^2/day, the maximum equilibrium water absorption rate is 72.2%.

Finally, the cytocompatibility of scaffold on HESF cells was studied. The results showed that the scaffolds have good biocompatibilities and no cellular toxicity through relative growth rate, AO/EB staining and cell apoptosis rate. Furthermore, cell adhesion and proliferation results suggested that the optimal weight ratio of SF/PLGA was 8/8 (Level: 1), which resulted in highest cell adhesion and proliferation.

Acknowledgements: Financial supports from NSFC (31000441 and 31170939), Natural Science Foundation of Fujian Province (2013J01189), Program for Prominent Young Talents in Fujian Province University (JA12004), China Postdoctoral Science Foundation (2014M551833), the Scientific Research Foundation for the Returned Overseas Chinese Scholars (State Education Ministry) and the Promotion Program for Young and Middle-aged Teacher in Science and Technology Research of Huaqiao University (ZQN-PY108) are gratefully acknowledged.

Organometal Halide Perovskite Solar Cells Featuring Inorganic Hole Conductors

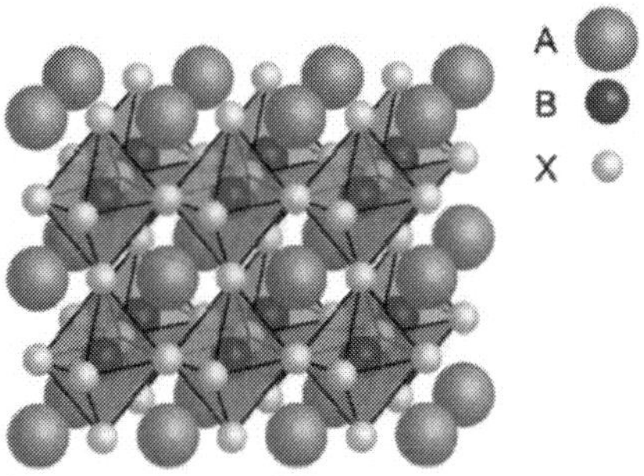

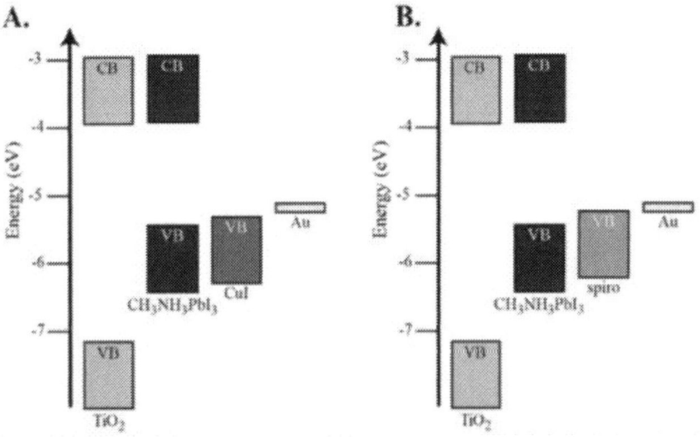

2014 AIChE Annual Meeting

Jeffrey A. Christians, and Prashant V. Kamat
11/18/2014

Dept. of Chemical and Biomolecular Engr.
Radiation Laboratory
University of Notre Dame

What is Perovskite?

- Prototypical compound is $CH_3NH_3PbI_3$
- Hybrid organic-inorganic materials
- Adopt a perovskite (ABX_3) crystal structure

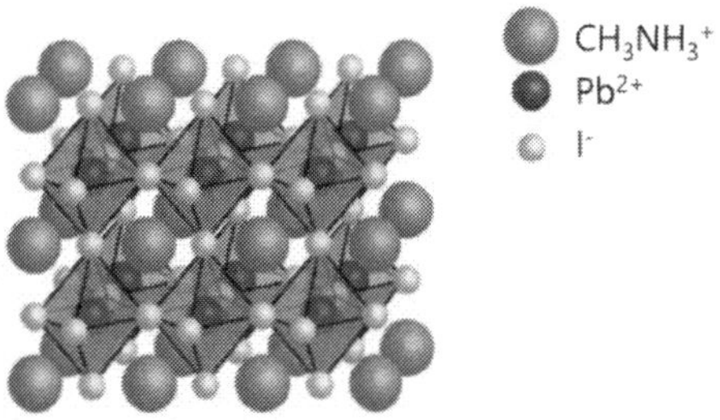

Advantages For Solar Cells

- Low temperature processing (<100 °C)
- Earth abundant materials
- High visible light absorption
- Long electron and hole diffusion lengths

2

Outline

- Motivation
- Solar Cell Fabrication
- Photovoltaic Efficiency
- Impedance Spectroscopy
- Conclusions

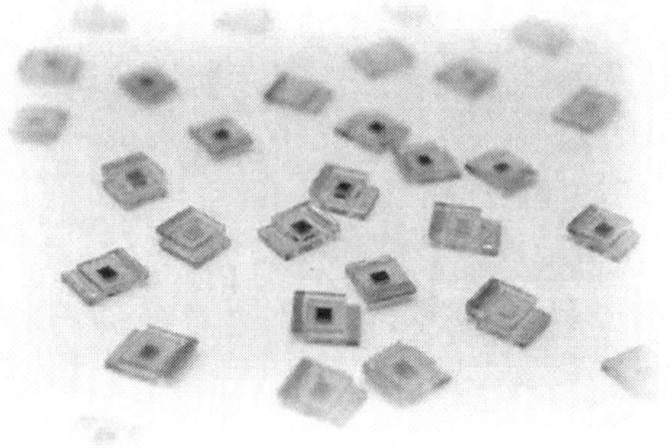

3

Design of Perovskite Solar Cells

Device Components

1. Optically Transparent Electrode
2. Electron Transport Material
3. ~~Perovskite Absorber~~
4. Hole Transport Material (HTM)
5. ~~Gold Counter Electrode~~

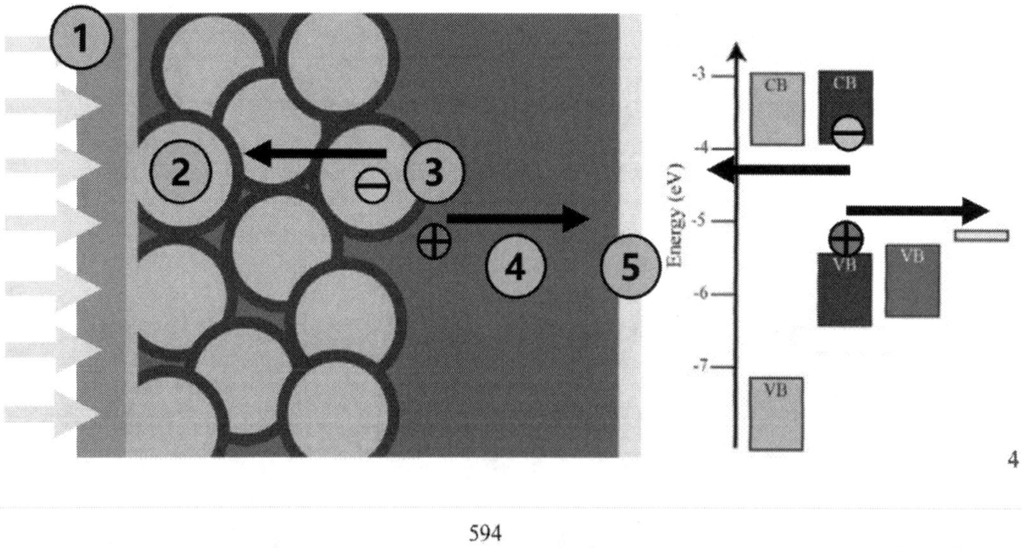

4

Employing Inorganic HTMs

- Record efficiency cells feature organic Hole Transport Materials (HTMs)
  - Expensive, lack stability
  - Sprio-OMeTAD market price > $200/g
- Inorganic HTMs → better stability and lower cost

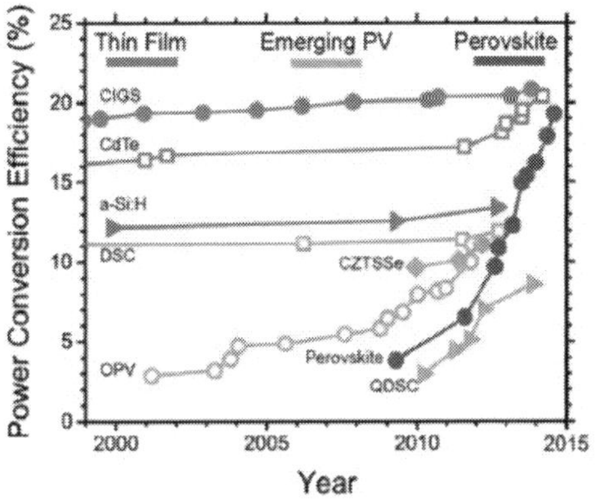

NREL, Research Cell Efficiency Records. url: *http://www.nrel.gov/ncpv/images/efficiency_chart.jpg*

5

Design Criteria for HTMs

- Solution processable
- Earth abundant, non-toxic
- Favorable valence band offset

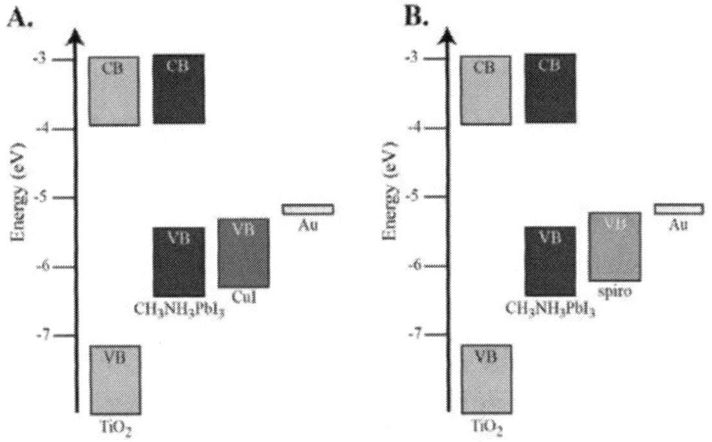

6

Solution-Processed CuI Films

- Deposit CuI by an automated drop casting technique
- 0.1 M CuI in 39:1 chlorobenzene:di-n-propyl sulfide by volume

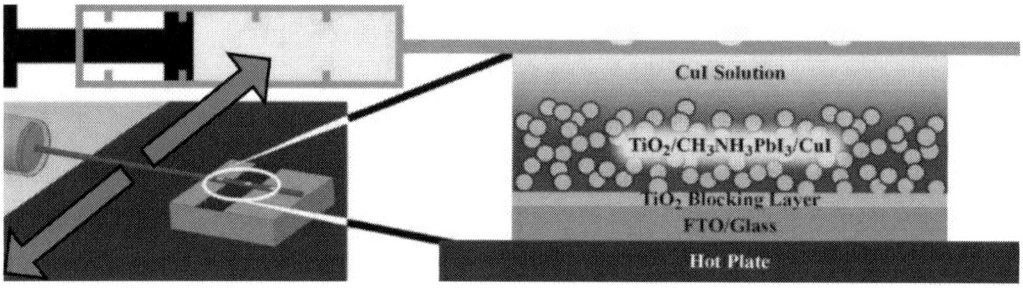

J. A. Christians; R. C. M. Fung, P. V. Kamat, *J. Am. Chem. Soc.* **2014**, 136, 758-764.

7

Solution-Processed CuI Films

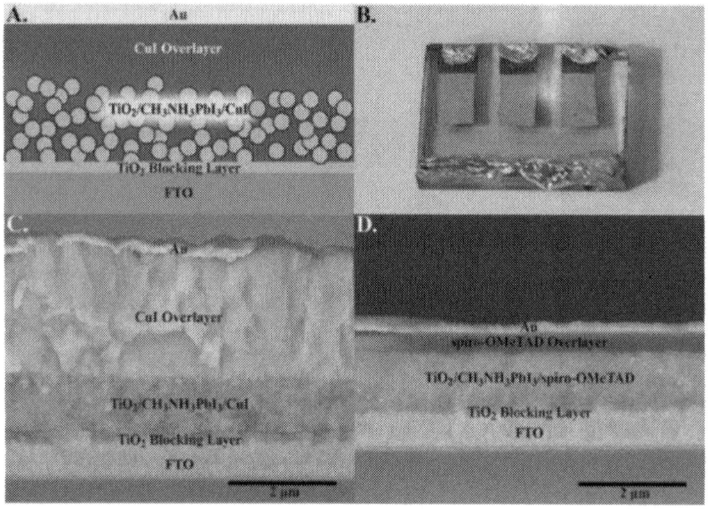

- Completed device architecture
  - $FTO/TiO_2/CH_3NH_3PbI_3/CuI/Au$

J. A. Christians; R. C. M. Fung, P. V. Kamat, *J. Am. Chem. Soc.* **2014**, 136, 758-764.

8

TiO2/CH3NH3PbI3/CuI Solar Cells

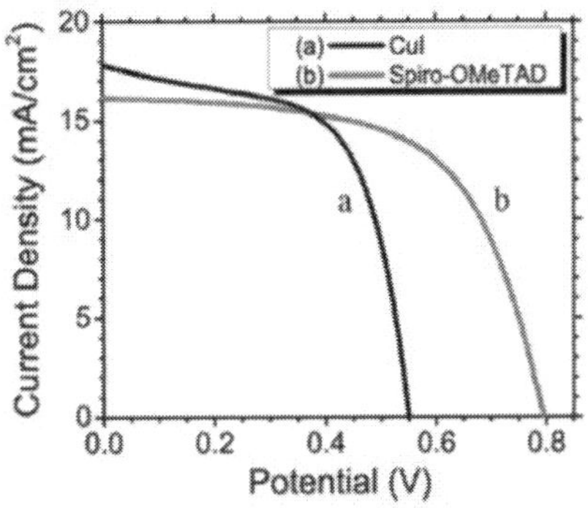

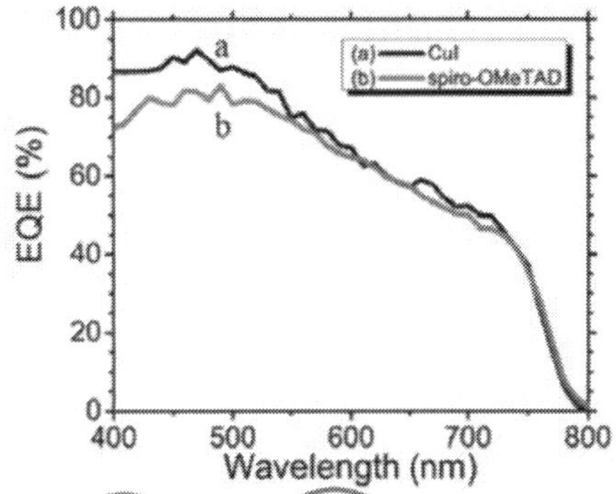

Hole Conductor		J_{SC} (mA/cm^2)	V_{OC} (V)	FF	η (%)
CuI	**Champion**	**17.8**	**0.55**	**0.62**	**6.0**
	Average$^{\alpha}$	12.1 ± 3.3	0.52 ± 0.06	0.60 ± 0.07	3.7 ± 1.1
	Maximum$^{\beta}$	18.9	0.62	0.71	--
Spiro-OMeTAD	**Champion**	**16.1**	**0.79**	**0.61**	**7.9**
	Average$^{\alpha}$	14.0 ± 1.7	0.78 ± 0.02	0.55 ± 0.07	6.0 ± 1.1
	Maximum$^{\beta}$	17.0	0.82	0.65	--

- Lower V_{OC} (avg. 0.52V vs. 0.78V)
- Higher fill factors (avg. 0.60 vs. 0.55)

J. A. Christians; R. C. M. Fung, P. V. Kamat, *J. Am. Chem. Soc.* **2014**, 136, 758-764.

9

Impedance Spectroscopy Analysis

- Sinusoidal voltage perturbation and measures current response
- Fit response to equivalent circuit model

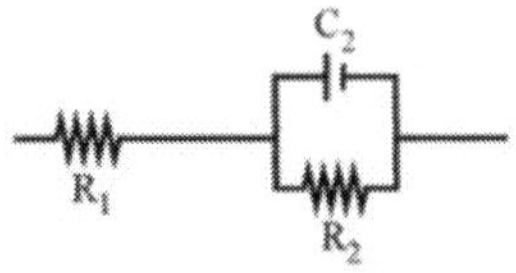

$Z(\omega) = I(\omega)/V(\omega)$

$Z = R_1 + R_2/(1+j\omega\tau_2)$

$Z = Z' + jZ''$

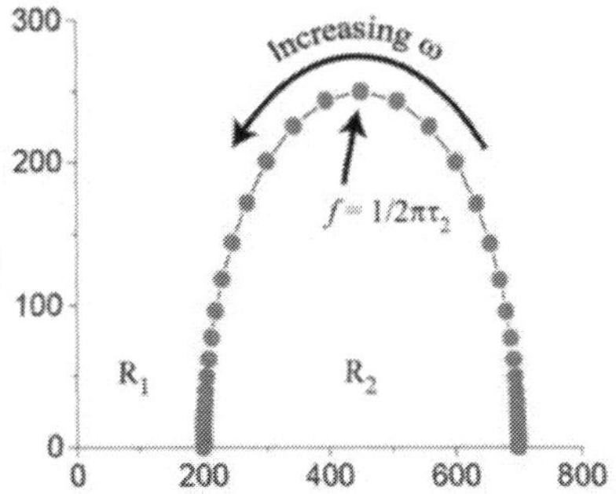

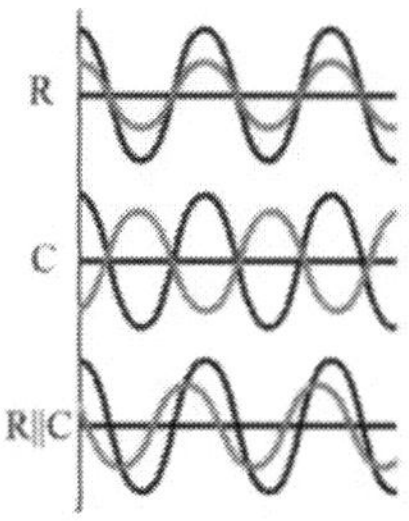

10

IS Spectra Features

(i) Large low frequency arc

(ii) Small high frequency arc (almost nonexistent with CuI

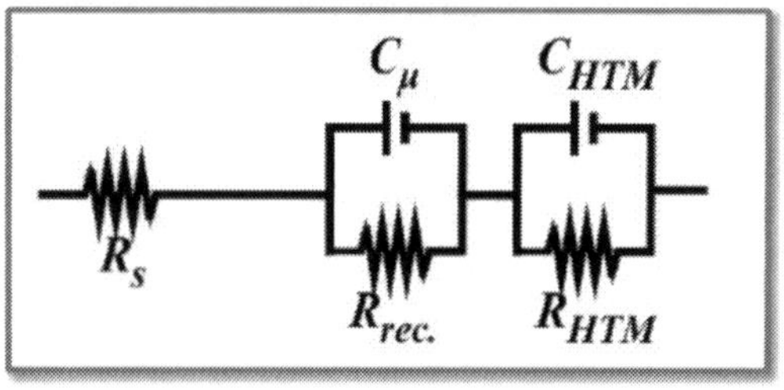

- $R_{rec.}$ = recombination resistance
- R_{HTM} = HTM resistance

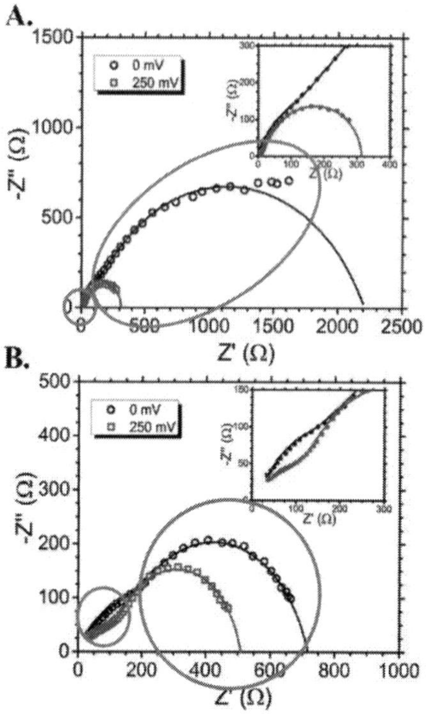

J. A. Christians; R. C. M. Fung, P. V. Kamat, *J. Am. Chem. Soc.* **2014**, 136, 758-764.

Impedance Spectroscopy Analysis

- Monitor equivalent circuit parameters with increasing voltage

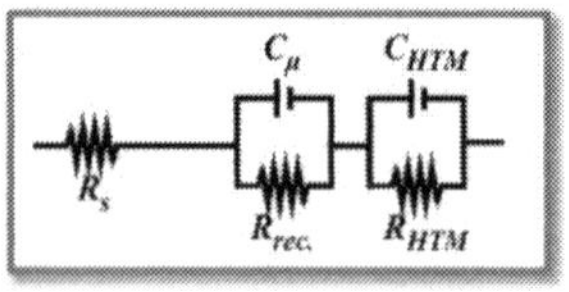

(i) Higher recombination at Perovskite/CuI interface

(ii) CuI displays ~ 2 orders of magnitude higher conductivity, σ_{HTM}

$$\sigma_{HTM} = L/R_{HTM}$$

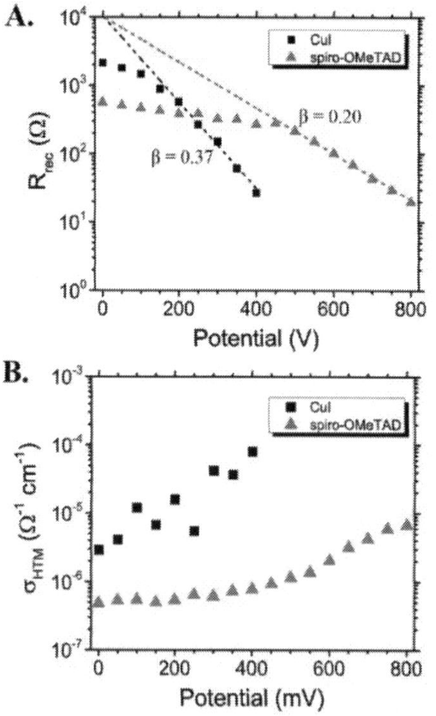

J. A. Christians; R. C. M. Fung, P. V. Kamat, *J. Am. Chem. Soc.* **2014**, 136, 758-764.

CuI Cell Stability

- Photocurrent stability
  - CuI-based > spiro-OMeTAD
- Efficiency decreased just 31% following unencapsulated storage for 1 year in ambient conditions

	J_{SC} (mA/cm^2)	V_{OC} (V)	FF	η (%)
Initial	17.8	0.55	0.62	6.0
1 year later	10.3	0.63	0.65	4.1

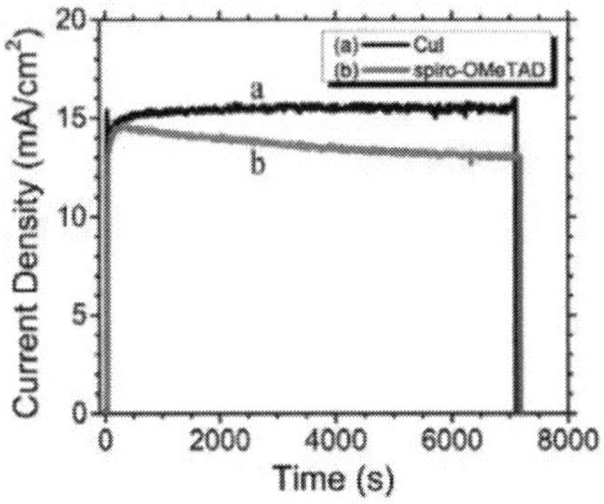

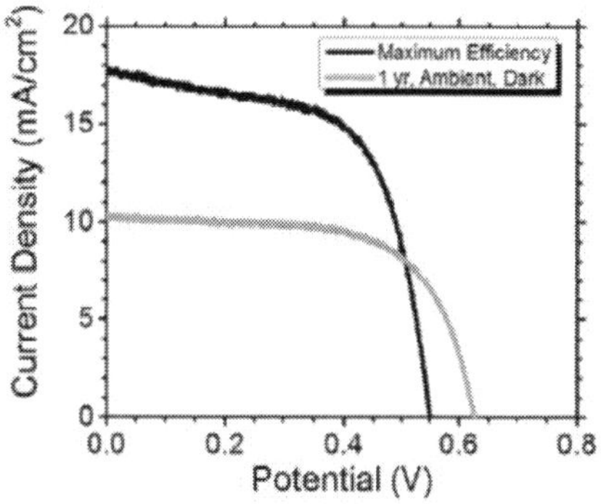

J. A. Christians; R. C. M. Fung, P. V. Kamat, *J. Am. Chem. Soc.* **2014**, 136, 758-764.

13

Conclusions

- CuI represents a promising new direction for perovskite solar cells – inorganic HTMs

Simple Fabrication
- Low-cost, solution-based
- Earth abundant materials
- Easily scalable

High Stability
- Stable photocurrent
- 31% decrease in performance over 1 year

Impedance Spectroscopy
- Low V_{OC} due to high recombination
- CuI shows very high p-type conductivity

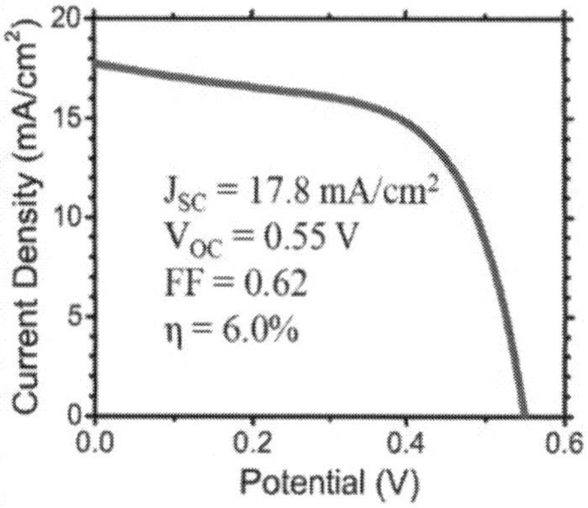

14

Acknowledgements

- Dr. Prashant Kamat
- Raymond Fung
- Kamat Lab Group Members
- Radiation Laboratory Staff

See more of our group's latest research and connect with us on the web.

Online: **KamatLab.com**
Facebook: **facebook.com/kamatlab**
Twitter: **@KamatLabND**
YouTube: **Kamat Lab**

Funding Support

- U.S. Department of Energy Office of Basic Energy Sciences
- Jana & Patrick Eilers Energy Research Fellowship
- CEST/Bayer Predoctoral Research Fellowship

15

389944 In Situ Formation of Submicrometer Pore Scattering Layer for Dye-Sensitized Solar Cells

Abstract:

In situ formation of submicrometer pore scattering layer for dye-sensitized solar cells

Su-Jin Ha, Jun Hyuk Moon

Abstract

Dye-sensitized solar cells (DSCs) are a promising substitute for conventional silicon solar cells. A scattering layer of submicrometer pores or particles has been widely introduced to achieve a high light-harvesting efficiency. However, many such fabrication processes require high temperatures and multiple steps to prepare the scattering layer. Here, a new fabrication approach of submicrometer pores scattering layer was demonstrated for dye-sensitized solar cells(DSSCs). The macropore was formed by the dissolution of polystyrene sphere in polystyrene/TiO_2 composite layer. Specifically, the in-situ macroporous scattering layer decreased the transmittance of the electrode film from 92% to below 1%. The DSCs using these macroporous scattering layers exhibited an increase in the efficiency of 22%. We believe that in-situ porous scattering layers show great promise for next-generation flexible DSCs.

All-Polymer Solar Cells: Structure-Property-Performance Relationships of New *n*-Type Polymer Semiconductors

Ye-Jin Hwang, Taeshik Earmme, Samson A. Jenekhe
University of Washington

To overcome the small photovoltage, high cost, poor photochemical stability, and other limitations of fullerene-based organic photovoltaics, there has recently been a remarkable growth of interest in developing new electron accepting (*n*-type) polymer semiconductors as replacements for fullerene-based acceptor materials. *n*-Type semiconducting polymers can provide better synthetic flexibility in tuning the electronic structure and optical, charge transport, and photovoltaic properties as well as the environmental stability and durability of the bulk heterojunction solar cells. However, the performance of all-polymer solar cells is still inferior compared to fullerene based solar cells. The major factors that limit the performance of all-polymer photovoltaic devices are the lack of suitable *n*-type semiconducting polymers, low photocurrent, and lack of approaches to optimize polymer/polymer blend morphology. To overcome these challenges, we have designed and synthesized a series of new naphthalene diimide (NDI)-selenophene based acceptor copolymers and investigated their optical absorption, electrochemical redox electronic structure, crystalline ordering, and charge transport properties.[1, 2] We investigated the photovoltaic properties and morphology of all-polymer blend solar cells composed of the new acceptor copolymers as the acceptor and a poly(3-hexylthiophene) (P3HT) or a thiazolothiazole-dithienosilole copolymer (PSEHTT) as the donor. The results reveal useful structure-property-performance relationships and fullerene-free all-polymer blend photovoltaic cells with power conversion efficiency of over 4 %.

1. Earmme, T.; Hwang, Y. -J.; Murari, N. M.; Subramaniyan, S.; S. A. Jenekhe, *J. Am. Chem. Soc.* **2013**, *135*, 14960.
2. Hwang, Y. -J.; Murari, N. M.; S. A. Jenekhe, *Polym. Chem.* **2013**, *4*, 3187.

High-efficiency graphene-based flexible organic solar cells

Sehoon Chang[1], Hyesung Park[2], Xiang Zhou[1], Jing Kong[2], Tomás Palacios[2], Silvija Gradečak[1]

[1]Department of Materials Science and Engineering, Massachusetts Institute of Technology, Cambridge, Massachusetts 02139

[2]Department of Electrical Engineering and Computer Science, Massachusetts Institute of Technology, Cambridge, Massachusetts 02139

Abstract

Flexible solar cell is one of the promising next generation optoelectronic devices, where the electrode materials need to be developed with good conductivity, transparency, and flexibility. Graphene arose as a promising flexible transparent electrode due to its good electrical conductivity and optical transparency along with mechanical and chemical robustness, and potentially low-cost processing. In this report, we have successfully designed both graphene anode- and cathode- based flexible polymer solar cells with thieno[3,4-b]thiophene/benzodithiophene (PTB7) and [6,6]-phenyl C71-butyric acid methyl ester (PC71BM) by resolving the issues occurring at the interface between graphene and charge transporting semiconducting materials such as molybdenum trioxide (MoO_3) or ZnO. We demonstrate high- efficiency graphene electrode-based flexible devices with power conversion efficiencies of 6.1% (anode) and 7.1% (cathode), respectively, which were achieved via thermal treatment of MoO_3 electron blocking layer and direct deposition of ZnO electron transporting layer on graphene surface. The advances accomplished in this work demonstrate graphene has great potential as a promising alternative transparent electrode system in a variety of flexible optoelectronic devices.

380288 Nanostructured Plasmonic Cathodes for Enhanced Performance of Organic Solar Cells

Abstract:

Organic solar cells (OSCs) offer significant potential to lower the manufacturing costs of solar energy harvesting devices and enable large area, lightweight, and highly flexible photovoltaics. A central challenge for OSCs is the trade-off between an active layer of sufficient film thickness to absorb as much light as possible while minimizing charge recombination. This balance between absorption and recombination typically results in using thin active layers that do not take full advantage of the incident light. Developing light manipulation techniques to maximize absorption is a key component for pushing OSCs to their peak potential. Integrating plasmonic nanostructures into these devices is one promising way to enhance absorption and is typically achieved by deploying Au or Ag plasmonic nanoparticles in various layers of an OSC. Additionally, replacing ITO as an electrode is desirable due to the high cost and limited supply of indium, as well as the lack of mechanical robustness and flexibility of ITO itself. In this work, we designed and fabricated plasmonic nanohole arrays to serve as the cathode for high efficiency OSCs replacing ITO and increasing forward light scattering and light trapping in the active layer while reducing reflectance. We used Finite-Difference Time-Domain (FDTD) electromagnetic simulations to design the nanohole size and pitch to maximize both the intensity of plasmon-induced electric fields and light absorption near the bandgap of the active layer. Once the most favorable nanostructure design was chosen, FDTD simulations and transfer matrix optical modeling were used to determine the optimal layer thicknesses of the active and interfacial layers. We used a blend of [6,6]-phenyl C71-butyric acid methyl ester ($PC_{71}BM$) and the low-bandgap semiconducting polymer thieno[3,4-b]thiophene/benzodithiophene (PTB7) to make the bulk heterojunction active layer and made both inverted and conventional devices. Plasmonically enhanced devices were fabricated on bare glass and on bare flexible PET and compared to control devices on ITO-coated glass and ITO-coated PET. The effects of these plasmonic electrodes to enhance light absorption, enable omnidirectional light harvesting, and increase robustness of the flexible devices will be discussed. These nanostructured plasmonic electrodes were fabricated using nanoimprint lithography in a process that is scalable to roll-to-roll manufacturing. To our knowledge, this is the first time nanopatterned plasmonic electrodes have been applied to high efficiency and flexible OSCs and the learnings presented here can be utilized to fabricate high performance electrodes for widespread photovoltaic systems.

381815 Comparing Electron Recombination Via Interfacial Modifications in Dye-Sensitized Solar Cells

Abstract:

Establishing a blocking layer between the interfaces of the photoanode is an effective approach to improve the performance of dye-sensitized solar cells (DSSCs). However, researchers have disagreed about the respective role of these blocking layers on electron transport processes at the transparent conductive oxide (TCO) or TiO_2 interface. In this work, HfO_2 blocking layers were deposited *via* atomic layer deposition (ALD) onto tin-doped indium oxide (ITO) and TiO_2. In both cases, addition of the blocking layer increased cell performance to an efficiency greater than 7%, corresponding to about a 10% improvement. The improved performance for a HfO_2 layer inserted between the ITO/TiO_2 interface is associated with an energy barrier that reduces electron recombination. If the blocking layer is too thick, cell efficiency degrades due to decreased tunneling probability of electrons. HfO_2 blocking layers between the TiO_2/dye interface showed more complex behavior and were more sensitive to the number of ALD cycles. For thin blocking layers on TiO_2, the improved device performance is attributed to the passivation of surface states in TiO_2, which immobilize electrons and impede interfacial electron transport. A distinct transition in dark current and electron lifetime were observed after about 2 ALD cycles. These changes to performance for thicker HfO_2 layers on TiO_2 indicate the formation of an energy barrier that reduces electron recombination but significantly impedes cell performance due to changes in the series resistance of the cell.

374791 Invited Talk: The Calorimetric Glass Transition Under Nanoconfinement

Abstract:

The behavior of materials confined at the nanoscale has been of considerable interest over the past two decades. Recent work in our laboratory has focused on the influence of nanoconfinement on the glass transition and associated kinetics, on melting and crystallization, and on polymerization kinetics and resulting properties. In this talk I will present recent results investigating the depression of the glass transition temperature for nanoconfined materials, including single polystyrene ultrathin films studied with ultrafast scanning chip calorimetry, glass-formers confined to nanopores, and glass-formers confined by crystalline domains. The single thin films are of particular interest because they show a T_g depression which can be reversed by film dewetting and thickening during high-temperature annealing. The results of materials confined in different geometries will be used to test the leading explanation for the T_g depression in nanoconfined glasses, i.e., that the enhanced mobility arises from free surface and interface affects.

Probing the Impact of Free Surfaces on the Glass Transition and Physical Aging in Ultrathin Polymer Films

Connie B. Roth and Justin E. Pye
Department of Physics, Emory University, Atlanta, GA, USA

Contact information:
Connie Roth: cbroth@emory.edu
Justin Pye: jepye@emory.edu

As technological applications are made smaller and smaller, the role of free surfaces on the properties of the material become increasingly important. Nanoscale materials, used in a variety of applications from microelectronics and nanolithography to gas separation membranes, exhibit material properties that can be strongly perturbed from their bulk values because of the increased surface-to-volume ratio of these materials. One of the least well-understood effects is how the presence of a free surface imparts excess mobility to the near surface region and is then propagated deeper into the material. To elucidate the underlying fundamental mechanisms of how this enhanced mobility near a free surface affects the glass transition of polymeric materials over length scales of tens of nanometers, we have performed glass transition temperature (T_g) and physical aging measurements of ultrathin free-standing polystyrene (PS) films.

Numerous studies have documented changes in T_g with decreasing thickness for thin polymer films less than ~100 nm in thickness. However, understanding of the fundamental mechanisms causing this phenomenon are still unknown. We have recently demonstrated that two separate mechanisms can act simultaneously to propagate enhanced mobility from the free surface deeper into the interior of the film resulting in two distinct reduced T_gs within single high molecular weight (MW) free-standing PS films [1]. These two transitions, which can be separated by up to 60 K, show qualitatively different behaviors. The lower transition is MW dependent and appears to be unique to high MW chains, while the upper transition is MW independent and appears to be common across multiple glass forming systems such as polymers, small molecules, and colloids. We believe this more universal mechanism induces a gradient in dynamics near the free surface whose length scale is related to cooperative motion [2].

To help determine the mechanisms of the two reduced T_gs observed in ultrathin high MW free-standing PS films, we performed physical aging measurements. For thick films (220-1800 nm) in which there are no T_g reductions, we find that the physical aging rate depends strongly on stress caused by thermal expansion mismatch between film and support. This stress, applied to the films as they are quenched into the glassy state, can nearly double the physical aging rate when changing the frame material from polycarbonate to silicon [3]. For much thinner, high MW PS films held at a temperature between or below the two reduced T_gs, we observe physical aging of the material consistent with both transitions leading to the kinetic arrest of some fraction of the film material. The corresponding thermal expansion coefficients of the film below, above, and in between the two transitions are also compared.

[1] J. E. Pye, C. B. Roth, *Phys. Rev. Lett.* **2011**, *107*, 235701.
[2] J. E. Pye, K. A. Rohald, E. A. Baker, C. B. Roth, *Macromolecules* **2010**, *43*, 8296-8303.
[3] J. E. Pye, C. B. Roth, *Macromolecules* **2013**, *46*, 9455-9463.

388848 Confinement Effects in Thin Films of Cyclic Polystyrene

Abstract:

The surface fluctuations of melt films of 6k cyclic polystyrene (CPS) and its linear analog were measured using X-ray photon correlation spectroscopy (XPCS) for films of various thicknesses. The surface fluctuations of the 6k linear PS melt films 17 nm and thicker and the 6k cyclic melt films 28 nm and thicker can be described using a hydrodynamic continuum theory (HCT) that assumes the film is characterized only by the bulk viscosity. When a film of CPS is 24 nm or thinner, the behavior can no longer be captured using the HCT with bulk viscosity. The surface fluctuations behave as though the film has an effective viscosity higher than the bulk value. There is no evidence of an effective modulus in the very thin films of cyclic chains. The thickness at which confinement effects are seen for the 6k CPS chains is larger than that for the linear analogs.

Acknowledgements: Use of the Advanced Photon Source at Argonne National Laboratory was supported by the DOE's Office of Science under Contract DE-AC02-06-CH11357. This work was supported by NSF Grants No CBET-0730692 and CBET-0731319.

367767 Rheology at the Nano-Scale and at Surfaces

Abstract:

There is considerable discussion in the literature concerning the dynamics of glass-forming materials, especially polymers, at the nanometer size scale. Of particular interest is the wide-spread view that increases in the mobility, especially as seen in reduced glass transition temperatures Tg, occur due to enhanced surface dynamics. One way of addressing the problem is to perform rheological measurements both on ultrathin films and on the film surface itself. Here we describe work in which ultrathin films are examined using both a novel nanobubble inflation test and a liquid dewetting experiment. We further look at surface dynamics using a spontaneous particle embedment method to extract surface rheological properties. Examination of the results shows while the thin films may show highly accelerated dynamics, their behaviors are surprisingly non-universal and, in some instances, contrary to what is observed for the surface dynamics. Furthermore, the surface dynamics themselves do not exhibit the typical Vogel-Fulcher-Tammann (VFT) or Williams, Landel and Ferry (WLF) behaviors expected for the temperature dependence of the dynamics of glass-forming materials, hence, suggesting that the surface either does not exhibit cooperative dynamics or it does not exhibit a glass transition at all.

In either case the surface, then, cannot be easily identified as the cause of observed reductions in Tg. We further remark that a great stiffening of the rubbery plateau in ultrathin films is observed and this may be consistent with observations that surface dynamics of polymers above the macroscopic glass transition seem to be slower than the macroscopic dynamics.

The authors thank the National Science Foundation under grants DMR-1207070 and CHE-1112416, the Office of Naval Research under project N00014-11-1-0424, and the John R. Bradford endowment at Texas Tech, each for partial support of this work.

378739 Dynamics of Confined Flexible and Unentangled Polymer Melts in Highly Adsorbing Cylindrical Pores

Abstract:

We performed coarse-grained molecular dynamics simulations to study the effect of the confinement of flexible and unentangled polymer chains in cylindrical pores. The results are compared with the recent neutron spin echo (NSE) experiments by Krutyeva and co-workers (Phys. Rev. Lett. 2013, 110, 108303), where anodized aluminum oxide (AAO) pores were infiltrated with polydimethyl-siloxane (PDMS) chains. The NSE experiments were designed to probe the interphase layer, which is hypothesized to have polymer chain dynamic properties between those of a glassy layer and the bulk. There is excellent agreement in the values obtained for the normalized coherent single chain dynamic structure factor, $S(Q,\Delta t)/S(Q,0)$, between experiments and simulations. The experimentally inaccessible mean squared displacement (MSD) of the confined monomers, calculated as a function of radial distance from the pore surface was obtained in the simulations and show a gradual increase of the MSD from the adsorbed but mobile layer to that similar to the bulk. This is akin to the hypothesized interphase layer. Additionally, the simulations provide information about the static polymer chain properties such as the mean-squared radius of gyration, $\langle R_g^2 \rangle$ and the average shape anisotropy, $\langle \kappa^2 \rangle$. These two properties indicate that the chains form a pancake-like conformation near the surface and a bulk-like conformation near the center of the confining cylinder. Direct visualization of the polymers in the simulation confirm the pancake-like conformation of the adsorbed chains and the presence of trains, loops and tails in the region between the adsorbed chains and the chains not in contact with the surface. Despite the presence of these different conformations, the average form factor of the confined chains still follows the Debye function which describes linear ideal chains and is in agreement with small angle neutron scattering (SANS) experiments.

A novel ellipsometry-based method was introduced to determine values of kinetic fragility in polymer films and to investigate the effect of nanoscale confinement on polymer fragility. Three polymer systems were studied: polystyrene (PS), polycarbonate (PC), and PS doped with small molecule diluents of 1,10-bis-(1-pyrene)decane (BPD). In bulk-like films, the fragility index measured by ellipsometry agreed very well with that by differential scanning calorimetry. With confinement, a dramatic decrease in fragility was observed in highly fragile PS and PC. In particular, the fragility decreased by 58% from 166 to 69 in PS and by 65% from 214 to 75 in PC as film thickness decreased from bulk to 27-28 nm; a substantially muted response was observed in the strongest of the three glass-forming polymer systems: PS + 2 wt% BPD, where the fragility decreased only 21% from 134 to 106 from a bulk film to a 27-nm-thick film. The larger fragility-confinement effect observed in more fragile polymers strongly correlates with a previous discovery regarding the T_g-confinement effect: the strength of the T_g-confinement effect increases with increasing fragility of the bulk polymer. This study indicates that bulk fragility, which is a parameter reflecting the packing efficiency of polymers, is associated with the susceptibility of polymers to effects of nanoscale confinement.

379884 Ordering and Properties of Confined Main Chain Liquid Crystalline Polymers

Abstract:

Ordering and Properties of Confined Main Chain Liquid Crystalline Polymers

Zhenpeng Li[1], Eric Baer[2], Donald R. Paul[1], Christopher J. Ellison[1]

[1]McKetta Department of Chemical Engineering, University of Texas at Austin, Austin, TX 78712

[2]Department of Macromolecular Science and Engineering, Case Western Reserve University, Cleveland, OH 44106

Abstract

Liquid crystalline polymers (LCPs) possess many attractive properties, including outstanding chemical resistance, good mechanical properties and extraordinary gas/liquid barrier properties. Such unique properties show promise for applications in barrier packaging materials. Previous studies have demonstrated that the structures and properties of crystalline polymers can be dramatically affected through nano-confinement. However, only several studies have been performed on the study of confined LCPs, few with confined main chain LCPs (MCLCPs). Therefore, it would be very interesting to understand the morphology and properties of confined MCLCPs. In this study, Vectra (a commercial grade MCLCP) thin films with 20nm - 2μm in thickness have been made by spin coating and its structure and ordering under nano-confinement was studied. It was found that the Tg of Vectra could be profoundly affected by the confinement and the interfacial interaction with the substrate. Additionally, Vectra polymer chains exhibited 'in-plane' orientation with their long axis aligned parallel to the substrate. Such 'in-plane' ordering could enhance the O_2 barrier properties of Vectra due to the increased tortuosity of the diffusion path way.

392487 Manipulating Polymers with Light Activated Chemistries

Abstract:

Light activated chemistries are a powerful tool for manipulating polymers with spatio-temporal control. In this talk, a few recent projects from our group that exploit such chemistries for patterning films and manufacturing fibers will be highlighted.

391402 Mixed-Linker Zeolitic Imidazolate Framework (ZIF) Materials and Scalable Membrane Processing on Polymeric Hollow Fibers

Abstract:

This talk will describe our recent progress in the synthesis and properties of ZIF-type MOF materials for molecular separation applications, and their subsequent use in the scalable fabrication of molecular sieving membranes on polymeric hollow fibers. New separations technology is key to increased energy efficiency in the chemical sector, as well as the efficient production of clean and renewable fuels. It is hypothesized that ZIF materials may find successful applications in this area. Towards this goal, we will first discuss approaches to controlling the pore structure and separation properties of ZIF materials via mixed-linker synthesis. We then discuss the processing and use of these MOF materials in the fabrication of membranes for gas and hydrocarbon separations. The transport properties of such membranes are discussed. We will discuss some key scientific and engineering questions whose resolution (via combined experimental and computational methods) may enable the successful application of ZIF membranes in separations processes.

392488 Microstructure and Stress Development in Particulate Coatings

Abstract:

A wide variety of products, including paints, adhesives, membranes, fuel cell components, printed electronic circuits and newsprint, are produced by liquid-applied coating and printing processes. A coating or printing process has two main steps: deposition of a liquid (e.g., solution, dispersion) onto a substrate and solidification, typically by drying or curing, into a functional coating or printed patch. This talk will focus on our efforts to understand and characterize the changes in structure that occur during drying and the impact of these changes on stress development. Two special techniques will be used: cryogenic SEM (cryoSEM) and stress measurement. CryoSEM involves freezing coatings at different stages of drying, fracturing to reveal cross-sections, subliming for contrast and imaging with an SEM equipped with a cryostage. CryoSEM reveals particle and water distribution through the thickness of a drying coating. Coating stress is measured during drying using a cantilever beam measurement. Walls built on the cantilever suppress lateral drying fronts, which simplifies the interpretation of stress development. Results from aqueous dispersions of silica particles will be presented. Parallel cryoSEM and stress measurement experiments were carried out to understand the origin of stress development. Additionally, using a variety of other coating dispersions, the effects of particle size, particle deformation and binder on particle distribution and stress will be presented.

392489 Molecular Elucidation and Engineering of the Stem Cell Fate Decisions

Abstract:

Elucidating the mechanisms that govern stem cell self-renewal and differentiation is critical for understanding the roles these cells play in organismal development and function as well as for harnessing stem cells to repair tissues damaged by disease or injury. It has become increasingly clear that stem cells are regulated not only by biochemical signals in the niche, but also by biophysical features in the way these signals are presented, though investigating the latter is challenged by experimental complexities in investigating and mimicking the complexity of the extracellular matrix (ECM), cell-cell interactions, and other niche components. Recent work has demonstrated that bioactive, synthetic materials can be harnessed to emulate and thereby study the effects of solid phase, biophysical cues on cell function, and by extension to develop enhanced stem cell culture systems. For example, activation of many cellular receptors involves the formation of oligomeric protein signaling complexes with ligands presented from the matrix, the surface of neighboring cells, and in some cases even from solution. We have developed multivalent ligands – polymers conjugated to signaling proteins to yield biomimetic signals with nanoscale spatial organization – which potently induce the differentiation of human pluripotent stem cells in vitro and neural stem cells in vitro and in vivo. In addition, these materials combined with optogenetics and super-resolution microscopy have yielded insights into signaling mechanisms that regulate the fate decisions of these cells. Finally, such biomimetic materials can be integrated into safe, scaleable, fully defined, robust culture systems for pluripotent stem cell expansion and differentiation.

IMPACT OF CELLULOSE ALLOMORPH ON ENZYMATIC SACCARIFICATION

Ting Cui[1, 2], *Jihong Li*[1, 2], *Zhipei Yan*[1, 2], *Shizhong Li*[1, 2*], *Menghui Yu*[1,2]
[1] *Institute of Nuclear and New Energy Technology, Tsinghua University, Beijing 100084, China*
[2] *Beijing Engineering Research Centre of Biofuels, Tsinghua University, Beijing 100084, China*

Abstract

Background

Bioconversion of cellulose into glucose or chemicals attracted extensive attentions for the sustainable development of human society in recent decades. The crystal allomorph of cellulose is a key factor to affect cellulose saccharification. However, due to the influence of lignin, hemicellulose and different characterizing methods of literatures, the effect of cellulose allomorph on cellulose saccharification is still confused. Thus, a systematic research on the effect of different cellulose allomorph on enzymatic saccharification is required.

Results

Multiple approaches, including ionic liquid (IL), ethylenediamine (EDA), glycerol and sodium hydroxide, were used to pretreat α-cellulose in this work. The properties of obtained cellulose (crystallinity index, lattice spacing, specific surface area and wettability) were characterized by X-ray diffraction, BET specific surface area analysis and water contact angle analysis, respectively. The distance of lattice spacing of cellulose III was larger than that of other cellulose samples. The crystallinity and water contact angles of all the cellulose samples were both in the following order: cellulose treated by IL < cellulose treated by NaOH < cellulose treated by EDA <α-cellulose < cellulose treated by glycerol. Cellulose treated by IL with a crystallinity index value of 20%, was very close to amorphous cellulose. After 72 h hydrolysis, the cellulose conversion rates ranged from 43% to 99%. Cellulose treated by IL exhibited a notably efficient hydrolysis profile, followed by cellulose treated by EDA.

Conclusion

Four cellulose allomorphs were prepared to investigate the effect of crystal form on the cellulose enzyme digestibility in this study. With the integrated experimental characterization, the most feasible enzymatic hydrolysis allomorph was cellulose III, followed by cellulose II, cellulose I_{α} and I_{β}. When the crystal type of cellulose was the same, the crystallinity index was the dominant factor. Low crystallinity index enhanced the enzymatic hydrolysis efficiency of cellulose. Water contact angle was also a feasible characterization to evaluate the enzymatic hydrolysis efficiency of cellulose except for cellulose III. High wettability of cellulose enhanced the enzymatic hydrolysis when crystal allomorph of cellulosic samples was the same.

364398 Bio-Inspired Magnetic Pollen Particles Tailored with Adherents for Tuning Adhesion Strength and Optical Properties

Abstract:

Nature provides remarkable examples of bio-microscale particles with well-optimized structures and/or chemistries for the design of novel functional materials with effective adhesion to a variety of porous or dense surfaces. Herein, we reported the development of bio-inspired magnetic pollen grains as novel multifunctional microparticles with tailored specificity in both adherence and optical activity. The adhesion strength of as-synthesized magnetic pollen grain is significantly enhanced through the synergic interactions of the short-range van der Waals (VDW) force//hydrogen bonding, the long-range magnetic force and capillary force into one particle. The long-range magnetic force is dependent on the amount of Fe_3O_4 particles loading in the core, while the VDW force and the capillary interaction are tuned through coating polystyrene (PS) nanoparticle, chitin nanofiber and a layer of viscous pollenkitt on the shell surface, respectively. For optical activity, the as-synthesized magnetic pollen grains exhibit varied light-reflective intensities with different surface modifications. The difference of reflective indices of surface coating materials determined the pollen grain light-reflective properties, which were enhanced with pollenkitt coating, and dramatically reduced with PS nano-particle coatings. This present strategy provides us with new insight on the exploring of the generation of multifunctional particles with significantly-enhanced taggants capabilities, and it could further extend to other pollen grains and their biomimetic-replica high-aspect-ratio particles with numerous different nanoscale surface features for adhesion purpose.

388676 Nanocomposites from Modified Lignin

Abstract:

Lignin comprises 25~30% of biomass, which is the second largest among all the materials extracted from wood. However, due to some limitations, such as amorphous formation, structure variation and low solubility both in water and organic solvents, most lignin are burnt for heat. To use lignin as high-valued chemicals and materials, lignin needs to be modified. In this research, lignin modification using chemical such as butyric anhydride and 2-Bromoisobutyryl bromide to change its solubility in different organic monomers were conducted. Modified lignin is able to dissolve in styrene and methyl methacrylate monomer and therefore, can participate in the polymerization of polymers. The mechanism of this reaction is the reaction between the –OH group in lignin and anhydride in butyric anhydride or bromide in 2-Bromoisobutyryl bromide. More works can be done based on the water-soluble lignin. In this research, 1HNMR and FTIR are used as characterization methods to justify the change of lignin structure. 31PNMR is used to determine content of –OH group. What's more, Zetasizer is used to measure the particle size of PMMA using modified lignin as additive and SEM is used to observe its morphology. Results obtained from this research can be used to develop more expanded area of lignin application.

Synthesis of pH-sensitive Core-shell Latex with Biobased Dendrimer via Miniemulsion Polymerization

Suguna Jairam [1], **Zhaohui Tong*[2],

[1, 2] Agricultural and Biological Engineering Department, University of Florida, Rogers hall, Museum Road, PO Box 110570 Gainesville, FL. 32611

*ztong@ufl.edu

Abstract: The pH-responsible nanostructures have gained vast interest in the areas of gene therapy, drug delivery, and sensitive coatings. In this study, we synthesized pH-responsive core shell latex with the biobased dendrimer encapsulated dye via miniemulsion polymerization. Glycerol, main component in the biodiesel waste stream, was used to synthesize a novel triacrylate monomer, which was crosslinked with ethylene diamine to produce dendrimeric structures. Next, the dendrimers were copolymerized with styrene in the presence of Solvent Blue 59 dye via miniemulsion polymerization to produce pH-sensitive core shell latex with dye encapsulation. Nuclear Magnetic Resonance (NMR) analysis of glycerol triacrylate, dendrimer and polymers (at varying pH) showed the successful synthesis of pure monomer from glycerol, completed reaction with dendrimer, and pH sensitivity in polymer; respectively. Transmission Electron Microscopy (TEM) images in combination with zetasizer analysis data indicate uniform size distribution of polymer particles, successful dye encapsulation and release upon pH variation.

389463 Biodegradable, Multifunctional Poly (amine-co-ester) with Ortho Ester in the Main Chain for the Delivery of Plasmid DNA and siRNA

Abstract:

Safe and effective delivery of genetic materials remains a most challenging problem. Viral vectors are commonly used and highly efficient, but there are many safety issues. Polymers are promising non-viral vectors because of their safety and high chemical versatility, but they usually suffer from low efficiency. Here, we report a family of biodegradable, multifunctional poly(amine-co-ester)s that include ortho ester groups in the main chain (PACEO). These polymers have low toxicity and high efficiency in the delivery of both plasmid DNA and siRNA. The best members of the PACEO family can deliver both plasmid DNA and siRNA with efficiency higher than or comparable to Lipofectamine, which is a standard for *in vitro* DNA/siRNA transfection.

The PACEOs were designed with several features: 1) biodegradability; 2) positive charge provided by tertiary amine in the main chain, which is used to bind DNA/RNA; 3) high hydrophobicity to assist in the complex formation; 4) acid-sensitivity provided by the ortho ester groups; 5) high tunability of the different functions, which enables the easy optimization of the polymer to accommodate different cargos. We have previously shown that high hydrophobicity allows stable gene binding with a greatly lowered positive charge density on the polymer, making the polymer less toxic. One major barrier for the delivery of genetic materials is endosomal escape. Polyplexes usually enter cells via endocytosis. Thus, the escape from intracellular vesicles is crucial since the high acidity (pH=4~6) in these vesicles can degrade the genes. It has been reported that acid sensitive polymeric vectors can facilitate the escape of the loaded genes from endosomes because they promptly break apart into low molecular weight products, increasing intra-endosomal osmotic pressure, and enhancing breakdown of the endosomes.

In this study, a novel lipase-catalyzed polymerization enables the synthesis of PACEOs with all these features. A series of novel, lipase-compatible ortho ester monomers were synthesized and copolymerized with several different monomers: diester to provide biodegradability, amino-diol to provide tertiary amine group, and lactone to provide high hydrophobicity. The lipase-catalyzed polymerization enables the fine tune of these functions simultaneously by adjusting the feed ratio of different monomers. Also, to our knowledge, this is the first use of ortho ester bearing monomers in lipase-catalyzed polymerization.

We have shown that PACEOs can complex with plasmid DNA or siRNA and form stable nanocomplexes with size ranging from 120nm to 200nm. The polymers also exhibit pH-dependent degradation: in a weight loss study, the representative PACEO showed less than 10% weight lost at pH7.4 during a 6-day incubation, but had a near complete weight loss at pH 4 within 4 hours. Also, these polymers have minimal toxicity: when tested on HEK293 cell line, PEI killed almost all cells after incubation with 50 ug/mL polymer for 48 hr, whereas cells exposed to PACEO were 100% viable, even up to 500ug/mL polymer.

Several PACEOs were highly efficient for *in vitro* transfection of plasmid DNA and siRNA. For plasmid DNA, the best PACEO showed an 8-times higher efficiency than Lipofectamine 2000 when delivering plasmid DNA encoding the luciferase gene to HEK293 cell line. When loaded with plasmid DNA encoding GFP, a single treatment with the best PACEO transfected over 60% of the treated U87 glioblastoma cells, while Lipofectamine2000 only transfected 30% of the cells. In the delivery of siRNA, the same PACEO showed a comparable efficiency to Lipofectamine RNAiMAX in knocking down luciferase expression in a HEK293 cell line stably transfected with Luciferase gene.

In summary, the PACEOs we report here are novel in chemistry, biodegradable, multifunctional, non-toxic, and highly efficient in the delivery of both plasmid DNA and siRNA. These polymers provide a potent and universal tool for gene therapy.

References:

Zhou, J., J. Liu, C.J. Cheng, T. Patel, C. Weller, J.M. Piepmeier, Z. Jiang, and W.M. Saltzman, *Biodegradable poly(amine-co-ester) terpolymers for targeted gene delivery*. Nature Materials, 2012. **11**(1): p. 82-90.

388654 DNA Modification of Collagen Scaffolds for Applications in Regenerative Medicine

Abstract:

The use of collagen-based biomaterials in regenerative medicine has rapidly increased over the past decade due to the biocompatibility, approved clinical usage, and broad importance of collagens in mammalian tissues. During this period, a variety of collagen modification techniques have been developed for the purpose of retaining and delivering therapeutics. Many techniques rely on chemical treatments, but a biomimetic strategy to modify collagen scaffolds offers compelling advantages. We have designed a collagen mimetic peptide (CMP)-based approach to create DNA-linked collagens for applications in therapeutic gene transfer. CMPs offer an enormously versatile (and reversible) method for modification of collagen with a range of materials. We created CMPs with a range of hybridization temperatures (T_m) on collagen I, and demonstrated the functionalization of DNA polyplexes with addressable CMPs at a range of densities. The retention of DNA polyplexes on collagen could be directly manipulated through variation of the CMP density on the polyplexes, and the CMP-based retention strategy also increased polyplex stability and sustained gene transfer over a period of several weeks in cultured murine fibroblasts. This novel collagen modification approach offers new opportunities to address key issues in therapeutic gene transfer for applications in regenerative medicine.

ID: 362986
Password: 589009

AIChE – Biomaterials – Biomaterials for Drug and Gene Delivery
Title: Topical Drug Delivery in Ulcerative Colitis Using an Inflammation-targeted Hydrogel
Authors: Sufeng Zhang, Joerg Ermann, Marc D. Succi, Allen Zhou, Jonathan N. Glickman, Praveen K. Vemula, Laurie H. Glimcher, Giovanni Traverso, Robert Langer, Jeffrey M. Karp

Ulcerative colitis (UC), one of the two main types of inflammatory bowel disease (IBD), is a chronic intestinal disorder characterized by mucosal inflammation limited to the colon [1]. A major hurdle in the treatment of UC is the lack of an effective drug delivery system to precisely deliver drugs to the sites of inflammation for higher local drug concentration, less systemic absorption, and therefore fewer side effects. To improve drug delivery in UC, we have developed an inflammation-targeted hydrogel system, made from ascorbyl palmitate listed as "Generally Recognized as Safe" (GRAS) by the Food and Drug Administration (FDA). The GRAS hydrogel was assembled as described previously with minor modifications [2]. Dexamethasone (Dex) was used as a model drug, and the release of Dex from the hydrogel *in vitro* was quantified in different release media. To demonstrate that the hydrogel preferentially adheres to the sites of inflammation, a fluorescent dye was encapsulated in the hydrogel for imaging purpose. Two murine colitis models, T-bet$^{-/-}$Rag2$^{-/-}$ Ulcerative Colitis (TRUC) [3,4] and Dextran Sodium Sulfate (DSS) induced colitis [5], were used for animal studies. Adhesion of the hydrogel to the inflamed colon was assessed *ex vivo* and *in vivo*. The therapeutic efficacy of Dex-loaded hydrogel administered to mice with colitis via enema was compared with Dex alone, hydrogel alone or untreated mice. Systemic Dex exposure after a single enema of Dex-loaded hydrogel or Dex alone to colitic mice was quantified using blood taken from mice at different time points, and the area under the curve (AUC) was analyzed. The self-assembled hydrogel demonstrated enzymatically triggered release of Dex and preferential adherence to the inflamed colon tissue. Dex-loaded hydrogel significantly inhibited colon inflammation in the TRUC model compared to controls. Dex quantification in the serum indicated an 8-15-fold lower Dex peak serum concentrations and 4-7-fold less systemic drug exposure (analyzed by AUC) after a single enema of Dex-loaded hydrogel in comparison with Dex alone. This topical drug delivery system using inflammation-targeted hydrogels enables local drug delivery to the sites of inflammation to lower systemic toxicity, and could potentially reduce the dosing frequency owing to the improved treatment efficacy. These studies establish a proof of concept in pre-clinical murine UC models for future studies for translation to the clinic.

References:
[1] Abraham C et al., *N Engl J Med,* 2009, 361(21): 2066-2078.
[2] Vemula PK, et al., *J Biomed Mater Res,* 2011, A 97(2): 103-110.
[3] Garrett WS, et al., *Cell,* 2007, 131(1): 33-45.
[4] Ermann J, et al., *Proc Natl Acad Sci USA*, 2011, 108(17): 7137-7141.
[5] Yan Y et al., *PLoS One*, 2009, 4, e6073.

pH-responsive nanoparticles for targeted and controlled drug release to disrupt oral biofilms

Benjamin Horev, [1,§] ***Marlise I. Klein,*** [2,‡,§]***, Geelsu Hwang,*** [3] ***Hyun Koo,*** [2,3,4]

and Danielle S.W. Benoit [1,5,6]

bhorev@ur.rochester.edu, mklein@foar.unesp.br, gellsuh@dental.upenn.edu, koohy@dental.upenn.edu, benoit@bme.rochester.edu

[1]Department of Biomedical Engineering, University of Rochester, NY 14627, United States, [2]Center for Oral Biology, University of Rochester, NY 14627, United States, [3]Biofilm Research Lab, Levy Center for Oral Health, University of Pennsylvania, PA 19104, United States, [4] Department of Orthodontics, School of Dental Medicine, University of Pennsylvania, PA 19104, United States, [5]Department of Chemical Engineering, University of Rochester, NY 14627, United States, [6]Center of Musculoskeletal Research, University of Rochester, NY 14627, United States, [‡]Current address: Department of Dental Materials and Prosthodontics, Araraquara Dental School, Univ Estadual Paulista, UNESP, Sao Paulo, Brazil.

Statement of Purpose:

In the US alone, $81 billion was spent on diagnosis, prevention, and treatment of oral biofilm-associated diseases in 2006.[1] Most infectious oral diseases result from dental biofilms. Bacteria within biofilms produce highly acidic microenvironments within biofilm matrix. Typical drug regimens are not effective in dental biofilm treatment, as topically introduced agents are not retained on dental surfaces. Therefore, treatment or prevention of dental biofilms requires localized delivery of high drug concentrations to dental surfaces, and may also benefit from pH-triggered drug release to deliver drug exactly when and where biofilms are developing. Thus, polymers and diblock copolymers composed of poly(2-dimethylamino)ethyl methacrylate (p(DMAEMA)) and pDMAEMA, butyl methacrylate (BMA), and 2-propylacrylic acid (PAA) (p(DMAEMA)-b-p(DMAEMA-co-BMA-co-PAA)), were explored for their ability to target the dental surfaces through electrostatic interactions and entrap and deliver the anti-biofilm drug, farnesol. Diblocks self-assemble into ~20 nm positively-charged nanoparticles with hydrophobic cores capable of drug loading.[2] Moreover, the diblocks have established pH-responsive behaviors[3] that may provide means to trigger drug delivery, as dental biofilm microenvironments exhibit greatly reduced pH (~4.5). We assessed the binding of diblock copolymer nanoparticles to tooth pellicle and biofilm exopolysacharide (EPS) surfaces that emulate clinically relevant tooth and dental biofilm surfaces. In addition, nanoparticle maximum drug encapsulation capacities and pH-responsive drug release were investigated.

Methods:

Diblock copolymers were synthesized by reversible-addition fragmentation chain transfer (RAFT) polymerization to achieve mono-disperse polymers (PDI<1.3) of ~20 kDa, with equivalent first and second block molecular weights. Reaction conditions were degrees of polymerization (DP) of 200 and 250 for 1st and 2nd blocks, respectively, and

1:10 initiator to chain transfer agent (CTA) ratios. Nanoparticles were labeled with Texas Red-isothiocyanate, to quantify adsorption to HA (hydroxyapatite), sHA (hydroxyapatite coated with saliva proteins), and gsHA (hydroxyapatite coated with saliva proteins and EPS glucans), surfaces that emulate the clinically relevant dental surfaces for biofilm targeting. Adsorption of the nanoparticles and several control polymers including p(DMAEMA) alone, nanoparticles with p(DMAEMA) coronas and hydrophobic p(BMA) cores, and nanoparticles with neutral p(PEGMA) coronas and pH-responsive p(DMAEMA-co-BMA-co-PAA) cores was performed. Finally, nanoparticles were loaded with farnesol through sonication. Drug loading capacities and pH-responsive release from the nanoparticles were measured by high performance liquid chromatography (HPLC). Additionally, efficacy of nanoparticle mediated drug delivery were demonstrated using *Streptococcus mutans (S. mutans)* in suspension as well as in biofilms grown on dental pellicle surface (sHA) analyzing bacterial colony forming units, biofilm mass, and strength of biofilms after clinically relevant treatment regimines.

Results:

The role of p(DMAEMA) in the adsorption of nanoparticles to dental surfaces was examined (Figure 1.A). For adsorption experiments performed using 1 μ M polymer, ~75% of p(DMAEMA) only (no core), ~55% of p(DMAEMA)-b-p(DMAEMA-co-BMA-co-PAA) and ~55% of p(DMAEMA)-b-p(BMA) nanoparticles adsorbed as compared to ~27% of nanoparticles formed from p(PEGMA)-b-p(DMAEMA-co-BMA-co-PAA). As they exhibited the greatest adsorption, have inherent pH-responsive behaviors,[3] and formed drug-loadable nanoparticles, more sophisticated adsorption experiments were performed on nanoparticles composed of p(DMAEMA)-b-p(DMAEMA-co-BMA-co-PAA) (Figure 1.B). According to Langmuir fits to the binding data, the average maximal binding capacity of nanoparticles on the dental surfaces was ~21 μmol/m^2 and the average binding affinity constants (K_a) were ~215 L*mmol^{-1}. These values did not differ statistically between the various dental surfaces. In addition, 0%, 26% and 36% binding of p(DMAEMA) to HA, sHA, and gsHA surfaces was observed at pH 10.5 respectively, which implies dental surfaces are primarily targeted through protonated tertiary amines within p(DMAEMA) coronas of NPC. Moreover, and owing to hydrophobic cores, NPC load farnesol, a hydrophobic antimicrobial agent, at up to ~22 wt%, which are >20 times greater drug amounts than the minimum inhibitory concentration of farnesol for *S. mutans*.

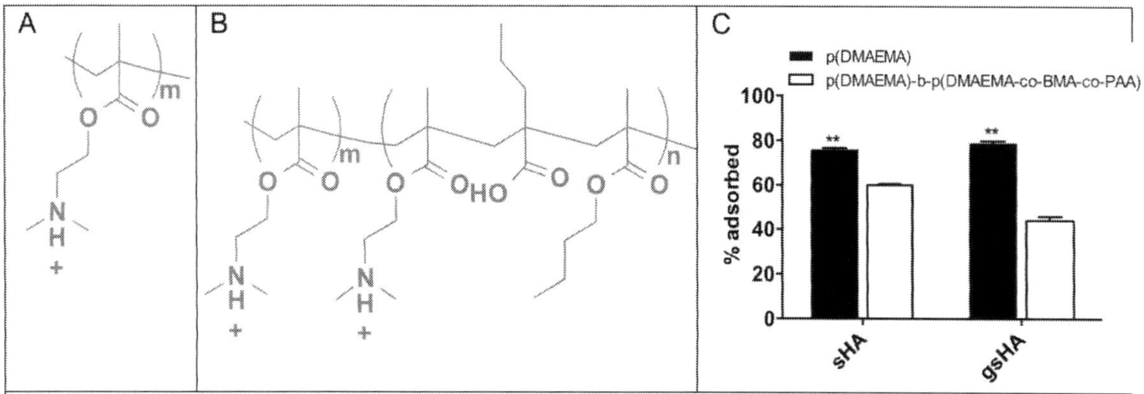

Figure 1. Structures of (A) p(DMAEMA) ((m=63, 9.3 kDa, PDI=1.3) and (B) p(DMAEMA)-b-p(DMAEMA-co-BMA-co-PAA) explored to target drug delivery to dental surfaces (n=92, 22 kDa, PDI=1.1). (C) Polymer binding to dental mimetic surfaces, saliva-coated hydroxyapatite (sHA) and glucans and saliva-coated hydroxyapatite (gsHA), was analyzed at 1 μM polymer concentrations.

Acidic pH (~4.5) within cariogenic biofilm microenvironments can be exploited for pH-triggered drug release *in situ*. Drug release experiments showed that at pH 4.5, ~37% of farnesol was released after 5 hours with ~90% released after 24 hours, as compared physiological pH, when ~24% release was achieved after 5 hours and ~90% after 48 hours (Figure 2A). The release was pH-dependent with $t_{1/2}$=7 and 15 h for release at pH 4.5 and 7.2, as the nanoparticles exhibit responsive behaviors that result in core destabilization at acidic pH, characteristic of cariogenic biofilm microenvironments. Encapsulated drug was biologically active as bacteria exposure to drug-loaded NPC resulted in a 3-log reduction in bacterial viability relative to free nanoparticles and PBS treated control groups. Additionally, 50% decrease in colony forming units per dry weight biofilm was observed in cariogenic biofilms treated with drug-loaded nanoparticles for 10 minutes, 5 times over 48 hours of biofilm growth (Figure 2B). Finally, treated biofilms exhibited 2-fold greater removal when exposed to shear forces compared with free farnesol, PBS, or nanoparticles alone.

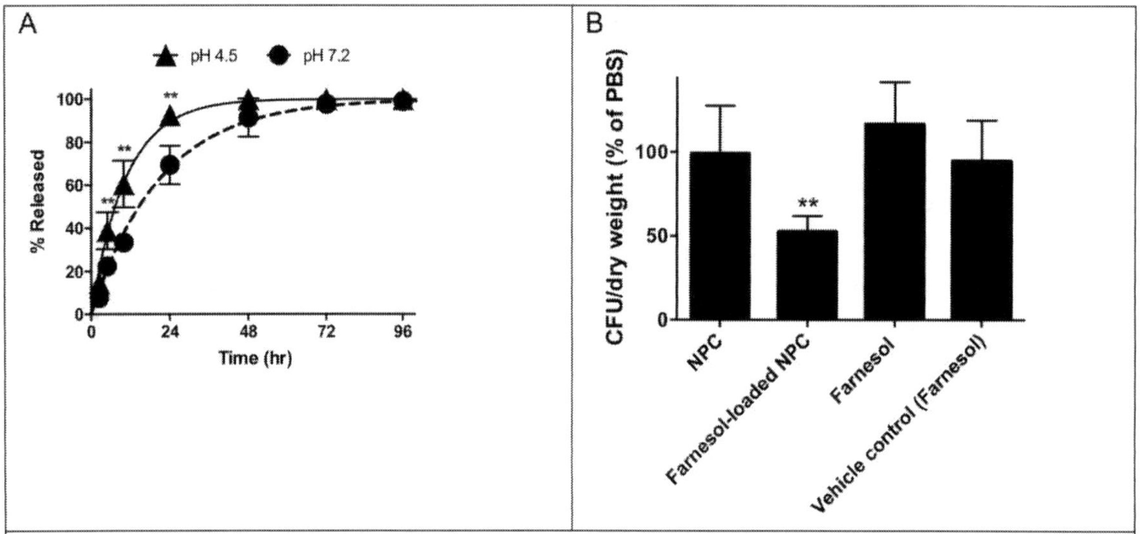

Figure 2. (A) Demonstration of pH-responsive farnesol release from diblock copolymer micelles (Fig. 1B) at pH 7.2 and 4.5. (B) 50% reduction in the number of colony forming units per dry weight were achieved in biofilms treated with farnesol-loaded NPC as compared to controls. Error bars represent standard error and asterisks denote significant difference as assessed by two-way ANOVA followed by Tukey's test for multiple comparisons ($p<0.01$).

Conclusions:

p(DMAEMA)-b-p(DMAEMA-co-BMA-co-PAA) nanoparticles strongly bind to dental pellicle and EPS surfaces with affinities ~20 times higher than the affinities of bisphosphonates to hydroxyapatite.[4] Nanoparticle affinity is likely due to positively-charged amine residues of p(DMAEMA) coronas that interact with the overall negative charge of dental surfaces,[5] as similar adsorption behaviors were observed for p(DMAEMA) alone, p(DMAEMA)-b-p(DMAEMA-co-BMA-co-PAA), and p(DMAEMA)-b-

(BMA). Farnesol, which is insoluble in water, can be loaded and delivered at >20-fold greater concentrations from nanoparticles than its minimal inhibitory concentration (MIC) for common oral bacteria such as *Streptococcus mutants* (*S. mutans*), thereby overwhelming biofilm resistance to regular drug concentrations.[6] Rapid and pH-responsive farnesol release from nanoparticles indicates that therapeutic delivery may be advantageous for the relatively short windows of dental treatment and low pH microenvironments of pathologic dental biofilms. Nanoparticle mediated drug delivery resulted in 50% reduction in viability of bacteria within treated biofilms. Therefore, diblocks reported herein have great potential to enhance efficacy of antibiofilm agents through targeting several dental surfaces simultaneously, localization of high drug amounts *in situ*, and rapid drug release in response to microenvironmental triggers.

References:

1. Flemmig TF, *Periodontol. 2000.* 2011;55(1):9–15; 2. Manganiello MJ, et al. *Biomaterials*. 2012;33(7):2301–9; 3. Convertine AJ and Benoit DSW, et al. *Control. Release*. 2009;133(3):221–9; 4. Al-Kattan A, et al. *Adv. Eng. Mater.* 2010;12(7):B224–B233; 5. Young A, et al. *Adv. Dent. Res.* 1997;11(4):560–565; 6. Koo H, et al. *Antimicrob. Agents Chemother.* 2002;46(5):1302–1309.

ENGINEERING OF LIPID BASED NANOPARTICLES AS A PLATFORM FOR THE DELIVERY OF NUCLEIC ACIDS TO TREAT INVASIVE BREAST CANCER

Steve L. Hayward and Srivatsan Kidambi

Department of Chemical and Biomolecular Engineering,
University of Nebraska-Lincoln, NE 68588

Breast cancer (BC) is the second leading cause of death for women in the United States with current projections including 232,670 new cases diagnosed and 40,000 deaths in 2014 alone. Depending on the specific subset and stage of BC, treatment options available include chemotherapy, radiation therapy, hormone therapy, biological therapy, and surgery. These treatment measures are invasive, induce an array of adverse side effects, or only work until the body develops a resistance. Gene delivery/ gene therapy is an innovative treatment option for BC that provides long-term disease management. Drug delivery through specialized nanocarriers is becoming more popular due to their ability to efficiently transport drug to specific tissues with minimal immune response or offsite effects. Established techniques for *in vitro* gene delivery include electroporation, lipofection, and viral vectors, but targeted delivery of active nucleic acids *in vivo* remains a significant hurdle to full scale implementation of targeted gene delivery. As a result, engineering of new delivery systems for *in vivo* applications are the focus of intense investigation by industrial and academic researchers. In this project, we propose using a highly customizable lipid based nanoparticle drug delivery vehicle for targeted nucleic acid delivery to treat invasive breast cancer. We have engineered lipid nanoparticles (LNPs) of various sizes, surface charges, and coating density of hydrophilic polymer using non-cationic lipid and cholesterol components. LNPs were characterized by both photon correlation spectroscopy to determine particle size distribution, hydrodynamic diameter, and polydispersity, and by zeta potential measurements to determine nanoparticle surface charge characteristics and to quantify colloidal stability. Our preliminary cell uptake studies using FITC tagged dextran (FD), a model hydrophilic drug which cannot permeate cell membranes without the aid of a nanocarrier, indicate the LNPs were internalized into the cytosol and provided confidence that this effect can be extended to nucleic acid molecules. A comparative study is underway to determine the delivery efficiency of nucleic acid molecules using LNPs to BC cell lines. We are investigating cellular response with optical microscopy, and probing biochemical changes by migration assays, western blot, and PCR. Our final goal is to utilize the LNP platform to perform gene therapy for the treatment of invasive breast cancer.

Characterization of pH-Sensitive Polymers for the Delivery of Hemophiliac Factors

Sarena D. Horava[1] and Nicholas A. Peppas[1,2,3]

[1]Department of Chemical Engineering, [2]Department of Biomedical Engineering, and [3]College of Pharmacy, The University of Texas at Austin, Austin, TX 78712

Oral delivery is a desirable route for a wide range of therapeutic agents for the treatment of systemic or topical diseases. Complexation hydrogels based on anionic monomers have been engineered to protect proteins from the harsh environment of the GI tract and delivery them to the small intestine due to pH-dependent swelling. We have successfully developed environmentally sensitive, input-responsive polymeric systems, notably poly(methacrylic acid) grafted with poly(ethylene glycol), denoted as P(MAA-g-EG), as vehicles for the delivery of insulin (5.8 kDa), calcitonin (3.4 kDa), and interferon alpha (23 kDa). Here we adopt this method to polymeric carrier systems for high molecular weight (HMW) proteins such as hemophilic factors VIII and IX. Hemophilia is a hereditary bleeding disorder characterized by the deficiency of clotting proteins: factor VIII (264 kDa) for type A hemophilia and factor IX (57 kDa) for type B. Here we focus on the optimization of P(MAA-g-EG) based hydrogel systems for the oral delivery of hemophiliac factor IX for hemophilia B treatment.

P(MAA-g-EG) films were synthesized by UV-initiated free radical polymerization using tetra ethylene glycol dimethacrylate and poly(ethylene glycol) (MW=1000) dimethacrylate as crosslinking agents with crosslinking densities of 0.8-1.7 mol % The ensuing polymer compositions were analyzed using FTIR and potentiometric titration. All formulations contained 96.5-96.9 mol % methacrylic acid. SEM micrographs showed a wide polydispersity of size and morphology of polymeric microparticles (45-75 μm). Based on results from dynamic swelling studies, all hydrogel formulations remained collapsed at low pH conditions (pH=1-4) and swelling increased above pH 5-6. Loading and release studies were performed with bovine serum albumin (BSA) (67 kDa), which serves as a size and isoelectric point analog for factor IX. The mass ratio of loaded protein to polymer ranged between 47.9-429.5 μg/mg. The encapsulated protein was retained within the collapsed polymeric network at low pH conditions relevant to that of the stomach and was released at physiologically relevant pH (pH 7.4) with release percentages of 1.1-10.0%. For microparticles loaded with fluorescein isothiocyanate (FITC)-conjugated BSA (FITC-BSA), the protein distribution was examined using confocal microscopy. The secondary structure of encapsulated protein can be determined by FTIR. *In vitro* studies using human colon adenocarcinoma Caco-2 cells showed that exposure (2h) to microparticles at physiological concentrations did not show appreciable cytotoxicity. Characterization of P(MAA-g-EG) networks is necessary for tailoring such systems for HWM protein delivery applications.

Acknowledgements: This work was supported in part by a grant from NIH (R01-EB-000246-20), an NSF-GRFP Fellowship to SDH, and the Fletcher S. Pratt Foundation. We also acknowledge the assistance of Jeremy Binagia.

Uniform chitosan microparticle prepared by microfluidics-aided double emulsion for controlled protein release

Abstract:

Emulsification has been widely used to synthesize spherical particles for drug delivery or food processing because of their high material encapsulation level and efficiency. However, emulsification by bulk mixing suffers from poor control of size and batch-to-batch variability. To improve on these aspects we developed a microfluidics-based W/O/W double emulsion system that is biocompatible and amenable to fine-tuning of the core microenvironment for various cargo encapsulation or studies of biological processes in a small compartment. Here, we produced uniform W/O/W double emulsions with controllable shell dimensions and number of inner droplets. We demonstrated formation of chitosan microparticles inside the droplets triggered by pH change and charge neutralization. Protein loading level and efficiency is significantly enhanced compared to bulk preparation, with additional advantage of well controlled particle size and narrow size distribution.

Surface coating with negatively charged polymers successfully prevented aggregation of microparticles. By modifying microparticle surface with different polymer (heparin, chondroitin sulfate, sodium sulfate), we could modify the loading level and release profile of encapsulated proteins. These coated chitosan microgels provided a biodegradable protective environment for protein delivery. A sustained release of the encapsulated protein mediated by both protein diffusion and hydrogel degradation is obtained for weeks. In summary, we demonstrate a highly controllable protein delivery system, with versatility of generalization for various therapeutic delivery.

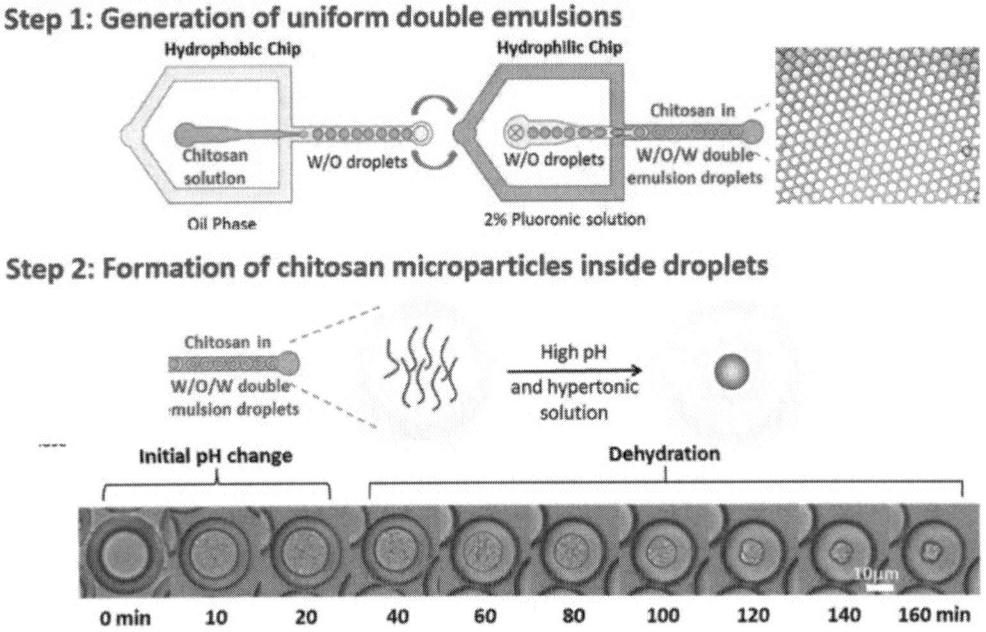

Figure 1. Schematic illustration of chitosan microparticle production by microfluidic systems. For protein encapsulation. Bovine Serum Albumin and other protein of interest is mixed with chitosan solution and co-flowed into the inner inlet for W/O droplet generation.

373603 A Nanofluidic Memristor Based on Anodic Oxide Formation on Silicon Microelectrode

Abstract:

We present a fast and robust fluid-based nonlinear ion memristor, based on a non-equilibrium anodic oxidation reaction of a silicon microelectrode in an aqueous solution, that can be used to synthesize nanofluidic logic circuits and fluid-based memory arrays with on-off logic components. The core components, with confined silicon microelectrodes and metal contacts, can be mass manufactured by simple silicon microfabrication method. Versatile dynamic scaling yields devices at different scales, with predictable performance features from simple geometric scaling, including nanodevices for nanofluidic operations with high on/off resistance ratios and high switching speeds. The memory of ions transporting across the silicon/solution interface is recorded by a continuous formation of discrete oxide layers, which produces a history dependent resistance. During this anodic oxide film formation, high field at the solid/solution interface (> 10^6~10^7 V/cm) is shown to split water after a few layers, giving rise to a large current signature, with a ½ diffusive scaling with respect to time, that originates from the diffusive advance of the hydroxyl ion front across the nanoscale oxide film. The enhanced hydroxyl ion flux from this Faradaic source rapidly oxidizes the silicon film until the entire silicon microelectrode is covered by the oxide layer. Under cathodic bias, hydroxyl ions are extracted from the oxide film and hydride formation decomposes the layer. This rectification phenomenon at different polarization and the memory effect due to continuous oxide film formation under anodic conditions are the key mechanisms behind this robust memristor with distinct on-off current states. The non-equilibrium hydroxyl ion transport in the oxide layer under anodic bias and its contribution to resistance switching is studied by X-ray photoelectron spectroscopy. The nanofluidic memristor is then designed based on this mechanism to achieve a high on/off resistance ratio larger than 10, and a short resistive state switching time of less than 100ms. We will report the use of this memristor in a nonlinear ion circuit, consisting of ionic NAND/NOR gates and latches, that can activate (code) and read (decode) multiple sensors with a binary sequence in a single-input-single-output design. Excitable, bistable and oscillatory ionic circuits that mimick neurons will also be reported.

Functionalized $Ag@SiO_2$ Core-Shell Nanoparticles for Photoluminescence Enhancement of Semiconductor Quantum Dots

Steven A. Warren[1], Jieqian Zhang[2], Trevor D. Ewers[3], Hong Yang[1]
[1]Department of Chemical and Biomolecular Engineering, University of Illinois at Urbana-Champaign, Urbana, IL
[2]Dow Chemical Company, Marlborough, MA
[3]Dow Chemical Company, Midland, MI

This presentation will discuss the synthesis and photoluminescent properties of plasmonic-coupled II-VI semiconductor quantum dot heterostructures consisting of a plasmonic silver nanoparticle core encased in a silica shell of controllable thickness that is surfaced modified for quantum dot adhesion. Plasmonic materials can be made with tunable plasmonic resonance profiles by controlling size, shape, and composition of the plasmonic metal nanoparticles. At specific wavelengths of incident light, metal nanoparticles exhibit a localized surface plasmon resonance (LSPR) in which coupling of the incident photon causes the coherent oscillation of free electrons on the metal's surface. At the LSPR frequencies, a strong electromagnetic field induced by the oscillating electrons extends from the metal surface. This electromagnetic field can lead to enhancement of the fluorescence from fluorophores near the metal surface through one of two proposed mechanisms. Excitation enhancement involves the increase in the number of excited fluorophores due to the enhanced electromagnetic field from LSPR. Emission enhancement involves an increase in the radiative decay rate and a decrease in the fluorescence lifetime, which increases the quantum yield. Previous work showed that silver-silica core shell structures can interact with CdSe/ZnS on the surface of the silica to slightly increase photoluminescence intensity when the silica shells are thick enough (>24 nm). This was suggested to be an example of excitation enhancement, as there was overlap between the silver extinction spectral peaks and the excitation wavelength for the photoluminescence.

In this work, functionalized $Ag@SiO_2$ nanoparticles are produced by depositing silica coating on Ag nanoparticles using a sol-gel approach. The silica shells are further functionalized with thiol groups to improve the attachment of quantum dots. CdSeS/ZnS quantum dots emitting at 490 nm are mixed in solution with these nanoparticles to generate $Ag@SiO_2@QD$ nanostructures. Shell thickness is shown to have a significant effect on the photoluminescence of the $Ag@SiO_2@QD$ solutions. Dispersions that contain silver particles with no shell or very thin shells (<5 nm) exhibit quenching behavior, presumably due to the direct, non-radiative energy transfer between the quantum dots and metal nanoparticles. Thick shells (10-20 nm) yield enhancement in the overall photoluminescence as compared to the quantum dots alone. As the shell thickness further increases, the enhancement decreases as the electromagnetic field effect weakens. In this system, there is spectral overlap between the silver particle extinction and the quantum dot emission, which leads to emission enhancement when the metal-QD spacing is optimized.

Stabilizing Colloidal Crystals by Leveraging Void Distributions

Nathan A. Mahynski
Athanassios Z. Panagiotopoulos

Dept. of Chemical & Biological Engineering
Princeton University

Dong Meng
Sanat K. Kumar

Dept. of Chemical Engineering
Columbia University

Colloidal Crystals

- Useful because they naturally diffract light when $\sigma < \lambda$
  - Potential in optical switches, filters, photonic band gap materials
  - Opals are polymorphic (FCC/HCP) ~ 300 nm Si spheres + H_2O
  - Light diffracts differently off of grain boundaries

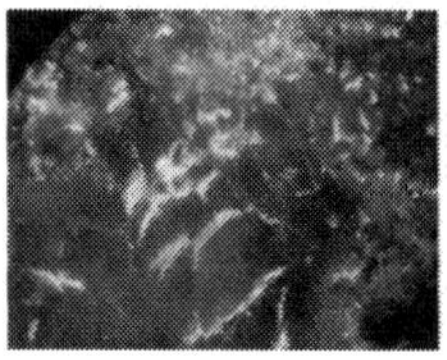

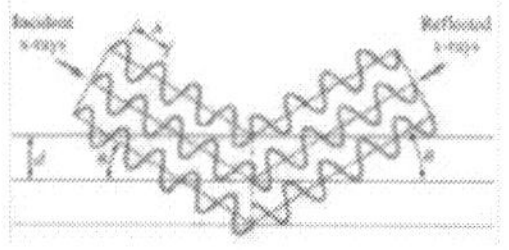

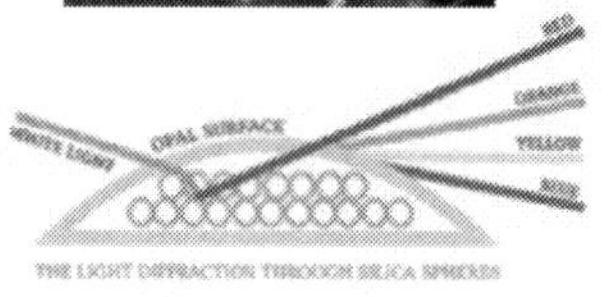

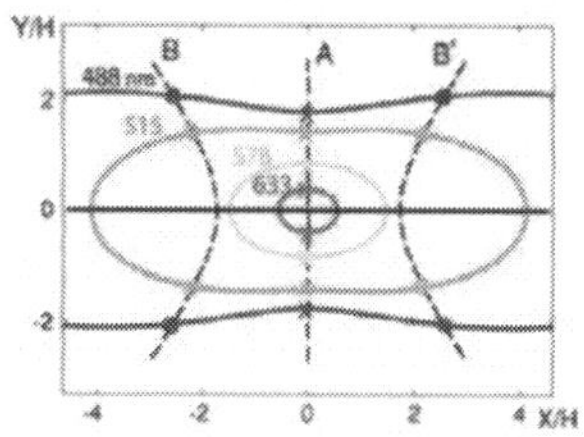

- However, most applications require <u>long range order</u>

Klein and Hurlbut, *Manual of Minerology* (1985)
http://www.minsocam.org/msa/collectors_corner/aam/opal.htm
Baryshev *et al.*, *PRB* **73** (2006)

11/19/14

AIChE Annual Meeting

2

The Dilemma

PRINCETON School of Engineering and Applied Science

- Polymorphs often separated by only marginal differences in free energy
- Many technologically relevant crystals exhibit cubic/ hexagonal polymorphism (*e.g.* diamond, tetrastack)

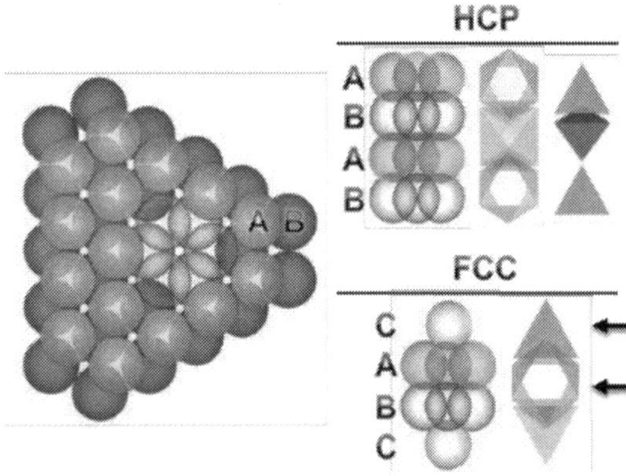

$\Delta F / k_B T \approx 0.001$

CN = 12
APF ~ 74%

$V_{ov} \sim 6V_{tv}$

Bolhuis *et al., Nature* **388** (1997)
Romano and Sciortino, *Nat. Comm.* **3** (2012)

11/19/14

AIChE Annual Meeting

3

The Dilemma

PRINCETON School of Engineering and Applied Science

- Polymorphs often separated by only marginal differences in free energy
- Many technologically relevant crystals exhibit cubic/ hexagonal polymorphism (*e.g.* diamond, tetrastack)

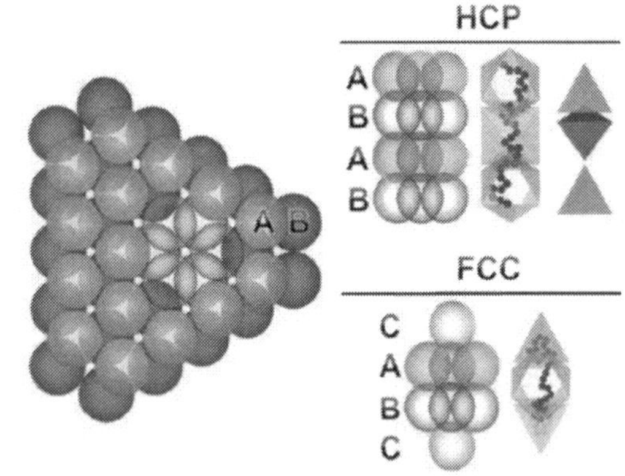

Can we amplify the free energy difference with an additive?

Bolhuis *et al., Nature* **388** (1997)
Romano and Sciortino, *Nat. Comm.* **3** (2012)

11/19/14 AIChE Annual Meeting 4

Depletion Drives Crystallization

PRINCETON

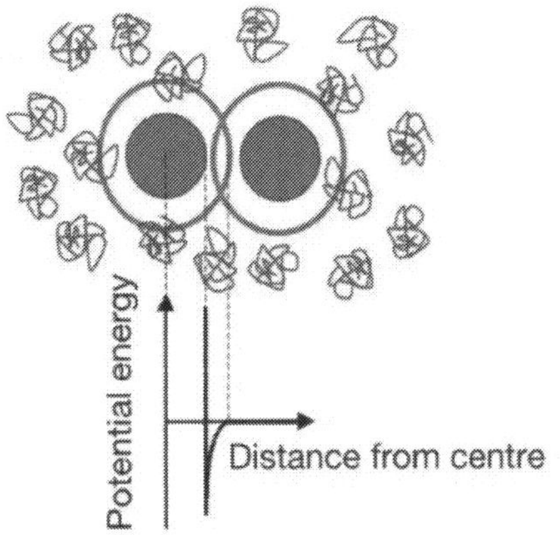

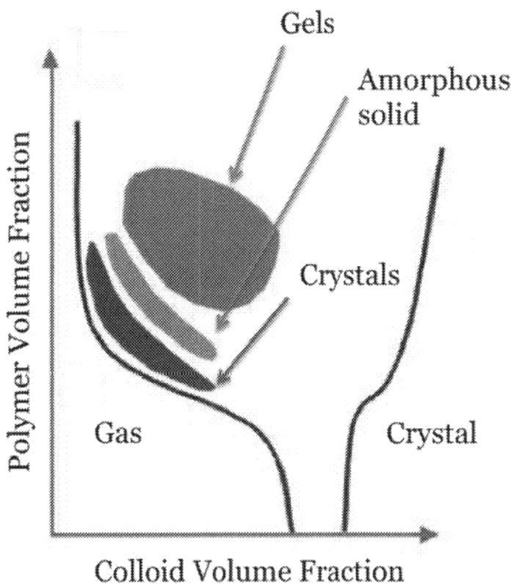

For nanoscale colloids, polymers are not just "blobs"

Anderson and Lekkerker, *Nature* **416** (2002)

11/19/14 AIChE Annual Meeting 5

Measuring Polymer Free Energy

PRINCETON

- Excess chemical potential of polymers can be measured "incrementally":

$$-\beta\mu_i^{ex} = \ln\langle \exp(-\beta U(r_i))\rangle$$

$$-\beta\mu^{ex} = \sum_i -\beta\mu_i^{ex}$$

6.45 : 1

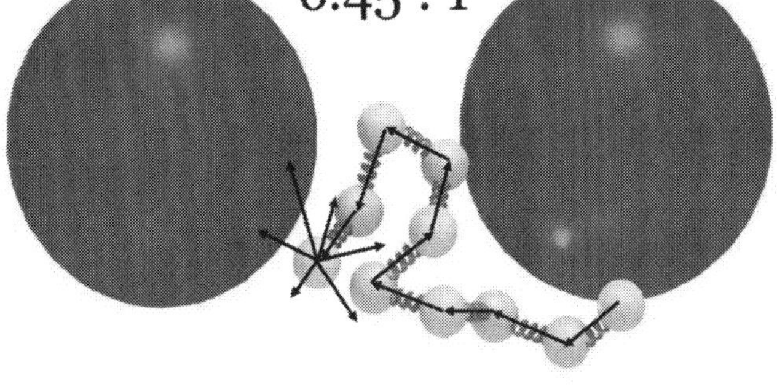

Kumar *et al.*, *PRL* **66** (1991)

11/19/14

AIChE Annual Meeting

6

Polymers Inside Crystals

PRINCETON School of Engineering and Applied Science

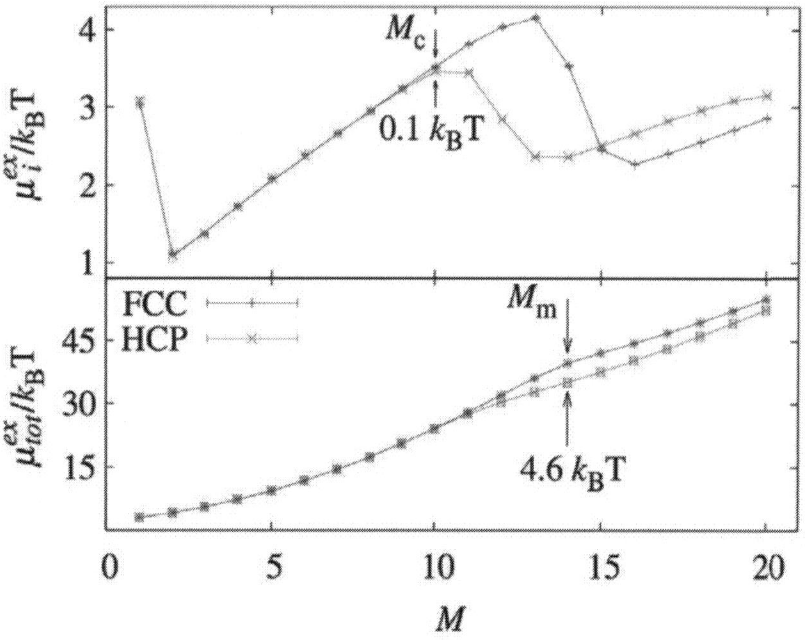

11/19/14

AIChE Annual Meeting

7

Importance of Void Distribution

PRINCETON School of Engineering and Applied Science

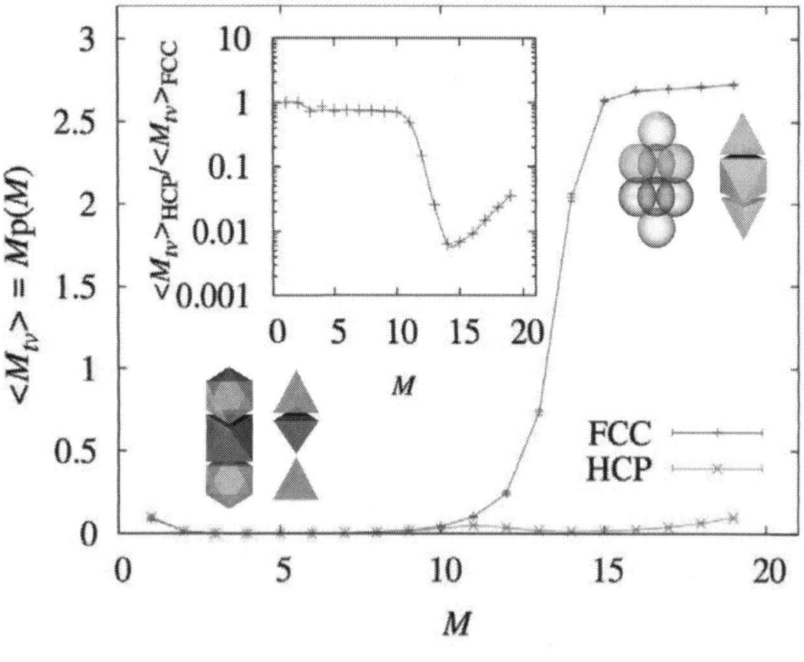

11/19/14

AIChE Annual Meeting

8

Persistence at Realistic Sizes

PRINCETON School of Engineering and Applied Science

$$M_m \sim (\Delta\mu_{tot}^{ex}/k_B T - 2)^{-1/2}$$

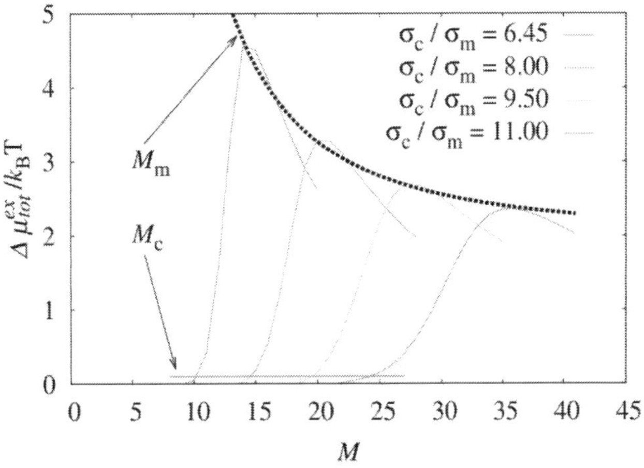

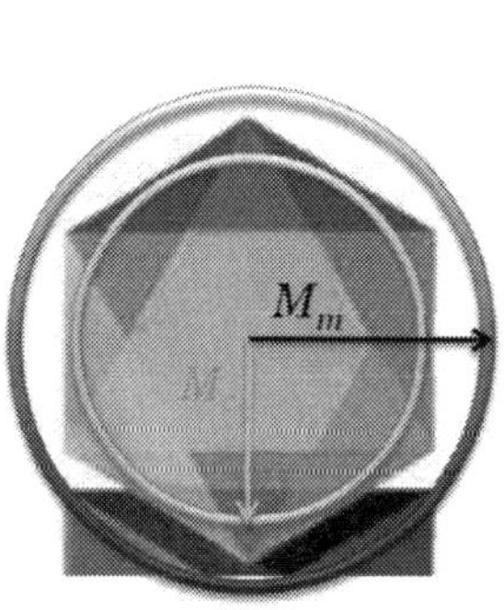

11/19/14

AIChE Annual Meeting

9

Macroscopic Scaling

PRINCETON School of Engineering and Applied Science

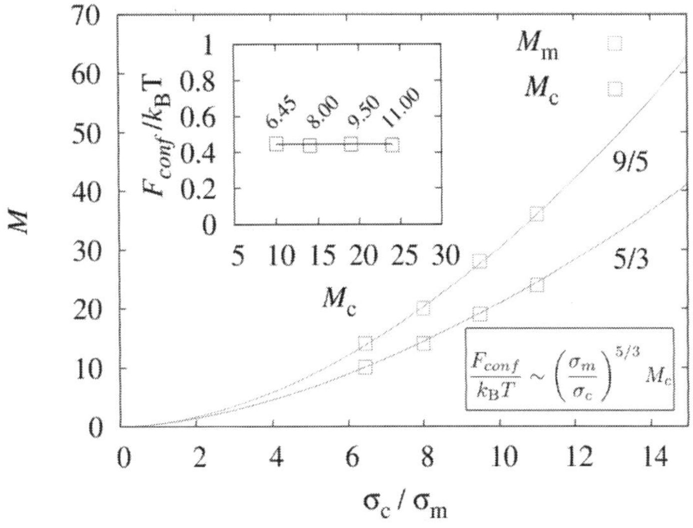

Implicates polymer confinement, not monomer (Kuhn segment) packing

Rubinstein and Colby, *Polymer Physics* (2012)

11/19/14 AIChE Annual Meeting 10

Molecular Dynamics

PRINCETON
School of Engineering and Applied Science

Q: What is the true phase behavior?
Can MC actually predict the most stable polymorph?

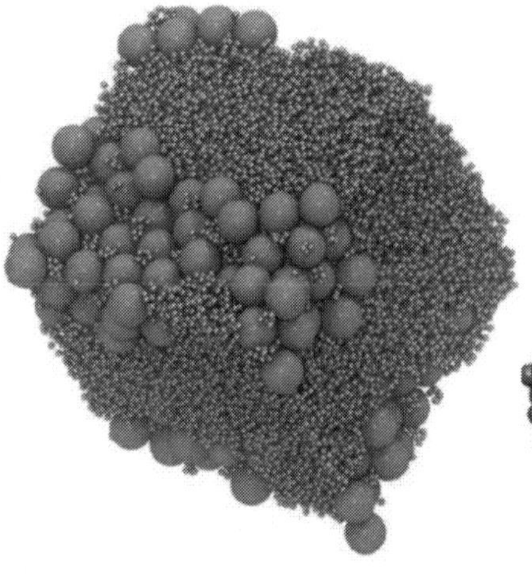

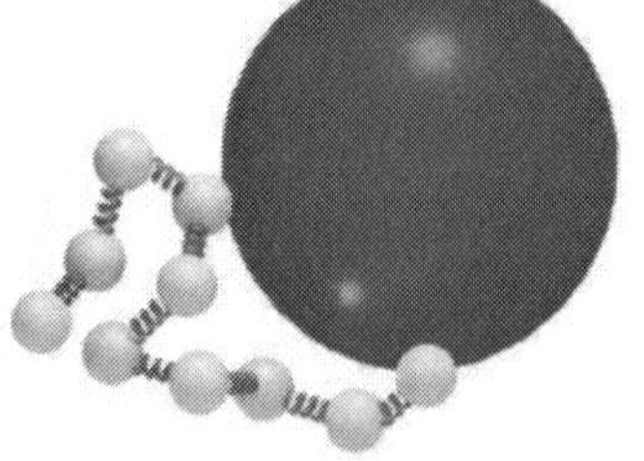

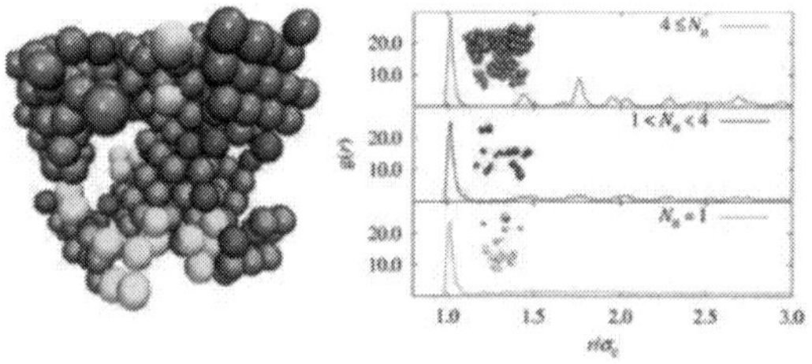

ten Wolde *et al.*, *JCP* **104** (1996)
HOOMD-blue http://codeblue.umich.edu/hoomd-blue (2013)

11/19/14 AIChE Annual Meeting 11

Linear Homopolymers, M = 10

PRINCETON
School of Engineering and Applied Science

GPU acceleration (HOOMD-blue)
7-10 days per 10^6 τ
Total wallclock time 1-2 months

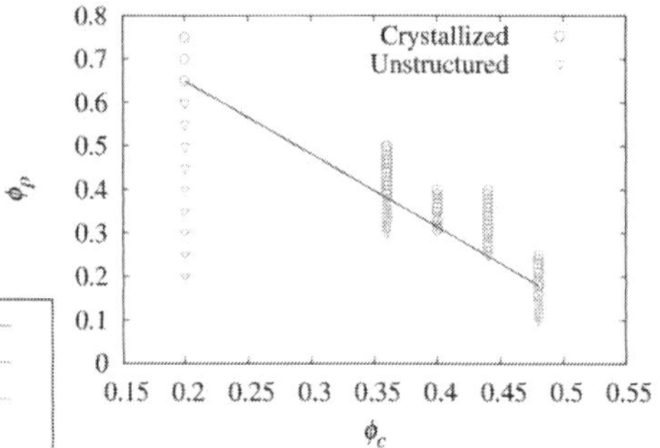

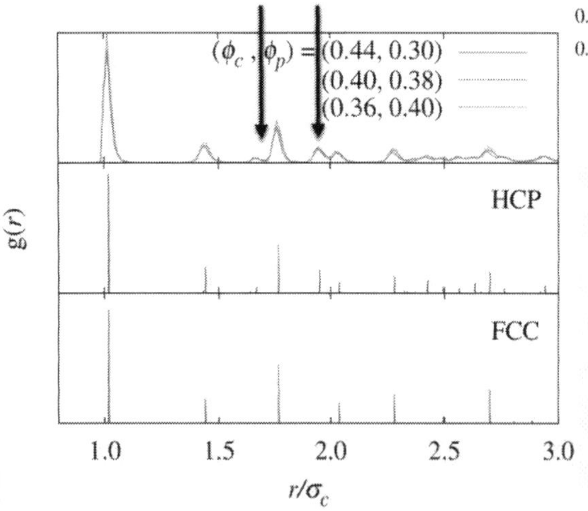

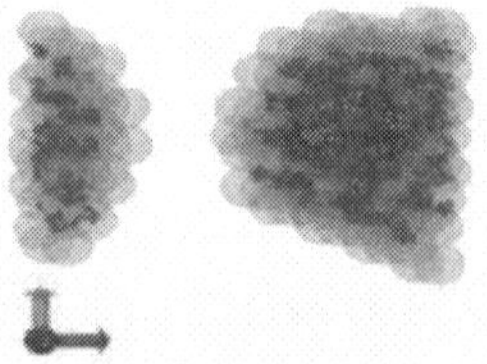

Linear Homopolymers, M = 10

PRINCETON School of Engineering and Applied Science

- Ostwald's Rule of Stages: phase transitions via intermediate metastable states. For hard spheres:
  - Fluid → BCC → HCP → → → → FCC (rHCP)
  - Contested, but the principle of "intermediate states" not disproven
- What about binary colloid + polymer systems?

“Melting”

$\phi_c = 0.2, \phi_p = 0.65\ (M = 10)$

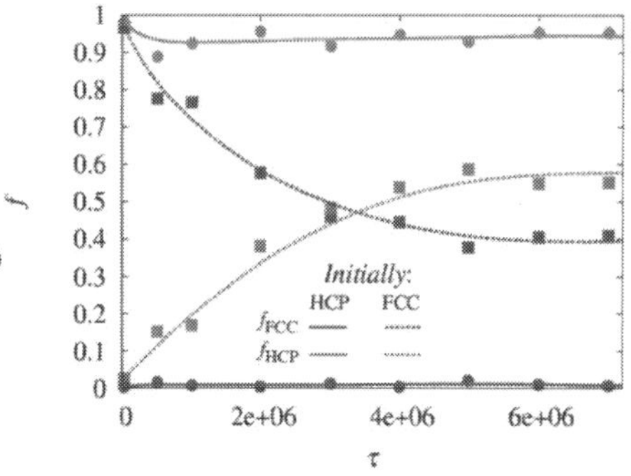

Anderson and Lekkerker, *Nature* **416** (2002)
Russo *et al.*, *Sci. Rep.* **2** (2012)

Block Copolymers

PRINCETON School of Engineering and Applied Science

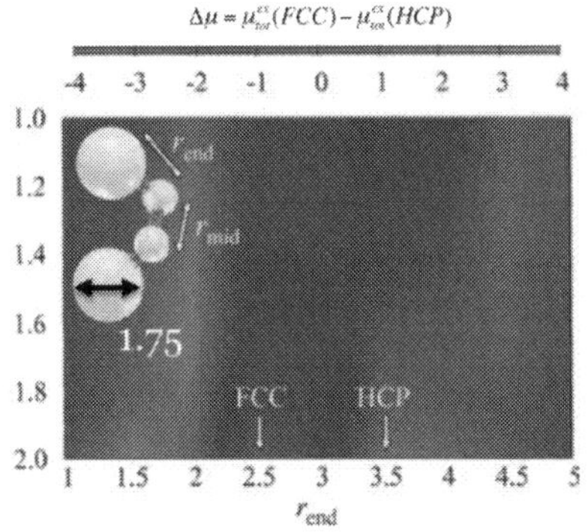

After ~5e7 τ (2 months of wallclock time)

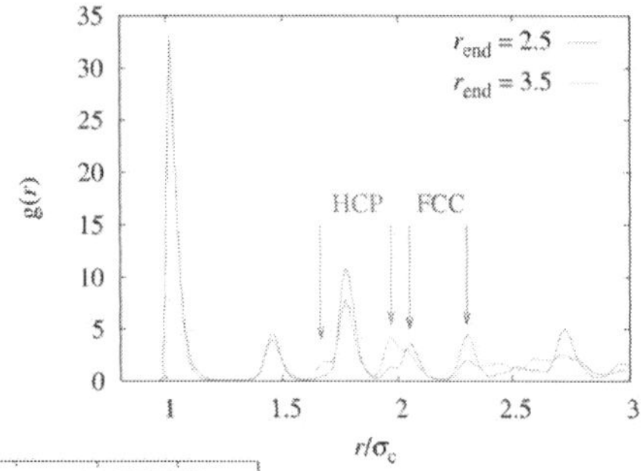

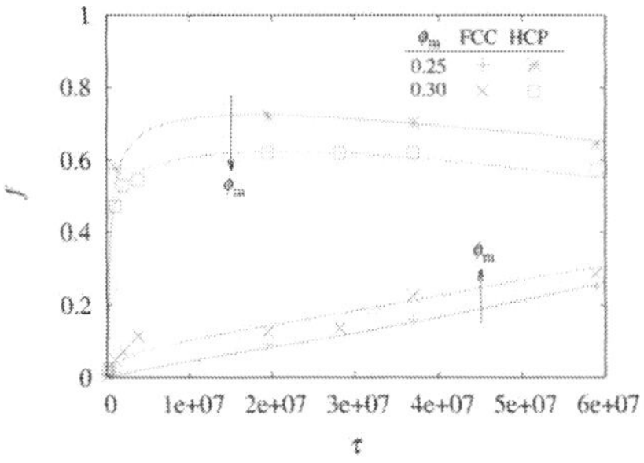

$\Delta\mu$ = -0.42 k_BT

Kinetics dependent upon polymer osmotic pressure

11/19/14

AIChE Annual Meeting

14

Conclusions

PRINCETON
School of Engineering and Applied Science

- In the nanoscale limit, polymer internal degrees of freedom become relevant
- Void-polymer interaction sets polymorph stability by amplifying the free energy difference between them in a binary system
- Polymer's architecture can be intelligently tuned to stabilize one close-packed polymorph over the other
- Purely entropic mechanism

11/19/14

AIChE Annual Meeting

15

Acknowledgements

PRINCETON
School of Engineering and Applied Science

Prof. A. Z. Panagiotopoulos

Prof. S. K. Kumar (Columbia Univ.)

Dr. Dong Meng (Columbia Univ.)

nature COMMUNICATIONS

ARTICLE

Received 9 Apr 2014 | Accepted 20 Jun 2014 | Published 21 Jul 2014

Stabilizing colloidal crystals by leveraging void distributions

Nathan A. Mahynski[1], Athanassios Z. Panagiotopoulos[1], Dong Meng[2] & Sanat K. Kumar[2]

NSF

Questions

PRINCETON
School of Engineering and Applied Science

11/19/14

AIChE Annual Meeting

16

378769 Elucidating the Chemical Origin Underlying Stable Semiconductor Nanowire Growth

Abstract:

Semiconductor nanowires are promising building blocks in the areas of energy conversion, electronics, and photonics due to their unique and tunable optoelectronic properties. Yet, significant challenges in heterostructure formation, uniform doping, and even simple axial growth remain, and stem from a poor understanding of the chemical phenomena governing bottom-up nanowire synthesis. Here, we use *in situ* infrared spectroscopy to quantitatively probe, for the first time, the chemical bonding underlying the vapor-liquid-solid mechanism and show that surface hydrogen atoms are vital for stable nanowire growth. We employ a novel method using infrared absorption spectra to extract the coverage of surface hydrogen, a reactive intermediate species, for Ge nanowires grown with Ge_2H_6 at a range of process conditions (260 – 330 °C, 0.5 – 1.5 × 10^{-4} Torr). Surface hydrogen coverage, which dictates the nanowire sidewall energy, is not a static value as commonly presumed, but instead varies dramatically over a narrow temperature and pressure range. Below a threshold coverage, the catalyst droplet unpins from atop the nanowire and growth abruptly terminates. To underscore the significance of the sidewall energy on vertical nanowire growth, we adsorb $-CH_3$ groups onto the sidewall which prevents droplet destabilization and enables growth beyond the process window of the Ge_2H_6 precursor. Our experiments elucidate a key chemical mechanism underlying nanowire growth and identify surface chemistry as a route to robustly control dopant profile and heterostructure formation, as well as synthesize previously inaccessible nanowire morphologies.

378987 Assembling Quasi-Quaternary Nanocrystal Superlattices for Enhanced Energy Transfer

Abstract:

Multifunctional materials are of great interest for their potential in a variety of fields such as sensing, catalysis, solar cell technology, and drug delivery. Binary nanocrystal superlattices (BNSLs) combine nanocrystals (NCs) with distinct physical properties by self-assembly of monodisperse components. The periodic ordering in BNSLs specifies stoichiometry and interparticle spacing with precision that are not achievable in glassy films or random mixtures and allows engineering of the interactions between the building blocks. Given the wide spectrum of crystalline and quasi-crystalline structures that have been obtained, the opportunities for exploration of these interactions in BNSLs are numerous. In such materials, external stimulation (e.g., light, magnetic fields) of a single component while examining interaction with another can be studied. To increase functionality, ternary NLSs and quasi-ternary NSLs have been demonstrated, but photoluminescent, plasmonic and magnetic components have not been combined in a single assembly.

In this contribution, we show the self-assembly of quasi-quaternary AB_{13}- and AB_2-type NSLs using exclusively core-shell particles as building blocks. The NSLs are made of Au/Fe_3O_4 nanostructures and CdSe/ZnS, CdSe/ZnS/CdS or PbSe/CdSe quantum dots (QDs). We investigated energy-transfer processes in the assemblies by measuring the photoluminescence (PL) of QDs in the final quasi-quaternary NSLs. We demonstrate that energy transfer is enhanced in the assemblies compared to random mixtures due to the maximized number of interactions between the components as a result of the ordering and the higher density of ordered phases.

383988 Enhancement of Visible Light Absorption of Ordered Mesoporous Titanium (Ti3+) and Nitrogen Doped Titanium Dioxide Thin Films for Solar Energy Conversion

Abstract:

The optical and electronic properties of TiO_2 thin films provide tremendous opportunities in several applications including photocatalysis, photovoltaics and photconductors for energy production. The primary challenge of using TiO_2 thin film for solar energy conversion is its inability to absorb the visible light of the solar spectrum due to its high band gap. In this study, visible light responsive cubic ordered mesoporous titanium (Ti^{3+}) and nitrogen codoped TiO_2 thin films (Ti^{3+}-N-TiO_2) are prepared by a surfactant templated sol-gel method. Hydrazine is used as a nitrogen source as well as a reducing agent to form Ti^{3+} in TiO_2. To prepare the films, cubic ordered mesoporous TiO_2 thin films are first prepared using Pluronic F127 as the template and $TiCl_4$ as the precursor. After aging and calcination, the films are dipped into hydrazine hydrate to induce nitrogen doping / reduction, followed by annealing at low temperature. After doping, the films are confirmed to have an accessible, well-ordered cubic mesopore structure by XRD and SEM analyses. XPS analysis shows that Ti^{3+} and both interstitial and substitutional nitrogen are present in the doped films. The Ti^{3+}-N-TiO_2 has a narrower band gap than pure TiO_2, and as a result, the Ti^{3+}-N-TiO_2 films show better visible-light photocatalytic activity than undoped TiO_2 for both the degradation of methylene blue in aqueous solution and hydrogen generation by water splitting. The effects of materials synthesis conditions on the doping process and photocatalytic activity will be discussed.

365480 Systematic Study of the Formation of Bio-Inspired Inorganic Nanoporous Membranes

Abstract:

Systematic Study of the Formation of Bio-inspired Inorganic Nanoporous Membranes

Silo Meoto, Marc-Olivier Coppens

Department of Chemical Engineering and EPSRC "Frontier Engineering" Centre for Nature Inspired Engineering, University College London, London, United Kingdom

Periodic mesoporous silica was synthesized inside the pores of a macroporous host template by acid-catalysed synthesis. Surfactants P123 and F127 were used as structure directing agents to form mesoporous silica with different mesostructures. Significant research effort has been expended on the fabrication of mesoporous silica structures using alumina membranes. However, obtaining a defect-free structure, in which the alumina channels are tightly filled with mesoporous silica, is not as trivial as extensively reported [1]. Additionally, knowledge about the mechanism of nucleation and growth of the silica structures within confined spaces is limited. Therefore, a systematic approach to experimental design has been adapted for the fabrication of silica nanocomposites with tuneable pores within alumina channels. Preliminary results reveal the effect of specific parameters on the development of a crack-free membrane.

Based on the underlying mechanism of the performance of biological nanopores, the diameter and surface chemistry of these hierarchical structures can be tailored by pore functionalization to enhance the selectivity of separations. The synthesized membranes were characterized by scanning electron microscopy (SEM), transmission electron microscopy (TEM), small-angle X-ray scattering (SAXS) and nitrogen adsorption/desorption measurements. The functionality of the membrane performance is ultimately demonstrated by its ability to separate solutes by size exclusion.

[1]. S. Meoto, M.-O. Coppens, *J. Mater. Chem. A*, **2014**, 2, 5640

374512 Current-Driven Evolution of Single-Layer Epitaxial Islands on Crystalline Solid Conductors

Abstract:

The driven assembly of confined quantum structures is of special importance to nanoelectronics and nanofabrication technologies. In this context, a particularly interesting problem is the current-driven dynamical response of single-layer adatom and vacancy clusters, i.e., islands and voids of single-layer thickness/depth, on surfaces of crystalline conducting or semiconducting substrates. In this presentation, we report theoretical and computational results on the current-driven morphological response of single-layer epitaxial islands on crystalline elastic substrates with periphery or edge diffusion being the dominant mode of mass transport.

We develop and validate a transport model for the current-driven dynamics of such single-layer epitaxial islands on crystalline substrates. Simulations based on the model show that the dependence of the stable steady island migration speed v_m on the inverse of the island size is not linear for larger-than-critical island sizes. In this nonlinear regime, we report morphological transitions, Hopf bifurcations, and instabilities for various substrate surface crystallographic orientations and island misfit strains; this strain is due to the lattice mismatch with the substrate for heteroepitaxial islands. A systematic parametric study is conducted to understand the effect on the individual island dynamics of edge diffusional anisotropy parameters, including the angle between a fast edge diffusion direction and the direction of the externally applied electric field. Proper rescaling of v_m accounting for the island morphology gives a universal linear relationship for its dependence on island size.

We also report numerical simulation results on an approach to *surface nanopatterning* based on the current-driven assembly of single-layer epitaxial islands on crystalline substrates. As a first step in obtaining a fundamental understanding of collective island dynamics, we focus on the evolution of pairs of different-size islands driven to coalescence and explore the effects of three key geometrical parameters: the sizes of the two islands of the pair and their center-to-center line misalignment with respect to the electric-field direction. Based on the understanding developed from the dynamics of island pairs, we extend this study to entire populations (distributions) of epitaxial islands on the substrates of interest. We discover various patterns ranging from equal- and different-size stable steady island-pair configurations to many-island patterns that can be tailored by controlling geometrical parameters of the initial island population and the duration of application of the electric field.

370596 Nanostructured Graphene Composite for Supercapacitor

Abstract:

Graphene materials show great potential for application in energy storage applications, in which every carbon atom is on the surface. When incorporated with appropriate transition metal oxides, those graphene materials can provide high pseudocapacitance, hence enhances their energy storage capability. However, the lower electric and ionic conductivities of transition oxide-based electrodes, hinders the charge transfer and consequently limited their energy storage performance. In this work, we show that by slight sulfurizing the oxides, we were able to significantly increase the specific capacitance of a graphene-based supercapacitor and achieve of the specific capacitance of ~1100F/g. Such high performance is mainly from the combined effect from improved electric conductivity provided by the nickel structures, along with improved ion diffusion by the porous electrodes structures. The approach developed in this method provides another venue enhancing the energy storage performance of transition metal oxide-based electrochemical supercapacitors.

385533 Invited Talk: Hybrid Electrodes for Electrochemical Energy Storage

Abstract:

Hybrid electrodes, which contain both electroactive polymers and inorganic materials, are promising for electrochemical energy storage. The electroactive polymer can bring both charge storage and conductivity to an otherwise insulating inorganic material. Further, the two components may synergistically perform better than either component alone. The overall challenge is to balance electrochemical performance (and in some cases mechanical properties) with composition and processing.

In this talk, several hybrid electrodes are presented, including polyaniline-graphene, polyaniline-V_2O_5, and poly(3-hexylthiophene)-block-poly(ethylene oxide) (P3HT-b-PEO)-V_2O_5 electrodes. Each are evaluated as cathodes in non-aqueous batteries. In the first case study, polyaniliine-graphene oxide electrodes are assembled via water-based dip and spray layer-by-layer assembly, followed by electrochemcial reduction of graphene oxide to graphene. In the second case study, polyaniline and V_2O_5 is assembled into layer-by-layer cathodes using layer-by-layer assembly. In the third case study, an electron- and ion-conducting block copolymer (P3HT-b-PEO) is combined with insulating V_2O_5 using a water-based micellization process, resulting in a flexible cathode. The basics of how polymers function in each of these hybrid electrodes will be discussed as well. The future of electroactive polymers in hybrid electrodes is promising because the offer conductivity, eletrochemical activity, and flexibility.

381012 Solid-State Siloxane Polymer Electrolyte for Lithium-Air (O2) Batteries

Abstract:

The need to reduce our dependence on fossil fuels and curb our contribution to global warming has increased interest in the development of electric vehicles. Current electric vehicles are mostly powered by lithium-ion batteries; the most energy dense batteries commercially available. However, lithium-ion batteries do not have the energy density needed for electric vehicles to compete effectively with gasoline-powered cars.[1-2] Therefore, newer battery chemistries are needed. One battery chemistry that can provide energy densities an order of magnitude greater than current lithium-ion batteries is lithium-air (O_2).[3]

Lithium-air (O_2) batteries utilize lithium metal as the anode, and a high surface area cathode (e.g., carbon) with oxygen serving as the active material. Oxygen reduction reactions (ORR) and oxygen evolution reactions (OER) are believed to occur during discharge and charge of a Li-O_2 cell respectively.[3] Despite the promises of high energy densities, several challenges such as electrolyte and electrode (e.g., carbon) instability,[4-6] low rate capability,[7] and poor round-trip efficiency,[6, 8] must be resolved before possible commercialization. In addition, the use of volatile, flammable liquid organic electrolytes pose severe safety concerns. To address some of these challenges, our group has focused on the development of solid polymer electrolytes for lithium-air.

We have utilized a solid polymer electrolyte with a siloxane (Si—O) backbone and alkoxy-like side chain that has a low glass transition temperature (T_g) and no melting transition. Although the synthesized polymer is viscous, it becomes a flexible solid when complexed with a lithium salt—a sign of physical crosslinking between the adjacent alkoxy chains and lithium ions. Ionic conductivity in polymer electrolytes is believed to occur through the segmental motion of the polymeric chain, and ion transport primarily in the amorphous region of the polymer.[9] Therefore, the lack of a melting transition, which corresponds to a lack of crystallinity, and the low T_g which is due to the highly flexible siloxane chain, supports ionic transport in this polymer electrolyte.

This polymer-salt complex was then incorporated in a Li-O_2 cell as a solid electrolyte and the discharge and charge performance of the Li-O_2 cell examined. In addition, the polymer electrolyte appears stable in the highly reactive lithium-air cell environment. Our work has shown that solid polymer electrolytes are a great alternative to flammable liquid electrolytes and can serve as the physical separator in the cell, can support the Li-O_2 electrochemistry, and with further optimization can move these novel batteries closer to commercialization.

1. Bruce, P. G.; Freunberger, S. A.; Hardwick, L. J.; Tarascon, J.-M., Li-O2 and Li-S batteries with high energy storage. *Nature materials* **2011,** *11* (1), 19-29.

2. Girishkumar, G.; McCloskey, B.; Luntz, A.; Swanson, S.; Wilcke, W., Lithium- air battery: promise and challenges. *The Journal of Physical Chemistry Letters* **2010,** *1* (14), 2193-2203.

3. Lu, Y.-C.; Gallant, B. M.; Kwabi, D. G.; Harding, J. R.; Mitchell, R. R.; Whittingham, M. S.; Shao-Horn, Y., Lithium–oxygen batteries: bridging mechanistic understanding and battery performance. *Energy & Environmental Science* **2013,** *6* (3), 750-768.

4. McCloskey, B.; Speidel, A.; Scheffler, R.; Miller, D.; Viswanathan, V.; Hummelshøj, J.; Nørskov, J.; Luntz, A., Twin problems of interfacial carbonate formation in nonaqueous Li–O2 batteries. *The Journal of Physical Chemistry Letters* **2012,** *3* (8), 997-1001.

5. Freunberger, S. A.; Chen, Y.; Drewett, N. E.; Hardwick, L. J.; Bardé, F.; Bruce, P. G., The Lithium–Oxygen Battery with Ether Based Electrolytes. *Angewandte Chemie International Edition* **2011,** *50* (37), 8609-8613.

6. Freunberger, S. A.; Chen, Y.; Peng, Z.; Griffin, J. M.; Hardwick, L. J.; Bardé, F.; Novák, P.; Bruce, P. G., Reactions in the rechargeable lithium–O2 battery with alkyl carbonate electrolytes. *Journal of the American Chemical Society* **2011,** *133* (20), 8040-8047.

7. Lu, Y.-C.; Kwabi, D. G.; Yao, K. P.; Harding, J. R.; Zhou, J.; Zuin, L.; Shao-Horn, Y., The discharge rate capability of rechargeable Li–O2 batteries. *Energy & Environmental Science* **2011,** *4* (8), 2999-3007.

8. Gallant, B. M.; Mitchell, R. R.; Kwabi, D. G.; Zhou, J.; Zuin, L.; Thompson, C. V.; Shao-Horn, Y., Chemical and morphological changes of Li–O2 battery electrodes upon cycling. *The Journal of Physical Chemistry C* **2012,** *116* (39), 20800-20805.

9. MacCallum, J. R.; Vincent, C. A., *Polymer Electrolyte Reviews*. Elsevier Applied Science: 1987.

388381 Controlled Self-Assembly and Ionic Conductivity Via Interfacial Modification of Lithium-Doped Block Polymers

Abstract:

Ion-conducting block polymers have drawn attention as electrolytes in lithium batteries due to their ability to self-assemble into ordered nanostructures. Such nanostructured block polymers permit the simultaneous control over ionic transport and mechanical strength, thus providing an enticing opportunity toward fabricating designer materials for polymer electrolytes. From the standpoint of increasing processability in polymer electrolytes, tapered block polymers provide an efficient way to decouple the processing temperatures from the polymer molecular weight by reducing interfacial interactions between pure blocks; in other words, the order–disorder transition temperatures (T_{ODT}) can be tuned independent of molecular weight using tapered interfaces. Recently, we successfully synthesized poly(styrene-*b*-oligo(oxyethylene) methacrylate) (P(S-OEM)), normal tapered P(S-SOEM-OEM), and inverse tapered P(S-OEMS-OEM) block copolymers with comparable molecular weights and compositions *via* atom transfer radical polymerization (ATRP) (note: the POEM is potentially superior to poly(ethylene oxide) due to its improved room temperature conductivity as a result of the lack of crystallization in POEM.). We found that the block copolymer nanostructures can be tuned by adjusting the taper volume fraction and the taper composition. Additionally, the effective interaction parameters (χ_{eff}) for the salt-doped copolymers were estimated to determine the overall influence of tapering on copolymer assembly. Moreover, the ion mobility in salt-doped materials was examined through Vogel-Fulcher-Tammann temperature dependence to explore the effects of interfacial modification on transport properties of salt-doped block polymer membranes. By developing better controls over interfacial interactions using tapered materials, we are able to design nanostructured membrane systems with increased conductivity, improved mechanical properties, and enhanced processability (reduced T_{ODT}) in ion transport devices.

Hydrated membranes with cocontinuous nanoscale hydrophilic and hydrophobic channels are needed to transport protons between the electrodes of hydrogen fuel cells and other electrochemical devices. In the present work, we study the water uptake and proton conductivity of a model fuel cell membrane comprising a triblock copolymer, polystyrenesulfonate-*block*-polyethylene-*block*-polystyrenesulfonate (S-SES), as a function of water activity in both humid air and liquid water. We demonstrate that the water uptake and proton conductivity of S-SES membranes equilibrated in liquid water are fundamentally different from values obtained when they were equilibrated in humid air. This phenomenon is commonly referred to as "Schroeder's Paradox". Using humidity-controlled synchrotron small angle X-ray scattering (SAXS) and cryogenic scanning transmission electron microscopy (STEM), we discover the morphological underpinnings of our observations. The discontinuous increase in water uptake and conductivity when nearly saturated humid air is replaced with liquid water coincides with the emergence of heterogeneity in the hydrated channels: a water-rich layer is sandwiched between two polymer-rich brushes. The majority of proton transport in water-equilibrated membranes thus occurs in the water-rich layer by Grotthuss mechanism. The sharp discontinuity in either water uptake or conductivity is, therefore, not necessarily paradoxical.

We also demonstrate a new strategy of fabricating block copolymer electrolyte membranes, specifically, by blending and extracting sacrificial homopolymer into and from block copolymer to induce voids, followed by sulfonation. By tuning the porosity of the membranes, the water uptake can be optimized for improved proton conductivities, in both humid air and liquid water. Furthermore, we discovered that the presence of voids also improved the dimensional stability membranes; i.e., they are more resistant to swelling.

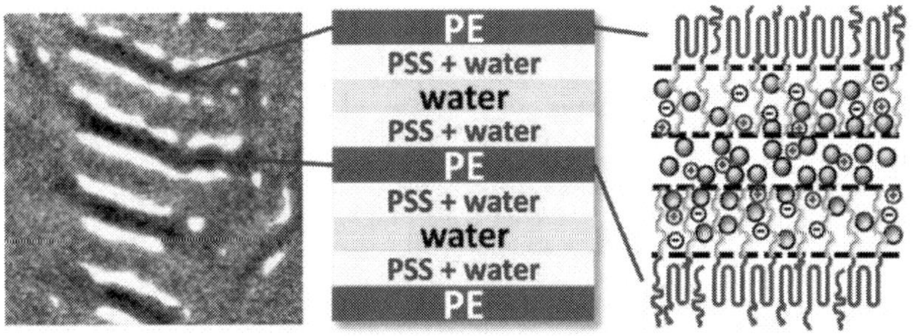

389309 Enabling High-Energy Density Supercapacitors with Conducting Polymer/Redox Biopolymer Composite Electrodes

Abstract:

Improvements in energy storage technologies are necessary to support the deployment of renewable energy generation systems (e.g. wind, solar), electric transportation, and power leveling systems. Future applications in these areas will demand increases in power and energy densities, as well as long term cycle-ability and scalable synthesis processes. Supercapacitors are emerging as promising devices to bridge the gap between conventional capacitors and batteries, and have already seen limited commercialization. A variety of materials can be used as supercapacitor electrodes, ranging from highly porous carbon nanomaterials to metal oxides and conducting polymers. Recent research has shown that redox polymers can be added to conducting polymers (CP) during the polymerization process to increase the Faradaic redox capacitance of the polymer film (e.g. polypyrrole, PPy). In this work, we present how redox polymer properties and process conditions can be utilized to increase the performance of mixed polymer electrodes comprising CP and redox materials for the development of high power and energy density hybrid supercapacitors.

Comprising about a third of the mass of plants and trees, lignin is one of the most abundant biorenewable polymers, yet it possesses very little value aside from its heat content. The random structure of lignin contains aromatic and aliphatic carbon subunits along with various forms of oxygen with phenolic groups that can undergo redox processes at a given electrochemical potential. While the sulfonated from of lignin (sodium lignosulfonate) has been used to enhance the redox capacitance in polymer electrodes, it makes up less than 1% of the available lignin. The widely available form of lignin (alkali lignin, AL) has limited solubility in aqueous acidic solutions thereby limiting the scope and cost advantages of lignin-polymer batteries. In this work, we describe how to overcome the solubility limitations of AL and synthesize PPy-AL electrodes with superior electrochemical performance compared to the sulfonated lignin form. Mixed polymer electrodes are prepared using PPy as the conductive element and various forms of lignin (type, MW, aromatic content) as the redox component. We describe how the film processing conditions (charge, solvent, monomer content, AL weight %) affect the physical and chemical structure of the polymer electrodes to develop processing-structure-electrochemical properties relationships. Optimized electrode compositions show a 2-3 fold increase in redox capacitance when using the alkali lignin relative to the sulfonated lignin. Hybrid supercapacitors comprised of these electrode materials validate the low-cost, high energy density advantages of electrodes containing abundant biorenewable polymers.

377515 Radical Polymers As Transparent Conductors in Organic Photovoltaic and Thermoelectric Applications

Abstract:

Conducting polymers have been studied extensively for their use in a range of energy conversion devices. Previously, π-conjugated polymers have dominated the research focus due to the high degree of electronic delocalization associated with their molecular structure; however many challenges regarding synthetic routes and the control of nanoscale structure continue to prevent their viability in widespread applications. To this end, we will discuss an emerging class of non-conjugated, electronically-active macromolecules, radical polymers, which have shown immense potential to transport charge despite being completely amorphous. These redox-active macromolecules have been used previously in electrolyte-supported applications (*e.g.*, flexible batteries). However, quantifying the ability of these non-conjugated macromolecules to conduct charge in the solid state has not been studied to the same degree.

Here, a model radical polymer, poly(2,2,6,6-tetramethylpiperidinyloxy methacrylate) (PTMA), was synthesized using the reversible addition-fragmentation chain transfer (RAFT) controlled polymerization mechanism, which produced polymers with readily-tunable molecular weights (5 kg mol^{-1} < M_n < 100 kg mol^{-1}) and narrow molecular weight distributions ($Đ \leq 1.2$). Importantly, we have demonstrated, for the first time, that the solid-state hole mobility (μ_h ~10^{-4} cm^2 V^{-1} s^{-1}) and conductivity (σ ~10^{-5} S cm^{-1}) values of these radical polymers are of the same order of many common conjugated polymers [*e.g.*, poly(3-hexylthiophene) (P3HT)]. Furthermore, because the polymer backbone is non-conjugated, these macromolecules are extremely transparent. Because of these promising optoelectronic properties, PTMA lends itself readily to serve as the hole-transporting layer in both regular and inverted architectures of organic photovoltaic (OPV) devices. We have shown that the addition of this radical polymer interlayer facilitated charge transport in both OPV geometries, and it allowed for relatively high power conversion efficiencies to be realized. Additionally, we have established that PTMA has a rather high thermopower value (S > 700 μV K^{-1}), which allowed it to perform at high levels in thermoelectric devices. Therefore, the combination of a well-controlled synthetic methodology with high electronic performance allows these radical polymers to be of great utility when they are incorporated into polymer-based electronic conversion devices.

387160 Membranes for Generation of Fuels from Sunlight

Abstract:

Polymer membranes are a vital component of solar-fuel generators. The requirements for the membranes include primarily low permeability of the gaseous or liquid products, good mechanical strength, and sufficient proton or hydroxide conductivity. The relative importance of these factors in the context of artificial photosynthesis is distinct from the requirements of fuel cell membranes such as Nafion, and present unique design opportunities. Current synthetic efforts in the group represent a multifaceted approach of incorporating new conductive motifs, novel structural elements, and exploring innovative concepts in membrane materials. In order to incorporate all of the necessary properties for a solar fuel membrane our attention has been concentrated around multicomponent block polymer membranes, which satisfy contradictory materials requirements through the fusion of chemically distinct functional building blocks. In this presentation, I will present an overview of membrane development efforts at the Joint Center for Artificial Photosynthesis: A Department of Energy (DOE) Energy Innovation Hub.

Highly Efficient All-Polymer Solar Cells : How to Control Polymer Blend Morphology

Taeshik Earmme, Ye-Jin Hwang and Samson A. Jenekhe
Department of Chemical Engineering
University of Washington, Seattle, WA 98195-1750

Organic photovoltaics (OPVs) based on conjugated polymers are promising for low cost, scalable solar energy conversion. Much progress has been made in the field of bulk heterojuction (BHJ) OPVs with polymer/fullerene derivative systems in the last decade, already reporting power conversion efficiencies (PCEs) approaching 10 %. In contrast, the performance of polymer/polymer (all-polymer) solar cells comprised of both donor and acceptor polymers and free of fullerenes, has been reported with very low efficiency. The development of highly efficient all-polymer solar cells has lagged significantly behind polymer/fullerene systems despite their potential advantages including tunable electronic structure, enhanced absorption coefficients, increased photovoltage, superior thermal/mechanical robustness, and facile control of solution viscosity for industrial coating process.

Here we report control of the blend morphology of all-polymer solar cells by tuning the solution processing which enables formation of nanophase-separated polymer domains in the few tens of nanometers length scale. The morphology of thiazolothiazole donor polymer and naphthalene-diimide based acceptor polymer blend was successfully tuned by co-solvent solution deposition and confirmed by atomic force microscopy (AFM). We show that the domain sizes of the polymers are significantly reduced by using chlorobenzene (CB):*o*-dichlorobenzene (DCB) mixture, leading to more than 20 % improvement in photovoltaic performance. Enhanced and balanced charge carrier mobilities in the nanostructured polymer blend morphology were also confirmed by space-charge-limited current (SCLC) measurements. The best photovoltaic parameter was observed in polymer/polymer solar cells, even exceeding similarly fabricated polymer/fullerene solar cells. These results suggest that control of polymer blend nanomorphology is a promising strategy to achieve high-performance polymer/polymer BHJ solar cells.

Reference

1. **Earmme, T.**; Hwang, Y.-J.; Murari, N. M.; Subramaniyan, S.; Jenekhe, S. A. "All-Polymer Solar Cells with 3.3% Efficiency Based on Naphthalene Diimide-Selenophene Copolymer Acceptor," *J. Am. Chem. Soc.*, **2013**, *135*, 14960-14963.

385941 Domain Compositions and Fullerene Aggregation Govern Charge Photogeneration in Polymer/Fullerene Solar Cells

Abstract:

The relationship between active layer morphology and photovoltaic device performance of organic semiconductor mixtures continues to be obscured by its complex microstructure. For example, in the case of polymer/fullerene solar cells the ubiquitous presence of mixed phases in the active layer creates multiple morphologically distinct interfaces which are capable of either charge recombination or separation. Here, we demonstrate that domain compositions and fullerene aggregation can strongly modulate charge photogeneration at ultrafast timescales through studies of a model system, mixtures of a low band-gap polymer, poly[(4,4'-bis(2-ethylhexyl)dithieno[3,2-b:2',3'-d]germole)-2,6-diyl-alt-(2,1,3-benzothia-diazole)-4,7-diyl], and [6,6]-phenyl-C_{71}-butyric acid methyl ester. We characterized the structure using energy-filtered transmission electron microscopy (EFTEM) and resonant soft X-ray scattering which show similar microstructures albeit the composition maps generated from EFTEM demonstrate that compositions of mixed domains in the microstructure vary significantly with overall film composition. Furthermore, the amount of polymer in the mixed domains is inversely correlated with device performance, which is corroborated by photo-induced absorption studies using ultrafast infrared spectroscopy that demonstrate that polaron concentrations are highest when mixed domains contain the least amount of polymer. Fullerene coherence lengths obtained from grazing-incidence X-ray scattering show correlation to polaron yields. Thus, for efficient charge photogeneration the purity of the mixed domains is critical as it modulates fullerene aggregation and electron delocalization.

392143 Invited Talk: Ionomer Design, Synthesis and Characterization for Ion-Conducting Energy Materials

Abstract:

For ionic actuators and battery separators, it is vital to utilize single-ion conductors that avoid the detrimental polarization of other ions; the commonly studied dual-ion conductors simply will not be used in the next generation of materials for these applications. *Ab initio* quantum chemistry calculations at 0 K in vacuum characterize ion interactions and ion solvation by various functional groups, allowing identification of constituents with weak interactions to be incorporated in ionomers for facile ion transport. Simple ideas for estimating the ion interactions and solvation at practical temperatures and dielectric constants are presented that indicate the rank ordering observed at 0 K in vacuum should be preserved. Hence, such *ab initio* calculations are useful for screening the plethora of combinations of polymer-ion, counterion and polar functional groups, to decide which are worthy of synthesis for new ionomers. Single-ion conducting ionomers are synthesized based on these calculations, with low glass transition temperatures (facile dynamics) to prepare ion-conducting membranes for ionic actuators and battery separators. Characterization by X-ray scattering, dielectric spectroscopy, NMR and linear viscoelasticity collectively develop a coherent picture of ionic aggregation and both counterion and polymer dynamics. Examples are shown of how *ab initio* calculations can be used to understand experimental observations of dielectric constant, glass transition temperature and conductivity of polymerized ionic liquids with counterions being either lithium, sodium, fluoride, hydroxide (for batteries) or bulky ionic liquids (for ionic actuators).

386801 Macroscopic Observation of Structure-Property Relationships in Bulk Poly(Ionic Liquid) Materials

Abstract:

Polymerized ionic liquids or "poly(ILs)" are of interest as conductive materials, gas separation membranes, novel surface coatings and a number of other thin film applications. However, we have found that poly(ILs) also exhibit unique and tunable behaviors and properties. Here, we focus on the macromolecular properties of poly(ILs) with various substituents appended to the imidazolium cation. Photopolymerization was employed to produce uniform material samples, which were then swelled in a variety of different organic solvents. Remarkably high swelling capacities were observed in certain solvents (e.g. > 200x in DMSO), indicating IL monomers produce long polymer chains that are loosely crosslinked. The relationships between IL structure, polymerization conditions and swelling solvent will be discussed. Furthermore, we have also observed self-healing and shape-memory properties.

Heavily doped poly(3,4-ethylenedioxythiophene) (PEDOT) thin films with high carrier mobility deposited using oxidative CVD (oCVD) for organic photovoltaic applications: conductivity stability and carrier transport

Sunghwan Lee and Karen K. Gleason

Conducting polymers based on PEDOT have received extra attention for potential use as hole transport layers and transparent electrodes (anode) in organic photovoltaic (OPV) applications for its ability to be deposited at lower temperature (RT-150 °C) in addition to high conductivity (>~500 S cm^{-1}) and optical transparency (>85 % in the visible regime).

Oxidative chemical vapor deposition (oCVD) PEDOT offers significant improvements in process and performance over conventional solution-based PEDOT:PSS. High purity gas phase process enables oCVD PEDOT to have the absence of dewetting defects causing non-uniformity in surface roughness and conductivity, and inherently conformal coating for the patterned substrates compared to liquid phase-processed PEDOT:PSS. These advantages make oCVD PEDOT suitable for large area, flexible and transparent devices on polymer and even paper substrates. Previous researchers reported[1,2] the electrical conductivity and structural properties of oCVD PEDOT. Also top- and bottom-illuminated organic photo voltaic (OPV) devices were demonstrated[3,4] using oCVD PEDOT electrodes on various substrates including papers which show the power conversion efficiency of ~2-3 %.

Achieving stable conductivity over time after low temperature annealing in air is the next challenge for the implementation of PEDOT electrodes for OPV applications as unexpected decrease in conductivity of PEDOT films has been reported[2,5-7]. Although the _ENREF_2decreases in PEDOT conductivity reported in the literature[2,5-7] lead to significant degradation of OPV device performance, the mechanisms for the conductivity instability have hitherto not been fully understood. The present study sheds new light on these mechanisms. There are four novel aspects of our work: (1) First, we characterize the variation in charge carrier density and carrier mobility during an accelerated aging experiment (i.e., annealing process at elevated temperatures) to establish the mechanism for the conductivity decrease. Previous researchers reported the conductivity decrease during annealing in air and speculated that this instability may be caused by dedoping. However, since the electrical conductivity is a function of carrier density and carrier mobility, further study is required to investigate the change in both these factors so that we can more precisely understand the mechanism for the conductivity decrease that occurs in vapor-phase polymerized PEDOT films. The present study, using a small current AC Hall Effect measurement system, reveals that a significant decrease in carrier density is the primary factor contributing to the decrease in conductivity during annealing in air and verifies the previous speculation on dedoping. (2) Also, we clearly identify for the first time the Fe-related dedoping phase of $Fe(OH)_2$ through X-ray diffraction analysis before and after the annealing process. We also establish the dedoping mechanisms, which at least partially operate in vapor-polymerized PEDOT films with typical $FeCl_3$ oxidant to produce the $Fe(OH)_2$ dedoping phase. (3) For carrier transport, a plot of carrier mobility vs. carrier density distinctly shows the trend that an increase in carrier density leads to lower mobility in the region of carrier density higher than ~10^{20} cm^{-3}. This trend is attributed to an increase in counter-anions that play roles as ionized scattering centers. The fit curve calculated by the well-known Brooks-Herring-Dingle (BHD) theory[8,9] strongly supports the ionized impurity scattering mechanism that governs the carrier transport in heavily-doped high carrier density PEDOT film. This is the first time that BHD theory, typically used for inorganic materials, has been applied to PEDOT. (4) Additionally, the electrical measurements show that 100% of the monomer units can be doped (~90% in the present study), which is much higher than the previous reports that describe that only 1/3 of the units can be doped[10].

The findings regarding the mechanisms for the PEDOT conductivity decrease and carrier transport behavior are important for OPV-device stability studies that show the strong effects of air exposure and low-temperature annealing on device performance degradation.

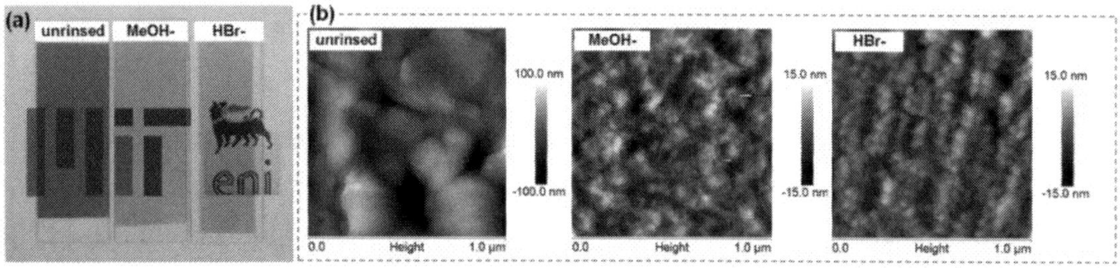

Figure 1. (a) Photographs and (b) AFM micrograph images of the oCVD PEDOT films of the unrinsed, MeOH-rinsed and HBr-rinsed samples in as-deposited state, all deposited at the same run.

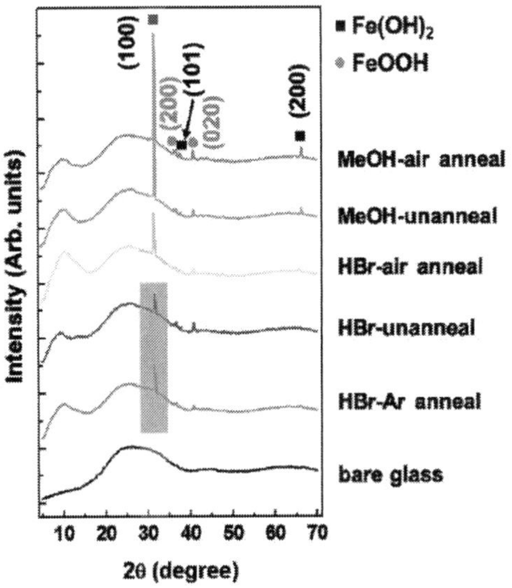

Figure 2. Bragg-Brentano x-ray diffraction patterns taken from the samples of HBr- and MeOH-rinsed PEDOT films before and after annealing in air and Ar ambient with a reference bare glass substrate.

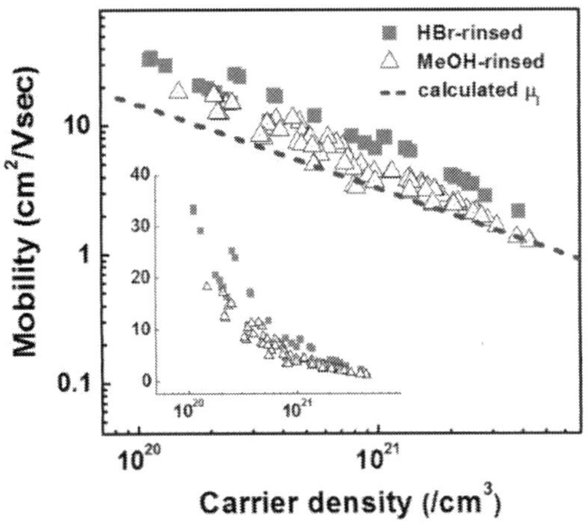

Figure 3. Carrier mobility as a function of carrier density in HBr- and MeOH-rinsed PEDOT films. The calculated mobility (μ_I) due to ionized charge centers is also shown by dotted line. At carrier density higher than ~10^{20} /cm^3, the mobility decreases with increasing carrier density which shows that the carrier transport is governed by the ionized impurity scattering.

1 S. G. Im, K. K. Gleason, and E. A. Olivetti, Applied Physics Letters **90** (2007).
2 R. M. Howden, E. D. McVay, and K. K. Gleason, Journal of Materials Chemistry A **1,** 1334 (2013).
3 M. C. Barr, J. A. Rowehl, R. R. Lunt, J. Xu, A. Wang, C. M. Boyce, S. G. Im, V. Bulovic, and

K. K. Gleason, Advanced Materials **23,** 3500 (2011).
[4] M. C. Barr, R. M. Howden, R. R. Lunt, V. Bulovic, and K. K. Gleason, Advanced Energy Materials **2,** 1404 (2012).
[5] H. Chelawat, S. Vaddiraju, and K. Gleason, Chemistry of Materials **22,** 2864 (2010).
[6] K. Kawano, R. Pacios, D. Poplavskyy, J. Nelson, D. D. C. Bradley, and J. R. Durrant, Solar Energy Materials and Solar Cells **90,** 3520 (2006).
[7] K. Norrman, M. V. Madsen, S. A. Gevorgyan, and F. C. Krebs, Journal of the American Chemical Society **132,** 16883 (2010).
[8] R. B. Dingle, Philosophical Magazine **46,** 831 (1955).
[9] H. Brooks, in *Advances in Electronics and Electron Physics*; *Vol. Volume 7*, edited by L. Marton (Academic Press, 1955), p. 85.
[10] A. Elschner, S. Kirchmeyer, W. Lovenich, U. Merker, and K. Reuter, *PEDOT: Principles and Applications of an Intrinsically Conductive Polymer* (Taylor & Francis, 2010).

387660 Self-Assembly and Mechanical Properties of Graphene/Acrylic Triblock Copolymer Gels

Abstract:

Polymeric gels are used in a wide variety of applications ranging from food materials to drug delivery to tissue engineering. Investigation of gel mechanical properties as a function of their microstructure is a significant research interest. This study presents the effect of graphene (or few-layer graphene) on the self-assembly and the mechanical properties of a thermoreversible gel consists of a physically cross-linked poly (methyl methacrylate)-poly (n-butyl acrylate)-poly (methyl methacrylate) [PMMA-PnBA-PMMA] triblock copolymer in 2-ethyl-1-hexanol, a midblock selective solvent. Graphene was obtained by sonicating exfoliated graphite in 2-ethyl-1-hexanol at various concentrations, 0.1-0.5 mg/mL. The level of exfoliation of graphene layers and the dimensions of individual graphene sheets were characterized using transmission electron microscopy. The dispersed graphene was then incorporated in a series of gels and the mechanical characterization of these gels was performed using a rheometer in oscillatory shear mode (small- and large-amplitude) and in compression mode. The effect of graphene on relaxation behavior of these gels was also studied. Small angle X-ray scattering (SAXS) was used to investigate the effect of graphene on the microstructure of these gels. The change in number of end blocks per junction zone and the average spacing between the junction zones were estimated from the SAXS data. The results indicate that the presence of graphene affects the self-assembly, gel points and the mechanical properties of the physical gels considered here.

371257 Nanomechanical Properties of Poly(*para*-phenylene vinylene) Determined Using Quasi-Static and Dynamic Nanoindentation

Abstract:

Mechanical properties of poly(*p*-phenylene vinylene) (PPV) conjugated polymer were determined by quasi-static and dynamic nanoindentation [1]. PPV films were prepared by thermolytic conversion of poly[*p*-phenylene (tetrahydrothiophenium)ethylene chloride] precursor films, at different temperatures. The mechanical properties of the films showed a dependence on the extent of conversion and chemical composition of the films. The storage modulus, E_r', and plasticity decreased with an increase in conversion, whereas the loss modulus, E_r'', showed the opposite trend. Both the precursor and the fully-converted PPV films were found to have significantly lower E_r'' than E_r', consistent with the glassy nature of the polymers at room temperature. The fully-converted PPV polymer was found to have room-temperature loss and storage moduli of 0.2 and 4.4 GPa, respectively. A comparison of the contact mechanical properties of the isotropic PPV films synthesized in this work, with the results of molecular dynamics simulations, and with previously reported values for stretch-oriented anisotropic films determined by conventional tensile testing methods, will also be discussed.

[1] Sreeram, A.; Patel, N. G.; Venkatanarayanan, R. I.; McLaughlin, J. B.; DeLuca, S. J.; Yuya, P. A.; Krishnan, S. Nanomechanical properties of poly(*para*-phenylene vinylene) determined using quasi-static and dynamic nanoindentation, *Polymer Testing*, In press. DOI: 10.1016/j.polymertesting.2014.04.012.

375647 Equilibrium Water Sorption in Poly(ethylene furanoate) Compared to Poly(ethylene terephthalate)

Abstract:

Recent introduction of poly(ethylene furanoate) (PEF), which is a bio-based, performance-enhanced furanic analog to poly(ethylene terephthalate) (PET), has sparked widespread investigation into previously un-reported fundamental properties of this polymer. The current work examines equilibrium water sorption properties in amorphous PEF and PET at 35°C over the entire water activity range. PEF exhibits a largely increased equilibrium water sorption uptake of ~1.7X compared to PET at unit activity, resulting from substitution of the non-polar phenyl ring in PET with the polar furan ring in PEF. Both polyesters exhibit dual-mode sorption up to ~0.6 activity, after which deviation is observed via a noticeable upturn in concentration vs. activity for both polyesters. Excellent agreement is found between three independent sorption measurement techniques, thereby providing a consistency check for the reported data. Supplementary characterization techniques are utilized to provide a more detailed understanding of the morphological changes induced by the increased sorption at unit activity.

379294 Structure and Dynamics of Star Polymer Melts Above Glass-Transition Temperature

Abstract:

Structural and dynamical properties of star-shaped polymer melts were investigated by performing molecular dynamics simulations based on a bead-spring model. Our model is based on an extension of Dobkowski's chain-length dependence correlation function [Eur. Polym. J. , 1982, 18 , 563]. It accounts for the fact that *f* chains of specific lengths are tethered to a branch point and it describes the properties under melt conditions and above the glass transition temperature, which, to date, are poorly understood. Although, simulations of the densities and isokinetic temperatures, based on the canonical definition of the laboratory glass-transition (T_g), are well described by the correlation function, a subtle behavior is manifested as the architecture becomes more complex. High functionality stars exhibit the opposite behavior from that of linear chain polymers and low functionality stars that is with increasing the arm length the density decreases rather than increasing. The difference between values of T_g and densities of low and high functionality stars is more pronounced for short arm lengths; for sufficiently long arm lengths, the behavior of stars of all functionalities is similar. Moreover, we find that increasing the arm length and the functionality leads results in a weaker dependence of the relaxations on temperature, i.e.: less fragile glass-forming behavior. Complementary results such as the specific volume and number of neighbors in contact provide further insights on the subtle relation between the structure and dynamics. The findings provide fundamentally important insights toward understanding the processing of polymers, colloids and nanocomposites.

Formation and distribution of rubbery phase in high impact polypropylene particles

K. Smolná[1], T. Gregor[2], J. Kosek[1]

[1]*Institute of Chemical Technology Prague, Technická 5, 166 28 Praha 6, Czech Republic, e-mail: juraj.kosek@vscht.cz;*

[2]*New Technologies – Research Center, University of West Bohemia, Univerzitní 8, 306 14 Plzeň, Czech Republic; e-mail: gregort@ntc.zcu.cz*

The detailed information about the distribution of pores and rubbery (EPR = ethylene propylene rubber) domains inside and on the surface of high impact polypropylene (hiPP) particles is needed both for the fundamental understanding of the process of EPR formation and for the optimization of the manufacturing process. To obtain the comprehensive information about the hiPP morphology, the set of hiPP particles originating from three catalysts was visualized by X-Ray micro-computed tomography (microCT) and atomic force microscopy (AFM). The combination of these techniques provides the overall evidence of the hiPP powder morphology with a good contrast between pores, PP and EPR phases. To discriminate between various theories explaining the EPR formation and evolution, the effect of the initial homopolymer porosity (originating in the used catalyst) and rubber content was studied for the statistically relevant number of particles. The explanation of the EPR layer formation on the particle surface is based on the systematic investigation of particle morphology in dependence on the homopolymer particle porosity, the EPR content and the antistatic agent which deactivates active centers of catalyst close to particle surface.

The principal findings of our research are:

- Contrary to a common belief we found that the copolymerization starting from homopolymer iPP particle of a low porosity (less than 10%) leads to a commercially viable products and that the most uniform EPR distribution is achieved.
- The fine EPR network was found in all studied samples with the rubber content from 10 to 50 wt.%. The detailed AFM images support the theory that EPR forms and advances along the iPP primary domains.
- Diffusion limitations in the high-impact PP are less significant than it was thought previously.
- EPR does not cover the entire particle surface but creates the rubber pools there. The quantitative characterization of these pools allows to address the problem of particle agglomeration.

Reference

Smolná K, Gregor T, and Kosek J. European Polymer Journal 2013;49(12):3966-3976.

Microstructure, Crystallization behavior and Surface Properties of Novel Fluorinated Copolymers Containing Short Perfluoroalkyl Chains

Qinghua Zhang, Xiaoli Zhan*, Fengqiu Chen

Department of Chemical and Biological Engineering, Zhejiang University, Hangzhou, 310027, P. R. China

*Corresponding author: xlzhan@zju.edu.cn

Fluorinated acrylates polymers with perfluoroalkyl side chain polymers have been widely explored because of their unique surface properties, such as the distinct oil and water repellence, low refractive index and high chemical resistance. Unfortunately, poly(fluoroalkyl acrylate)s with long fluoroalkyl side chain, such as perfluorooctyl or perfluorooctanesulfonyl groups, have been confirmed to produce perfluorooctanoic acid (PFOA) and perfluorooctanesulfonate (PFOS) by hydrolyzation and oxidative degradation. Developing novel environment friendly fluorinated monomers and polymer materials is attracting the attention of the researchers.

The aim of the present work is developing an efficient synthetic approach to nontoxic and nonbiopersisten fluorinated acrylate containing a sulfonamide group as a space link between the fluorinated chain and acrylic group. The reactivity of the novel fluorinated monomers were investigated, and the microstructure, crystallization behavior and surface properties of the fluorinated copolymers were explored in detail. The reactivity ratios and the Q-e values were reported for the first time. In addition, the microphase segregation, crystallization structures and the surface wetting properties of the fluorinated homopolymers and copolymers were studied by FTIR, ^{1}H-NMR, dynamic contact angle analysis, differential scanning calorimetry (DSC) and wide-angle X-ray diffraction (WAXD).

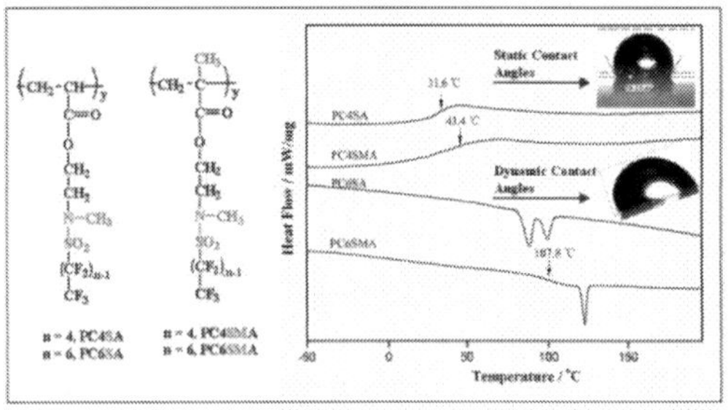

Fig.1 Chemical structure and thermal behavior of novel fluorinated polymers

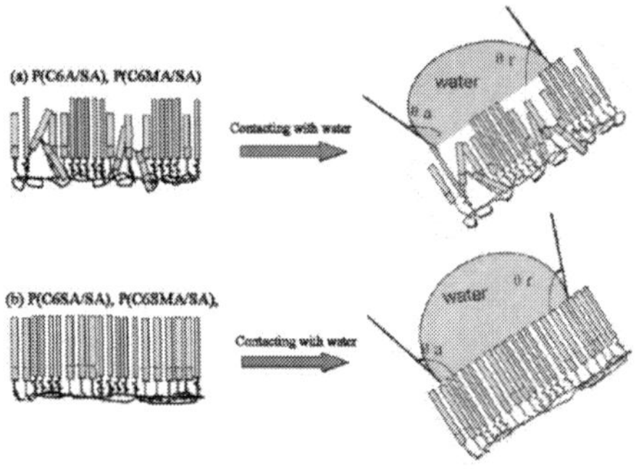

Fig.2 Models for water repellency of fluorinated copolymers

The results showed that the fluorinated side chains interrupted and impeded the crystallizable side chains of SA from forming complete crystals. As a result, the target polymers showed extreme low surface energies and excellent dynamic water repellency, which mainly caused by the synergistic interaction between the fluorinated and nonfluorinated crystalline pendant groups. The research work reported here will help in application of these fluorinated monomers in designing and synthesizing novel environment friendly fluoropolymer materials.

Natalie Girouard
School of Chemical & Biomolecular Engineering, Georgia Institute of Technology
Dr. Shanhong Xu[1], Dr. J. Carson Meredith[1], and Dr. Meisha L. Shofner[2]

[1] *School of Chemical & Biomolecular Engineering*
[2] *School of Materials Science and Engineering*

Abstract

Thermal Stability and Mechanical Reinforcement of Cellulose Nanocrystal/Waterborne Epoxy Composites

One of the building blocks of cellulose structures found in natural materials are individual cellulose crystallites known as cellulose nanocrystals (CNCs). CNCs are reinforcing fillers of emerging interest for polymer composites due to their high modulus and potential for sustainable production. In this study, CNC-based composites with a waterborne epoxy resin matrix were prepared using two different processing strategies and characterized to determine morphology, water content, and thermal and mechanical properties. A major challenge with composite materials is achieving homogeneous nanoparticle dispersion. CNCs have strong intermolecular interactions with each other through hydrogen bonding from hydroxyl groups present on the CNC surface. The surface hydroxyl groups influence the not only the dispersion levels that may be achieved in CNC composites, but also the physical and chemical interactions with the polymer matrix. This study explored the effect of processing by simply altering the addition of mixing on cellulose nanoparticle dispersion and the resulting impact on composite performance. While the level of dispersion achieved was different for the two cases examined in this study, all composites retained the transparency inherent to the neat matrix material. Mechanical and thermal property analyses were correlated with polarized light microscopy in order to explain macroscopic structure-property relationships. Preliminary results have shown that improved nanoparticle dispersion can be achieved through altering the order of addition and without compromising other important variables surrounding composite synthesis. This affords a unique opportunity to compare two systems with varied degrees of nanoparticle dispersion and to assess this impact on bulk composite performance. The results of this study provide evidence that CNCs can improve thermomechanical performance of thermosetting polymers, as observed through changes in the glass transition temperature and modulus values. Overall, the results of this research suggest that CNCs are promising fillers for waterborne epoxy matrices with applications in structural materials and coatings.

370301 Surface Functionalization of Nanocrystalline Cellulose

Abstract:

The chemical modification of nanocrystalline cellulose (NCC) mainly occurs on the surface by introducing functional groups to meet the needs of different applications. The chemical reactions modify the hydroxyl groups of cellulose primarily by esterification, silylation, oxidation and etherification. TEMPO (2,2,6,6-tetramethylpiperidine-1-oxyl) mediated oxidation selectively converts hydroxymethyl group into carboxylic group on the surface of cellulose, which results in a better dispersion of HCl-hydrolyzed NCC. Most of previous efforts on surface functionalization have been focused on the chemical modification of hydroxyl groups of cellulose and improve NCC dispersion in polar organic solvents or its specific mechanical applications. Few researchers have focused on the chemical modification or polymer grafting of NCC to be dispersed in non-polar solvents.

In this study, we will use polyisobutylene succinic anhydride (PIBSA) to graft a hydrophobic tail on the surface of NCC. The succinic anhydride can covalently bond with hydroxyl groups through esterification. In the process, NCC will react with polyisobutylene succinic anhydride (Mw ~950-1300) in toluene at 100°C for 5-30 min. Finally, the resulting nanocellulose-PIBSA will be collected via centrifugation. The grafted NCC will be dispersed in non-polar solvents by mechanical and ultrasonic mixing with the disperser homogenizer. In addition, (2-dodecen-1-yl) succinic anhydride with lower molecular weight will also be used in the polymer grafting to examine the effect of short alkyl chain on the dispersion of NCC.

Injectable hydrogel encapsulated with pH-responsive and amphiphilic copolymer for controlled drug release

Nusheng Chen[1], **Zhaohui Tong*[2],

[1, 2] Agricultural and Biological Engineering Department, University of Florida, Frazier Rogers hall, Museum Road, PO Box 110570 Gainesville, FL. 32611

*ztong@ufl.edu

Abstract: Injectable hydrogel can work as an effective carrier for drug delivery because of its biocompatibility, minimal invasion and excellent permeability. Furthermore, the entrapment of nanoscopic or microscopic structures inside hydrogel can broaden its application and prolong the delivery of target drugs. In this project, a pH-sensitive copolymer: poly ethylene oxide-block-poly dimethylaminoethyl methacrylate (PEO-b-PDMAEMA) was constructed and embedded into a cellulose derived injectable hydrogel in the form of aqueous micelles. The micelle structure can facilitate the encapsulation of hydrophobic substance and thus, enhance the loading ability for the hydrogel. In addition, the release of inclusion could be triggered by the external stimuli due to pH-sensitive of PDMAEMA. Hydrogel and micelles will then work together as double barriers avoiding the burst release of drugs. In this work, PEO was first modified to become macro initiator and the following ATRP (atom transfer radical polymerization) reaction led to the formation of copolymer PEO-b-PDMAEMA. After dialysis, the corresponding aqueous micelles were generated, which was employed for a hydrophobic dye (drug model) encapsulation. Injectable gel was fabricated following the procedure below. Carboxymethyl cellulose (CMC) was oxidized to induce aldehyde groups and modified to add amine groups respectively. Then, Schiff base reaction rendered the formation of gel by simple mixture of cellulose derivatives under mild condition. The properties of injectable hydrogel were characterized in the aspects of gelation time, elasticity, or swelling ratio. The results showed that the gel could be generated in a very short period after blending. Besides, pH-triggered release characteristic of micelles and its injectable gel complex were studied by fluorescent spectroscopy. The results indicated that pH-sensitive and amphiphilic copolymer not only facilitates the loading of hydrophobic substances, but also provides the capacity of pH triggered release.

389396 Chemical Surface Modification of Cellulose Nanofibrils and Their Reinforcement Effect in Polystyrene Matrix

Abstract:

Cellulose nanofibrils (CNFs) have been of a great interest as reinforcing fillers due to their low weight, high aspect ratio, and excellent mechanical strength properties. The advantages of CNFs, however, are often limited by their aggregation tendency and interfacial compatibility with hydrophobic polymer matrices. Chemical surface modification of CNFs by polymer grafting is introduced to limit the intermolecular hydrogen bonding between nanofibrils by functionalizing surface hydroxyl groups of cellulose and to improve wettability with hydrophobic material while maintain its biodegradability and reinforcing capability. In this study, maleated styrene block copolymers were successfully grafted on the surface of CNFs through esterification. The evidences of grafting were identified through changes in chemical functional groups of CNFs. The influences of polymer grafting on the morphological, surface wetting, thermal stability, crystallinity, and solvent redispersion characteristics were evaluated and compared with those of neat CNFs. The reinforcing capability of modified CNFs as reinforcing fillers in polystyrene matrix was assessed. The study also explored the mechanical performance of different nanocomposite processing methods utilizing neat CNFs. The results of this study would provide indication on the effects of CNFs surface modification as reinforcing fillers in hydrophobic polymers.

RESEARCH ON CRYSTALLINE CELLULOSE HYDROGEN BONDS NETWORK

Ting Cui[1, 2], *Jihong Li*[1, 2], *Menghui Yu*[1,2], *Shizhong Li*[1, 2*], *Zhipei Yan*[1, 2]
[1] *Institute of Nuclear and New Energy Technology, Tsinghua University, Beijing 100084, China*
[2] *Beijing Engineering Research Centre of Biofuels, Tsinghua University, Beijing 100084, China*

Abstract

Background

Hydrogen bonds play an important role in crystalline cellulose's stability. However, crystalline cellulose bioconversion is partly inefficient due to the deleterious impact of cellulose multiple hydrogen bonds on enzymatic accessibility. The type of hydrogen bond (interchain and intrachain hydrogen bond) and the total number of hydrogen bond differ a lot among different cellulose allomorphs. However, there was no systematic report about hydrogen bond network of different crystalline celluloses. Thus, in this study, the hydrogen bond network of crystalline cellulose I, II, III was investigated through Fourier transform infrared spectroscopy (FTIR). The relationship between hydrogen bond network and cellulose enzymatic digestibility was studied.

Results

The results of FTIR indicated that the intrachain/interchain hydrogen bond ratio of crystalline cellulose was strongly correlated with cellulose allomorph. The transformation of crystalline allomorph from native cellulose I to cellulose II, III would led to an increase in the number of interchain hydrogen bonds and a decrease in the number of cellulose intrachain hydrogen bonds. Cellulose enzymatic hydrolysis rate had strongly positive correlation with cellulose interchain/intrachain hydrogen bond ratio. The order of crystalline cellulose interchain/intrachain hydrogen bond ratio was as followed: cellulose III > cellulose II > cellulose I.

Conclusion

In this study, through the experimental method, the results revealed that compared to cellulose interchain hydrogen bond, the intrachain hydrogen bond of crystalline cellulose seemed to be more difficult for cellulase to break. Our experimental efforts may provide a new insight for further improvements of lignocellulose pretreatment method.

379982 Shape-Dependent Modulation of Immune Response to Pathogens

Abstract:

The immune system has evolved remarkably to recognize and respond to a wide variety of pathogens and produce distinct immune responses against diverse pathogenic structures. Despite remarkable immunological advancements, the generalized mechanism by which the immune system differentiates distinct pathogens is yet to be elucidated. This is due to vast variations in physical (for example, size, shape, surface charge, functional groups, and hydrophobicity) and chemical (for example, surface antigen) properties of pathogens. In particular, pathogen shape has gained wide attention in recent years for its ability to influence pathogen interaction with key components of the immune system, such as macrophages, dendritic cells, complement proteins and T-cells. However, the significance of pathogen shape in immune response modulation is not yet explored. This is primarily due to the limited availability of safe pathogens, which display similar physicochemical properties, but varying shapes. Moreover, it is technically quite challenging to modify the shapes of biological disease-causing pathogens. A potential solution to these limitations includes the use of safe, non-biological pathogen mimicking structures, which mimic pathogen induced immune responses without the associated adverse effects of a pathogen. To address this issue, we report herein the importance of shapes in mimicking pathogenic structures and their role in modulating the immune response. In the current study, we report hydrophobic polystyrene particles of various sizes and shapes, displaying a model antigen (ovalbumin) on their surface, as pathogen-mimicking structures, and evaluated their role in generating distinct shape-dependent immune responses against the antigen. The results revealed that spherical (200 nm in diameter) particles produced a Th1-biased (cellular) immune response that is characteristic of viruses and rod-shaped (1500 nm in length) particles produced a Th2-biased (humoral) immune response that is characteristic of bacterias. Among different pathogenic structures, smaller spherical particles (200 nm in diameter) produced stronger immune response as compared to larger spherical (500 nm in diameter) or rod-shaped (450 nm and 1500 nm in length) particles. These results provide evidence of shape-dependent modulation of immune system against diverse pathogenic structures. This knowledge can be used to rationally design "pathogen-mimicking" materials and develop next generation vaccines that can induce appropriate immune responses to protect against a wide range of pathogens.

383128 Understanding Protein Nanoparticle Vaccine Adjuvancy through Dendritic Cell Antigen Presentation

Abstract:

Highly conserved pathogen proteins are essential for broadly cross-protective vaccines, but tend to be poorly immunogenic. We have previously demonstrated that protein nanoparticle vaccines made from conserved pathogen proteins trigger specific, adaptive immune responses that soluble protein vaccines cannot[1]. Without excipients or adjuvants, protein nanoparticles eliminate the possibility of off-target immune responses, and their abiotic nature makes them amenable to cold chain-independent storage and use. The mechanisms by which protein nanoparticles enhance component protein immunogenicity are still not well understood. It is hypothesized that dendritic cells, the most potent antigen-presenting cells, and the bridge between innate and adaptive immunity, are responsible for mediating this effect. In this work, we investigate the role of size and surface properties of model ovalbumin (OVA) protein nanoparticles on particle uptake, maturation factor expression and antigen presentation in the JAWS II dendritic cell (DC) line.

Pure OVA nanoparticles of 250, 500 and 750 nm have been synthesized by a modified desolvation procedure. By incorporating Alexa Fluor 488-conjugated OVA into our particles, we demonstrate via flow cytometry that nanoparticles enhance DC protein delivery compared to soluble protein administration. However, enhanced delivery alone does not correlate to a greater inflammatory response. Ensuring that the proteins on the surface of the nanoparticles have retained their secondary structure is also critical and amplifies the DC TNF-α response. Optimal particle formulations are determined by a combination of uptake, TNF-α expression indicating an inflammatory response, and CD86 upregulation indicating dendritic cell maturation. Additional cytokine expression and surface marker upregulation is also investigated to determine the nature of the costimulatory signals induced by protein nanoparticles. Knowledge of critical particle properties identified here will be used for fabricating nanoparticles from recombinantly-expressed influenza proteins, and the relevant biomarkers identified in this study will have their downstream correlates assessed in an *in vivo* mouse vaccination model to further develop protein nanoparticles as an antigen-specific vaccination strategy.

1. Wang, L, Hess, A, Chang, TZ et al. "Nanoclusters self-assembled from conformation-stabilized influenza M2e as influenza vaccines." *Nanomedicine* 2014 Feb;10(2):473-82.

386380 Design of Well-Defined Vaccines from Polyelectrolyte Multilayer Capsules

Abstract:

Challenging diseases such as cancer, HIV, and malaria could benefit from new prophylactic and therapeutic treatments that provide control over the type of immune responses that are elicited. However, vaccines are often formulated with mixtures of adjuvants, antigens, and excipients which may each contribute to the type of immune response that arises. In particular, many recent studies demonstrate that common biomaterials such as PLGA and polystyrene exhibit intrinsic immune activity. This characteristic creates a hurdle for more rational design of future vaccines because the role of the carrier itself may alter how other immune signals (e.g., antigens, adjuvants) are received by the immune system.

To overcome this challenge, we have developed a fully-defined vaccine composed entirely from self-assembled antigens and immune signals. Using a layer-by-layer process, immunogenic components were assembled onto sacrificial microparticles, followed by core removal to create a new type of polyelectrolyte multilayer (PEM) capsule. The resulting microcapsules exhibit loading levels proportional to the number of layers deposited and offer properties (e.g., size, loading, stability) that can be tuned by altering pH and ionic strength during synthesis and core removal. Capsules are efficiently taken up by primary dendritic cells (DCs) (bone marrow-derived) in a dose-dependent manner at levels up to 60%. Observation by laser confocal microscopy revealed capsules distributed throughout the interior of the cell as punctate structures. Corresponding flow cytometry studies demonstrate that capsules assembled from peptide antigen and immune signals efficiently activate DCs, as indicated by upregulation of activation and co-stimulatory surface markers. Ongoing studies are focused on presentation and binding affinity of peptide presented by treated DCs and the subsequent induction of antigen specific T cells *in vitro* and *in vivo*. The approach described here could lead to a platform for design of well-defined vaccines that allow better, more predictable induction of immunity.

383907 Surface Properties of Nanoparticle Vaccines for Potent Pulmonary Mucosal Immunity

Abstract:

Direct pulmonary immunization can provide enhanced local humoral and cell mediated mucosal protection, which is critical for vaccination against lung-specific pathogens such as influenza or tuberculosis. Nanoparticle formulations for pulmonary vaccines offer potential solutions to overcome biological barriers of the lung, inherently targeting APCs and penetrating the mucosal layer. A wide variety of particle formulations have been investigated in preclinical studies for pulmonary vaccine development, yet the role of surface charge on downstream immune responses remains poorly understood. We developed a pair of nanoparticles (NP) with varied surface charge which were covalently attached to a model antigen ovalbumin (OVA) that are otherwise identical in size, shape and antigen loading. OVA-conjugated cationic NPs were found to induce more potent antigen presenting cells (APCs) than anionic NPs both *in vitro* and *in vivo*. Following treatment with cationic NP-OVA, dendritic cells (DCs) induced strong $CD4^+$ OTII T cell proliferation in co-culture at ~10 fold lower concentrations than anionic NP-OVA, which was attributed to increased expression of surface receptors and key cytokines/chemokines. *In vivo*, pulmonary instillation of cationic NPs resulted in enhanced systemic and lung antibody titers, stemming from sufficient particle association with lung DC populations and active particle transport to draining medistinal lymph nodes. Our results indicate that particle surface charge is a main variable in driving T-cell dependent antibody responses *in vivo* and suggest the continued role of cationic NP platforms for engendering potent mucosal responses.

Polyanhydride nanovaccine platform for viral antigens

Julia Vela Ramirez[1], Lorraine Tygrett[2], Habtom Habte[3], Rajarshi Roychoudhury[4], Jihua Hao[5], Neil Greenspan[5], Nicola Pohl[4], Michael Cho[3], Thomas Waldschmidt[2], and Balaji Narasimhan[1]

[1]Department of Chemical and Biological Engineering, Iowa State University, Ames, IA 50011

[2]Department of Pathology, University of Iowa, Iowa City, IA 52242

[3]Department of Biomedical Sciences, Iowa State University, Ames, IA 50011

[4]Department of Chemistry, Indiana University, Bloomington, IN 47405

[5]Department of Pathology, Case Western Reserve University, Cleveland, OH 44106

HIV (Human Immunodeficiency Virus) has been a major challenge for researchers around the world for the past three decades. According to the World Health Organization, there were 35.3 million people infected with HIV in 2012 and in the same year 2.3 million new infections and 1.6 million deaths occurred from the resulting disease, AIDS (Acquired Immunodeficiency Syndrome). The high mutation and recombination rates of the virus, structural features of the spike glycoproteins, and infection and functional subversion of key immune system cells, have made production of a vaccine capable of eliciting effective neutralizing antibody responses for this virus a difficult challenge to overcome with traditional vaccination approaches. In the development of efficacious HIV vaccines, the strategies that have received significant attention include induction of virus-neutralizing antibodies, antigenic stability, and improved patient compliance. Among these strategies, the use of novel adjuvants that can be double as delivery vehicles has been studied. In this context, the use of a platform technology based on polyanhydride nanoparticles for antigen delivery offers advantages in comparison to more standard methods, such as preservation of immunogenicity, sustained antigen release, activation of antigen presenting cells, and immunomodulatory behavior.

Our work is focused on the gp41-54Q-GHC immunogen, which is based on the membrane proximal external region of an HIV-1 glycoprotein (gp41) that elicits broadly neutralizing antibodies. In these studies, the goal was to design a biodegradable nanoparticle-based delivery platform for HIV antigens that could provide an antigenicity-preserving environment for these subunit antigens and elicit a robust immune response. The polyanhydride system studied was based on 1,6-bis(*p*-carboxyphenoxy) hexane (CPH) and 1,8-bis(*p*-carboxyphenoxy)-3,6-dioxaoctane (CPTEG). CPTEG:CPH copolymer nanoparticles encapsulating gp41 proteins were synthesized using anti-solvent nanoencapsulation. In the design of these particle-based vaccine formulations, several approaches were used, including the co-encapsulation of TLR agonists with the gp41 proteins, surface modification of the nanoparticles with targeting moieties such as di-mannose and glycolic acid, and the use of surfactants in the synthesis process. The release kinetics and the structural and functional stability of the immunogens upon release were studied. The particles were administered to BALB/c and C57BL/6 mice using footpad injections and germinal center formation and isotype switching analysis were performed. The results indicated that the chemistry of the formulation played an important role in stabilizing HIV antigens, controlling their release kinetics, and in the formation of germinal centers following different immunization protocols. The studies described herein will aid in the rational design of the polyanhydride nanovaccine platform for future *in vivo* applications.

PAMAM dendrimer as a Platform for Peptide Subunit Vaccine Delivery: Application in the Prevention of *Chlamydia trachomatis* Infection

Ingrid S. Ganda[1,2], Qian Zhong[2], Mirabela Hali[3], Mariene R. Amorim[1], Izabel K. Carvalho[1], Ricardo L. C. A. Junior[4], Francine F. Padilha[1], Judith Whittum-Hudson[3], Sandro R. P. da Rocha[1,2*]

[1]Industrial Biotechnology, Tiradentes University, Aracaju, SE, 49032-490 Brazil.
[2]Department of Chemical Engineering and Materials Science, Wayne State University, Detroit, MI 48201, USA.
[3]Department of Immunology and Microbiology, School of Medicine, Wayne State University, Detroit, MI 48201, USA.
[4]Morphology and Structural Biology, Technology and Research Institute, Tiradentes University, Aracaju, SE, 49032-490 Brazil.

Chlamydia trachomatis is an obligate intracellular bacteria causal agent of many important diseases. Chlamydiae are the most prevalent sexually transmitted bacteria, and cause trachoma, the leading cause of preventable blindness. Chlamydia genital infection is asymptomatic especially in women, and if untreated it can cause chronic pelvic pain, pelvic inflammatory disease (PID) and infertility. The first infection does not provide protection against reinfection and antibiotic treatment can result in persistent chlamydial infection. Vaccination is considered the best strategy to reduce the prevalence of chlamydia, but there is no commercial vaccine available, despite many efforts.

Chlamydial glycolipid exoantigen (GLXA) has been studied as vaccine candidate due to its presence in different species of *Chlamydia* sp. and in all stages of its life cycle, amplifying the possibilities of action of chlamydia-antibodies and has the potential to protect against other *Chlamydia* related diseases (trachoma, pneumonia, rheumatoid arthritis). Peptides that mimic GLXA epitope were generated and patented as potential vaccine candidates. However, stability *in vivo* and uptake by immune cells are some of the challenges in using peptide subunit vaccines.

Nanocarriers can be utilized to encapsulate or conjugate antigens allowing their delivery to specific tissues and cell populations that may improve peptide uptake by immune cells and generate more efficient immune response. Polyamidoamine (PAMAM) dendrimer nanocarriers (DNCs) are monodisperse, tree-like, 3D polymeric systems with multifunctional valences that allow for the conjugation of multiple copies of peptide sub-unit vaccines. In this work we developed a novel peptide-based vaccine platform that was applied to deliver a mimic of a chlamydial antigen – a peptide called Pep4. PAMAM DNCs of generation 4, with –OH terminal groups (G4OH) were modified by the addition of $–NH_2$ terminal groups and the antigenic peptides conjugated through a pH sensitive bond for intracellular targeted delivery, while avoiding release in the high pH physiological environment. This synthetic vaccine was chemically characterized, and its effectiveness evaluated by *Chlamydia* genital infection mouse model.

Four groups of animals (n=8) were immunized with: i) phosphate buffered saline (PBS), ii) Pep4 with aluminum hydroxide (adjuvant), iii) G4OH-Pep4 (Pep4 conjugated to G4OH) and iv) G4OH+Pep4 (Pep4 only mixed with G4OH). After 3 subcutaneous immunizations (2 weeks interval), animals were bleed and sera evaluated by immunostaining. G4OH-Pep4 and G4OH+Pep4 immunized groups generated more chlamydia-specific antibodies than PBS and Pep4+$AlOH_3$ groups. Additionally, G4OH formulations could keep Pep4 properties once the antibodies produced were able to recognize different *Chlamydia* species (*C. trachomatis*, *C. muridarum*, *C. abortus* and *C. pecorum*). After immunization, animals were challenged with *C. trachomatis* infectious form intravaginally. Every week vaginal swabs were collected from all animals. The proposed G4OH-Pep4 vaccine formulation significantly protected animals from *Chlamydia* genital infection by reducing infectious loads. Thirty-five days post-infection, genital tract were collected from all animals and histology slides prepared to evaluated tissue damage. G4OH formulations immunized animals presented less fibrosis and other tissue damages after challenge. No signs of toxicity were noticed during the entire experiment. Notably, G4OH+Pep4 group also presented positive results demonstrating a possible adjuvant effect of G4OH.

In conclusion, G4OH was used for the design of an antigenic peptide subunit vaccine delivery with a pH-sensitive bond for intracellular targeting. G4OH was successful in maintaining the immunogenic activity of the peptides, and was shown to increase its efficiency compared to the free peptide. The G4OH-Pep4 conjugates were shown to enhance the immune system and generate protection against *Chlamydia* infection. Additionally, G4OH was also observed to work as an adjuvant helping the immune system to recognize the peptide. The relevance of this work stems from the fact that the proposed system is a potential platform for the development of novel vaccines for a broad range of infectious diseases.

Keywords: subunit vaccine, vaccine nanocarrier, polyamidoamine

389206 Synthesis and Characterization of Chemically Modified Immunostimulatory Polysaccharide Serving Dual Function As Adjuvant and Protein Antigen Delivery Vehicle

Abstract:

Aluminum salt-based (alum) adjuvants are currently the only Food and Drug Administration (FDA)-approved adjuvants for subunit vaccines. However, alum-based formulations have several drawbacks. Antigens with certain surface charges and hydrophobicity are required to bind to alum, which limits its applicability, and protein adsorption to alum can lead to undesired protein denaturation. Alum typically results in the production of antibodies against the antigen and polarization towards a cluster of differentiation 4 (CD4+) Type 2 helper T cell (Th2) response, whereas a more balanced response that also includes a CD8+ and Type 1 helper T cell (Th1) response is desired. Effective vaccines should create long-term adaptive immunity against the target pathogen through a combined humoral and cell-mediated response that incorporates all of these responses. This multifaceted response is preferred because many viruses and bacteria exist intracellularly, rendering an antibody (i.e. humoral) response unable to fully prevent or clear an infection. Furthermore, long-term storage of antigens often requires certain cold-chain conditions, but alum is not structurally stable enough to be easily frozen or lyophilized. Structural damage may disrupt its adjuvanticity and therefore limits the long-term storage options of alum-containing vaccines. Other adjuvants that elicit a Th1 response have been considered but are often too cytotoxic to use clinically (e.g. Freund's complete adjuvant and cholera toxin). An alternative approach to using alum or toxic Th1 adjuvants is an immunostimulatory polysaccharide. Here, we have chemically modified a naturally occurring polysaccharide, which has recognized immunostimulatory properties, that fulfills a dual function as both the adjuvant and acid-sensitive delivery vehicle. The vehicle has been prepared to passively target antigen presenting cells, and in vivo immunization with an encapsulated model antigen has led to the production of antibody levels comparable to an alum-based formulation. Moreover, the polysaccharide has previously illustrated cryoprotectant properties, which could facilitate extended storage outside the cold-chain. Our work demonstrates the potential for this modified immunostimulatory polysaccharide to serve a dual function as a subunit vaccine adjuvant and antigen delivery vehicle.

387896 Single-Cell Analysis of Specific B Cell Binding and Uptake of Peptide-Targeted Liposomes for Vaccine Formulations

Abstract:

Lipid-based nanoparticles have been widely characterized for the delivery of small molecules, peptides and proteins. Liposomes containing surface-bound antigens are similarly being developed as vaccines and have been shown to induce strong immune responses when compared to soluble antigens by facilitating cross-presentation and enhancement of humoral responses. Yet so far no studies have focused on how specific liposome design considerations modulate immune responses at the single cell level to enhance CD4+ T cell help and antibody production through interaction and activation of specific B cells. Using an antigen-specific 3-83 Ig transgenic mouse model, we evaluated the role of liposome composition on the direct binding and internalization of liposomes with surface-displayed peptide antigens to antigen-specific B cells. Initially we focused on the role of surface PEGylation and antigen tethering. We hypothesized that while surface PEGylation has been shown to improve liposome stability in serum and enhance lymph node targeting in vivo, it may hinder binding and internalization through specific B cell receptors. Our studies showed that unilamellar PEGylated liposomes containing maleimide-functionalized PEG-lipids conjugated to cysteine-containing antigenic peptides (p31 and p0 peptides with high and no binding affinities, respectively) bound with varying intensities to B cells as a function of PEGylation (10-0% PEG): 5% PEG liposomes bound with the highest intensity to B cells, while 0% PEG liposomes containing peptide antigen anchored directly to the bilayer surface had significantly reduced binding as a result of steric hindrance. These conditions as well as a third condition (5% PEG with surface anchored peptide antigen which resulted in even lesser binding compared to 0% PEG liposomes) were evaluated for their ability to activate antigen-specific B cells in vitro. B cell activation was assessed at the single-cell level using flow cytometry to quantify the up-regulation of early B cell activation markers, such as CD69, CD86, CD40, and MHCII. As predicted, 5% PEG liposomes with tethered p31 resulted in the strongest antigen-specific B cell activation, while the weakest binding 5% PEG liposomes with surface anchored p31 resulted in the lowest levels of activation comparable to unstimulated cells. The incorporation of TLR4 agonist, monophosphoryl lipid A (MPLA), into the liposome bilayer significantly enhanced early activation (day 1) and helped sustain activation and B cell viability at later times (up to 4 days) in an antigen-specific manner. However, MPLA incorporation into liposomes interestingly provided no enhancement on B cell binding despite the presence of TLR4 on the surface of B cells, signifying MPLA-triggered activation is dependent on specific antigen-mediated uptake of liposomes by B cells. Finally, we evaluated the effectiveness of these 3 liposome conditions to induce antigen-specific humoral responses in naïve mice. Antigen-specific antibody titers in mice immunized with antigen-bearing liposomes containing MPLA were found to correlate directly to B cell binding profiles, with 5% PEG liposomes with tethered p31 resulting in titers ~$\log_{10}$-fold higher than 5% PEG liposomes with surface anchored p31. Current studies are focused on understanding the relationship between B cell binding and activation and the ability of these B cells to present helper peptides to CD4+ T cells, as well as how these liposomes can be utilized to probe and expand low-frequency antigen-specific B cells in spleens and lymph nodes post-immunization. We believe the results obtained in these studies will further the development and understanding of effective particle-based peptide vaccines.

SYNTHESIS AND PROPERTIES OF AMBERLITE®-SILICA COMPOSITES

Gifty Osei-Prempeh and James Ingles
West Virginia University Institute of Technology, Montgomery, WV 25136

Ceramic-polymer composites can be optimized based on surface functionalization for specific applications. Incorporating nanoporous silica into polymer matrices to form polymer-silica composites provides potential access to the high surface area in porous silica and the functionality in the silica.

Amberlites® are commercially available polymeric adsorbents used for various separation and catalytic processes. These materials possess fairly large pores which can be tailored for applications in areas such as CO_2 capture, water purification, immobilized enzyme catalysis etc. Amberlite®-silica composites were made from two different amberlite materials XAD7HP and IRA-900. These materials show high silica incorporation. Amine functionalization of the composites resulted in good functional group incorporation and materials with fairly large pores for easy access to the functional groups. The surface area, pore size and volume, thermal properties and functional group incorporation and access of these composites will be compared. Application of these materials in CO_2 capture and contaminant removal from water will be presented.

Modification of Chitosan/MWCNTs nanocomposite for Lead (II) and Cadmium (II) Detection in Application of Disposable Electrochemical Sensor

Yu-Cheng Liu[a], Young Ku[a]* and Kuan-Jung Chen[b]

[a]*Department of Chemical Engineering, National Taiwan University of Science and Technology, Taipei City 106, Taiwan.*

[b]*M-Right Biomedical Inc., New Taipei City 231, Taiwan.*

Abstract

This study was investigated an ongoing research on the development of rapid, low-cost and friendly-use techniques with modification of chitosan and multi-wall carbon nanotubes (MWCNTs) nanocomposite for trace heavy metals, such as lead (II) and cadmium (II) detection by square wave stripping voltammetry. It was conducted using glassy carbon electrode (GCE), while a highly porous surface with an orderly three-dimensional network sensing layer (chitosan-MWCNTs) was fabricated by electro-deposition. The results shows that the peak potential of the reoxidation of Cd is -0.74 V, and of Pb -0.51 V vs. Ag/AgCl. The correlation coefficient (R^2) of the calibration curves was 0.996 (for Cd) and 0.995 (for Pb) with the same linear range (from 0.5 to 2 ppm).

It was especially electrochemical methods, are widely recognized as a powerful tool for the simultaneous determination of multiple types of metal ion, because of the combination of an effective pre-concentration step with advanced electrochemical measurements of the accumulated analystes. The chitosan-MWCNTs/GCE modified electrode could be applied on the analysis of drinking water and mineral water.

Keywords: Electrochemical Sensor; chitosan; multi-wall carbon nanotubes lead; cadmium

LAMAR UNIVERSITY
A Member of The Texas State University System

Anticorrosive Polyurethane Sensing Nanocomposites

Daowei Ding, **Huige Wei, Dr. Hongbo Gu, Dr. Suying Wei, Dr. Zhanhu Guo**
Integrated Composites Lab (ICL)
Dan F Smith Department of Chemical Engineering
Lamar University, Beaumont, Texas 77710
http://dept.lamar.edu/zhanhu/guo.html
Nov 19th, 2014

Outline

- **Introduction**
- **Experiment**
- **Results and Discussion**
- **Conclusions**
- **Acknowledgements**

Introduction

(Steel in the concrete)

(Plant tanks)

(Gas & oil pipes)

(Electronics)

Introduction (cont.)

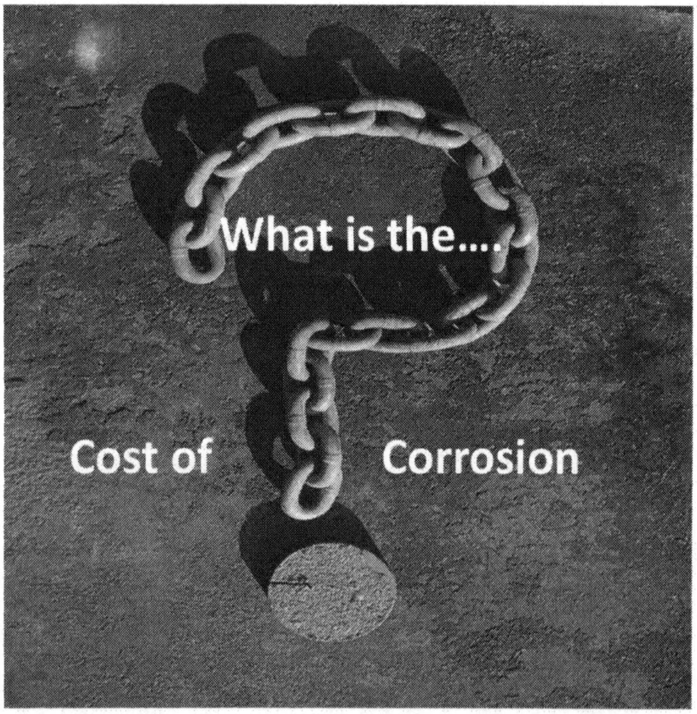

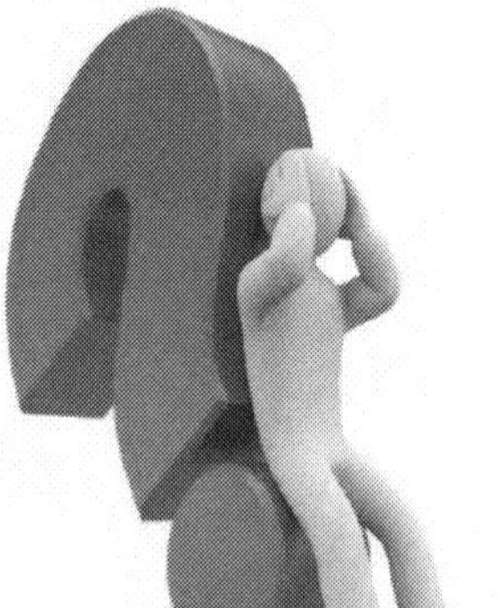

Introduction (cont.)

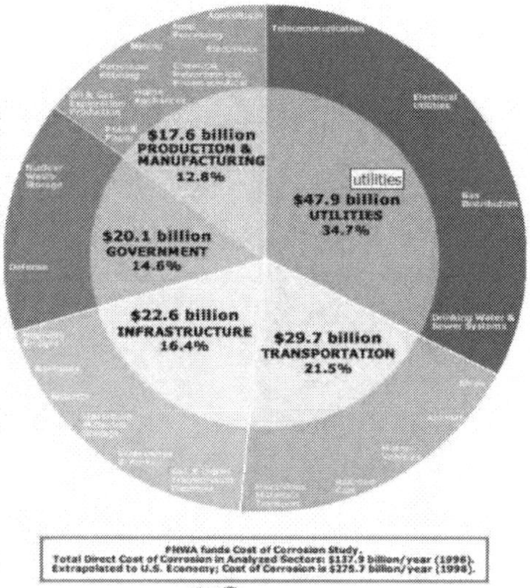

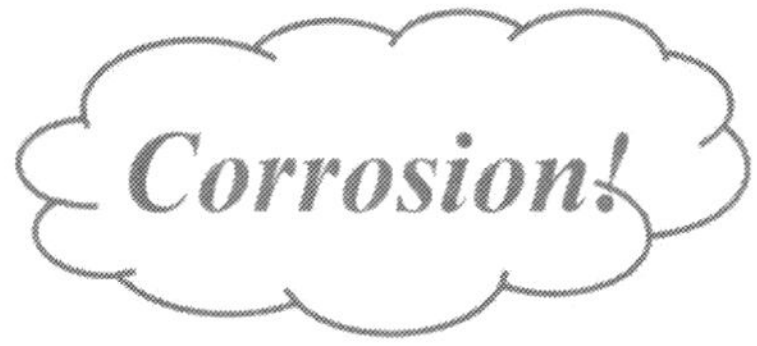

5

Introduction (cont.)

***How to prevent corrosion*?**

Introduction (cont.)

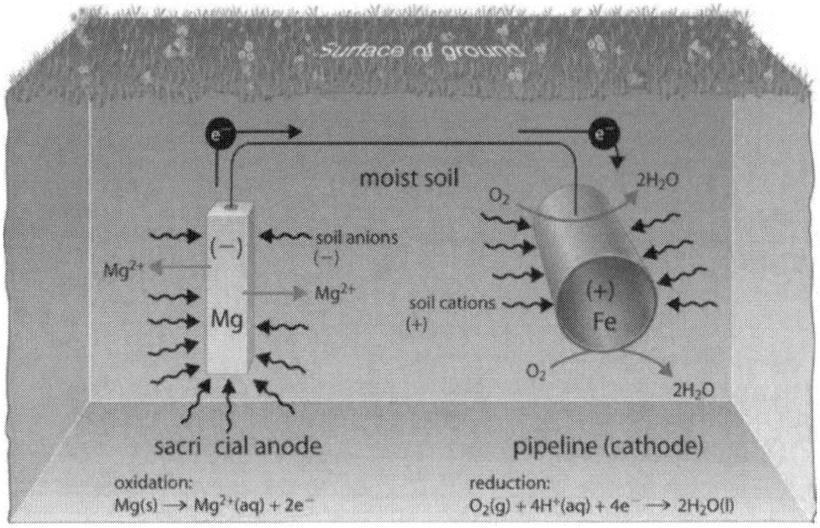

Cathodic protection

Introduction (cont.)

Anodic protection

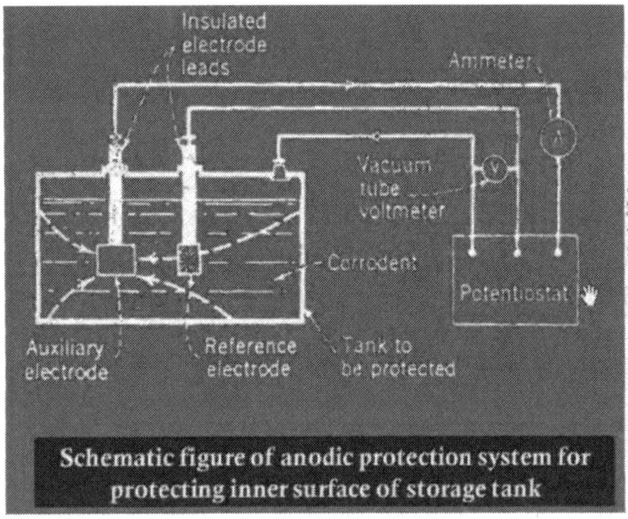

Schematic figure of anodic protection system for protecting inner surface of storage tank

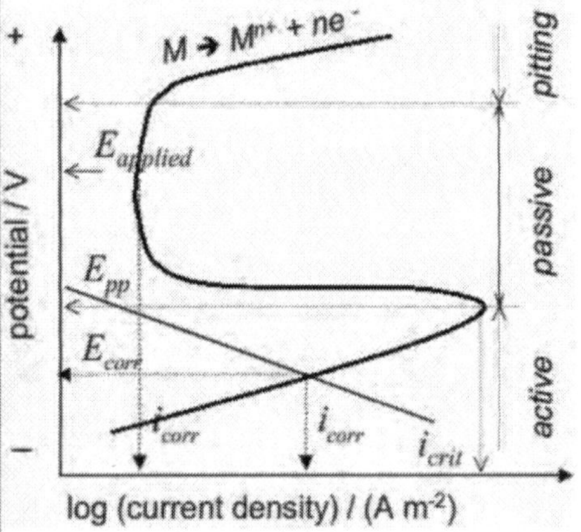

Introduction (cont.)

Anodic/Cathodic inhibitors

- **Anodic inhibitors**

  reducing oxidation reaction rate

  ✓ Nitrite (NO_2), Chromate (CrO_4^{2-})

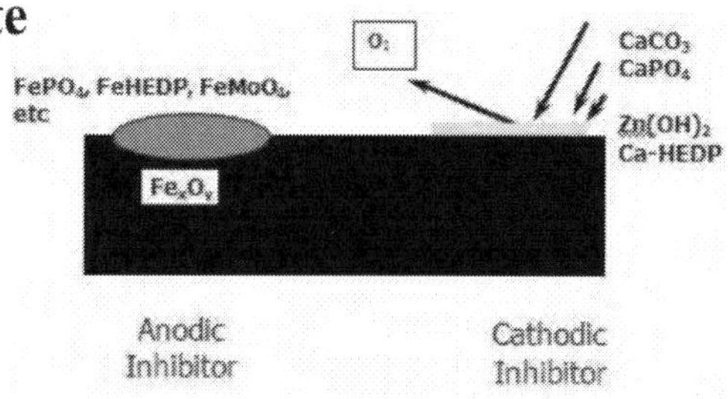

- **Cathodic inhibitors**

  reducing reduction reaction rate

  ✓ Zinc Oxide, Volatile amines

Introduction (cont.)

PPolypyrrole (**PPy**)

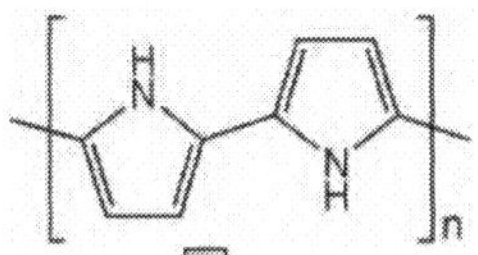

Polyaniline (**PANI**)

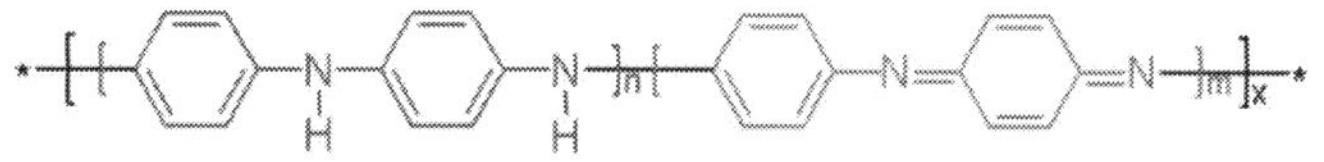

Poly(3,4-ethylenedioxythiophene, **PEDOT**)

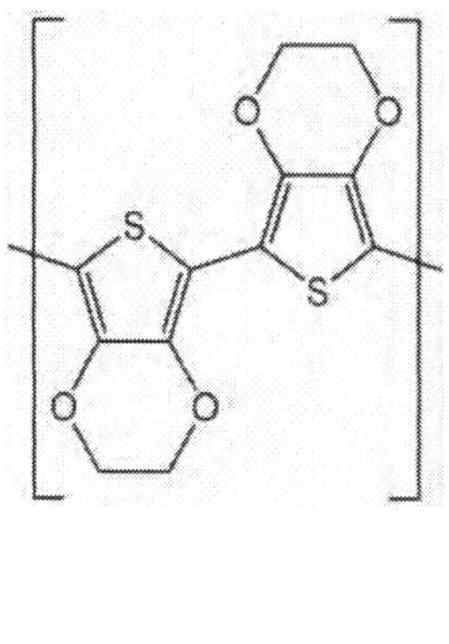

Conductive polymers (CPs)

Introduction (cont.)

CPs can be applied in the following formations..

- **Coatings in pure or composite form**
- **Primer coating with conventional topcoat**
- **Blended with conventional coating**
- **Additive to modify conventional coatings**

Introduction (cont.)

However……

- **Dissolution** of active metal substrates

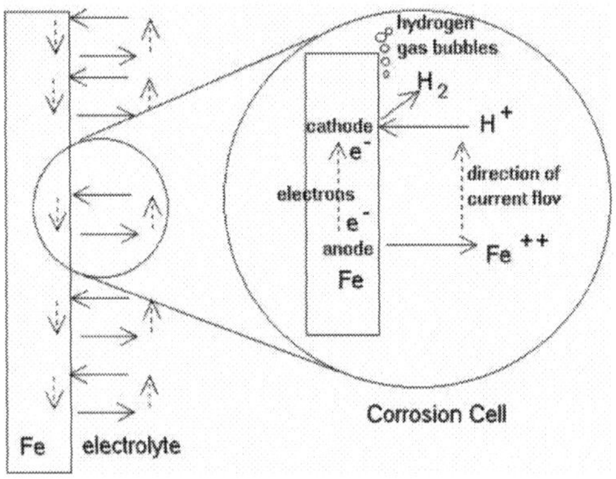

- Polymerization process **blocked** by the metal oxide formed on the metal substrate

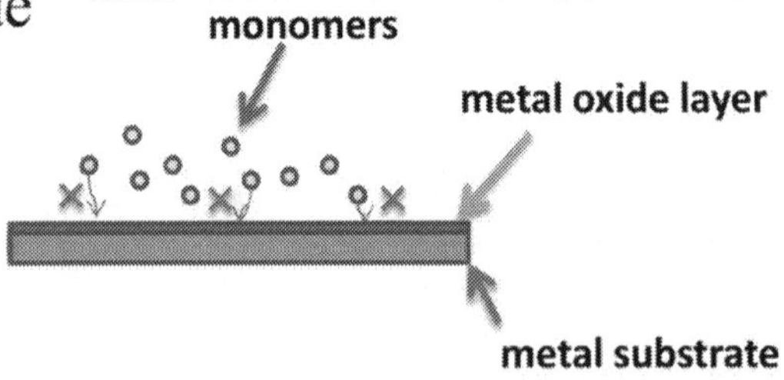

Introduction (cont.)

Epoxy | **Polyvinyl acetate** | **Polyvinyl butyral**

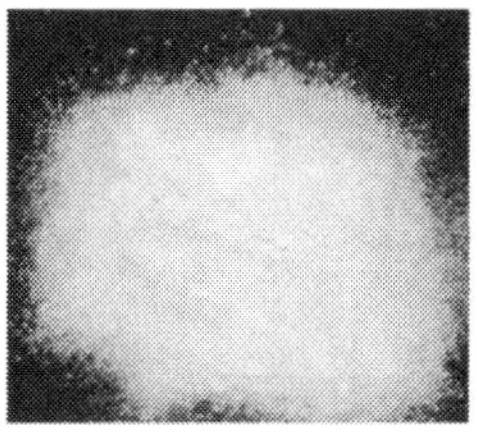

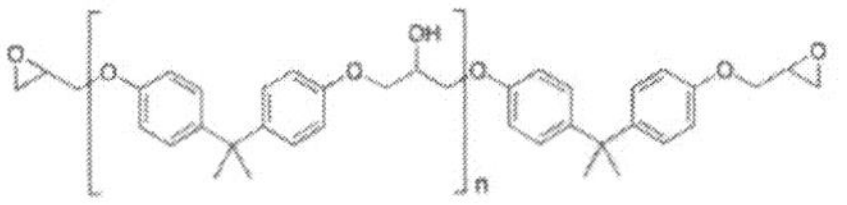

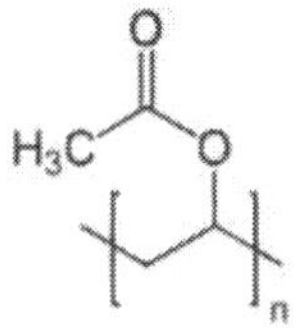

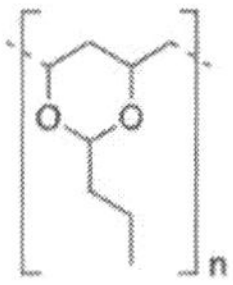

Conventional coating merits[1]
- Low cost
- Ease of application
- Aesthetic functionality

1, Buchheit, Rudolph G., et al. *Progress in Organic Coatings* 47.3 (2003): 174-182.

Introduction (cont.)

Advantages of Polyurethane (PU):

- *High extendable strength*
- *High anti-corrosion properties*
- *Good adhesive properties*

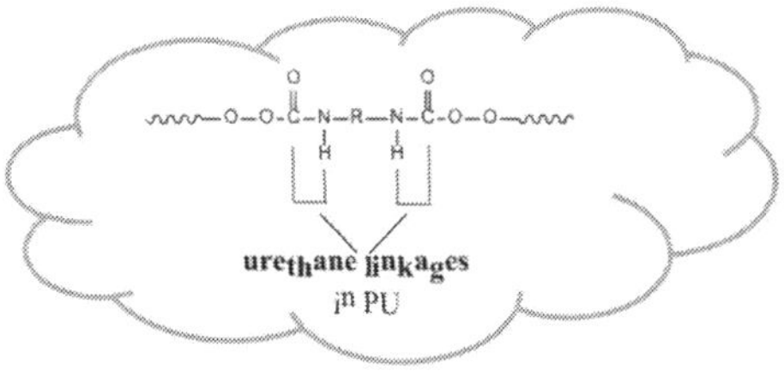

Sørensen, Per Aggerholm, et al. *Journal of Coatings Technology and Research* 6.2 (2009): 135-176

Introduction (cont.)

Advantages of Carbon Nanotube (CNTs):

- *High electrical conductivity*
- *High chemical stability*
- *Good stretchable property*
- *Compatible to many function groups*

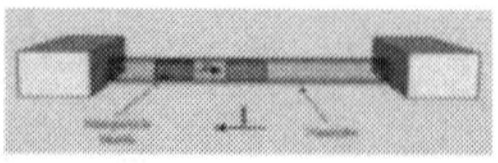

Memory chip

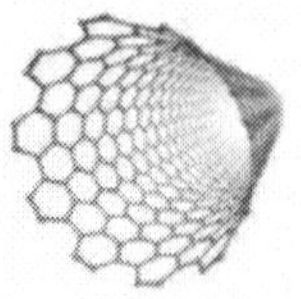

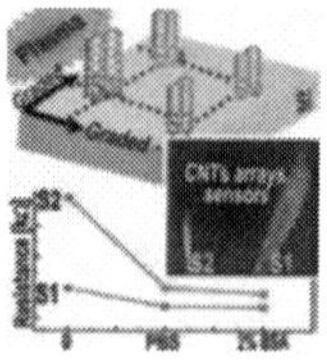

CNTs-based Super capacitor

Hydrogen storage

Baughman, Ray H., Anvar A. Zakhidov, and Walt A. de Heer. *Science* 297.5582 (2002): 787-792.

Introduction (cont.)

Primer coating

PU

+

CNTs

Antistatic Reinforcement

Anticorrosive Conductive CNTs/PU composites

Experiment

Synthesis

Surface-Initiated-Polymerization (SIP) method

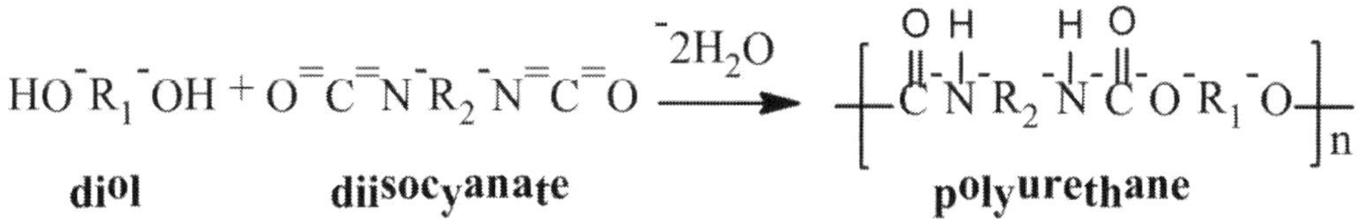

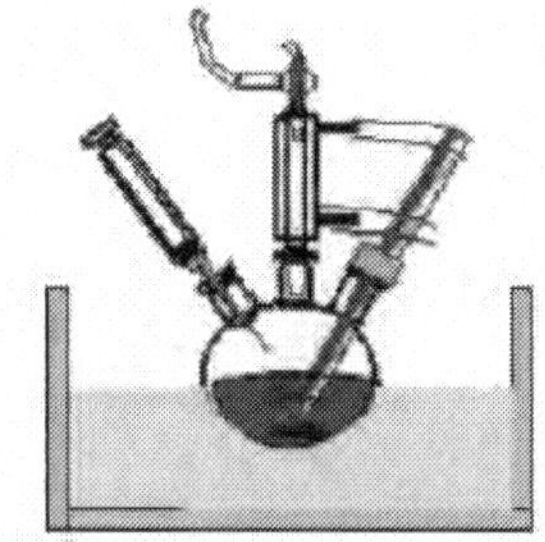

Fabrication of PU and CNTs/PU composite by **SIP** method.

Experiment (cont.)

In situ resistance in cyclic strain test

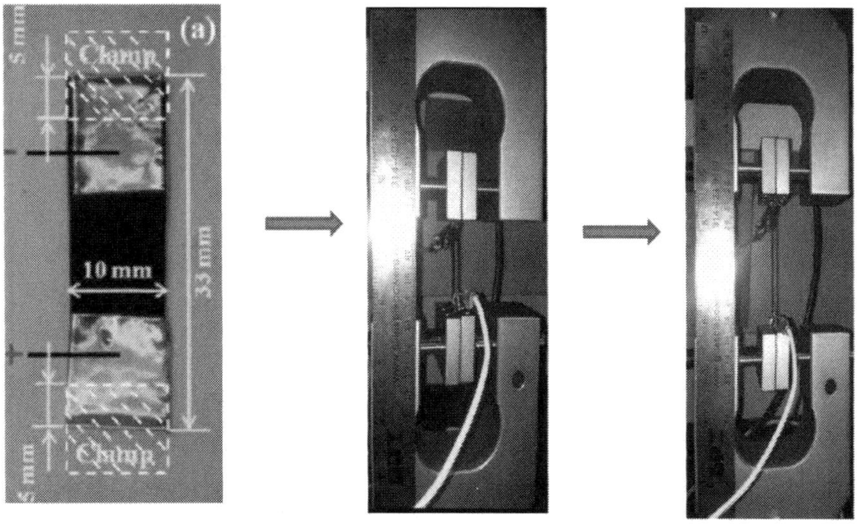

(a) Sample dimensions and (b-c) at various of elongation

Experiment (cont.)

Anticorrosive properties evaluation set up

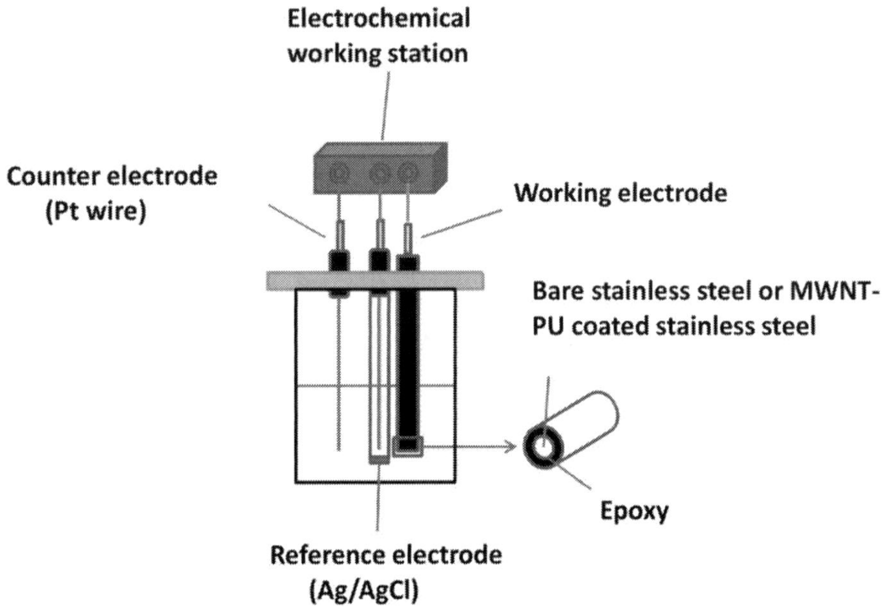

Results and Discussion

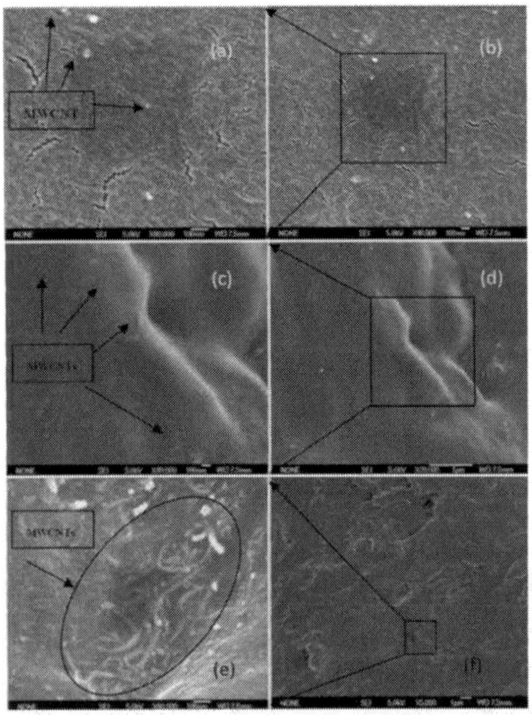

SEM Images

- (a and b) 1.0 % MWNTS/PU,
- (c and d) 3% MWNTS/PU,
- (e and f) 5% MWNTS/PU.

Results and Discussion (cont.)

FT-IR Characterization

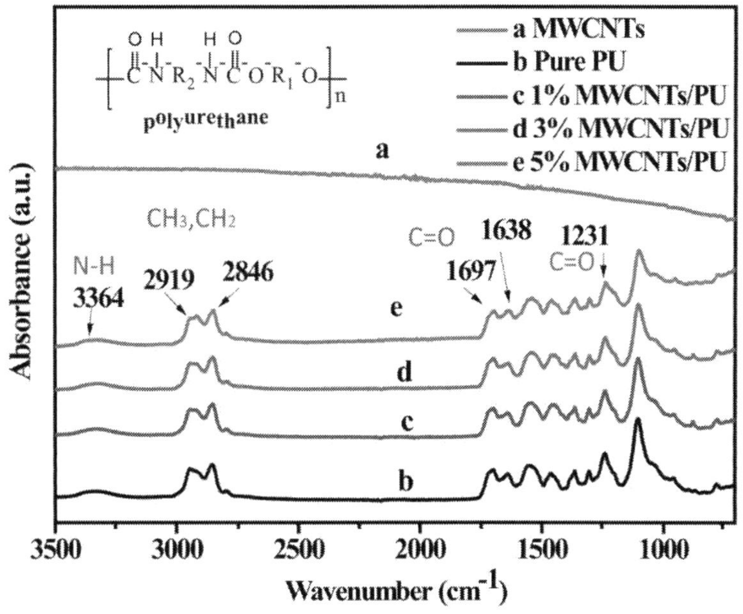

Results and Discussion (cont.)

Thermal Gravimetric Analysis

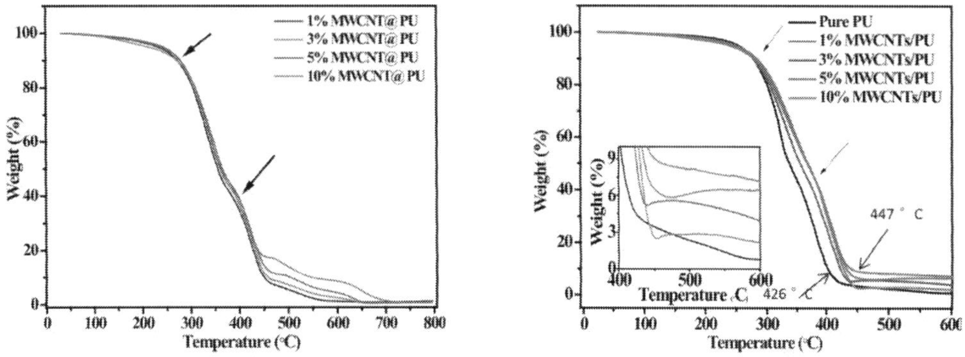

TGA graph of MWNTS/PU under (A) air atmosphere and (B) nitrogen condition

Results and Discussion (cont.)

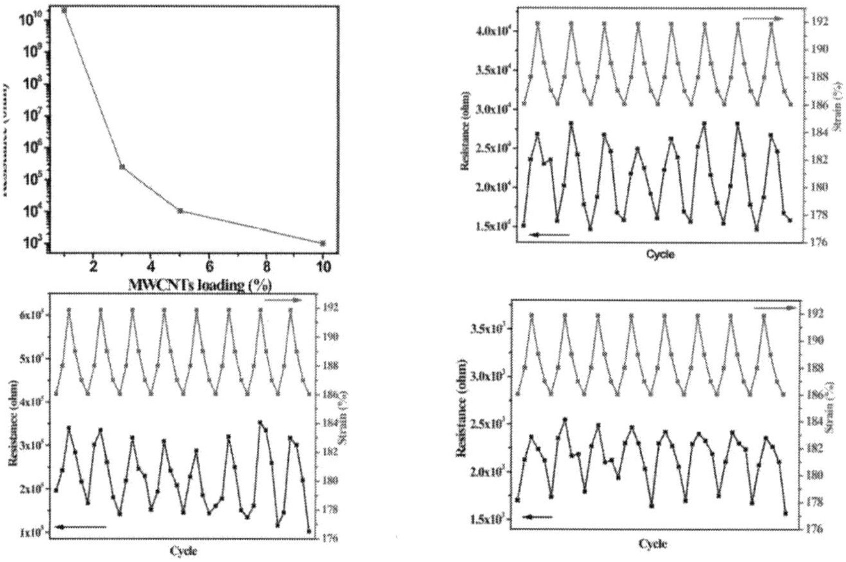

CNTs/PU strain sensoring property. Quick response for the physic-change

Results and Discussion (cont.)

E_{ocp}-time curves

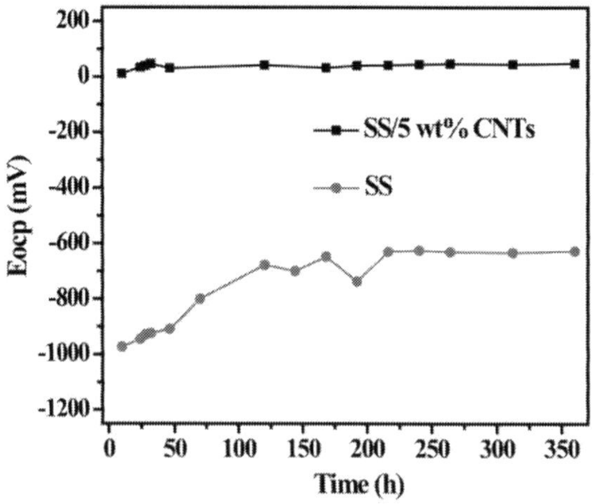

***Eocp*-time** curves for steel and CNTs/PU coated-steel in 3.0 wt% NaCl solution using a two-electrode cell.

Ag/AgCl and steel or CNTs/PU coated-steel serves as the reference electrode and working electrode, respectively.

Results and Discussion (cont.)

Eocp of the steel and CNTs/PU coated steel at various exposure time in 3% NaCl solution.

Time (h)	10	24	28	32	46	120	168	192	216	240	264	312	360
SS	-973	-945	-929	-924	-909	-678	-648	-738	-630	-625	-631	-633	-627
SS/PU-CNTs	11	34	40	47	31	43	33	41	33	45	48	46	49

-200 mV for PANI coated stainless steel in the literature[1]

1. A. Özyılmaz and M. Erbil, *Current Applied Physics*, 2006, **6**, 1-9.

Results and Discussion (cont.)

Tafel Curve

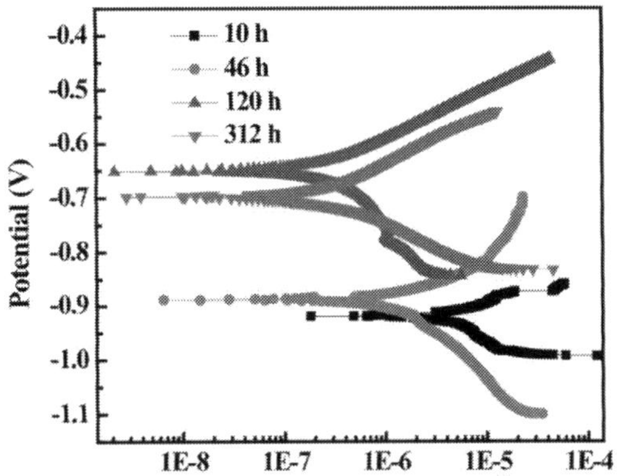

Bare stainless steel after different immersion time in 3.0 wt% NaCl solution.

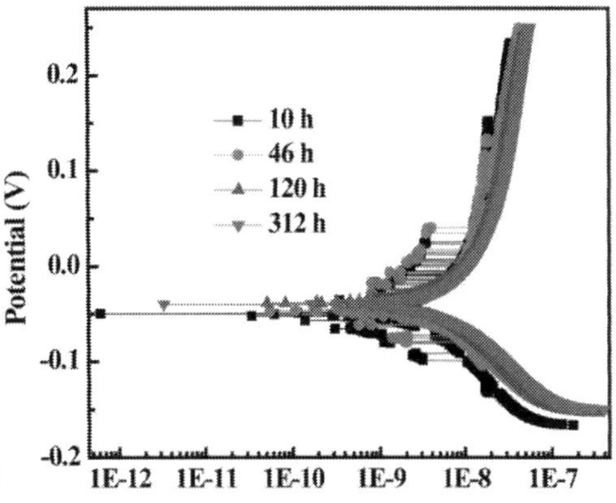

CNTs/PU coated-steel after different immersion time in 3.0 wt% NaCl solution.

higher E_{corr} and **lower** i_{corr} indicate **better** anticorrosion properties

Results and Discussion (cont.)

E_{corr} and I_{corr} of the **steel** and **CNTs/PU coated-steel** obtained from the fitting data of potentiodynamic polarization curves.

	Time (h)	10	24	28	32	46	120	168	192	216	240	264	312
E_{corr}/mV	SS	-919	-939	-899	-880	-888	-763	-650	-682	-634	-671	-648	-698
	SS/PU-CNTs	-53	-60	-49	-46	-49	-47	-48	-37	-35	-31	-39	-44
I_{corr}/nA	SS	67925	1862	3382	1208	2407	92	858	1151	289	449	188	286
	SS/PU-CNTs	4.4	3.8	10.6	3.5	5.7	11.0	9.6	11.1	11.0	12.2	13.0	13.5

Compared with improved -**140** mV in the E_{corr} for PANI coated carbon steel[2]

2. B. Yao, G. Wang, J. Ye and X. Li, *Materials Letters*, 2008, **62**, 1775-1778.

Results and Discussion (cont.)

EIS

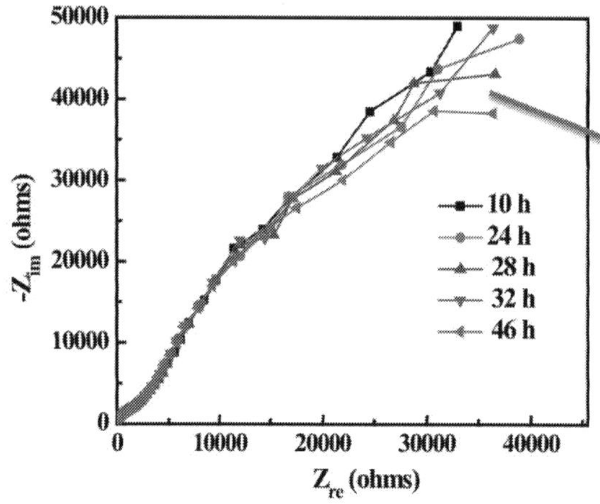

No obvious enhanced impedance observed, indicating no **metal oxide** formed on the CNTs/PU composite-coated steel

EIS of the **CNTs/PU coated-steel** after different immersion time in 3 wt% NaCl solution in a three-electrode cell. The frequency ranges from 100, 000 to 0.01 Hz at a 10 mV amplitude referring to the open circuit potential.

Results and Discussion (cont.)

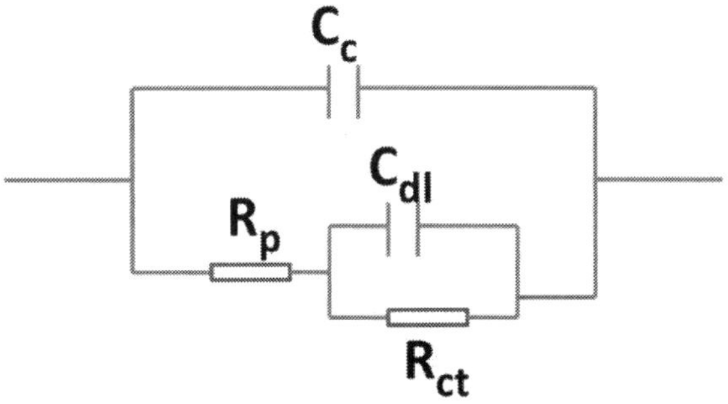

Equivalent circuit **for** CNTs/PU coated-steel **composite coating**

C_c — coating capacitance;
R_p — pore resistance;
C_{dl} — double layer capacitance;
R_{ct} — charge transfer resistance.

Uniform, defect free coating gives a **higher** **R_{ct}** and **R_p**, **lower** **C_c** and **C_{dl}**.

Results and Discussion (cont.)

Fitting parameter values from the **equivalent circuit** of the EIS obtained from aqueous 3.0 wt% NaCl solution.

time (h)	R_p (Ω)	C_c (F)	C_{dl} (F)	R_{ct} (Ω)
10	4974	2.234×10^{-9}	1.346×10^{-8}	8.22×10^{4}
24	4709	2.352×10^{-9}	1.332×10^{-8}	7.751×10^{4}
28	4786	2.364×10^{-9}	1.337×10^{-8}	7.459×10^{4}
32	4641	2.405×10^{-9}	1.347×10^{-8}	7.735×10^{4}
46	4499	2.479×10^{-9}	1.365×10^{-8}	6.749×10^{4}

R_p ~10^2, C_c ~10^{-6} to 10^{-8}, C_{dl} ~10^{-5} to 10^{-7}, Rct ~10^4 [3].

3. M. Shabani-Nooshabadi, S. Ghoreishi and M. Behpour, Electrochimica Acta, 2009, 54, 6989-6995.

Results and Discussion (cont.)

Anticorrosion mechanism

$$Fe \longrightarrow Fe^{2+} + 2e^-$$

$$Fe^{2+} \longrightarrow Fe^{3+} + e^-$$

$$O_2(g) + 2H_2O + 4e^- \longrightarrow 4OH^-$$

$$2Fe^{2+}(aq) + O_2(g) + 2H_2O \longrightarrow 2FeOOH + 2H^+$$

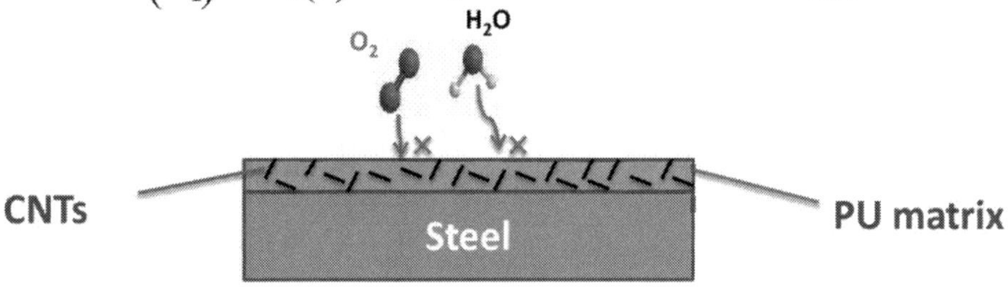

Conclusion

- **Conductive** CNTs/PU composite has been fabricated successfully via SIP method for anticorrosion and strain sensing purpose

- CNTs/PU composite exhibits **enhanced** mechanical properties and thermal stabilities compared to pure PU

- Both of the visual observation and the measurement confirm the **excellent** anticorrosive properties of the CNTs/PU coating

"Strain Sensitive Polyurethane Nanocomposites Reinforced with Multiwalled Carbon Nanotubes", D. Ding, H. Wei, J. Zhu, Q. He, S. Wei* and Z. Guo*
Energy and Environment Focus, 2014

"Anticorrosive Conductive Polyurethane Multiwalled Carbon Nanotubes Nanocomposites" H. Wei, D. Ding, S. Wei* and Z. Guo*
Journal of Materials Chemistry A, 1, 10805-10813 (2013)

Acknowledgements

CMMI-1030755

Integrated Composites Laboratory

Thank you for your attention !

Comments? Questions?

Atomistic Modeling of Cross-linked Polyamide/Graphene and Polyamide/Graphene Oxide Composites Reverse Osmosis Membrane

Heng Ma[1], Haiyang Zhao[1,2], Lin Zhang[2], Tao Wei[1]

1. Dan F. Smith Department of Chemical Engineering, Lamar University, Beaumont, TX, 77710, United States
2. Key Laboratory of Biomass Chemical Engineering of MOE, Department of Chemical and Biological Engineering, Zhejiang University, Hangzhou 310027, China

Polyamide reverse osmosis (RO) membrane is one of the most successfully commercialized membranes for desalination and water purification. The introduction of inorganic nano-particles into polyamide (PA) thin layer is expected to enhance water permeation, the separation performance and operation stability of RO membrane. In this work, atomic models of cross-linked polyamide membrane and their composites of graphene (PA/GS) and graphene oxide (PA/GO) were established. The heterogeneous pore structures and size distribution were characterized. The activation energy barriers for water permeation were estimated to compare between the pure PA membrane and the composites membranes using the efficient free energy computations. Ions diffusion and hydrations inside nano-pores are also investigated. This fundamental study will rationalize the experimental development of PA/GS and PA/GO composites RO membrane.

380889 TiO2/Fe3O4 Nanoparticles Embedded Thermo Responsive Composite Gel Beads for Water Treatment

Abstract:

Because of its environment-friendly properties, TiO_2 nanoparticles can be used for wastewater treatment. However, although the reaction rate is high, separation and recovery process of TiO_2 nanoparticles from the treated water is required after the reaction. To overcome the disadvantage, here we propose to prepare both TiO_2 and Fe_3O_4 nanoparticles embedded thermo-responsive composite gel beads. In the composite, poly(N-isopropylacrylamide) (PNIPAM) and calcium alginate were used as thermo-responsive polymer and structure support polymer, respectively. Here, Fe_3O_4 nanoparticles will be used to achieve localized heating. If AC magnetic field is applied for the composite intermittently, the composite beads will swell and shrink repeatedly and it will achieve the promoted diffusion of substrates and products through the composite.

The composites were prepared using single tube nozzle by modifying the simple process to prepare thermo-responsive capsule reported in our previous report. The obtained samples were characterized by examining their thermo-responsive property, photocatalytic activity and localized heating property. The experimental results showed that localized heating of thermo-responsive bead containing Fe_3O_4 was also achieved without raising a bulk solution temperature by applying alternative current (AC) magnetic field. The simulation based on the characterization experiments on thermo-responsive property, photocatalytic activity and localized heating property proved the effectiveness of thermo-responsive pumping to improve the phtocatalytic activity of TiO_2 nanoparticle embedded in the composite gel.

390009 Cellulose Derived Mesoporous Magnetic Carbon Nanocomposites with Enhanced Cr(VI) Removal

Abstract:

Cr(VI) pollution is increasingly serious in surface water and groundwater due to its highly toxic for human health [1]. The cost effective and sustainable materials are urgently needed for the Cr(VI) removal. Cellulose is considered as a good adsorbent for for Cr(VI) removal due to its sustainability and abound active sites. But the low adsorption capacity and slow removal rate limit its application due to its small surface area and the lack of porosity. In this study, the newly magnetic carbons based on cellulose were fabricated through a simple calcination method.[2] The properties of magnetic carbons synthesized by using the Fe_3O_4 nanoparticle and $Fe(NO_3)_3$ solution as the processors (MC-O and MC-N, respectively) were investigated. The materials were then used as the adsorbents for Cr(VI) removal from aqueous solution. Both Fe_3O_4 and $Fe(NO_3)_3$ were reduced to zero-valent Fe (ZVI) after calcination with cellulose at 800 ^{o}C. The MC-N exerts a higher proportion of ZVI and a bigger surface area and exhibits a better resistance against oxygen and acid than MC-O due to its smaller pore size. Both MC-O and MC-N present great performances on Cr(VI) removal. 4.0 mg/L Cr(VI) can be completely removed from neutral solution by both MC-O and MC-N within 10 min. Cr(VI) removal is highly pH-dependant, 1000 mg/L Cr(VI) can be completely removed in 10 min at pH 1.0. MC-O performs better than MC-N at pH 7.0, while MC-N presents a higher capacity than MC-O at pH 1.0. The magnetic carbons can be easily separated from solution by using permanent magnet after treated with Cr(VI). Finally, the Cr(VI) removal mechanisms including the Cr(VI) reduction and precipitation were discussed as well.

[1] Xu C., Qiu B., Gu H., *et al.* Synergistic Interactions between Activated Carbon Fabric and Toxic Hexavalent Chromium. *ECS J. Solid State Sci. Technol.,* **2014**, 3, M1-M9.

[2] Zhu J., Gu H., Guo J., *et al.* Mesoporous Magnetic Carbon Nanocomposite Fabrics towards Highly Efficient Cr(VI) Removal. *J. Mater. Chem. A*, **2014**, 2, (7), 2256-2265.

383243 Invited Talk: Tail State Distribution of Polymer Donor Dictates Whether Vertical Phase Separation Affects Device Characteristics in Bulk-Heterojunction Solar Cells

Abstract:

With soft-contact lamination and delamination schemes, we have been able to elucidate how the presence of an electron-donor wetting layer at the electron-collecting interface of solar cells affects device performance. We constructed conventional bulk-heterojunction solar cells comprising three model electron donors: poly(3-hexyl thiophene), or P3HT; poly[2,6-(4,4-bis-(2-ethylhexyl)-4*H*-cyclopenta-[2,1-*b*;3,4-*b*′]dithiophene)-*alt*-4,7(2,1,3 benzothiadiazole)], abbreviated PCPDTBT; and 7,7′-(4,4-bis(2-ethylhexyl)-4H-silolo[3,2-b:4,5-b′]dithiophene-2,6-diyl)bis(6-fluoro-4-(5′-hexyl-[2,2′-bithiophen]-5-yl)benzo[*c*][1,2,5]thiadiazole), abbreviated p-DTS$(FBTTh_2)_2$, with fullerene derivatives. To exaggerate the influence of vertical phase separation, we laminated in each case a thin layer of electron donor on the bulk-heterojunction active layer prior to cathode deposition. While devices with PCPDTBT and p-DTS$(FBTTh_2)_2$ exhibit drastically reduced short-circuit current densities compared to devices without the additional electron-donor wetting layer, devices with P3HT only exhibit a marginal drop in short-circuit current density compared to devices without a thin P3HT wetting layer at the electron-collecting interface. We ascribe this difference in the reduction of short-circuit current density to intrinsic differences in the ionization potential and tail state distribution of the electron donors. Relative to PCPDTBT and p-DTS$(FBTTh_2)_2$, P3HT has a shallower HOMO energy level (closer to vacuum level) and a substantially broader tail state distribution. Against the energy levels of fullerene, both of these factors can support electron-hole recombination, followed by electron injection from the cathode into the tail states of P3HT when the solar cell is under operation. As such, the presence of a P3HT wetting layer does not substantially affect device performance. In comparison, PCPDTBT and p-DTS$(FBTTh_2)_2$ have deeper HOMO energy levels and narrower tail state distributions compared to P3HT, the combination of which cannot support electron-hole recombination and hole injection into the electron donor from the cathode. The presence of thin wetting layers of PCPDTBT and p-DTS$(FBTTh_2)_2$ at the electron-collecting interface thus adversely affects device performance. These observations implicate the relative importance of vertical phase separation in the bulk-heterojunction active layers of these materials pairs in dictating solar cell device performance.

386073 Structural Characterization of Exceptionally Aligned Regioregular CDT-PT Based Copolymer with High Field-Effect Mobility

Abstract:

Polymer-based field-effect transistors typically have a charge mobility of less the 1 cm^2/Vs. The mobility can be further improved through alignment of the conjugated polymer chains in the direction of charge transport. For our study, we use a regioregular conjugated copolymer (PCDTPT) based on cyclopenta[2,1-b:3,4-b′]dithiophene (CDT) and pyridal[2,1,3]thiadiazole (PT) structural units, which has a hole mobility around 0.6 cm^2/Vs for isotropic spin-coated films. When the regioregular PCDTPT is casted on an unidirectional nano-grooved substrate, the hole mobility has been measured to be greater than 20 cm^2/Vs in the direction of the nano-grooves. To provide further insight on the significance of the alignment in enhancing the mobility, this talk will focus on the detailed characterization of the crystalline orientation of the aligned regioregular PCDTPT films. The characterization techniques include a combination of transmission electron microscopy (TEM), grazing incidence wide angle X-ray scattering (GIWAXS), and near edge x-ray absorption fine structure (NEXAFS) spectroscopy. The results indicate elongated polymer fibers/chains along the alignment direction (*i.e.* parallel to nano-grooves) allowing for efficient intrachain transport between source-drain contacts. Furthermore, the conjugated polymer backbone has preferential edge-on orientation resulting in pi-pi stacking in the plane of the film, thus allowing for efficient interchain charge transport. Such efficient routes for charge transport between the source-drain contacts is responsible for the measured high mobility in the aligned films.

380198 Fluidic-Directed Assembly of Highly Crystalline Semiconductor Supramolecular Structures with Enhanced Charge Transport

Abstract:

Continuous-flow microreactors have emerged as a new tool for both synthetic and process chemists, which provide several advantages compared to traditional batch reactors. For example, enhanced heat- and mass-transfer characteristics, safety of operation when using highly exothermic, explosive or toxic reagents, precise control over residence (reaction) time, isolation of sensitive reactions from air and moisture, high surface-to-volume ratio, the possibility of automation and the ease of scale-up or ability to operate several devices in parallel (numbering up). In addition, these microreactors allow for integration of several reaction steps into one single streamlined process, which results in a significant time-gain compared to traditional batch processes. Further, laminar flow is an important advantage for microfluidic technology, due to the possibility of tuning materials properties. Self-assembly of semiconducting polymers into ordered supramolecular structures commensurate with efficient charge transport has been achieved by tuning a range of processing parameters (e.g. film formation methods, different solvents (low or high boiling point), polymer-dielectric interface treatment, and solution treatment such as ultrasonication or UV irradiation). However, these strategies present limitations for achieving highly crystalline semiconducting polymers, including uniform, rapid processing, and continuous fabrication of devices.

Here, fluidic-directed assembly of highly crystalline semiconductor supramolecular polymer structures with enhanced charge transport characteristics is demonstrated. Fluid flow provides for enhanced intramolecular ordering of solubilized polymer chains, and thereby effects formation of anisotropic supramolecular polymer assemblies via favorable π–π stacking (intermolecular interaction). Molecular ordering is thus dramatically enhanced with concomitant, enhanced charge transport characteristics of corresponding films. Furthermore, we combine the microfluidic treatment strategy with a blade coating method to achieve continuous and large scale preparation of FET (field-effect transistor) devices with enhanced performance. The methodology exhibits great potential for application in flexible semiconductor device manufacturing.

A Molecular-Scale Understanding of Cohesion/Fracture in Conjugated Polymer:Fullerene Mixtures

Naga Rajesh Tummala, Chad Risko, and Jean-Luc Brédas
School of Chemistry and Biochemistry &
Center for Organic Photonics and Electronics,
Georgia Institute of Technology, Atlanta, Georgia 30332-0400

Abstract

Active-layer mechanical flexibility is a characteristic feature of organics-based electronic devices, which is considered as a main technological advantage over traditional inorganic electronics. Recent experimental studies quantifying the strength of materials derived from π-conjugated organic molecules and polymers, however, reveal that limited fracture toughness is a serious drawback, especially for multi-component polymer:fullerene blends used in photovoltaic applications.[1,2] Here, through a series of molecular dynamics simulations, we investigate the cohesion, mechanical properties, and fracture mechanisms of pure and blended films derived from poly-(3-hexyl thiophene) [P3HT] and chemically modified fullerenes. Comparison to experimental data reveals how differences in terms of the fullerene chemical substitution, P3HT molecular weight, and P3HT:fullerene derivative interactions influence these characteristics. The molecular-scale insights derived from the results provide an understanding that can be used to refine design principles and enhance material robustness and reliability.

(1) Dupont, S. R.; Oliver, M.; Krebs, F. C.; Dauskardt, R. H. *Sol. Energy Mater. Sol. Cells* **2012**, *97*, 171.
(2) Savagatrup, S.; Makaram, A. S.; Burke, D. J.; Lipomi, D. J. *Adv. Funct. Mater.* **2014**, *24*, 1169.

384271 Device Performance Is Independent of Crystallite Texture in Conjugated Block Copolymer Photovoltaics

Abstract:

Conjugated block copolymers have the potential to tune donor/acceptor interfaces and the mesoscale structure within the active layer of organic photovoltaic devices. The ability to control and modify the micro-phase separation of the copolymer can offer a useful platform in understand the relationship between chemical structure, nanoscale morphology, and photovoltaic device performance. We have demonstrated that utilizing poly(3-hexylthiophene)–*block*–poly-((9,9-dioctylfluorene)-2,7-diyl-alt-[4,7-bis(thiophen-5-yl)-2,1,3-benzothiadiazole]-2′,2″-diyl) (P3HT-*b*-PFTBT) as the active layer of solar cells leads to power conversion efficiencies near 3% and remarkable 1.2 V open-circuit voltages without the use of a fullerene acceptor. We also find that we can control the P3HT crystallite orientation from "face-on" (π-stacking predominantly out-of-plane) to "edge-on" (π-stacking mainly in-plane) by altering the solvent casting conditions. Nevertheless, our device performance is independent of the crystallite orientation, which is contrary to our current understanding – face-on crystal orientations are expected to maximize charge transport in solar cells. We hypothesize that the ubiquitous broad distribution of crystallite textures always provides pathways for charge extraction, regardless of the average orientation of P3HT crystals. This phenomenon is explored in model materials, where we have made perturbations to the P3HT-*b*-PFTBT structure by adding side chains to the TBT unit. The combination of energy-filtered TEM, grazing-incidence wide angle X-ray scattering and soft X-ray scattering provides a comprehensive picture of the role of the molecular structure on the morphology.

Bandgap tunable cross-linked PEDOT copolymers with superior optical property via oxidative chemical vapor deposition (oCVD)

Sunghwan Lee and Karen K. Gleason

High optical transmittance conjugated polymers with electrical conductivity are garnering much attention for the applications in organic optoelectronic devices including organic field effect transistors and organic photovoltaic solar cells. Polymers based on poly(3,4-ethylenedioxythiophene) (PEDOT) are particularly promising candidates with high conductivity, uniform surface planarity and excellent ductility. In this work, homopolymer PEDOT films deposited using oxidative chemical vapor deposition (oCVD) show the maximum conductivity as high as ~3500 S cm^{-1} at a deposition rate <0.5 nm min^{-1}. However, their utility is limited due to the relatively low transmittance and abrupt decrease near the red edge in the visible regime.

Here, we report on the significantly improved optical properties achieved via tuning the band gap of cross-linked PEDOT copolymers using oCVD, offering a single-step process for the synthesis and deposition of copolymer films. The cross-linking monomers of biphenyl or anthracene are simultaneously evaporated with EDOT monomer and an oxidant of $FeCl_3$ during the deposition. Poly(anthracene-co-EDOT) [p(ANTH-co-EDOT)] shows the superior transmittance of ~ 93% to homopolymer PEDOT (~80%) and poly(biphenyl-co-EDOT [p(BPH- co-EDOT)] (~88%) at the wavelength of 550 nm. In addition, copolymer films show no significant transmission decay in the red edge regime of the light unlike homopolymer PEDOT that presents an abrupt transmission falloff. An improvement in optical transmittance is in agreement with an increase in band gap of materials (p(ANTH- co-EDOT), ~2.3 eV vs. PEDOT, ~1.8 eV). oCVD-processed band gap tunable PEDOT copolymers with enhanced transmittance may, therefore, have application in organic optoelectronic devices that require the high optical transparency.

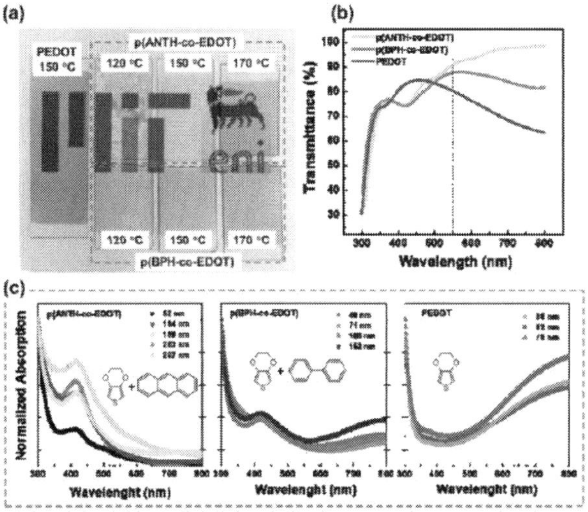

Figure 1. Optical properties of homopolymer and copolymer PEDOT films (a) photographs of p(ANTH-co-EDOT) and p(BPH-co-EDOT) copolymers with typical homopolymer PEDOT deposited on 25x75 mm glass slides, (b) optical transmission spectra from ~55 nm-thick samples investigated in the visible regime from 300 to 800 nm. The dotted line indicates the standard light wavelength of 550 nm, and (c) normalized absorption spectra of oCVD polymers as a function of film thickness. Each inset shows the monomer(s) used.

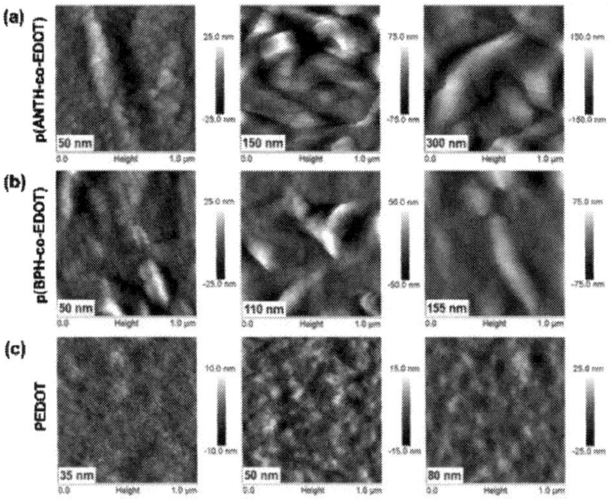

Figure 2. AFM surface topographic images obtained via non-contact height mode from oCVD films of (a) p(ANTH-co-EDOT), (b) p(BPH-co-EDOT) copolymers and (c) homopolymer PEDOT in different thickness.

378241 Understanding the Effects of Physical and Chemical Features of Additives on the Morphology of Blends of Conjugated Polymers and Fullerene Derivatives Using Molecular Simulations

Abstract:

The morphology of blends of conjugated polymers (electron donors) and fullerene derivative molecules (electron acceptors) strongly affects the efficiency of organic solar cells. Recently, minority component solvent additives such as diiodooctane (DIO) and chloronapthalene (CN) have been used in acceptor-donor blends to alter blend morphology and to significantly improve device efficiency. In this work we present coarse-grained molecular dynamics simulations of conjugated polymer and fullerene derivative blends with additives of varying length, chemical functionalization, and volume fraction to understand how these additives affect morphology and to elucidate what physical and chemical features of solvent additives are important for tuning morphology.

This work introduces oxidative Molecular Layer Deposition (oMLD) as a chemical route to synthesize highly conductive and conformal poly(3,4-ethylenedioxythiophene) (PEDOT) thin films via sequential vapor exposures of molybdenum(V) chloride ($MoCl_5$, oxidant) and ethylene dioxythiophene (EDOT, monomer) precursors. The growth temperature strongly affects PEDOT's crystalline structure and electronic conductivity. Films deposited at ~150°C exhibit a highly textured crystalline structure, with {010} planes aligned parallel with the substrate. Electrical conductivity of these textured films is routinely above 1000 S cm^{-1}, with the most conductive films exceeding 3000 S cm^{-1}. At lower temperatures (~100°C) the films exhibit a random polycrystalline structure and display smaller conductivities. Post-polymerization acid rinses aid in removing spent oxidant and facilitate dopant ion exchange, leading to enhanced conductivity. This sequential-reaction synthesis method produces highly conformal coatings over high aspect ratio structures making it attractive for many device applications. In this work, oMLD was invesitgated as a method to deposit PEDOT as an electrocatalyst for solar cells. PEDOT-coated nano-ITO particles on FTO glass had a conversion efficiency of 7.18%, close to that of Pt at 7.26%.

389035 Rapid Assembly of Mesoporous Electrodes of Polyaniline Nanofibers and Multi-Walled Carbon Nanotubes

Abstract:

The design and assembly of nanoscale materials is critical for developing high performance mesoporous electrodes for energy storage devices that can be scale-up for manufacturing. To address this challenge, we have designed a modified layer-by-layer (LbL) nanofabrication technique based on electrostatic self-assembly with vacuum assisted filtration (VF-LbL). Electrostatic interactions with vacuum force allow fabrication of thick electrodes (4 - 50 microns) of conjugated polyaniline (PANi) nanofibers and oxygen functionalized multiwalled carbon nanotubes (MWNT) in short timeframes. Controlled thermal treatment of the electrode sample improves the electronic conductivity and mechanical stability that also shows high surface area with interpenetrating networks of polymeric nanofibers and carbon nanotubes. Electrochemical measurements reveal high specific capacity originating from the MWNTs and redox active PANi nanofibers that store charge through double layer charge storage and faradaic mechanism with excellent charge/discharge stability over 10,000 cycles. I will discuss on the strategies to improve the physicochemical properties and scale up of the nanostructured electrodes that could be extended to other nanomaterials for high performance batteries and supercapcitors.

368822 Water Content and Solute Partitioning in Hybrid Silicone-Hydrogels

Abstract:

With roughly six times the amount of corneal oxygen supply compared to HEMA-based soft contact lenses, silicone-hydrogel (SiHy) soft contact lenses (SCLs) provide comfort for extended wear, while maintaining ocular health [1-3]. Despite their commercial importance, the structure and morphology of SiHy SCLs is not well understood. It is, however, widely accepted that silicone-based SCLs should contain co-continuous hydrophilic and hydrophobic polymer phases [3]. The hybrid structure of SiHys allows for sufficient salt transport through the water phase [4] localized in hydrophilic domains, as well as, sufficient oxygen permeability through the silicone domains [2,3]. Additionally, partitioning of care-solution and wetting agents in SCLs for improved on-eye lens performance [5, 6] is dictated largely by the water content [7] and, thus, the SiHy microstructure. To our knowledge, no studies attempt prediction of water uptake or solute partitioning in silicone-based hydrogels.

We investigate equilibrium water and solute uptake in SiHys both experimentally and theoretically. Specifically, we focus on quantitatively predicting how initial SiHy formulation design affects equilibrium water content and partitioning of aqueous and oleophilic solutes of similar size (~100-300 Da). The SiHy materials studied were synthesized with free radical polymerization and consist of varying hydrophilic fraction (HEMA, MAA) and silicone fraction (Tris, mPDMS) with a block-copolymer macromer to ensure a co-continuous structure. Equilibrium water content was measured gravimetrically. UV/vis spectrophotometry and fluorescence confocal laser scanning microscopy were employed to measure partition coefficients for aqueous and oleophilic solutes respectively. Water content is predicted assuming that water is confined to the hydrophilic phase in the gel. Likewise, aqueous solutes have a high affinity for water, and dominantly partition in the hydrophilic phase as well. Conversely, oleophilic solutes predominantly partition into the silicone phase due to their low water solubility. To account for aqueous-solute exclusion and oloephilic-solute-adsorption in the silicone phase, extended Enhancement Factor Partitioning Theory [7] is adopted. Comparison of theory for water content and solute partitioning to the experimental data is excellent.

[1] Tighe B. Silicone hydrogels: structure properties and behavior. In: *Silicone Hydrogels: Continuous Wear Contact Lenses*; Sweeney DF, Ed. Butterworth-Heinemann: Oxford, 2004; pp.1-27

[2] Stapleton F, et al. Silicone hydrogel contact lenses and the ocular surface. *The Ocular surface*. 2006; 4; 24.

[3] Nicolson PC, Vogt J. Soft contact lens polymers: an evolution. *Biomaterials*. 2001; 22; 3273.

[4] Guan L, et al. Permeability and partition coefficient of aqueous sodium chloride in soft contact lenses. Journal of Applied Polymer Science. 2011; 122; 1457.

[5] Jones L, et al. The TFOS international workshop on contact lens discomfort: report of the contact lens materials, design, and care subcommittee. Invest Ophthalmol Vis Sci 2013; 54:TFOS37-70.

[6] Ketelson HA, Meadows DL, Stone RP. Dynamic wettability properties of a soft contact lens hydrogel. Colloid Surface B. 2005; 50; 1.

[7] Dursch TJ, et al. Water-soluble drug partitioning and adsorption in HEMA/MAA hydrogels. Biomaterials. 2014; 35; 620.

387551 Polymer Directed Self-Assembly of Targeted Theragnostic Nanoparticles

Abstract:

Multicomponent nanoparticles that combine diagnostic and therapeutic functionality would be versatile biomaterials in the development of personalized medicine, pharmacogenomics, and optimized drug selection. In this work, we produce nanoparticles encapsulating supramolecular complexes of antioxidants (therapeutic) and metal salts (diagnostic) via Flash NanoPrecipitation (FNP). In this process, amphiphilic block copolymers direct self-assembly of biocompatible nanoparticles with high concentrations of encapsulated components. The precipitation process is controlled by carefully tuning the time scales of micromixing, self-assembly, complexation as well as nucleation and growth. We have incorporated metal salts such as iron and copper for potential PET imaging as well as gold for potential X-ray contrast. Targeted delivery of the hybrid nanoparticle can be achieved by tailoring the surface chemistry of the nanoparticle; therefore, functionalization of the block copolymer with a targeting ligand such as an antibody will also be discussed.

382702 Inhibiting Nosocomial Infections and Foreign Body Reaction with Engineered Hydrogel Scaffolds

Abstract:

Implantable medical devices, such as sensors, drug releasing materials, etc, are playing an important role in medical treatments. However, two major challenges are remained impeding the successful application of medical implants. First, bacterial infections caused by the implants have become one of the leading causes of death in hospitals. Second, foreign body reaction causes implants to become encapsulated in a dense collagenous capsule that blocks mass/signal transport. Here, we present the design of polymeric hydrogel scaffolds that can resist nosocomial both infections and the foreign body reaction. First, a non-antibiotic antimicrobial strategy was used in this work, giving the scaffold an excellent ability to resist the growth of both gram-positive and gram-negative bacteria. Second, the engineered porous structure of the scaffold greatly mitigated the formation of a capsule and facilitated mass transport. Third, the scaffold is fully degradable in vivo, making it a safe platform for many biomedical applications.

370545 Dual-Stages Biomimetic Adhesives

Abstract:

Mussel adhesive proteins (MAPs) are ideal candidates for effective tissue adhesives and orthopedic sealants. They are natural products from marine organisms which are friendly to the environment with high curing rate, biocompatibility, and outstanding mechanical property. In our work, a predominant functional chemical in mussel adhesive proteins, the dopamine, was introduced to poly(acrylic acid) (PAA) to form a straightforward biomimetic material, PAADopa. Under suitable metal-chelating method, new phenomenon of coacervate adhesives and the subsequent hydrogel adhesives during the cross-linking process were found. The coacervate adhesives were formed instantly by mixing PAADopa and zinc liquid, which could be easily cured near body-temperature and show robust mechanical property compared to popular commercial glue under both dry and wet conditions. With pH trigger, coacervates could be changed into hydrogel which showed perfect self-healing property and adjustable modulus. The obtained water-born coacervate and hydrogel adhesives should be ideal candidates for the clinical adhesives and biomaterials.

Superior Dispersion and Property Enhancements in Nanocrystalline Cellulose-Polyolefin Biocomposites Prepared via Solid-State Shear Pulverization

Krishnan Iyer[1], Gregory Schueneman[3] and John Torkelson[1,2]
[1]Department of Chemical and Biological Engineering, Evanston, IL 60208, USA
[2]Department of Materials Science and Engineering, Evanston, IL 60208, USA
[3]Forest Products Laboratory, US Forest Service, Madison, WI, USA
Contact information:
Krishnan Iyer – kai@u.northwestern.edu
Gregory Schueneman - gtschueneman@fs.fed.us
John Torkelson – j-torkelson@northwestern.edu

Cellulose, the main component of plant cell walls, constitutes the most abundant natural polymer resource available worldwide. It is the basic building block for textiles and for paper. Due to its remarkable strength with stiffness values exceeding 150 GPa, cellulose has garnered a lot of interest as a filler for polymers. Ever since the discovery of colloidal cellulose particles by Ranby in 1950's, the process of producing defect-free, rod-like crystalline nanocellulose has undergone considerable changes. Such cellulosic nanocrystals are known by different names such as nanowhiskers, nanocrystals, nanofibrils and nanocrystallites. Here, we refer to them as nanocrystalline cellulose (NCC). Such nanocrystals have large aspect ratios, being 3-5 nm in diameter and 100-300 nm in length. The unique physical properties of NCC have thus inspired several literature studies to explore its application as novel green nano-reinforcement in the production of polymer nanocomposites. Green nanofillers offer several sustainable benefits such as low cost, light weight, less energy for processing and reduced wear to instruments.

As a green filler, NCC has been incorporated into a variety of polymer matrices. Due to its hydrophilic nature, NCC is easily dispersed in water. Therefore, many studies have attempted to produce polymer-NCC composites with water-soluble polymers such as poly(vinyl alcohol). Favier et al. (Polym Adv Technol 1995;6:351–5) reported the first production of styrene-butadiene rubber (SBR)-NCC composites with water-based emulsion systems. In addition, NCC has been dispersed in several organic solvents, producing composites with enhanced mechanical properties. Compared to solution processing, relatively few studies have investigated the possibility of conventional melt-processing techniques such as twin-screw extrusion (TSE) for producing NCC-based nanocomposites. Achieving good dispersion of NCC in non-polar polyolefin matrices is a major challenge with TSE because of the hydroxyl groups on the filler surface and the often-agglomerated nature of NCC. Furthermore, NCC has poor thermal stability with degradation temperature as low as ~180 °C. This poses issues associated with filler degradation during melt processing for the production of composites.

Poor filler dispersion and degradation of NCC are detrimental to mechanical performance of polyolefin-NCC composites. Ljungberg et al. (Polymer 2006;47:6285–92) and Bahar et al. (J Appl Polym Sci 2012;125:2882–9) produced composites of PP with NCC using solution-processing techniques observing minor to no property enhancements. Durfesne and coworkers (Polymer 2009;50:4552–63) employed TSE to produce polyethylene-NCC composites with only modest increment in mechanical properties due to large filler agglomeration during processing. Surface treatment with organic acid chloride aliphatic chains of different size did not result in any further improvement in properties. Oksman and coworkers (Compos Sci Technol 2006;66:2776–84) employed a novel liquid feeding technique in which a NCC dispersion in water with plasticizer (poly(vinyl alcohol)) is pumped into molten poly(lactic acid) during extrusion. Such techniques still cause severe filler aggregation with several added challenges in removing moisture and in suppressing polymer degradation. Recently, Lin et al. (Macromolecules 2013;46:5570–83) used a wrapping technique in which NCCs were coated with low molecular weight polyethylene oxide to suppress NCC degradation during processing. However, filler agglomeration prevails due to incompatibility and inability of melt-mixing to provide sufficiently large shear forces to break up filler agglomerates. Therefore, there is a need for a solvent-free, continuous and industrially scalable processing technique for the production of such green polymer biocomposites.

Solid-state shear pulverization (SSSP) is a novel processing technique, which employs a modified twin-screw extruder where materials are processed under near-ambient temperature conditions. During SSSP, polymers are processed in their solid state, and large compressional and shear forces are imparted to the material. The repeated fragmentation and fusion of the material result in effective break-up of filler agglomerates. Near-ambient conditions in SSSP provide intimate mixing between polymer and nanofiller without regard to viscosity mismatch or thermodynamic limitations. Additionally, such conditions can prevent filler degradation of cellulosic materials, with effective particle break-up and dispersion, all of which are advantageous in the production of polyolefin-NCC biocomposites. In the past, these unique advantages of SSSP processing have been employed for the successful production of the following:

1. *Polymer blends:*
SSSP processing has been successfully used for the compatiblization of immiscible polymer blends with in-situ formation of block copolymers as a result of inter-radical coupling during the mechanochemical treatment.

2. *Nanocomposites:*
Fine dispersion and exfoliation of a variety of fillers ranging from platelet such as graphite and clay to agglomerated carbon nanotubes have been observed with dramatic mechanical property enhancements.

3. *Green composites:*
Recently, SSSP has been employed for the production of green polymer composites based on renewable waste materials such as rice husk ash with major enhancements in mechanical properties. Well-dispersed rice husk ash acts as a substituent for commercially available silica commonly employed as filler for plastics.

4. *Polar functionalization of polyolefins:*
Taking advantage of the unique radical chemistries associated with near-ambient temperature processing conditions in SSSP, polar functionalization of polyolefins (e.g., polypropylene) with ester or maleic anhydride moieties can be achieved with dramatically suppressed molecular weight reduction compared to melt processed systems.

Here we demonstrate the effective dispersion and major suppression of NCC degradation in green polyolefin-NCC biocomposites via SSSP with up to 10 wt% filler content.

Excellent dispersion and retention of pristine cellulose color are observed in both low-density polyethylene (LDPE) and PP matrices through optical microscopy and scanning electron microscopy. Synergistic property enhancements due to effective dispersion are demonstrated via a variety of characterization techniques such as differential scanning calorimetry (DSC) and tensile testing. Well-dispersed NCC acts as an excellent nucleating agent for PP crystallization with dramatically reduced crystallization half-times. In short, the superior filler dispersion and suppressed filler degradation give rise to synergistic enhancements in Young's modulus and excellent retention of elongation at break for 10 wt% NCC loadings in both PP and LDPE. Previously, poor outcomes were reported by Menez et al. when incorporating unmodified NCC in LDPE produced by TSE (Polymer 2009;50:4552–63). They observed only a modest 14 % increase in Young's modulus and a 85 % decrease in elongation at break relative to the neat polymer. In comparison, SSSP processed LDPE-NCC composite with 10 wt% NCC show up to 69 % increase in Young's modulus and no decrease in elongation at break relative to that of the neat polymer. To the best of our knowledge, the mechanical property enhancements shown by SSSP-processed polyolefin-NCC composites are superior to those reported in the research literature for similar composites made by melt processing. The degradation behavior of NCC, LDPE-NCC and PP-NCC systems is analyzed under nitrogen and air atmospheres using a thermogravimetric analyzer (TGA). Interestingly, LDPE-NCC composites show enhanced thermal stability relative to neat LDPE due to char formation from well-dispersed NCC. This study thus demonstrates the promise of SSSP processing in exploiting the full potential of this novel green nanofiller.

MFC-based composite films for gas barrier applications with improved water resistance

Çağlar Meriçer[1], *Luca Ansaloni*[1], *Matteo Minelli*[1,2], *Marco Giacinti Baschetti*[1,2]

[1]Dipartimento di Ingegneria Civile Chimica, Ambientale e dei Materiali (DICAM)
[2]Centro Interdipartimentale per Ricerca Industriale - Meccanica Avanzata e Materiali (CIRI-MAM)
Alma Mater Studiorum - Università di Bologna, Bologna, Italy

Nano-sized cellulose fibers are highly attractive candidates for the production of bio-composite materials, due to their low weight coupled with high strength, their large availability and good biodegradability. Among the possible uses, microfibrillated cellulose (MFC) has been demonstrated to present excellent gas barrier properties when prepared in self-standing films, likely related to the crystalline domains present in the structure and to the ability of molecules to create strong inter-fibrillar bonds forming a dense network. MFC represents thus an appealing biodegradable material for packaging applications[1]. Nevertheless, the use of MFC films at the industrial scale is still limited due to the high hydrophilic nature of the fibers, which can significantly alter the barrier properties. Indeed, self-standing MFC layers showed a marked water uptake (up to 10-15 wt%, depending on the fibers fabrication technique), able to produce a relevant swelling of the film matrix, leading to a drastic increase of the O_2 transport rate. Consequently, the permeability value increases of two orders of magnitude, even at low relative humidity (R.H.)[2,3].

To overcome this limitation, a composite bilayer is proposed, which couples the barrier properties of MFC with the hydrophobic nature of poly(lactic acid) (PLA), a biopolymer largely used as compostable packaging material. Thin coatings of two different MFC generations (obtained by carboxymethylation, MFC-G1, or through enzymatic pre-treatment, MFC-G2) are deposited on a PLA layer, used as substrate. The developed deposition protocol, based only on a physical pretreatment of PLA films, allowed the achievement of a very good adhesion between the two phases. In addition, the coating improves the mechanical resistance of MFC, with respect to the self-standing film properties, making the composite material suitable for a wider range of packaging applications.

Dry oxygen permeation tests reveal a significant improvement of the barrier properties of PLA[4], whose O_2 transfer rate (O.T.R.) decreased about two orders of magnitude. In addition, the permeability of the sole coating is conveniently evaluated from the comparison of the pure PLA datum and that of the multilayer films. The permeability of the coating layer is in good agreement with the value obtained from self-standing thick films (both in case of MFC-G1 and MFC-G2), demonstrating the capability of the present material to preserve the original properties of the two phases. Interestingly, when the R.H. of the system is increased up to 25%, the oxygen transfer rate across the film is barely affected by the presence of water, and at 50% R.H. the O.T.R. increases only to a limited extent, compared to the neat MFC film, that in the same condition already showed a marked increase. Only when high water content in the feed stream are reached (R.H. > 75%), the O_2 permeability tends to the PLA value in dry conditions, suggesting that no further transport resistance is given by MFC.

The present study reveals that the novel composite material preserves the advantageous properties of the two neat materials, the hydrophobic behavior and the oxygen barrier, providing a fully renewable and biodegradable packaging material able to compete with the conventional oil-based products available in the market.

References:

1. Siró & Plackett, *Cellulose* **17,** 459 (2010).
2. Minelli *et al.*, *J. Memb. Sci.* **358,** 67 (2010).
3. Aulin *et al.*, *Cellulose* **17,** 559 (2010).
4. Auras *et al.*, *J. Appl. Polym. Sci.* **92,** 1790 (2004).

Biopolymer Composites: Mechanical, Thermal, Rheological and Biodegradation Properties

Ali U Chaudhary, Khalid Al Zaabi, Vikas Mittal
Chemical Engineering Department,
The Petroleum Institute, Abu Dhabi, UAE
vmittal@pi.ac.ae

Introduction

- Biodegradable polymers (biopolymers) have gained significant research focus in the recent years due to increasing consciousness towards the use of more environmentally friendly materials.
- In a recent study, a significant compound annual growth rate of 22% for biodegradable polymers during the 5 year period starting from 2012 has been estimated.
- Incorporation of a large variety of fillers in the conventional polymers has been observed to cause significant enhancements in their mechanical, thermal, rheological and gas barrier properties. In order to generate high value materials from biopolymers, similar composites using fillers like layered silicates, nanotubes etc. have also been reported.

2

Aim/Objective

- Due to the increasing commercial importance of biopolymers, it is needed to further explore their composites and applications.

- In the presentation, composite systems of different biopolymers (PBAT, PLA, chitosan, PBS and PA) with silicate, graphene and silica reinforcement phases are presented. The objective was to study the interfacial interactions of the polymer chains with the surface of filler particles as well as to correlate the structural and morphological features of the composites with their thermal, rheological and mechanical properties. In addition, the use of bio-filler based on date seed powder to generate 'true bio-composites' has been explored.

3

Date Seed Powder Composites

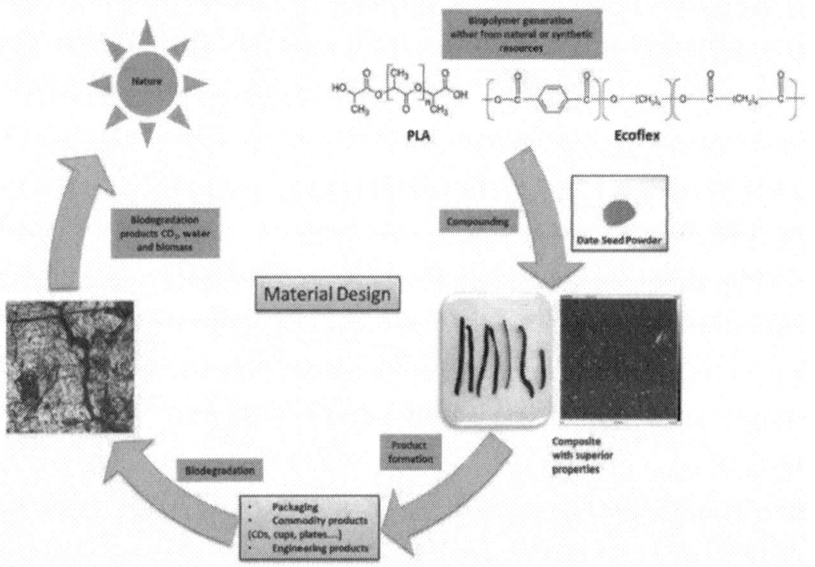

Date Seed Powder

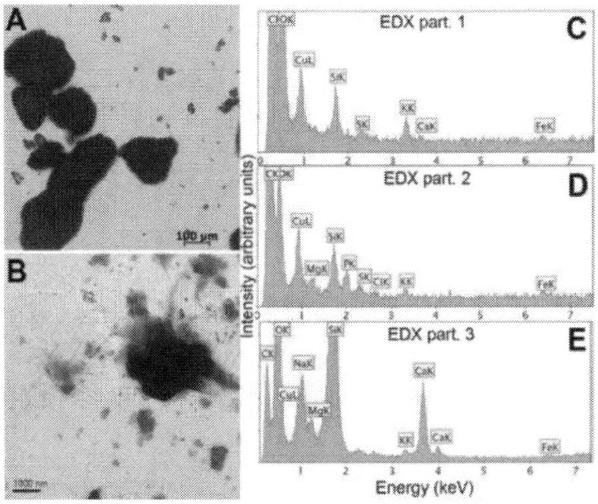

Source: various web resources

Date Seed Powder

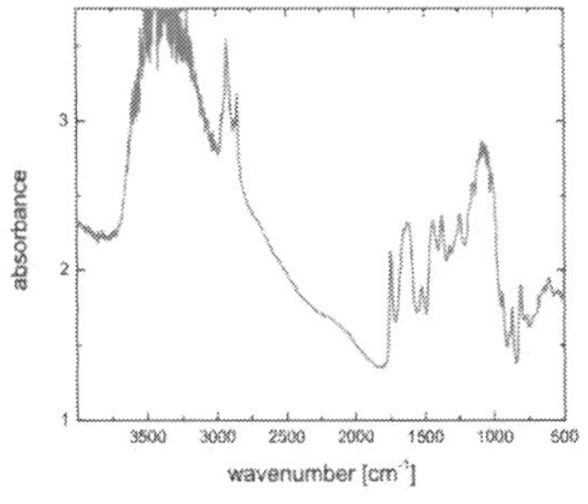

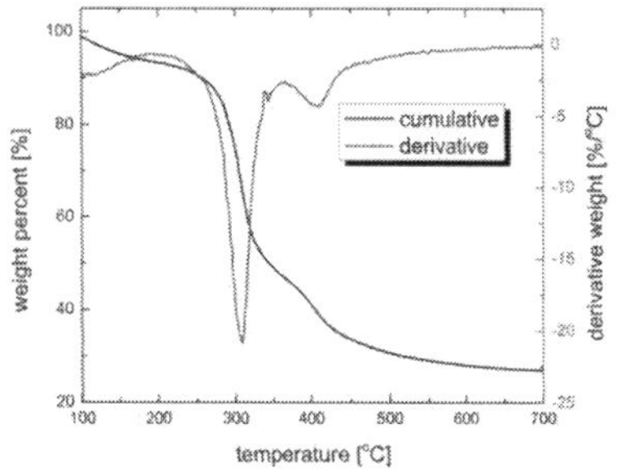

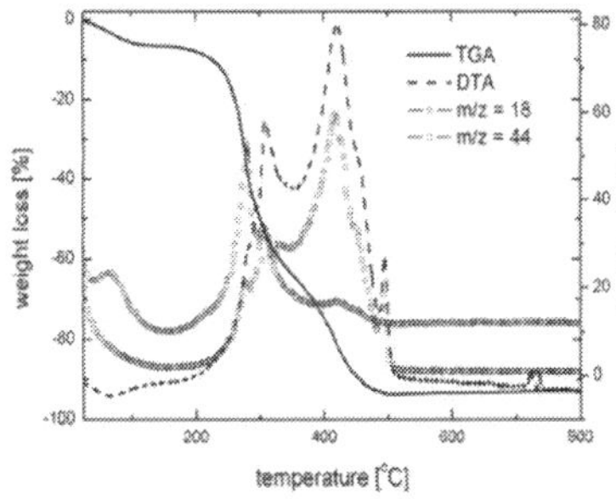

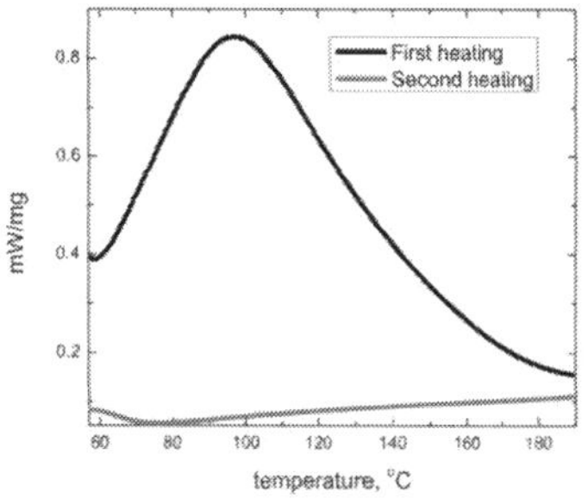

Mechanical Properties

PBAT composites

Composites	Tensile modulus, MPa	Yield stress, MPa	Yield strain, %	Stress at break, MPa	Elongation, mm
0%	82.1	9.1	13.1	20.9	88.7
10%	116.7	9.0	14.1	14.8	83.4
20%	149.0	9.7	9.7	11.8	50.9
30%	197.1	9.3	7.4	8.5	22.5
40%	256.7	10.7	5.9	8.2	13.3

Mechanical Properties

PLA composites

Composites	Tensile modulus, MPa	Yield stress, MPa	Yield strain, %	Stress at break, MPa
0	3376	70.3	3	63.7
10%	3528	53.5	2.2	43.7
20%	3508	44.9	2	28.1
30%	2495	7.3	0.3	16.3
40%	2216	5.1	0.3	18.8

Thermal properties

PBAT

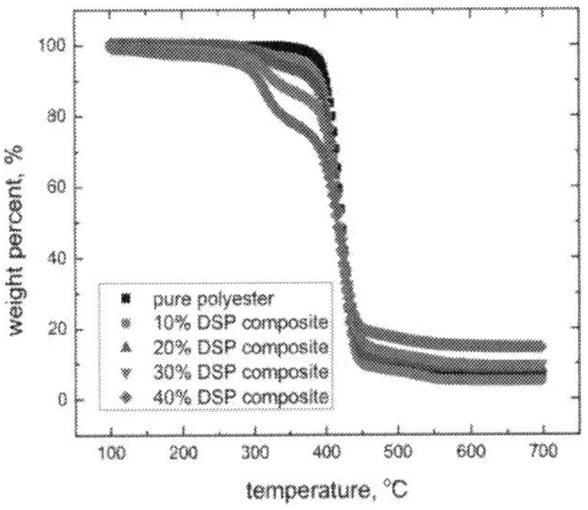

PLA

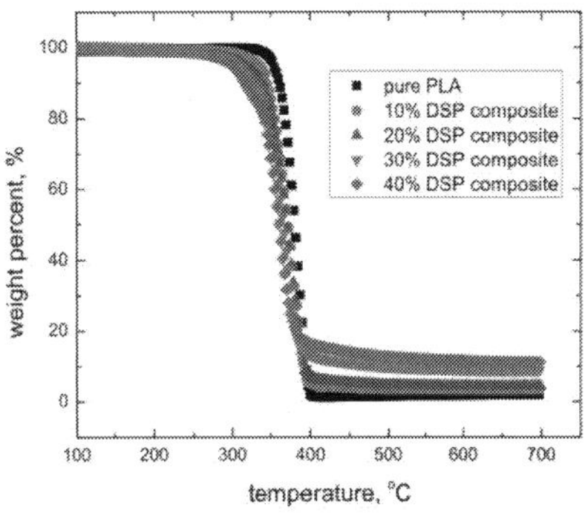

The filler peaks are visible individually

Thermal properties

PBAT

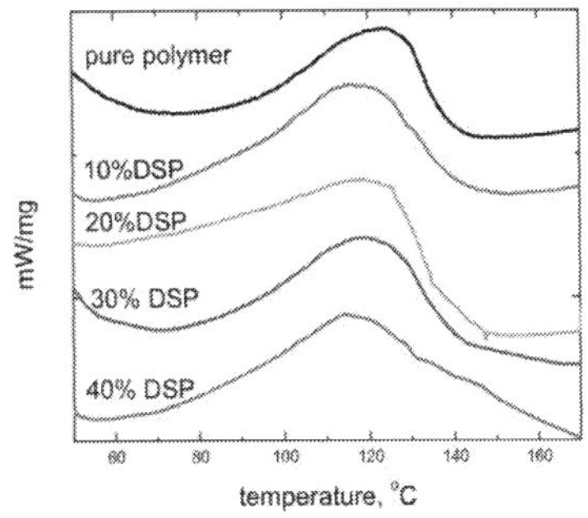

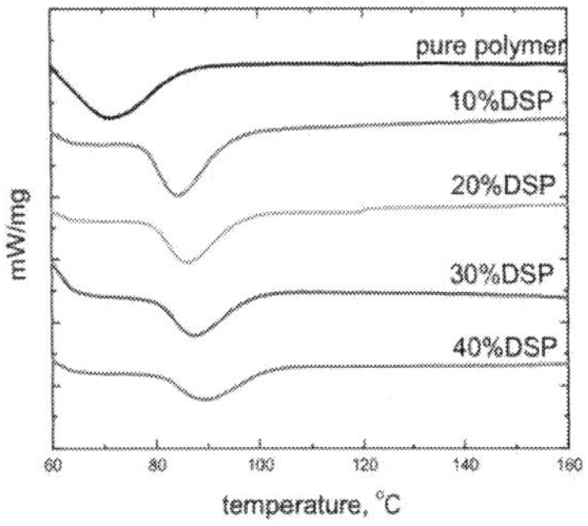

PLA

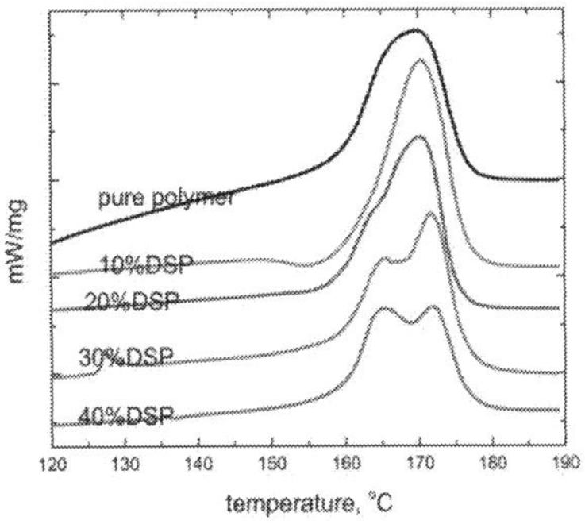

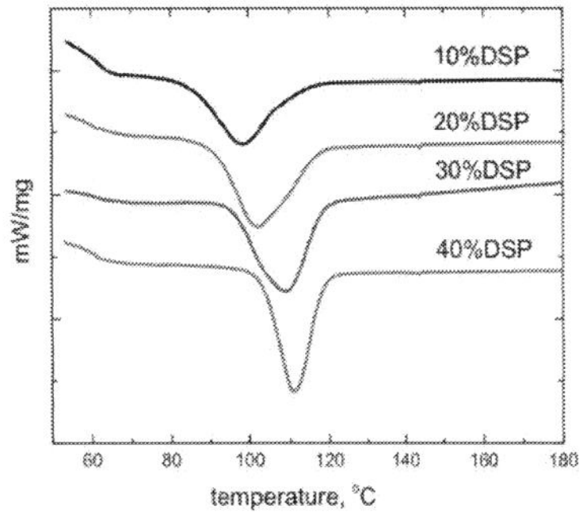

Rheological Properties

PBAT

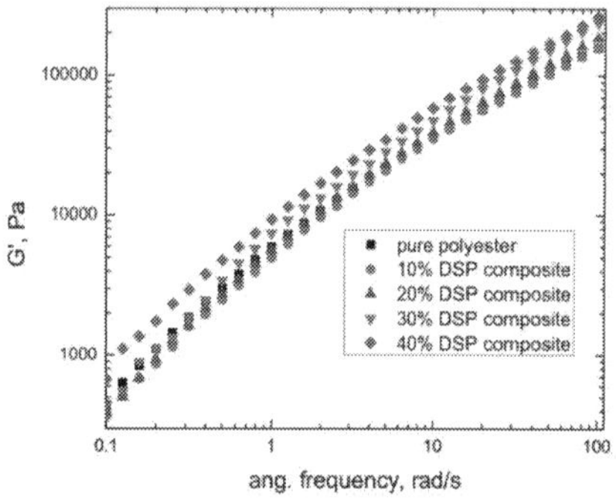

PLA

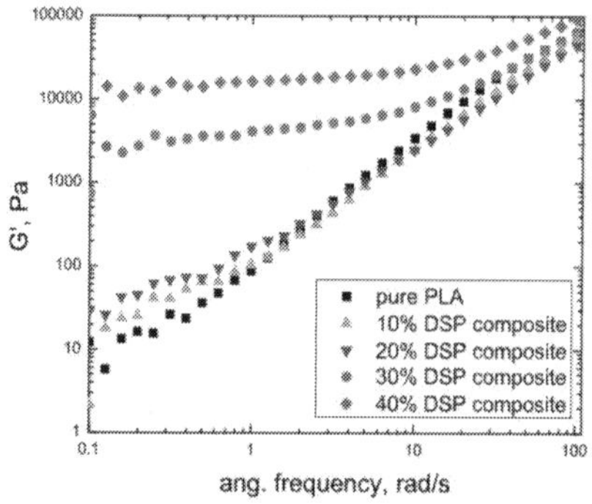

Morphology

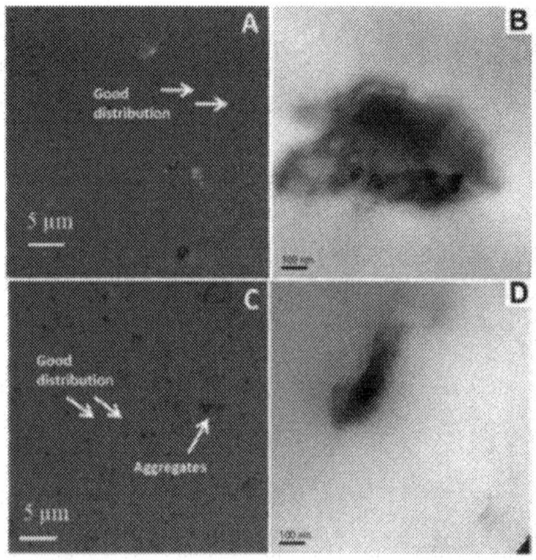

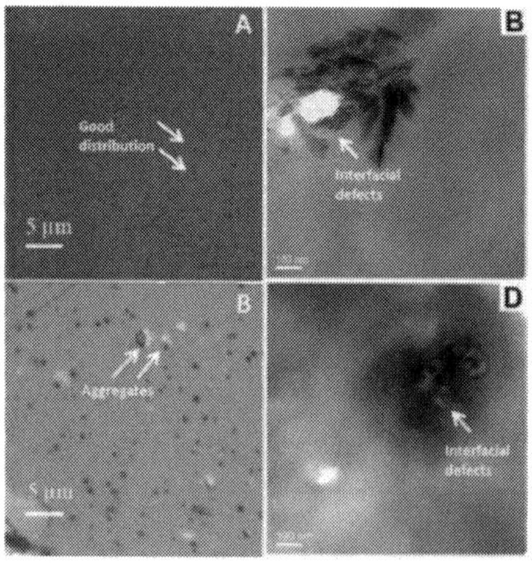

(A) Light microscopy and (B) TEM images of PBAT composite with 10% DSP content. (C) and (D) correspond to light microscopy and TEM images of PBAT composite with 30% DSP content.

(A) Light microscopy and (B) TEM images of PLA composite with 10% DSP content. (C) anc (D) correspond to light microscopy and TEM images of PLA composite with 30% DSP content.

12

Morphology

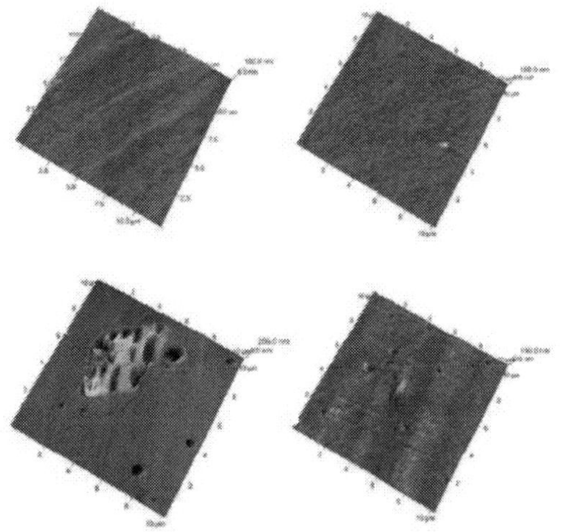

AFM height micrographs of PBAT composites with (top left) 10% and (bottom left) 30% DSP content; micrographs of the PLA composites with (top right) 10% and (bottom right) 30% DSP content.

13

Biodegradation

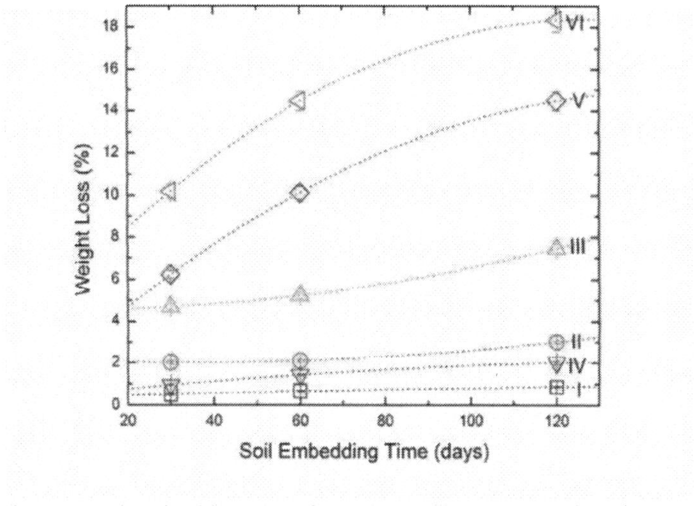

Weight loss vs. soil embedding time for PLA, PBAT and composites. I: PLA; II: PLA+20%DSP; III: PLA+40%DSP; IV: PBAT; V: PBAT+20%DSP; VI: PBAT+40%DSP.

14

Biodegradation

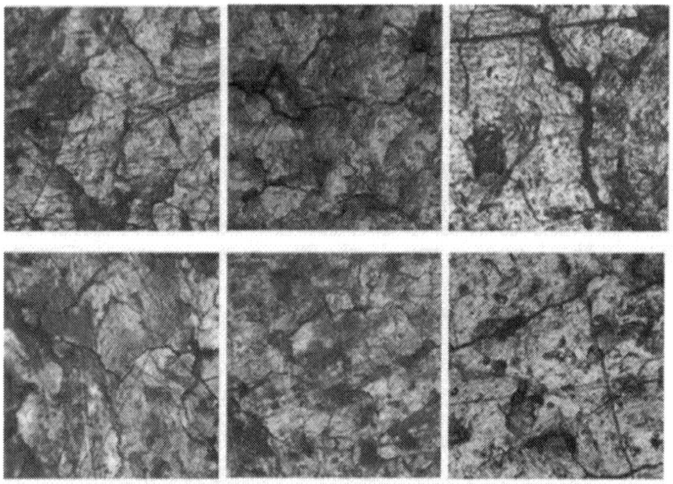

Pure polymer 10% DSP composite 30% DSP composite

Optical micrographs of (upper row) PBAT and (lower row) PLA and their composites after embedding into natural soil for 120 days. The width of images reads 100 μm.

15

Chitosan/PBS Blend Based Composites

Mixing ratio of PBS:CS	Temperature (°C)	Screw speed (rpm)	Young's modulus (N/mm²)	Stress @yield (N/mm²)	Stress @peak (N/mm²)	Strain @yield (%)
100:00	170	100	500	17	37	1.40
50:50	160	80	1848	17	25	0.66
50:50	160	100	2064	16	33	0.36
50:50	170	100	2025	16	33	0.5
60:40	170	100	1691	16	32	0.52
70:30	170	100	1282	12	28	0.43

16

Mechanical Properties

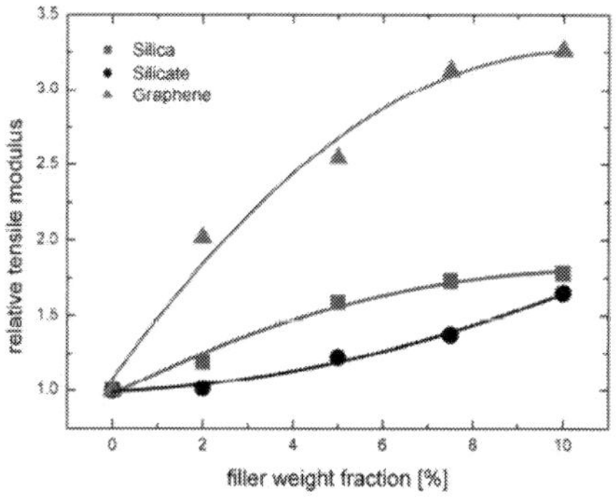

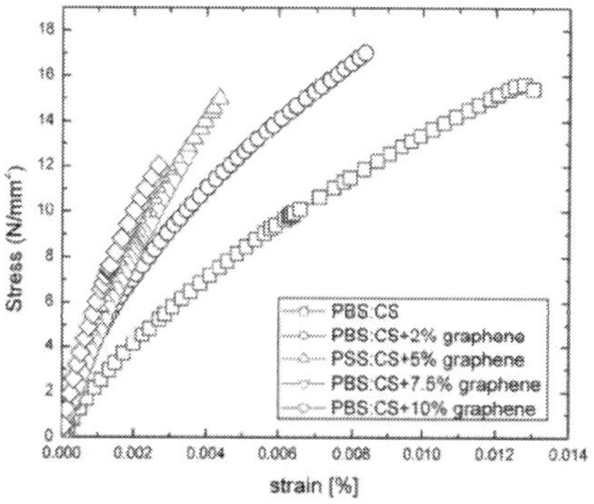

Relative tensile modulus of the nanocomposites as a function of filler weight fraction and stress-strain curves of nanocomposites with graphene

17

Mechanical Properties

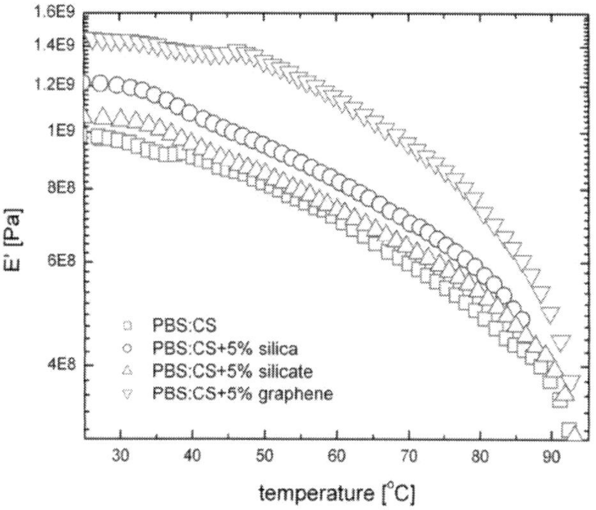

Storage modulus of nanocomposites with 5% filler content, obtained from DMA analysis.

18

Rheological Properties

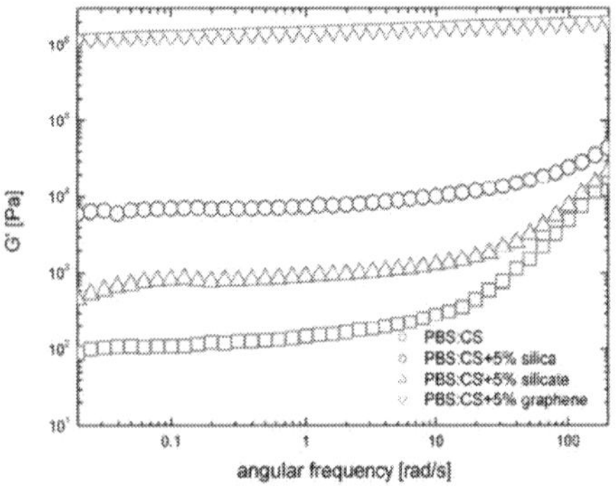

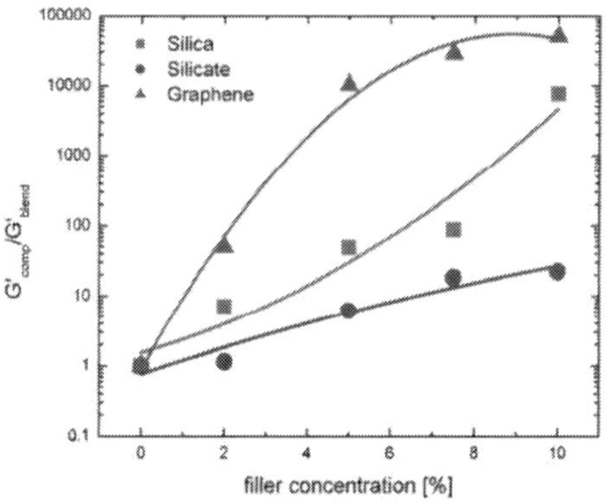

Shear rheology analysis to generate storage modulus of the composites (with 5% filler content) as a function of angular frequency; relative G' of composites as a function of filler concentration (1 rad/s angular frequency)

19

Thermal Properties

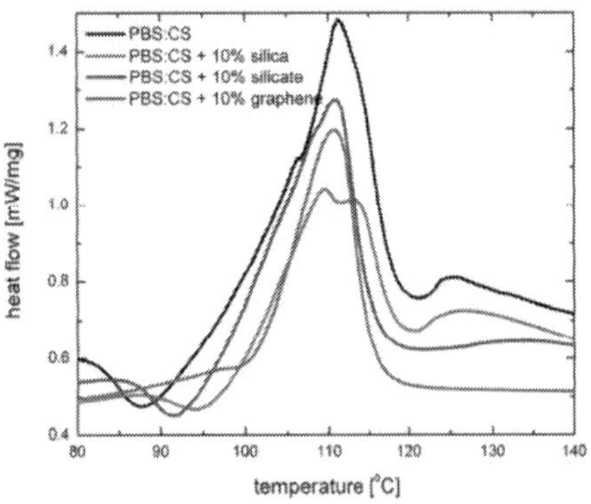

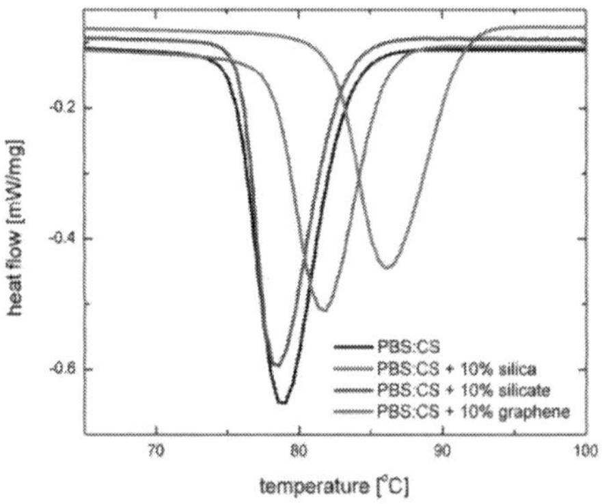

DSC (a) melting and (b) crystallization thermograms of PBS:CS nanocomposites with 10% filler content.

20

Thermal Properties

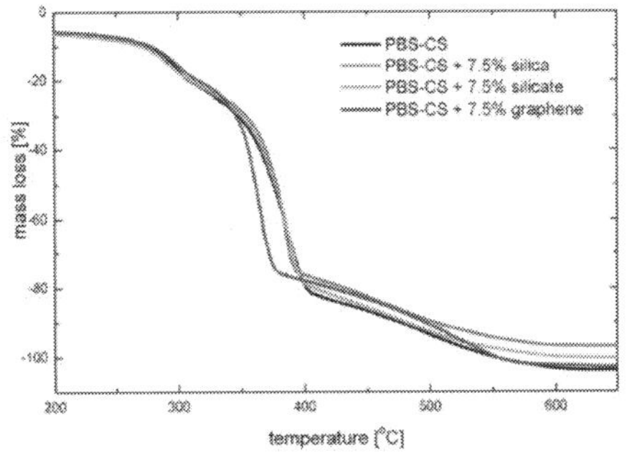

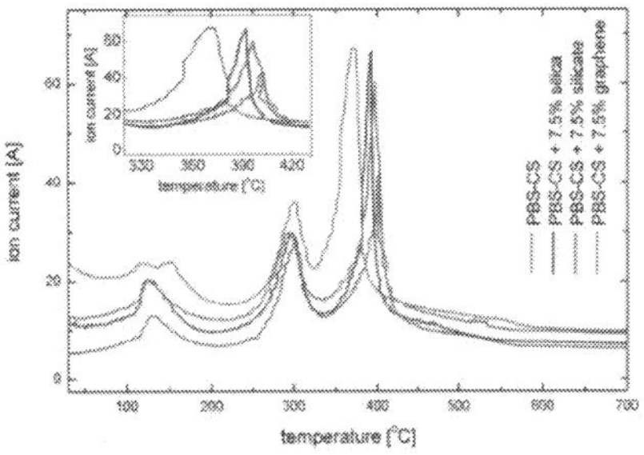

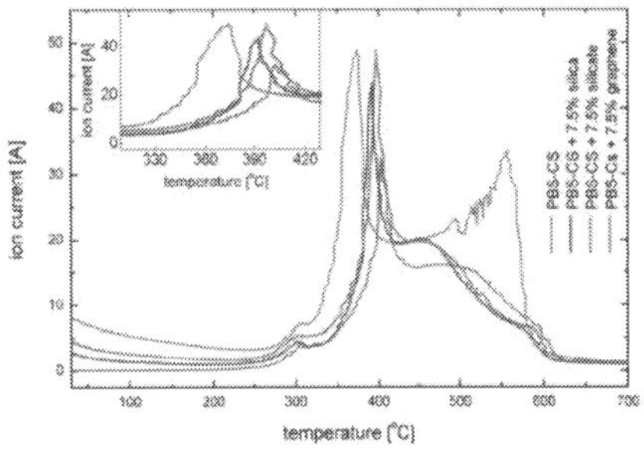

TGA-MS (in air) analysis of the polymer blend and nanocomposites with 7.5% filler content: (a) cumulative mass loss, (b) m/z = 44 and (c) m/z = 18

Morphology

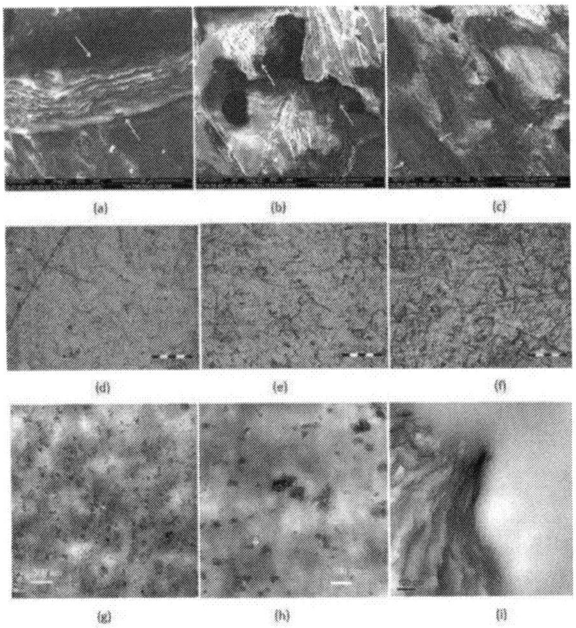

(a)-(c) SEM micrographs of PBS:CS blend representing sheathing of chitosan particles and PBS fibrils (shown by arrows); (d)-(f) optical images of nanocomposites with 5% silica, silicate and graphene respectively; (g)-(i) TEM images of nanocomposites with 5% silica, silicate and graphene respectively.

22

Biodegradation

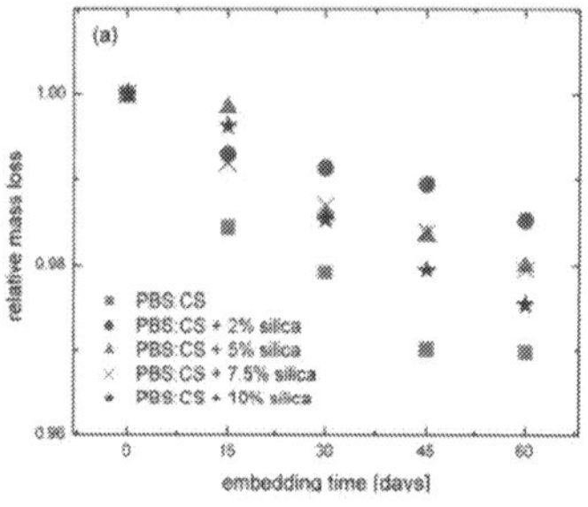

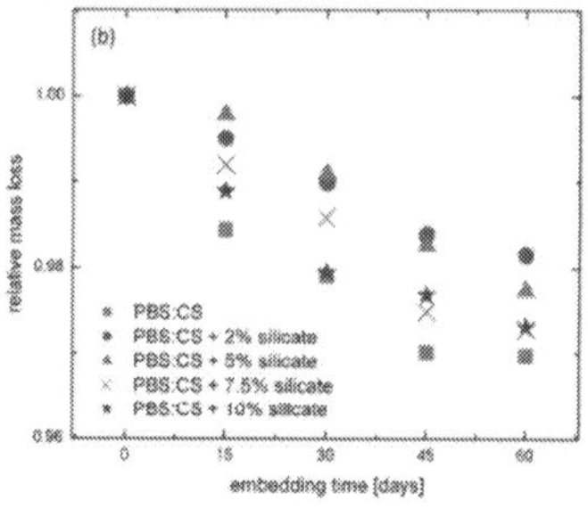

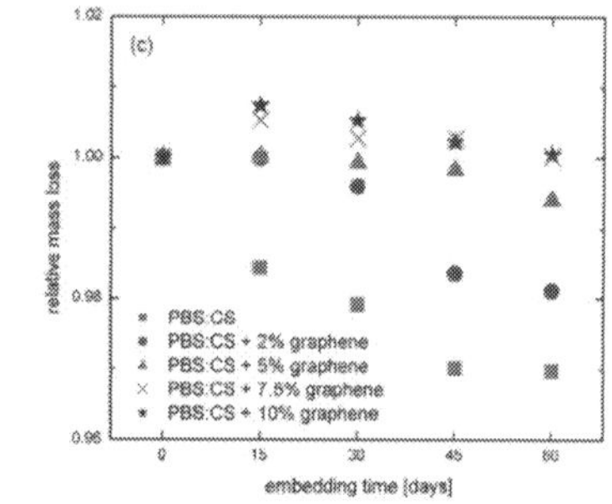

Mass loss of the samples as a function of embedding time; (a) silica, (b) silicate and (c) graphene composites

Biodegradation

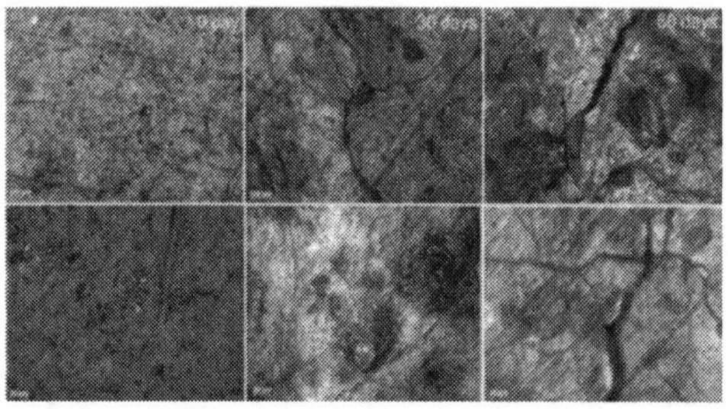

Optical images of the PBS:CS+10% silicate composite as a function of time; (first column) 0 days, (middle column) 30 days and (third column) 60 days.

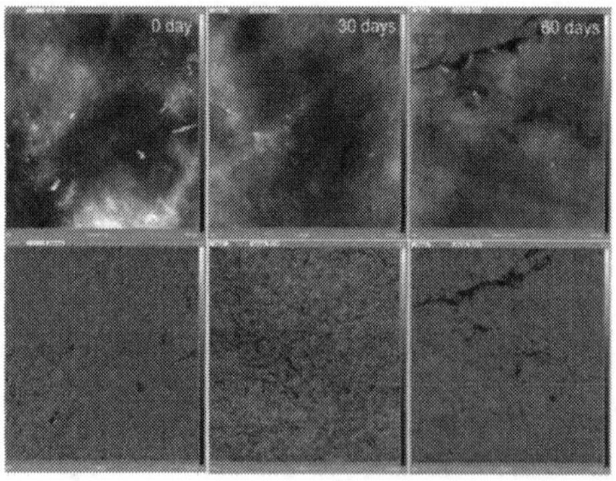

AFM images of the PBS:CS+10% silica composite system.

24

Biodegradation

Optical images of the PBS:CS+10% silicate composites showing the cross-section in relation with the embedding time; (first column) 0 days, (middle column) 15 days and (third column) 30 days.

25

Biodegradation

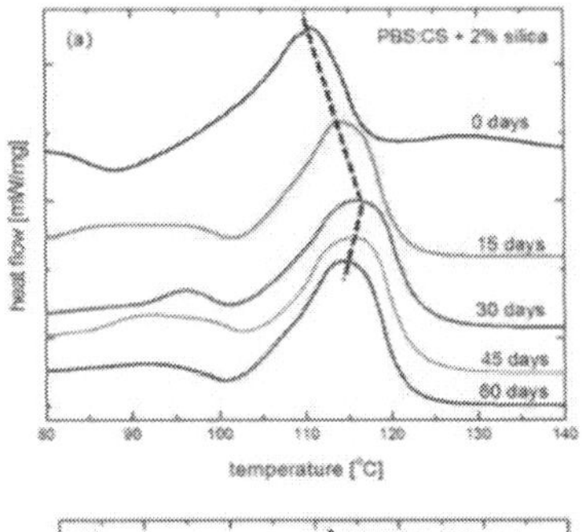

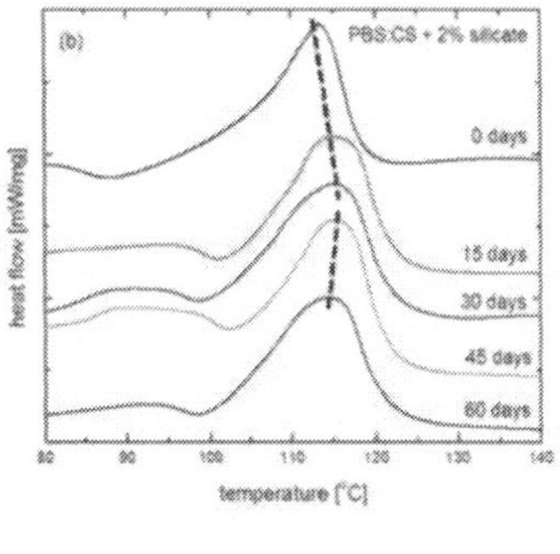

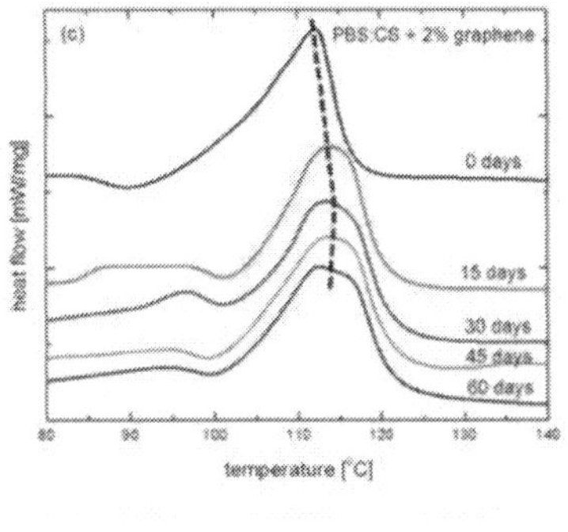

DSC melting thermograms of PBS:CS+2% filler composites as a function of embedding time in soil; (a) silica, (b) silicate and (c) graphene

26

388096 Invited Talk: Photochemical Reactions for Replicating and Aligning Block Copolymer Thin Film Patterns

Abstract:

New methodologies for patterning micro- and nano- scale features in polymer thin films are desired because of their high technological relevance to a range of applications, including microelectronics fabrication. Here recent research will be described where light is used to photochemically replicate the patterns on the surface of block copolymer thin films. In general, directed self-assembly (DSA) of block copolymers (BCPs) could potentially complement or even supplant existing photolithographic techniques, but requires high-resolution guiding patterns to direct the block copolymer self-assembly into device-oriented structures over large areas. A facile replication procedure of the guiding patterns could increase the overall throughput of DSA given that making the original chemical guiding patterns can be relatively slow. Our strategy involves placing benzophenone-containing liquid compositions between the top surface of a block copolymer thin film and a transparent substrate. Upon irradiation the liquid composition solidifies and covalently binds to the BCP, thus creating a mirror-image copy of the original pattern on the transparent substrate. This replicated pattern serves to direct the assembly of a new BCP thin film, while the patterned wetting characteristics of the original substrate are recovered for use in further replication cycles. The process is scalable to large areas, takes less than 1 h, and occurs below the glass transition of the BCP.

372896 Azobenzene Containing Main-Chain Liquid Crystalline Elastomer for Fully Reversible Photo- and Thermal Actuation

Abstract:

A series of main-chain liquid crystalline elastomers (LCE) containing novel azobenzene mesogens, containing two, three, four phenyl rings respectively, were synthesized via hydrosilyzation addition reaction under ambient condition. Their structures and properties of the elastomer films were studied in detail by a combination of nuclear magnetic resonance (NMR), differential scanning calorimetry (DSC), polarized optical microscopy (POM), RSA solid dynamic mechanical analyzer and in-situ two-dimensional X-ray scattering. With large uniaxial stretch at elevated temperature, followed by annealing procedure, a very perfect reorientation monodomain Smectic-C liquid crystalline phase was obtained and confirmed by the X-ray diffraction pattern. The prepared monodomain films showed strong opto-mechanical responses upon exposure to ultraviolet (UV) light at 365 nm, as a result of an optically induced isomerization of azobenzene moieties in the backbone. Compared to the previously reported side-chain and copolymer LCEs, the new end-on main-chain LCEs presented much more pronounced mechanical responses due to the large backbone molecular motion during the trans-cis transition of the azobenzene moieties upon UV irradiation. The UV light induced stress responses that obtained with the mesogenic groups reached over 800 kPa, which provide a promising prospect for artificial muscle, actuation and sensor applications.

Controlling the Physical Aging of Polymer Glasses by Vitrifying Under Stress

Laura A. G. Gray, Justin E. Pye, and Connie B. Roth
Department of Physics, Emory University, Atlanta GA, USA

Contact information:
Laura Gray: lgolick@emory.edu
Justin Pye: jepye@emory.edu
Connie Roth: cbroth@emory.edu

Under many processing conditions, polymers naturally form a glass. Their stereo-irregularity and sluggish relaxation times hinder crystallization readily trapping the polymer molecules into an amorphous glassy state. Unfortunately, the non-equilibrium nature of this glassy state results in an unstable material that slowly densifies over time in a process referred to as physical aging that is typified by decreases in permeability and increases in brittleness. The speed of these glassy dynamics and the resulting stability of the material depend on how far the current state of the system is from equilibrium, which is determined from thermal and deformation histories. As such, different processing conditions can significantly influence the resulting properties of polymer glasses, an area with limited understanding currently at a fundamental molecular level.

During common processing conditions leading to glassification, polymers are often subjected to large volume contractions either through temperature changes or during solvent removal when casting. However, in many cases, this size change of the polymeric component can be restricted by the surrounding support or mold leading to significant stresses. A simple example of this process where the stress imparted to the polymer can be readily calculated involves cooling polymer films while supported on rigid frames. The thermal expansion differences between the polymer film and underlying support, easily over an order of magnitude, can result in stresses of several MPa. We have recently demonstrated that the physical aging rate of polystyrene (PS) films is strongly affected by the type of material used for its underlying support or frame. We find that the physical aging rate correlates strongly with the stress caused by the thermal expansion mismatch between film and support [1]. This stress, applied to the films as they are thermally quenched into the glassy state, can nearly double the physical aging rate when changing the frame material from polycarbonate to silicon.

How stress and mechanical deformation impart mobility to polymer glasses has been a long standing issue related to the underlying causes of the glass transition. To date, most such studies have primarily focused on materials where the glassy state was formed stress free. In contrast, we are interested in the stability of polymer glasses when stress is applied during the formation of the glassy state (vitrification). To investigate how stress imparted to polymer glasses on cooling alter the resulting physical aging behavior, we have constructed a unique jig to apply a known stress to free-standing films during the thermal quench. Ellipsometry is used to measure the physical aging rate of PS films transferred onto silicon wafers by quantifying the time-dependent decrease in film thickness that results from an increase in average film density during aging. We find that stress values above a threshold value result in less stable polymer glasses with faster physical aging rates [2]. We hypothesize that increased stresses may trap the glassy state into higher, less stable potential energy minima resulting in faster aging rates. Initial measurements of the rubbery plateau creep compliance also indicate that these thin films are stiffer than bulk by two orders of magnitude, consistent with other studies in the literature. However, our results appear to be independent of film thickness over the range studied (150-700 nm). Current efforts are now focused on computer-controlled application of stress and strain during the quench to investigate these unusual material properties in thin films.

[1] Pye, J. E.; Roth, C. B. *Macromolecules* **2013**, *46*, 9455–9463.
[2] Gray, L. A. G.; Roth, C. B. *Soft Matter* **2014**, *10*, 1572 – 1578.

An Examination of Post-Processing Orientation in Coextruded poly (ε-caprolactone) Fibers

Alex Michael Jordan

Traditional methods of producing synthetic polymeric fibers include wet and dry spinning, melt spinning, and electrospinning. By utilizing melt coextrusion, typically used to produce discrete layers, the melt flow can be manipulated to produce a fiber/matrix composite in a continuous and solvent-free fashion. [1] Unlike traditional fiber processing techniques, it is possible to introduce post-processing orientation utilizing melt coextrusion, which provides a unique avenue to tune the properties of the fiber system. In this work, the fiber/matrix composite consisted of rectangular fibers of poly (ε-caprolactone) (PCL) in a poly (ethylene oxide) (PEO) matrix, which were chosen for their widespread use in the bioengineering field as tissue engineering scaffolds and for ease of surface modification on PCL fibers. Following coextrusion, the composite was washed using water and methanol to produce a nearly pure (99.2% by weight) PCL fiber system. After melt processing and composite washing, the Hermans orientation factor of the PCL fibers was ~0.01, which indicated a random crystal orientation. The fibers were then drawn at a temperature of 50 °C in order to develop an oriented crystal system. After fiber drawing (draw ratio, DR = 11), the Hermans orientation factor reached ~0.94, which indicated a highly aligned crystal phase along the fiber axis (drawing direction). [2] This drastic shift in orientation during the drawing process provides a platform to investigate *in-situ* morphology during drawing and the resulting shift in mechanical properties.

The macro/microscale and nanoscale structure of the fibers was evaluated using scanning electron microscopy (SEM) and detailed analysis of *in-situ* wide-angle X-ray scattering (WAXS). On the macroscale, it was observed that following the washing process a system of highly aligned rectangular fibers is produced. On the microscale, the dimensions (both width and thickness) of individual fibers were measured as a function of draw ratio. In the as-extruded state, the width was measured to be 2.6 ± 0.6 μm, while the thickness was measured to be 1.6 ± 0.4 μm. Following the drawing process, the fiber width was measured to be 0.31 ± 0.05 μm and the thickness was measured to be 0.13 ± 0.02 μm. It was observed that a linear relationship exists between the fiber dimensions and draw ratio, suggesting that the fibers undergo affine deformation and demonstrating the ability to control fiber dimensions through post-processing orientation. Scherrer WAXS analysis revealed that PCL crystallites are fragmented into smaller crystallites during the drawing process. [3] Following the orientation step, the fibers were cooled down to room temperature and in the process, pass through the crystallization temperature of PCL, and the crystallinity increased from roughly 40% in the as-extruded state, to almost 60% in the oriented state. Based on these investigations, a complete picture of the macroscopic, microscopic, and nanoscopic hierarchical structure formed within the coextruded PCL fibers was developed using advanced X-ray scattering techniques and microscopy imaging.

In the second part of the study, we explored the thermo-mechanical properties of the as-extruded fibers and the oriented fibers to correlate the observed structure to the properties developed. A comparison of the melting endotherms of the first heating differential scanning calorimetry (DSC) thermograms revealed that the melting enthalpy of the oriented fibers was substantially higher than that of the as-extruded fibers, which agrees with the higher crystallinity estimates obtained from WAXS. Additionally, the melting endotherm of the oriented fibers was more well-defined than that of the as-extruded fibers, shifting to a slightly lower melting temperature, which is consistent with smaller individual crystallites that are also observed in

WAXS. The dynamic mechanical properties and tensile properties of bundles of as-extruded and oriented fibers were also examined, revealing a 10x increase in storage modulus (dynamic mechanical analysis measurements) as well as a 10x increase in tensile strength and 30x increase in tensile modulus (uniaxial tensile measurements). The extreme enhancement in the mechanical properties of the oriented fiber bundles was attributed to the shift from large, randomly oriented crystallites to smaller, highly ordered crystallites.

In order to assess the viability of replacing traditional electrospun fibers with extruded fibers, electrospun fibers of comparable diameter to those of the extruded fibers were produced and the thermal and mechanical properties of both systems were investigated. As-extruded, rectangular PCL fibers compared well to electrospun, circular PCL fibers with similar mechanical behavior, while having a surface area roughly 30x higher than conventionally produced electrospun fiber mats. The oriented PCL fibers significantly outperformed electrospun fibers, exhibiting higher storage modulus, tensile modulus, and tensile strength (Figure 1). Finally, a facile method of solvent and PEO recovery was developed to reduce waste generated during fiber production.

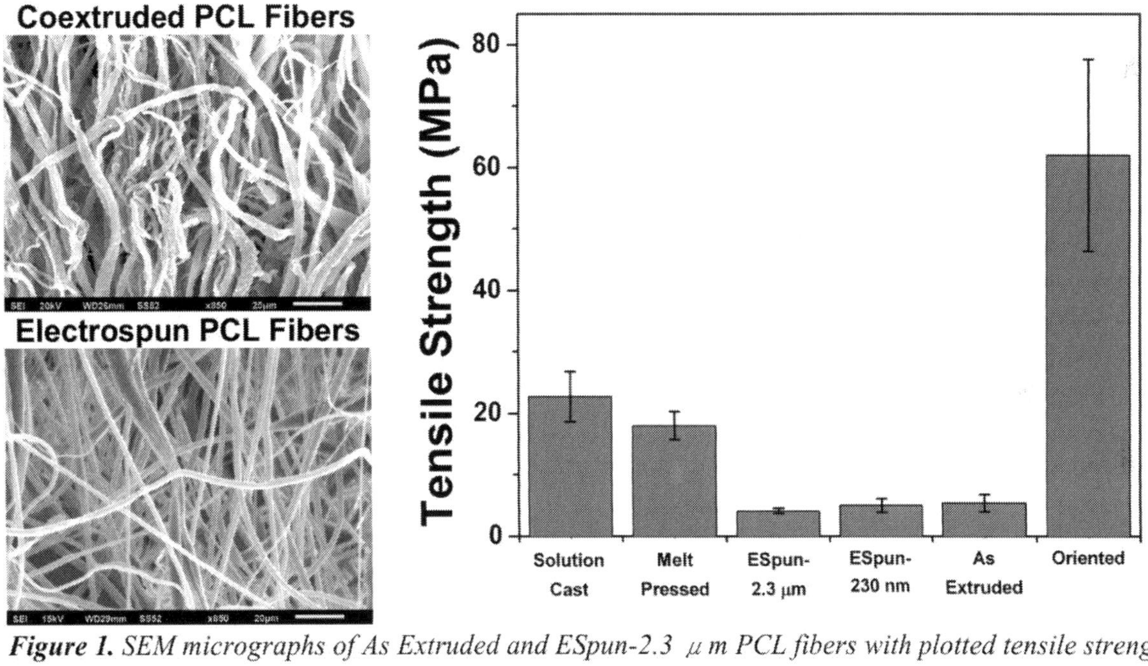

Figure 1. *SEM micrographs of As Extruded and ESpun-2.3 μm PCL fibers with plotted tensile strength of solution cast and melt pressed PCL mats, electrospun, as extruded and oriented PCL fibers.*

References

1. Wang, J.; Deepak, L.; Ponting, M.; Wnek, G.; Korley, L. T. J.; Baer, E. Manufacturing of Polymer Continuous Nanofibers using a Novel Co-extrusion and Multiplication Technique. *Polymer* **2014,** *55* (2), 673-685.
2. Kolbuk, D.; Sajkiewicz, P.; Kowalewski, T. A. Optical birefringence and molecular orientation of electrospun polycaprolactone fibers by polarizing-interference microscopy. *Eur. Polym. J.* **2012,** *48* (2), 275-283.
3. Scherrer, P. Bestimmung der Grosse und der inneren Struktur von Kolloidteilchen mittels Rontgensrahlen (Determination of the size and internal structure of coilloidal particles using X-rays). *Math-Phys KI* **1918,** 98-100.

368405 Mechanistic Function of Costabilizers in Peroxide-Induced Crosslinking of Industrial Polyethylene Blends

Abstract:

Polyethylene compounds for industrial applications are typically crosslinked via a radical pathway by means of a radical initiator, such as dicumyl peroxide. The compositions also contain additives designed to improve heat-aging performance. However, due to interactions with the peroxide, some of these additives can negatively affect peroxide stability in the polymer compound product. Using dodecane as the model for polyethylene, the current work examines the kinetics and product distribution associated with the thermal decomposition of dicumyl peroxide in the presence of several costabilizers in air and under a nitrogen atmosphere. Mechanistic explanations based on the experimental observations will be presented.

AIChE
The Global Home of Chemical Engineers

14AIChE
Annual Meeting, Atlanta, GA

浙江大学化学工程与生物工程学院
College of Chemical and Biological Engineering, Zhejiang University

Toughening of Epoxy Thermosets by Poly[(styrene-*alt*-(maleic anhydride)]-*block*-Polystyrene-*block*-Poly(n-butyl acrylate)-*block*-Polystyrene Tetrablock Copolymer (SMA-SBAS) Prepared *via* RAFT Mini-Emulsion Polymerization

2014 AIChE, Nov. 16-21, Atlanta

Ren He
Professor, Xiaoli Zhan
Associate Professor, Qinghua Zhang

Toughening of Epoxy Thermosets

14AIChE
Annual Meeting, Atlanta, GA

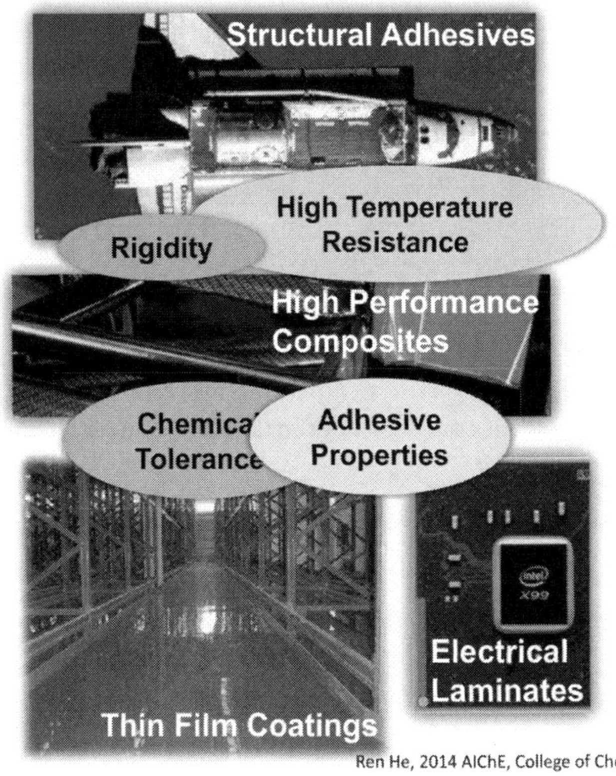

Highly Crosslinked Network Structure

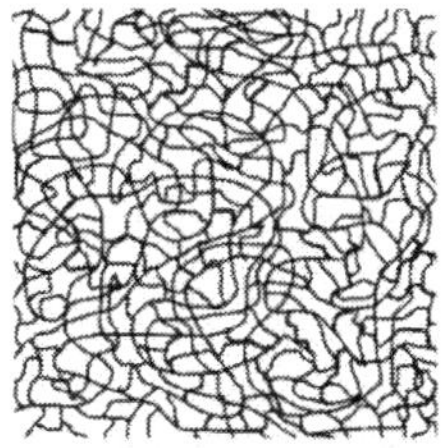

Toughness Modification

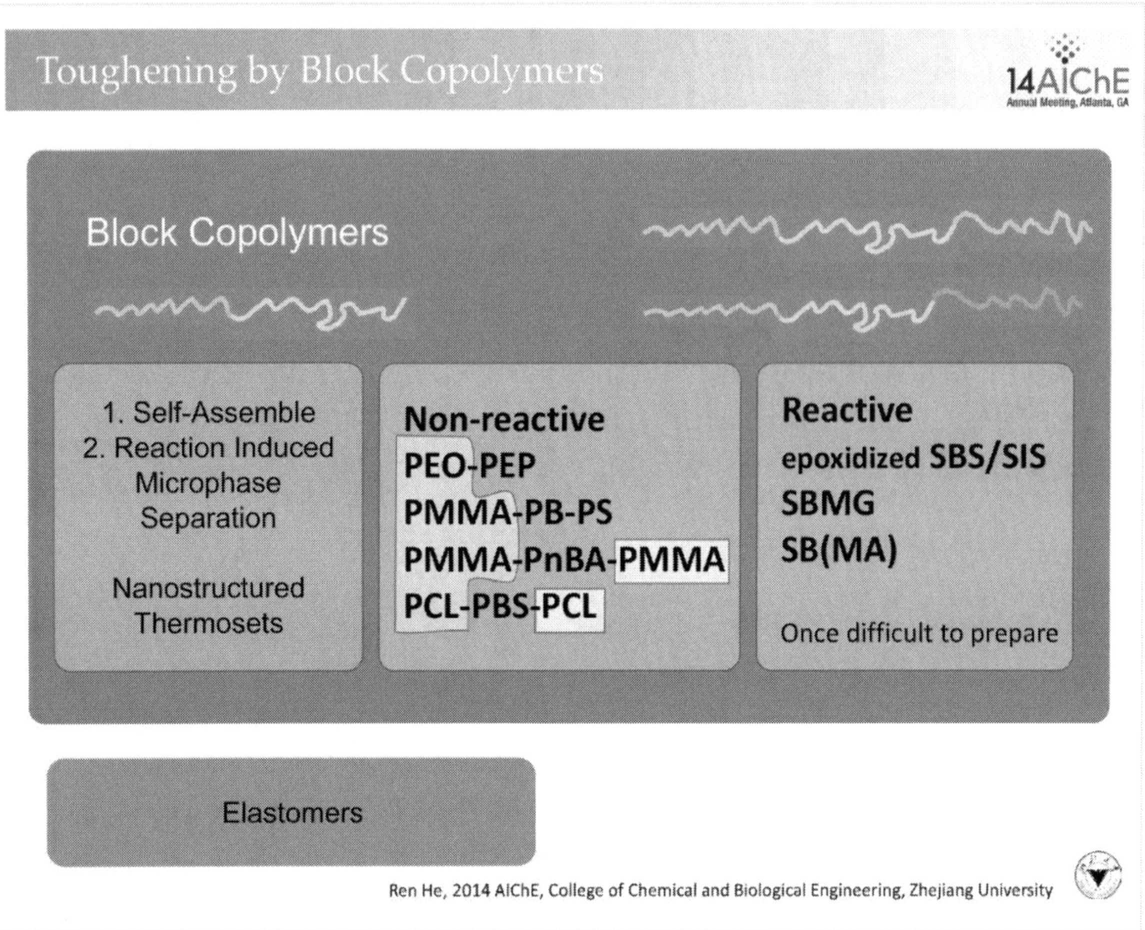
Toughening by Block Copolymers
14AIChE
Annual Meeting, Atlanta, GA
Block Copolymers
1. Self-Assemble
2. Reaction Induced Microphase Separation
Nanostructured Thermosets
Non-reactive
PEO-PEP
PMMA-PB-PS
PMMA-PnBA-PMMA
PCL-PBS-PCL
Reactive
epoxidized SBS/SIS
SBMG
SB(MA)
Once difficult to prepare
Elastomers
Ren He, 2014 AIChE, College of Chemical and Biological Engineering, Zhejiang University

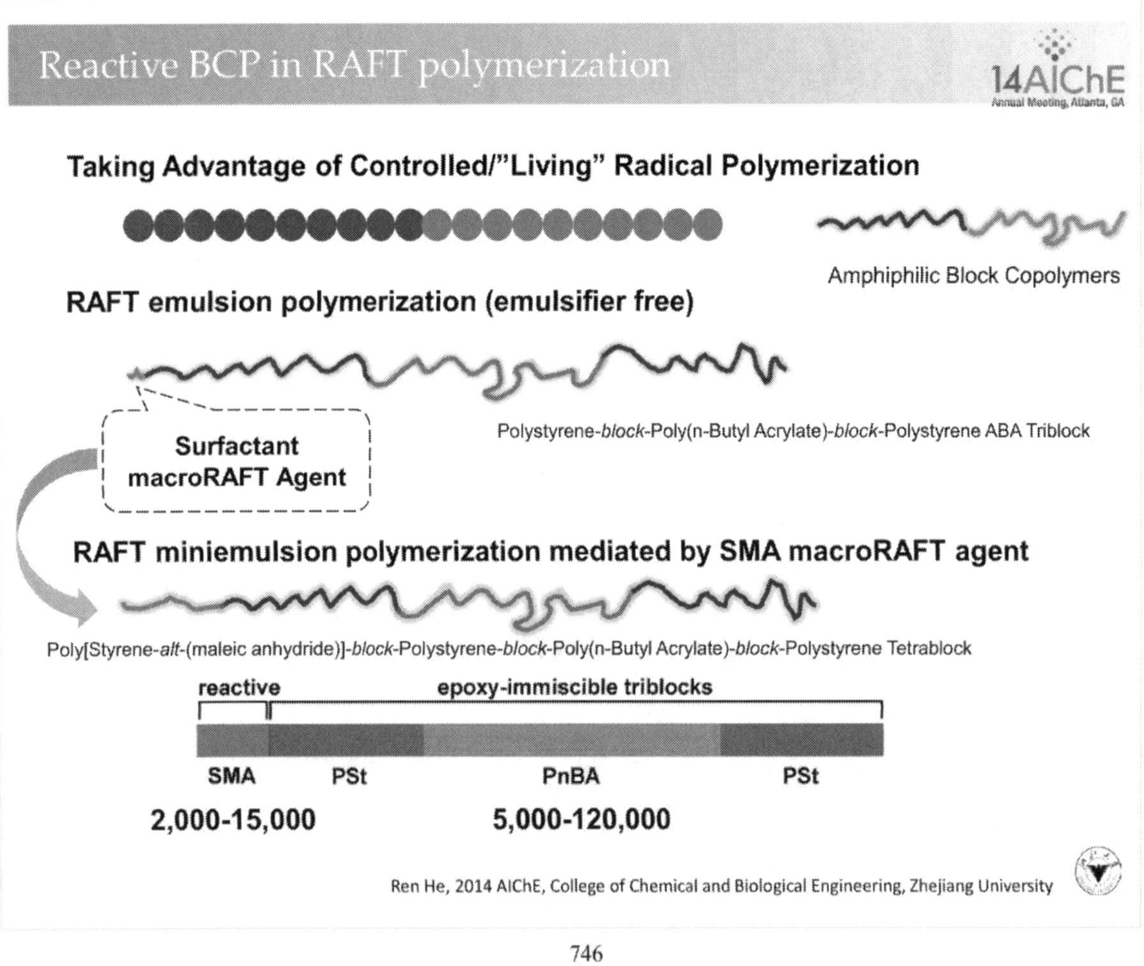
Reactive BCP in RAFT polymerization
14AIChE
Annual Meeting, Atlanta, GA
Taking Advantage of Controlled/"Living" Radical Polymerization
Amphiphilic Block Copolymers
RAFT emulsion polymerization (emulsifier free)
Polystyrene-block-Poly(n-Butyl Acrylate)-block-Polystyrene ABA Triblock
Surfactant macroRAFT Agent
RAFT miniemulsion polymerization mediated by SMA macroRAFT agent
Poly[Styrene-alt-(maleic anhydride)]-block-Polystyrene-block-Poly(n-Butyl Acrylate)-block-Polystyrene Tetrablock
reactive
epoxy-immiscible triblocks
SMA
PSt
PnBA
PSt
2,000-15,000
5,000-120,000
Ren He, 2014 AIChE, College of Chemical and Biological Engineering, Zhejiang University

Polymerization

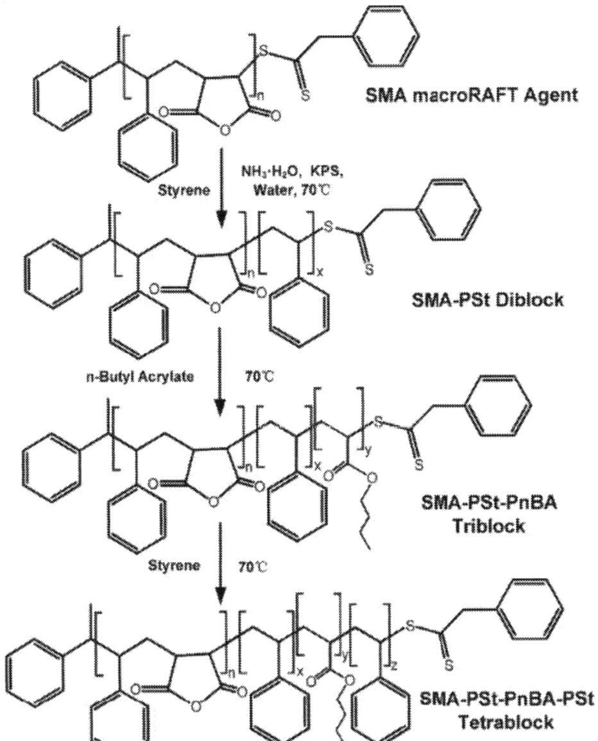

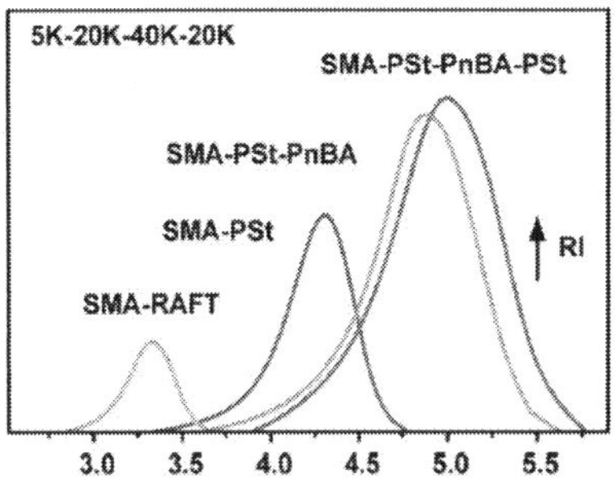

▲**Fig.1** GPC chromatogram variations of before and after chain extensions during the synthesis of sample 5K-20K-40K-20K.

◀**Sc.1** Synthesis of the amphiphilic tetrablock copolymers via RAFT miniemulsion polymerization mediated by SMA-RAFT macroRAFT agent.

DGEBA/SMA-PSt-PnBA-PSt/MDA Blends

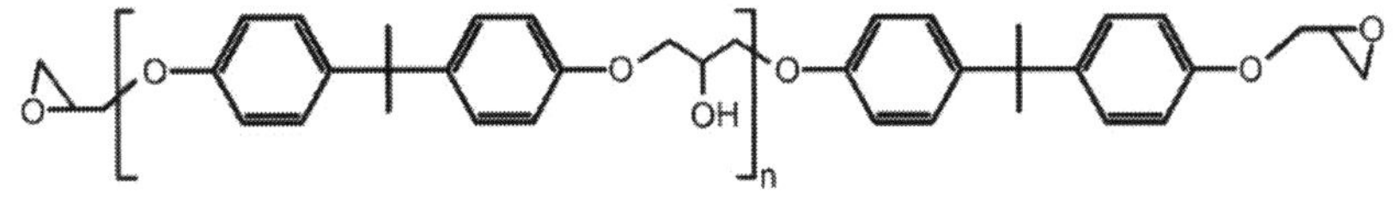

DGEBA, E51, equivalent weight 196g/eq.

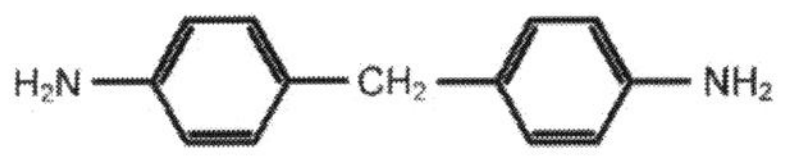

Hardener: MDA

▲ **Sc.3** Chemical Structures of DGEBA and MDA used in this investigation.

1. Covalently linked to DGEBA

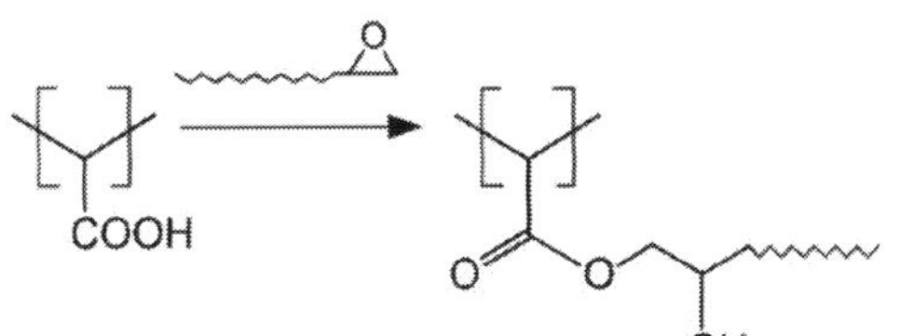

2. Covalently linked to amine hardener

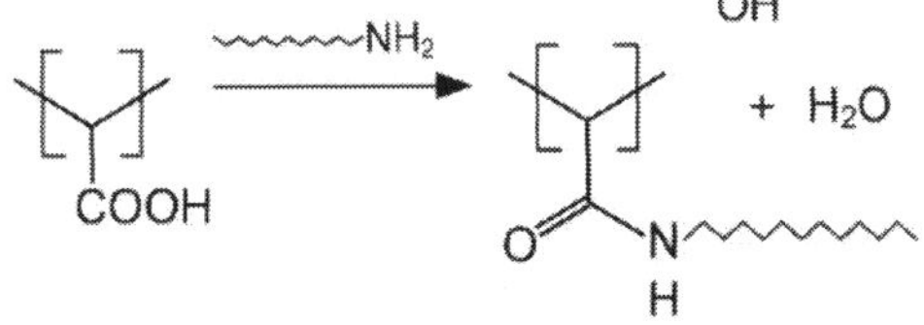

▲ **Sc.4** Two modes of incorporation of SMA-SBAS in the epoxy-amine system.

Optical Transparency

14AIChE
Annual Meeting, Atlanta, GA

Neat epoxy **10% SMA-SBAS** **10% SBAS**

▲**Fig.3** Optical transparency of cured neat epoxy (Left) ,10wt% SMA-PSt-PnBA-PSt modified epoxy (Middle), 10wt% PSt-PnBA-PSt modified epoxy (Right).

Thermal behavior and morphology

14AIChE
Annual Meeting, Atlanta, GA

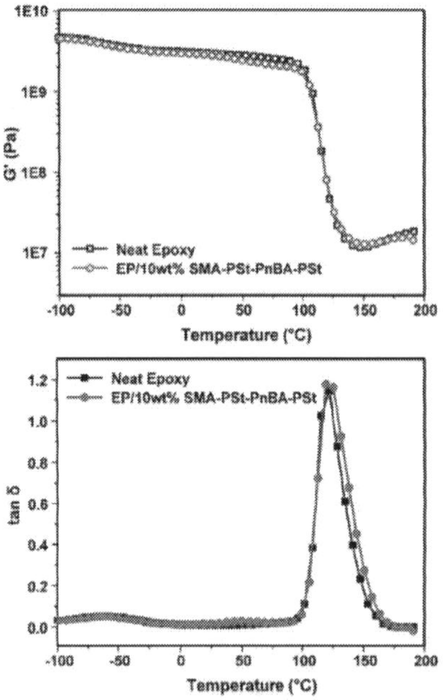

▲ **Fig.4** DMA Plots of Neat Epoxy and EP/10wt% SMA-PSt-PnBA-PSt/MDA. Storage Modulus (G') and tan δ curves as a function of temperature are presented.

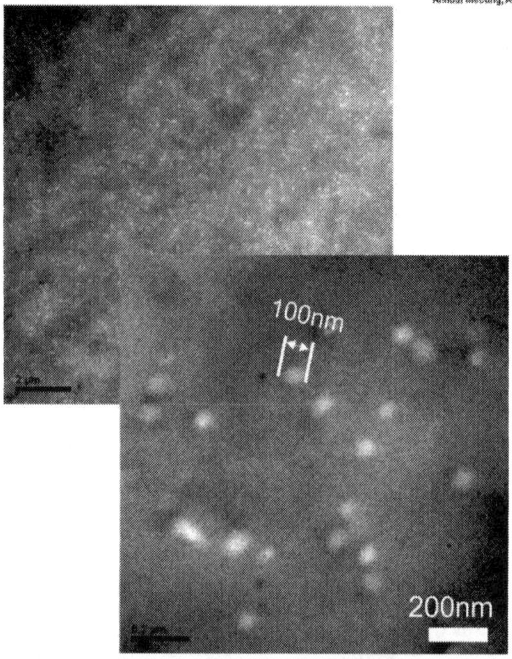

▲ **Fig.5** TEM photographs of DGEBA/MDA blend modified with 10wt% SMA-PSt-PnBA-PSt. Left scale bar, 2μm, Right scale bar, 200nm.

Toughness

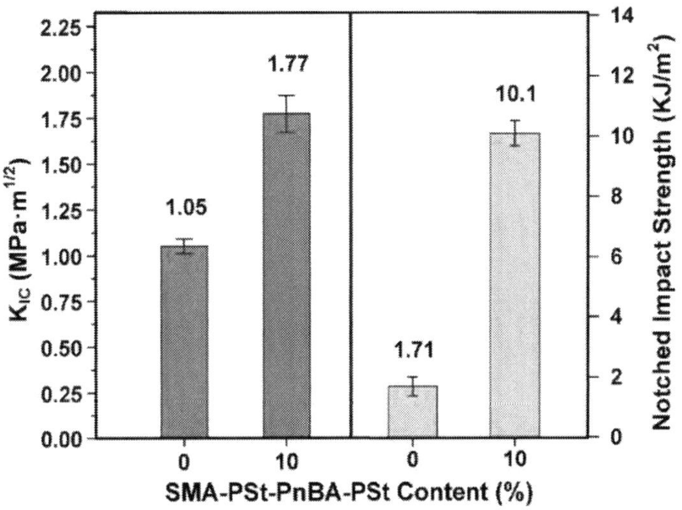

▲ **Fig.6** Fracture toughness values (23°C, ASTM D5045) and Notched Izod impact strength (23°C, GB/T1843) of neat epoxy and 10wt% of SMA-PSt-PnBA-PSt modified DGEBA/MDA thermosets.

70% increase in critical stress intensity factor (K_{IC}) and 5 folds increase in impact strength were observed. And we are currently making efforts to reveal the mechanisms.

Conclusions

Features

1. Achieved nanostructure by extending linear non-miscible triblock copolymer elastomer with a reactive block;
2. Unexpected Toughening Effects;
3. Balance of Properties: Limited loss of T_g and Modulus;
4. Ease of designing and preparing various block copolymer structures;
5. Cosolvent is need due to high molecular weight;
6. Increased viscosity.

Acknowledgement

14AIChE
Annual Meeting, Atlanta, GA

Thank you!

AIChE
The Global Home of Chemical Engineers

National Natural Science Foundation of China

www.cmce.zju.edu.cn

384051 Morphology and Mesoscale Characterization of Self-Assembled Poly (ethylene glycol)-Poly(MTC-benzyl ester) Triblock Copolymer Hydrogels

Abstract:

A novel class of supramolecular hydrogels was produced by synthesizing a triblock polycarbonate material using organocatalytic ring-opening polymerization (ROP) of a poly(ethylene glycol) PEG macroinitiator with a cyclic monomer (MTC-OBn), then subsequently physically cross-linking triblock copolymers by hydrophobic collapse to form hydrogels in water. Controlling the molecular properties of triblock copolymer materials, primarily poly(carbonate) and poly(ethylene glycol) segment lengths, resulted in a viable platform for probing mesoscale material properties using small angle X-ray scattering (SAXS). Furthermore, the macroscale mechanical properties of these supramolecular networks were characterized using oscillatory shear rheology. SAXS spectra show a strong scattering peak, indicating a molecular domain size of 20 - 25 nm. At higher concentrations of the triblock hydrogelator a structural transition from isolated hydrophobic clusters to a lamellar assembly was observed. Modeling of poly(carbonate) and poly(ethylene glycol) backbone size demonstrated that the polymers in this structure behaved as a 2D SAW. SAXS and rheological characterization indicates that the molecular weight of the polycarbonate segment affects the gel structure on both the mesoscale and macroscale.

379093 Surface Mechanical Behavior of Biocompatible Poly((D,L-lactic acid-ran-glycolic acid)-Block-Ethylene Glycol) (PLGA-PEG) Block Copolymers at the Air-Water Interface

Abstract:

The surface pressure-area isotherms of poly((D,L-lactic acid-ran-glycolic acid)-block-ethylene glycol) (PLGA-PEG) monolayers at the air-water interface exhibit mixed patterns of behavior that seemingly represent both the glass transition of the condensed PLGA film and the lateral repulsion of the PEG brush chains that could occur simultaneously under lateral compression of the monolayer. A study was conducted to investigate which one of the above two mechanisms is more responsible for the observed PLGA-PEG isotherm behavior. Particularly, this study is focused on the isotherm behavior of PLGA-PEG monolayers at small monolayer areas. In this regime, the PLGA-PEG isotherms show a surface pressure upturn during continuous monolayer compression, similarly to the isotherms of PLGA homopolymers. However, unlike PLGA, the surface pressure vs. area behavior of the PLGA-PEG diblock copolymer exhibits little dependence on the compression rate. Also, the high surface pressure generated due to the monolayer compression and the molecular morphology of the monolayer developed under high compression were seen to relax very little over time. These observations suggest that in the diblock situation, the high monolayer surface pressure observed at high surface polymer concentrations is mainly due to the lateral repulsion of the PEG chains in the brush layer. These experimental results also compare favorably with predictions of the self-consistent field polymer brush theoretical model. A comparison of the interfacial rheological properties of the air-water monolayers formed by the PLGA-PEG diblock copolymer vs. a non-glassy analogue of this diblock copolymer, poly((D,L-lactic acid-ran-glycolic acid-ran-caprolactone)-block-ethylene glycol) (PLGACL-PEG) also support this conclusion. PLGA-PEG monolayers exhibit protein resistance and also reasonable chemical stability at the air-water interface, and therefore seem to have the potential to be used in lung surfactant applications.

UPGRADING LIGNIN ANTIOXIDANT PROPERTIES BY SELECTIVE DISSOCIATION OF WHEAT STRAW LIGNIN LINKAGES

Mohammadali Azadfar and Shulin Chen

Department of Biological Systems Engineering, Washington State University, Pullman, WA 99164-6120, USA

Funding Sources: Funding for this research was provided by Washington State University

ABSTRACT

Production of value-added products from lignin obtained after pretreatment of biomass requires advanced characterization and modification techniques to illustrate the lignin structure in detail and also to increase the lignin's usability. The lignin modification methods which have been proposed so far endeavor to decompose the natural structure of lignin for producing the value-added products. One of the attractive applications of raw lignin obtained after pretreatment is for use as an antioxidant. In order to preserve the intrinsic structure of lignin for producing antioxidant, the lignin linkages should be dissociated utilizing the ingenious methods. This project aims to modify the lignin structure employing the benefits of the effect of alkaline and acidic conditions on ester and ether lignin linkages without destruction of valuable original structure of wheat straw lignin.

The characterization of wheat straw lignin obtained from sequential ozone and soaking aqueous ammonium pretreatment revealed the presence of phenolic derived-compound with comparable antioxidant activity with other synthetic antioxidants. The characterization of lignin structure demonstrated that the good antioxidant activity of the lignin is due to the predominant presence of hydroxycinnamic acids and G-lignin derived-compounds such as ferulic acid and guaiacol during the alkaline pretreatment (S/G ratio of 0.35). Alkaline pretreatment (soaking aqueous ammonium in this project) of wheat straw is mainly responsible for breaking down of the ester linkages. The dissociation of ester linkages open this opportunity that the hydroxycinnamic acids derivatives can be appeared with applied antioxidant properties including the presence of hydroxyl group, *ortho*-methoxy substitution on the phenolic ring, and low molecular mass (<1000 Mn).

Regarding the molecular mass of lignin's precursors and characterization of obtained lignin structure, it is possible to determine the number of linked lignin monomers (G, S, H lignin units). Consideration of the lignin macromolecule structure opens this opportunity to characterize the potential linkages among lignin monomers. The ether linkages have predominant capability to link the lignin's monomers in order to show the obtained lignin's molecular weight. Based on the selective destruction effect of acid on ether bonds, it has been proposed that by utilizing the acidic condition the majority of ether linkages in lignin structure will be dissociated. Thus, lignin-derived compounds with a substantial lower molecular weight and also containing new active functional groups will be obtained. The prepared lignin derivatives including aforementioned properties will be indicated excellent antioxidant activity.

Keywords: Lignin linkages; Selective dissociation; Characterization, Modification; Antioxidant

Engineered Kraft Lignin Particles for Foam Stabilization

Stephanie Lam, Andrew Tibbits, Orlin D. Velev

Department of Chemical & Biomolecular Engineering, North Carolina State University, Raleigh, North Carolina 27695-7905

Lignin, which exists in the cell wall of woody plants, is one of the most abundant biopolymers found in nature and has many potential uses as a stabilizer, encapsulation agent, and as a component of composite materials. In this work, we present the results of a process where KL particulates, which are themselves emulsifiers, are reformed into drastically more efficient foam stabilizers using a water-based shear process. The flash-precipitation procedure is environmentally-friendly and does not involve the use of organic solvents or harsh chemicals. We characterized the effect of this reformation process on the surface chemistry and structure of the nanoscale voluminous lignin particles produced. The ζ-potential of the KL particles was found to change with pH, and became slightly lower following the reforming process. The effective volume of the particles was found to increase by ~30 times following the reformation procedure, which resulted in an enhanced ability of reformed KL to sterically stabilize foams and emulsions. We also evaluated the stability of foams made with reformed particles at various pH conditions and found enhanced foam stability to correlate with KL reformation at low pH (~2), which corresponds to a surface potential of approximately -20 mV and increased particle hydrophobicity. Thus, we have developed an efficient, environmentally-friendly method by which to expand the effective volume of lignin particles, which after the reformation process have shown enhanced properties as foam and potentially emulsion stabilizers.

2014 AIChE Annual Meeting
Division 17: Forest and Plant Bioproducts Division
17004 Biobased Materials III: Lignin-Based Materials
November 16-21, 2014
Atlanta, GA

Charaterization of Biochar from Loblolly Pine: Changes in Carbon Structure and Consequent Effects on Applications as Sorbents for Aromatic Contaminants

Junyeong Park [1,*], Kwang Hun Lim [2], Martin A. Hubbe [1], Orlando J. Rojas [1], Sunkyu Park [1]

[1] Department of Forest Biomaterials, North Carolina State University, Raleigh, NC 27695
[2] Department of Chemistry, East Carolina University, Greenville, NC 27858
* Presenting author: jpark7@ncsu.edu

Thermally-treated biomass has been generally considered as lower-valued solid fuel or by-products from various thermal conversion processes for lignocellulosics. Recently, however, these solid products such as biochars and their activated counterparts, are getting wide attentions, due to their possibilities for sustainable and versatile applications, including soil amendment, carbon sequestration, and sorbents for various environmental contaminants. It has been reported in existing studies that the structural characteristics of thermally-treated biomass have correlations with required properties for those applications. However, the details about what is going on inside lignocellulosic biomass during thermal treatment have not been fully understood yet. In this study, the changes in structural characteristics, such as aromatic structures and sizes of carbon clusters, were monitored using several complementary analytical tools, including solid-state NMR spectroscopy, compositional analysis, and other characterization methods. After the thermal treatment of loblolly pine at 700 °C under inert atmosphere, the fraction for aromatic carbons inside was increased to 69 % out of total carbons in solids, and even higher to 76% after the subsequent activation process. When biomass was treated under slightly oxidative atmosphere, the development of aromatic cluster was promoted substantially. Surface analysis revealed that the properties like BET surface area and pore volume were mainly dependent on the structural characteristics of the feedstock before activation. It was shown that structural and surface characteristics of thermally-treated biomass were largely governed by process conditions for thermal treatment, and this can make it possible to design solid products for specific applications with enhanced properties.

390387 Phase Behavior of Kraft Liqnins with a Tunable, Renewable Solvent System

Abstract:

Lignin is unique among biopolymers in having significant aromatic character, which makes it potentially attractive for a wide range of uses, including polymeric applications and carbon fibers. Unfortunately, most of the commercial-grade lignins available today (i.e., the Kraft lignins) are severely limited in their applications because of their high metals (primarily sodium) content.

However, recently Clemson researchers have discovered a powerful, versatile, and renewable solvent system that creates a phase behavior when contacted with lignin such that the sodium content can be reduced 10-fold with each phase split; furthermore, this phase separation also separates the lignin into fractions with significantly different molecular weights and chemical properties. Thermodynamic variables such as temperature, solvent composition and amount, and lignin type (i.e., hardwood vs. softwood) have been studied to determine their effect on the location of phase boundaries, the metals selectivities between phases, and the molecular weight distribution of the lignin within each phase.

Pseudo-ternary phase diagrams will be presented for different types of feed lignins with the solvent system at several different temperatures, and the global phase-behavior characteristics of this lignin-solvent system will be discussed, along with the implications of the observed phase equilibria for the development of a viable purification and fractionation process for lignin.

392452 Production and Evaluation of Bioplastic Blends from Phbv, PCL, Epoxidized Natural Rubber and Lignin

Abstract:

Biodegradable polymers, from biological sources such as as poly(3-hydroxybutyrate-co-3-hydroxyvalerate) (PHBV) and petroleum based sources such as poly(ε-caprolactone) (PCL), have been shown to be effective in solving many of the problems with current commercial plastics, including greenhouse gas emissions, continued stability of fossil fuels and concerns related to end of life. Lignin, a by-product of the paper and cellulosic ethanol industries, has recently received attention in biomaterials and chemical applications due to its favorable polyphenolic structure and large quantity of available functional groups. Because of the widespread nature of the pulp and paper industry, the most commonly available form of lignin is the kraft lignin, a byproduct refined from the black liquor stage of the pulping process. The main challenge associated with composites of biopolymers and lignin the appearance brittle behavior caused by the presence of lignin in the polymer matrix. In this study small quantities of epoxidized natural rubber (ENR) have been compounded with the bioplastic blends to operate as an impact modifying agent. Characterization of these bioplastic blends was performed through mechanical testing (tensile, flexural test and impact tests) and thermal properties of the bioplastic blends. The emphasis of this study was improving the impact resistance of lignin-biopolymer blends.

The Researchers acknowledge the financial support from the Natural Sciences and Engineering Research Council (NSERC), Biomaterials and Chemicals Network.

376814 Biomimetic Coacervate Environments for Protein Analysis

Abstract:

Living cells have evolved sophisticated intracellular organization strategies that are challenging to reproduce synthetically. Biomolecular function depends on both the structure of the molecule itself and the properties of the surrounding medium. The ability to simulate the *in vivo* environment and isolate biological networks for study in an artificial milieu without sacrificing the crowding, structure, and compartmentalization of a cellular environment, represent engineering challenges with tremendous potential to impact both biological studies and biomedical applications.

A major challenge in designing synthetic organelles and reconstituted *in vivo* microenvironments is maintaining both crowding and compartmentalization while controlling the available intermolecular interactions. Emerging experience has shown that complex coacervates (electrostatically-driven liquid-liquid phase separation) utilizing biomolecules produces an effective biomimetic microenvironment. Initial efforts are focused on understanding (i) the incorporation of various model proteins and biomolecules into coacervate phases, (ii) the stability of the encapsulated proteins, and (iii) how environmental factors regulate activity, as in the case of the dynamic self-assembly of F-actin.

*This work is supported by the University of Chicago and the U.S. Department of Energy Office of Science program in Basic Energy Sciences and the Materials Sciences and Engineering Division, and the University of Chicago NSF-MRSEC.

Biomimetic Concentric Microgrooved Structures for Regulating Bone Cellular Behavior

Shanfeng Wang

Department of Materials Science and Engineering, The University of Tennessee, Knoxville, TN 37996

Here I present three methods used in our research group to prepare biomimetic concentric microgrooved structures for regulating bone cellular behavior. These three methods are polymer banded spherulites, polymer replica-molding from microfabricated silicon chips, and crystallization of calcium carbonate ($CaCO_3$). In the first study, mouse pre-osteoblastic MC3T3-E1 cell attachment, proliferation, and nuclear localization were assessed on the hot-compressed flat disks and spherulites-roughened films of the high-molecular-weight PCL and one of its homo-blends. The number of attached MC3T3-E1 cells and the proliferation rate were greater on the rougher surfaces than those on the flat ones. Cell nuclei were preferentially, though not absolutely, located in or close to the valleys of the banded spherulites. In the second study, microgrooved substrates with four groove widths of 7.5, 16.1, 44.2, and 91.2 μm and three groove depths of 0.2, 1, and 10 μm were prepared through replica molding, i.e., photo-crosslinking PCLTA on micro-fabricated silicon wafers with pre-designed concentric groove patterns. MC3T3-E1 cell attachment and proliferation could be better supported by the stiffer substrates while not significantly influenced by the microgrooves. Microgroove dimensions could regulate MC3T3-E1 cell alignment, nuclear shape and distribution, mineralization, and gene expression. Among the microgrooves with a fixed depth of 10 μm, the smallest width of 7.5 μm could align and elongate the cytoskeleton and nuclei most efficiently. Strikingly, higher mineral deposition and upregulation of osteocalcin gene expression were found in the narrower microgrooves when the groove depth was 10 μm. In the third study, $CaCO_3$ concentric microgrooves with groove widths of 5.0 and 10 μm were fabricated through simply controlling incubation temperature. Mouse pre-osteoblastic MC3T3-E1 cells were cultured on flat and microgrooved substrates of $CaCO_3$ and their adhesion, spreading, proliferation, alkaline phosphatase activity, and calcium content were remarkably enhanced by the microgrooves, in particular, the narrower ones. Furthermore, focal adhesions and actin filaments of MC3T3-E1 cells could be aligned on both 5.0-μm and 10-μm-wide $CaCO_3$ grooves. Compared with the original round nuclei on the flat substrates and expanded round nuclei on the narrower microgrooves, the MC3T3-E1 cell nuclei on 10-μm-wide $CaCO_3$ grooves demonstrated preferred entrapment in the grooves and significant alignment with a smaller area after two-day culture.

389580 Biomimetic Growth of a Pathologic Biomineral in Hydrogel Networks

Abstract:

Biominerals found in living organisms have diverse biological functions and unique properties arising from the particular structure, orientation, and morphology of the constituent minerals. A systematic investigation of the morphological evolution of biominerals grown in organic matrices, and the interaction of such matrices with the inorganic crystals are essential for understanding biomineralization processes and for developing bioinspired materials. We employ here hydrogels as a model organic matrices and a 'double-diffusion' method to mimic the formation of calcium oxalate, the major constituent of kidney stones. Gel-mediated crystallization provides a great way to vary the local supersaturation through controlling the diffusion rate of the reacting ions. We study the nucleation, growth, type of polymorph and orientation of the calcium oxalate crystals as affected by the gel density, reservoir concentrations, molar stock solution ratio, and additives. The obtained crystals are characterized for their morphology and crystal structure using Scanning Electron Microscopy (SEM) and X-ray Diffraction (XRD), respectively. We also investigate the possible occlusion of gelatin in calcium oxalate through Thermogravimetric Analysis (TGA). These findings help our understanding of the mechanism of gelatin incorporation in calcium oxalate crystals, and provide insights into the formation of organic-inorganic composite materials.

388477 Peptoid-Modified Bicelles As Surrogate Cell Membranes for Membrane Protein Sensors and Analytics

Abstract:

The study of membrane-affiliated interactions is significant to the understanding of cell function, detecting biomarkers to diagnose disease, and testing the efficiency of new therapeutic targets. The amphiphilic character of membrane proteins makes them difficult to produce, leading to low availability and stability. Surrogate cell membranes can be used to stabilize membrane protein structure and maintain native activity. One type of surrogate cell membranes that we are interested in are bicelles, which are disc-shaped lipid bilayers that are created by combining long- and short-chain phospholipid pairs. Our goal is to modify the edges of bicelles by incorporating peptoids with different functional groups. For example, thiol groups can be included so that the peptoid-modified bicelles can be anchored to a gold surface. Peptoids are ideal for this application since a large diversity of side chains can be easily incorporated in a sequence-specific manner. Bicelles with different peptoid compositions are imaged using SEM, TEM and AFM. The presence of peptoids in the bicelles is confirmed by fluorescence microscopy and attachment of gold nanoparticles.

388496 Self-Assembled Protein-Inorganic Hybrid Supraparticles for Robust Protein Immobilization

Abstract:

Biomineralization is a process during which mineral clusters and biomacromoecules self-assemble into nanostructured composite materials. Proteins serve dual roles in these hybrid materials, first as templating agents to regulate structural and physical properties of the material, and second as ligands for biological communication. Inspired by this natural process, we developed a method to fabricate self-assembled and porous hybrid supraparticles with specific affinity recognition for robust protein immobilization. First, we constructed a recombinant fusion protein that contains naturally derived leucine zipper coiled coil motifs, which can enable non-covalent conjugation of proteins via a high affinity interaction between the zippers. In the fusion protein, two identical leucine zipper motifs were combined with a random-coil midblock. Second, the engineered protein directs formation of flower-like nanohybrids that are further self-organized into porous supraparticles. When calcium phosphate was precipitated in the presence of the fusion protein, it provided nucleation sites and directed the growth of calcium phosphate petals into flower-like nanohybrids. The van der Waals interactions between the flower-shaped nanostructures resulted in their colloidal assembly into interlocked chains, which were packed into hierarchically structured and porous supraparticles on the micrometer length scale. The protein-inorganic supraparticles are biocompatible, and provide large surface on which leucine zipper motifs, the high-affinity conjugation sites, are available. Globular proteins fused with the leucine zippers can be immobilized on the supraparticles via simple incubation in aqueous solution. In particular, immobilization of enzymes can enhance their stability and catalytic activity, thus providing benefits for many applications. Thus, our approach provides a versatile method to fabricate complex biomaterials by mimicking both specific biomolecular recognition and the natural biomineralization process for formation of hybrid nanostructures.

389233 Engineering Antiphagocytic Drug Carriers Using CD47-Streptavidin Fusion Protein

Abstract:

Despite the remarkable advances and successes in the efficacy of the synthetic carriers such as polymer and lipid particles, they often struggle to meet clinical expectations such as fast clearance from blood circulation. Natural particulates, which range from pathogens to mammalian cells, are therefore worth examining in more depth, as they are highly optimized for their specific functions *in vivo* and possess features that are often desired in drug delivery carriers. To extend the blood circulation time of drug delivery carriers, many strategies have been used in particle surface modifications to avoid clearance by mononuclear phagocyte system. Here we report a biomimetic approach of functionalizing synthetic particulates with CD47 self-marker proteins to endow drug delivery carriers with phagocytosis-resistant features. CD47-streptavidin fusion protein was constructed and attached to the biotinylated particles. Our results revealed that CD47 could reduce macrophage-mediated clearance of particles via its interaction with signal regulatory protein alpha on macrophages.

387352 Applications of Proteolipobeads in 3D Ligand Display and Cellular Interactions

Abstract:

Biomaterials and biomedical research are often focused on the re-creation of complexities found in cells and cellular microenvironments. We hypothesize that biomembrane-microsphere assemblies are a viable biomimetic means to present ligands/bound factors to stem cells and mimic cellular communication in the stem cell niche. Our platform, termed proteolipobeads (PLBs), features laterally-mobile proteins and peptides positioned at the interface of a synthetic biomembrane. The presentation of ligands/bound factors involved in cell-cell interactions (e.g. cadherin) has begun in 2D culture systems by surface patterning but has not yet been well established in 3D stem cell culture systems.

We present results on three aspects of this study: 1) the fabrication and characterization of functionalized microspheres that contain supported lipid bilayers; 2) the construction and characterization of biocompatible hydrogel/microsphere hybrid scaffolds that integrate lipid bilayer/microsphere assemblies and 3) preliminary experiments of human mesenchymal stem cells (hMSCs) microencapsulated into matrix/PLB hybrid scaffolds. Special attention was given to the viability of hMSCs within this platform, as well as the frequency and extent of hMSC-PLB interactions. A primary feature of the PLB construct is a membrane-embedded synthetic peptide, which serves as a conjugation point for both ligands displayed at the PLB exterior and for biomembrane-microsphere tethering at the PLB interior. Moreover, the phyics of lipid-peptide interactions permits spatial control of the peptide, and consequently, ligand display.

Confocal microscopy studies were conducted to assess the diffusivity of laterally mobile PLB constituents (via FRAP) and to visualize N-cadherin-mediated PLB interactions with hMSCs in situ. Immunohistochemical staining was used to detect N-Cadherin within the hybrid matrices relative to counterstained supported lipid bilayers. Finally, we used flow cytometry to detect the level and homogeneity of specific fluorescently labeled proteins, peptides, and lipid bilayers included in the PLB assemblies. This work constitutes a new method for displaying a wide range of complex membrane proteins and signaling molecules to live cells within 3D matrices.

390275 Antifouling Surfaces Inspired By Nature: Nuclear Pore Complex

Abstract:

Membrane fouling, the process in which proteins adhere to the external and internal surface of synthetic membranes, is considered a major limitation in membrane filtration of biofluids. This causes permeation flux and changes in selectivity, with the overall effect of reducing membrane performance and increasing process cost. In the past 30 years researchers have focused on how to reduce membrane fouling. The Whitesides group has published a set of criteria for obtaining protein-resistant chemistries, viz. the surface of such membranes or materials should (i) be hydrophilic, (ii) include hydrogen bond acceptors, (iii) do not include hydrogen bond donors, and (iv) be net electrically neutral. An alternative approach to producing synthetic surfaces based on these criteria is to observe Nature's method of producing protein-resistant surfaces. In particular, we focused on the Nuclear Pore Complex (NPC), the sole mediator of exchange between the nucleus and the cytoplasm in all eukaryotic cells. Protein transport across the NPC is fast, energy-dependent (to give directionality), often receptor-mediated and most important it occurs in the absence of unwanted fouling. For this reason we focused on "NPC-proteins" and covalently attached them on a sensor surface to test their antifouling properties. A series of different proteins, including BSA, were used to test the efficacy of this new biomimetic surface. For comparison (control), we also tested surfaces modified with other inert proteins and PEG, the gold standard with respect to protein-resistance. A quartz crystal microbalance with dissipation was used in all these experiments.

378201 Invited Talk: Simulations of Ionic Aggregate Morphology and Dynamics in Ionomer Melts

Abstract:

Ionomers are attractive as single-ion conducting polymer electrolytes, which would be safer and have higher efficiency than currently-used liquid electrolytes. However, ionomers do not have sufficiently high conductivities for practical application. An understanding of the relationships between ionomer chemistry, morphology, and ion transport is needed to design ionomers with improved conductivity. To investigate the morphology formed by aggregates of the ions, we have performed molecular dynamics simulations of a series of polyethylene-based model ionomer melts, in which the spacing between functional groups is precisely controlled. The simulations provide new insights into the shape, size, and composition of ionic aggregates. In particular, we observe a wide variety of aggregate morphologies, ranging from small spherical aggregates to string-like shapes and large percolated networks. The structure factors calculated from simulation agree well with X-ray scattering data. Depending on the morphology, the simulation and experimental scattering curves can be well fit with either a modified hard sphere or a newly-developed modified hard cylinder model. To reach long enough simulation times to follow ion transport in these systems typically requires coarse-grained models. We apply a static electric field to the system to extract the ion mobility and conductivity as a function of ionomer morphology. The coarse-grained simulations show that systems with percolated aggregate morphologies have higher ion mobilities.

Sandia National Laboratories is a multi-program laboratory managed and operated by Sandia Corporation, a wholly owned subsidiary of Lockheed Martin Corporation, for the U.S. Department of Energy's National Nuclear Security Administration under contract DE-AC04-94AL85000.

385258 pH-Mediated Control over Polyelectrolyte Complex Shape and Internal Structure

Abstract:

When oppositely charged polyelectrolytes are mixed in water they form polyelectrolyte complexes (PECs), whose size and morphology typically reflect their formation kinetics. This process has an electrostatic driving force and (when weak polyelectrolytes are used) depends strongly on pH. Using poly(styrene sulphonate) (PSS) as the polyanion and poly(allylamine) (PAH) as the weak (pH-sensitive) polycation, we have exploited this sensitivity to generate PECs with custom-designed shapes and internal structures. PAH and PSS were mixed at high pH (far above the pKa of PAH), where the PAH amine groups were largely uncharged and the PAH/PSS mixtures formed single-phase solutions. PECs were then prepared from these solutions by reducing the pH, thus ionizing the PAH and triggering complexation. This was achieved through two strategies. The first strategy, which was used to form PEC beads with solid internal structures, relied on the dropwise addition of PAH/PSS mixtures to acid baths. The diffusion of acid into the droplets led to rapid PAH ionization and, consequently, conversion of the PAH/PSS solution droplets into PEC beads, which (unlike the solvent-filled PEC capsules prepared through traditional methods) had solid internal morphologies. Furthermore, to extend PEC formation to more-intricate structural motifs, the pH-sensitivity of PAH was exploited to form custom-designed shapes through photopatterning. In this second strategy, a photoacid generator (PAG), which lowered the pH when exposed to UV light, was added to the PAH/PSS mixtures. The localized protonation of PAH amine groups then led to the formation of PECs, whose custom-designed shapes reflected the photoirradiation patterns. These distinct PEC shapes and (in the case of the beads) internal morphologies suggest pH-triggered PEC formation as an effective approach to directing PEC structure.

388820 Interfacial Properties of Polyelectrolyte Coacervates

Abstract:

Under favorable conditions, mixing aqueous solutions of cationic and anionic polymers can lead to a type of liquid-liquid phase separation known as coacervation, where one phase is rich in polymers (coacervate). These coacervates have a wide range of applications, including as underwater adhesives, in food processing, drug delivery, or advanced coatings. The key to understanding the formation of coacervates and fine-tuning their performance is to develop a model of their interfacial properties, capable of describing the width of the interface, ion distributions, and polymer concentration gradients. We have developed such a model by combining the Cahn-Hilliard theory with the modified Debye-Hueckel treatment on electrostatic interactions. We investigated the coacervation phase diagram, interfacial density profile, and ion distribution, by comparing the predictions of our model with results from direct molecular simulations. We further investigated the effects of dielectric inhomogeneity and chain connectivity.

390382 Rheological Properties and Electrospinning of Chitosan-Oil Solutions

Abstract:

Despite recent progress in electrospinning, the parameters that govern the ability to spin a water-immiscible phase is limited. Previously, seven studies that have focused on spinning solutions that contain two immiscible phases, which have all utilized a surfactant to stabilize drops of one phase against coalescence. In this work, we explore the effect that chemically modifying chitosan, a "green" hydrophilic polymer has on its ability to stabilize oils and electrospin them into nanofiber mats. Two oils were investigated, cinnamaldehyde (CIN), which can chemically react with chitosan, as well as an oil that did not react with chitosan, hydrocinnamyl alcohol (H-CIN). Proton nuclear magnetic resonance was used to confirm the degree of deacetylation of the chitosans and their solution-based reactions to the oils. Stability testing and solution rheology were then correlated to the spinnability of chitosan/poly(ethylene oxide) (PEO) precursor solutions.

360028 Coacervate Driven Assemblies Using α-Helical Polypeptides

Abstract:

A variety of materials with diverse structures and properties can form as a result of electrostatic interactions between oppositely charged macromolecules. Under defined conditions, complexation can lead to a phase separation phenomenon, referred to as complex coacervation. Using polypeptides, derived from amino acids, as a model system we identified the external parameters that affect coacervation,[1] explored the thermodynamics of coacervate formation,[2] and studied the rheological[3] and interfacial properties[4] of polypeptide coacervates. Here, we focus on the self-assembly of a different type of polypeptides. These water-soluble ultra-stable α-helical polypeptides are produced by elongating the charged side chains from the polypeptide backbone. Under similar conditions, mixing of these helical polypeptides with oppositely charged polyelectrolytes leads to the formation of liquid complexes (complex coacervates). Using the same strategy, coacervate core micelles are formed when the helical polypeptides are linked to a neutral hydrophilic block and are mixed with homopolymers of the opposite charge. The effects of salt, chirality and block length on both self-assembly structures will be discussed.

1. Priftis D., Tirrell M., *Soft Mater* **2012**, *8*, 9396-9405.

2. Priftis D., Laugel N., Tirrell M., *Langmuir* **2012,** *28*, 15947-15957.

3. Priftis D., Megley K., Laugel N., Tirrell M., *J. of Colloid and Interface Sci.* **2013,** 398, 39-50.

4. Priftis D., Farina R., Tirrell M., *Langmuir* **2012,** *28*, 8721-8729.

369327 Segmented Imidazolium Ionenes: Solution Rheology, Thermomechanical Properties, and Electrospinning

Abstract:

Segmented imidazolium ionenes with poly(tetramethylene oxide) (PTMO) soft segments and ionic, bis-imidazolium hard segments formed nonwoven fibrous mats upon electrospinning. The segmented imidazolium ionenes exhibited high thermal stability, excellent mechanical properties, and two glass transition temperatures indicative of microphase separation. Tensile analysis revealed strain-induced crystallization for ionenes with a 2 kg/mol PTMO soft segment and increased stress at break with increasing soft segment molecular weight. Solution rheology enabled the determination of solvent quality, overlap and entanglement concentrations, and the influence of ion content on solution properties. Fiber diameters for polymeric fibers electrospun from acetonitrile suggested solution polyelectrolyte behavior that was independent of the PTMO molecular weight. This work reports the successful electrospinning of segmented imidazolium-containing ionenes and an in-depth analysis of their thermomechanical and solution properties.

369616 Ion and Cooperativity Effects in Complex Coacervate Structure

Abstract:

Complex coacervates are solutions of oppositely-charged polyelectrolytes that can undergo liquid-liquid phase separation to form polymer-rich and polymer-deficient phases. The physics governing this system is relevant to a whole host of biopolymeric and synthetic systems with applications ranging from drug encapsulation to layer-by-layer assembly. Widely-used Voorn-Overbeek theory is often used to explain the coacervation process, and is typically reasonable in many of its predictions. Nevertheless, it is based upon classical Debye-Huckel theory that is known to be insufficient in the conditions realized in typical coacervation experiments. We instead use the Polymer Reference Interaction Site Model (PRISM), which is an extension of Liquid State theory ideas, to gain insight into the local organization of chains and ions such that we can elucidate the physical driving forces in complex coacervation. We demonstrate a rich array of behaviors such as cooperative interactions between the polyelectrolytes, finite salt size effects, the effects of valency, and discuss how Voorn-Overbeek based models tend to provide reasonable experimental estimates despite being based on incorrect assumptions. We show how these results relate to experimental data suggesting the presence of ion-specific effects, and correspondingly demonstrate an expanded set of parameters in which complex coacervates may be tuned by the design of both the polymer and salt components.

C.E.S. acknowledge support from the International Institute for Nanotechnology, S.L.P. and M.T. acknowledge support from the U.S. Department of Energy Office of Science program in Basic Energy Sciences and the Material Sciences and Engineering Division.

373806 Thermal Transitions in Layer-By-Layer Assemblies Observed Using Electrochemical Impedance Spectroscopy

Abstract:

Layer-by-layer (LbL) assemblies have been of great interest due to their versatile functionality and ease of fabrication. The sensitivity of an LbL film to assembly conditions as well as external stimuli can be attributed to the weak nature of non-covalent interactions (electrostatic, hydrogen bonding, van der Waals) existing between the two adsorbing species. As it will be shown here, temperature is a particularly intriguing parameter in that some LbL films respond dramatically and others do not in aqueous media. The reasoning behind this phenomenon is not well understood, and is explored herein. We will present the structure and transport properties of LbL films of poly(diallyldimethylammonium chloride) (PDAC) and poly(styrene sulfonate) (PSS) as a function of temperature using electrochemical impedance spectroscopy and a redox-active probe. The effects of thickness, salt concentration, and the identity of outermost layer on the thermal transition temperature are investigated. It will be shown that equivalent circuit modeling can be applied to assess a film's transition temperature, and results are in good agreement with calorimetry and quartz crystal microbalance with dissipation.

Layer-by-Layer (LbL) assembly of oppositely charged polyelectrolytes (PE) involves a complicated set of physical processes. Despite the numerous applications of PE multilayer films (PEM), there is still no quantitative model that describes salient features of this process such as chain complexation, and diffusion of polyions, salts, and water, as functions of time and position in the film during deposition and rinsing. Previous models are either only thermodynamic in nature or otherwise based on severe assumptions about dynamics and thus fail to predict some of experimentally observed trends. A continuum description of LbL assembly, accounting for PE chain diffusion, complexation, network relaxation is reported here. By formulating the model as a moving boundary problem, the height of the PEM can be continuously tracked over time. PE complexation is incorporated into model by considering the ion pairing reactions whereby repeat units of oppositely charged PEs lose their counterions to join the PEM. Using a combination of Flory-Huggins and Flory-Rehner free energy model and Poisson equation to solve for the electrostatic potential profiles, we observe that diffusion of PE chains even into a neutral PEM network is markedly different from diffusion of neutral polymers into a network. As PE chains diffuse, counterions readjust themselves to minimize the net local charge, but fail to do so completely as they would have to pay a significant entropic penalty. Diffusion of PE chains predominantly driven by the electrostatic field induced by the entropy of counterions is characterized by pulse-like PE composition profiles. Even without considering chain complexation, we demonstrate that it is possible to at least qualitatively explain the non-monotonic variation of PEM growth kinetics versus salt concentration, observed experimentally.

*** Formatting Issue - Best Available Paper/Graphic ***

A facile method for N-doped grpahene and the application in supercapacitors

Jin, Hong[1]; Wang, Xiaomin[1] Gu, Zhengrong[1]*;

1 (Affiliation): Agricultural and Biosystems Engineering Department

South Dakota State University, P.O. Box 2120

1400 North Campus Drive, AgE building, SAE 221, Brookings SD 57007

Phone 605-688-5372

* Corresponding author Email: Zhengrong.gu@sdstate.edu

1.1 Introduction

One of the great challenges we are facing today is energy conversion and storage (Lee et al., 2012). It is crucial that high efficiency, low-cost and environmentally friendly systems can be developed for energy conversion and storage; therefore, the energy storage topic has been becoming one of the hottest research fields (Huang et al., 2012b; Jiang et al., 2013a; Jiang et al., 2012; Raymundo-Piñero et al., 2006; Wang et al., 2012a; Zheng et al., 2011). Particularly, supercapacitors have been becoming attractive in terms of the large power density, long cycling life, modest energy density (Hao et al., 2013b; Huang et al., 2008; Lu et al., 2012; Wang et al., 2014; Wu et al., 2012). The performance of supercapacitors is mainly based on the electrode materials. Because of the high surface area, high conductivity and chemical stability, carbon materials are thought as the ideal candidate for electrode of supercapacitors (Hao et al., 2013a; Qie et al., 2013).

Carbon xerogels have been studied in terms of the non-binding electrode; however, the limitation is that the original materials (organic xerogels) are not cost effective and not environmental friendly (Perez-Cadenas et al., 2009). Activated carbon has been used commercially because the low cost of its preparation and high surface area; however, it is not a good choice for high quality supercapacitor (Li et al., 2013). The reason is that activated carbon has a long diffusion distance and a high ion-transfer resistance, which will limit the application in high current density. Carbon nanotubes have also been studied as the electrode for supercapacitors based on the excellent conductivity and high current density performance; nevertheless, the low specific capacitance and high cost as the critical drawbacks impede the application as good electrode for supercapacitors (Portet et al., 2007).

Graphene is considered as the attractive material in the area of nanocomposites (Young et al., 2012), transparent conduction films (Wang et al., 2010), sensors (Zeng et al., 2011), nanoelectronics (Roche, 2011), and energy storage devices (Fan et al., 2012; Gong et al., 2013) since it's been discovered in 2004. The theoretical capacitance of single layer graphene is 550 F g^{-1}, which is much higher than the carbon materials we were studying; therefore, it becomes more attractive than other carbon candidates for the researchers who are working on the supercapacitors(Han et al., 2014; Zhu et al., 2011). Several methods, such as mechanical exfoliation (Pang et al., 2010), chemical exfoliation

(Hernandez et al., 2008), chemical synthesis (Eigler et al., 2013), chemical vapor deposition (Huang et al., 2012a) and electrochemical synthesis (Parvez et al., 2014), etc. have been explored to synthesis high quality graphene. However, these synthesis methods are somehow not easy to scale up and the re-stacking of graphene is difficult to avoid and resulting in some extent of decreasing of specific capacitance. Chemical modification can improve the performance of graphene in the application of supercapacitors, specifically, nitrogen doped grpahene has been considered as the good choice (Cao et al., 2013; Wen et al., 2012).

Various methods were used to synthesis the N-doped grpahene. The electron beam deposition (Zhang et al., 2011), arc discharge (Li et al., 2010a), nitrogen plasma (Moon et al., 2014; Park et al., 2014) were used to produce N-dope grpahene, with the assistance of ammonia, hydrazine or pyridine as nitrogen source. However, the high requirement of reaction condition, the low products yields, the high energy consumption and environmental issue due to the precursor of nitrogen sources are prompting researchers to find a facile way for getting high quality N-dope grpahene.

Our approach for N-dope grpahene is innovative and cost effective. The protein rich (25% to 35% protein) distillers dried grains with soluble (DDGS) was selected as the N-dope grpahene precursor. Before dried, DDGS was immersed by $Ni(NO_3)_2{\cdot}6H_2O$ solution. During this process, Nickel element can be dispersed onto the surface of DDGS. After dried at 400 °C, the mixture was immersed into KOH solution. The N-dope grpahene was obtained by graphitizated activation at 700 °C. The most important thing is that the as-prepared N-dope grpahene presented excellent electrochemical properties (high specific capacitance, low inner resistance and extremely good rectangular shape of cyclic voltammetry) as the electrode for supercapacitors.

1.2 Experimental

1.2.1 Preparation of N-dope grpahene

The protein rich DDGS was used as the feeding material for generating N-dope grpahene. 1.45 gram of $Ni(NO_3)_2 \cdot 6H_2O$ was dissolved in 20 gram of water, and then the solution was transferred into a crucible where 20 gram of DDGS was loaded as well. The mixture was placed for 1 hour at room temperature, and then put into muffle furnace. These soaked mixture was then dried at 400 °C in muffle furnace (chamber is 15*15*22 cm) in N_2 atmosphere (N_2 flow is 500 ml min^{-1}) for 20 minutes to decompose $Ni(NO_3)_2 \cdot 6H_2O$. The as-prepared product was cooled down and soaked into 20 ml 7.5 mol/L potassium hydroxide solution for 1 hour. This mixture was then put into oven and dried at 105 °C for 24 hours. After that the graphitizated activation was carried out at 600, 700, 800, 900 and 1000°C separately for 1 hour with the protection of 500 ml min^{-1} N_2 flow. The obtained samples were cooled down in muffle furnace in the same N_2 atmosphere. N-dope grpahene samples were further purified in the 1.0 mol L^{-1} HCl at 150 °C within the hydrothermal reactor, and then washed with deionized water to pH value at 7 and dried at 105 °C overnight under vacuum. The samples were listed as N-600, N-700, N-800, N-900 and N-1000 based on the reaction temperature.

1.2.2 Characterization of N-dope grpahene

Physisorption analyses with N_2 were carried out at 77K (liquid nitrogen bath), using Tristar 3000 Micropore analyzer. The specific surface areas were calculated using the Brunauer–Emmett–Teller (BET) equation. The total pore volumes were obtained at the relative pressure P/P_0=0.978. The micropore volume, mesopore volume and the pore size distribution were determined by the NLDFT analysis for carbon with slit pore model (Micromeritics Inc.) based on the N_2 isotherm adsorption data. Spectrum (Horiba LABRam confocal Raman microscope) was used to characterize N-dope grpahene, while the excitation wavelength was set at 532 nm from a diode pumped solid-state laser. The micro-morphology was detected by SEM (HITACHI S-3400N). The XPS measurements were performed on an SSX-100 system (Surface Science Laboratories, Inc.) equipped with a monochromated Al K_α X-ray source, a hemispherical sector analyzer (HSA) and a resistive anode detector. The base pressure of the XPS system was 6.0 x 10^{-10} Torr. During the data collection, the pressure was ca. 1.0 x 10^{-8} Torr. JEM-2100F was used to characterize the crystal and porous structure of N-doped graphene structure.

1.2.3 Electrode preparation and electrochemical measurements

The supercapacitors were assembled as the general. For all of the electrochemical characterization, the surface area of the electrode was 1 cm^2. Meanwhile, the electrolytic medium was 6mol L^{-1} KOH. The electrodes were prepared by pressing the slurry of N-doped graphene, carbon black (conducting material from Fisher scientific) and PTFE (binder) with a mass ration of 8:1:1 on nickel foam. Two-electrode sandwich-type cells were built using a stainless coin cell (2032) with a microporous PP separator celgard-3501 between the electrodes.

Cyclic voltammetry (scan rate from 5 to 200 mV s^{-1}) and galvanostatic charge/discharge cycling (current density load from 0.1 to 10.0 A g^{-1}) were performed using a SP-150 multichannel potentiostat–galvanostat-EIS (Biologic, France). The electrical conductivity of supercapacitors was measured through the electrochemical impedance spectroscopy (i.e. frequency response analysis yields over the frequency ranging from 0.1 Hz to 200,000 Hz with potential amplitude of 10 mV) with a SP-150 multichannel potentiostat–galvanostat-EIS (Biologic, France). The values of specific capacitance was calculated from the galvanostatic discharge characteristics at current density of 0.1 to 5.0 A g^{-1}, and expressed in Farad per gram (farad per square meter was depend on the surface area of the activated carbon in the electrode) of electrode active material.

1.3 Results and discussion

1.3.1 Physiochemical characterization of N-dope grpahene

*** **Formatting Issue - Best Available Paper/Graphic** ***

The chemical composition of N-doped graphene samples were characterized by the X-ray photoelectron spectroscopy (XPS). The results are shown in Figure 7.1.

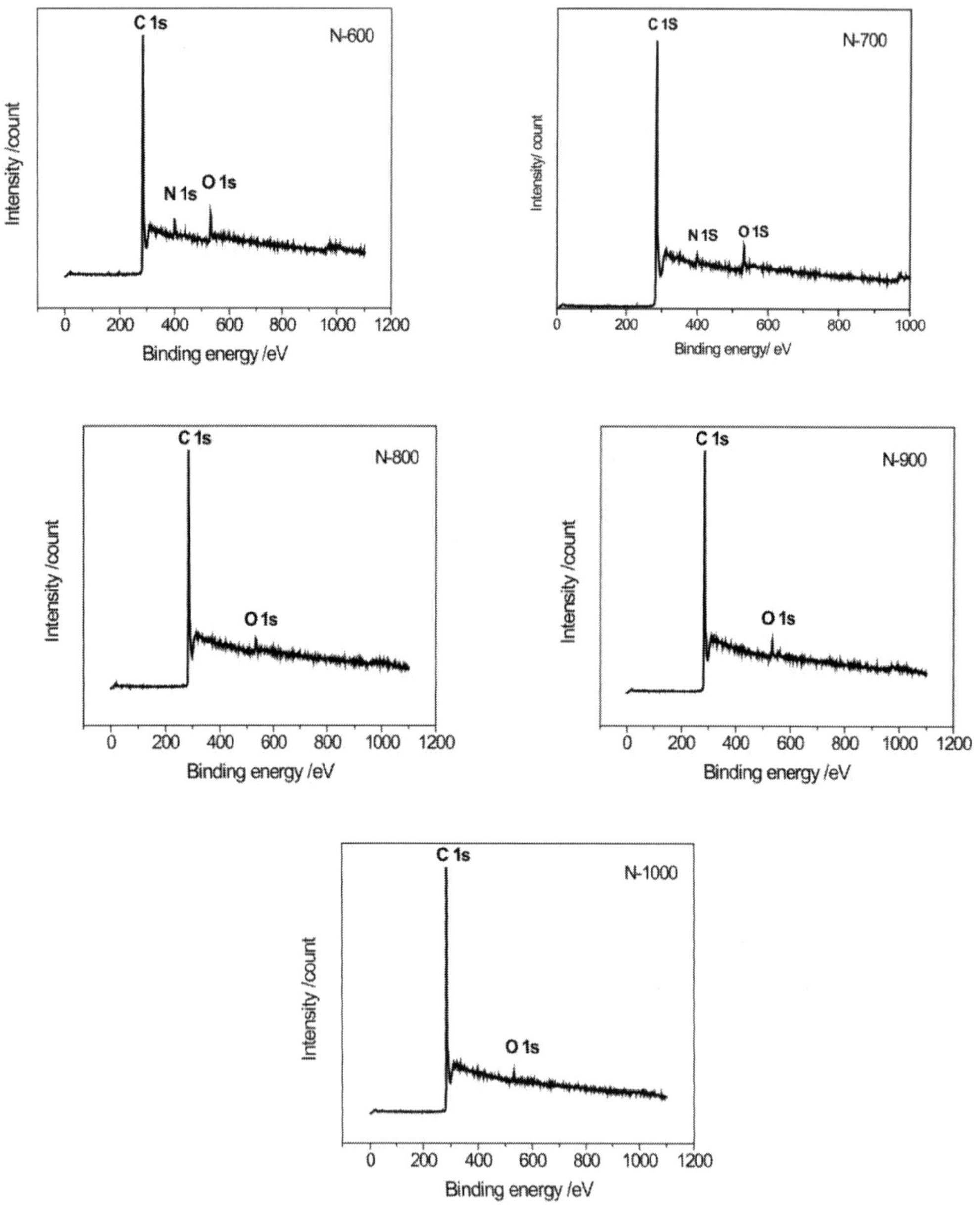

Figure 7.1. X-ray photoelectron spectroscopy of as prepared samples

*** **Formatting Issue - Best Available Paper/Graphic** ***

The N-600 and N-700 samples show the N 1s peak, while it is not obvious for the N-800 and there is no N 1s peak for N-900 and N-1000. The elemental contents were characterized by XPS as well, and they are shown in Table 7.1.

Table 7.1. The elemental contents of as prepared samples

Samples	C (Group %)	N (Group %)	O (Group %)	C (Atom %)	N (Atom %)	O (Atom %)
N-600	80.59	7.19	12.22	89.77	4.77	5.47
N-700	84.77	4.06	11.17	92.47	2.64	4.89
N-800	91.49	1.18	7.33	96.17	0.74	3.09
N-900	93.10	0	6.90	97.11	0	2.89
N-1000	95.37	0	4.63	98.9	0	1.91

Besides, for the samples of N-900, N-1000, there are no nitrogen elements left. These results indicate that the N-doped graphene structures were firstly formed at the lower temperature, and with the temperature increase, the N-doped carbon would thermodynamically decompose.

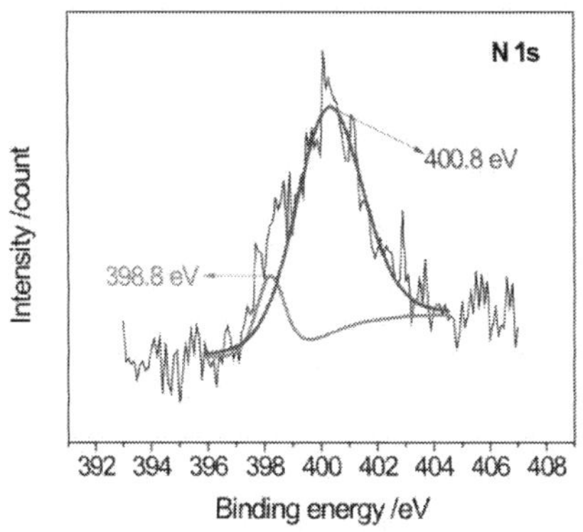

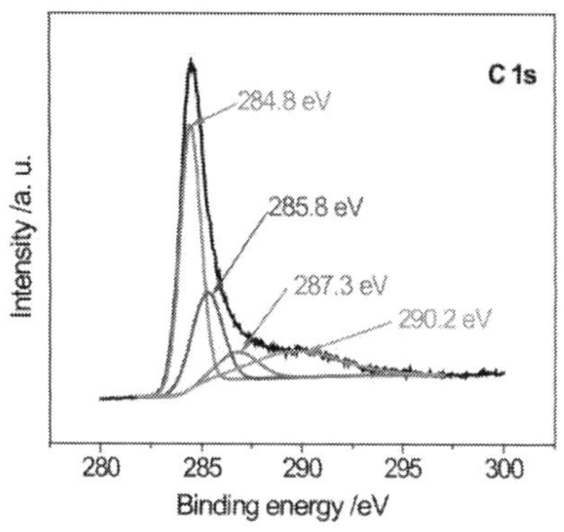

*** **Formatting Issue - Best Available Paper/Graphic** ***

Figure 7.2. Detailed C 1s and N 1s XPS peaks of N-700

The detailed C 1s peaks of N-700 are shown in Figure 7.2. Two components centered at 398.8 eV and 400.8eV are detected in the N 1s peak of N-700, which are corresponded to the nitrogen forms of pyridinic-N and pyrrolic-N, respectively. The pyrrolic-N is the dominant nitrogen form in N-600, N-700 and N-800. The pyrrolic-N ratios of these three samples are 70.2%, 87.3% and 100% separately. The total nitrogen ratios of these three samples are 4.77%, 2.64% and 0.74%. The main peak is located at 284.8 eV, which is the same as the graphite-like sp^2 carbon. While the N- sp^2 carbon and N- sp^3 carbon are reflected by the peaks at 285.8 eV and 287.3 eV. In addition, the peak at 290.2 eV is because of the carbon-oxygen bond.

*** Formatting Issue - Best Available Paper/Graphic ***

***** Formatting Issue - Best Available Paper/Graphic *****

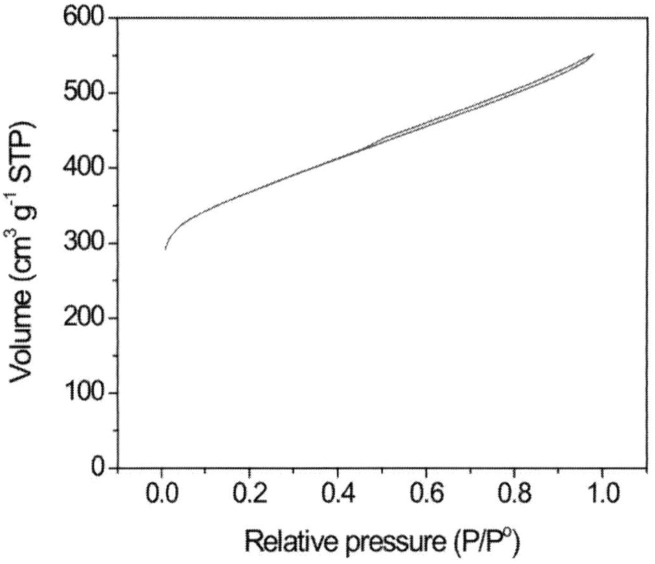

*** Formatting Issue - Best Available Paper/Graphic ***

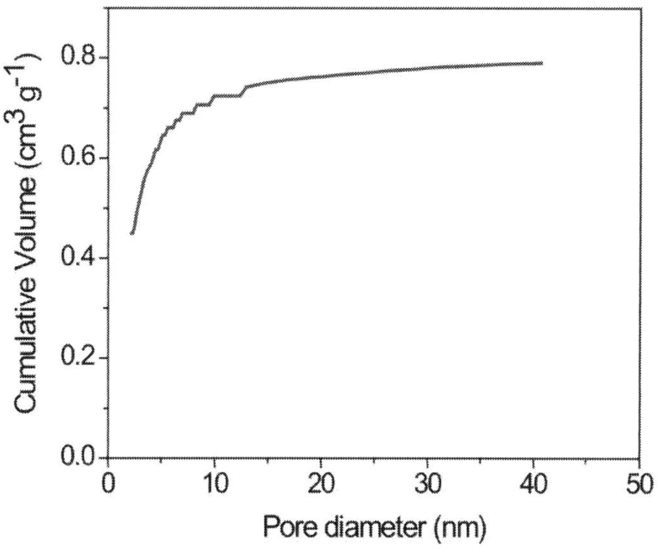

The morphology and microstructure of N-700 were characterized by scanning electron microscopy (SEM), transmission electron microscopy (TEM) and Brunauer-Emmett-Teller (BET) surface area analysis. The distinctive microstructures and the complex three-dimensional porous structure are shown in Figure 7.3A. The wrinkled nanostructures, nanoporous structures and the few-layer graphene structures are shown in Figure 7.3 (B-E). Most of the sample is composed with graphene layers less than 10-layer and the length of the graphene structures are in the nanoscale. Very few parts of graphene structure are about 10-30 layers. This nano structures were futher proved by nitrogen adsorption-desorption isotherm curve and the pore size distribution curve. As Figure F and G show, the small hysteresis loop at the relative pressure P/P^0 between 0.45 and 1.0 indicates that N-700 contains a small amount of mesopores. And the pore size distribution curve is further used to analyze N-700, while the data are shown in Figure 7.3 G and Table 7.2. The result shows the total pore volume is 0.85 cm^3 g^{-1}, and 0.45 cm^3

*** Formatting Issue - Best Available Paper/Graphic ***

g^{-1} pores are less than 2 nm, while 0.29 cm^3 g^{-1} pores are ranging from 2 to 10 nm and 0.11 cm^3 g^{-1} pores are above 10nm.

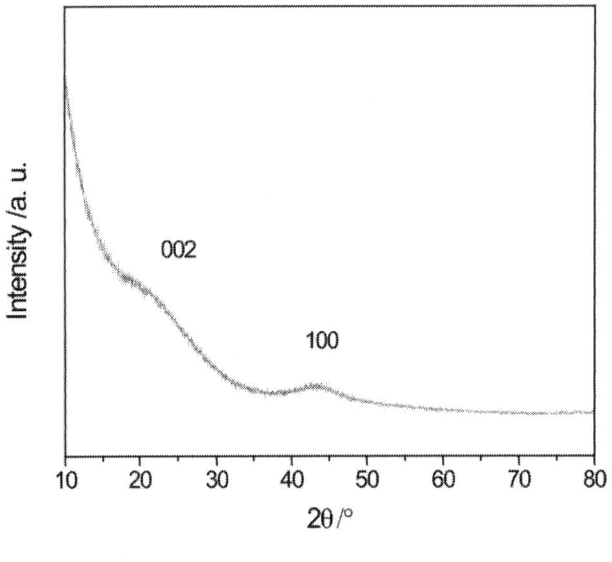

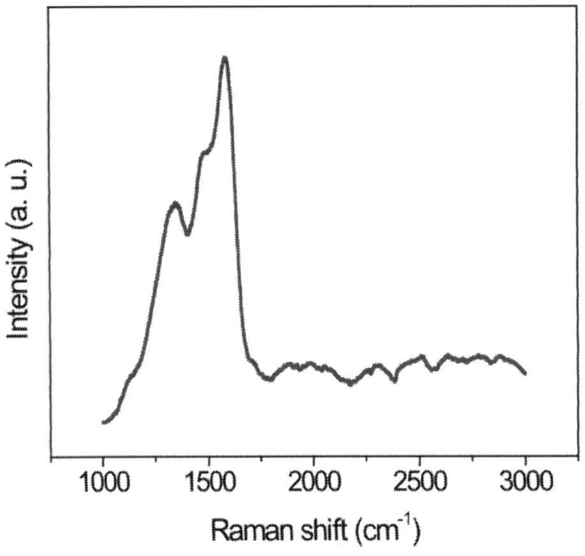

The crystalline property of N-700 was also characterized by Raman spectroscopy. The Raman spectrum shows N-700 is a typical partly disordered graphene system. The intensity ratio of G band versus D band (I_G/I_D) is 1.71, which means that N-700 contains more graphtized structures than disordered carbon. The compressed 2D band is due to the defective structure of graphene, namely it is caused by the nitrogen doping affection. The X-ray diffraction (XRD) pattern of N-700 is shown in Figure 7.4 B. It is clear that N-700 presents the peak 100, while the 002 peak is not obvious. This result is in accordance with the TEM characterization, which means that N-700 is composed by more single layer or few layer N-doped graphene sheets but not graphite chunk.

*** Formatting Issue - Best Available Paper/Graphic ***

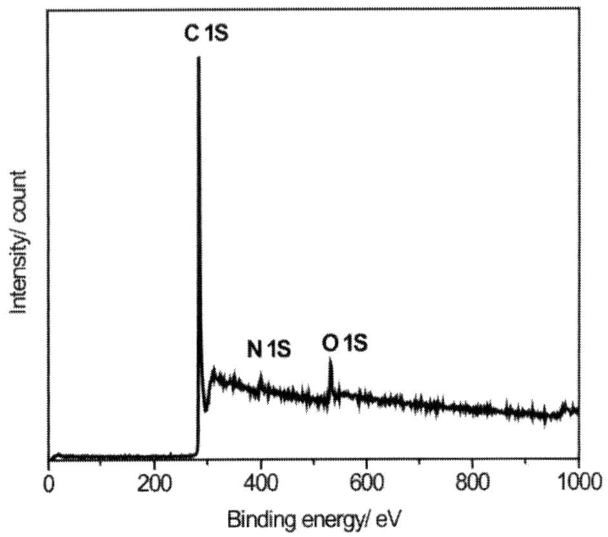

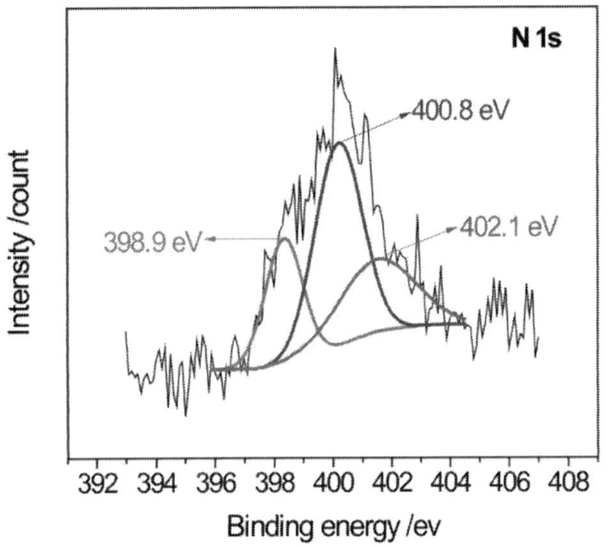

*** Formatting Issue - Best Available Paper/Graphic ***

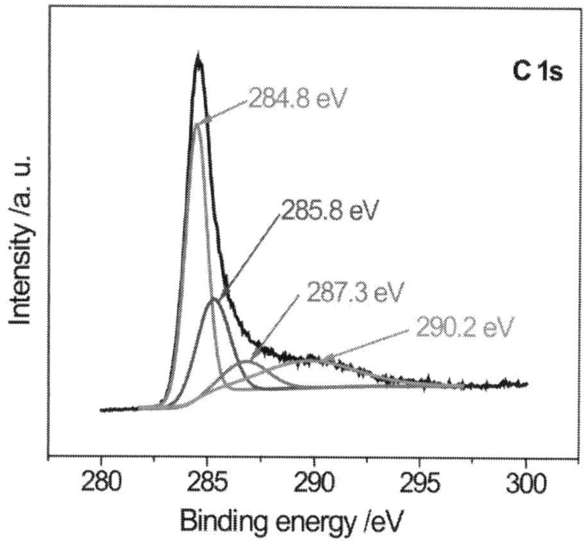

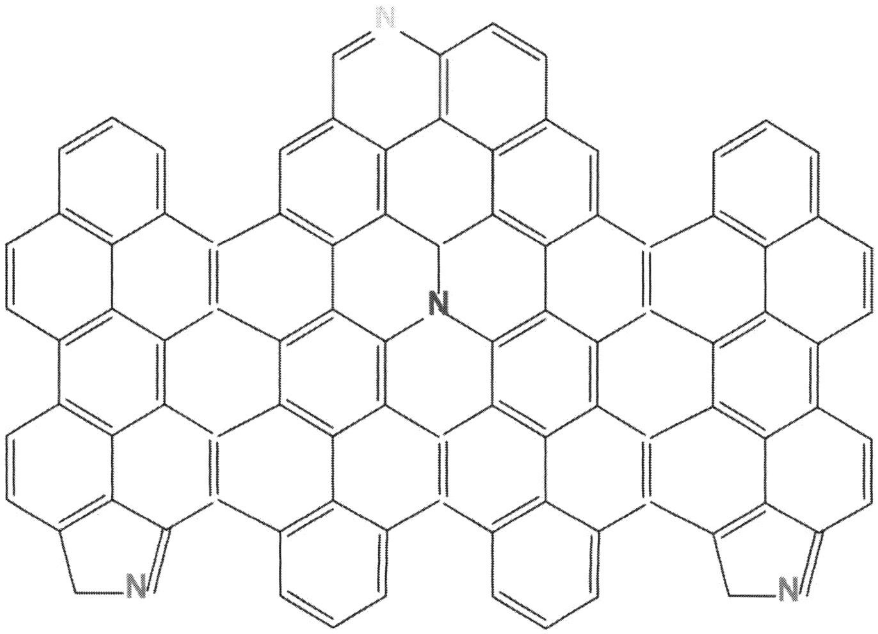

1.3.2 Electrochemical performances of activated carbons

The supercapacitor properties of the N-700 were studied through a set of symmetrical two-electrode 2032 coin-type system in a 6 M potassium hydroxide

*** **Formatting Issue - Best Available Paper/Graphic** ***

electrolyte. The cyclic voltammetry curves (CVs) of N-700 supercapacitor at various scan rates are shown in Figure 7.5A. The typical rectangular shape of all of the scan rates ranging from 5 to 200 mV s^{-1} indicates pure electric double layer capacitive features of N-700. The specific capacitances at different scan rates are shown in Figure 7.5B. The high capacitance of 293 F g^{-1} can be obtained at the scan rate of 5 mV s^{-1}, while it is shown high capacitance of 236 F g^{-1} at the high scan rate of 200 mV s^{-1}.

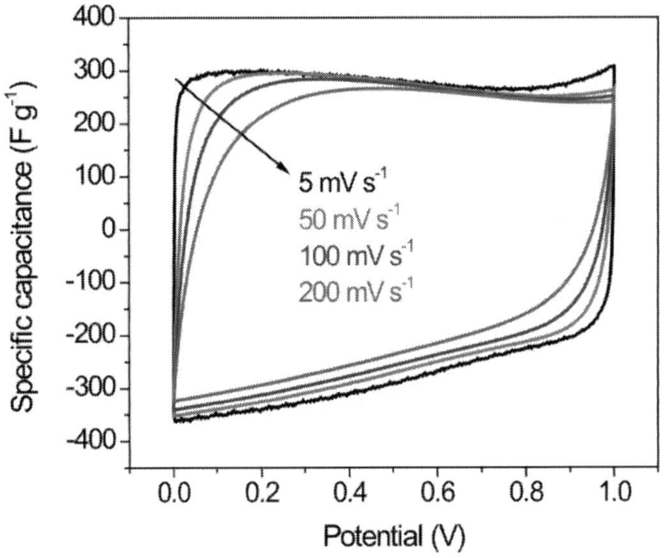

*** Formatting Issue - Best Available Paper/Graphic ***

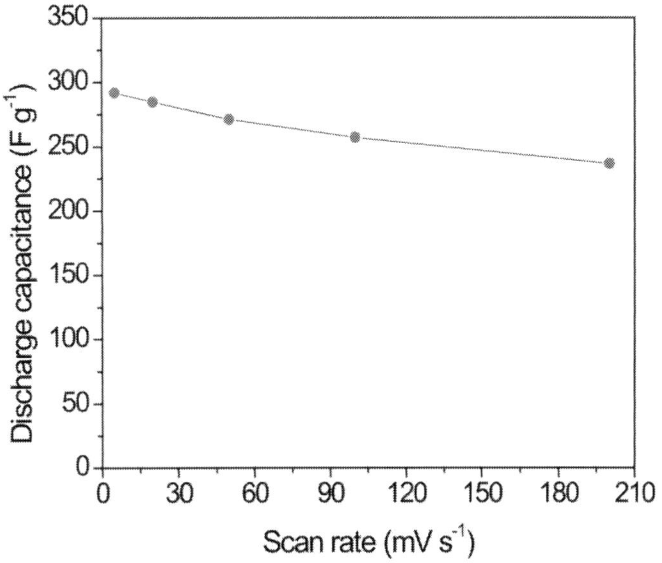

The galvanostatic charge/discharge properties are shown in Figure 7.6. As can be observed, the charging plots are mainly symmetric to the discharging plots at the current densities, especially at the current density of 0.1 and 1.0 A g^{-1}. This indicates N-700 is a high capacitive reversible material for the application as the electrode of supercapacitors. The specific capacitances are shown in Figure 7.6B. The specific capacitance is as high as 330 F g^{-1} at the current density of 0.1 A g^{-1}, which is higher than most of the reported materials. More importantly, at the high current density of 10.0 A g^{-1}, the specific capacitance can still achieve 270 F g^{-1}.

*** Formatting Issue - Best Available Paper/Graphic ***

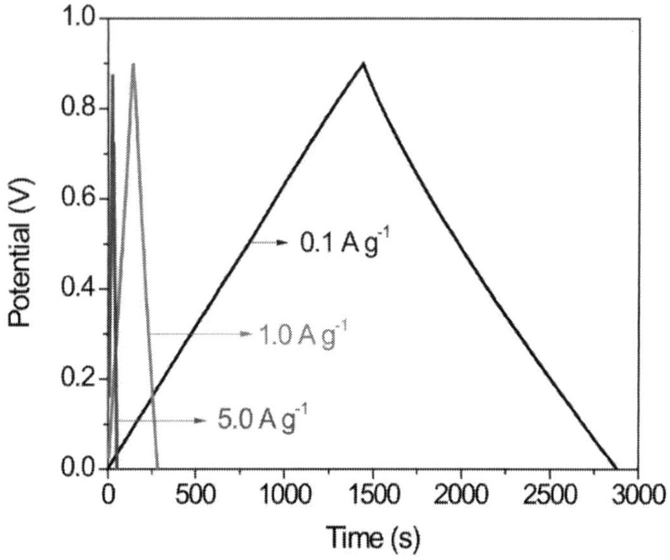

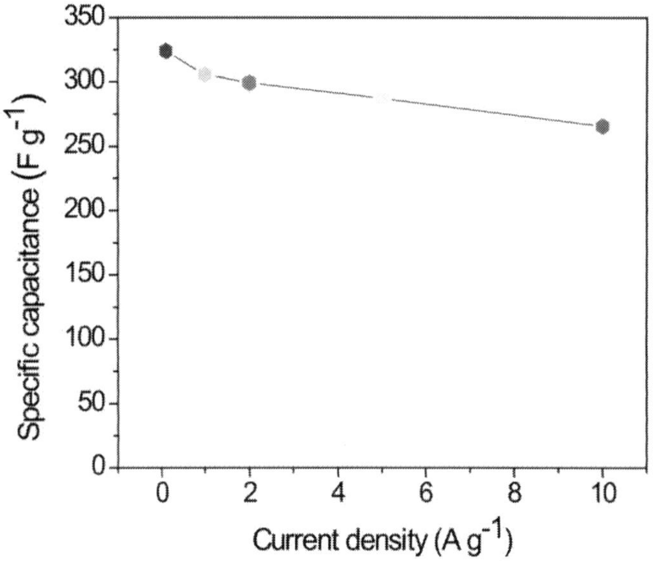

The electrochemical properties of the N-700 were further characterized by electrochemical impedance spectroscopy (EIS). As shown in Figure 7.7, the Nyquist plot of N-700 electrode was performed in a frequency ranging from 20 kHz to 10 mHz. The inner resistance (equivalent series resistance) is around 0.1 ohm, which indicates the excellent conductivity of the sample N-700. And the vertical linear part in the lower frequencies indicates the as prepared supercapacitor is an ideal double-layer supercapacitor. The inset in Figure 7.7 shows properties of the medium frequency region. As is shown here, there is a 45° phase shift after the semicircle in the high frequency, which is caused by the porous structure of the N-doped grpahene. The transport of ions from bulk electrolyte solution through the porous structure of the electrode to the surface of the electrode formed a diffusion inner resistance. And the diffusion inner resistance is very small in the supercapacitor, which is because the ions transport is promoted by the hierarchical porous structures.

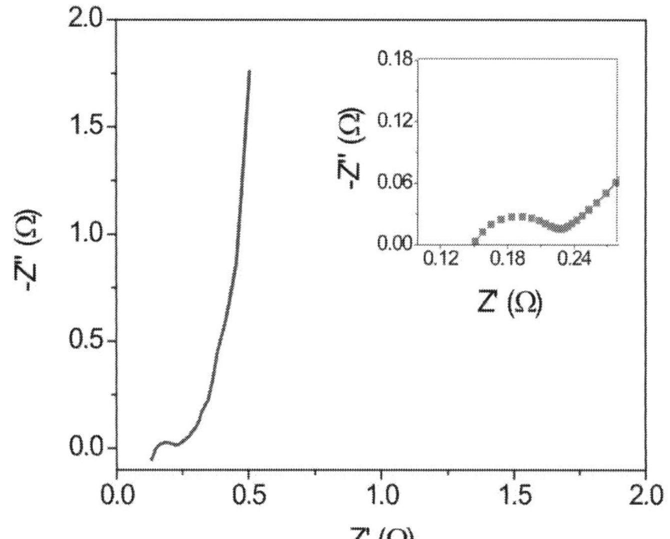

1.4 Conclusion

According to the SEM, TEM, XRD, Raman spectroscopy, BET, and XPS results, the N-700 sample is the crumpled nanoporous few-layer N-doped grpahene. The high performance N-doped graphene has been developed by a facile method with the DDGS as the carbon and nitrogen precursor. The carbonization temperature is found as the key factor to the properties of the N-doped graphene. The most important observation is that the N-700 sample presented high specific capacitance (330 F g^{-1}) and extremely low inner resistance (0.1 ohms) while it is used as the electrode for supercapacitors. The synthesis method is extremely attractive, since the N-doped graphene can be produced economically and effectively. This method is easy to scale-up and the precursor is easy to get and extremely cheap; nevertheless, it will be an attractive way to synthesis N-doped graphene for the application of supercapacitors in industry.

378918 Lignin Derived Mesoporous Carbons for Ultracapacitor Electrode Material

Abstract:

Mesoporous carbons were synthesized from pre-crosslinked lignin based materials by surfactant templating with pluronic F127 followed by physical and chemical activation to enhance the pore texture properties. The BET surface areas of the mesoporous carbons were in the range of 185 to 1148 m^2/g with total pore volume of 0.28 to 1.0 cm^3/g. Besides pore textural properties, these materials were also characterized with electron microscopy, electron energy loss spectra (EELS), small-angle x-ray scattering (SAXS) and x-ray photoelectron spectroscopy (XPS). The ultracapacitor electrodes were fabricated with three samples with varying porosity and aqueous KOH as electrolyte. Majority of the plots in the cyclic voltammetry exhibited rectangular shape exhibiting ideal double layer capacitance. The trend in mass and surface area based specific capacitance did not follow the pore textural properties and that can be attributed to the electrode-electrolyte interactions, surface functionality and various pore geometries. The capacitive characteristics of those materials are investigated with impedance spectra in 10 mHz and 0.2 MHz frequency and a vertical tail in the low frequency domain demonstrated the good capacitive behavior. The electrochemical capacitive characteristics were also investigated with Bode plot, RC characteristics and equivalent electrical circuit by employing the constant phase elements (CPEs).

384619 Ultracapacitor and Its Applications in Rapid Energy Storage and Conversion

Abstract:

Liquid fuels such as gasoline fossil fuels and FTS non-fossil fuels have high energy density and as a result are desirable for onboard energy storage through chemical bonds. To increase fuel efficiency, kinetic energy from braking, conventionally thought of as unusable, can in reality be recovered and electrochemically converted into reusable energy that can charge Ni-MH or Li-ion battery packs through chemical reactions. However, due to the effects of severe weather and the consequences of thermal and mechanical stresses during charge and discharge, the expected cycle lifetime of the energy storage system is diminished, reducing its efficacy. An alternative to using battery packs is to use the ultracapacitors, which store electrostatic charges in Helmholtz layers. An additional advantage using an ultracapacitor is that it has higher energy density than the conventional dielectric capacitor. Due to the absence of chemical reactions, it also has a longer cycle lifetime, albeit a higher discharge rate than a system using batteries to store energy. This work examined the viability of using an ultracapacitor module as a power capturing device. In order to understand the physical processes inside the capacitor power module, an ultracapacitor module with 300 F at an operating window from 4.0 to 14.5 volts was tested and evaluated through an ac impendence measurement via 5 mV small *ac* currents that were used to disturb the energy storage system. The capacitance values were determined using *ac* impendence technique and also calculated using *V-t* curves at different constant currents. The further effort likely extends the lifetime for the onboard energy storage system. The advancement of the system structure and design for energy storage and conversion gives valuable support to efficient power capturing for sustainable energy needs.

Comparison of Argon and Oxygen Plasma Activation of Biochar for Supercapacitors

Rakesh Kumar Gupta[1], Mukul Dubey, Zhengrong Gu[2]*, Qi Hua Fan[1]*

[1] Department of Electrical Engineering and Computer Science
[2] Department of Agricultural and Biosystems Engineering
South Dakota State University

* Author for correspondence, zhengrong.gu@sdstate.edu, qihua.fan@sdstate.edu,

Abstract

Biochar, a byproduct of the pyrolysis process, is biomass-derived carbon that has been traditionally used for soil amendment. As a low-cost and environmentally friendly material, biochar has shown the potential to replace the synthesized carbon nanomaterials (e.g. carbon nanotubes) for future supercapacitors. To achieve high capacity for charge storage, biochar needs to be properly activated. Conventional activation involves high-temperature baking (>900°C) for hours using a strong base mixed with biochar. The process is lengthy and energy intensive. Recently, a few groups reported using plasma to treat carbon nanomaterials and demonstrated improved capacitance. The mechanisms are unclear and the materials microstructures of plasma treated biochar are indefinite. This work intends to reveal the mechanisms of plasma activation of biochar. Two source gases (argon and oxygen), which yield highly active and inert species in the plasmas, were used to activate yellow pine biochar. Preliminary study demonstrated significant effects of plasma treatment. The specific capacitance of supercapacitors made with biochar treated by oxygen plasma for 5 minutes was 171 F/g, by Ar plasma treatment for 30 minutes was 80 F/g, and untreated was 60 F/g. This paper investigates the surface evolution and microstructure change of yellow pine biochar upon the plasma reactions. Possible reasons accounting for the three times increase in the capacitance after the oxygen plasma treatment were discussed.

Key words: biochar, supercapacitor, plasma

390312 Polypyrrole Modified Cobalt Oxide for Electrochemical Energy Storage

Abstract:

Electrochemical capacitors have attracted significant attention for their promising potential applications ranging from portable electronic devices to hybrid electrical vehicles and large industrial equipments.[1-5] Two cobalt oxides with three-dimensional flower- and ball-like morphologies, respectively, were prepared by a facile solvothermal method for electrochemical capacitor applications. Binder free electrodes were obtained by drop casting cobalt oxide suspensions in the ethanol solvent onto carbon paper to evaluate the electrochemical performances. The flower-like cobalt oxide exhibits better capacitive performances due to the unique mesoporous structure compared to the non-porous ball-like cobalt oxide, as confirmed by SEM and TEM images. Furthermore, conductive polymer, polypyrrole (PPy), was in situ electropolymerized to the flower-like cobalt oxide via the cyclic voltammetry techniques. A significant enhancement in the performances in terms of both energy and power density has been witnessed by the PPy coated flower-like cobalt oxide. Reduced charge transfer resistance is inferred to be responsible for the superior performances as confirmed by the electrochemical empedance spectroscopy.

References

(1) Wei, H.; Gu, H.; Guo, J.; Wei, S.; Guo, Z. *ECS Journal of Solid State Science and Technology* 2013, *2*, M3008.

(2) Wei, H.; Gu, H.; Guo, J.; Wei, S.; Guo, Z. *Journal of The Electrochemical Society* 2013, *160*, G3038.

(3) Zhu, J.; Chen, M.; Qu, H.; Zhang, X.; Wei, H.; Luo, Z.; Colorado, H. A.; Wei, S.; Guo, Z. *Polymer* 2012, *53*, 5953.

(4) Wei, H.; Yan, X.; Wu, S.; Luo, Z.; Wei, S.; Guo, Z. *The Journal of Physical Chemistry C* 2012, *116*, 25052.

(5) Wei, H.; Zhu, J.; Wu, S.; Wei, S.; Guo, Z. *Polymer* 2013, *54*, 1820.

370784 Metal Nanoparticles Directed NiCo2O4 Nanostructure Growth on Carbon Substrates with Large Capacitance

Abstract:

The rapid development of portable flexible electronics requires significant advancement of energy storage materials in terms of energy and power density, mechanical strength, flexibility and durability. Current techniques focus on developing conductive carbon substrate and subsequently coating pseudoactive nanostructures on carbon to achieve high energy/power density. However, most of the decorated pseudoactive nanostructures meet interfacial failure during long term cycling tests and results in significant energy drop after certain cycles. Here, we developed a novel two-step approach to manufacture electroactive nanocomposites with excellent interfacial properties. Nanoparticles were introduced to promote carbon graphitization in the first annealing process and further served as "seeding" material for $NiCo_2O_4$ nanostructure growth in the second step. With the "bridging" effect of the metal nanoparticles, excellent interface between $NiCo_2O_4$ and graphitized carbon fabric are obtained. The electrodes show large capacitance of >700 F/g. More important, excellent cycling performance with >90% retention over 10000 cycles are achieved. This two-step method provides a general and promosing approach to synthesize hierarchical nanostructures for advanced energy storage devices.

376623 In-Situ Atomic Force Microscopy Study on Solid-Electrolyte Interphase Formation of Li-Ion Batteries

Abstract:

Solid-electrolyte interphase (SEI) forms on the electrode during the first several charge-discharge cycles. This thin layer on Li-ion electrode is believed to provide a voltage slop at the interface of the electrode and the electrolyte to prevent further electrolyte decomposition, while still provide significant ionic conductivity for Li ions insertion and extraction during cycling. The SEI grows during cycling and develops into a thicker insulate layer to Li ions at the end. Many groups have reported the composition, evolution and thickness of the SEI layer on the electrodes of Li ion batteries. With the application of in-situ AFM, more precise scanning of the morphology can be achieved without the interference of the electrolyte residues and the effects of the oxidation in atmosphere. However, the effect of decomposition products on lithium ion insertion, the thickness changes of the SEI layer and the compositions of the SEI during cycling have not been fully elucidated. A better understanding of the formation and evolution of SEI will provide valuable data to explain the performance, self-discharge and failure mechanism of Li-ion batteries. In this report, the SEI formation was investigated a step-by-step by in-situ AFM. A dynamic competition mechanism of the SEI formation process is proposed based on the experimental results.

360493 Template-Free Synthesis of Conducting Polymer Microstructures for Supercapacitor Electrodes

Abstract:

Interest in energy storage technologies, especially to support transportation, portable electronics and renewable energy generation, has increased at a tremendous rate over the past decade. The development of new technologies faces many challenges, such as material accessibility, efficiency, charge storage capacity and cycle-ability. Emerging devices, such as supercapacitors, have the potential to provide high energy devices with fast discharge rates to bridge the gap between traditional batteries and high-power capacitors. Materials ranging from high surface area, inert carbon nanomaterials to Faradaic metal oxides and conducting polymers have been used to achieve a range of performance properties in supercapacitors. Porous carbon nanomaterials utilize electrical double layer capacitance, where charges are stored physically at the electrode interface, allowing devices to achieve high discharge rates (high power density) at the cost of lower energy density. Electroactive conductive polymers (ECPs) are promising alternatives, due to their conductivity, increased energy storage capacity through Faradaic charge transfer, and their amenability to low cost, large area synthesis methods. In this work, we demonstrate a simple approach to preparing large quantities of conducting polymer microtubes without the need for a solution or substrate template. We will discuss how to control the polymer assembly and microtube synthesis on stainless steel mesh substrates, and how the microstructure affects the electrochemical performance.

Microtubes of polypyrrole were electrochemically polymerized on stainless steel mesh substrates (variable mesh sizes) using a range of growth conditions, including current density and monomer concentration and overall mass. Electrodes were studied to understand the growth mechanism and microtube formation, and how the electrode structure affects the electrochemical properties. The microtube's formation and growth mechanism were studied using scanning electron microscopy (SEM) on samples prepared through different stages of growth (1C up to 30C). The tube structures evolve from an initial nucleus, which forms at the intersection of two metal wires in the mesh, to the final tube. We found that changing the mesh size (mesh diameter and spacing) provided an easy tool to manipulate the microtube size (100μm up to 1 mm), shape (cone-like or tube-like structure) and microtube density on the electrode. Finer mesh substrates (200x200 and 400x400) resulted in the highest density (about 350 microtubes/cm^2), which also exhibited the highest electrode capacitance (200 F/g). This microstructure synthesis approach also applied to large substrates, which showed similar physical and electrochemical performance. The results show the effects of the substrate, current density and monomer concentration on the template-free assembly and electrochemical performance of polypyrrole, as well as new ways to manipulate the physical structure of redox materials for supercapacitor electrodes. Importantly, this approach is amenable for large-scale syntheses of micro and nanostructured electrode systems.

388073 Vesicular Protein Suprastructures Via Temperature-Responsive Self-Assembly of Elastin-like Polypeptide Fusion Proteins

Abstract:

Elastin-like polypeptides (ELPs) are artificial peptide repeats derived from tropoelastin. A unique property of ELPs is their inverse transition from the soluble to insoluble phase in aqueous solution during an elevation of temperature. Thus, ELPs can self-assemble into coacervate particles in response to thermal triggers. More interestingly, it is possible to construct self-assembled suprastructures using amphiphilic block copolypeptides of ELP and hydrophilic fragments.[1,2] Here, we describe temperature-responsive self-assembly of vesicular suprastructures using ELP-based fusion proteins that also contain fully folded globular proteins. In our approach, an ELP fragment was fused with a hydrophilic coiled coil protein motif (C1), and served as a hydrophobic fragment of the di-block fusion protein (C1-ELP). For the second building block, we used two fluorescent proteins mCherry and EGFP as model proteins, and constructed di-block fusion proteins, mCherry-C2 and EGFP-C2. Because of high affinity between C1 and its peptide binding partner C2, fusion proteins containing C2 can bind C1-ELP by simply mixing in aqueous solution. When mixture solutions of C1-ELP and mCherry-C2 (or C1-ELP and EGFP-C2) prepared at 4°C were placed at room temperature for an hour, the solutions became turbid, indicating inverse phase transition of ELP motifs and formation of suprastructures. Using confocal microscopy, the resulting protein vesicles were characterized and indicated that the globular fluorescent proteins were homogenously incorporated in the self-assembled layers. Ionic strength of protein solutions played an important role in formation of vesicular suprastructures. We hypothesize that modulation of surface tension by ion concentration allows segregation of ELP fragments in the presence of different types globular proteins, mCherry and EGFP. Our system allows temperature-responsive programmed assembly of engineered recombinant proteins in aqueous environment, without use of any organic solvents, through exploitation of the inverse phase transition behavior of ELP. Incorporation of various types of folded globular proteins including enzymes, growth factors, therapeutic ligands can lead to significant benefits in applications such as biocatalysts, drug delivery, and molecular imaging.

References

1. T. A. T. Lee, A. Cooper, R. P. Apkarian, V. P. Conticello, Thermo-reversible self-assembly of nanoparticles derived from elastin-mimetic polypeptides. *Adv. Mater.* **2000,** 12, 1105 – 1110

2. M. R. Dreher, A. J. Simnick, K. Fischer, R. J. Smith, A. Patel, M. Schmidt, A. Chilkoti, Temperature triggered self-assembly of polypeptides into multivalent spherical micelles. *J. Am. Chem. Soc.* **2008**, 130, 687 – 694

385737 Sequence-Dependent Self-Assembly of Peptide Amphiphiles Via Molecular Simulations

Abstract:

Stimulus-responsive biomaterials have been of great interest for a wide range of therapeutic applications including the design of biomimetic tissue scaffolds and drug delivery systems. Peptide amphiphiles (PAs), which are comprised of a bioactive peptide sequence that is chemically conjugated to a hydrophobic alkyl tail, have been shown to undergo self-assembly to form nanostructures (e.g., cylindrical nanofibers, spherical micelles) whose morphology is strongly dependent upon the solvent condition. In this study, we investigate a series of PAs to examine the effect of peptide sequence on self-assembly behavior and morphological properties of stimuli-responsive nanostructures via molecular simulations. Implementation of a newly extended intermediate-resolution protein model coupled with discrete molecular dynamics simulation technique enables the simulation of large PA systems to observe the whole self-assembly process. This allows us to elucidate kinetic mechanisms of self-assembly to form cylindrical nanofibers and spherical micelles as a function of the pH condition. Moreover, placing nonpolar peptide residues adjacent to the alkyl tail causes them to experience a nanofiber-to-micelle transition as a function of decreased pH values due to the close proximity of charged residues. By manipulating the placement of peptide residues, the intermolecular interactions involved can ultimately be controlled and used as an effective parameter to design smart biomaterials.

389050 Molecular Description of LCST Behavior of Elastin-like Peptides Poly(VPGVG) and Poly(VGPVG)

Abstract:

The elastin-like peptide (ELP) is temperature sensitive biopolymer which undergoes conformational transition at critical temperature. Elastin-like polypeptides (ELPs) with the repeat sequence of VPGVG are widely used as a model system for investigation of lower critical solution temperature (LCST) transition behavior. Recent experimental observations show that the order of amino acids in the ELP sequence also plays an important role since $(VGPVG)_n$ and $(VPGVG)_n$ peptides show the differences in thermo-transition behavior. In order to elucidate the factors driving the conformational transitions of ELPs we performed molecular dynamics simulations of $(VPGVG)_n$ and $(VGPVG)_n$ peptides with different length and at a range of temperatures. Comprehensive analyses of the hydrophobicity, shape, size, hydration and dynamics of $(VPGVG)_n$ and $(VGPVG)_n$ suggest that the transition behavior includes the thermal disruption of water network, loss of hydration, and increase in peptide hydrophobicity. Our simulations show that single poly(VGPVG) is more hydrophilic than poly(VPGVG) at high temperatures while similar hydrophobicity of two types of ELPs at low temperatures is observed. Simulations of the peptide-peptide association demonstrated that the aggregates formed by double poly(VGPVG) chains are more densely associated than poly(VPGVG) at high temperature. Even more compact association of double poly(VGPVG) was observed upon cooling the aggregates obtained from the double-peptide simulation. In the contrary, poly(VPGVG) aggregates dissociate upon cooling which is consistent with the reversibility exhibited in experiments. The results from our study provide an atomic-level description of the thermo-responsive conformational properties of elastin-like peptides and explain experimental observations. Support for this research was provided by the NSF's Research Triangle MRSEC (DMR-1121107).

389774 Thermo-Responsive Poly(N-isopropylacrylamide) Grafted on Polydopamine-Coated Surface for Cell Harvesting

Abstract:

Thermo-responsive Poly(N-isopropylacrylamide) (PNIPAAm) grafted surfaces have been widely used in the area of tissue engineering and drug release due to the reversible phase transition at its lower critical solution temperature around 32°C. The aim of this research is to develop a simple and cost-effective method to produce PNIPAAm using polydopamine-coated surface as a platform. To do so, dopamine was first conjugated with polymerization initiator, and then a layer of functionalized polydopamine film was formed onto a glass substrate when the pH of the solution was adjusted to 8.5. Surface-initiated PNIPAAm was obtained after standard polymerization procedure. The PNIPAAm-grafted polydopamine-coated surface was characterized by analytic tools such as FTIR and contact angle. Anchorage-dependent cells cultured on the thermo-responsive surface were easily detached without trypsin at ambient temperature.

383747 Structure and Thermodynamics of Single- and Double-Stranded DNA Oligomers Near Hydrophilic and Hydrophobic Functionalized Surfaces

Abstract:

Understanding the influence of functionalized surfaces on the behavior of DNA will accelerate the development of emerging biotechnologies that rely on the self-assembly of single-stranded DNA (ssDNA) molecules into double-stranded DNA (dsDNA) duplexes. Using atomistic molecular dynamics simulations and enhanced sampling techniques, we study the effects of model hydrophobic and hydrophilic surfaces, and the effects of temperature, on the conformations and stability of ssDNA and dsDNA oligomers. These simulations capture the expected interactions between ssDNA and the two surfaces (e.g. hydrogen bonds, hydrophobic interactions), and we quantify the thermodynamics governing the conformations of ssDNA oligomers, including contributions from counterions and water molecules. Then, we study the interactions of dsDNA oligomers with the hydrophilic and hydrophobic surfaces to examine how DNA-surface interactions may influence the stability of nanostructures constructed from hybridized DNA strands.

387120 Invited Talk: Networked Structure and Properties of Lignin-Based Thermoplastics

Abstract:

Our recent study demonstrates a significant performance enhancement in formulations of lignin-based thermoplastics. Lignin, a biomass-derived polyphenolic material, is highly branched in structure and amorphous in nature. Its chemical structure depends on biomass type and method of extraction. In this work we have investigated network formation in lignin or its plasticized derivatives to formulate various melt-processible plastics. Understanding network formation in lignin polyblends is critical for tailoring their mechanical properties. In this work, the lignin phase was crosslinked by a reactive melt-mixing route both in the presence and absence of plasticizers. It was observed that the reactivity, processibility, and mechanical properties of lignin-modified polymer resins depend on the chemical architecture of lignin. The lignin-derived polymer formulations obtained had ultimate elongation >100% and mechanical strength >15 MPa. Depending on the melt-mixing temperature and sequence, the plasticizer and reinforcing agents offer synergistic compatibility in the polyblend. Successful use of lignin bioplastics can significantly enhance biorefinery and pulping industry economics.

380378 Crosslinked Fiber Production Via Simultaneous Centrifugal Spinning and Thiol-Ene Photopolymerization

Abstract:

Current synthetic fiber manufacturing technologies typically use either solvent or heat to transform a solid preformed polymer into a liquid before applying a force to draw the liquid into fiber. While the use of solvent poses concerns regarding process safety, environmental impact, and solvent recovery, the use of heat may lead to polymer degradation and excessive energy consumption. To address these critical problems, here we present an alternative fiber manufacturing method that encompasses extruding a monomer solution using centrifugal forcespinning while simultaneously polymerizing the monomers into solid fiber using UV initiated thiol-ene chemistry. This method not only negates the use of both heat and solvent, but also produces fibers that are highly crosslinked, mechanically robust, and thermally stable. In this process, the balance between curing kinetics, fiber flight time, and solution viscoelasticity is essential for fiber formation. An operating diagram is developed to show how the intricate interplay of these key factors lead to the formation of smooth fibers and other undesirable fiber defects such as beads-on-string, fused fibers, non-uniform fibers, and spray. In addition, the influence of processing parameters such as solution viscosity, spin speed, and orifice diameter on fiber diameter will be addressed.

384219 Effect of Network Structure on the Mechanical Properties of Poly(dicyclopentadiene): A Molecular Simulation Study

Abstract:

Dicyclopentadiene (DCPD) has attracted attention as an inexpensive industrial byproduct that can be catalytically cross-linked into a robust polymer network that is resistant to corrosion and impact. Depending on the type of catalyst and on reaction conditions, the mechanical and thermal properties of cross-linked DCPD vary, likely due to differences in the nanoscale structure of the network (e.g., cross-link density, ratio of cis/trans isomers, cross-linking mechanism, local heterogeneity). Using atomistic molecular dynamics simulations, we investigate how several characteristics of the nanoscale network structure affect the macroscale thermal and mechanical properties of cross-linked poly(DCPD).

389957 Structure and Transport Properties of Highly Resilient Networks

Abstract:

Polymer networks are critically important for numerous applications including soft biomaterials, adhesives, coatings, elastomers, and gel-based materials for energy storage. One long-standing challenge these materials present lies in understanding the role of network defects, such as dangling ends and loops, developed during cross-linking. These defects can negatively impact the physical, mechanical, and transport properties of the gel. Here we report the structure and transport properties of chemically cross-linked poly(ethylene glycol) (PEG) and PEG-PDMS gels formed through a unique cross-linking scheme designed to minimize defects in the network. Small-angle neutron scattering (SANS) was utilized to investigate the network structures of gels in both D_2O and d-DMF. The highly resilient mechanical properties of these systems, results from SANS, and diffusivity of small molecules in these gels all suggest that the gels have a low incidence of defects.

Strain Recovery in Dual Cross-linked Polymer Grafted Nanoparticle Networks With Slip and Catch Bonds

Balaji V.S. Iyer, Victor V. Yashin, Matthew J. Hamer and Anna C. Balazs
Swanson School of Engineering, University of Pittsburgh, Pittsburgh, PA-15261

Via computer simulations, we investigate the effect of mechanochemical reactions on the strain-recovery in networks of dual cross-linked polymer grafted nanoparticles. The individual nanoparticles are composed of a rigid core and a corona of grafted polymers that encompass reactive end groups. With the overlap of the coronas on adjacent particles, the reactive end groups form the labile bonds, and some number of permanent bonds is present in the system. The labile bonds are assumed to be of either "slip" or "catch" type that differ in dependence of the rate of rupture on force. In slip and permanent bonds, the rupture rate is accelerated by the applied force and is described by the Bell model. In contrast, the rupture rate of the catch bonds may both increase and decrease depending on the rate of deformation and the force exerted on the bond. We capture the catch bond behavior by considering bond rupture from two bound states and force-induced transitions between the states. We model the tensile deformation and strain recovery during relaxation after release of the tensile force. The existing labile bonds break and new labile bonds form in the course of deformation and hence, a damaged material is "resuscitated" both in terms of strain recovery and network re-formation through relaxation over long time. We show that the force-dependent behavior of catch bonds can be harnessed to obtain tough networks with enhanced recovery and reduced hysteresis.

Reactive Extrusion process and elasticity in sol gel polymers

S.K. Ahuja

Abstract:

Styrene acrylate and polyesters were reactive extruded in twin screw extruder Werner Pfleiderer 30mm Twin Screw Extruder, Type ZSK 30 and in Werner Pfleiderer ZSK 70, with organic peroxide as cross-linker. The extrudates were crushed into small particles and they were compressed into discs which were characterized for dynamic viscoelastic modulus and complex viscosity using TA instrument of Rheometrics Spectrometer. The strain amplitude was kept at or below 1%. Screw speeds, feed rate and barrel temperature were varied as processing variables while material variable was the variation in peroxide concentration that resulted in different concentration of gel in the composite. The results show that as the gel concentration is increased, Weissenberg number (Wi) increases Threshold in screw speed and peroxide concentration was investigated finding conditions when mixing was poor, had lower melt flow rate and when plug flow occurred.

Introduction

Twin screw extruder (TSE) is a versatile melt mixing equipment for polymer processing especially in reactive extrusion, compounding, and in controlling sol gel process. The screw elements contribute to conveying and kneading enables optimizing dispersive and distributive mixing for reactive extrusion. The residence Time distribution (RTD), residence volume distribution (RVD) and residence rotation distribution (RRD) are powerful considerations in characterizing extruder performance. Residence Time Distribution can be predicted from input variables such as moisture content, screw speed, nozzle diameter and barrel temperature. Increase of rotation speed results in significant increase of filler dispersion, whereas an increase of throughput resulted in worse dispersion. Beside these machine parameters, the design of the screw can further promote the filler dispersion especially when using distributive screw configurations containing mixing elements. Viscoelastic nature of the polymer must be taken into account in analyzing extrusion flows. The characteristic relaxation time λ of the polymer has proved to be the key quantity. When λ is small and viscous effects dominate, the RVD model approach is applicable. When λ is large, the elastic effects of the polymer may dominate under typical flow conditions(1)..

Co-rotating intermeshing twin screw extruders are the most widely used in reactive extrusion, because it is possible to build the screw profile and the barrel configuration. The feed rate is controlled by volumetric or gravimetric feeder(s). Consequently, the screws work mostly partially filled, which entails some additional benefits: i) it is possible to insert along the barrel different feeding ports, either for liquids or for solids, ii) devolatilization is easy in zones with a low filling ratio, iii) the screw profile can be divided into successive independent sections, with specialized functions (polymer feeding and melting, injection of reagents, mixing, reaction development, devolatilization, pumping and shaping···) and iv) the residence time and the level of thermomechanical stresses are relatively easy to control; they have superior mixing capacities, both in distributive and dispersive terms. Mixing can be controlled by the geometry of certain screw sections (in general, blocks of staggered kneading discs and by the operating conditions; the screws can rotate at high screw speeds (up to more than 1500 rpm, thus promoting high dispersion levels whilst guaranteeing high production rate. moisture content, While in some experiments feed rate and not screw speed while in others screw speed, barrel temperature and nozzle diameter had significant effects on RTD (2)

Experimental.The twin screw is allowed to rotate at about 300 revolutions per minute, the temperature along barrel 5 is maintained at about 70° C. in the first barrel section, 120° C. upstream, and 90° C.downstream, and the die pressure is about 50 pounds per square inch. When crosslinking in a batch internal melt mixing device, the residence time is preferably in the range of about 10 seconds to about 5 minutes. The rotational speed of a rotor in the device is preferably about 10 to about 500 revolutions per minute. Dry mixed powders of styrene acrylate copolymer with divinyl benzene or with unsaturated polyester containing cross-linker benzoyl peroxide were put in co-rotating Twin Screw Extruder and the rod extrudates were crushed and micronized into small particles. These particles were made into discs and were characterized by subjecting them into sinusoidal deformation in a rheometer. Details of the extrusion process are given in patents (3,4).

RTDs for the extrusion process were determined by a reflectance method, Reflectance probes were placed past the tips of the screws just before the die and mixing zones. Premixed pellets containing titanium dioxide were added to the feed and the response can be considered an impulse response. The data were collected at a sampling rate of different frequencies from 60 Hz to 20 Hz. Average of three replicates were taken filtered, and normalized. Shear rate in the extruder is defined as (4)

$$\dot{\gamma} = \frac{\pi D N}{h}$$

---[1]

where D is the screw diameter, N is the screw speed and h is the channel depth. Weissenberg number is the product of relaxation time and shear and reduces to

$$Wi = \lambda\dot{\gamma} = \left(\lambda\frac{\pi D}{h}\right)N$$

---[2]

Discussion and Results

The effect of the mass throughput and screw speed on the scaled-up operation is found that when the lower mass throughput is used for the smaller extruder keeping a constant screw speed, the predicted residence times of extrusion for the larger extruders are lower, than those corresponding to the reference device, and a converse situation occurs for the higher implemented value of the mass throughput. Also, in general terms, the higher increase of the reaction temperature on the scaled-up operation corresponds to the lower mass throughputs and higher screw speeds specified for the reference extruder. For more restrictive flow conditions, the final molecular weights and polydispersity are lower in the case of the screw element of the larger extruder.

It is found that the distributive mixing ability of the reactive flow is mainly related to the specified mass throughput and almost independent of the specified peroxide concentration for a particular extruder size. For the analyzed screw elements, the conveying element corresponding to the small size extruder shows a slightly better distributive mixing performance. The special type of screw element consists of screws rotating at different speeds which have different cross sections. In this case, the outer and inner diameters of both the special and the conventional type of screw elements are specified to be the same. The distributive mixing capabilities is found to be independent of the specified peroxide concentrations but dependent on the mass flow rate. It is speculated from the simulation results, from both the transient- as well as the steady-state flow conditions, that the screw element with the special design would yield lower final values of the polydispersity and molecular weight. Depending on the configuration of screw element improved distributive mixing capabilities as well as a wider RTD is found (5). In Figure,1, the coefficient of expansion and free volume increases up to 20% gel and as the gel concentration is increased beyond 20%,both free volume and coefficient of expansion decrease from densification. Our experiments show indicated in Figures 2 and 3 that the effect of gel results in greater changes in elastic modulus than viscous modulus resulting in lowering of tan delta as the sinusoidal oscillating shearing frequency is low.

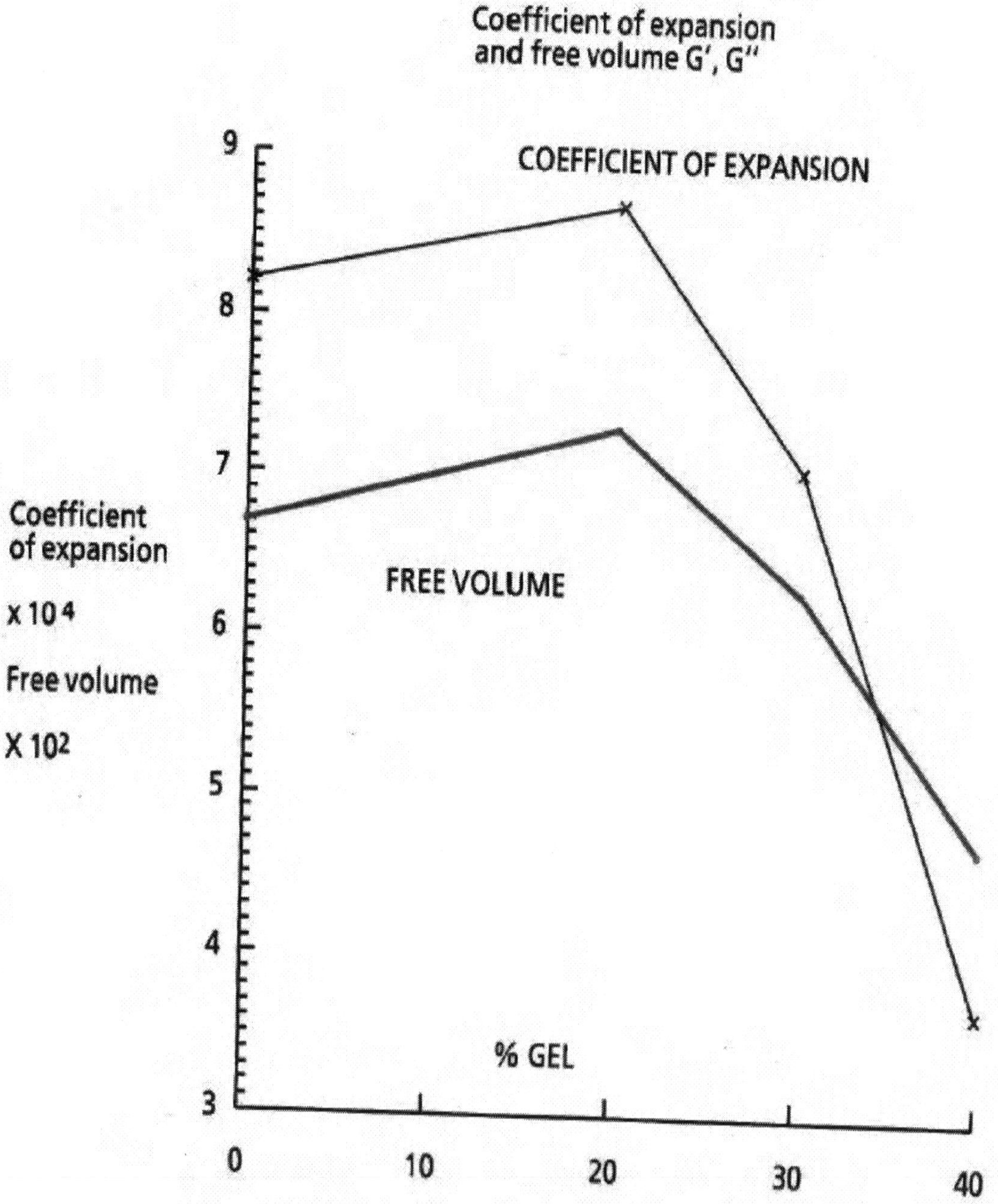

Fig 1 Free volume and Coefficient of expansion as function of percentage of gel

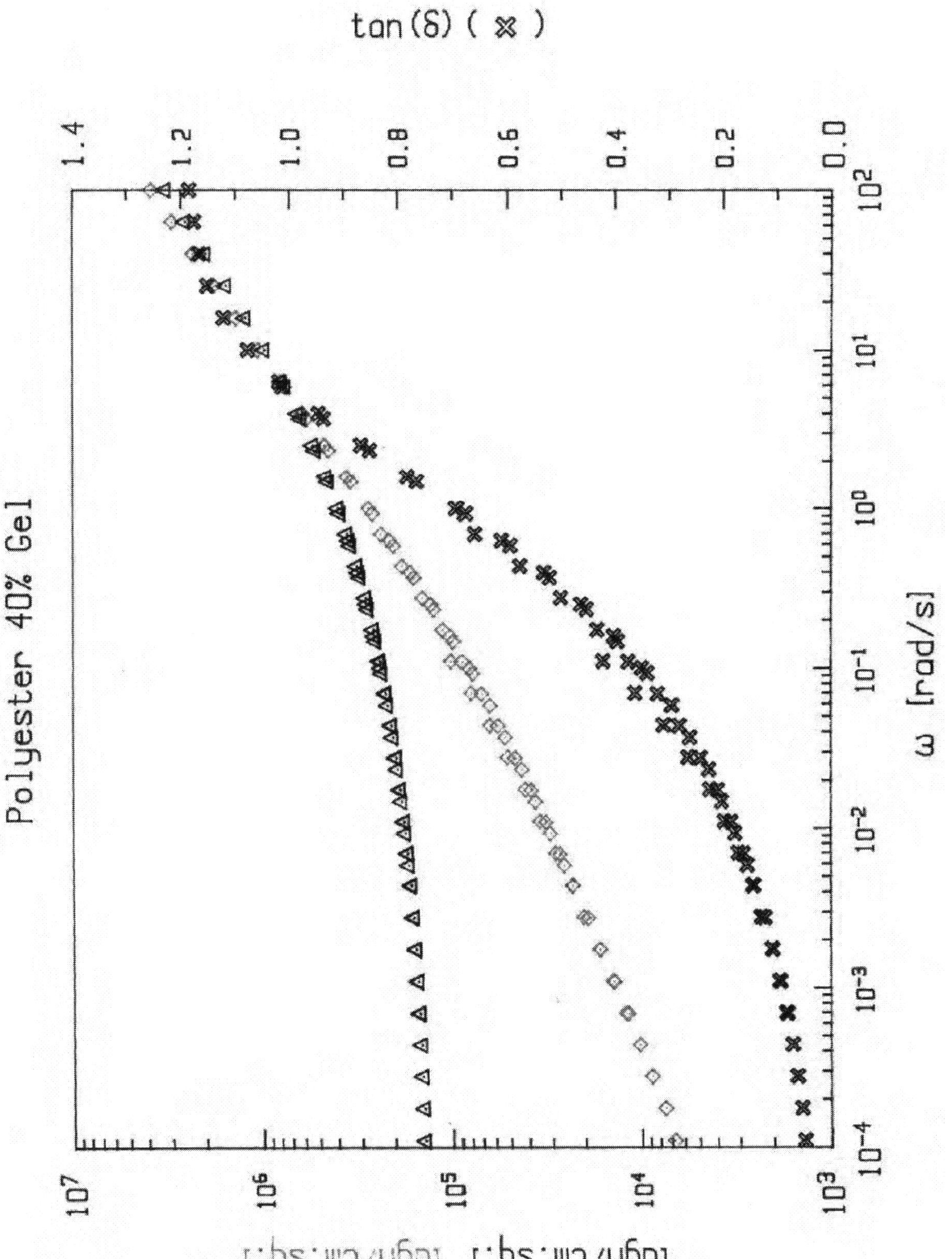
Polyester 40% Gel
tan(δ) (✱)
ω [rad/s]
G' (∇) [dyn/cm.sq.]
G" (◇) [dyn/cm.sq.]

Figure 2 Time temperature superposition shifted curves of polyester gel at 120C.

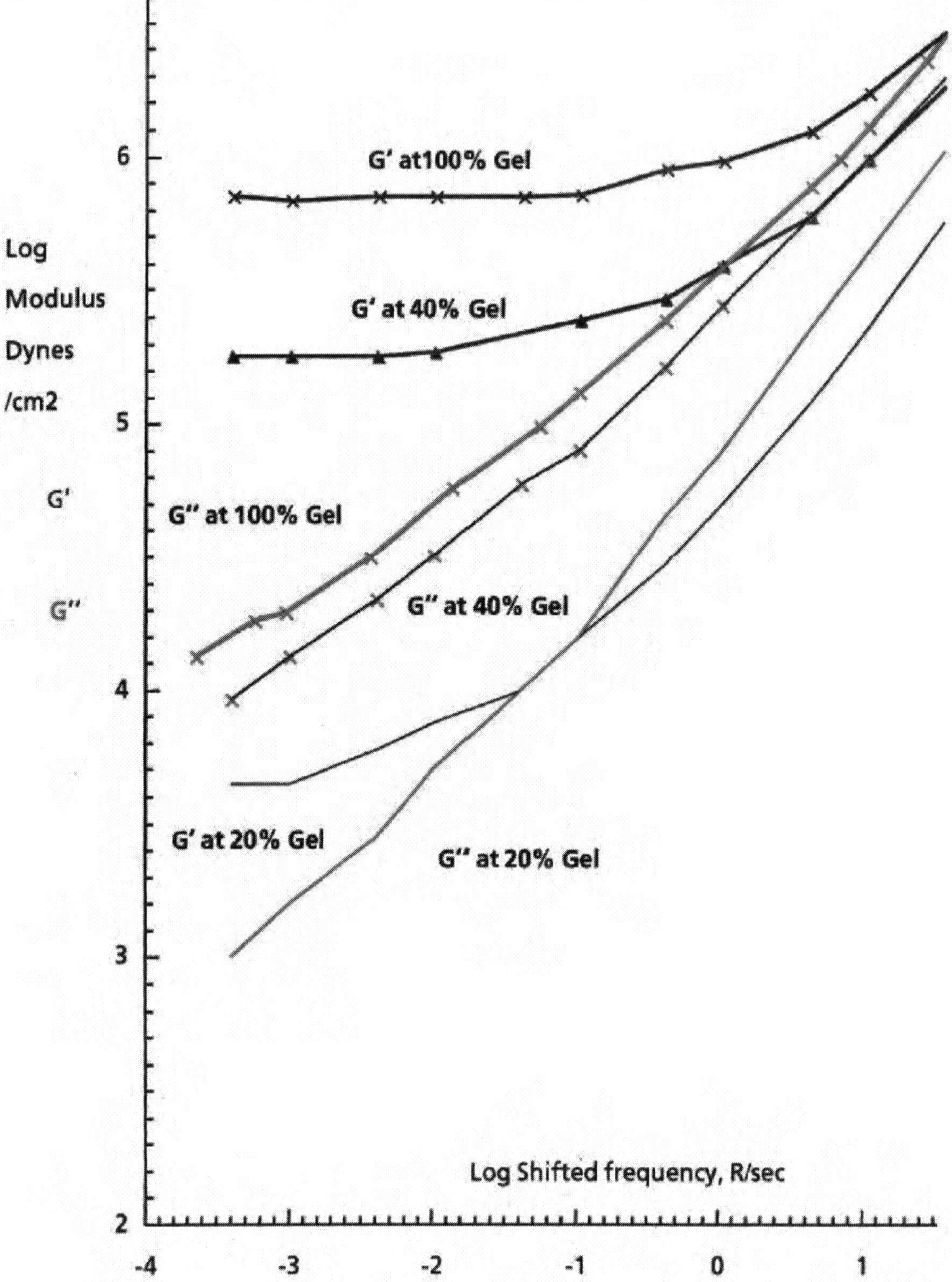

Figure 3 Elastic and viscous modulus as function of shifted frequency for different gel concentrations

In ZSK 30,the effect of increasing screw speed from 100 RPM to 300RPM was to increase the reaction temperature which increased gel concentration. Beyond 300RPM, excessive frictional heating took place resuling in inhomogeneous sol gel structures and frequent plug flow.In scale up from ZSK30 to ZSK 70,temperature,a combination of different levels of peroxide concentrations and feed rates at a constant screw speed were tried. In the second factor, peroxide level was kept constant but screw speed and feed rate was varied. In order to get to the same gel concentration, screw speed had to decreased from 300RPM to 200RPM to correspond to same thermal time.

Conclusions

Weissenberg number(elasticity) appears to be a better indicator than viscosity in relating to processing parameters such as Residual Time Distribution, feed rate and screw configuration., Scale up requires several considerations besides residence time via factorial design i.e., thermal time.

Refrences:

1. Gao J, Walsh G, Bigio D, Briber R. Wetzel M. Residence-time distribution model for twin-screw extruders. AIChE J. 1999;45:2541-2549.
2. Puaux, J.P., Bozga, G., Ainser, A., 2000. Residence time distribution in a corotating screw extruder. Chemical Engineering Science, 55, 1641-1651.

3 Mahabadi et al. United States Patent: 5,571,655, (1996)

4 Odell et al. Reative Melt Mixing Processes United States Patent 6, 114,076(2000)

5 Estanislao O. Rodríguez, Numerical Simulations of Reactive Extrusion in Twin Screw Extruders, (2009) Ph. .D, Thesis

384562 Investigation of Mechanical Properties of a Triblock Copolymer Gel Using Cavitation Rheology and Laos

Abstract:

Soft materials often display non-linear mechanical behavior at large deformation, i.e., these materials either stiffen or soften with the application of strain. Here, we use cavitation rheology, an experimental technique for measuring mechanical properties of soft materials, to investigate the deformation behavior of a swollen, physically cross-linked polymer gel. The gel was obtained by dissolving poly methylmethacrylate-poly n-butylacrylate-poly methylmethacrylate (PMMA-PnBA-PMMA) in a mid-block selective solvent, 2-ethyl 1-hexanol. Different polymer volume fractions (5%, 7%, and 10% v/v) were selected to study the effect of crosslink density on the critical pressure of cavitation. Cavitation is a snap-through instability caused by the pressurization of a defect inside a gel. Similar mode of instability was observed for the cavitation experiments conducted at 6 °C and 22 °C. The experiments were also conducted at different compression rates and a decrease of the critical pressure was observed at a low compression rate likely caused by the relaxation of the polymer chains. Both neo-Hookean and Gent models were applied to capture the elastic contribution in pressure at different expansion ratios before and at instability point. In addition to cavitation experiments, large amplitude oscillatory shear (LAOS) was used to characterize the gel deformation behavior. The Fourier coefficients and the Chebyshev coefficients were calculated from the stress responses to elucidate the strain stiffening behavior of these gels.

Probing the impact of heterogeneous structure on mechanics and transport in PEG hydrogels

Paula Malo de Molina, Sahger Lad and Matthew E. Helgeson

Department of Chemical Engineering, University of California Santa Barbara, Santa Barbara, CA 93106

The two most important properties of polymer hydrogels for most common applications are their mechanical strength and transport of molecular and biological species. Typically, these properties depend critically on the crosslink density or "mesh size" of the polymer network, as extracted from macroscopic measurements assuming a homogeneous polymer structure. However, most polymer hydrogel networks are structurally heterogeneous due to polymer architecture and chemical heterogeneity. Although heterogeneity in hydrogels is widely recognized, its specific role in determining mechanical and transport properties is not well understood.

To fill this gap, we present detailed studies of structure, mechanics, and diffusion in two types of model poly(ethylene glycol) (PEG) hydrogels: heterogeneous PEG diacrylate networks and homogeneous tetra-PEG networks. Small angle neutron scattering (SANS) measurements on PEG hydrogels with varying precursor molecular weight and concentration allow us to differentiate the molecular-scale structure of the gels. In the case of PEG diacrylate gels, we present a new model whereby the heterogeneous structure is comprised of a fractal network of star-like polymer blobs, giving a more accurate description of the than the often-used Ornstein-Zernike model. This allows for better insight into the impact of inhomogeneity on the mechanical and transport properties of the gels. Specifically, fluorescence recovery after photobleaching (FRAP) and differential dynamic microscopy (DDM) measurements allow us to quantify the size-dependent diffusivity of both molecular and colloidal probes, and mechanical measurements are used to determine the scaling behavior of elastic properties with the gel composition. In both cases, we find that non-homogeneous PEG hydrogels exhibit qualitatively different behavior than their homogeneous counterparts. Our results have significant implications for the characterization and design of hydrogels for many applications in biotechnology and separations.

386078 Modulating the Thermal and Light Responsive Behavior of Spirobenzopyran-Based Poly(N-isopropylacrylamide) Hydrogels

Abstract:

Spirobenzopyrans are an interesting class of compounds which have recently attracted significant attention in the field of "smart" material development, owing to their diverse stimuli behavior, e.g. light, temperature, stress and pH. However, the functionality of such material alone is often limited, and thus the integration of these with other functional polymers is desired. Here we demonstrate the capability of synthesizing light and temperature responsive spirobenzopyran-functionalized poly(N-isopropylacrylamide) (PNIPAAm) hydrogels. To improve the light responsive behavior and broaden the spectra of light absorbed by the hydrogel, particularly pertaining to the visible range, different derivatives of sprirobenzopyran were integrated in the hydrogel along with a pore forming agent. The thermoresponsive behavior of these hydrogels was studied with differential scanning calorimetry and then compared to traditional PNIPAAm hydrogels. The effect of surfactants and solvents on the transition temperature of such hydrogels was also investigated. The integration of such multi-functional compounds in hydrogels could have interesting implications for the development of novel "smart" devices.

392454 Advanced Biorefinery Towards a Sustainable Bioeconomy: Value-Added Biobased Materials

Abstract:

In both the biochemical and thermo-chemical bioconversion process we generate a considerable amount of downstream products. The value-added uses of co-products and bio-products from biofuel industries are termed as "Advanced Biorefinery" in this presentation. Dry milling corn ethanol industries produce distillers' grains and carbon dioxide whereas lingo-cellulosic ethanol industries produce lignin and carbon dioxide as their downstream products. The thermo-chemical conversion of biomass produces bio-oil, syngas and bio-chars. The value-added biobased materials from co-products/byproducts will not only help in enhancing the profitable return of the related renewable biofuel industries but also can help in substituting certain petroleum-based products in the manufacturing sectors. These new biobased materials can be cost-competitive with added advantages of eco-friendliness. The recent developments of such biobased materials for potential uses in consumer products, packaging and interior automotive parts will be discussed.

Acknowledgement: The financial support from the Natural Sciences and Engineering Research Council (NSERC) Canada for the Discovery Grant; NSERC NCE AUTO21; the Ontario Research Fund (ORF) Research Excellence (RE) Round-4 from the Ontario Ministry of Economic Development and Innovation (MEDI); the Ontario Ministry of Agriculture and Food (OMAF) and Ministry of Rural Affairs (MRA) - University of Guelph Bioeconomy-industrial uses research program; and OMAF-MRA New Directions Research Program to carry out this research is gratefully acknowledged

368896 Selective Conversion of Biorefinery Lignin to Dicarboxylic Acids

Abstract:

Lignin is a ubiquitous component in almost all plant biomass. Large quantities of industrial lignin are already produced annually as a waste product of the pulp and paper industry, where the vast majority is burned as a low cost fuel to provide energy for the papermaking process. The emerging biomass refinery industry will further introduce an enormous amount of lignin. Thus, there is an urgent need to develop technologies that can create new applications for biorefinery lignin. Lignin is the largest source of renewable material with an aromatic skeleton.

Majority of previous in area have been focused on depolymerizing lignin to low molecular weight phenolic compounds (LMWPC). However, one potential group of chemicals has been overlooked among oxidative conversion products: dicarboxylic acids (DCA). DCA such as muconic, maleic, and succinic acids are important and highly valuable industrial chemicals and intermediates used in many industries including the biopolymer, pharmaceutical, and food additives industries. Current commercial dicarboxylic acids are all produced from petroleum based feedstock. Developing a green route to producing DCA from renewable biomass lignin will be of a prime interest to both chemical and biomass conversion industry. Oxidative aromatic ring cleavage to yield muconic acid and its derivatives is a well-recognized reaction. Although DCA and their derivatives have been detected in the reaction mixture from oxidative delignification of wood pulp, there has not been successful attempt to produce DCA from biomass lignin. Discovering a selective route to convert lignin to DCA will pave a new avenue towards lignin utilization.

This paper demonstrates a novel route to creating dicarboxylic acids from biorefinery lignin via ring cleavage by chalcopyrite catalyzed oxidation in a highly selective process. We reported a method of selective production of DCA via chalcopyrite ($CuFeS_2$) catalysed oxidation of biorefinery lignins in the presence of hydrogen peroxide (H_2O_2) under mild reaction conditions. A clear two stage reaction mechanisms is proposed, including depolymerization and oxidative aromatic ring cleavage, as explaination of the products profiles through lignin and model compounds study. Up to 95% selectivity of stable, saturated dicarboxylic acids was obtained from several types of biorefinery lignin and model compounds under mild, environmentally friendly reaction conditions.

Making activated carbon from low/un-hydrolyzed biomass residue

Chen Li and Sandeep Kumar

Department of Civil & Environmental Engineering, Old Dominion University, Norfolk, VA, USA

Corresponding author: skumar@odu.edu

Bioethanol has become one of the most promising biofuels for short to medium alternative to fossil transportation fuels. During production of bioethanol, a large amount of un-hydrolyzed solids (UHS) is generated. The UHS resulting from enzymatic hydrolysis during cellulosic ethanol production is a wet and carbon-rich residue. It mainly contains lignin and unconverted sugars depending upon the types of biomass pretreatment used in the earlier stages. Several tons of UHS are expected to be generated during cellulosic ethanol/butanol production from lignocellulosic biomass. The efficient utilization of this residue will have a positive impact in making the biofuels cost-competitive. In recent few decades, adsorption has become a well-established technique to remove pollutants from water and air. Activated carbon is becoming the prevailing adsorbent for the purification of water with low pollutant concentration due to their advantage in high internal surface areas and active surfaces. However, the major drawbacks of commercially available carbons are expensive, so more efficient and low-cost ones now are needed for water industry[1].

Recently, there have been lots of efforts on research in developing recycling processes to produce activated carbon from biomass waste, for instance, thermal conversion of pyrolysis and gasification processes [2]. There are already many kinds of biomass waste such as apricot stone shells, bean pods, and corn stover are being used to produce activated carbon, [3]. Under certain process conditions several activated carbons were produced from agricultural residues having BET surface areas ranging between 250 and 2410 m^2/g and pore volumes of 0.022 and 91.4 cm^3 /g [4].

In this study, the UHS generated after enzymatic scarification of corn stover is used as a feedstock for activated carbon production. Hydrothermal processing of biomass under subcritical water condition is another promising green technology to extract useful sugar from biomass. Large amount of solid residue is recovered after the hydrothermal treatment. The other feedstock including corn stover and pinewood are first subjected to hydrothermal hydrolysis to extract sugars and the residues are used for activated carbon production. Our objective is to hydrolyze maximum sugar in aqueous phase and use the low sugars containing solid residue for activated carbon production. In this study, solid residues after hydrothermal hydrolysis are named low-hydrolyzed biomass residue (LHR).

Two-step experiment is designed. The first step is focused on maximizing hydrothermal sugar extraction result and producing LHR. During the hydrothermal hydrolysis, reaction time between 3 mins to 8 mins, reaction temperature between 180°C to 260 ^{O}C, and sulfuric acid concentration (upto 1 v/v%) are studied. The second step is focused on making activated carbon form UHS and LHR. The high temperature chemical activation is our primary method to make activated carbon. The effect of reaction time (30 mins to 120 mins) and temperature (750°C to 800 ^{O}C) on the BET surface area are studied. Different chemicals such as $ZnCl_2$, H_3PO_4 and K_2CO_3 are tested. The quality evaluation matrix of activated carbon included the analysis of BET surface areas, pore volumes, iodine value, and burn off percentage. One of the experiments result showed that the surface area of activated carbon from LHR was 658m^2/g and the pore volume was 0.13cc/g. The objective of the study is to optimize the process for high quality activated carbon production from LHR and UHS.

References

1. Budinova, T., et al., *Biomass waste-derived activated carbon for the removal of arsenic and manganese ions from aqueous solutions.* Applied Surface Science, 2009. **255**(8): p. 4650-4657.
2. Daifullah, A. and B. Girgis, *Removal of some substituted phenols by activated carbon obtained from agricultural waste.* Water research, 1998. **32**(4): p. 1169-1177.
3. Tay, T., S. Ucar, and S. Karagöz, *Preparation and characterization of activated carbon from waste biomass.* Journal of Hazardous Materials, 2009. **165**(1): p. 481-485.
4. Ioannidou, O. and A. Zabaniotou, *Agricultural residues as precursors for activated carbon production—a review.* Renewable and Sustainable Energy Reviews, 2007. **11**(9): p. 1966-2005.

392455 Reactive Blending of Protein Rich Meals and Biodegradable Polymers for Green Packaging

Abstract:

Plastics are highly valued materials because of their low cost and extraordinary versatility and they constitute the second largest application of petroleum next to energy. Efforts to replace commodity plastics based packaging materials with biodegradable counterparts are usually hampered by strong completion with cheaper petroleum-based polymers and limited performance. Blending of relatively expensive biodegradable polymers such as poly (butylene succinate) (PBS), poly (butylene adipate co-terephthalate) (PBAT), and polycaprolactone (PCL) with low value and cheap agricultural feedstock can reduce the cost of the resulting materials while maintaining their biodegradability. For example, the use of less expensive protein rich meals such as soybean meal, canola meal, corn meal etc. obtained from the food and biodiesel industry as co-reactants of biodegradable polymers in green packaging applications is expected to reduce the cost competitiveness barrier. Reactive extrusion processing could be one of the potential techniques with a lot of promise for future development of biodegradable plastics. This is because, it is a continuous and cost effective polymer processing techniques that allows melt blending and various other chemical reactions including (co) polymerization, compatibilizing, grafting, branching and functionalization without the need for solvents. The direct use of relatively cheap soy protein meal in the reactive blending, and study of compatibilization chemistry to develop a cost competitive green packaging films are the main objectives of this research. The soy meal underwent through three types of fermentations to de-structurize the ingredients of the soy meal including simple sugars and cellulose that hinders film making behavior. Post-fermentation, the protein and pectin concentrations in the soy meal have significantly increased. Also, plastic films made out of the blending of the fermented soymeal and PBAT exhibited a significant performance improvement.

Acknowledgement: The financial supports from 1) the Ontario Ministry of Agriculture and Food (OMAF) and Ministry of Rural Affairs (MRA) - University of Guelph Bioeconomy-industrial uses research program; 2) OMAF-MRA New Directions Research Program; and 3) the Ontario Research Fund (ORF) Research Excellence (RE) Round-4 from the Ontario Ministry of Economic Development and Innovation (MEDI) and Natural Sciences and Engineering Research Council (NSERC) Discovery grant individual (to Mohanty) to carry out this research are gratefully acknowledged.

387114 Separate Hydrolysis and Fermentation of Untreated and Pretreated Miscanthus Cake to Produce Ethanol

Abstract:

The use of lignocellulosic biomass to produce biofuel will add value to land and reduce emissions of greenhouse gases by replacing petroleum products. Valuable co-products derived from fractionation of miscanthus give the resulting fibrous fraction an economic advantage as a feedtock for ethanol production. Freshly harvestedmiscanthus from the North Carolina A and T farm was dewatered using press, whereby miscanthus is separated into a fiber-rich cake and a nutrient-rich juice. The production of cellulosic ethanol from miscanthus fibers with or without alkali(NaOH) pretreatment were investigated by this work using separate hydrolysis and fermentation (SHF). The objective was to compare the effect of alkali pretreatment on the miscanthus samples with regards to ethanol yield.

The pretreatment of miscanthus with NaOH was carried out to assess the effect of various NaOH concentrations on the enzymatic digestibility of the wet miscanthus cake. The concentrations of NaOH investigated were 0, 1, 4 and 7%, Enzymatic hydrolysis of the miscanthus cake was carried out using a cocktail of enzymes comprising of cellulase (NS 50013), β-glucosidase (NS 50010) and hemicellulose (NS 22002) at a temperature of 50oC for 96 hours. This was followed by fermentation using Escherichia coli (E. coli) at a temperature of 32oC, also for 96 hours.

The hydrolysis products investigated using High Performance Liquid Chromatography (HPLC) included glucose, xylose, arabinose and uronic acids. Samples were further analyzed for fermentation products including ethanol, glucose, xylose, arabinose and acetic acid. The results show a steady increase in fermentable sugars from the enzymatic hydrolysis over the time period, and also an increase in ethanol produced from the pretreated miscanthus as compared to the untreated miscanthus with the overall effectiveness of the alkali pretreatment being a function of the NaOH concentration. This suggests effective delignification leading to enhanced ethanol production.

Chemical activation of biomass-derived hydrochar for environmental applications: Influence of processing conditions during hydrothermal treatment

Akshay Jain[a], *Rajasekhar Balasubramanian*[b] & *M.P. Srinivasan**[a]

[a] Department of Chemical and Biomolecular Engineering, National University of Singapore, Singapore 117585, Singapore.

[b] Department of Civil and Environmental Engineering, National University of Singapore, Singapore 117585, Singapore

Abstract

Coconut shell-derived hydrochar has been used as precursors to produce mesoporous carbonaceous adsorbents with high mesopore surface area (570- 1150 m^2/g). The textural properties of the activated carbon can be easily tuned by modifying the concentration of reactants (coconut shell, $ZnCl_2$ and water). The tuning of the stoichiometry of reactants governed the formation of oxygenated functional groups on hydrochar and $ZnCl_2$ solubility and thus its availability at the time of activation. Up to 36% increase in mesopore area and 18% increase in BET area could be achieved by the use of hydrothermal pre-treatment compared to conventional soaking. A combination of hydrothermal pre- treatment using H_2O_2 followed by the hydrothermal treatment using $ZnCl_2$ at different concentrations resulted in a significant improvement in the mesopore surface area (from 570 to 1150 m^2/g) and BET surface area (from 1720 to 2440 m^2/g) of activated carbon. Increase in mesopore area is validated by the increased adsorption capacity of organic dyes.

Use of simple and resource-/energy- efficient pre-treatment processes will be the key to designing high-surface area mesoporous adsorbents for environmental applications such as wastewater treatment and gas separation.

The use of Amazonian biomass for small biorefineries serving small communities in the rainforest

Angela A. González, Juan C. Higuita, Paola A. Moreno, Carlos A. Cardona
agonzalezvil@unal.edu.co , jchiguitav@unal.edu.co, palmorenonu@unal.edu.co , ccardonaal@unal.edu.co

Instituto de Biotecnología y Agroindustria
Universidad Nacional de Colombia
Manizales - Colombia

14AIChE
Annual Meeting, Atlanta, GA

Amazon Region

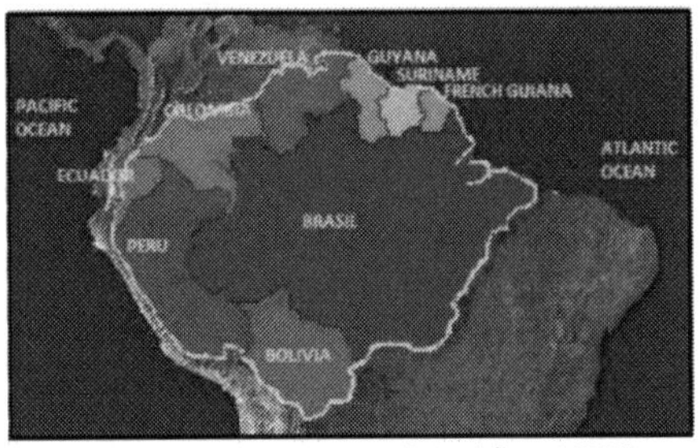

Fig 1. Amazon Region. Adapted from WWF

Table 1. The Amazon Region area per country

Country	Area (km^2)	%
Bolivia	724.000	9,76
Brazil	5,034,740	67,91
Colombia	477,274	6,43
Ecuador	115,613	1,56
Guyana	214,960	2,89
Perú	651,440	8,79
Suriname	142,800	1,93
Venezuela	53,000	0,73
Total	7,413,827	100

Source from: Palacios (2011)

- 56% of the world rainforest.
- 28 million people (4.5 inhabitants / km^2).
- 385 indigenous groups.

It represents the 40% South America territory.

Colombian Amazon Region

Economy: Natural resources exploitation.

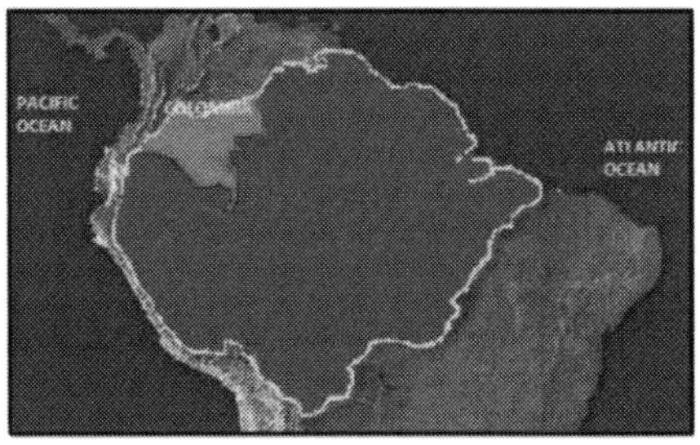

Fig 2. Colombian Amazon Region. Adapted from WWF

Colombian Amazon Region: 483.164 km² (41% of the Country area)

By 2011 the region recorded poverty rates of 35,6%

Source from: https://www.dane.gov.co/files/investigaciones/condiciones_vida/pobreza/cp_pobreza_2011.pdf

The economy of Colombian Amazon Region

Natural resources exploitation.

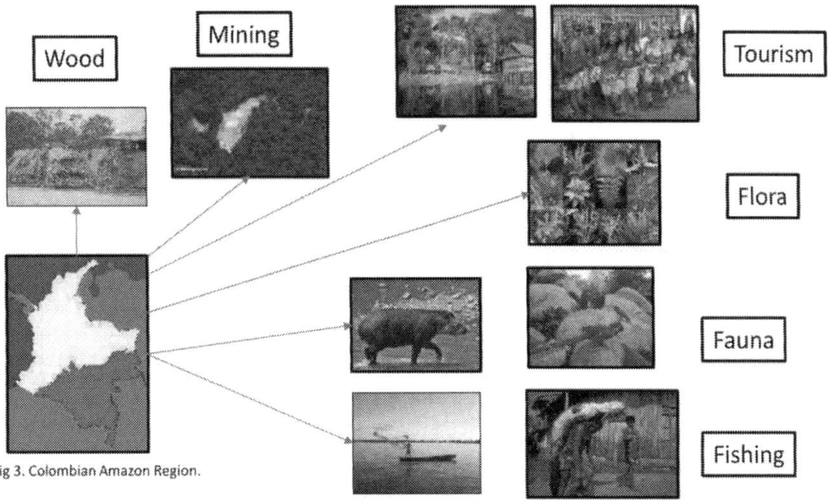

Fig 3. Colombian Amazon Region.

Source from: www.lugaresde viaje com, http://geografiadelamazonas.blogspot.com.es /, www.keywordpictures.com, www.andigena.org, http://es.mongabay.com/, www.imeditores.com

Entering to the Colombian Amazon

Fig 3. Amazonas - Colombia.

- Amazonas: Provence at south of Colombia.
- Connectivity limited.
- Air and water communication. NO roads.

Area: 109,665 km² (equivalent to Cuba)
9,6% of the Continental Territory
PIB participation: 0,1%
90,000 inhabitants (0,8 inhabitant/km²)

Entering to the Colombian Amazon

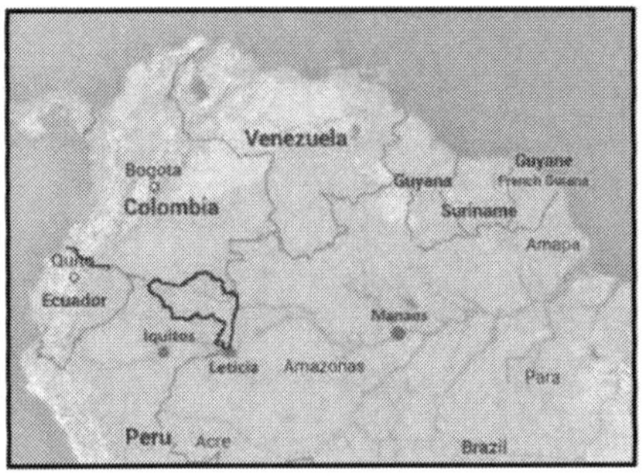

Fig 3. Amazonas - Colombia.

- Potential of production:
  - Agro: fruits, medicinal plants.
  - Agroindustry: cacao, juices, pulps, vitamins...
  - Wood: cedar, mahogany.
  - Services: tourism, environmental.

The study case

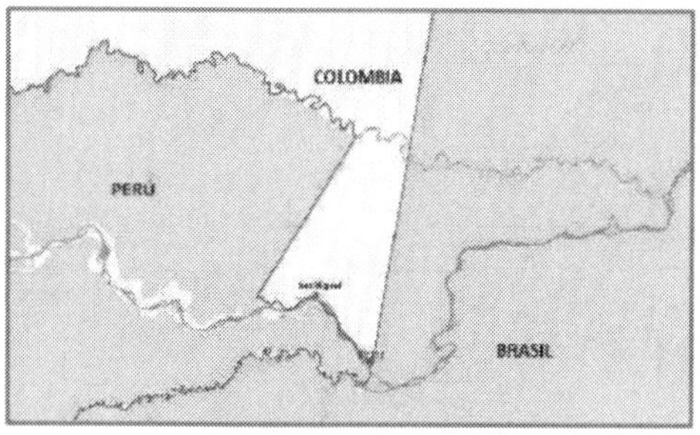

Fig 4. San Miguel Amazonas Colombia

- San Miguel.
- Small community on the Amazon River two hours from Leticia.
- Fluvial communication.
- 300 inhabitants.

Local initiative of production

Cacaoteros del Amazonas is an association of ten natives families (Tikuna) that they have been producing cacao since 4 years ago.

Currently they market handmade cocoa.

Cocoa production

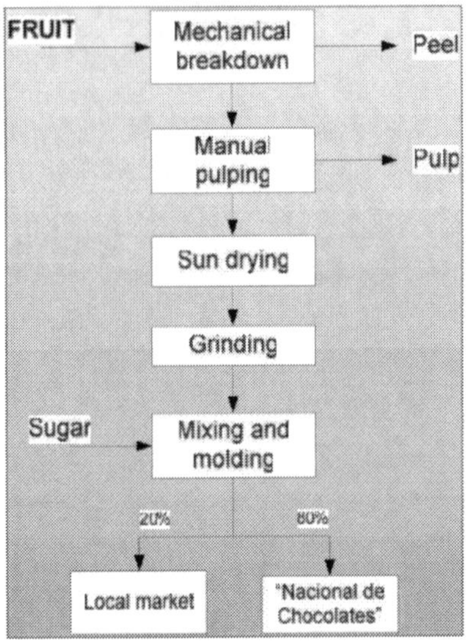

Sale Price:
2,81 US/kg product cocoa

Fig 6. Production process flow sheet according to the Cacaoteros

"Amazonian Cocoa, an economic alternative for Amazonas"

Research Group in Chemical, Catalytic and Biotechnological Processes
Universidad Nacional de Colombia Manizales Campus 9

Objective

To evaluate technical, economic and environmental the process for obtaining products with add-value from *Theobroma cacao* in the Colombian Amazon context.

Research Group in Chemical, Catalytic and Biotechnological Processes
Universidad Nacional de Colombia Manizales Campus 10

Cocoa – *Theobroma cacao*

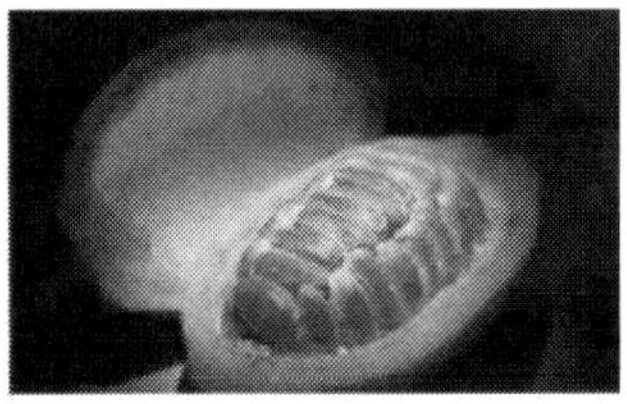

Table 2. Cocoa composition used for simulation process

Feature	Peel[1]	Pulp[2]	Seed[3]
Percentage	75,0	4,0	21,0
Moisture (%)	8,5	80,0	4,6
Ash (%)	6,7	0,5	3,6
Protein (%)	8,6	0,1	20,1
Lipid (%)	1,5	N.A.	54,5
Carbohydrates (%)	32,3	13,0	11,4
Lignin (%)	21,4	N.A.	5,8
Fiber (%)	36,6	1,1	N.A.

1 Vriesmann et. al. (2011).
2 Liendo et. al. (1997).
3 Sánchez V.A. (2007) .

Methodology

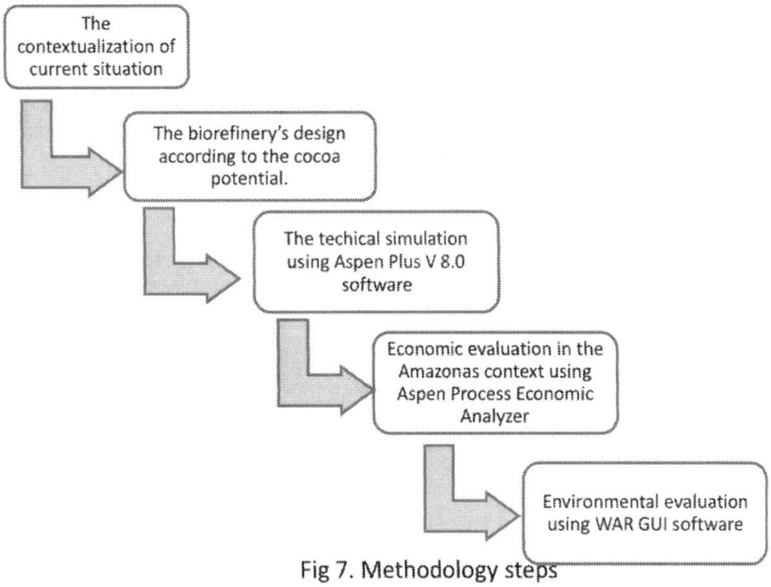

Fig 7. Methodology steps

Research Group in Chemical, Catalytic and Biotechnological Processes
Universidad Nacional de Colombia Manizales Campus

Results

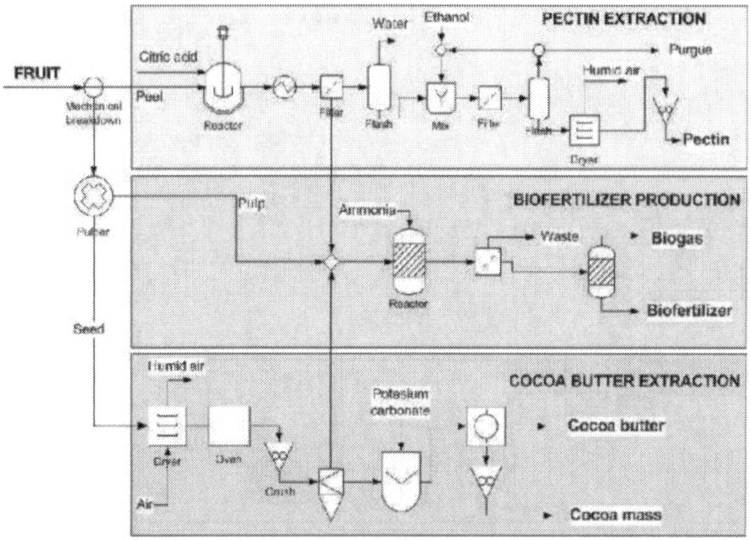

Fig. 8 Biorefinery flow sheet

Scenarios

According to agricultural production capacity of San Miguel.

Table 3. Proposed scenarios

SCENARIO	RAW MATERIAL (TON/DAY)
1	0,5
2	1
3	2

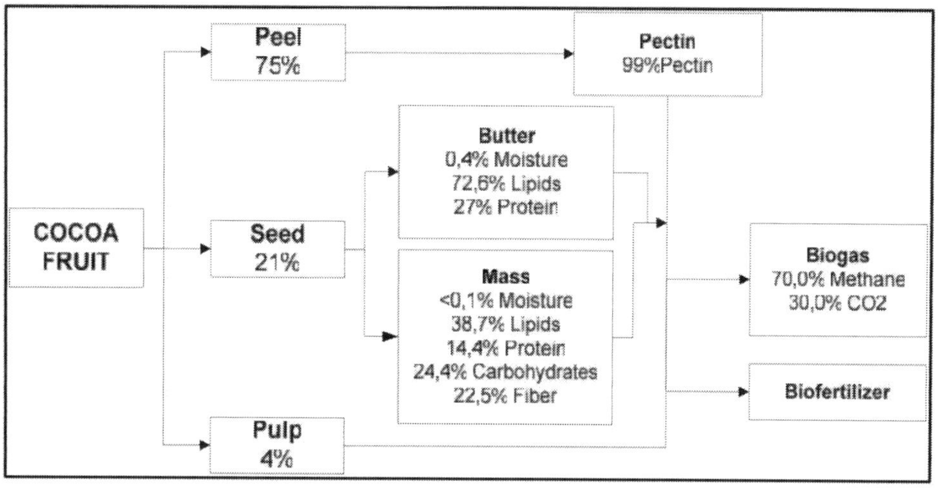

Fig. 9 Quality of cocoa biorefinery products

Research Group in Chemical, Catalytic and Biotechnological Processes
Universidad Nacional de Colombia Manizales Campus 15

Distribution cost

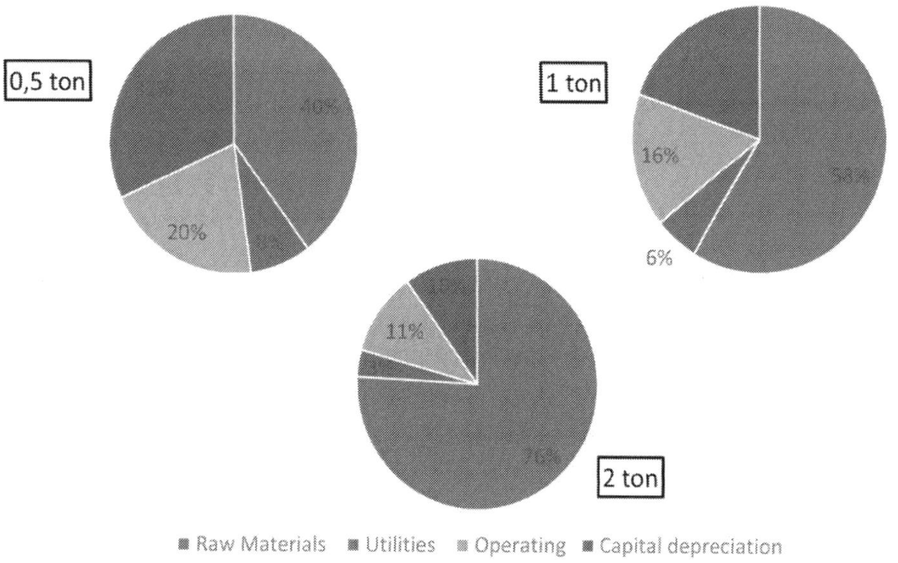

Fig. 10 Total distribution cost of Cacao Biorefinery

Research Group in Chemical, Catalytic and Biotechnological Processes
Universidad Nacional de Colombia Manizales Campus 16

Distribution cost

Table 4. Cost of production per product

SCENARIO	PECTIN	BUTTER	MASS	BIOGAS	BIOFERTILIZER
	USD/kg	USD/kg	USD/kg	USD/kg	USD/kg
0,5 ton	37,47	0,48	0,60	13,10	2,12
1 ton	28,77	0,20	0,25	9,99	1,62
2 ton	27,87	0,15	0,19	8,42	1,37

Environmental analysis

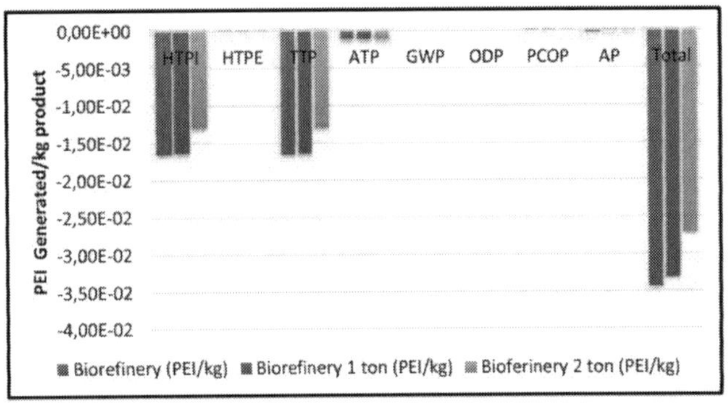

Fig. 11 Biorefinery Potential Environmental Impact generated

HTPI: Human Toxicity Potential by Ingestion.
HTPE: Human Toxicity Potential by Exposure.
TTP: Terrestrial Toxicity Potential.
ATP: Aquatic Toxicity Potential.
GWP: Global Warming Potential.
ODP: Ozone Depletion Potential.
PCOP: Smog Formation Potential.
AP: Acidification Potential.

Conclusions

According this study it was possible conclude:

- The production under biorefinery concept is a possibility to take in count in the Amazonas Provence.

- The simulation methodology can help in the research of productive Amazon initiatives because it reduces the pressure over ecosystems and allows to evaluate previously the technical, economic and environmental feasibility of the project.

REFERENCES

- Palacios, Pablo. 2011. "Devenir Amazónico. La Transformación Del Paisaje Amazónico." In *Amazonia Colombiana: Imaginarios Y Realidades. Cátedra Jorge Eliécer Gaitán*, edited by Catalina Pérez Niño Juan Alvaro Echeverri. Bogotá.
- Vriesmann L.L., de Mello A., R.D, de Oliveira P. C.L. (2011) *Cacao pod husk (Theobroma cacao L.): Composition and hot-water-soluble pectins.* Industrial Crops and Products 34 1173-1181.
- Barazarte, H., Sangronis, E., Unai, E. (2008) *La cáscara de cacao (Theobroma cacao L.): una posible fuente comercial.* Archivos latinoamericanos de nutrición.Vol. 58 N°1.
- Redgwell, R., Trovato, V., Merinat, S., Curti, D., Hediger, S., Manez, A., (2003) *Dietary fibre in cocoa Shell: characterisation of component polysaccharides.* Food chemistry 81, 103-112.
- Liendo R., Padilla F.C., Quintana A., (1997) *Characterization of cocoa butter extracted from Criollo cultivars of Theobroma cacao L.* Food Research International. Vol. 30. No. 9 727 – 731.
- Sánchez C., V.A., (2007) *Caracterización organoléptica del cacao (Theobroma cacao L.), para la selección de árboles con perfiles de sabor de interés comercial.* Tesis de grado. Universidad Técnica estatal de Quevedo.
- Cardona, C.A. & Cerón, I.X., (2012) Conceptual design of sustainable biorefineries - Metabolites and bioenergy production in the Amazoon. The World Congress on Industrial Biotechnology and Bioprocessing, Orlando, Florida.
- Moncada, J., El-Halwagi, M. M., & Cardona, C. A. (2013). *Bioresource Technology, 135*(0), 533-543.
- Rincón, L., Becerra, L., Moncada, J., & Cardona, C. (2014). *Waste and Biomass Valorization, 5*(2), 189-198.
- Rincón, L. E., Hernández, V., & Cardona, C. A. (2014). *Process Biochemistry, 49*(3), 474-489.

Corresponding Autor: Carlos Ariel Cardona Alzate.
Km 9 Vía al Aeropuerto La Nubia, Manizales, Caldas
tel. +57 68879400 ext 55354
e-mail: ccardonaal@unal.edu.co
Universidad Nacional de Colombia. Sede Manizales

Research Group in Chemical, Catalytic and Biotechnological Processes
Universidad Nacional de Colombia Manizales Campus

371010 Understanding Dissolution As a Mode of Capacity Fade in Lithium-Ion Batteries

Abstract:

$LiNi_{1/3}Co_{1/3}Mn_{1/3}O_2$ (NCM) positive electrode material is commonly used in lithium ion batteries because of its favorable characteristics including high reversible capacity, power capability, and thermal stability. However, dissolution of transition metals from the positive electrodes is a concern in lithium ion batteries containing NCM positive electrode as it has been implicated in a rise in impedance and capacity fade of cells. In a lithium ion cell, there are many competing effects that contribute to the growth of the solid electrolyte interphase (SEI) and capacity fade. In this work, the effects of metal dissolution on full cells employing NCM positive electrode and graphite negative electrode was studied by accelerating the degradation due to dissolution process. The effects of metal dissolution on cell capacity, cyclic stability, and impedance were investigated by using electrochemical techniques. Postmortem analyses were carried out on electrodes to study the morphological changes due to the transition metal.

376894 Nitrogen-Doped Graphene Paper As Electrodes for High-Performance Lithium/Dissolved Polysulfide Batteries

Abstract:

Graphene sheets with adsorbed polydiallyldimethylammonium chloride can be transformed into nitrogen-doped, free-standing paper (NGP) by high temperature treatment. The NGP are conducting, and can be used as binder-free electrodes for the state-of-the-art Li/dissolved polysulfide batteries. When tested in the catholyte configuration of soluble polysulfides in a liquid electrolyte, the NGP showed rather stable cycling behaviour and high charge storage capacities, achieving specific capacities of ~1000 mAh/g (based on sulfur) after 100 cycles and coulombic efficiencies >98%. These values were significantly better than undoped graphene papers (GP). The NGP material and the electrochemical cells were characterized with XRD, XPS, TGA, AFM, electron microscopy, and Raman and impedance spectroscopy before and after cycling. The results showed stronger binding of sulfur to NGP relative to GP, which were consistent with modelling results from DFT calculations, which further showed that the pyrrolic and pyridinic N were more effective than quaternary N in interaction with lithium polysulfides. The modelling results were substantiated experimentally by thermal conversion of pyrrolic and pyridinic N into quarternary N with accompanying degradation in battery performance. Reasons for the improved performance of nitrogen-doped material were proposed.

382209 High Capacity, Stable Silicon/Carbon Anodes for Lithium-Ion Batteries Prepared Using Emulsion-Templated Directed Assembly

Abstract:

Silicon (Si) is a promising candidate for lithium ion battery anodes because of its high theoretical capacity. However, the large volume changes during lithiation/delithiation cycles result in pulverization of Si, leading to rapid fading of capacity. Here, we report a simple fabrication technique that is designed to overcome many of the limitations that deter more widespread adoption of Si based anodes. We confine Si nanoparticles in the oil phase of an oil-in-water emulsion stabilized by carbon black (CB). These CB nanoparticles are both oil- and water-wettable. The hydrophilic/hydrophobic balance for the CB nanoparticles also causes them to form a network in the continuous aqueous phase. Upon drying this emulsion on a current collector, the CB particles located at the surfaces of the emulsion droplets form mesoporous cages that loosely encapsulate the Si particles that were in the oil. The CB particles that were in the aqueous phase form a conducting network connected to the CB cages. The space within the cages allows for Si particle expansion without transmitting stresses to the surrounding carbon network. Half-cell experiments using this Si/CB anode architecture show a specifi c capacity of 1300 mAh/g Si + C and a Coulombic efficiency of 97.4% after 50 cycles. Emulsion-templating is a simple, inexpensive processing strategy that directs Si and conducts CB particles to desired spatial locations for superior performance of anodes in lithium ion batteries.

Nanostructured $LiMnPO_4$ Matrix as Flexible High-rate Cathode Materials

Feng Jiao

Center for Catalytic Science & Technology, Department of Chemical Engineering, University of Delaware, Newark, DE, 19716 USA. Fax: +1-302-831-1048; Tel: +1-302-831-3679; E-mail: jiao@udel.edu

Electronic systems that can cover large areas on flexible substrates have received increasing attention in the past few years, because they enable classes of applications that lie outside those easily-addressed with wafer-based electronics[1-3]. As the critical component for flexible electronics, a flexible energy storage device is crucial because all the functionalities onboard rely on the power supply. The performance of the current generation is greatly limited by the electrode materials, and many efforts have been devoted to development of high performance flexible batteries[4-7]_ENREF_1. Recently, compliant materials on curvilinear surfaces, such as carbon nanotubes[8], carbon nanofibers[9], graphene[10], metal oxide-based nanowires[11], and slurry-type mixtures of nanostructured active materials[12], have been explored as flexible battery electrodes. However, these studies have mainly focused on anode electrode materials, and there has been very limited progress on cathode materials for flexible batteries. It is critical to explore cathode electrode materials with the potential to couple with the developed flexible anodes for device applications.

One-dimensional (1D) electrode materials have attracted much attention recently because of their potential application in flexible battery technology. Many 1D anode materials based on carbon and metal oxides have been synthesized for flexible batteries, but only limited studies on the cathode side have been conducted. In this talk, we will present the synthesis of high-rate cathode electrodes based on Mg-modified $LiMnPO_4$ nanofibers. The nanofibers are embedded inside a nanostructured conducting carbon matrix to enhance their electronic conductivity and structural integrity while retaining flexibility. As a result, a 50% increase in capacity is obtained, achieving an outstanding performance of 135 mAh g^{-1} at a C/10 rate (15 mA g^{-1}). This nanostructured Mg-modified $LiMnPO_4$ matrix also exhibited superior rate capability and much better cycleability compared to its $LiMnPO_4$ counterpart. Even at a high charge/discharge rate of 5C (750 mA g^{-1}), 80% of the capacity (107 mAh g^{-1}) is still retained, representing, to the best of our knowledge, the best rate performance for $LiMnPO_4$-based electrodes. More importantly, such a superior rate capability is achieved with an excellent cycleability (no capacity fading for 200 deep cycles).

Reference:

1. Lu, Q., Hutchings, G. S., Zhou, Y., Xin, H., Zheng, H., & Jiao, F. Nanostructured flexible Mg-modified $LiMnPO_4$ matrix as high-rate cathode materials for Li-ion batteries. J. Mater. Chem. A, 2014, 2, 6368-6373.

384806 Performance of Ternary Ionic Liquid–Lithium Salt Mixtures for Li-O2 Batteries

Abstract:

Identification of stable electrolyte materials which can operate in an oxygen rich atmosphere is a major challenge for the development of the Li-O_2 battery. Conceivably, a hydrophobic room temperature ionic liquid (IL) electrolyte, because of its low volatility, can prevent the hydrolysis reaction at the Li anode and drying of the O_2 electrode during cycling. Pyrrolidinium IL can create a passivation layer on the Li anode and is stable against highly reactive superoxide O_2^-. However, it has a high viscosity and low O_2 solubility that is detrimental to cell performance. Ternary IL mixtures comprised of both high ionically conductive and less viscous imidazolium (i.e., 1-butyl-3-methylimidazolium bis(trifluoromethylsulfonyl imide (BMIMTFSI)) and pyrrolidinium (i.e., *N*-butyl-*N*-methylpyrrolidinium bis(trifluoromethanesulfonyl) imide (PYR_{14}TFSI)) based ILs are tested as the electrolyte for Li-O_2 batteries. 4:1 (80/20 v/v%) BMIMTFSI : PYR_{14}TFSI based ternary mixtures show excellent cyclability with 330 $mAhg^{-1}$ over 50 cycles at a current density of 0.1 $mA.cm^{-2}$, much better than the pure BMIMTFSI, PYR_{14}TFSI and TEGDME electrolytes. The results clearly show the beneficial synergic effect of the two ionic liquids on the electrochemical properties of the mixtures.

389884 The Effect of Solid-Electrolyte Interphase on Lithium/Sulfur Liquid Battery Performance

Abstract:

The Effect of Solid-Electrolyte Interphase on Lithium/Sulfur Liquid Battery Performance

Negar Mosavati, Steven O. Salley and K. Y. Simon Ng

Department of Chemical Engineering and Materials Science, Wayne State University, Detroit, MI 48202

Lithium-Sulfur batteries (Li-S) are one of the most attractive power sources and promising candidates to replace the conventional state-of-the-art lithium ion batteries. Sulfur as a cathode active material is a by-product of petroleum and has the advantages of high natural abundance and low cost. Although a Li-S battery can deliver almost 5 fold higher theoretical capacity than lithium ion batteries (1675 mAh/g sulfur) with a theoretical specific energy (~2600 Wh/kg), it suffers from several drawbacks. The major barrier is the redox shuttle reactions in which lithium polysulfides are reduced by the Li anode and fail to reoxidize back completely to elemental sulfur at the cathode side upon charging. This leads to low active material utilization, poor capacity retention, high self-discharge and a short cycle life.

In current research, we have introduced a dense ionic conductive passivation solid electrolyte interface (SEI) layer on the Li anode, which effectively prevents soluble polysulfide penetration into the Li bulk and subsequent parasitic reactions. As a result, the redox shuttle reaction and the sulfur active material loss on the Li anode surface decreases. Consequently, the overall Li-S battery electrochemical performance and capacity retention improved substantially. In order to accomplish a profound understanding of the chemical and electrochemical aspects that influence SEI layer formation and overall Li-S battery performance, X-ray diffraction (XRD), X-ray photoelectron spectroscopy (XPS), scanning electron microscopy (SEM), electrochemical impedance spectroscopy (EIS), cyclic voltammetry (CV) and charge-discharge measurements have been conducted under controlled conditions.

Facile Synthesis of Ultrathin-shell Graphene Hollow Spheres for High-performance Lithium-ion Batteries

Dandan Cai, Liang-Xin Ding, Suqing Wang, Haihui Wang**

School of Chemistry & Chemical Engineering, South China University of Technology, Wushan Road, Guangzhou 510640, China

Lithium-ion batteries (LIBs) have been widely used in portable electronic devices due to their high energy and power density and long cycle life. As the further application in hybrid electric vehicles and electric vehicles, their performance urgently needs to be further improved.[1-3] Herein, we report a facile and efficient template assisted method without surfactant to fabricate ultrathin-shell and interconnected graphene hollow spheres as an advanced anode material for LIBs. The designed hollow sphere structure can act as an effectively barrier to suppress agglomeration of graphene nanosheets. Compared with other conventional hollow carbon spheres, the ultrathin-shell and interconnected graphene hollow spheres not only retain the superior performance of graphene nanosheets but also show the following additional advantages: i) ultra-thin graphene shell can significantly reduce the diffusion length for lithium ions during insertion/extraction process; ii) the superior electronic conductivity and interconnected network structure of graphene hollow spheres can provide a highway network for fast electron transport; iii) the wrinkled and porous structure of graphene shell can endow abundant lithium storage sites. As expected, the as-prepared graphene hollow spheres exhibit very high reversible lithium storage capacity (824.2 mA h g^{-1} at 100 mA g^{-1}) and excellent rate capability (249.3 mA h g^{-1} at 5000 mA g^{-1}), showing a great promising anode material for the next generation high-performance LIBs.

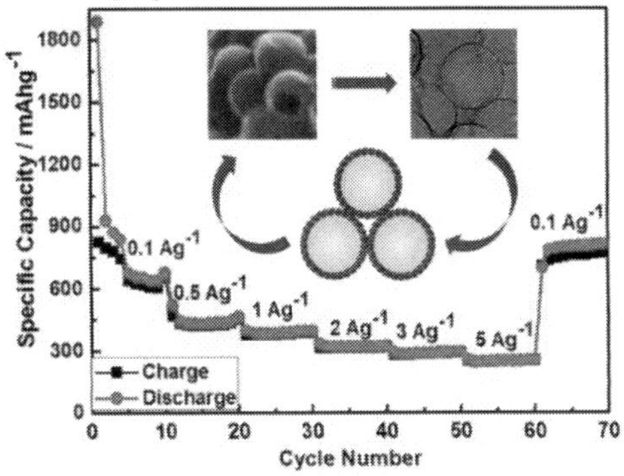

Figure 1. Rate capabilities of the graphene hollow spheres at various current densities from 100 mA g^{-1} to 5000 mA g^{-1}.

References

1. Aricò, A. S.; Bruce, P.; Scrosati, B.; Tarascon, J.-M.; Van Schalkwijk, W. Nanostructured Materials for Advanced Energy Conversion and Storage Devices. *Nat. Mater.* **2005**, *4*, 366-377.
2. Choi, N. S.; Chen, Z.; Freunberger, S. A.; Ji, X.; Sun, Y. K.; Amine, K.; Yushin, G.; Nazar, L. F.; Cho, J.; Bruce, P. G. Challenges Facing Lithium Batteries and Electrical Double-Layer Capacitors. *Angew. Chem. Int. Ed.* **2012**, *51*, 2-33.
3. Kaskhedikar, N. A.; Maier, J. Lithium Storage in Carbon Nanostructures. *Adv. Mater.* **2009**, *21*, 2664-2680.

***Corresponding Author**: lxding@scut.edu.cn, hhwang@scut.edu.cn. Tel.:+86-20-87110131.

363570 Solvothermal Route Based in Situ Carbonization to Fe3O4@C As Anode Material for Lithium Ion Battery

Abstract:

Magnetite (Fe_3O_4) is a very promising candidate as an anode material in rechargeable lithium ion battery (LIBs) because of its ability to store up to eight Li per formula unit via reversible reactions, resulting in a high theoretical capacity of 924 mAh g^{-1}. Moreover, advantages such as low cost, ease of synthesis and environmental friendliness make it a promising candidate for large scale commercial applications for LIBs. However, Fe_3O_4 suffers from poor conductivity, large volume change and voltage hysteresis during the electrochemical reaction, which strongly limits its practical application. For the fact that carbon in its own feature exhibits several interesting properties, including electronic conductivity, high surface area, tunable porous structure, etc., the combination of Fe_3O_4nanostructures and different forms of carbon materials show promising potential in the preparation of composite anode materials.

In this work, a high pressure and temperature based solvothermal route was developed for the synthesis of Fe_3O_4 with carbon as a composite anode for LIBs. The carbon in products was a result of the in situ carbonization of organic components under high pressure (24.0 MPa) and temperature (350 °C). Composites with different amounts of carbon were prepared by annealing the solvothermal products at different temperature. Taking advantage of the high theoretical capacity of Fe_3O_4 and favorable characteristics of carbon, a capacity of about 610 mAh g^{-1} after 100 cycles was achieved for the composite with 54.6% carbon. The carbon amount depended electrochemical performance was also investigated. The Fe_3O_4@C composite can be used as an alternative anode material and the introduced synthetic strategy may provide further insights into the preparation of inorganic oxides coupled with carbon via in situ carbonization of organic components.

390498 High Capacity Silicon Nitride-Based Composite Anode for Lithium Ion Batteries

Abstract:

One of the busiest areas of research in Li ion batteries is the development of stable high energy anode materials. The main material of choice is Si, due to its high theoretical specific capacity (4,200 mAh/g), which is more than ten times the value of current anode material, graphite (370 mAh/g). One of the main challenges for Si as an anode is its huge volume expansion (~400%) upon the formation of silicon-lithium alloys, resulting in the pulverization of electrodes and performance degradation. In this paper, we report the synthesis and performance of a Si nitride (SiN_x) based composite anode material that demonstrates a two-step reaction mechanism during cycling: reversible conversion reaction then alloying reaction. These steps are necessary to divert the full alloying reaction to follow reaction pathways that improve the cycling characteristics of the anode material. Elucidation of the resulting performance was performed using spectroscopy, microscopy and electrochemical techniques. The potential of this material that displays better volume stress management and thus improves the overall electrochemical cycling stability is presented.

Gradually Softening Hydrogels for Understanding Cell Behavior During Fibrosis Regression

Steven R. Caliari[1], Gi yun Lee[1], Maryna Perepelyuk[2], Rebecca G. Wells[2], Jason A. Burdick[1]
[1]Department of Bioengineering, University of Pennsylvania
[2]Department of Medicine, University of Pennsylvania

Introduction: There is a critical unmet need for hydrogel systems that can recapitulate salient features of native tissues during processes such as morphogenesis, aging, and disease. The extracellular matrix (ECM) of tissues is dynamic, presenting an evolving set of mechanical and compositional cues to resident cells. *In vivo* studies suggest that during the regression of liver fibrosis hepatic stellate cells (HSCs), the primary source of fibrogenic myofibroblasts, either undergo apoptosis or acquire a phenotype that appears similar to healthy quiescent HSCs but is characterized by abnormal gene expression and unusually rapid myofibroblastic differentiation after a repeat insult [1]. However, the role of liver mechanics in mediating this process is unknown. This work introduces a methacrylated hyaluronic acid (MeHA) hydrogel system containing a combination of stable and hydrolytically degradable crosslinks that presents a gradually softening (but not fully degradable) milieu to mimic physiological softening events such as fibrosis regression.

Results: MeHA hydrogels (3 wt%) were crosslinked with a combination of thiol-based crosslinkers via Michael-type addition. Stable crosslinks were introduced with dithiothreitol (DTT) while PEG-based thiolated crosslinkers containing ester groups [2] susceptible to hydrolysis were also incorporated. These gels displayed relatively high initial stiffness (~15-20 kPa) and gradually softened over the course of 14 days to a stable stiffness of ~2-3 kPa. These values compare favorably to fibrotic and healthy liver respectively. Rat HSCs that were mechanically primed on tissue culture plastic initially remained spread on stiff gels (both non-degradable and degradable variants) with significantly larger cell area, loss of lipid droplets, and expression of the myofibroblast marker α-SMA. However, following 14 days of culture HSCs on softening gels showed significantly reduced cell area and more lipid droplets compared to HSCs on stiff gels. Human mesenchymal stem cells (hMSCs) showed a similar response when cultured on softening gels, displaying gradually decreasing cell area (Day 1: $3110 \pm 400\ \mu m^2$, Day 7: $842 \pm 126\ \mu m^2$) and reaching a similar area to hMSCs cultured on soft static gels after 7 days ($743 \pm 86\ \mu m^2$).

Conclusions: We describe a softening hydrogel system to explore the influence of dynamic changes in mechanical properties on cell phenotype during fibrosis regression. Ongoing work is developing gel formulations with a wider range of degradation profiles and evaluating HSC and hMSC phenotype in response to different softening dynamics.

References:

1. Kisseleva, T., et al., *Myofibroblasts revert to an inactive phenotype during regression of liver fibrosis.* Proc Natl Acad Sci U S A, 2012. **109**(24): p. 9448-53.
2. Zustiak, S.P. and J.B. Leach, *Hydrolytically degradable poly(ethylene glycol) hydrogel scaffolds with tunable degradation and mechanical properties.* Biomacromolecules, 2010. **11**(5): p. 1348-57.

386120 Tunable Hydrogels to Understand the Role of the Microenvironment in Regulating Breast Cancer Dormancy and Recurrence

Abstract:

Breast cancer reoccurs in approximately 20 percent of all patients between 5 and 10 years after successful treatment of the primary tumor.[1] Many of these recurrences are believed to arise from tumor cells at distant metastatic sites that have reactivated after long periods of dormancy. Remodeling of the extracellular matrix (ECM) at these sites over time is hypothesized to trigger the release of the tumor cells from dormancy.[2] Two-dimensional (2D) and three-dimensional (3D) *in vitro* culture models have been developed to recreate some of the interactions between cancer cells and their microenvironment.[3,4] Lack of mechanical property control and differences between biochemical composition of these model systems and the native ECM make it challenging to identify key ECM signals that promote dormancy or activation. Synthetic, hydrogel-based biomaterials are well suited for the creation of a new culture model to examine cancer recurrence, allowing user-directed control over the presentation of biomechanical and biochemical cues to direct cell behavior *in vitro*.[5, 6] Further, light-mediated hydrogel formation and modification permit spatiotemporal control over the presentation of biomechanical and biochemical cues to mimic the dynamic, nonhomogeneous *in vivo* microenvironment.[7]

We are designing a hydrogel-based 3D culture platform to mimic critical mechanical and biochemical properties of different metastatic site tissues where dormancy occurs, including the bone marrow, liver, and lung. With this approach, we aim to identify cues that play a role in dormancy and reactivation of tumor cells toward the development of new therapeutics for recurrent breast cancer. Specifically, poly(ethylene glycol)(PEG)-based hydrogels have been polymerized by a radically-initiated thiol−ene chemistry using PEG-4-thiol, prepared by modification of a four-arm PEG-OH monomer with thiols,[8] and peptides modified with alloxycarbonyl-protected lysine residues to supply a biologically-orthogonal reactive functional group (−ene) for the formation of a step-growth hydrogel with ideal network structure. Hydrogels formed within 5 minutes after the application of cytocompatible doses of UV light (10 mW/cm^2, 365 nm) in the presence of the lithium acylphosphinate photoinitiator.[9] Mechanical properties of the hydrogels were measured by rheometry, and the elastic moduli of soft metastatic site tissues were achieved (~0.5-5 kPa).[7] Further, biomimetic peptides have been spatially patterned within the matrix.

To establish matrix compositions that promote cancer dormancy or activation, we first compared the effects of soft and stiff substrates coated with relevant whole ECM proteins on tumor cell dormancy, including fibronectin (FN), laminin (LN), and collagen I and IV (Col I, Col IV) that are commonly found within metastatic tissue sites. Cells were cultured on top of these soft and stiff substrates in the presence of various growth factors, including basic fibroblast growth factor (FGF2), and their continued activation or dormancy was evaluated by immunostaining for a marker of proliferation (Ki-67). Finally, moving to 3D culture, human breast cancer cells (MCF7s) and human mesenchymal stem cells (hMSCs) were encapsulated within hydrogels decorated with integrin-binding peptides and remained viable post encapsulation as assessed by a membrane integrity assay (LIVE/DEAD) and metabolic activity assay (CellTiter96). These results demonstrate the creation of a hydrogel-based synthetic ECM to investigate the role of the microenvironment in regulating cancer dormancy and recurrence.

1 A Brewster, et al. *J Natl Cancer Inst.* 2008, 100, 1179-1183.

2 J Townson, et al. *Cell Cycle*. 2006, 5, 1744-1750.

3 J Barrios, et al. *Cancer Microenviron.* 2009, 2, 33-47.

4 D Barkan, et al. *Cancer Res.* 2008, 68, 6241-6250.

5 A Kloxin, et al. *Adv Mater.* 2010, 22, 3484-3494.

6 M Lutolf, et al. *PNAS*. 2003, 100, 5413-5418.

7 C DeForest, et al. *Chem Mater.* 2010, 22, 4783-4790.

8 B Fairbanks, et al. *Macromolecules.* 2011, 44, 2444-2450.

9 B Fairbanks, et al. *Biomaterials.* 2009, 30, 6702-6707.

PEG-based Tumor Millibeads for Three-dimensional Cancer Cell Culture

Shantanu Pradhan, Jacob M Clary, Chloe S Chaudhury, Elizabeth A Lipke

Abstract:

Models that mimic the pathophysiology of human cancers are critical to advancing cancer research and drug discovery. In this research, we have developed a novel cancer model (called, 'tumor millibead model') through the use of poly(ethylene glycol)(PEG)-based hydrogel biomaterial for three-dimensional (3D) cancer cell culture, which can be potentially used in the drug discovery and development process. Whereas most current cancer research relies on two-dimensional (2D) cultured cells or animal models, the use of biomaterials to create tissue-engineered cancer models provides advantages in replicating the tumor microenvironment. Animal and 2D cell culture models have some inherent limitations in their ability to mimic the native human cancer pathophysiology. 2D models lack the complexity of the 3D extracellular microenvironment, an important player in the tumorigenic progression. Animal models are physiologically different from the microenvironmental conditions of the human body and, hence, frequently fail to provide accurate information about the future clinical efficacy of the tested drugs. Therefore, there is a clear need for the development of cancer models which can provide physiologically and clinically relevant information about drug action and efficacy. In order to address this problem, the emerging field of cancer tissue engineering is focused on incorporating the principles of cancer biology in tissue engineering to create 3D *in vitro* engineered cancer tissue through the use of hydrogel biomaterials, facilitating the investigation of tumorigenic phenomena and testing efficacy of new anti-cancer drugs.

In order to create the tumor millibead model reported in the study, we have developed a novel method for the fabrication of millimeter-scale hydrogel spheres (called 'hydrogel millibeads') using a surface-tension based, liquid-liquid dual phase system. This technique allows the creation of highly uniform, PEG-based hydrogel spheroids, which are significantly larger in size compared to PEG-based hydrogel microspheres obtained by previous fabrication methods. In addition, the ability to culture cancer cells within the hydrogel millibeads provides a unique cancer model which can mimic the millimeter-sized tumors in the human body.

Briefly, poly(ethylene glycol diacrylate) (PEGDA) precursor solution of a fixed volume (2, 5 or 10 µl) was pipetted on the surface of an oil phase solution in a Petri dish, where the precursor droplet floated at the air-oil interface due to surface tension differences between air, oil and aqueous phase. The droplets were crosslinked in presence of light of wavelength 365-700 nm for 10 seconds to form hydrogel millibeads. They were subsequently harvested and washed with DMEM to remove all oil. The hydrogel millibeads obtained were larger in size, more uniform in shape with better ease of size control and consistency between fabrication batches. The swelling behavior and diffusion characteristics of hydrogel millibeads were correlated to their initial fabrication size. Human breast cancer MCF7 cells were encapsulated within PEGDA millibeads to generate 'tumor millibeads', which were maintained in long-term 3D culture. Fluorescence imaging and SEM imaging revealed the spherical morphology and uniform distribution of cells in tumor millibeads. Live/Dead staining revealed the presence of a necrotic core of cells surrounded by healthy viable cell layers at the periphery of the millibeads, which is a prominent feature observed in many native and *in vivo* tumors.

Overall, in this study, we have demonstrated a novel technique for the fabrication of hydrogel millibeads which can support long-term 3D cancer cell culture. Important characteristics of the native tumor microenvironment, including tumor heterogeneity, necrotic core formation and 3D cancer cell morphology, are closely simulated within the tumor millibeads. In addition, use of the PEG-based hydrogel system allows for the biochemical and mechanical modifications of the model engineered tissue to match the properties of native cancer tissue in the human body. Biochemical modifications include the use of PEG-peptide or PEG-protein conjugates that allow cellular adhesion, degradation and remodeling of the engineered tissue. Mechanical modifications include tuning of the matrix stiffness to match the native stiffness of cancer tissue by controlling relative PEG concentration, crosslinking density and chain length. This engineered tumor model could be potentially used to obtain physiologically and clinically relevant information from future drug studies. The millibead fabrication technique can be scaled up to create engineered 3D cancer tissue models for high-throughput drug-testing applications. Hydrogel millibeads can also be potentially used to encapsulate suitable biomolecules or therapeutics for controlled delivery at the site of interest. Hence, PEG-based hydrogel biomaterials can significantly contribute to the creation of useful cancer models, and thereby advance cancer research, and drug discovery and development.

373647 Injectable Thermo-Sensitive Hydrogel As an Adjuvant: In Situ Modulation of Dendritic Cells for Cancer Vaccine

Abstract:

Attempts to develop cell-based cancer vaccines have shown limited efficacy due to the failure of transplanted dendritic cells (DCs) to survive long enough to reach the lymph nodes. The development of materials that can modulate DCs *in situ* to enhance antigen uptake and presentation has emerged as a novel method toward developing high quality cancer vaccines. Here, we propose a two-step hybrid strategy to produce a more robust cell-based cancer vaccine *in situ*. First, a significant number of DCs are recruited to an injectable thermo-sensitive mPEG-PLGA hydrogel through sustained release of chemo-attractants, like granulocyte-macrophage colony-stimulating factor (GM-CSF). Then, these resident DCs can be loaded with cancer antigens through the use of viral or non-viral vectors. We demonstrate that GM-CSF releasing mPEG-PLGA hydrogels successfully recruit and house DCs and macrophages, allowing subsequent introduction of antigens by vectors to activate the resident cells, thus initiating antigen presentation and boosting immune response. Moreover, this two-step hybrid strategy generates a high level of specific anti-tumor immunity demonstrated in both prophylactic and therapeutic models of melanoma. This injectable thermo-sensitive hydrogel shows great promise as an adjuvant for cancer vaccines, potentially providing a new approach for cell therapies through *in situ* modulation of cells.

377110 Novel 2.5D Cell Culture Platforms to Investigate the Role of Stiffness Gradients on Cell Migration

Abstract:

Although it is now well understood that migrating cells *in vivo* often encounter spatial variations in extracellular matrix stiffness, current research investigating cell migration focus on the use of culture platforms with static material parameters. Therefore, we developed a novel 2.5D culture platform where the cells are placed at the interface of stiff tissue culture plastic and a soft hydrogel matrix, to mimic the transitions in microenvironmental stiffness experienced by cells *in vivo*. Alginate was chosen as the hydrogel matrix because it facilitated independent investigation of both matrix stiffness gradients and cell-matrix adhesions on migration. Under these conditions, we observed cell migration from the underlying stiff substrate into alginate. Observation of migration into alginate in the presence of integrin inhibition as well as qualitative microscopic analyses suggested an adhesion-independent cell migration mode. Additionally, the observed migration was dependent on alginate matrix stiffness as well as the RhoA-ROCK-myosin-II pathway; inhibitors specifically targeting ROCK and myosin-II arrested cell migration. These studies will potentially facilitate the development of relevant *in vitro* platforms to discover therapeutic strategies against tumor cell motility and invasion.

Theoretical and Experimental Evaluation of the Influence of Material Properties on Hydrogels for Use as an Artificial Pancreas

Kaitlin M. Bratlie[1,2] and Scott Beckman[1]

[1]Department of Material Science & Engineering, Iowa State University, Ames, IA, USA
[2] Department of Chemical & Biological Engineering, Iowa State University, Ames, IA, USA

According to the American Diabetes Association, diabetes affects over 20.8 million people in the United States or about 7% of the population. Current treatments involve daily injection of insulin from recombinant human or animal sources. Controlling diabetes has proven difficult since regulation of insulin secreted by the β-cells of the pancreatic islets in response to blood glucose is a highly dynamic process. We are particularly interested in developing materials for reducing host-responses for polymer encapsulated insulin producing β -cells, which has potential as a type I diabetes therapeutic. Encapsulating β -cells in polymers to evade host-responses was first proposed by Lim & Sun[1] using alginate, a natural polymer derived from seaweed. Elliott *et al.*[2] found that alginate encapsulated porcine islets remained viable 9.5 years after transplantation. However, they were encapsulated in fibrous tissue, resulting in poor diffusion of insulin into the patient's blood stream. In this work, we have focused on altering the phenotype of macrophages (MΦs) through interactions with our hydrogel such that they release less tumor necrosis factor-α (TNF-α). We are also interested in using these materials to prevent transport of cytotoxic molecules while allowing insulin to freely diffuse.

MΦs were polarized to the M1 and M2 phenotypes and markers of the Th1 response – TNF-α – and the Th2 response – interleukin-10 – were monitored upon exposure to polymers modified with various functional groups. These materials were able to reduce the amount of TNF-α produced by the cells up to 50%. The ability of these materials to increase the permselectivity of the hydrogels was also explored. Finite element analysis, Monte Carlo, and analytic theoretic methods were employed to develop a fundamental understanding of how these materials improved permselectivity. Finally, we examined the material properties to uncover correlations between altering the phenotype of MΦ and the transport properties of the materials through principal component and partial least squares analytics.

References:

1. Lim, F. & Sun, A. M. Microencapsulated islets as bioartificial endocrine pancreas. *Science (80-.).* **210,** 908–910 (1980).

2. Elliott, R. B. *et al.* Live encapsulated porcine islets from a type 1 diabetic patient 9.5 yr after xenotransplantation. *Xenotransplantation* **14,** 157–61 (2007).

Visible-Light-Induced Synthesis of pH Responsive Hydrogels with Crosslink Density Gradients

Ozlem Cevik, Seda Kizilel

Chemical and Biological Engineering, Koc University, 34450 Sariyer, Istanbul, Turkey

Stimuli responsive hydrogels mimic biological systems and sense changes in environmental cues, such as pH. Previous studies on pH sensitive hydrogels mainly focused on the synthesis under UV light, where bulk polymerization was considered. Here, we developed the synthesis of pH responsive composite hydrogel networks via surface initiated photopolymerization. The significant differences between the pH responsive hydrogels synthesized here and previous UV light photopolymerized pH sensitive hydrogel networks are: (i) the use of xanthene dye instead of Irgacure in previous systems, (ii) incorporation of nanodomains within the network, (iii) presence of gradients of permeability within the structure. Formation of the pH responsive layer is based on methacrylic acid grafted with poly(ethylene glycol) tethers P(MAA-g-EG) as macromer, eosin Y as photoinitiator, and triethanolamine (TEA) as co-initiator. The physical properties of the hydrogels including swelling ratio and pore structure were investigated. Swelling behavior of the hydrogels was depended on the environmental pH changes (Figure 1), and SEM images demonstrated porous structure within the network. In addition, in vitro cellular responses were assessed, which suggested desirable biocompatibility for these gels. This is first instance for the synthesis of a pH responsive hydrogel formed under visible light by surface initiated photopolymerization. These hydrogels will be useful for controlled drug/ligand delivery, tissue engineering or various biomedical applications where visible light exposure and gradients in permeability is desirable.

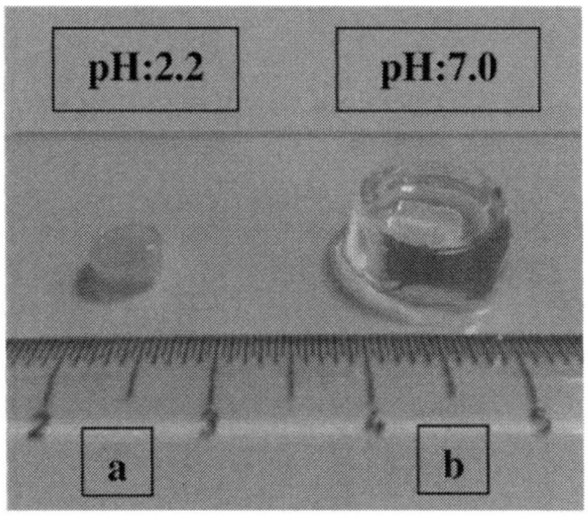

Figure 1. pH responsive P(MAA-g-EG) hydrogels formed with visible ligh. Hydrogel is (a) collapsed at pH:2.2 and (b) swollen at pH:7.0. solutions.

385293 A High Throughput Combinatorial ECM Screening Platform for Optimizing 3D Microenvironment for Tissue Engineering Applications

Abstract:

Development of optimal biomaterials by modulating cell microenvironment in 3D matrices remains a challenging task in tissue engineering. Cell behavior is directed by cell-cell and cell-ECM interactions which are controlled by composition and concentration of ECM and growth factors. Identification of synergistic ECM combinations by rational and combinatorial screening of the scaffold, ECM proteins and growth factors will lead to development of optimal biomaterials. Current macro-scale combinatorial screening methods to study the effect of microenvironmental cues are cost-prohibitive, effort intensive and limited by low throughput. To address this challenge, we have developed a high throughput screening platform which utilizes a combinatorial hydrogel printer for generating combinatorial libraries from base scaffold, ECM proteins and growth factors. High throughput combinatorial printer employs a set of syringe pumps for controlling the composition and concentration of components, static mixer system for uniform mixing of materials and translational stage for spatially addressable dispensing of materials into a hydrogel array. A high throughput confocal flatbed scanner was used for imaging hydrogel arrays and cellular responses of proliferation and morphology can be quantified. We studied the endothelial cell responses in composite Collagen I and hyaluronic acid scaffolds. Our results show endothelial cell responses is directed by collagen and hyaluronic acid concentration. Increase in collagen density showed a marked reduction in cell number and addition of hyaluronic acid to collagen does not show marked differences in endothelial cell number. Doping of ECM rich matrigel into collagen scaffolds showed increased cell proliferation and endothelial cell spreading compared to composite collagen I hyaluronic acid scaffolds. The high throughput screening platform is an efficient tool for discovery of synergistic interactions of ECM proteins and soluble factors for optimizing biomaterials for tissue engineering applications.

375805 UV-Assisted Stabilization of Modified Softwood Kraft Lignin Fibers

Abstract:

Production of high strength carbon fibers from bio-derived precursors is of topical interest. Carbon fibers were produced from a partially acetylated softwood kraft lignin, but the thermo-oxidative stabilization required a low heating rate below 0.2°C/min to prevent the fibers from being heated too rapidly and becoming soft / tacky. Therefore, in this study, stretched as-spun fibers were UV irradiated for 15 min. The fibers underwent sufficient crosslinking on the surface such that they could be subsequently thermally stabilized at heating rate as high as 1.4°C/min. The extent of reaction was studied using FTIR, and the glass transition temperature evolution was assessed by DSC. Stabilized fibers were successfully carbonized at 1000°C, and resulting carbon fibers displayed a tensile strength as high as 1 GPa. These results establish that UV irradiation can effectively shorten the total stabilization time for lignin-based carbon fibers.

384364 Effect of a Natural Cactus Based-Mucilage Dispersant on the Surface Tension and Droplet Size of Dispersed Crude Oil

Abstract:

The purpose of this work is to evaluate the effect of the cactus plant-based mucilage on the surface tension and droplet size of dispersed crude oil in oil in water emulsions. The mucilage is extracted from the cactus plant, Opuntia ficus-indica, commonly known as Nopal or Prickly pear cactus. The extraction process yields two types of mucilage: the non-gelling extract (NE) and the gelling extract (GE). Cactus mucilage extracts and conventional dispersants were evaluated under different conditions and concentrations, including two concentrations of oil (3, 6% v/v) and several dispersant to oil ratios (1:1,1:3, 1:20, 1:40) at room temperature. Synthetic seawater and surrogate oil from Britsh Petroleum (properties similar to the Macondo rig) were used. The results of the dispersion obtaned using mucilage extracts were compared with results using conventional dispersants (Corexit EC9500A). The average droplet size was smaller in the systems with cactus mucilage (1.5% NE generated 5 μm droplets) when compared with the systems using the conventional dispersant Corexit EC9500A (1.5% EC9500A formed 6.2 μm droplets). The surface tensions of Oil/Water emulsion (3, 6% v/v) were reduced from around 50 mN/m to 30 mN/m by adding the cactus mucilage extracts. A standard EPA toxicity test using Daphnia magna colonies was exposed to both NE and GE mucilage extracts in concentrations ranging from 0 to 5000 mg/L showed that mucilage can be considered non-toxic to the evaluated species with a median lethal concentration (LC_{50}) above 2000 mg/L. Cactus mucilage is an alternative technology to mitigate the damage that oil may cause to the aquatic ecosystem and minimize undesired effects associated with the use of synthetic dispersants in oil spills.

388378 Artificially Engineered Protein Gels Derived from Nucleoporins

Abstract:

The nuclear pore, a tiny channel 45 nm in diameter, is filled with a polypeptide-based hydrogel capable of selectively filtering < 0.1% of all proteins in a eukaryotic cell while translocating over 1,000 molecules per second. The responsible polypeptides are nucleoporins, commonly containing "FG" repeat sequences, which are known to associate with one another to construct physical gels and give rise to the hydrogel's selective filtering property. Major obstacles in utilizing nucleoporin-based hydrogels for separation technology include low biosynthetic yield and long gel processing time, which typically takes several hours to days.

In order to overcome these drawbacks, we developed artificially engineered protein building blocks that attempt to mimic the selective permeability of nuclear pore hydrogels, are produced in high yields, and possess reduced gelation times. A well-investigated nucleoporin, Nsp1, was used as a template sequence for its well-defined repeat domains. To mimic Nsp1's transport properties, nucleoporin-like polypeptides (NLPs) were designed from consensus Nsp1 repeat sequences. To facilitate hydrogel formation, NLP sequences were flanked with coiled-coil domains which create three dimensional polymer networks. Here, we demonstrate that artificially engineered biopolymer building blocks can be expressed in high yield and can construct a physically crosslinked hydrogel within a few seconds. Using a combination of selective permeability fluorescence assays, shear rheology, and Raman spectroscopy, the mechanism of artificially engineered NLP hydrogel function is studied.

379734 Enhancing Oxygen Permeability in Hydrogel Wound Dressing By Cyanobacterial Gas Vesicles

Abstract:

Hydrogel wound dressing is commonly used nowadays. However, for some wounds the use of hydrogel wound dressing may incur insufficient oxygen availability for proper healing. Oxygen has an important role in angiogenesis, cell motility, and extracellular matrix formation. Oxygen limitation during the wound-healing process is one of the major causes that can lead to formation of chronic wounds. Cyanobacterial gas vesicles are hollow, protein-walled, gas-filled structures in cylindrical shapes with conical caps. Gas vesicle wall is impermeable to liquid water but permeable to gases with very low mass transfer resistance. Because oxygen diffusivity in a gas phase is much higher than that in liquid and solid phases, introducing gas vesicles into the hydrogel has the possibility of effectively enhancing oxygen permeation through the gel. Nonetheless, we have shown earlier that if dispersed randomly, gas vesicles need to be provided in high volume fractions to be effective in enhancing oxygen transfer, because the oxygen transferring rapidly through the gas vesicles is bottlenecked by the slow transfer across the liquid/gel continuous phase that separates the dispersed vesicles. We have thus taken the approach of attaching gas vesicles as continuous layers on the surface of fibers that extend across the whole hydrogel thickness. For example, PMMA coated, highly porous, and large pored polyurethane foam can be introduced as a core porous structure in the hydrogel wound dressing. After the coated PMMA surface is gently modified by acid hydrolysis, gas vesicles collected from the *Anabaena flos-aqua* culture are chemically attached onto the modified surface. This foam structure creates connected gas vesicle layers or channels that allow fast oxygen transfer through the hydrogel wound dressing. The PMMA surface modification has been optimized and tested. AFM pictures confirmed the attachment of gas vesicles on the modified surface. Much faster oxygen permeation through attached gas vesicle layers has been visualized by the localized color change in resazurin solution immediately adjacent to the gas vesicle-attached surface of a PMMA strip placed in the originally colorless anaerobic solution. We are currently measuring the oxygen permeability to provide more quantitative information on the oxygen transfer enhancement offered by different methods and amounts of gas vesicle attachment on surfaces. Results of cyanobacterial gas vesicle production and harvesting as well as the oxygen transfer enhancement achieved through the use of gas vesicle will be reported and discussed in the presentation.

386251 Adhesive Elastin-Based Proteins As Soft Tissue Glues

Abstract:

Each year, millions of major surgeries are performed worldwide. Many of these major surgeries involve suturing of tissue, which results in additional damage to the tissue and slows the healing process. Surgical adhesives are a promising alternative to sutures. To be a successful surgical adhesive, a product must be: biocompatible, set in a wet environment, and have proper adhesive and cohesive properties. In addition, a soft tissue surgical adhesive must also be flexible to prevent excessive stress on surrounding tissue. Unfortunately, most modern adhesives lack at least one of these characteristics.

The goal of this project is to utilize protein engineering to develop and characterize a biomimetic soft tissue surgical adhesive. The protein-based adhesive will contain two domains: a structural domain that provides flexibility and an adhesive domain that provides the ability to adhere in wet environments. The structural domain is inspired by elastin, a protein that provides elastic properties to tissues such as skin, blood vessels, and ligaments. The adhesive domain is inspired by mussel adhesive proteins (MAPs), which have the ability to bond underwater to nearly any material. MAPs contain numerous 3,4-dihydroxyphenyl-L-alanine (DOPA) residues, a chemical moiety that provides adhesion and cohesion through a number of mechanisms. Together, the elastin-based domain and the mussel-inspired domain will provide the characteristics necessary for an effective surgical adhesive.

Using standard molecular biology techniques, a family of proteins has been designed. Each protein contains elastin-like pentapeptides in combination with numerous tyrosine residues. Following expression in *E. coli* and subsequent purification, the proteins will undergo enzymatic conversion to produce the adhesive DOPA residues from tyrosines. To be successful soft-tissue surgical adhesives, the proteins must be biocompatible, have strong bulk adhesive strength, and a low elastic modulus. A LIVE/DEAD fluorescence assay will be utilized to assess protein cytotoxicity. Lap shear adhesion testing will be used to examine bulk adhesive strength. Finally, uniaxial tensile testing of crosslinked protein hydrogels will be performed to assess the elastic modulus of the adhesive. Initial results show high cytocompatibility (>98% viability) and significant bulk adhesive strength on aluminum (approximately 1 MPa). In summary, this new family of flexible adhesive proteins combines the mechanical properties of elastin and the adhesive properties of MAPs. Its unique properties are ideal for application as a soft tissue surgical adhesive, as well as coatings for cell attachment or scaffolds for tissue engineering.

Simulation and experimental investigation of osteogenic activity of hydrogel-conjugated BMP-2 peptide

Seyedsina Moeinzadeh, Esmaiel Jabbari

Department of Chemical Engineering, University of South Carolina, Columbia, SC, USA 29208

Introduction: The Peptide derived from the active domain of osteoinductive rhBMP-2 protein increased osteogenic differentiation and mineralization of the bone marrow stromal cells. Peptides can be conjugated to a macromer and crosslinked into a gel to immobilize the peptide at the site of regeneration. However, peptide aggregation and nanoscale structure formation in the conjugated state in aqueous solution can influence morphogenesis of the encapsulated mesenchymal stem cells (MSCs). In this work, we simulated nanostructure formation in the precursor solutions of polyethylene glycol (PEG) based hydrogels containing BMP2 peptide, BMP2 pre-conjugated to PEG macromers and BMP2 pre-conjugated to amphiphilic PEG macromers.

Materials and methods: The atomistic structure of the peptide and macromers were coarse-grained into different beads (set of atoms). There were totally 16 bead types for peptide back bone and amino acid side chains, ethylene oxide repeating unit (EO), acrylate group (Ac) and lactide repeating unit (L) (Fig 1a). Simulations were performed under NVT ensemble dynamics using the MARTINI force filed.[3] A Lennard-Jones (12-6) function was used for Van der Waals interactions. A Coulombic field with a cutoff distance of 12 Å was employed for electrostatic interactions. A harmonic function for bond length, a cosine harmonic function for bond angles and a dihedral function for torsions were used to account for interaction between bonded beads. A time step of 10-fs was employed and total simulation time was 1-μs. The temperature was held at 37°C using the Nose thermostat. All of the simulations were performed in 150×150×150 Å boxes with 3D periodic boundary conditions. The Mesocite module of Materials Studio (v5.5, Accelrys) was used for the mesoscale simulations. The osteogenic activity of the conjugated peptide was assessed with human mesenchymal stem cells by measuring alkaline phosphatase activity (ALP) and the extent of mineralization with incubation time.

Results: The simulation results showed that acrylated BMP2 peptides in the PEG aqueous solution formed aggregates with a distribution of aggregation numbers nagg (number of peptides per aggregate) ranging from 1 to 6. The average Nagg did not change significantly with conjugation of BMP2 peptide to the PEG macromer but the distribution of Nagg was narrower. The number of aggregates decreased and the average size of each aggregate increased when the peptide was pre-conjugated to the macromers containing degradable lactide segments. The average Nagg increased from 2.2 to 3.1, 4.2 and 5.8 and the number of aggregates decreased from 11 to 8, 5 and 4 when the number of lactide monomers per macromer increased from 0 to 2, 4 and 6, respectively. This was attributed to the contribution of hydrophobic lactide units to the formation of micelle like aggregates. Moreover, the hydrophobic side chains of the peptide (I, V, L) formed the center of the aggregate and the charged residues (K and E) located on the surface of the aggregates. Furthermore, the Ac-Ac integration number (INAc-Ac) which indicates the proximity of reactive acrylate groups in PEG solutions increased from 1.9 to 2.3, 2.6 and 3.3 with increasing the number of lactide monomers per peptide-macromer conjugate from 0 to 2, 4 and 6. The increase in INAc-Ac resulted in an increase in the conjugation efficiency of BMP2 peptide. Conjugation of the BMP-2 peptide to PEG increased critical micelle concentration (CMC) and osteoinductive potential of the peptide whereas conjugation to lactide-capped PEG reduced CMC and osteoinductive potential of the peptide. Experimental and simulation results revealed that osteoinductive potential of the BMP-2 peptide is correlated with its CMC and the free peptide concentration in aqueous medium and not the total concentration.

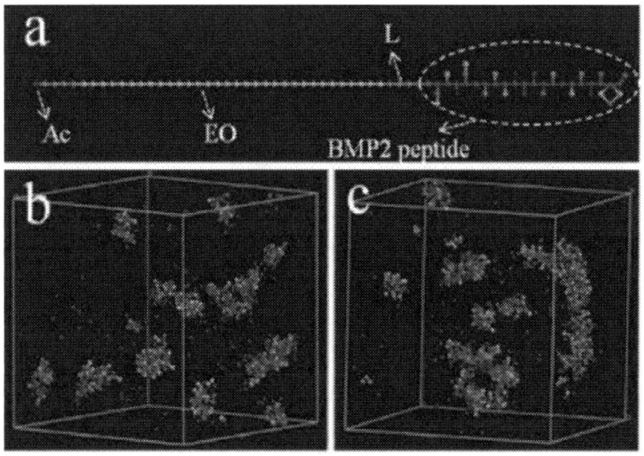

Figure 1. (a) Coarse-grained structure of BMP2 peptide conjugates; aggregate formation in BMP2-PEG (b) and BMP2-PEG-Lactide conjugates (c).

370147 Designing Multi-Component Nanostructured Soft Biomaterials Interacting with Charged Nanoparticles

Abstract:

We develop a computational model to design and characterize multi-component nanostructured soft biomaterials and their interactions with functionalized and charged nanoparticles using the Molecular Dynamics simulation technique. Our objective is to generate a stable vesicle through the self-assembly of amphiphilic lipid molecules using implicit solvent coarse-grained molecular models in order to investigate biologically important physiological processes occurring on the mesoscopic spatio-temporal scales. The amphiphilic lipid molecules are composed of a hydrophilic head group and two hydrophobic tails. In addition, we use two component lipid mixtures to study the formation of hybrid lipid vesicles through the self-assembly. We find that the degree of phase segregation between the two lipid species is tunable via their distinct effective chemical specificity and molecular architecture. Furthermore, we have extended the model to introduce the charged lipid molecules into the vesicle by using long-range electrostatic interactions between charged molecules. We investigate the interactions of various functionalized nanoparticles with the charged lipid species and effects of these interactions on the dynamics and morphologies of the lipid vesicles. Our findings can be used to design new soft biomaterials for various applications in medicine, sensing and energy.

375919 Invited Talk: Generation of Stable Non-Spherical Capsules with Controlled Interfacial Coverage of Surface-Active Particles

Abstract:

The complex interfaces of particle-stabilized drops and bubbles have a significant effect on the rheology of emulsions and foams. Common high-shear emulsification methods offer no direct control over the rate of droplet generation, the final droplet size distribution, or the composition of the interface, all of which directly impact the rheology of the final emulsion. In this talk, a simple microfluidic device is used to generate monodisperse droplets with independent control of droplet size, production rate, and the degree of particle loading on the interface. Droplets are formed on demand at a T-junction via pressure control of each fluid phase. The freshly formed droplets travel along the axis of a circular capillary containing a surface-active particle suspension. The residence time of the droplets within the capillary and the transport of particles in the thin films and fluid volumes between droplets determine the coverage of particles on the interface. The flow along the axis of the droplet sweeps adsorbed particles to the rear of the droplet, leading to a highly concentrated region of the interface that does not relax upon exiting the confined geometry of the microchannel. The length of the concentrated region increases with particle coverage, producing capsules with shapes from pears to maracas to cylinders. The capsules are monodisperse, can be produced at high rates, and remain stable for tens of hours after formation. Measurements of the dilatational and shear interfacial rheology help rationalize the observations of highly stable non-spherical shapes. Thus, careful control of transport in microscale devices allows for production of emulsions with tunable morphology, interfacial rheology, and bulk rheology.

Preparation and Ultralow Interfacial Tension Properties of Polystyrene Sulfonate-based Amphiphile

Varun Shenoy Gangoli[1], Sarah S. Munday[1], Jason Mann[2], Hadi Shamsijazeyi[1], Maura Puerto[1], Rafael Verduzco[1], Clarence A. Miller[1], George J. Hirasaki[1], James M. Tour[2], Michael S. Wong[1,2]

[1]*Department of Chemical and Biomolecular Engineering, Rice University, Houston, TX, USA 77005*
[2]*Department of Chemistry, Rice University, Houston, TX, USA 77005*

The demand for oil in the US far exceeds the current supply from existing oil fields. Retrieval of oil from reservoirs is not very efficient and thus enhanced oil recovery (EOR) techniques have been employed by industries. We have developed a novel material that is synthesized by controlled desulfonation of polystyrene sulfonate (PSS) in water at low pH as seen in Figure 1. The desulfonation introduces hydrophobicity into the polymer and the nature as well as extent of desulfonation controls the surface activity of the product. We have achieved ultra-low interfacial tension (~10^{-3} mN/m) when measured against oils such as toluene. This material is also very stable at high temperature and salinity- characteristic of oil reservoirs. Lastly, the synthesis method is inexpensive, reliable and can be scaled up easily for real world applications.

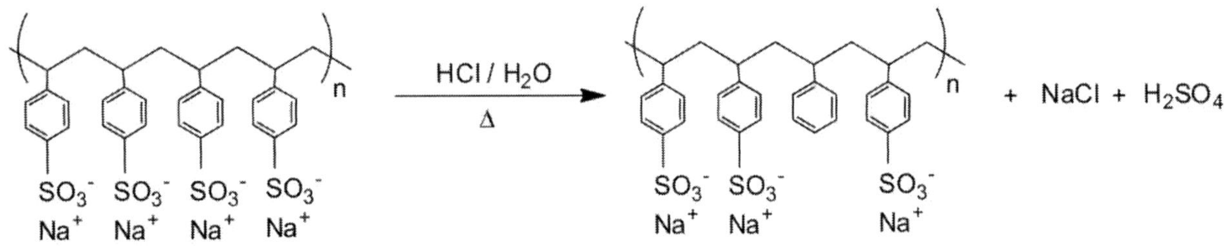

Figure 1. The desulfonation reaction shown for 1 PSS monomer unit

VT

Squeeze Flow for Enhanced Strain Induced Orientation in Thermotropic Liquid Crystalline Polymers: Applications to H2 Storage

Craig D. Mansfield, Chen Qian, Donald G. Baird
Virginia Tech Department of Chemical Engineering
November 20, 2014

VT

Motivation

- Emerging hydrogen economy
  - New natural gas deposits, hydrogen from steam reforming and water gas shift reactions
  - Water electrolysis
- Hydrogen is an efficient multipurpose fuel

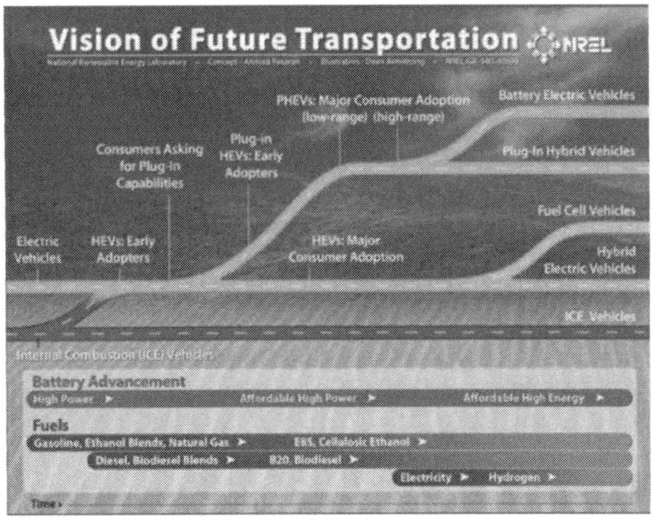

2

Design Requirements for H_2 Tanks

- Cryogenic Storage
  - Low temperature (20 K)[2]
  - High pressure (10,000 psia)[2]
- Tank Physics
  - Mechanical
    - High modulus
    - High strength
  - Thermal
    - Low thermal expansion coefficient
  - Transport
    - Low permeability
  - Duty cycle
    - Large mechanical/thermal range
    - Variable frequency of use
- Polymer Processing
  - Low melt viscosity
  - High melt strength
- Process Economics
  - Common stock materials
  - High stored mass to tank mass ratio
  - Low overall cost
  - High market residence time
  - Recyclable w/ minimal attrition

3

Common TLCPs: Composition

- HX series polyesters: HX3000, HX6000, HX8000

Terephthalic acid | Hydroquinone | Hydroquinone derivatives | p-hydroxybenzoic acid

- Vectra series polyesteramides: Vectra B950

6-hydroxyl-2-naphthoic acid, 60mol% | Terephthalic acid, 20mol% | p-aminophenol, 20mol%

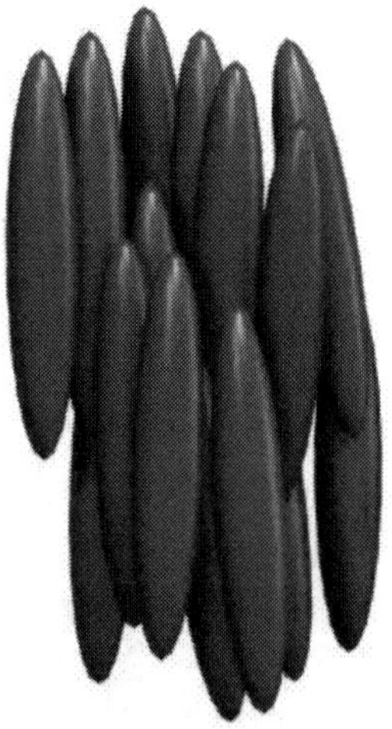

- Aromatic monomers act as rigid rods
- TLCPs shown form a nematic phase

Image of nematic phase by reference 12

4

Common TLCPs: Properties

Materials	Injection molded parts[a]	
	Modulus /GPa	Strength /MPa
Vectra A950	6~10[2-4]	100~140[2-4]
Vectra B950	10~22[2-4]	100~140[2-4]
HX series	12~20[5]	100~130[5]

Materials	Fiber	
	Maximum modulus /Gpa	Maximum strength /Mpa
Vectra A950	150[9]	1500[9]
Vectran M®[c]	80[10]	1200[10]
Vectran HS®[c]	80[10]	2000[10]
302 Stainless Steel	190[11]	1300[11]

Materials	H_2 permeability /mol •(m^{-1} sec^{-1} Pa^{-1})[b]
TLCP[6]	1.6×10^{-20}
HDPE[7,8]	$\sim 9 \times 10^{-16}$

a. injection molded parts include both end-gated plaques and tensile bars, for end-gate plaque, properties listed in the table are in the flow direction.
b. b. H_2 permeabilities were measured at the temperature of 298K. The TLCP reported in reference 7 has similar structure as Vectra A950.
c. Vectran is the brand name of commercially available LCP fiber, currently produced by Kuraray America Inc. (previously known as Celanese Advanced Materials Inc.).

5

Extrusion Blow Molding

1) Parison is extruded from an annular die

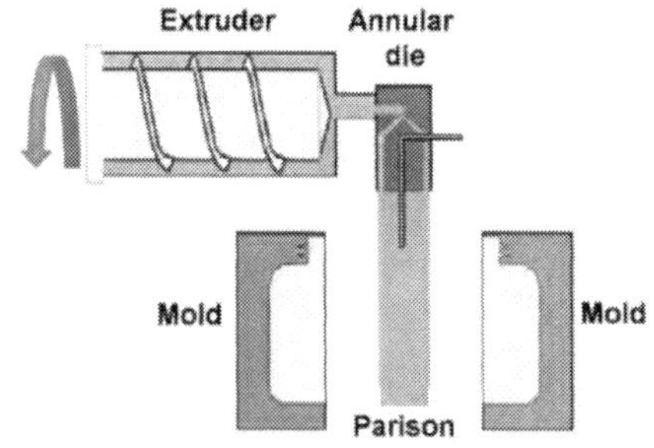

2) Mold is closed, sealing ends of parison

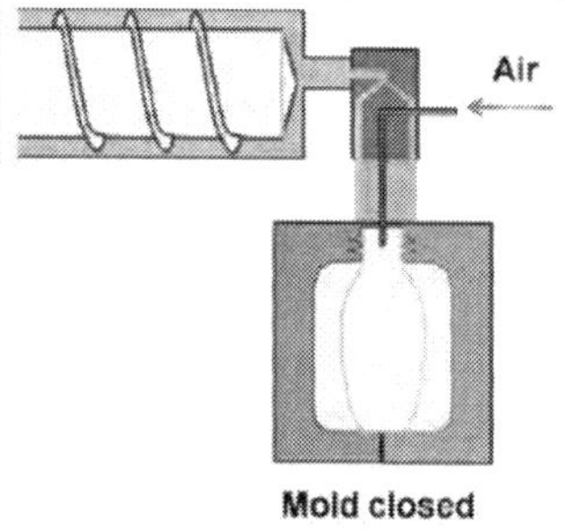

3) Parison is expanded into mold, forming a bottle

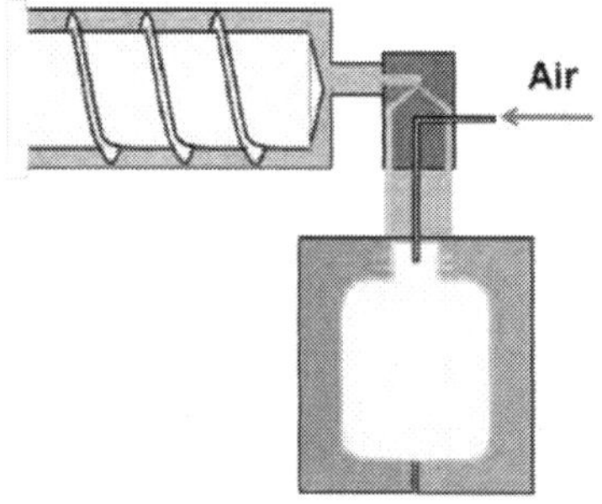

4) Mold is opened to release the bottle

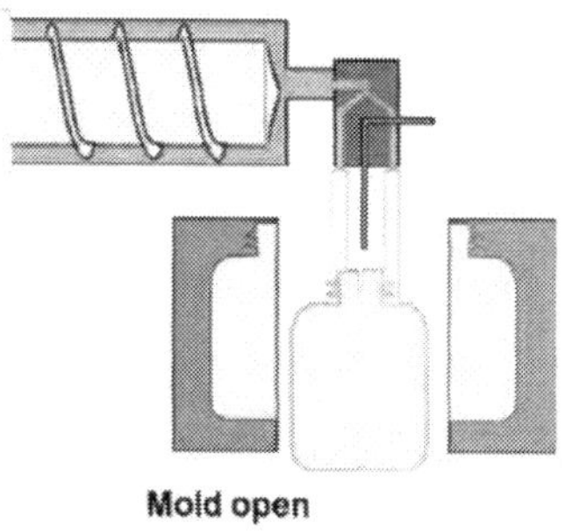

Images by Chen Qian

6

Subcooling of HX-3000

- HX-3000 is processed below melting point in the die during extrusion of parisons

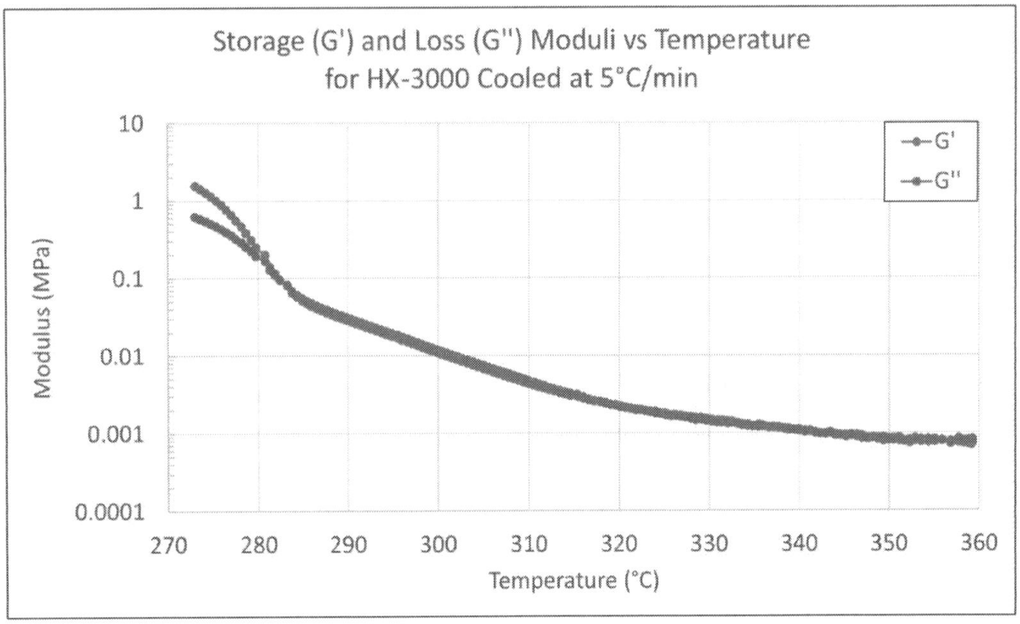

7

Model Tank Geometry and Orientation

- Blue: Machine (axial) direction
- Red: Hoop direction

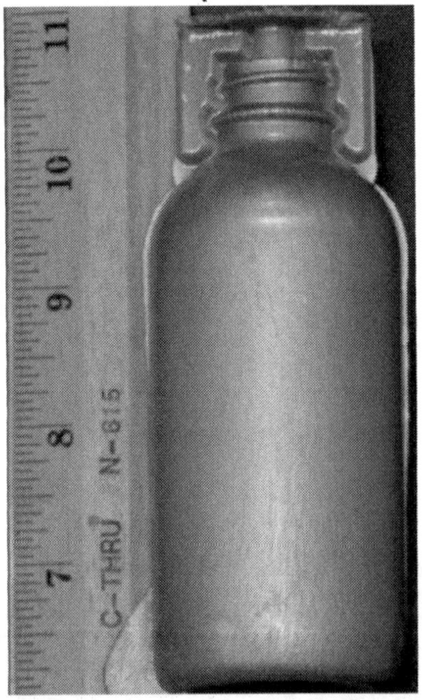

8

Enhanced Mechanical Properties in the Machine Direction

- Bottles obtained from blow molding HX-3000 parisons with different amount of strain in the machine direction.
  - Observed increased tensile modulus in machine direction
  - Observed decreased tensile strength in hoop direction

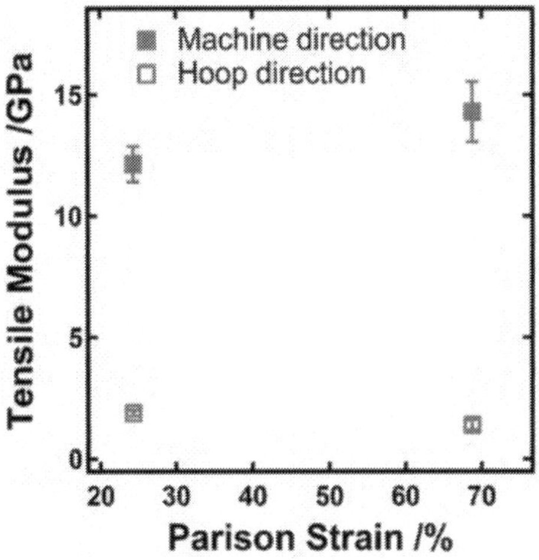

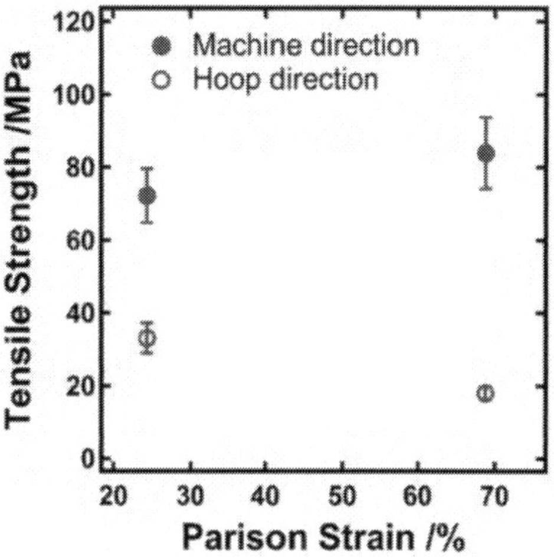

9

Relating Squeeze Flow to Extrusion Blow Molding

- Squeeze Flow
  - Planar extension

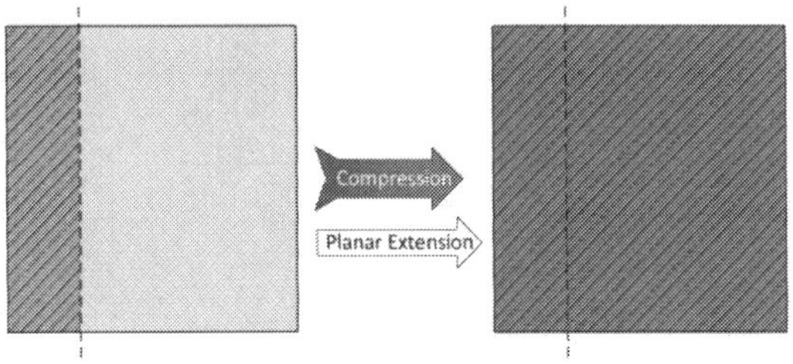

- Extrusion Blow Molding
  - Extrusion and stretching

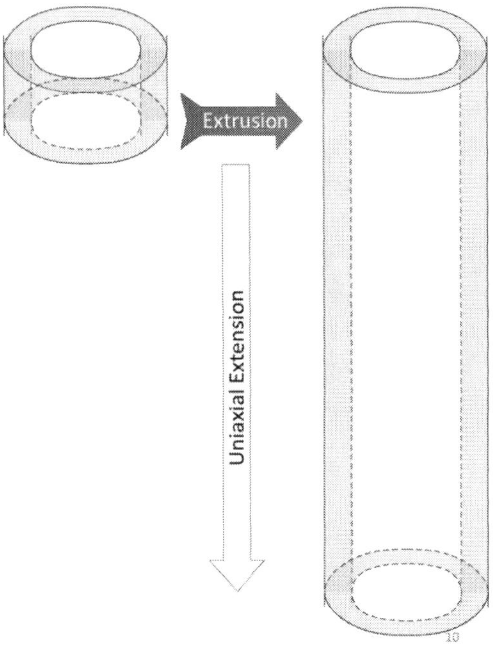

10

Relating Squeeze Flow to Extrusion Blow Molding

- Squeeze Flow
  - Biaxial extension

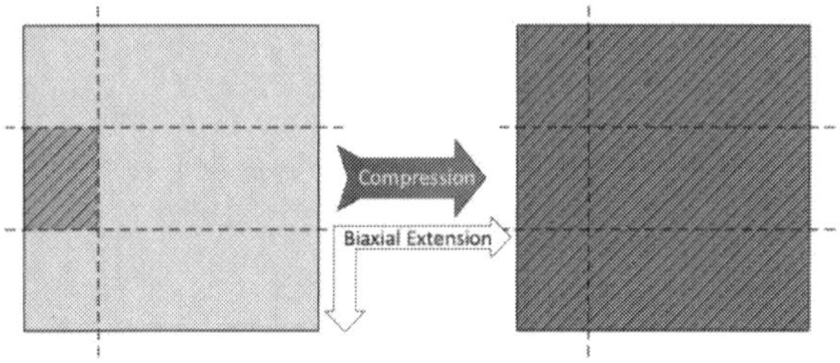

- Extrusion Blow Molding
  - Expansion

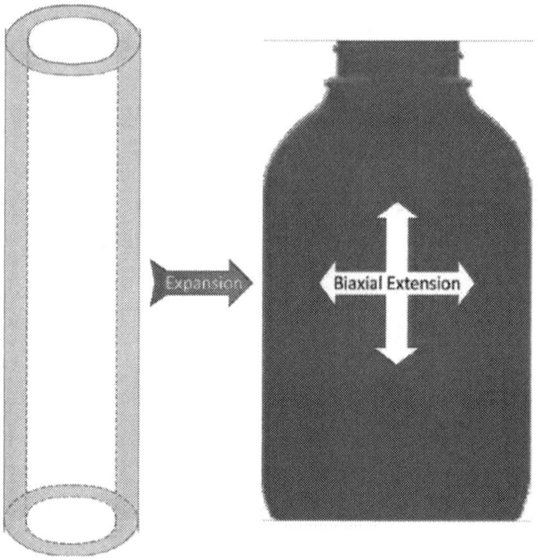

Squeeze Flow via Compression Molding

- Used sequential constant force compressions to impart strain

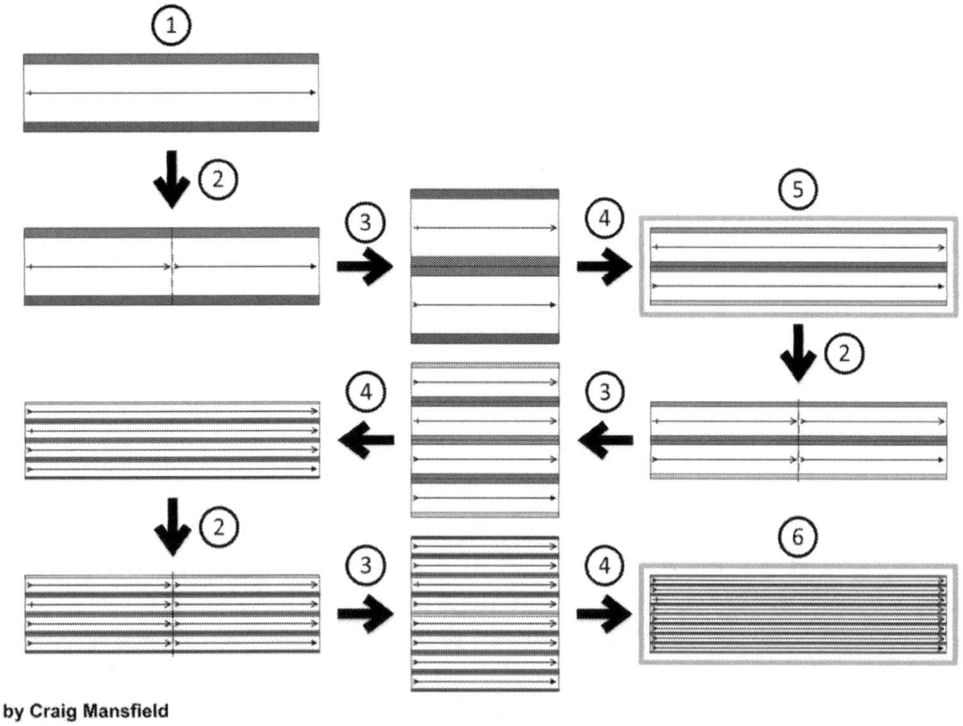

Image by Craig Mansfield

Compressions at T = 330°C

- No observed trend in tensile strength

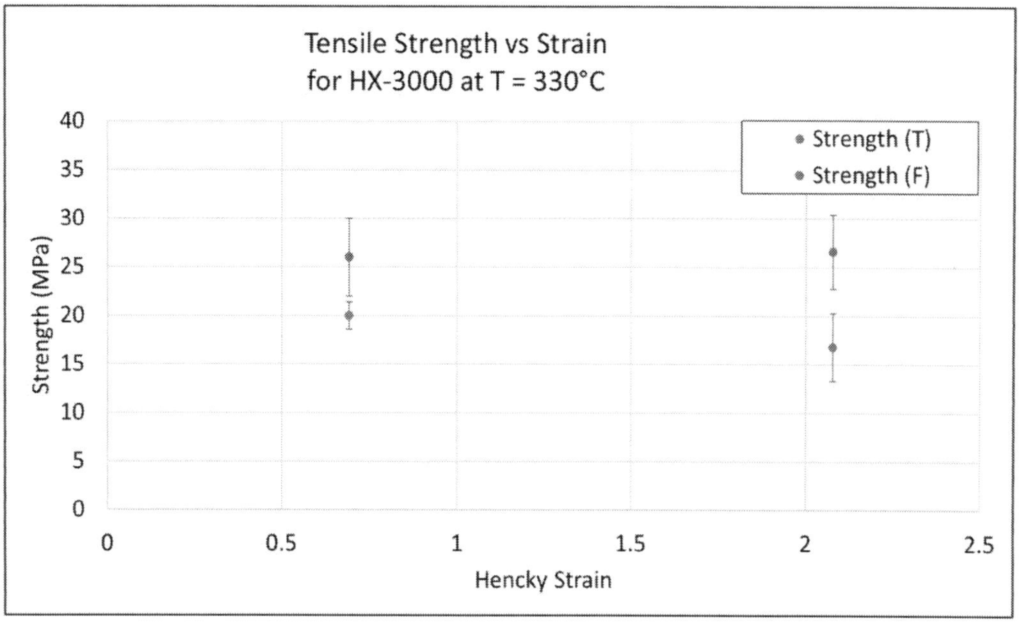

13

Compressions at T = 330°C

- Tensile modulus decreased with strain in the flow direction

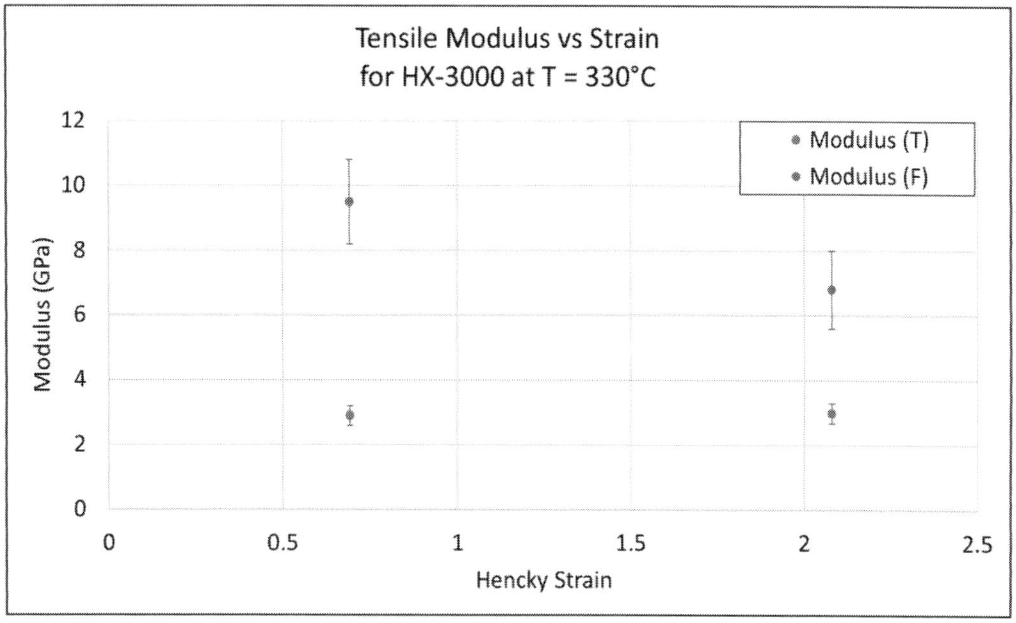

14

Compressions at T = 350°C

- Tensile strength increased with strain in the flow direction

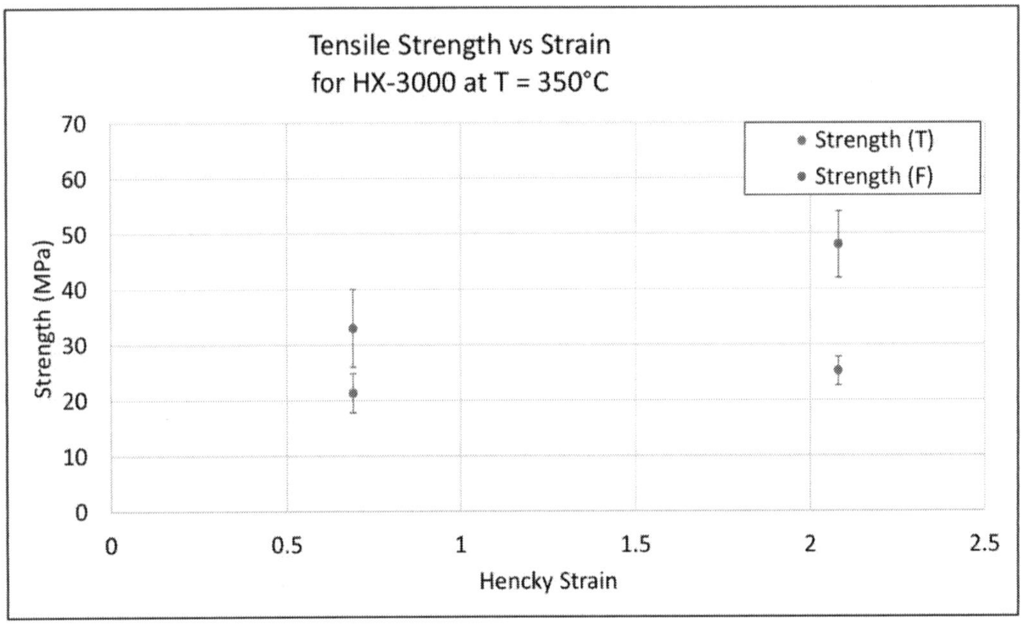

15

Compressions at T = 350°C

- Tensile modulus decreased with strain in the transverse direction

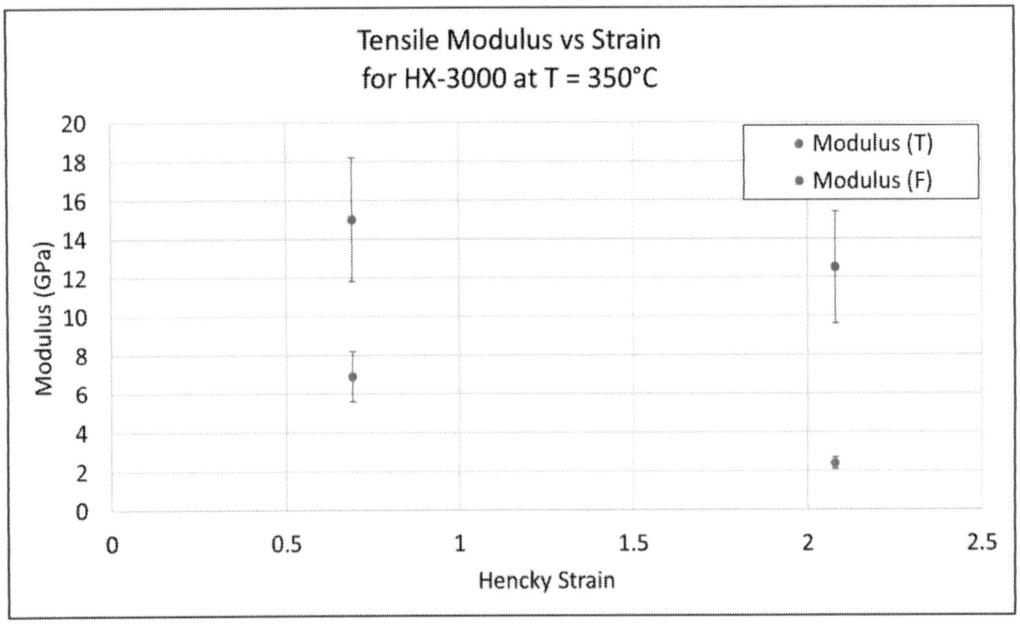

16

Modified Measure of Strain

- Modified strain to account for time spent at elevated temperature

$$\epsilon_T = \epsilon_h \frac{t_{preheat} T_{melt}}{t_{compression} T_{compression}}$$

- Modified strain (ϵ_T) represents ratio of competing driving forces on orientation: flow alignment and Brownian motion.

17

Effect of Modified Strain

- Observed positive trends in flow and transverse directions for tensile strength with modified strain

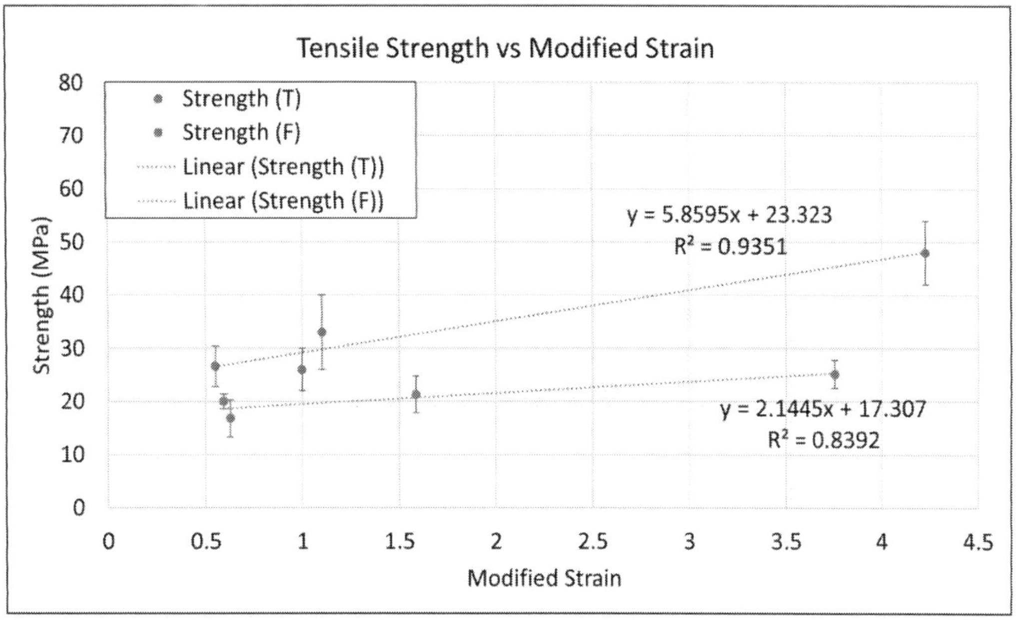

18

Effect of Modified Strain

- No observed trends in tensile modulus with modified strain

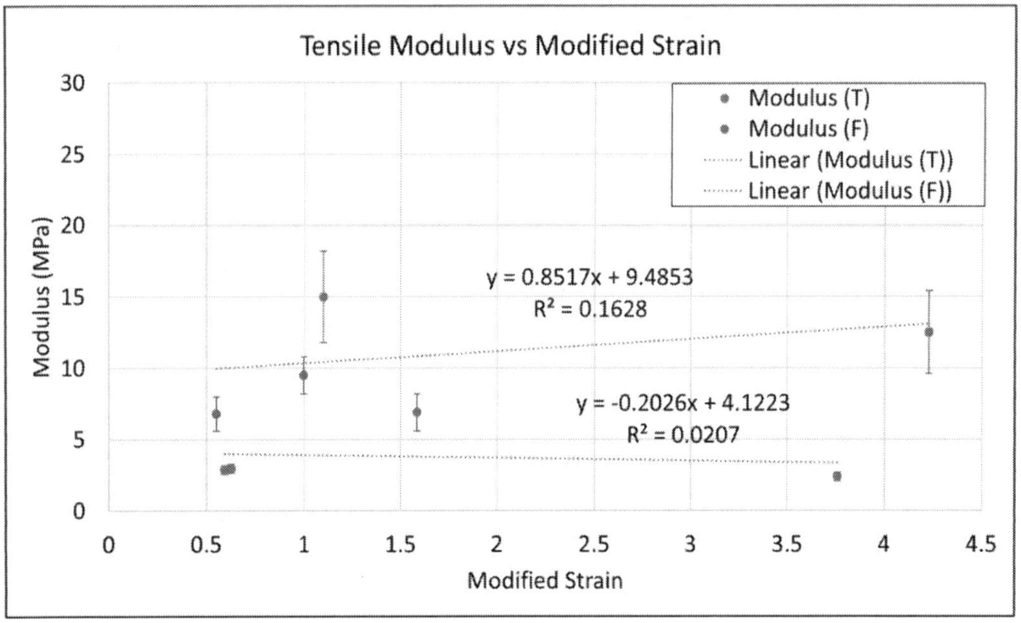

19

Conclusions and Future Work

- Conclusions
  - Tensile strength is positively correlated with strain in the flow direction above the melting point
  - Tensile strength is positively correlated with modified strain
- Future Work
  - Repeat present work to decrease uncertainty of observed trends
  - Vary applied force during compressions
  - Impart strain via biaxial extension
  - Correlate results from squeeze flow with properties of blow molded bottles

20

Acknowledgements

- Funding sources
  - Savannah River National Laboratory (SRNL)
  - U.S. Department of Energy (DOE)
- Research group

21

Questions?

22

References

Works Cited

1. R. D. McCarty, J. Hord, and H. M. Roder, "Selected properties of hydrogen (engineering design data)," National Engineering Lab.(NBS), Boulder, CO (USA), 1981.
2. Schott, R. *et al. Physical properties of injection molded liquid crystal polymers and high temperature engineering polymers*, 1993.
3. Datta, A.; Baird, D. G. *Polymer* **1995**, *36*, 505.
4. Datta, A.; Chen, H. H.; Baird, D. G. *Polymer* **1993**, *34*, 759.
5. Results from experimental work by Chen Qian.
6. Chiou, J. S. *et al. Journal of Polymer Science Part B: Polymer Physics* **1987**, *25*, 1699.
7. Humpenoider, J. *Cryogenics* **1998**, *38*, 143.
8. Schultheiss, D. Permeation barrier for lightweight liquid hydrogen tanks **2007**, page 121.
9. Kalfon-Cohen, E. *et al. Polymer* **2009**, *50*, 1797.
10. Pegoretti, A. *et al. Composites Science and Technology* **2006**, *66*, 1970.
11. MatWeb, http://www.matweb.com/search/DataSheet.aspx?MatGUID=c49afdac596a4ebe9011951468b284cd
12. "File:LiquidCrystal-MesogenOrder-Nematic.jpg," *Wikipedia, the free encyclopedia*.

Further Reading

1. G. S. Lee, J. S. Cho, J. C. Kim, T.-H. Yoon, and S. T. Shin, "Direct confirmation of biaxiality in a bent-core mesogen through the measurement of electro-optic characteristics," *Journal of Applied Physics*, vol. 105, no. 9, p. 094509, 2009.
2. S. Sircar and Q. Wang, "Dynamics and rheology of biaxial liquid crystal polymers in shear flows," *Journal of Rheology*, vol. 53, no. 4, p. 819, 2009.
3. A. G. Vanakaras and D. J. Photinos, "Thermotropic biaxial nematic liquid crystals: Spontaneous or field stabilized?," *The Journal of Chemical Physics*, vol. 128, no. 15, p. 154512, 2008.
4. W. Wang, P. Zhang, and Z. Zhang, "Rigorous derivation from Landau-de Gennes theory to Ericksen-Leslie theory," *arXiv preprint arXiv:1307.0986*, 2013.
5. D. Mucci and L. Nicolodi, "On the Elastic Energy Density of Constrained Q-Tensor Models for Biaxial Nematics," *Archive for Rational Mechanics and Analysis*, vol. 206, no. 3, pp. 853–884, Jul. 2012.
6. V. M. Gun'ko, V. I. Zarko, E. V. Goncharuk, L. S. Andriyko, V. V. Turov, Y. M. Nychiporuk, R. Leboda, J. Skubiszewska-Zięba, A. L. Gabchak, V. D. Osovskii, Y. G. Ptushinskii, G. R. Yurchenko, O. A. Mishchuk, P. P. Gorbik, P. Pissis, and J. P. Blitz, "TSDC spectroscopy of relaxational and interfacial phenomena," *Advances in Colloid and Interface Science*, vol. 131, no. 1–2, pp. 1–89, Feb. 2007.
7. S. Bensaad, B. Jasse, and C. Noël, "Orientation development in injection moulded plaque of liquid crystal copolyesteramide using normal incidence specular reflection," *Polymer*, vol. 34, no. 8, pp. 1602–1605, 1993.
8. S. Bensaad, B. Jasse, and C. Noel, "Orientation measurements from the surfaces of injection moulded plaques of thermotropic liquid crystalline polymer using normal incidence specular reflection," *Polymer*, vol. 40, no. 26, pp. 7295–7301, 1999.
9. Theory of Polymer Dynamics, Chapter 10, M. Doi and S. F. Edwards, 1986.
10. The Structure and Rheology of Complex Fluids, R. G. Larson, 1999.
11. D. S. Done and D.G. Baird, "Solidification Behavior and Recovery Kinetics of Liquid Crystalline Polymers," Polym. Engr. Sci., 30:16, 985989 (1990).
12. P. G. de Gennes and J. Prost, *The physics of liquid crystals*. Oxford; New York: Clarendon Press ; Oxford University Press, 1993.
13. André M. Sonnet and Epifanio G. Virga, *Dissipative Ordered Fluids: Theories for Liquid Crystals*. Springer, 2012.

23

387755 Highly Conductive Polyolefin Nanocomposites: Controlled Dispersion of the Mixed Solid Nanofillers

Abstract:

Carbon nanotube is commonly used as conductive nanofiller to improve the electric conductivity of hosting polymer matrix.[1] However, the conductivity of the polymer nanocomposite depends on the net work formation of carbon nanotube in the polymer matrix, and high CNTs content was required to accomplish the conduction network in polymer matrix. To enhance the conductivity, additional treatments including chemical functionalization[2] and heating were have been reported to improve the dispersion of CNTs in polymer matrix. However, these treatments would cause negative effect on the conductivity of the pristine CNTs. In order to make a more efficient network formation of CNTs in polymer matrix, the conductive polypropylene (PP) nanocomposites were prepared by innovatively coating carbon nanotubes (CNTs) on the surface of gelated/swollen soft polypropylene (PP) pellets. The brand new processing method ensures the CNTs network formation in the polyer matrix and the electrical conductivity (σ) studies revealed a percolation value of only 0.3 wt% CNTs. At lower processing temperature, the CNTs formed network structure more easily in the polymer matrix resulting in a higher σ. Compared with the mixed α and γ crystal phases for pure PP, more γ phase PP (0.2533 at 160 °C) was observed in the polymer nanocomposites (PNCs) and the fraction of γ phase increased with increasing the pressing temperature. In addition, the CNTs at lower loading (0.1 wt%) served as the nucleating sites and promoted the crystallization of PP by reducing the surface free energy barrier towards nucleation. The CNTs were observed to favor the shear thinning behavior of the polymer matrix and at high shear rates, the CNTs serving as the branches of polymer chains favored the disentanglement of polymer chains and thus caused an even lower viscosity of PNCs than that of pure PP. The calculated optical band gap of CNTs from Tauc plot was observed to increase with increasing the processing temperature, i.e., 1.55 eV for PNCs prepared at 120 °C and 1.70 eV for PNCs prepared at 160 and 180 °C. Both Drude model and interband transition phenomenon have been used for theoretical analysis of the observed real permittivity variation from negative to positive (also called plasma frequency) and from positive to negative, respectively.

1. Y. Li, J. Zhu, S. Wei, J. Ryu, Q. Wang, L. Sun and Z. Guo, *Macromolecular Chemistry and Physics*, 2011, **212**, 2429-2438.

2. H. Gu, S. Tadakamalla, X. Zhang, Y.-D. Huang, Y. Jiang, H. A. Colorado, Z. Luo, S. Wei and Z. Guo, *Journal of Materials Chemistry C*, 2013, **1**, 729-743

388377 In-Situ Observation of Phase Separation and Morphology of Polymeric Membrane Dopes in Shear and Elongational Flow

Abstract:

In hollow fiber membrane spinning, the effects of shear and elongational stresses on membrane morphology are often convoluted due to the complexity of the spinning process. In this study, we systematically investigate the effect of shear and elongational stresses on phase separation and membrane morphology utilizing microfluidics, video-microscopy and electron microscopy (SEM) of the formed fibers. Polymer solutions were extruded through microfluidic channels into non-solvent, and the *in situ* processes of phase separation and defect formation were observed through video-microscopy and compared with the resulting cross-sectional images of the fibers. The rheological properties of dope solutions and the resulting fiber morphology as a function of shear rate/elongational rate were combined to construct flow regime maps with clearly identifiable characteristic features. Each of the observed regimes was shown to have limiting physical process that determines fiber formation and morphology: at low extrusions rates mass transfer is the dominant process, the inherent thermodynamics of the polymer solution plays a key role at intermediate extrusion rates, and shear and elongational stresses are critical factors at high extrusion rates. Defect-free regimes were found in particular regimes and most importantly, high elongational stresses can eliminate the defect formation in membrane fabrication.

376243 Carbon Fiber Derived from PAN/Lignin Bi-Component Precursor

Abstract:

Carbon Fiber Derived From PAN/Lignin Bi-component Precursor

Jing Jin, Amod A Ogale

Department of Chemical and Biomolecular Engineering

Center for Advanced Engineering Fibers & Films (CAEFF)

Clemson University, Clemson, SC 29634

Abstract

Bio-derived precursors are of increasing interest in the production of carbon fibers to reduce the reliance on synthetically-derived polyacrylonitrile (PAN) precursor, which is primarily used in current commercial carbon fibers. Lignin, an abundant biomass, is currently recycled back into paper production process. However, there are renewed efforts to find value-added applications for lignin, including that as a carbon fiber precursor. Several studies have been reported on the conversion of chemically-modified lignin into carbon fibers. However, the tensile strength of such fibers has been limited to about 1 GPa. Therefore, the current study investigates the processing and properties of carbon fibers derived from bi-component lignin and PAN precursors.

PAN and softwood kraft lignin were wet-spun into precursor fibers. Hydroxyl groups in lignin enable the solubility in DMSO (solvent) for wet spinning and higher rate of thermal stabilization. Carbon fibers were successfully produced by heat-treatment in inert environment at 1000°C. The properties of such fibers are much lower than those derived from conventional PAN fibers. This observation is consistent with other limited studies on PAN/lignin polymer blends. Essentially, the lignin component was not found to contribute to the strength of resulting carbon fibers. Therefore, ongoing studies are dealing with alternative routes for processing PAN/lignin, including two different layers to form a bi-component precursor fiber.

Key Words: Carbon fiber, Lignin, Wet spinning

INFLUENCE OF DIE DESIGNS ON THE PRODUCED NONWOVEN WEBS IN MELT BLOWING PROCESS

Dawud H. Tan, William J. Kopecky, James C. Briester, James P. Endle, 3M Company, St. Paul, MN, USA.

Introduction

A nonwoven is a web of randomly oriented fibers bonded by physical entanglements or adhesion but not woven or knitted[1]. Nonwoven applications range from disposable wipes to medical apparel to filtration media [2]. One common process to produce nonwovens is melt blowing. Polymer melt is extruded through a collection of small orifices while simultaneously blows hot air jets to attenuate the extrudates and form microfibers, which are subsequently collected as a nonwoven web.

The earliest concept of melt blowing was documented in Hall's patent where melt blown vinyl resin was claimed for its use as storage battery retainer [3]. Approximately a decade later, Wente [4] published his effort to use melt blowing process in making sub-micron fibers to trap radioactive particles in the upper atmosphere. His work is usually referred to as the first documented effort of using melt blowing process. In the next two decades, various melt blowing patents belonging to industrial companies (most notably Freudenberg [5], Kimberly-Clark [6], Exxon [7], [8], [9]) were published. The melt blowing process claimed in these patents all used the same die design where one row of polymer orifices is impinged by two air jets, which now is commonly known as Exxon die.

Two die design concepts that deviate from Exxon die were first documented in 1980s. Lau [10] showed the design idea of utilizing a central row of air being impinged by two rows of polymer orifices. Mende [11] documented another design in which multiple co-axial polymer/air orifices are arranged in a rectangular array. Mende's idea spawned other publications with similar concept, such as those proposed by Schwarz [12], Torobin & Findlow [13], James et al. [14], Brown et al. [15], Gerking [16]. Yet another idea was proposed by Fabbricante et al. [17] where stacked-shims concept was used to pack polymer orifices and air slots in a rectangular array.

The present work centers on two novel die designs, cf. Figure 1, where both have two-dimensional (non-linear) orifice arrangements that lead to a higher orifice density and potentially improved productivity. These designs increase the orifice density without significantly reducing the orifice size and spacing as reductions in orifice size and spacing have been shown to correlate with reduction in web uniformity (i.e. web contains more shot, a irregularly shape polymer blob that's much larger than the fibers). Various modifications and alterations of the designs shown in Figure 1 will be apparent to those skilled in the art without departing from the scope and spirit of this study. This study should not be restricted to that which has been set forth herein only for illustrative purposes.

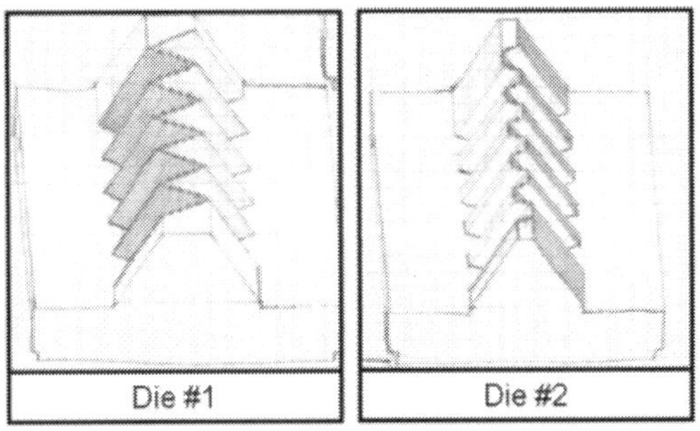

Figure 1: Two new melt blowing die designs that were investigated for the purpose of improving process throughput without compromising the resulting web's uniformity.

The two-dimensional nature of these new designs is expected to produce a different aerodynamics. Figure 2 shows that Die #1 results in rotational aerodynamics that is new and different than that of Exxon die, while Die #2 produces aerodynamics that is a 50:50 mixture of the Exxon die and Die #1. The rotational aerodynamics is expected to result in new air-fiber interaction in melt blowing.

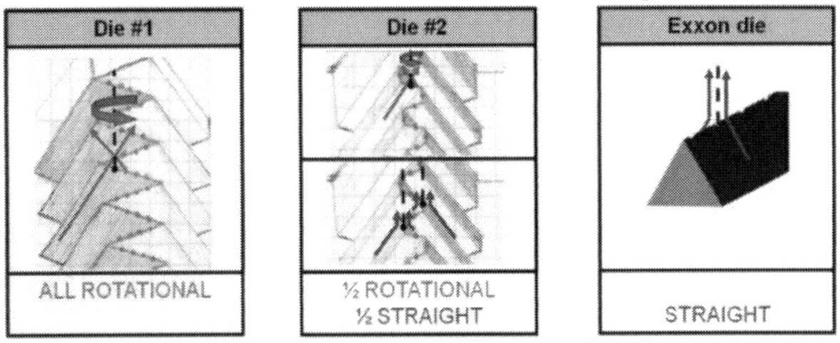

Figure 2: Schematics detailing the different aerodynamics observed in Die #1 and Die #2 in comparison with the aerodynamics of Exxon die. Each die have its corresponding air knives but they are not shown here for clarity.

Experiment

Melt Blowing Equipment & Material

Experiments were done on a melt blowing line with 2" wide die at the 3M Corporate Process Research Laboratory (St. Paul, MN). The die was oriented horizontally and heated with PID-controlled cartridge heaters. The air was heated with an in-line heater (Sylvania SureHeatMax, 10 kW, 240 V) that was set at 20-25 °C higher than the die temperature. The air flow rate was measured with a rotameter (Omega Engineering Inc. FL-2072) located upstream of the heater. A single-screw extruder (Brabender, 16 mm screw diameter) was used to feed polymer into the die. Polymer flow rate was measured by weighing the meltblown web within a given time. The produced webs were collected on a porous, woven, thermoplastic conveyor belt (50 ft/min maximum speed) located 20-24 inch away from die face. Three dies were investigated in this study. Table 1 shows that in comparison to Exxon die, the new designs double the orifice density without changing the orifice spacing and with slight reduction in orifice diameter. In all cases, the die tips were set to protrude out from the air gap set by the air knives (known as positive setback). Unlike in Exxon die where the air gap is equal along the die width, the new designs produced air gap that varies along the die width. Table 2 shows the two polypropylene (PP) materials that were used in this study. They were utilized due to their difference in viscosity, which is quantified with melt flow index (MFI).

Table 1: Three die designs investigated in this study. The orifice density is defined as the total number of orifice divided by the linear width of die tip. The die tip setback is positive when the die tip protrudes out from the air gap.

Die	Orifice density (hpi)	Orifice Dia.(inch)	Orifice Spacing (inch)	Die Tip Setback (inch)	Air Gap Width (mils)
Exxon	25	0.015	0.040	+ 0.020	0.030
Die #1	50	0.012	0.040	+ 0.020	0.015 - 0.045
Die #2	50	0.012	0.040	+ 0.010	0.017 - 0.030

Table 2: Two polypropylene materials investigated in this study. The melt flow index and melting temperature values are provided by the resin manufacturers. The melt flow index values were measured following D-1238 ASTM method (2.16 kg – 230 °C). The melting temperature (T_m) values were measured with differential scanning calorimetry (DSC).

	Trade Name	Melt Flow Index (g/10 min)	Tm (°C)
PP-low MFI	Total 3860X	100	165
PP-high MFI	Basell Metocene MF650Y	1800	154

Fiber Diameter & Shot Analysis

The average fiber diameter was measured during the experiments using pressure drop-diameter correlation and after the experiments using scanning electron microscopy (SEM). The effective fiber diameter (EFD) calculated during melt blowing experiment was based on an empirical equation set forth by Davies [18]. EFD was measured by forcing air to flow through the web (face velocity = 13.7 cm/s) and measuring the web's pressure drop, thickness, and weight. The average fiber diameter was determined after the experiment by recording fiber images with a desktop SEM (Phenom Pure). Typically 100-150 fiber diameters were measured from the images, and statistical analysis software package (Microsoft Excel) was used to determine the average fiber diameter. The amount of shot in the web was quantified by measuring the area covered by shots in the images using image analysis software (Olympus). Subsequently, shot percentage was defined as the percentage of area covered by the shot(s) to the total area in each image and an average was taken from all images.

High Speed Visualization

Figure 3 shows the high speed visualization setup used herein to better understand the air-polymer interaction. The high speed camera (FastCam APX RS with 200 mm macro lens) focused on an area indicated by the red box, which was illuminated with a high intensity light source (ARRI Pocket Lite 400W HMI). Figure 3 shows an example where the camera focused at the die tip where the polymer first exits the die and interacts with the air jets. The camera field of view was set at 6 x 12 mm. The camera recording speed was set to 20,000 frames per second (fps).

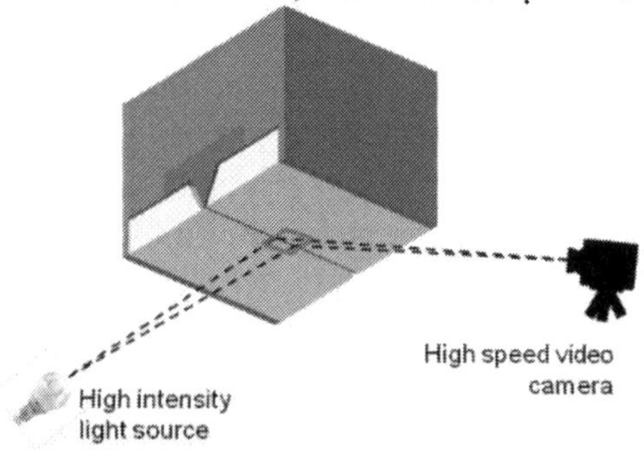

Figure 3: An illustration of high speed visualization setup used herein to visualize air-polymer interaction.

Results & Analysis

Die #1

Table 3 lists two melt blowing experiments performed with Die #1 in comparison with Exxon die while using the two PP materials. The results show that PP-high MFI is better than PP-low MFI in terms of shot formation (5.10 & 13.62%) but they are both worse than the Exxon dies (0.35 & 3.19%). This might be due to the new twisting aerodynamics inherent in Die #1. High speed visualization setup illustrated in Figure 3 was used to investigate the higher shot formation with Die #1. Figure 4 shows chronological snapshots from two videos that were recorded with Die #1 at two different locations: a) at the die tip, and b) 20 mm from the die tip. Figure 4a) points to fiber twirling motion in addition to fiber whipping motion regularly-observed in Exxon die, which was studied in detail by Breese & Ko [19]. The twirling motion is suspected to enlarge the radius of the fiber whipping cone and cause the fiber to touch neighboring fibers sooner, and thus produces more shots. This is further confirmed with another video recorded at ~20 mm away from the die tip, snapshots of which are shown in Figure 4b). Unlike in Exxon die where continuous fiber streams are typically seen, Figure 4b) shows more shots intermixed with the continuous fiber streams moving away from the die tip.

Table 3: Melt blowing experiments with Die #1 with two PP materials while monitoring shot formation. Exxon die have a hole-density of 25 hole-per-inch, hole diameter of 0.015", setback of +0.020", and air gap width of 0.030". Die #1 have a hole-density of 50 hole-per-inch, hole diameter of 0.012", setback of +0.020", and air gap width of 0.015"-0.045". The Q_{air} in Exxon die was set lower in order to match the fiber diameter of the resulting webs.

			Die #1				Exxon Die			
Material	T_p (°C)	Q_p (g/hole.min)	Q_{air} (SCFM)	SEM d_{av} (µm)	EFD (µm)	Shots (%)	Q_{air} (SCFM)	SEM d_{av} (µm)	EFD (µm)	Shots (%)
PP-low MFI	320	0.24	43	1.50	12.30	13.62	31	3.15	9.90	3.19
PP-high MFI	260	0.10	41	2.15	7.20	5.10	30	1.60	5.60	0.35

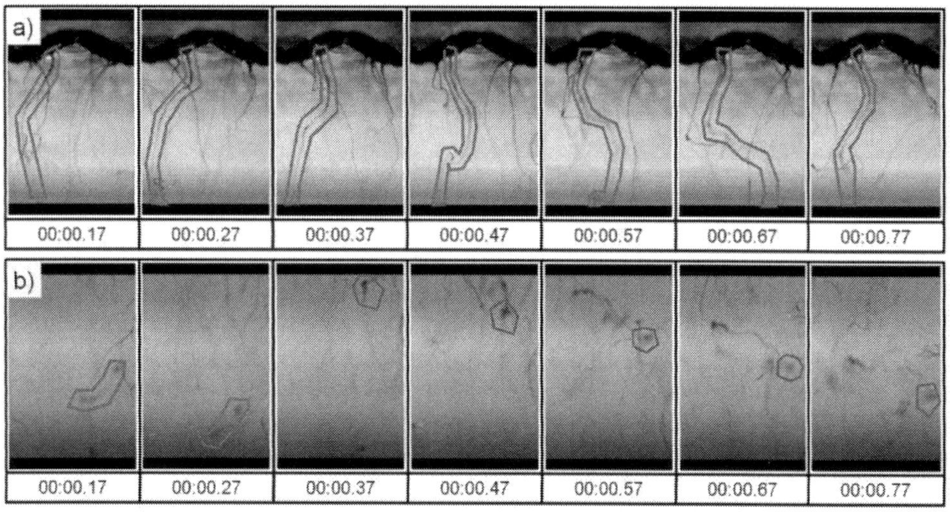

Figure 4: Chronological snapshots from high-speed videos of PP-low MFI interacting with air jets with Die #1 at 320°C, 0.24 g/hole.min and 43 SCFM. The videos were recorded: a) at the die tip where the movement of one fiber strand is highlighted by the red box, b) 20 mm away from the die tip where the movement of two different shots is highlighted by the red and green boxes. For clarity, the video field of view was post-reduced from 6x12 mm^2 to 6x4 mm^2. The time stamp of each snapshot is listed underneath each image. The videos were recorded at 20,000 fps and displayed at 10 fps.

SEM images of webs produced with Die #1 shown in Figure 6 highlight the presence of a small portion of helical fibers that are not usually seen in webs produced with Exxon die. This is attributed to the novel aerodynamics in Die #1. This is new and unlike the other known methods to generate helical fibers, which is achieved by either using two distinct materials with different coefficient of thermal expansion (e.g. Terakawa [20], Marmon & Midkiff [21], Yokozawa et al. [22]), differential temperature of air jets (e.g. Brown et al. [23], Tsubata [24]), or simply mechanically crimping the fibers.

Die #1 was first compared with Exxon die to produce meltblown webs with PP-high MFI at less benign processing conditions (typically high Q_{air} and $T_{polymer}$) as shown in Table 4. Approximately 10 SEM images were taken from each nonwoven web and the amount of helical fibers in each picture was counted. The average concentration of helical fiber ($C_{helical,avg}$) was defined as the amount of helical fibers found in all SEM images divided by the total number of SEM images. For fair comparison, each SEM image was taken at different magnification levels to guarantee that the total number of fiber in each image is approximately constant (~500).

Table 4 shows two experiments with the Exxon die where the $Q_{polymer}$ values (0.09-0.15) bracket the $Q_{polymer}$ of Die #1 (0.11). Furthermore, Die #1 was operated at a higher Q_{air} (40) than Exxon die (30) to account for the larger air slot area in Die #1. Table 4 also shows another run with Die #1 where the air flow rate was the same as Exxon die (~30). Despite all the variations, Table 4 shows that Die #1 always produces web with a higher concentration of helical fiber than Exxon die (~ 3x). Figure 6 compares the webs produced with Die #1 and Exxon die and it can be seen that the helical fibers in web made with Die #1 are usually longer and wound tighter than those seen in web made with Exxon die.

Table 4: Less benign processing conditions that were used to compare Die #1 and Exxon die with PP-high MFI as a starting material. $C_{helical,avg}$ is a measure of helical fibers concentration in each sample.

Die	$T_{polymer}$ (°C)	T_{air} (°C)	$Q_{polymer}$ (g/hole.min)	Q_{air} (SCFM)	$C_{helical,avg}$ (fibers/picture)
Die #1	260	280	0.11	33	3.33
	260	280	0.11	40	3.50
Exxon	260	280	0.09	30	0.44
	260	280	0.15	30	0.60

The helical fibers in Exxon die webs are likely caused by the higher frequency of fiber breakage due to stronger turbulence at less benign processing conditions. Turbulent eddies likely catch and twist the fiber's end, and produce a fiber that is partially helical along its length. This is confirmed by the observation of some helical fibers with one broken end. The higher $C_{helical,avg}$ in Die #1 is likely due to the synergistic combination of: 1) fiber breakage at less benign processing condition (as seen in Exxon die) and, 2) the larger fiber whipping cone radius in Die #1. Larger whipping cone radius increases the chance of fiber to interact with its neighbors and break more often. The same rotational aerodynamics is also hypothesized to cause the longer and tighter-wound helical fibers seen in Die #1 webs. The rotational aerodynamics produces a stronger turbulent eddies, which twists the fiber more after it breaks.

Comparison of Die #1 and Exxon die were also done at more benign processing conditions (lower Q_{air} and $T_{polymer}$), which should reduce/suppress fiber breakage and thus helical fibers in Exxon die webs. In contrast, helical fibers are still expected in Die #1 webs because more benign operating conditions weaken the turbulence but not the rotational aerodynamics. Table 5 shows that Exxon die webs now do not contain helical fibers. On the other hand, Die #1 webs still have some helical fibers. The measured $C_{helical,avg}$ values are lower than those measured at less benign conditions, which is expected. Figure 7 shows helical fibers in Die #1, but not Exxon die, webs.

Table 5: Comparing Die #1 and Exxon die with PP-high MFI as a starting material at a more benign processing condition. $C_{helical,avg}$ is a measure of helical fibers concentration in each sample.

Die	$T_{polymer}$ (°C)	T_{air} (°C)	$Q_{polymer}$ (g/hole.min)	Q_{air} (SCFM)	$C_{helical,avg}$ (fibers/picture)
Die #1	220	220	0.06	21	1.50
	220	220	0.13	21	1.25
Exxon	220	210	0.06	16	0.00
	220	210	0.14	16	0.00

Fiber-fiber interaction in Die #1 was manipulated by plugging 80% of the holes as shown in Figure 5. Other processing conditions, except $Q_{polymer}$, remained constant. This should allow each fiber to whip and twirl without touching its neighbors too soon and consequently reduce the amount of helical fibers in Die #1 webs. Table 6 shows that the 80% plugged cases result in lower amount of helical fibers. It should be noted that the higher frequency of fiber breakage observed in Die #1 causes the produced webs to always have a higher shot content that those made with Exxon die.

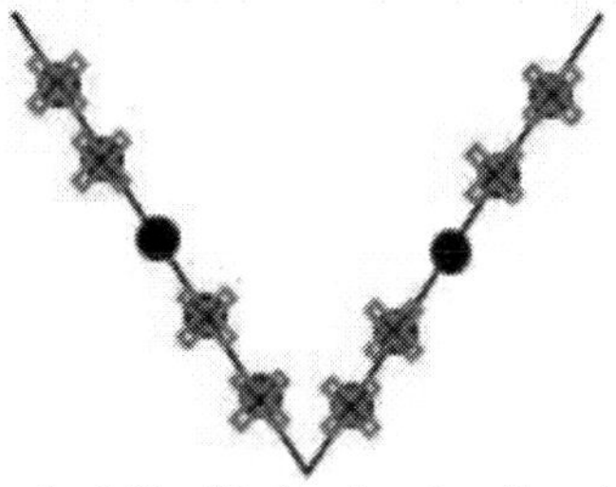

Figure 5: A top-view illustration of a segment of Die #1 tip showing the chosen configuration while plugging 80% of the holes to allow fibers to whip and twirl without significantly interact with its neighbors.

Table 6: Comparing a regular Die #1 with the same Die #1 but whose holes are 80% plugged. PP-high MFI was the starting material and a more benign processing condition was used. $C_{helical,avg}$ is a measure of helical fibers concentration in each sample.

Die	$T_{polymer}$ (°C)	T_{air} (°C)	$Q_{polymer}$ (g/hole.min)	Q_{air} (SCFM)	$C_{helical,avg}$ (fibers/picture)
Die #1	220	220	0.06	21.0	1.50
	220	220	0.13	21.0	1.25
Die #1	220	210	0.05	21.0	0.89
(80%plugged)	220	210	0.13	21.0	0.33

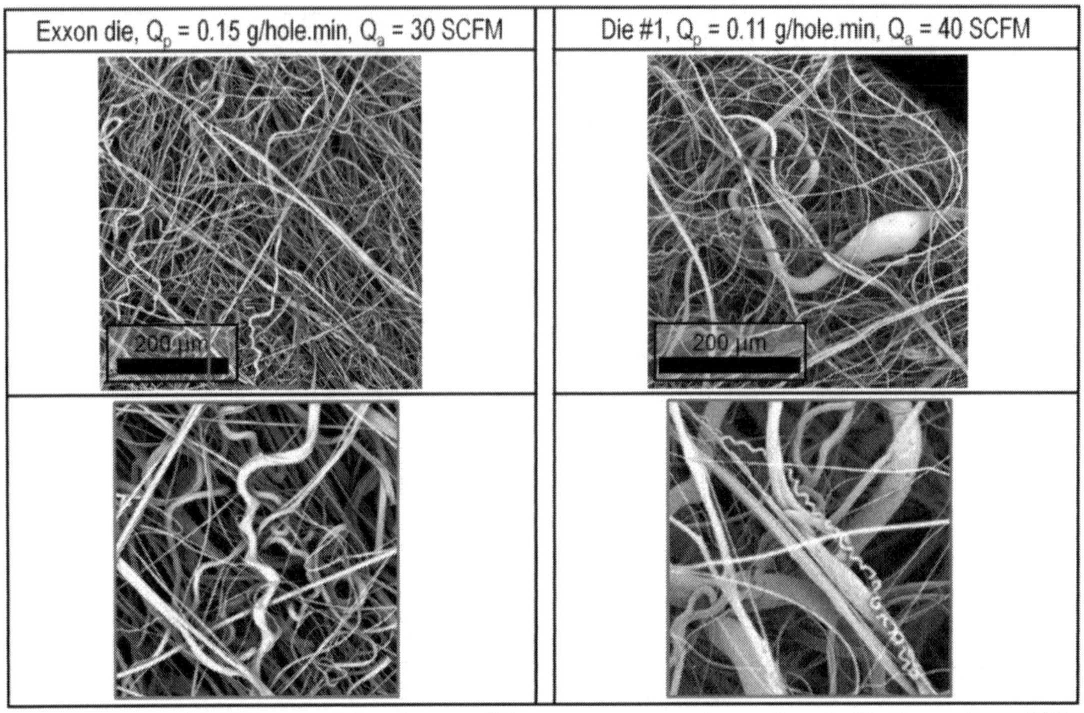

Figure 6: A collection of SEM images taken from meltblown nonwoven webs produced with Exxon die and Die #1 at $T_{polymer}$ = 260 °C and T_{air} = 280 °C. A closer inspection of these SEM images (magnified red-boxed images) shows that the helical fibers produced with Die #1 are usually longer and wound tighter.

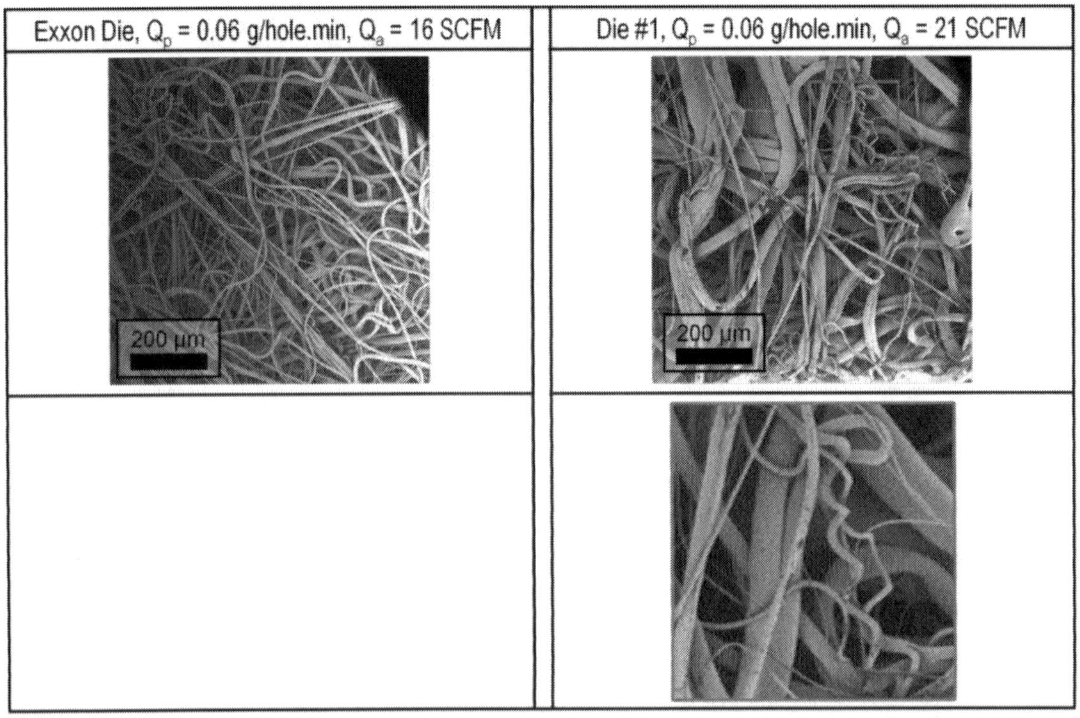

Figure 7: A collection of SEM images taken from meltblown nonwoven webs produced with Exxon die and Die #1 at $T_{polymer}$ = 220 °C. A closer inspection of these SEM images (magnified red-boxed image) shows that the helical fibers are still observed in Die #1 webs but completely absent from Exxon die webs when a more benign processing condition was used.

Die #2

Table 7 shows two experiments done with Die #2 in comparison to Exxon die while using two PP materials. As seen with Die #1, PP-high MFI is better than PP-low MFI (4.56 & 9.74%) in terms of shot formation but both are worse than Exxon die (2.46 & 0.95%). Videos of both PP materials were recorded as it came out from the die tip and representative snapshots are shown in Figure 8a) and b). Both materials behaved very differently as they came out from the die tip. Unlike in Exxon die where the polymer is instantaneously stretched after exiting the die tip, polymer builds up on the tip of Die #2 and forms a bulbous mass before it eventually got stretched and/or blown away from the die tip. These droplets are likely responsible for the higher shot formation observed with Die #2.

Table 7: Experiments comparing Die #2 with Exxon die using two PP materials while monitoring shot formation. Exxon die have a hole-density of 25 hole-per-inch, hole diameter of 0.015", setback of +0.020", and air gap width of 0.030". Die #2 have a hole-density of 50 hole-per-inch, hole diameter of 0.012", setback of +0.010", and air gap width of 0.017"-0.030". The Q_{air} in Exxon die was set lower in order to match the fiber diameter of the resulting webs.

			Die #2				Exxon Die			
Material	$T_{polymer}$ (°C)	$Q_{polymer}$ (g/hole.min)	Q_{air} (SCFM)	SEM d_{av} (µm)	EFD (µm)	Shots (%)	Q_{air} (SCFM)	SEM d_{av} (µm)	EFD (µm)	Shots (%)
PP-lo MFI	300	0.20	42	1.75	13.90	9.74	30	2.40	10.30	2.46
PP-hi MFI	260	0.20	42	2.05	5.40	4.56	30	1.40	5.90	0.95

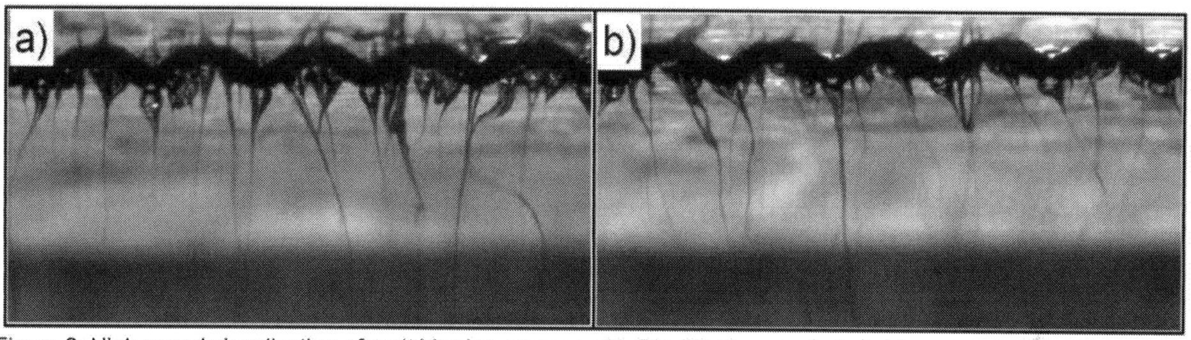

Figure 8: High speed visualization of melt blowing process with Die #2: a) snap shot of video recorded for PP-low MFI at 300°C, b) snap shot of video recorded for PP-high MFI at 260°C. Both were run at 0.20 g/hole.min and 42 SCFM. The video field of view was 6x12 mm^2. They were recorded at 20,000 fps and displayed at 30 fps.

The tip's shape of Die #2 possibly makes it easier for polymer to build-up on the corners and/or compromises the jet aerodynamics such that there's inadequate force to attenuate the polymer. The die tip setback of Die #2 was changed from positive to negative (i.e. tip is hidden behind air gap), to improve the aerodynamics and attenuation force. Table 8 shows changing the setback does not help with shot formation and if any, negative setback appears to worsen shot formation in Die #2, i.e. from 8.09 to 13.84%, and from 4.12 to 5.74%.

Table 8: Experiments to investigate the effect of setback in Die #2 with PP-low MFI while monitoring shot formation. Die #2 have a hole-density of 50 hole-per-inch, hole diameter of 0.012", and air gap width of 0.017"-0.030".

			Setback = +0.010"			Setback = -0.010"		
$T_{polymer}$ (°C)	$Q_{polymer}$ (g/hole.min)	Q_{air} (SCFM)	SEM d_{av} (µm)	EFD (µm)	Shots (%)	SEM d_{av} (µm)	EFD (µm)	Shots (%)
320	0.08	42	1.29	6.90	8.09	2.40	9.20	13.84
320	0.14	42	1.28	7.10	4.12	1.60	8.00	5.74

Summary

Two new melt blowing die designs were investigated, all with two-dimensional (non-linear) hole arrays. In comparison to Exxon die, Die #1 produced a novel aerodynamics that resulted in unique air-fiber interaction as evidenced by high speed visualization. The new aerodynamics also created nonwoven fabrics containing small amount of helical fibers with diameter similar to meltblown fibers. Die #2 was challenged by polymer build-up on die tip. In comparison to Exxon die, both designs undesirably created nonwoven webs with more shots.

References

[1] I. M. Hutten, "Processes for Nonwoven Filter Media," in *Handbook of Nonwoven Filter Media*, Burlington, MA, Elsevier/Butterworth-Heinemann: Oxford, 2007, pp. V/195-244.

[2] I. Butler, Worldwide Outlook for the Nonwovens Industry: 2007 - 2012, Cary, NC: INDA, 2008.

[3] C. A. Hall, "Storage battery retainer". US Patent 2335757, 30 November 1943.

[4] V. A. Wente, "Superfine Thermoplastic Fibers," *Journal of Industrial and Engineering Chemistry,* vol. 48, no. 1, pp. 1342-1346, 1956.

[5] L. Hartmann, W. Riedmuller and G. Muller, "Apparatus and process for production of filaments". US Patent 3379811, 23 April 1968.

[6] E. Schwarz, "Apparatus for producing strong and highly opaque random fibrous webs". US Patent 3806289, 23 April 1974.

[7] D. Lohkamp and J. Keller, "Melt-blowing die using capillary tubes". US Patent 3825379, 23 July 1974.

[8] J. Harding, J. Keller and R. Buntin, "Melt-blowing die for producing nonwoven mats". US Patent 3825380, 23 July 1974.

[9] R. Buntin, J. Harding and J. Keller, "Non-woven mats by melt blowing". US Patent 3849241, 19 November 1974.

[10] J. C. Lau, "Extrusion process using a central air". US Patent 4818464, 4 April 1989.

[11] T. Mende, "Melt blow die". US Patent 4826415, 2 May 1989.

[12] E. C. Schwarz, "Process and apparatus for producing non-woven webs of strong filaments". US Patent 6013223, 11 January 2000.

[13] L. Torobin and R. C. Findlow, "Method and apparatus for producing high efficiency fibrous media incorporating discontinuous sub-micron diameter fibers, and web media formed thereby". US Patent 6315806, 13 November 2001.

[14] M. D. James, S. R. Jackson, S. Aydore, H. Eroglu, D. E. Ensign, P. D. Trokhan and

E. A. Stewart, "Method for forming fibers". US Patent 7939010, 10 May 2011.

[15] D. B. Brown, J. D. Stark, C. A. Granato and D. K. Zacharias, "Apparatus for extruding cellulose fibers". US Patent 8029260, 4 October 2011.

[16] L. Gerking, "Spinneret for spinning threads, spinning device for spinning threads and method for spinning threads". US Patent 20130217290, 22 August 2013.

[17] T. J. Fabbricante, A. S. Fabbricante and G. F. Ward, "Micro-denier nonwoven materials made using modular die units". US Patent 6114017, 5 September 2000.

[18] C. N. Davies, "The Separation of Airborne Dust and Particles," in *Institute of Mechanical Engineers Proceedings 1B*, London, 1952.

[19] R. R. Bresee and W.-C. Ko, "Fiber formation during melt blowing," *International Nonwoven Journals,* vol. 12, no. 2, pp. 21-28, 2003.

[20] T. Terakawa, "Continuous fiber nonwoven and the method for producing it". US Patent 6518208, 11 February 2003.

[21] S. E. Marmon and D. G. Midkiff, "Multilobal conjugate fibers and fabrics". US Patent 5707735, 13 January 1998.

[22] M. Yokozawa, R. Kinoshita, S. Okuda and K. Tanaka, "Polyester conjugate fiber". JP Patent 2655557, 8 March 1989.

[23] B. K, C. Quinn, J. Shelley and S. Sykes, "Helically crimped, shaped, single polymer fibers and articles made therefrom". US Patent 20030104748, 5 June 2003.

[24] M. Tsubata, "Nonwoven fabric made of crimping continuous filament and its production". JP Patent 04100963, 2 April 1992.

365779 Modeling of Wet Spinning

Abstract:

Wet spinning is an important processing operation used in the production of fibers from systems such as polyacrylonitrile and the cellulosic and vinyl-based polymers, as well as in the fabrication of hollow fibers for membrane separation and drug delivery applications. In wet spinning, an initially homogeneous solution is extruded directly into a liquid bath where mass, momentum, and heat transfer processes take place, resulting in liquid demixing and gelation to form a solid, structured fiber. Modeling involves coupling the simultaneous transport and phase separation processes with constitutive models that account for viscoelastic effects. Important effects that should be captured include predictions of the locking-in of fiber diameter, stresses, and internal structure along the spin line, since these, to a large extent, fix the as-spun fiber properties. The text by Ziabicki (1) reviews wet spin modeling up to about 1976; however, since then, there have been relatively few new studies reported. The most complete recent studies we are aware of are those of Kalabin et al. (2) that combine Newtonian constitutive modeling with a two-phase binary diffusion model that incorporates thermally-induced phase separation/gelation along the spinline. Our recent studies follow a different approach built on our successful dry spinning modeling (3) that incorporates a two-component constitutive model based on parallel combination of Newtonian and Giesekus equations. This is coupled with a 2-d pseudo-ternary diffusion formalism that utilizes separate diffusion equations for the fiber gel and solution phases. These are combined with mass transfer boundary conditions at the fiber-bath interface and incorporation of the ternary phase diagram based on Flory-Huggins theory to quantify the gelation dynamics along the spinline. Model results illustrating predictions of the spinline fiber profiles and structuring for a variety of systems and processing conditions will be discussed along with comparisons of literature data for the wet spinning of PAN fibers.

1. A. Ziabicki, *Fundamentals of Fiber Formation*, John Wiley and Sons, NY, 1976.
2. A.L. Kalabin, *Fibre Chemistry,* **37**, 141 (2005).
3. Z. Gou and A.J. McHugh, *J. Non-Newt. Fl. Mech.* **118**, 121 (2004).

362246 A Modelling Pathway and Software Tool Connecting Molecular Structure to Linear Viscoelastic Properties and Melt Flow Index for Polyolefins

Abstract:

Polymer rheological properties are crucial for the processing behaviour as well as the end-use properties of polyolefins. Moreover, they are widely used for polymer quality control. Many different experimental methods exist in order to describe the complexity of polymer rheological behaviour. In industrial practice, frequency sweep and melt index measurement are mainly used. Frequency sweep measurement is typically employed for viscoelastic properties description providing polymer melt viscosity versus frequency diagram (flow curve). On the other hand, melt flow index measurement (MI) is the one that has gained great popularity in the field defined by the mass of polymer, in grams, flowing in ten minutes through a capillary under specified conditions. MI offers a fast, easy to perform and low cost polymer rheological behaviour characterization connected to end-use customer specifications.

Polymer micro-structure is affecting the rheological behaviour of the polymer. The term micro-structure describes fundamental properties of the polymer including molecular weight, co-polymer composition and chain sequence distribution. A clear example describing this connection is the following: high molecular weight polyolefins exhibit improved end-use mechanical properties (e.g., toughness, strength, impact resistance and environmental stress cracking resistance, etc.). At the same time, they have higher melt viscosities and, therefore, it is more difficult to process them. Usually, a balance is sought; a broad or bimodal MWD exhibits improved mechanical properties and good processability levels as well. In this way, rheology acts as a link between molecular structure and final properties of a polymer (Touloupidis V., **2011**, 'Mathematical Modelling and Simulation of an Industrial α-olefins Catalytic Slurry Phase Loop-reactor Series', *PhD thesis*, Aristotle University of Thessaloniki). Thus, fundamental understanding of polymer rheology and having a method for the prediction of viscoelastic polymer properties on base of the underlying molecular micro-structure offers a valuable tool for more efficient polymer product development and reverse engineering.

In this work, a modelling pathway and software tool is presented for linking entangled linear polymer molecular properties to linear viscoelasticity and MI values. A two-step approach is followed: (i) a model developed under Matlab environment links molecular properties with flow curves, and then, (ii) a Polyflow model calculates MI values based on the flow curves predicted. The method applies for linear homo-polyethylene grades while it can be additionally extended for 'slightly' non-linear co-polymers.

The connection between the polymer molecular micro-structure and the viscoelasticity is achieved through a model combining reptation and Rouse relaxation concepts and basic rheology theory found in literature (van Ruymbeke E., Keunings R., Stephenne V., Hagenaars A. and Bailly C., **2002**, 'Evaluation of Reptation Models for Predicting the Linear Viscoelastic Properties of Entangled Linear Polymers', *Macromolecules* 2002, 35, Kiparissides C., Pladis P. and Moen O., **2006**, 'From Polyethylene curves to Molecular Weight Distributions', *Multiscale Modelling of Polymer Properties*, Elsevier). Based on molecular weight distribution measurements obtained via size exclusion chromatography (SEC), the developed rheology model is able to predict viscoelastic properties (including relaxation modulus, storage and loss moduli, flow curve, etc.) of linear polymers (e.g., polyethylene, PE, polystyrene, PS, polycarbonate, PC). The strong point of this method is that the predicted flow curve includes the polydispersity effect and the actual shape of the molecular weight distribution as this information. A software tool was developed under Matlab programming language based on the method described as a user-friendly stand-alone application, entitled 'BorVisc', presented via a user-friendly graphical user interface environment.

Furthermore, on the basis of the commercial Finite Elements CFD software package Ansys®Polyflow a model was developed computing the MI value based on the flow curve predicted by BorVisc. The simulation model takes into account geometry details of the MI instrument as well as the specific loading of measurements (e.g., 2.16, 5, 10, 21Kg). In addition to MI values, the model is able to predict in detail pressure, velocity and shear rate profiles of measurements (i.e., barrel and capillary region). Polyflow model together with Borvisc offer the pathway to move from purely molecular micro-structure to the other end of MI application-oriented bulk properties.

The method was thoroughly tested and validated for different low- and high-density polyethylene grades exhibiting great performance. Accuracy levels in the range of ±10% on average is exhibited considering both steps of computation (i) MWD to flow curves and (ii) flow curves to MI. Taking into account that MWD, flow curve and MI experimental measurement deviations are inevitably present, one could note that the model is close to resolution of measurements. These promising results offer a solid ground for using the methodology presented (i.e., BorVisc software and Polyflow model) as a valuable tool to enhance product development towards the direction of end-use polymer bulk properties prediction. Moreover, the method offers also high potential when applied for reverse engineering, computing a molecular microstructure based on desired rheological properties.

SYNTHESIS OF POLYLACTIDE-GRAFT-MALEIC ANHYDRIDE VIA FREE RADICAL INITIATION IN SUPERCRITICAL CARBON DIOXIDE

Barbara A. Wheelden, Amber R. Tupper, Sunggyu Lee
Ohio University, Athens, OH

One of the most popular bio-based polymers on the market today is polylactide (PLA). Although PLA is bio-based, biodegradable, produced on a large scale, and able to be processed by traditional polymer processing equipment, its widespread use is limited by its relatively poor mechanical properties. PLA possesses a high tensile modulus, but low elongation and toughness. It is an inherently brittle material and therefore a poor choice for flexible packaging applications in its virgin form. PLA can be melt blended with more flexible polymers to increase its suitability for a wider range of applications, but due to their high molecular weight and viscosity, polymers are rarely miscible with each other. Immiscible polymer blends cause premature separation of polymeric phases, discontinuity of mechanical and thermal properties, and shortened service life. Graft copolymers can serve as compatibilizers between immiscible polymer blends to improve their miscibility and subsequent mechanical properties.

In this work, polylactide-graft-maleic anhydride (PLA-g-MA) will be synthesized with the ultimate objective of incorporating this graft copolymer as a compatibilzer into bio-based polymer blends with PLA as the main phase. PLA-g-MA will be synthesized in a stainless steel batch reactor in the presence of supercritical carbon dioxide ($scCO_2$). ScCO2 is an effective medium for polymerization because while most polymers do not dissolve, they tend to swell in scCO2, allowing for diffusion of reactive monomer into the swollen polymer. In addition, the density of the CO2 can be easily tuned by changing pressure and/or temperature, and the reaction can take place under ideal conditions without the harsh organic solvents utilized in solution grafting or high temperatures employed in molten state grafting.

In this study, the amount of initiator and MA will be kept constant at 2.5 wt% and 10 wt%, respectively, and the scCO2 density will be held at 300 kg/m^3. Azobisisobutyronitrile (AIBN) and Benzoyl Peroxide (BPO) will be used as the free radical initiators, and reactions will be carried out at three different temperatures for each of two initiators studied.

After the grafting reaction has taken place, the samples will be purified using Soxhlet extraction in methanol to remove excess monomer and spent initiator and Fourier Transform Infrared (FTIR) and Proton (^1H) Nuclear Magnetic Resonance (NMR) spectroscopy will be utilized to ensure that grafting has taken place and to determine the grafting level. The optimum reaction temperature and initiator will be determined based on these results.

387512 Ductility Improvement of Film Products Made of Nano-Fibrillated Saccharification Residues of Wood

Abstract:

The objective of this study was to investigate the effects of polymer incorporation on the tensile properties of films made of nano-fibrillated saccharification residues of wood. In earlier work, we demonstrated that it is technically feasible to utilize these lignin-rich residues for making plastic-like films, but the resulted films are brittle. In this study, we addressed the issue by incorporating a ductile thermoplastic at relatively low mass fractions, in contrast to the conventional approach (low fiber content) with nanofiber-polymer matrix composites. To implement this research, aspen wood flour was first pre-treated with an alkaline solution and then enzymatically saccharified to attain 80% glucose release from glucan. The solid residuals were mechanically fibrillated in aqueous media to fibrous fragments of micron and nanometer size. Some pre-treated wood flour (without saccharification) was also subjected to mechanical fibrillation. Fibrillated fragments from both pre-treated and saccharified biomass were blended at a solid mass ratio of 50:50. An appropriate amount of poly (vinyl alcohol) (PVA) solution was added to the mixture, and the slurry was immediately decanted to form into mats and subsequently dried to films. Results showed that PVA incorporation at 1.2% and 21% mass fraction of the films slightly improved tensile strain at break, but the films remained brittle. Increasing the PVA content to 37% mass fraction increased the strain to three folds, without compromising tensile strength of the films. From 24h-soaking study, incorporation of PVA also has the advantage of maintaining the film integrity in water with negligible weight loss (~1.0%). Other results such as wet strength and moisture adsorption will be discussed in the light of application potential. Overall, we successfully improved ductility while sustaining the tensile strength of nanofiber films. This outcome contrasts with the use plasticizers which usually increase ductility at the expense of strength (in addition to stiffness). By improving ductility of films containing saccharification residues, this study contributes to expanding the application window and value of the film products, thereby advancing the possibility of coproducing sugars and value-added materials from bioconversion of woody biomass.

Chemistry and chemical engineering process for making PET from bio based monomers.

Damian Adrian Salazar Hernandez

Advisor: Dr. Saleh A. Jabarin

Polyesters are one of the most important and versatile polymers in use today. They are produced by the polycondensation reaction of a glycol or dialcohol with a difunctional carboxylic acid. Polyesters are continuously used for several purposes ranging from bottles for water, carbonated drinks, health care products to fibers or carpet.

Poly ethylene terephthalate (PET) is polyester formed by step-growth polymerization from ethylene glycol (EG) and terephthalic acid (PTA). EG is traditionally obtained from the oxidation of ethylene over metal catalyst. PTA is obtained from liquid phase oxidation of p-xylene. The synthesis of PET requires a two stage process: esterification and polycondensation. PET raw material EG and PTA are currently obtained through reforming process of oil and natural gas. There are alternative chemical routes identified on how to successfully transform the sugars obtained from biomass material into terephthalic acid and ethylene glycol. Yield of final product obtained is still in the range of low molecular weight material and effective separation represents a challenging step for researchers. In this project we produce PET using EG and PTA obtained from oil and also with both raw materials obtained from a bio based source. Polymer produced with each set of raw material is characterized and compared to identify similarities or differences between them. In our approach we separate PTA precursor (p-xylene) from a bio-based aromatic mixture. This mixture can come from any of the following chemical routes: 1) bio based oil (BTX) provided to us, 2) product from Diels Alder reaction of 2,5-dimethylfuran (DMF) and ethylene, 3) product effluent from pyrolysis reaction of 2-methylfuran (2MF) and propylene. It is anticipated that in any of the separations there might be impurities (other organic compounds) that may affect the purity of the final product obtained; therefore our main objectives of the project are to determine the effect of these low molecular weight compounds (impurities) on the polymerization and characteristics of PET and identify critical areas on the separation train of precursor.

Thermochemical transformation of biomass produces furanic compounds by dehydration of sugar molecules present in the cellulose and hemicellulose. These compounds can be reacted with olefins via Diels-Alder reaction to produce aromatics. Using CDS Analytical Pyroprobe 2-methylfuran and propylene react at 600 °C over zeolite catalyst. Product obtained contains p-xylene along with other high carbon aromatics. This experiment was done to prove the concept on the formation of desired chemicals. The next step is to do pyrolysis in a fixed bed reactor to collect the product effluent and do separation of p-xylene. As extension of the work with furanic compounds, 2-5 dimethylfuran is reacted with ethylene over zeolite Y catalyst in a 4590 Micro Bench Top Parr Reactor. This reaction follows a cyclo addition mechanism producing aromatics with mono, di and tri substituted groups.

Bio oil and liquid samples obtained from separation train of experiments in the research are characterized using Shimadzu GC2010 with flame ionization detector, Bruker GC450 with mass spectrometer and Shimadzu VP HPLC with UV detector. In all cases PTA precursor is separated from each of the mixture using fractional distillation and crystallization. For the first step, we use a packed bed and a plate column along with a reflux distillation head to separate streams containing aromatics with different boiling point. The stream containing xylene isomers is cooled down to temperatures below -50 °C, this allows us to crystallize p-xylene and separate it via vacuum filtration. P-xylene, cobalt acetate, acetic acid and air are fed to Parr reactor to do oxidation. The product obtained is a low grade crude terephthalic acid that contains different impurities, i.e. 4-carboxylbenzaldehyde (4-CBA). The subsequent step is the purification, the crude acid is mixed with Pd over carbon and deionized water to create a slurry of 13% w acid. This mixture is placed in the same Parr reactor and hydrogen is used to pressurize the system. A hydrogenation reaction takes place, transforming the 4-CBA into toluic acid. The solid/liquid product is separated by vacuum filtration obtaining purified terephthalic acid in solid form. This bio based acid and bio based ethylene glycol (provided to us) are used to do polymerization on Parr reactor and produce PET. To make a comparison with traditional process, EG and PTA are obtained from industry and used to make polymer with the same procedure and reactor as the bio based.

Table 1 show the results obtained from analysis of bio oil. Benzene, toluene and xylenes account for 76% w of the total mixture, p-xylene in specific is present at 10% w. At the current stage of the research we are able to separate meta-, ortho- and para-xylene as a combined stream with 78% purity; we are working on crystallization to obtain p-xylene as a purified product.

Table 1 – Identified components in Bio oil.

A	Benzene	M	Benzene, 1,2,4,5-tetramethyl-
B	Toluene	N	1-propenal, 3-phenyl-
C	Ethylbenzene, m-p-xylene	O	Benzofuran, 2-methyl-
D	o-xylene	P	Benzene, 4-ethenyl-1,2-dimethyl-
E	Benzene-propyl	Q	2-methylindene
F	Benzene, 1-ethyl-3-methyl-	R	1H-Indene, 1-methyl-
G	Benzene, (1-methylethyl)-	S	Naphthalene
H	Benzene, 1,2,4-trimethyl-	T	Naphthalene, 1,2-dihydro-3-methyl
I	Benzene, 1,3,5-trimethyl-	U	Naphthalene, 1,2-dihydro-4-methyl
J	Indane	W	Naphthalene,2-methyl-
K	Indene	X	Naphthalene,2-ethyl-
L	Benzene, (2-methyl-1-1propenyl)-	Y	Naphthalene,1,2-dimethyl-

On the thermochemical approach, we followed a stage test to determine the effect of catalyst and propylene on the formation of aromatic material. 2MF is tested on pyroprobe unit; the effluent of the reaction is send to Bruker GCMS for characterization. Once the chromatogram is obtained, the area under the curve for the desired peaks is measured. Table 2 show the ratio of the area under the curve considering the pyrolysis of sole 2MF as the reference value. We can see an increase in the formation of aromatics by adding catalyst to the system. Then when we add propylene to the reaction media there is a dramatic increase in the formation of xylenes. This proves the idea that the formation of aromatics follows Diels Alder process and the feed of an olefin boost reaction rate. In each of the chemical routes we have promising results towards obtaining polymer precursor from bio derived source.

Table 2 – Area ratio for selected aromatics in 2MF pyrolysis.

	Catalyst Location			Area Ratio			
Material	Tube	Reactor	Gas	Benzene	Toluene	EB + m/p-xylene	o-xylene
2-MF	X		He	1.0	1.0	1.0	1.0
2-MF		X	He	8.1	7.5	13.4	12.9
2-MF		X	HePro	11.3	22.1	65.3	73.0

392458 Co-Injection Moulding Bioplastics: Investigating Procedure and Performance

Abstract:

To help combat environmental issues like climate change and landfill waste, bioplastics are being developed to provide renewable and/or biodegradable polymer options. Compostable, corn-based polylactic acid (PLA) is strong and stiff, but brittle. Polycaprolactone (PCL), a petroleum-based compostable polymer, is very tough but has poor strength. Combining these plastics by melt blending is an established method to create a new material that balances their complementary properties; however, alternative processing methods have the potential to be more successful. Co-injection moulding is a manufacturing technique that combines polymers in a layered structure; it can help to produce more aesthetically pleasing products, reduce processing steps, and may result in improved mechanical performance. Determining operating parameters for optimal co-injection conditions for PLA and PCL will be investigated and final products will be compared to similar melt blended samples. Emphasis will be placed on proper part formation and impact resistance in evaluating a successful co-injected material.

Acknowledgement: The financial support from the Natural Sciences and Engineering Research Council (NSERC), Canada for the Discovery grants individual (to Mohanty) and the Ontario Ministry of Agriculture & Food and Ontario Ministry of Rural Affairs (OMAF-MRA) to carry out this research is gratefully acknowledged.

Ecological Glue – Towards a Sustainable Industry?

Merten Morales[a,*], Reto Bättig[a], Stavros Papadokonstantakis[a], Konrad Hungerbühler[a]

[a]Institute for Chemical and Bioengineering, ETH Zurich, CH-8093 Zurich, Switzerland

merten.morales@chem.ethz.ch

1. Background + Aims

The introduction of a new generation of biorefineries into the market presents a multi complex problem, where questions like environmental sustainability, economic feasibility and conceptual exploration need to be answered beforehand. The biorefineries of the second generation need to tackle the objective of how chemicals can be produced from biomass, and not compete with the food production. Lignocellulosic biomass presents a raw material with a high potential to replace corn or sugarcane. Innovative production routes from wood to chemicals have to be included within the biorefinery concept. The quest of finding the optimal biorefinery concept is becoming even more complex by considering besides the conventional also innovative processes for the production of chemicals. The approach for finding the optimal concept is that generic optimization algorithm, developed by EPFL in Lausanne, is able to integrate different processes in order to find a bio refinery concept under the objective of profit maximization and environmental impact minimization.

A major part of the research project is the modelling of the different production paths to produce either platform chemicals (e.g. Glucose, Ethylene, Succinic acid) or high value added chemicals (Polyurethanes, Polyesters, Vinyl polymers). For the production routes conventional concepts as well as innovative concepts are considered for finding the optimal production routes. The concepts are evaluated through environmental indicators (Global Warming Potential, Ecoindicator 99, Cumulative Energy Demand) and economic indicators (Profit, Operational costs, Investment cost).

2. Methods + Results

An important platform chemical, which needs to be considered for a biorefinery, is acetic acid. The main application of acetic acid is the production of vinyl acetate, which is used to produce polyvinyl acetate (PVA). PVA is a common chemical for glue or latex emulsions (e.g. application paper coating, textile treatment) [1]. Nowadays acetic acid is mainly produced from petrochemical resources. The most important synthetic route over the last 25 years to produce acetic acid is the carbonylation of methanol. For introducing a feasible route for acetic acid in a biorefinery concept we modelled and optimized the process of Wacker Chemie [2]. The Wacker process is converting ethanol to acetic acid by the gas phase oxidation of ethanol. The major aim of this research project is to point out the major environmental and economic cost in the production of PVA from lingocellulosic biomass. The results of the project show the economic and environmental impact of different steps for the production of PVA.

1. Converting Wood to Glucose
2. Fermentation of Glucose to Ethanol
3. Dehydration of Ethanol to Ethylene
4. Oxidation of Ethanol to Acetic acid
5. Addition of Ethanol and Acetic acid to Vinyl acetate
6. Polymerization of Vinyl acetate to PVA

3. Summary/Conclusion

Through the assessment based on different rigorous models a conclusion can be drawn if PVA produced from the renewable resource wood can be produced more environmental friendly or is more economical attractive than PVA coming petrochemical industry.

References

[1] Ullmann`s Enycylopedia of Industrial Chemistry, Acetic acid, ISBN: 9783527306732

[2] Wacker Chemie AG, Method for producing Acetic and Formic acid by the gas phase oxidation of ethanol, WO 2008/110468 A1

366034 Chitosan from Solid-State Fermentation of Soybean Residues By Filamentous Fungi As Biobased Paperboard Coating Additives for HVAC Applications

Abstract:

The HVAC (Heating, Ventilation, and Air-conditioning) industry markets phenolic resin coated-paperboard media for use in the air-cooling of agricultural livestock pens and enclosures. Phenolic resins such as phenol formaldehyde have been shown to improve dry stiffness and water resistance of paper base sheets. However, in HVAC applications where good wet strength and water absorption properties are required, phenol formaldehyde-coated paperboards unfortunately have reduced wet stiffness levels compared with uncoated paperboards, thus resulting into premature failure of the air-cooling media. Furthermore, the use of phenolic resins may reduce the recyclability or biodegradability of the used coated paperboards. Chitosan, a versatile biopolymer commonly derived from crustacean shells in seafood wastes, has found numerous applications across multiple disciplines due to its unique and interesting properties. Some of these properties include its antimicrobial quality, chelating properties, and the ability to form hydrogels, fibers, and films. Chitosan has also found potential applications in the pulp and paper industry, mainly as surface coating or sizing additive for paper and paperboard products with enhanced or functionalized properties. These properties include water resistance, air barrier, wet and/or dry strength, and antimicrobial properties. Thus, chitosan could serve as a bio-based alternative to conventional phenolic resin as paperboard coating materials. However, the current process for chitosan production involves costly, labor-intensive, and harsh chemical treatments of seafood waste and also suffers from raw material supply inconsistencies and geographical limitation. Fortunately, chitosan can also be found as cell wall components of filamentous fungi. These fungi can be grown in fermentation cultures using agricultural and food processing residues as substrates to produce high-quality chitosan continuously and sustainably, and with expectedly lower separation and recovery costs.

The current study investigated the use of soybean processing residues such as meal and hulls as substrate for the growth and chitosan production of fungi using solid-state fermentation (SSF). Selected fungal strains with documented high chitosan production capabilities were screened for direct growth and utilization of solid soybean residue fermentation substrate. Results show that three out of the six screened fungi strains exhibited expansive mycelial growth on soymeal substrate while the rest only showed liquefaction of the soymeal substrate. The fungal strains exhibiting growth on the soymeal substrate were selected for fermentation kinetics and optimization studies to determine optimal SSF conditions for growth and chitosan production. The product chitosan is intended for application as coating additives for paper-based evaporative cooling media requiring wet stiffness and water absorption/wicking characteristics. Economic analysis indicate a target chitosan yield of 20 % (w/w of substrate) in order to match the price of the conventional phenolic resin coatings of $3.00 per lb. Further cost reductions are possible with lower chitosan application weights or coating techniques for optimum wet stiffness and wicking performance. Parallel investigations were conducted using commercially available chitosan to determine the optimum levels of chitosan concentration in carrier solution, coating weight, and other coating process variables for improved coated paperboard dry and wet stiffness and water absorption/wicking. Chitosan-coated paper basesheets, along with the uncoated and phenolic resin-coated sheets were characterized and compared for Wet Taber Stiffness and Wicking Height according to the manufacturer's methods and specifications. The target critical-to-quality (CTQ) parameters and their levels for application as cooling pads are 1600 mg wet Taber stiffness and 2 inches of wicking height over 10 minutes. The economics of paperboard coating using chitosan compared with phenolic resins based on coating amounts on weight or area basis, unit costs, and product performance will be further discussed.

380663 Invited Talk: Dynamics of Artificially Engineered Protein Gels Based on Coiled-Coil Associating Groups

Abstract:

Artificially engineered protein hydrogels provide an attractive platform for biomedical materials due to their similarity to components of the native extracellular matrix; however, a better understanding of the polymer science underlying their mechanical response can help to extend the range of mechanical properties achievable with such systems. One particular area of interest is improving the toughness of protein gels. Herein, we demonstrate an approach to biomaterial toughening through engineering both topological entanglements and physical associations into the gel to produce two distinct length and timescales of network interaction. Oxidation or coupling of cysteine residues in proteins enables chain extension through protein end coupling, producing polymer gelators of high molecular weight that are not easily achievable directly by protein expression. When the molecules are processed to induce a chain extended state, the plateau modulus of the network remains largely unchanged, but significant increases in low frequency modulus and elastic recovery after loading are observed. In addition, the resulting gels can be extended to over 3,000% strain before break. These materials serve as models for comparison with theories of sticky reptation. Using responsive coupling chemistries allows entanglements to be produced dynamically within the material, enabling responsive enhancement of the network.

When the materials are prepared with a molar mass below the entanglement molar mass, they can be compared to sticky Rouse theories of polymer dynamics. Clear deviations from single mode relaxation behavior are observed in shear rheology, and superdiffusive behavior can be identified in the materials. Comparisons between multi-sticker and simple telechelic proteins provide insight into the effects of molecular structure on the observed behavior. Finally, we report the development of a simple associative network model capable of calculating chain end-to-end distance distributions under flow as a first step to understanding the specific behaviors observed in these systems. Interconversion between dangling, looped, and bridging chains is modeled with a Smoluchowski equation using the dumbbell model and capturing chain end association and flow terms. This model shows for the first time the effect of looped chains on stress in gels and the presence of flow instabilities. Furthermore, by tracking the entire chain end distribution, this model is able to quantify the effects of chain tumbling on stress in the polymer networks.

Microrheological investigation of a biofilm and its constituting extracellular polymers

Mahesh Ganesan [1], John G. Younger [2] and Michael J. Solomon [1]

[1] Department of Chemical Engineering, University of Michigan, Ann Arbor, MI 48109
[2] Department of Emergency Medicine, University of Michigan, MI 48109

We utilize passive microrheology to understand the origin of viscoelasticity in a bacterial biofilm by measuring the mechanical properties of its individual constituents using thermally driven microspheres as probes. Biofilms are a biological soft matter consisting of bacterial aggregates embedded within an extracellular polymeric hydrogel. Owing to their composite microstructure, the bulk mechanical properties are likely due to a synergistic response of cells and polymers, which occur at very short temporal and spatial scales, thus requiring the use of microrheology to access this feature. Using a model bacterial biofilm formed by *Staphylococcus epidermidis*, we show that the biofilm's creep compliance can be viewed differently at short and long time scales. The short time response was found to be governed by a composite polymer system consisting of the biofilm's extracellular polymers. Microrheology measurements on biofilms show that the long time plateau in creep is due to the presence of large bacterial clusters. Comparison of these studies with recently published data obtained from conventional bulk measurements show good agreement at long times, thus confirming our approach. Furthermore, the viscous modulus of the biofilm was found to be due to the extracellular polymers while the origin of elasticity couldn't be evinced clearly. Due to the detrimental prevalence of biofilms in a variety of environments, studies in this direction is well suited to understand their sustenance, yielding and subsequent fracture under different shear conditions.

383524 Hybrid Chemical-Physical Crosslinking in Artificial Protein Hydrogels

Abstract:

Artificial proteins are an important class of macromolecules for the development of engineered hydrogels in which structural and functional information can be encoded at the level of the protein sequence. In this work, we have designed triblock artificial proteins capable of both chemical crosslinking through reactive amino acid residues at the protein termini and physical crosslinking through associative coiled-coil midblock domains. Under the appropriate conditions, the resulting hydrogel is a hybrid network that contains both chemical and physical junctions and behaves as a viscoelastic solid. While the chemical network is essentially permanent, the coiled-coil physical crosslinkers are able to reversibly dissociate, allowing some of the stress that is stored upon network deformation to relax. The level of stress relaxation can be tuned by adjusting the ratio of chemical and physical crosslinking. Swelling the hydrogels in buffer containing denaturant or introducing a point mutation at one of several key residues disrupts coiled-coil formation and results in a purely elastic, rather than viscoelastic, response to an applied strain. This work demonstrates how bulk hydrogel properties such as viscoelasticity can be programed through variation of the protein sequence. Chemical-physical hybrid networks have potential applications as cell culture platforms to study cellular mechanosensing in a viscoelastic environment, as components to sensors or stimuli responsive materials, and for dissipating mechanical energy during stretching to improve hydrogel toughness.

390355 Equilibrium Structure and Dynamics of Peptide-Based Molecular Gels

Abstract:

Aromatic short peptide derivative molecules can form rigid gels ($G' > 10^5$ Pa) at very low volume fractions (< 1%) due to highly specific anisotropic interactions that result in the self-assembly of molecules into space-filling fibrous networks, featuring characteristic gel-like mechanical properties. These systems are seeing potential applications in cell culture, tissue engineering, and drug delivery, but differ from typical biopolymer networks that are commonly used in these areas, as the molecules interact predominantly through hydrogen bonding and π-π stacking, and in an absence of covalent bonds. We explore a model molecular gel, formed by mixing a solution of Fmoc-diphenylalanine (Fmoc-FF) in dimethyl sulfoxide (DMSO) with water. Using optical, scattering, and rheological techniques, we study the gel structure, as well as the kinetics and dynamics of gel formation and aging. We find that the aging process of these gels is associated with a tendency of the system to reach a steady state through the structure evolving into an increasingly uniform network, which is correlated with an increase in relaxation times and growth of the elastic modulus to a plateau value. Further, we demonstrate that the gels are mechanically and thermally reversible, and hypothesize that the system is in equilibrium in its fibrous network state. The observations provide experimental evidence for simulations of aggregation of valence-limited "patchy" particles published in the literature, which suggest that an equilibrium gel can be formed at low volume fractions without an intervening phase separation.

389231 Synthesis and Characterization of Novel Stimuli-Responsive Hydrogels Based on Polyglycerol

Abstract:

The present study was focused on the synthesis and characterization of a hydrogel material from glycerol. The synthesized hydrogel absorbs up more than 7 times its own weight. The hydrogel absorption capacity changes in response to temperature and pH external stimuli. A hydrogel is a high-valued added product of glycerol, a co-product of the transesterification of vegetable oils. This work is part of the effort to develop new technologies for using glycerol as a building block for valuable chemicals that may contribute to transform the actual biodiesel industry into a bio-refinery [1]. The studied hydrogel was synthetized using as building block polyglycerol obtained by the direct polymerization of glycerol. Polyglycerol has been used as starting material for drug delivery [2, 3] temperature-responsive copolymers [4], dyes adsorption from aqueous media [5], among many other applications [6-9]. Recently, Salehpour and coworkers have been reported the synthesis of polyglycerol – based hydrogels using poly (ethylene glycol) diglycidyl ether (PEGDE) as crosslinking agent [10]. In this work, polyglycerol was cross-linked with biodegradable acids such as citric and oleic acids. Chemical reactions involved in the functionalization of polyglycerol were esterification and acid-catalyzed hydration of alkenes. The effect of the proportion of cross-linked agents, citric, and oleic acid, in the final physical properties of the hydrogel were studied. It was found that hydrogel properties depend on the content of cross-linked agent, the ratio between citric and oleic acid, the functional groups present in these acids as well as their molecular length. The characterized hydrogel has heterogeneous porous with diameters in a range of 5-50. The relation between the synthetized hydrogel glass transition temperature and its response to external stimuli is discussed. Physical and chemical properties were analyzed by Scanning Electron Microscopy (SEM), Differential Scanning Calorimetric (DSC), Thermogravimetric Analysis (TGA), and Fourier Transform Infrared Spectroscopy (FTIR).

REFERENCES

1. Yuguo Zheng, X.C., and Yinchu Shen, *Commodity chemicals derived from glycerol, an important biorefinery feedstock.* Chemical Reviews, 2010. **108**: p. 5253–5277.

2. Steinhilber, D., et al., *Surfactant free preparation of biodegradable dendritic polyglycerol nanogels by inverse nanoprecipitation for encapsulation and release of pharmaceutical biomacromolecules.* J Control Release, 2012.

3. Calderon, M., et al., *Development of efficient acid cleavable multifunctional prodrugs derived from dendritic polyglycerol with a poly(ethylene glycol) shell.* J Control Release, 2011. **151**(3): p. 295-301.

4. Taylor, D.K., et al., *Temperature-Responsive Biocompatible Copolymers Incorporating Hyperbranched Polyglycerols for Adjustable Functionality.* J Funct Biomater, 2011. **2**(3): p. 173-194.

5. Chen, Z., et al., *Multicarboxylic hyperbranched polyglycerol modified SBA-15 for the adsorption of cationic dyes and copper ions from aqueous media.* Applied Surface Science, 2012. **258**(13): p. 5291-5298.

6. Li, M., et al., *Methotrexate-conjugated and hyperbranched polyglycerol-grafted Fe(3)O(4) magnetic nanoparticles for targeted anticancer effects.* Eur J Pharm Sci, 2013. **48**(1-2): p. 111-20.

7. Ahn, C.H., J.-D. Jeon, and S.-Y. Kwak, *Photoelectrochemical effects of hyperbranched polyglycerol in gel electrolytes on the performance of dye-sensitized solar cells.* Journal of Industrial and Engineering Chemistry, 2012. **18**(6): p. 2184-2190.

8. Mamiński, M.Ł., et al., *Fast-curing polyurethane adhesives derived from environmentally friendly hyperbranched polyglycerols – The effect of macromonomer structure.* Biomass and Bioenergy, 2011. **35**(10): p. 4461-4468.

9. Mamiński, M.ł., M. Czarzasta, and P. Parzuchowski, *Wood adhesives derived from hyperbranched polyglycerol cross-linked with hexamethoxymethyl melamines.* International Journal of Adhesion and Adhesives, 2011. **31**(7): p. 704-707.

10. Salehpour, S., C.J. Zuliani, and M.A. Dubé, *Synthesis of novel stimuli-responsive polyglycerol-based hydrogels.* European Journal of Lipid Science and Technology, 2012. **114**(1): p. 92-99.

387599 Kinetics Study on Hyperbranched Thermoplastic Elastomer from Soybean Oil Via RAFT

Abstract:

In this contribution we show how hyberbranching and gelation may be controlled in the Reversible Addition-Fragmentation Chain Transfer (RAFT) polymerization of multifunctional acrylated epoxidized soybean oil (AESO). Under certain conditions, RAFT, a type of controlled radical polymerization delays the onset of gelation to high conversion, enabling the production of biobased thermoplastic elastomers. We have explored a number of factors that influence gelation, *e.g.* number of functionality per monomer, monomer concentration, initiator concentration, etc... Kinetic data, as well as light scattering data will be provided, which show how and why RAFT delays gelation. Dynamic shear rheology of polymer with different degree of branching s will also be discussed, which shows how chain architecture affects the growth of hyperbranched system and the viscoelasticity of the final product.

382444 Rheological and Microstructural Characterization of Native Lung Mucus

Abstract:

Mucus is a heterogeneous, gel-like material critical to life that coats different organs, including the respiratory, GI, and reproductive tracts, and performs multiple tasks including protection from bacteria, virus, toxins, and foreign particles. The major components of mucus are water and mucin *o*-linked glycoproteins with different monomer/oligomer configurations. Mucin glycoproteins and other mucus components contribute significantly to the viscoelastic properties of mucus. In terms of modeling the viscoelastic behavior of mucus for application to simulations or production of a mucus substitute, there are relatively few studies in the literature conducted on native mucus samples. We established an experimental protocol to harvest pulmonary mucus from different mammals and here we report the mechanical properties of native lung mucus from a postmorterm horse. Rheological characterization were performed using oscillatory shear (small- and large- amplitude) and steady shear experiments. The native mucus displayed nonlinear characteristics with a strain-stiffening behavior followed by a strain-softening behavior. Using optical microscopy and atomic force microscopy, linkages were examined between the microstructure behavior of mucus and the observed rheological data. These results provide new insights for describing the viscoelastic behavior of mammalian lung mucus. Especially important is developing a model to accurately describe both the viscoelastic behavior over a range of strain rates, as key physiological processes occur at both ends of the range from ciliary clearance to coughing and sneezing. These processes are of key to understanding and modeling the clearance of mucus and particulates, and impacts occupational and environmental regulations of particulates, drug delivery methods and uptake, and changes in lung mechanisms in lung diseases.

388234 Polyacrylamide Hydrogels Prepared Using Templates: Synthesis, Characterization and Efficiency of Separation of Macromolecules

Abstract:

Nanoporous polymer hydrogels offer a desirable combination of mechanical, optical, and transport characteristics that have placed them at the core of a variety of biomedical technologies including engineered tissue scaffolds, substrates for controlled release of pharmaceutical compounds, and sieving matrices for electrophoretic separation of DNA and proteins. All these applications are closely related to the nanoporous morphology of the matrix, especially the mean pore size and its distribution.

Crosslinked polyacrylamide is a widely used gel matrix for protein and DNA electrophoresis because it offers a number of attractive properties including excellent separation resolution, optical/UV transparency, and electro-neutrality. In order to rationally engineer gels with properties that are optimally suitable for any of these applications, it is essential to understand the nature of the gel pore structure and parameters through which it can be manipulated.

In this work we study the efficiency of separation of macromolecules through a porous structure of polyacrylamide hydrogel obtained by two methods of polymerization: 1) through a more common protocol to develop a separation in a capillary gel electrophoresis system and 2) through photo-initiated polymerization. The porous structure was formed using nanotemplates based in Sodium Dodecyl Sulfate. Different concentrations of SDS were studied in order to assess the influence of the size and distribution of the pores on the separation process of macromolecules. The porous structure was characterized using novel techniques in which the hydrogel samples are prepared at room temperature for direct visualization of nanoporous morphology using transmission electron microscopy (TEM). Additionally, other techniques to characterize the porous structure were explored such as: Thermoporometry, Dynamic Mechanical Analysis and Computed Tomography Imaging Spectrometer. The results regarding the efficiency of separation of macromolecules and the characterization of the porous structure are compared and discussed for each type of polymerization. Directions for future research are also included.

387837 Diffusion and Binding Dynamics in Viscoelastic Protein Hydrogels

Abstract:

Proteins perform a wide range of biological functions, but are built from a small set of amino acids. What if we could outfit proteins with new amino acids, and tailor them for applications to biology and beyond? Our laboratory has pioneered strategies for incorporating dozens of new amino acids into proteins, and has begun integrating these proteins into functional materials. This talk will describe chemical and biophysical approaches that we have taken to explore the dynamic properties of a novel protein copolymer, which self-assembles into physically crosslinked, viscoelastic hydrogels. Special emphasis will be placed on fluorescence recovery after photobleaching (FRAP), which has been used to characterize chain diffusion and reversible binding in these gels. This technique has revealed that the dynamic properties of gel-associated chains are highly sensitive to changes in protein structure. Additionally, photobleaching has enabled us to control an unusual feature of pattern formation during a volume phase transition associated with these gels.

386590 Solar Fuel Generation at Near-Neutral pH Conditions: Operational Advantages and Disadvantages

Abstract:

An efficient, long lasting, and cheap solar fuel generator is an alluring goal to support future energy demands. Solar fuel generator is a photoelectrochemical cell (PEC) where a photovoltaic (PV) cell is in direct contact with the water-splitting electrochemical cell. The available materials for the PV cells and catalysts can make a solar fuel generator that can simultaneously satisfy two out of the three characteristics mentioned above. For example, either efficiency or durability can be compromised to make a less expensive PEC. The efficiency and durability of the device are strongly coupled and dependent on i) the interaction of photoelectrodes with the liquid electrolyte and ii) the overall ionic resistances in the electrolyte. The photoelectrodes can corrode in extreme pH conditions and are only stable in a finite pH range. Therefore, the stability and longevity of the photoelectrodes require PEC operation in near neutral pH solutions.

The near-neutral pH electrolysis requires more energy to compensate for potential losses due to pH gradients, electrodialysis and ohmic resistance. The ohmic losses and concentration polarization can be minimized by increasing ionic strength and buffer capacity of the electrolyte, respectively. The supporting electrolyte in polyprotic acid and/or buffer solutions shows a positive effect on limiting current densities, whereas it has a negative effect on limiting current densities in monoprotic acid solutions. The major source of solution losses in neutral pH devices is due to pH swings at the electrodes which can be as high as 200 mV at 20 mA/cm^2 current density. Another major contributor to overall potential losses is the OER overpotential which can be > 320 mV for most OER catalysts. This talk will also discuss the effect of convection and mixing on potential losses and gas crossover in the device.

390208 Intensity Modulated Photocurrent Spectroscopy (IMPS) Used to Detect Photoactive Intermediates during Niw Electrodeposition

Abstract:

Intensity modulated photocurrent spectroscopy (IMPS) is a technique conventionally used to characterize the semiconductor/electrolyte interface. It is used here in a new way, to verify the presence of adsorbed, photoactive intermediates during the electrodeposition process, using NiW alloy electrodeposition as an example. Interest in NiW alloys are related to their high corrosion resistance, ability to catalyze the hydrogen evolution reaction from water, and superior wear resistance. The NiW deposition behavior from aqueous solutions is described as a couple reaction mechanism and better understanding of the mechanism can help to control the deposit composition which in turn is related to the resulting properties. To this end, IMPS was used to examine codeposition using an ammonia free citrate-boric acid electrolyte. The IMPS spectra was generated for different applied potentials. A significant current response to the incident modulated UV light was detected where steady state NiW codeposition takes place. The results confirm the presence of photoactive intermediates during codeposition, and is the first use of this technique in the electrodeposition process resulting in a metallic product.

First-Principle Modeling Approach Towards Molecular Design of Electrode Materials Candidates for Lithium Ion Battery Applications

Ki Chul Kim,[1] Seung Woo Lee,[2] and Seung Soon Jang[1]

1 School of Materials Science and Engineering, Georgia Institute of Technology, 771 Ferst Drive NW, Atlanta, GA 30332-0245, USA

2 School of Mechanical Engineering, Georgia Institute of Technology, 801 Ferst Drive NW, Atlanta, GA 30332-0405, USA

Abstract

Energy storage technology is a critical factor for sustainable energy supply that will meet our growing energy demand from portable electrical devices to large electrical grids. To date, efforts to develop applicable energy storage systems have mostly focused on electrochemical devices, particularly lithium ion batteries. Despites their high energies and charge storage densities, they have been suffered from their poor power densities. It has been recently reported that carbon nanotube (CNT) based electrode materials in molecular levels could improve their power densities arising from their characteristics such as high electron conductivity, high electrochemically active surface areas. In addition, it is expected that the transport of lithium ions can be enhanced due to relatively short diffusion paths. Although CNT based electrode materials have therefore tremendously attracted as promising electrode candidates, systematic investigation of the performance of CNT based electrode materials has not been reported so far. Herein, we have employed the First-Principles Modeling approach such as density functional theory (DFT) method to study the redox potentials of CNT based electrode material candidates. Our investigation has aimed to identify the relationship of the structure electrode materials with the performance.

389416 Unifying the Hydraulic, Electro-Osmotic, and Diffusive Perspectives on Water Transport in Ionomer Membranes

Abstract:

Unifying the Hydraulic, Electro-Osmotic, and Diffusive Perspectives on Water Transport in Ionomer Membranes

Charles W. Monroe

Department of Chemical Engineering, University of Michigan

Modern studies of water transport in ionomer membranes for PEM fuel cells usually take into account the traditional diffusive mode of transport, in which water-activity gradients drive water flux, and the pressure-diffusive transport mode, in which gradients in the local pressure induce chemical-potential gradients that drive flux [1]. Some models also treat the ionomer gel as a porous medium, through which water flows via hydraulic permeation [2,3]. These two perspectives are seemingly contradictory, in that one set of phenomena is diffusive, and governed by material (and charge) balances, while the other is controlled by the momentum balance. In fact, a general transport model can be constructed in which the diffusional and hydraulic transport modes are taken to occur in parallel, and are coupled.

A general transport model will be presented that is applicable to swollen ionomers under isothermal, non-isobaric conditions in general. The case of a Nafion membrane (for which most transport properties are readily available) will be treated in particular detail. The basis of the multicomponent, momentum-coupled mass- and charge-transport model in irreversible thermodynamics will be addressed.

In addition to water diffusivity, ionic conductivity, and the electro-osmotic coefficient, the model also incorporates the membrane compressibility, alongside other thermodynamic factors. The theoretical construct will be used to illustrate a number of effects on transport that arise when water-swelled ionomers are constrained spatially, or subjected to external compressive forces. Theoretical analysis will be used to predict several effects observed in experiments, such as stresses exerted by constrained membranes subjected to water flux, and constant-current diffusion potentials that appear to change with the applied mechanical load. The model will be useful to describe practical PEM fuel-cell membranes constrained within reactor assemblies.

References

[1] D.M. Bernardi and M. W. Verbrugge, *J. Electrochem. Soc.* **142**(1992) 2477.

[2] C. Fabiani, G. Scibona, and B. Scuppa, *J. Membrane Sci.* **16**(1983) 51.

[3] Q. Duan, H. Wang, and J. Benziger, *J. Membrane Sci.* **392-393** (2012) 88.

390620 Influence of Electrolyte Composition and Positive Electrode Materials on the Mg-Containing Type Nimh Batteries

Abstract:

Nickelemetal hydride (NiMH) batteries have been widely used in today's power tools and portable applications due to their advantages such as flexible design, excellent power, long cycle life etc. Traditionally, KOH solutions with high concentration are widely used as electrolyte in NiMH batteries. However, KOH solution is highly corrosive and not environmental friendly, especially for some active materials that contained in metal hydride electrode, such as Mg, Al metals. KOH solutions with high concentration can easily react with Mg, Al, and greatly decrease the electrochemical performance of cathode electrode. Some MgNi2, Mg2Ni base alloys have been reported having high discharge capacity and high energy density. However, because KOH reacts with Mg in alloy and changed the surface structure of anode electrode material, NiMgH cells show a very short cyclic life and commercialization becomes impossible. In this research some additives are used to minimize the negative effect of KOH electrolyte in NiMgH. A series of Cl^- and F^- containing salts are proved to efficiently decrease the corrosion of KOH solutions and promote the electrochemical performance of electrolytes, such as LiCl and KF. New electrolytes are prepared and characterized in NiMgH system. In this study, a series of modified Ni(OH)2 are prepared and used as positive electrode materials. The effects of new electrolytes on these modified Ni(OH)2 are investigated. It shows that α-Ni(OH)2 phase in the positive electrode materials is stabilized in new electrolytes, which greatly prolong the cyclic life of NiMgH cells.

Computational Study of C14 Laves Phase AB_2 Alloy Structure and PCT Hysteresis in Nickel-Metal Hydride Batteries

Diana F. Wong [a,b], Kwo Young [a], and K. Y. Simon Ng [b]

[a] BASF Battery Materials-Ovonic, 2983 Waterview Drive, Rochester Hills, MI 48309, USA

[b] Department of Chemical Engineering and Materials Science, Wayne State University, Detroit, MI 48202, USA

Abstract

Computational tools offer fast and flexible screening capabilities to facilitate the design of battery materials, in this case, the anode materials for nickel-metal hydride (Ni-MH) batteries. Laves phase (C14 and C15) based AB_2 metal hydride alloys are good candidates to replace the rare earth (RE) element based AB_5 alloys currently used in Ni-MH batteries. C14 and C15 phase alloys offer relatively higher energy density and lower cost in a volatile RE market; however, these phases are not equal and contribute to different electrochemical properties: C14 phase tends to improve capacity and cycle life while C15 phase offers better rate capability.[1,2] AB_2 metal hydrides currently used for anode materials in Ni-MH batteries are mainly C14-predominant. The pulverization rate of C14 phase alloys is highly correlated to pressure-concentration-temperature (PCT) isotherm hysteresis, which is experimentally reported to be related to the *a/c* lattice parameter ratio of the crystal unit cell.[3] Calculation results of the C14/C15 predominating phase determination of a series of pseudo-binary AB_2 alloys using density functional theory are presented in this study, showing the correlation between the predominant phase structures to the average electron density. Geometry optimizations of the C14-predominant compositions yield the theoretical *a* and *c* lattice parameters that are used to predict battery cycle life performance. The effects of different chemical modifiers on the phase selection and cycle stability will be reported and compared to experimental results. With 6 million hybrid electric vehicles on the road worldwide powered by Ni-MH batteries, this screening method can help the optimization of cycle life with high electrochemical capacity while keeping anode material costs low.

References

1. Young, K.; Ouchi, T.; Huang, B.; Chao, B.; Fetcenko, M. A.; Bendersky, L. A.; Wang, K.; Chiu, C. *J. Alloys Compd.* **2010**, *506*, 841.
2. Young, K.; Nei, J.; Ouchi, T.; Fetcenko, M. A. *J. Alloys Compd.* **2011**, *509*, 2277.
3. Young, K.; Ouchi, T.; Fetcenko, M. *J. Alloys Compd.* **2009**, *480*, 428.

Sodium di-vacancy patterns and staging onset in layered TM-oxides

Masoud Aryanpour and Young-Gyoon Ryu
Advanced Materials Lab, SAIT-America
1st Cambridge Center, Cambridge, MA-02142

In recent years, electrode materials for Na-ion batteries have been the focus of extensive research. If successful, that would potentially lower the price of advanced batteries in a wide range of applications from mobile devices up to electric vehicles.

Designing Na-based electrode materials cannot be achieved by simply replacing Li with Na in known materials for lithium-ion batteries. Experiment and theory point to the requirement of redesign and re-optimization endeavors to reach similar or better performances.

An important class of materials is layered compounds containing alkaline atoms intercalated in between layers of transition metal oxides (NaMO2). These compounds have been the topic of research for a few decades [2,3], yet their complex electrochemical behavior remains much to be fully understood.

De-intercalation of sodium as the cathode material being charged is often accompanied by several phase changes that in turn prevent reaching high capacity and energy density that are offered theoretically by layered materials. In a previous work, we studied the effect of Na mono-vacancy in electrode materials [4]. In this work, we consider the effect of two atomic vacancies in one or two sodium layers the first few transition metal compounds NaMO2 (M=Sc, Ti, V, Cr, Mn). This study can elucidate dominant vacancy superstructures, as well as the inclination of these materials toward staging as the charging proceeds.

[1] BL Ellis and LF Nazar. "Sodium and sodium-ion energy storage batteries", Current Opinion in Solid State and Materials Science, 16(4):168-177, 2012.

[2] C. Didier, M. Guignard, C. Denage, O. Szajwaj, S. Ito, I. Saadoune, J. Darriet, and C. Delmas, "Electrochemical Na-Deintercalation from NaVO2", Electrochemical Solid-State Letters, 14(5), A75-A78, 2011

[3] M.M. Doeff, M.Y. Peng, Y. Ma, and L.C. De Jonghe, "Orthorhombic NaxMnO2 as Cathode Material for Secondary Sodium and Lithium Polymer Batteries", Journal of Electrochemical Society, 141 (11), L145-L147

[4] M. Aryanpour, H.-S Lee, "Atomic Re-arrangements and Energetics of Selected TM-oxides Caused by an Na-vacancy", 223rd Electrochemical Society Meeting, May2013

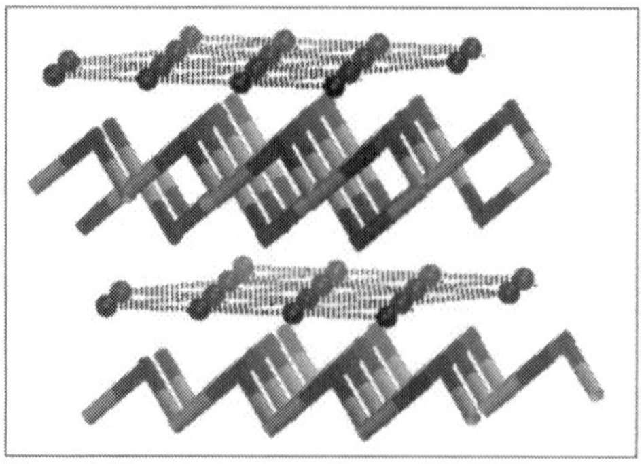

Figure 1 Crystal structure of layered NaMO2 materials used in Na-ion batteries

370256 Plasmon-Enhanced Upconversion Emission of Hexagonal Er3+, Yb3+ Co-Doped Lutetium Oxyfluoride Using Silver Nanoparticles

Abstract:

In this work, Ag@LuOF: Er^{3+}, Yb^{3+} micro-hexagonal composites were prepared. X-ray diffraction, energy dispersive spectroscopy, and high-resolution transmission electron microscopy results confirm that Ag nanoparticles with size around 30 nm have been uniformly immobilized on the surface of LuOF: Er^{3+}, Yb^{3+} hexagons. Upconversion spectra under near-infrared laser excitation suggest that the emission intensity of LuOF: Er^{3+}, Yb^{3+} is significantly enhanced by combining with Ag nanoparticles. Possible mechanisms for the plasmon-enhancement on upconversion have been proposed based on the analysis of absorption spectra and decay lifetime curves. Because of the near-infrared excitation, high upconversion intensity, and pure red emission, the Ag@LuOF: Er^{3+}, Yb^{3+} composites exhibit promising applications in solid-state lighting, biological labeling, and imaging.

Measuring How the Local Glass Transition Temperature Shifts Across a Glassy-Rubbery Polymer-Polymer Interface

Roman R. Baglay and Connie B. Roth
Department of Physics, Emory University, Atlanta, GA, USA

Contact information:
Roman Baglay: rbaglay@emory.edu
Connie Roth: cbroth@emory.edu

Recent advances in block copolymer synthesis have led to new routes to forming nanostructured polymer blends putting rubbery, glassy, and crystalline phases in intimate contact. These new multicomponent block copolymers can be designed to self-assemble into a range of morphologies including unique intertwined co-continuous phases with domain sizes easily less than 100 nm in size. Efforts to understand the unique properties of these new nanostructured materials have been limited to measurements of macroscopic properties correlated with the observed microstructure. Here, we use individually assembled multilayer films with specific layer thicknesses to act as a model system mimicking the local morphology of such nanostructured blends. A fluorescence technique is used to directly measure the local glass transition temperature (T_g) profile as a function of distance from a glassy-rubbery polymer-polymer interface.

We focus on a polystyrene (PS) / poly(*n*-butyl methacrylate) (PnBMA) weakly immiscible system with a 7 nm interfacial width [1] that creates an 80 K T_g difference across the interface from one bulk T_g value to another. Multilayer films of high molecular weight PS and PnBMA are constructed where an ultrathin (10-15 nm thick) pyrene-labeled layer is inserted into the multilayer structure at different positions (z) from the glassy-rubbery interface. Temperature-dependent fluorescence intensity is collected to determine the local $T_g(z)$ at a given position z from the interface. Using a series of different samples, we are able to map the $T_g(z)$ profile across this glassy-rubbery interface. Our work reveals a gradient in the local $T_g(z)$ that spans ~400 nm across the interface from one bulk T_g value ($T_g^{PnBMA} = 20$ °C) to another ($T_g^{PS} = 100$ °C). The results demonstrate that local properties can be highly coupled across polymer-polymer domains within nanostructured blends, easily spanning entire domain sizes. This is in contrast to how the glass transition of polymer blends are traditionally treated, assumed to depend only on the local composition within a few nanometers [2], such that neighboring domains, even in nanostructured polymer blends, are presumed to have independent T_gs.

Surprisingly we find that the local $T_g(z)$ gradient across the glassy-rubbery PS/PnBMA interface is highly asymmetric, extending much further into the glass PS side. This observation is consistent with recent theoretical predictions from a limited mobility model for glassy dynamics in confined systems [3]. For decades T_g in confined systems has been studied with the hopes of uncovering the length scales that impact the glass transition. In ultrathin confined films, free surface and substrate perturbations induce gradients in local T_g spanning several tens of nanometers indicating that cooperative dynamics are coupled in some fashion across multiple cooperatively rearranging regions (CRRs) [4,5]. However, understanding length scales of local T_g gradients near a free surface has been hampered by limitations of how to treat the enhanced mobility at the free surface theoretically. In our study, a glassy-rubbery polymer-polymer interface is used to establish an 80 K local T_g gradient from one well-defined T_g value to another, thereby introducing a well-defined system with T_g asymptotics on both sides of the glassy and rubbery polymer-polymer interface.

[1] Rauscher, P. M.; Pye, J. E.; Baglay, R. R.; Roth, C. B. *Macromolecules* **2013**, *46*, 9806–9817.
[2] Lodge, T. P.; McLeish, T. C. B. *Macromolecules* **2000**, *33*, 5278–5284.
[3] Tito, N. B.; Lipson, J. E. G.; Milner, S. T. *Soft Matter* **2013**, *9*, 9403–9413.
[4] Ellison, C. J.; Torkelson, J. M. *Nature Materials* **2003**, *2*, 695–700.
[5] Pye, J. E.; Rohald, K. A.; Baker, E. A.; Roth, C. B. *Macromolecules* **2010**, *43*, 8296–8303.

383404 Tuning the Wall Thickness of Templated Polystyrene Nanotubes Produced from Melt Infiltration

Abstract:

In this work, we demonstrate methods of varying the wall thickness of polystyrene nanotubes produced from template synthesis by utilizing classical theories of liquids spreading on flat and curved substrates. We find that factors which affect the surface energy of the polymer and substrate such as the annealing temperature, annealing atmosphere and polymer molecular weight influence the thickness of the nanotubes which are achieved. Results from this study are in qualitative agreement with what is predicted from classical theories. Nanotubes with wall thicknesses ranging from 18 to 40 nm were achieved by infiltrating different molecular weight polystyrene. Higher molecular weight samples possessed larger wall thicknesses compared to lower molecular weight samples. Thermal characterization with differential scanning calorimetry measurements revealed significant Tg-confinement behavior as a function of wall thickness. Post infiltration processing such as annealing was also found to affect the thickness. Supported tubes annealed at high temperatures possessed thinner wall diameters compared to those annealed at lower temperatures. Additional results suggest that atmosphere can have an effect on the wall thickness. Alteration of the surface energy through changes to factors such as the temperature, polymer molecular weight and atmosphere affects the equilibrium tube thickness and may allow for dimensional control over the nanostructures produced from template synthesis.

The ability to tune the dimensions of nanomaterials is critical for applications which rely upon the unique properties which arise due to the greater effect of free surfaces, interfacial interactions and small diffusion lengths at the nanoscale. Confinement in one or more dimensions can lead to enhancements in the mechanical strength, electrical, optical, thermal properties of materials and benefit areas involving drug delivery, medical devices, interconnects, sensing, catalysis, filtration and nanocomposites. The ability to adjust material properties using the length scale of these structures imparts additional requirements e.g. the ability to achieve uniform size and shapes and the means by which one can tune their dimensions.

Template synthesis is often employed because it allows for dimensional tunability and can be used to infiltrate a wide range of organic and inorganic materials. Anodic aluminum oxide templates (AAO) can be produced using a variety of different electrolytes and conditions and yield pores tens to several hundred nanometers in size. Infiltration into the templates results in negative replication of the size, shape and length of the pore morphology. This is useful for creating nanostructures with custom exterior dimensions and shapes; however, the ability to tune the characteristics of internal features such as the wall thickness of nanotubes is very limited. Control over the wall thickness of nanotubes is typically carried out using solution based techniques to infiltrate polymer into the membrane. The nanotube wall thickness can be varied by changing the molecular weight of the polymer, solution concentration, type of solvent and size of the template used. However, the uniformity and the reproducibility of nanostructures produced using solution based approaches can be very unpredictable. Holes or undulations are often observed as a result of the solvent evaporation process or the evolution of Rayleigh instabilities. Furthermore, extensive drying of the samples must be carried out because residual solvent may influence the morphology of the nanostructures or influence their properties. Melt infiltration generally yields more uniform and reproducible nanostructures and is simpler because it is solvent-less but little is known or has been done regarding the wall thickness tunability of nanotubes produced in this manner. Classical theories regarding the spreading of low energy fluids on flat and curved substrates and recent experimental work suggest that the thickness of polymer films or tubes supported in AAO templates can be modified by annealing under different conditions.

According to classical wetting theories the propensity for a low energy liquid (e.g. a polymer melt) to spread across a high energy surface is characterized by the spreading coefficient which accounts for the difference in energy between an exposed substrate versus one that has been wetted by a liquid film. A polymer liquid will wet the substrate when the total system energy can be lowered be minimized. This process can occur very rapidly and is often described as a "precursor film". When exposed to a template membrane under complete wetting conditions, e.g. high temperatures, the polymer coats the interior and forms nanotubes. The thickness of the precursor film reaches an equilibrium value which is determined by the balance between long range Van der Waals forces and surface energies. This prevents the film from spreading without bounds into an infinitely thin layer of material. Changes to the surface energy can shift the equilibrium value and result in changes to the equilibrium thickness. Factors such as the temperature and atmosphere are known to influence the surface energy of a polymer melt and can be used to modify the wall thickness of template nanotubes. Likewise, the molecular weight of the polymer or the functionality of the substrate can also affect the surface energy of the system.

Various molecular weights of polystyrene were melt infiltrated into AAO templates under the same thermal conditions. Nanotubes with wall thicknesses ranging from 18 to 40 nm were achieved. Thicker nanotubes were created by infiltrating higher molecular weight samples whereas thinner tubes were obtained from lower molecular weights. This is in agreement with classical theory; the surface energy of a polymer increases with molecular weight and increases the equilibrium thickness. The infiltrated templates were also measured using differential scanning calorimetry to determine the change in Tg-confinement behavior as a function of thickness. Tg reductions as large as 20 °C from the bulk Tg were observed for the thinnest tubes whereas bulk-like Tg behavior was obtained for the thickest tubes. It was also discovered that wall thickness can be modified by annealing at different temperatures after infiltration. For the same molecular weight polymer, nanotubes annealed at 160 °C possessed a lower wall thickness than those annealed at 130 °C. These results suggest that it is possible to tailor the wall thickness of a polymer sample by modifying the annealing conditions without having to use different molecular weight polymers or modifying the surface energies of the substrate. These results further suggest that other factors such as the atmosphere and the surface energy of the substrate may affect the thickness and that factors such as the amount of polymer available for infiltration will not based upon what is predicted from classical spreading behavior.

362494 Depletion, Bridging, and Ordering in Bare and Grafted Nanorod-Polymer Mixtures

Abstract:

We present a detailed study of polymer-nanorod composites. Using integral equation theory we have determined the structure and phase behavior non-interacting, attractive, and grafted nanorods in polymer melt. Depletion-induced phase separation is predicted between polymer and rods of nearly all aspect ratios. Miscibility is achieved by introducing a small amount of attractive interaction between polymer and nanoparticles. At large polymer-nanoparticle attraction strengths a few polymer molecules bridge several nanoparticles giving rise to another phase separated regime. Large diameter nanorods tend to be bridged my more than a single polymer layer (telebridging). This creates a finite miscibility window below the depletion temperature and above the bridging temperature.

Small changes in nanorod aspect ratio induce a drastic narrowing of the miscibility window, particularly for thin nanorods. However, as the rod radius of gyration approaches the radius of gyration of the polymer, the phase boundaries are nearly independent of aspect ratio. Our work confirms previous experimental findings that depletion-driven phase separation is enhanced with increasing size of nanospheres. However, we predict that nanorods display a subtle, and non-montonic dependence of the depletion line on the rod thickness. In the bridging regime, larger nanospheres show enhanced bridging while nanorods display a non-monotonic behavior. The width of the miscibility window is, therefore, a non-monotonic function of the nanorod thickness. Close to both bridging-driven, and depletion-driven phase separation, the attractive force between rods increases with rod thickness, while remaining independent of rod length. However, this attractive force operates at different inter-particle separations in the two regimes. Our work with grafted nanorods (with various graft lengths) illustrates that even short polymer grafts at moderate grafting densities can prevent the entropic depletion-driven phase separation. On the other hand, grafted nanoparticles are likely to form ordered, microphase separated structures within the polymer melt.

384265 Ordered Polymer-Grafted Nanoparticle Networks for Ionic Conductivity

Abstract:

Ionic liquids have been used effectively to improve ionic conductivity of diblock copolymers in past works. We suggest that polymer-grafted nanoparticles with sulfonated end groups can be used as a new way of controlling particle dispersion and ion aggregation within ionic liquids. In this talk, the structures of synthesized ionically functionalized polymer-grafted iron oxide nanoparticles will be examined and discussed through SAXS and TEM analysis.

390373 Preparation of Conductive Polymer Nanofibers with a Coaxial Nozzle Electrospinning Method

Abstract:

Conductive polymer is a polymeric compound which shows electrical conductivity, and is applied to various fields because of its high electrical conductivity, high flexibility, and lightweight characteristics. Further, nonwoven microfabrics of the conductive polymers are expected to be utilized for the highly sensitive sensors in medical and environmental fields. However, because of their infusible and insoluble nature, process of the conductive polymer nonwoven microfabrics must become a multistep. In recent years, the electrospinning of microfibers has been employed due to its simplicity, versatility, and scale-up opportunities. In the electrospinning, coaxial nozzles are often used as the discharge port for the formation of the fine structured materials. In this study, we developed a single step processing method for the formation of nonwoven conductive polymer fabrics which uses electrospinning method with a coaxial two fluid nozzle of a monomer and an oxidant solution.

3,4-ethylenedioxythiophene (EDOT) was chosen as a monomer due to the facts that poly-EDOT (PEDOT) has a good conductivity, environmental stability, and harmless to the human body. Iron( ) *p*-toluenesulfonatehexahydrate (FeTS), one of the most efficient oxidants for EDOT polymerization and known to produce intrinsically doped and electrically conductive PEDOT, was chosen as a oxidant. Before the electrospinning experiments, appropriate solvent for the synthesizing the PEDOT was examined with a bulk polymerization of EDOT, and acetonitrile was chosen for the solvent. Cellulose acetate (CA) and Poly(methyl methacrylate) (PMMA) were used as matrices for the conductive polymer microfibers. Effects of spinning conditions on the properties of conductive polymer were examined.

The monomer solution was prepared by dissolving the CA powder and the liquid EDOT into acetone or acetonitrile. The oxidant solution was prepared by dissolving the PMMA powder and the FeTS powder into acetone or acetonitrile. Both solutions were filled into grass syringes, and placed in automatic syringe pumps. The grass syringe filled with the monomer solution was connected to the outer tube of the coaxial nozzle, while that of the oxidant solution was connected to the inner tube. Both solutions were fed simultaneously to the coaxial nozzle, and electrospunned to the aluminum foil counter electrode to obtain the PEDOT nonwoven microfabrics. The PEDOT microfibers were characterized with scanning electron microscopy (SEM), mass spectrometric analysis (MALDI-TOF-MS) and four-point probes conductivity measurements.

With the coaxial nozzle electrospinning, nonwoven microfabrics were successfully obtained. Average diameter of PEDOT microfibers were varied from 490.6 nm (PMMA 15 wt%, CA 1 wt%) to 800 nm (PMMA 15wt%, CA 10 wt%) with the change in the CA concentration. This fact indicates that the fiber diameter could be controlled by the concentration of the matrix solution. Conductivity of thus formed microfabrics was the order of 10^{-3} S/cm, which is the same with the PEDOTS obtained by the bulk polymerization. We are now working on the improvement of the conductivity.

383834 Novel Pressure-Responsive Shape Memory Polymers

Abstract:

Shape memory polymers (SMPs) are stimuli-responsive materials that can change their shape upon application of an external stimulus (e.g., heat and light). They have been extensively exploited for a wide spectrum of technological applications ranging from smart biomedical devices to aerospace morphing structures. Most of the currently available SMPs are thermoresponsive and they suffer from slow response speed and heat-demanding programming and recovery steps. Although pressure is an easily adjustable process variable like temperature, pressure-responsive SMPs are largely unexplored. Here we demonstrate a new class of pressure-sensitive SMPs that enable the instantaneous recovery of the memorized shapes by applying a very small pressure. By combining these novel SMPs with self-assembled colloidal crystals, we have developed tunable macroporous photonic crystal films showing large optical stop band shifts triggered by a small pressure change. Rewritable photonic crystal microstructures, such as optical waveguides, can be easily printed on the pressure-responsive SMP films. We have also demonstrated that nanoporous smart coatings exhibiting tunable anti-glare properties can be fabricated by using the new SMPs. These pressure-responsive SMPs could find important applications in anti-counterfeiting and biometric recognition markets.

370773 Synthesis of Polymer Brush in Oil-Phase through Photo-Emulsion Polymerization

Abstract:

Oil-dispersed spherical polymer brush consisting of a hydrophilic core and lipophilic shell was synthesized through a novel "grafting from" procedure. Poly(acrylic acid) (PAA) nanoparticles with an average diameter of 100 nm were firstly prepared in toluene by inverse emulsion polymerization. At the end of polymerization, photo initiator 2[p-(2-hydroxy-2-methylpropiophenone)]-ethylene glycol- methacrylate (HMEM) was slowly added to copolymerize at the surface of PAA nanoparticles. After the addition of N-vinyl carbazole monomer, poly(N-vinyl carbazole) (PVK) chains finally generated from the surface of PAA core to form a brush-like structure via photo-emulsion polymerization under the irradiation of UV light. Dynamic light scattering (DLS) results display an increase in hydrodynamic diameters after photo-emulsion polymerization. In addition, the characteristic peaks of PVK were clearly observed in UV-vis spectra and photoluminescence spectra, indicating PVK chains were grafted onto PAA matrix. Transmission electron microscopy (TEM) shows the spherical shape of PAA-PVK brushes with relatively narrow size distribution. Compared with pure PAA core, the stability of PVK grafted PAA brush was significantly improved in organic solvent. This kind of novel core-shell structure is promising in the field of fluorescence labeling carrier and protein encapsulation in non-aqueous media.

387876 Thermodynamic Interactions and Tunable Properties of Thermoplastic Elastomers Derived from Vegetable Oils

Abstract:

Linear ABA triblock copolymers, such as poly(styrene-*b*-butadiene-*b*-styrene) and poly(styrene-*b*-isoprene-*b*-styrene), are among the most important and widely used thermoplastic elastomers (TPEs). However, those TPEs are derived from fossil fuels. Due to the finite availability of fossil fuels and the environmental impact of fossil fuels manufacturing, there has recently been an urgent need to develop alternative polymeric materials from sustainable sources. In this study, lauryl acrylate (LAc) and stearyl acrylate (SAc), derived from vegetable oils, were polymerized with reversible addition-fragmentation chain transfer polymerization, and subsequently extended for the preparation of poly(styrene-*b*-(LAc-*co*-SAc)-*b*-styrene) triblock copolymers. The thermal and mechanical properties of polyacrylate random copolymers were readily tuned by variation of the acrylate composition and resulting distribution of side-chain lengths. The order-disorder transition temperatures of the triblock copolymers were surprisingly independent of the mid-block composition, implying the Flory-Huggins interaction parameter is also independent of the acrylate side-chain length. Structural analysis revealed the non-equilibrium spherical morphologies of the triblock copolymers, which transformed to cylindrical microstructures under large amplitude oscillatory shear with the cylinders aligned transverse to the shear flow direction. Tensile testing experiments indicated that these triblock copolymers exhibit elastomeric behavior at room temperature.

379939 Invited Talk: Using Light to Locally Tune the Properties of Polymers

Abstract:

This talk will describe two related projects that utilize generic light sources to induce localized changes to the physical and mechanical properties of polymers in a strategic manner. The first project utilizes photopolymerization to change the properties of the walls of microfluidic channels. Poly(dimethylsiloxane) is one of the most popular materials for microfluidics, but swells in the presence of most organic solvents and is therefore limited to handling aqueous fluids. We show that it is possible to diffuse photoinitiator into the walls of the channel to localize photopolymerization that results in a thin shell that is impervious to organic solvents due to crosslinking. The remainder of the polymer that composes the microchannel retains the desirable properties of siloxanes. The second project utilizes uniform light from an IR heat lamp to induce self-folding of polymer sheets. Regions of black ink patterned on sheets of pre-strained polymer absorb the light and heat up due to the photothermal effect. The heat causes the polymer beneath the ink to relax and therefore shrink, which can induce self-folding. Self-folding is similar to origami but is done without direct human intervention. The appeal of our approach is that it utilizes two dimensional patterns to create a variety of complex three dimensional objects. In essence, it allows shape memory polymers to be programmed using a variety of two dimensional patterning techniques and triggered using only light. The talk will discuss the fundamental science that enables both of these projects, which are united by the importance of localized changes in polymer properties initiated by exposure to light.

385435 Improved Photopolymerization Kinetics of Acrylate and Amide Monomers in the Presence of Bulky Organic Salts

Abstract:

As radical reactions are commonly used in the manufacture of polymers and polymer products, interest has increased in new methods of radical polymerization. Among these techniques is the process of photopolymerization in which the reaction proceeds through the production of radicals from a photoinitiator. Previously, ionic liquids (ILs) have been found to increase the polymerization rate of multiple vinyl monomers including styrene and methacrylates. In our study, coordinated ionic liquids were synthesized containing the anion bis(trifluoromethylsulfonyl)imide (Tf_2N) and Li^+ cations coordinated to the polar vinyl monomers methyl methacrylate, hydroxyethyl methacrylate, methyl acrylate, methacrylic acid, acrylic acid, and acrylamide. The resulting substances were photopolymerized and the monomer conversion measured using real-time Fourier transform infrared spectroscopy (FTIR). In addition, salt extraction tests were performed on the resulting polymers and the material properties were examined using differential scanning calorimetry (DSC).

pH-dependent Degradation Kinetics of Polylactic Acid

STEFANO LAZZARI, FABIO CODARI, DAVIDE MOSCATELLI[1], GIUSEPPE STORTI, MASSIMO MORBIDELLI

Institute for Chemical and Bioengineering, Department of Chemistry and Applied Biosciences, ETH Zurich, Valdimir-Prelog-Weg 1, CH-8093 Zurich (Switzerland)
[1] *Dipartimento di Chimica, Materiali e Ingegneria Chimica "Giulio Natta", Politecnico di Milano, I-20131 Milano (Italy)*

Abstract

Polylactic acid (PLA) is a biodegradable and biocompatible polymer. Due to its properties it raised significant interest in the biomedical field, where it is often employed in drug delivery applications. The knowledge of the degradation kinetics of PLA has a key role when designing such drug releasing devices, as it allows modulating their release profiles. Among the different parameters affecting the PLA degradation kinetics, pH is one of the most relevant ones. While previous studies clarified the pH-dependent degradation mechanisms occurring (acid-, base- and non-catalyzed), the values of the corresponding kinetic constants were actually provided only at specific pH values.[1-3]

In this frame, the aim of this work is to study the degradation of lactic acid oligomers as a function of pH in a wide range of values of interest for biomedical applications. In particular, the degradation of oligomers with chain length varying from 2 to 8 repeating units was followed at 37°C for pH values equal to 1, 3.5, 5.7 and 7.4. These experimental data, along with those previously reported at pH = 2[1], are used to develop a detailed kinetic scheme and evaluate the corresponding kinetic constants. The resulting deterministic, pH-dependent model is able to describe the experimental data with reasonable accuracy and an average error of about 5%. The following values of the kinetic constants have been estimated: $k_D^{\alpha} = 4.44\times10^{-3}$ $L^2mol^{-2}h^{-1}$, $k_D^{\beta} = 9.29\times10^{-4}$ $L^2mol^{-2}h^{-1}$ (acid-catalyzed), $k_{RH} = 1.56\times10^{3}$ $L^2mol^{-2}h^{-1}$, $k_{BB} = 2.88\times10^{6}$ $Lmol^{-1}h^{-1}$ (base-catalyzed) and $k_W = 2.49\times10^{-5}$ $Lmol^{-1}h^{-1}$ (non-catalyzed). Notably, these kinetic constants can be used to study also the degradation of PLA polymer chains, as the values have been found to be chain-length independent from the tetramer on. Taking advantage of the results of this kinetic analysis and accounting for the diffusion limitations typical of PLA biomedical devices (e.g. nano- and microparticles), a predictive modeling of the degradation behavior becomes possible for a wide range of pH conditions.

[1] F. Codari, S. Lazzari, M. Soos, G. Storti, M. Morbidelli, D. Moscatelli, *Polym Degrad Stabil* **2012**, *97*, 2460.
[2] S. J. de Jong, E. R. Arias, D. T. S. Rijkers, C. F. van Nostrum, J. J. Kettenes-van den Bosch, W. E. Hennink, *Polymer* **2001**, *42*, 2795.
[3] C. F. van Nostrum, T. F. J. Veldhuis, G. W. Bos, W. E. Hennink, *Polymer* **2004**, *45*, 6779.

Tandem Catalysis: A Prospective Method for Producing Ethylene/α-Olefin Copolymers from Ethylene Stock

Song Guo [1], Hong Fan [1], Bo-Geng Li [1,*] and Shiping Zhu [2]

[1]State Key Laboratory of Chemical Engineering, Zhejiang University

[2]Department of Chemical Engineering, McMaster University, Hamilton, Canada L8S 4L7

1

Introduction of polyethylene

Polyethylene (PE)
- Homopolymers
  HDPE LDPE
- Copolymers (propylene, 1-butene, 1-hexene, 1-octene. <20%)
  LLDPE, POE

Simplest Petroleum-Based Polymer

World production of polyolefins in 2010

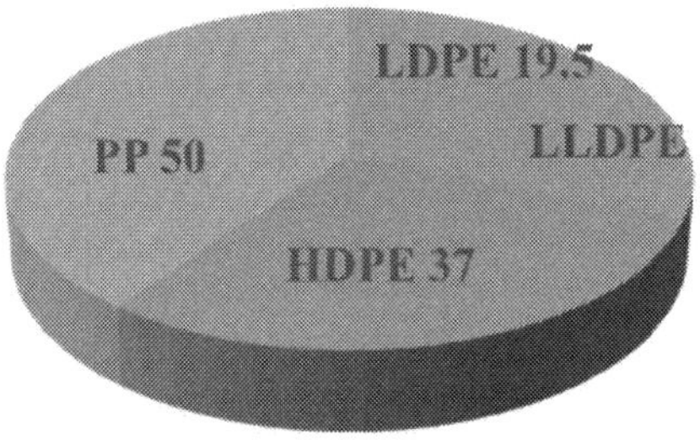

Unit: million tons

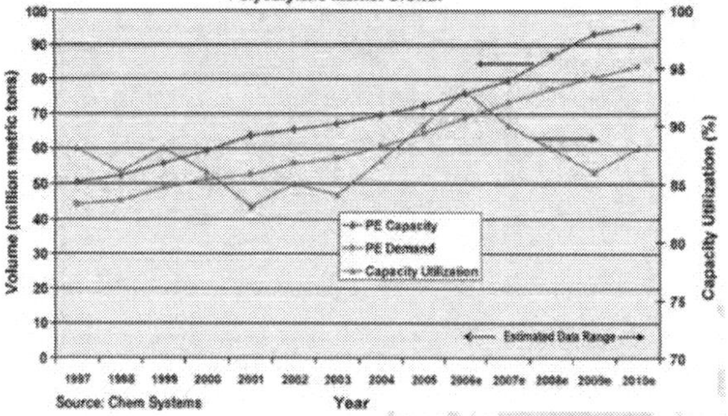

Figure 1. Market growth of polyethylene from 1997.

P. Chum, K. Swogger. Prog. Polym. Sci. 2008, 33, 797–819.

2

Introduction of polyethylene

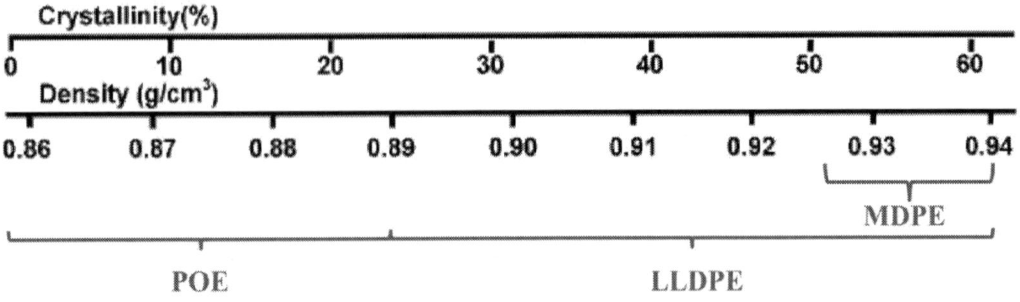

APPLICATIONS	film, sheet, fiber, foams, coatings, pipes
PROCESS	Gas Phase: UCC, BP Solution Phase: DuPont, Dow, DSM Slurry Phase: Philips, Solvay, Mistui
COMONOMER	1-butene, 1-hexene, 1-octene
PRODUCER Using 1-Hexene	ExxonMobil, Eastman, Equistar, Chevron Phillips, Dow, Basell, Innovene, Samsung Total

3

Introduction of Tandem Catalysis for Preparation of PE

Tandem Catalysis

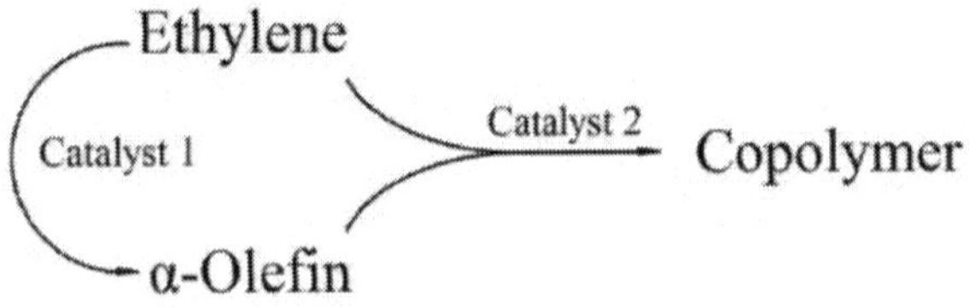

- Using ethylene-only stock in single reactor
- Cost Save in comonomer production, transportation, and storage

1980s-2000, ethylene oligomerization was mainly focused on dimerization.

SNS & PNP Catalysts for Ethylene Trimerization

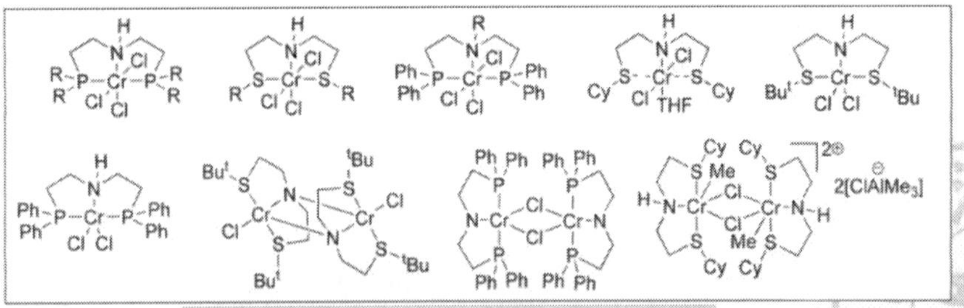

High activity: > 10^6 g/(mol Cr × h)

High Selectivity: > 98% for 1-hexene

D. S. McGuinness. *Chem. Rev.* 2011, 111, 2321–2341.

4

Linear low density polyethylene (LLDPE)

- T_m 120~125 °C, Comonomer content 5~10 wt%.
- Gas phase (1-C4, Unipol, UCC)
- Slurry phase (1-C4, Philips; 1-C6, Solvay)
- 1-C6 and 1-C8 better mechanical properties than 1-C4

Slurry Phase vs. Gas Phase

Cost in production	**Slurry Phase > Gas Phase, but comparable**
Cost in trademark switch	Slurry Phase is lowest

Preparation of EH LLDPE via Tandem Catalysts

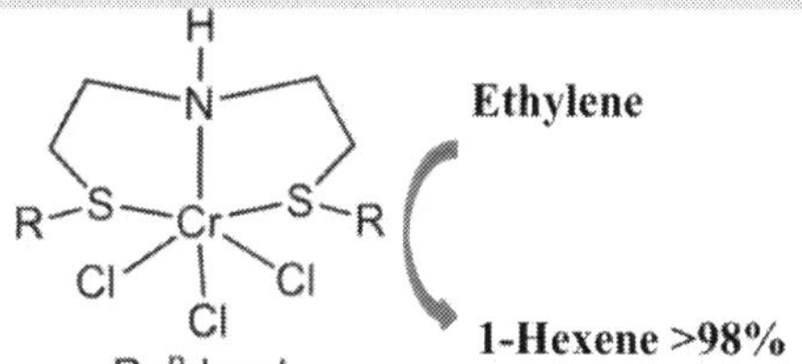

R=ⁿdecyl

SNS-Cr

+

Supported Z-N catalyst

SLC-S donated
Mg: 6.66 wt%
Ti: 2.24 wt%
THF: 28.58 wt%
Average size: 12.40μm

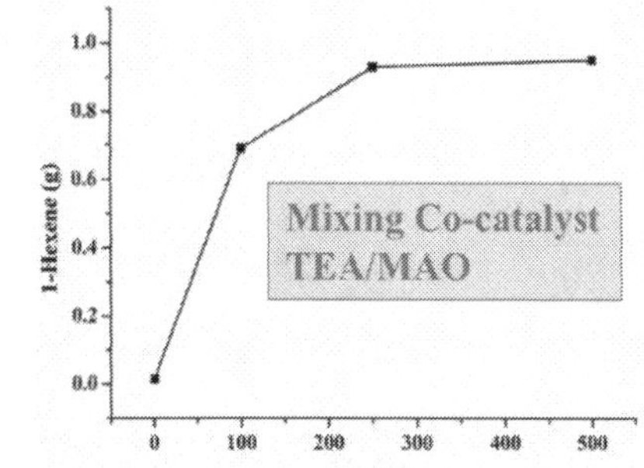

Figure 2. Ethylene trimerization results with different MAO/Cr ratios where TEA/Cr=200.

Advantages

- **Cost in MAO: low level**
- **PE byproduct: trace**
- **Optimal temperature of 70-90 °C: suitable for slurry phase**
- **Morphology of resulting polymer: spherical particle**

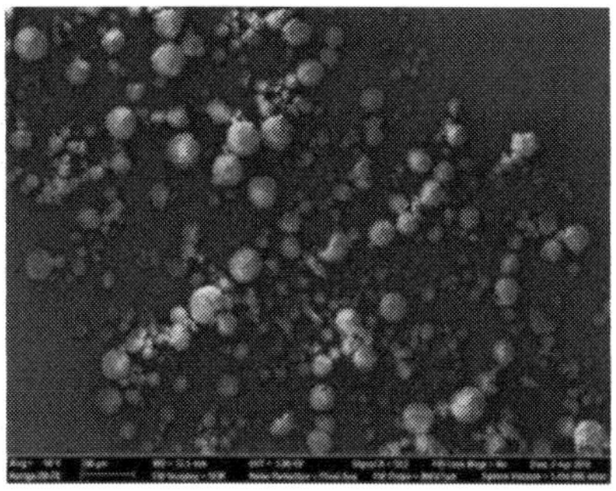

Ethylene homopolymer prepared with SLC-S.
T_m=131.8 °C, M_w=238 kg/mol, PDI=17.7

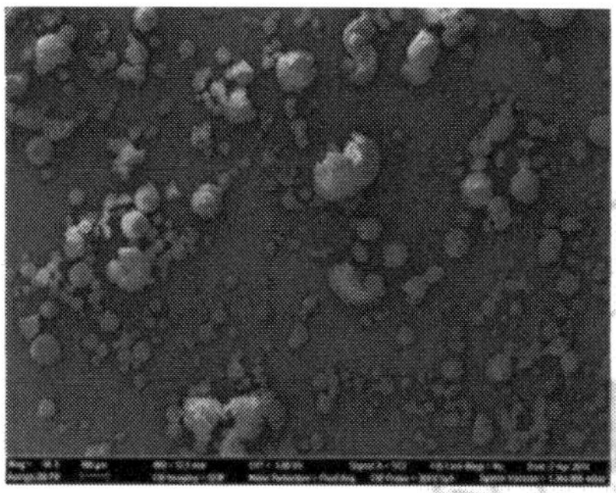

EH LLDPE prepared with SNS-Cr & SLC-S.
T_m=125.0 °C, M_w=149kg/mol, PDI=18.3

ZHEJIANG UNIVERSITY

Tandem Catalysis for Preparation of Polyethylene

High performance polyolefin elastomer (POE)

- **High comonomer incorporation (>8 mol%)**
- **1-C6 and 1-C8 better than 1-C4, but cost more**
- **Solution Process, such as INSITE™ of DOW**

High Temperature is The Challenge for Tandem Catalyst System

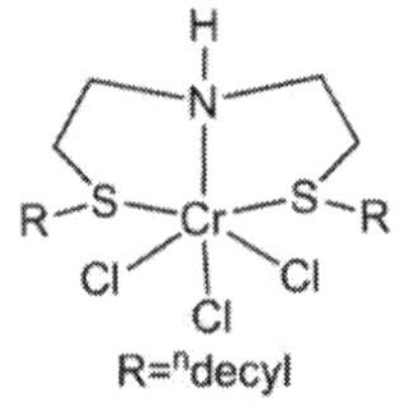

R=^{n}decyl

SNS-Cr

+

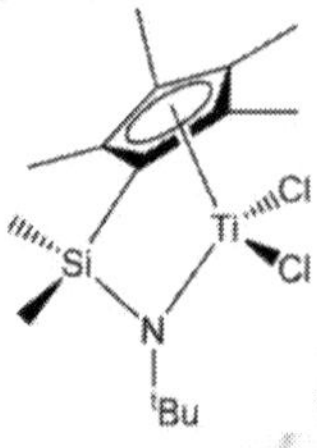

CGC

- High trimerization selectivity > 99%
- High 1-hexene selectivity > 99%
- Can be active over 110 °C

- High Copolymerization Capability
- Can be active over 140 °C
- Commercially used by Dow

ZHEJIANG UNIVERSITY

Synthesis of Ethylene-1-Hexene Copolymer via Tandem Catalysis System of SNS and CGC

75 °C & Atmospheric pressure

Range: Semi-crystallite LLDPE ~ Completely amorphous PE

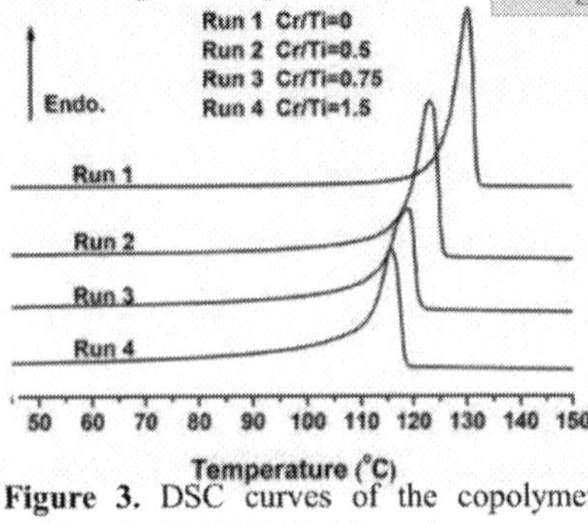

Figure 3. DSC curves of the copolymers produced with CGC-Ti=10 μmol.

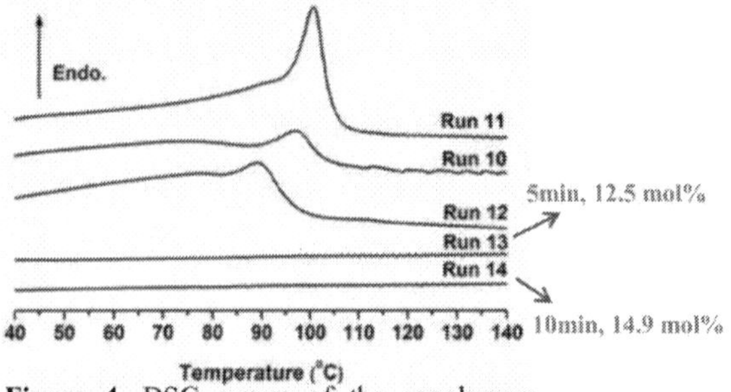

Figure 4. DSC curves of the copolymers produced with pre-trimerization.

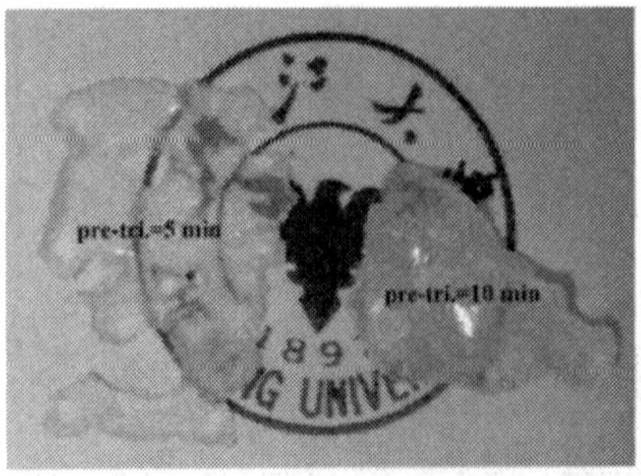

Figure 5. Photos of the products obtained in Run 13 and 14.

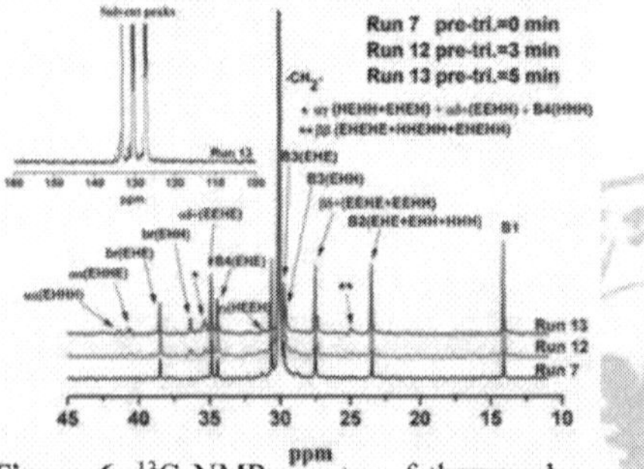

Figure 6. ^{13}C NMR spectra of the copolymers produced with different pre-trimerization times.

S. Guo, et al. ***Macromol. Chem. Phys.*** 2014, 215, 1661-1667.

High Temperature (110 °C) High Pressure (31 bar)

Table 1. Results of ethylene trimerization with SNS at high pressure[a)]

Run	T (°C)	1-C6 (g)	Productivity (10^6 g/mol Cr h)	1-C6 in C6 (%)	PE Product: Yield (g)	T_m (°C)	X_c[b)] (%)	Mw (kg/mol)	PDI
O1	75	50.5	6.05	99.8	0.05	125.6	40.2	539	18.3
O2	90	39.7	4.75	99.5	0.09	125.5	44.2	388	6.6
O3	110	28.3	3.39	99.3	0.04	125.8	49.0	284	8.6

[a)]Polymerization condition: SNS = 16.7μmol, Al/Cr = 1000, ethylene pressure = 31 bar, toluene volume = 250 ml, reaction time = 30 min; [b)]The heat of fusion of completely crystalline PE is 290 J/g.

Table 2. Ethylene polymerization with CGC and SNS at high pressure and high temperature[a)]

Run	CGC (μmol)	SNS (μmol)	P (bar)	Pre-tri.[b)] (min)	Productivity (10^6 g/mol Ti h)	F_2[c)] (mol %)	T_m (°C)	Xc (%)	M_w (kg/mol)	PDI
T1	5	10	31	0	3.94	1.5	101.0 / 110.4	34.3	124	3.0
T2	10	10	31	1	7.94	3.0	105.0	22.4	80.2	4.0
T3	10	20	31	5	6.48	3.1	103.8	21.9	64.2	2.5
T4	10	20	23	1	5.12	2.5	103.6	24.2	54.6	2.7

[a)]polymerization condition: copolymerization time = 30 min, reactor set temperature = 110 °C, 600 rpm, MAO = 15 mmol, 250 ml toluene; [b)]Trimerization time for SNS before introduction of CGC; [c)]Comonomer incorporation in copolymer.

S. Guo, et al. *Macromol. React. Eng.* 2014, DOI: 10.1002/mren.201400027.

9

Sample	T_m (°C)	Xc (%)	Mw (kg/mol)	PDI
CPOE	97.1	23.7	97.1	1.8

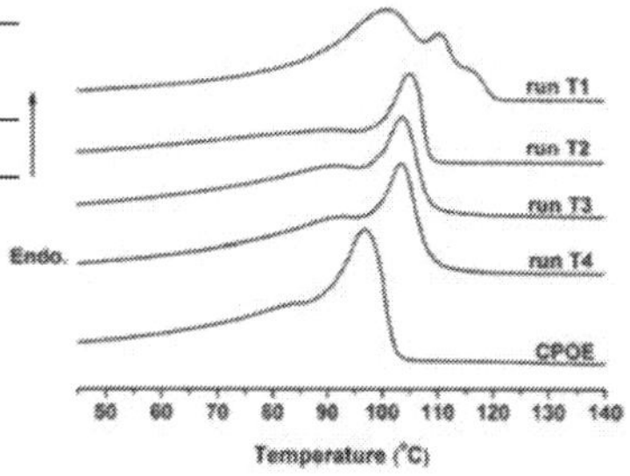

Figure 7. DSC curves of the samples prepared from tandem polymerization of SNS and CGC at high temperature and high ethylene pressure.

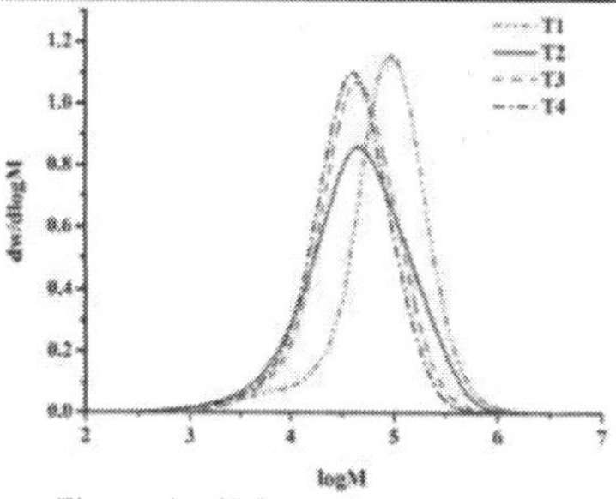

Figure 8. GPC curves of the samples prepared from tandem polymerization of SNS and CGC at high temperature and high ethylene pressure.

Comparable mechanical properties with commercial POE!

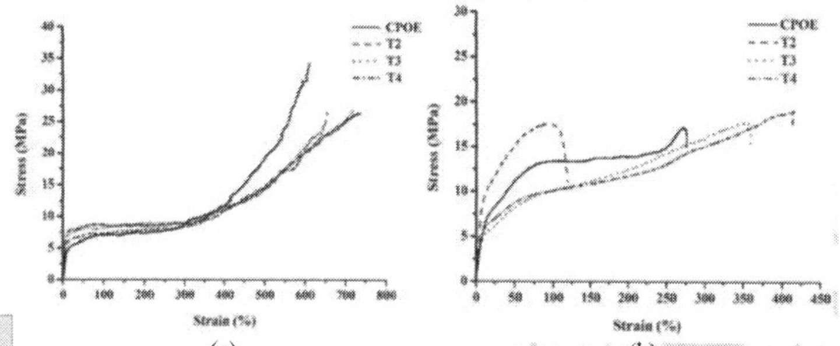

(a) (b)

Figure 9. Engineering stress-strain curves of Sample T2-T4 and CPOE. The tensile specimens were prepared by (a) compression molding and (b) injection molding.

10

Mechanism and Employing Moment Method to Build Kinetics Model

Trimerization mechanisms	
Active-site formation	$C_1 \xrightarrow{k_{a1}} C_1^*$
Ethylene chelation	$C_1^* + 2M_A \xrightarrow{k_1} C_1M_AM_A^*$
Ethylene insertion	$C_1M_AM_A^* + M_A \longrightarrow C_1M_AM_AM_A^*$
1-Hexene elimination	$C_1M_AM_AM_A^* \longrightarrow M_B + C_1^*$
Deactivation	$C_1^* \xrightarrow{k_d} C_{d1}$

Copolymerization mechanisms	
Active-site formation	$C_2 \xrightarrow{k_{a2}} C_2^*$
Active-site initiation	$C_2^* + M_j \xrightarrow{k_{ij}} P_{1,j}^*, \quad j = A, B$
Chain Propagation	$P_{n,i}^* + M_j \xrightarrow{k_{pij}} P_{n+1,j}^*, \quad i = A,B \quad j = A,B$
Transfer to monomer	$P_{n,i}^* + M_B \xrightarrow{k_t} D_n + P_{1,B}^*$
β-hydrogen elimination	$P_{n,i}^* \xrightarrow{k_\beta} D_n + C_2^*$
Spontaneous deactivation	$P_{n,i}^* \xrightarrow{k_{d2}} D_n + C_{d2}$

Semi-Batch Experiments to Obtain Kinetics Parameters

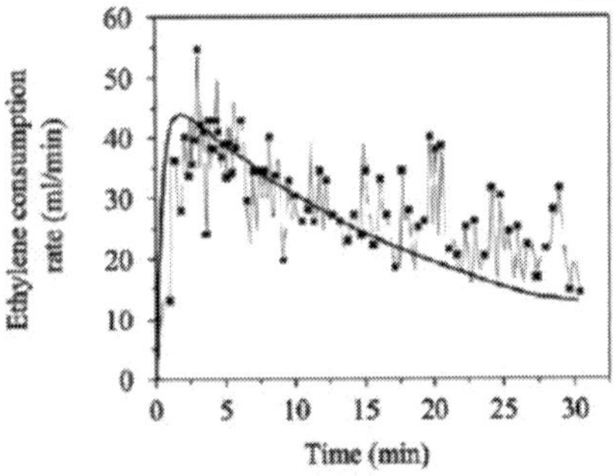

Trimerization

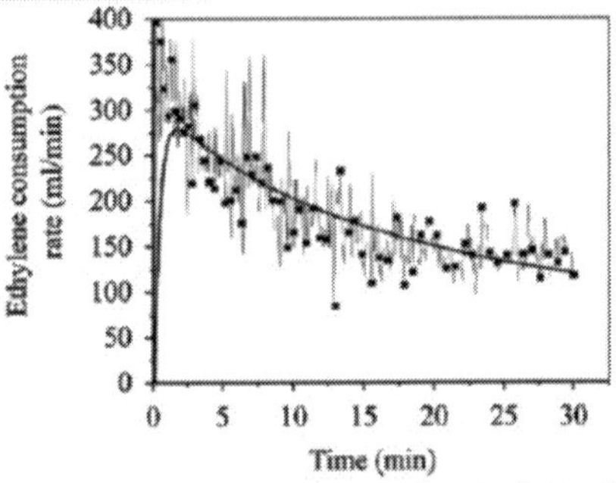

Tandem Polymerization

B -G Li, *et al. Chem. Eng. Sci.* 2008, *63*, 2057 – 2065.

Simulation in CSTR Steady state will be approached when switching process conditions

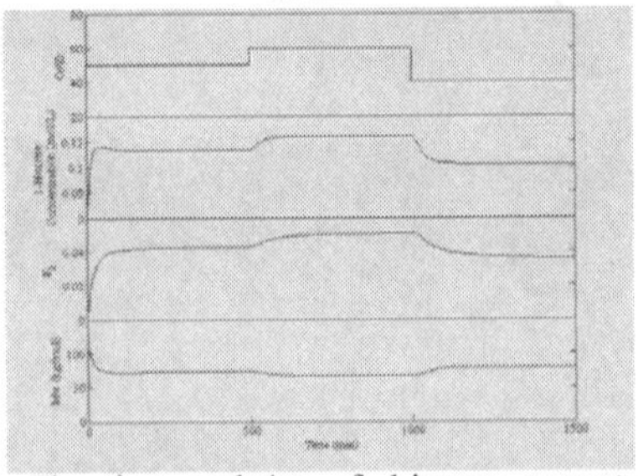

Figure 10. The time evolution of 1-hexene concentration, copolymer composition, and Mw for continuous copolymerization carried out with switching $[Cr]_{in}/[Zr]_{in}$.

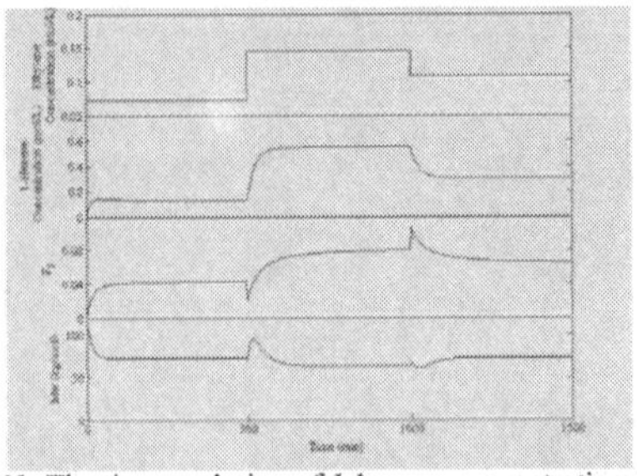

Figure 11. The time evolution of 1-hexene concentration, copolymer composition, and Mw with ethylene concentration switched from 0.0723 mol/L to 0.1446 mol/L, and to 0.1085 mol/L.

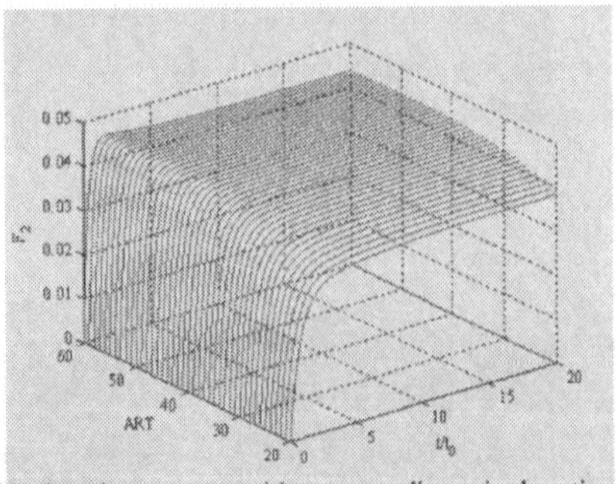

Figure 12. Copolymer composition versus dimensionless time for continuous copolymerization carried out with different ARTs.

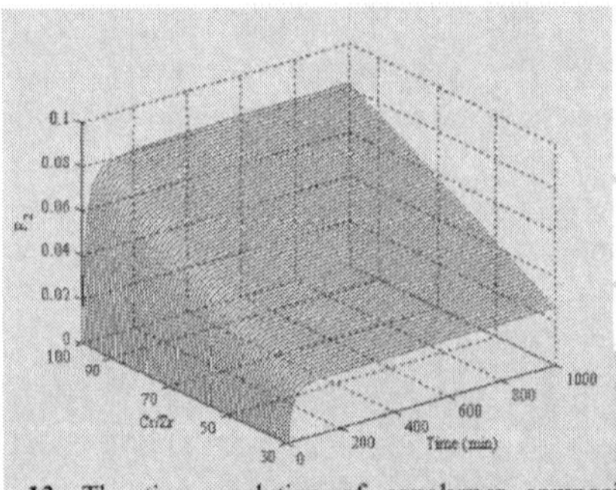

Figure 13. The time evolution of copolymer composition for continuous copolymerization carried out with different $[Cr]_{in}/[Zr]_{in}$.

浙江大学 ZHEJIANG UNIVERSITY

Summary

• The trial of preparing 1-hexene based LLDPE via tandem polymerization of SNS and supported Z-N catalyst appears promising.

• Metallocenes cooperate well with SNS complexes in the presence of MAO as co-catalyst, under 75 °C and atmospheric pressure.

• Ethylene-1-hexene copolymers with comonomer content up to 14.9 mol% was obtained by the tandem polymerization of SNS and CGC.

• Above 110°C, the resulting copolymers having comparable mechanical properties with commercial POE.

• The above results indicate the tandem polymerization carried out in CSTR under high-temperature high-pressure will be inspiring.

• According to the simulation result, higher Cr/Ti ratio and shorter residence time should be considered when applied in CSTR system under high-temperature high-pressure.

15

385176 Dispersed Microenvironments for Chemical Evolution

Abstract:

Biopolymers in current living organisms are usually synthesized with the help of enzymes and cell compartmentalization. Such mechanism was unlikely to exist in the prebiotic earth. Therefore, the first biopolymers for the evolution of early life forms should come from simpler systems with abiotic reaction mechanisms. Miller-Urey experiments have demonstrated the possibility of the formation of amino acids and hydroxy acids in the prebiotic world.[1] Those materials can be used as building blocks for biopolymers. However, even though the source of monomers is understood, the formation of polymers in the prebiotic ocean still faces a number of challenges. The high water concentration in the ocean can hamper the polycondensation reaction by shifting the equilibrium to the reactant side. This enhances the difficulty of forming long chain polymers and shortens their life time.

In this work, an oil/water biphasic reaction media is designed to solve this problem. It has been shown that some water immiscible organic solvents can exist in the prebiotic world.[2] Because of the lower water concentration in those oil solvents, they can facilitate the polymerization reaction by shifting the equilibrium to the product side and protect the synthesized polymers from hydrolysis by water. Also, the increased hydrophobicity of the growing polymer chain can further concentrate those polymers in the oil phase and speed up the evolutionary process. The oil phase can also be dispersed and stabilized by surfactants to form an emulsion. The dispersed oil phase has much higher interfacial area and can facilitate the absorption of building blocks from the prebiotic ocean.

The present work first focuses on the polymerization of lactic acid in the oil/water biphasic system. Polyesters have been proposed as one possible ancestor for polypeptides[3] and lactic acid is a significant product from the Miller-Urey experiments. Also, polylactic acid (PLA) is a biodegradable polymer that has been well characterized and can be used in drug delivery applications.

Two different abiotic reaction pathways have been studied in this work. The first one is the polycondensation reaction of lactic acid with dicyclohexylcarbodiimide as condensing agent and 4-dimethylaminopyridine as catalyst. It has proved to be an efficient condensing agent for the polymerization of lactic acid when organic solvent is used alone.[4] In our work, the interfacial reaction between the hydrophobic carbodiimide and hydrophilic lactic acid is able to produce long chain PLA. The second reaction pathway is starting from the ring-opening polymerization of lactide, the cyclic dimer of lactic acid. [5] Lactide has higher solubility in organic solvents than lactic acid. Therefore, the reaction can proceed directly in the oil phase. The ring-opening polymerization can be catalyzed by 4-dimethylaminopyridine which is a weak organic base and its distribution between oil and aqueous phase can be controlled by the pH of water. Polyester with high yield can be obtained and preserved in the oil phase under near neutral pH and room temperature.

The findings in this work provide another methodology to solve the mystery of the origin of life. The presence of oil phase successfully facilitates the polymerization and extends the life time of the synthesized polyester. The same biphasic reaction system can also be applied to the reaction and evolution of other biopolymers in the future.

Reference

1. Miller, S. L.; Urey, H. C., Organic compound synthesis on the primitive earth: Several questions about the origin of life have been answered, but much remains to be studied. *Science* **1959,** *130* (3370), 245-251.

2. Lasaga, A. C.; Holland, H. D.; Dwyer, M. J., Primordial oil slick. *Science* **1971,** *174* (4004), 53-55.

3. Orgel, L., Some consequences of the RNA world hypothesis. *Origins Of Life And Evolution Of The Biosphere* **2003,** *33* (2), 211-218.

4. Akutsu, F., Inoki, M., Uei, H., Sueyoshi, H., Kasashima, Y., Naruchi, K., Yamaguchi Y., Sunahara, M., Synthesis of poly(lactic acid) by direct polycondensation of lactic acid using 1,1'-carbonyldiimidazole, N,N,N',N'-tetramethylchloroformamidinium chloride and N,N'-dicyclohexylcarbodiimide as condensing agents. *Polym Journal* **1998,** *30*, 421.

5. Coulembier, O., Dubois, P., 4-dimethylaminopyridine-based organoactivation: From simple esterification to lactide ring-opening "Living" polymerization. *Journal of Polymer Science Part A: Polymer Chemistry* **2012,** *50* (9), 1672-1680.

14AIChE Annual Meeting, Atlanta, GA

Non-aqueous suspension polycondensation for the preparation of poly(*p*-phenylene terephthalamide)

Peijian Wang, Kai Wang, Jisong Zhang, Guangsheng Luo*

The State Key Laboratory of Chemical Engineering, Department of Chemical Engineering, Tsinghua University, Beijing, 100084, China

wpjqhu@gmail.com, 86-10-62797493

Sklochе

Outline

- Background
- Research Contents
- Conclusions

Background of poly(*p*-phenylene terephthalamide) (PPTA)

High performance properties :

- Tensile strength
- Corrosion resistance
- Roughness
- Weight and thermal stability

Wildly used in:

- Composite material
- Bulletproof material
- Rubber reinforced material
- Asbestos replacement material

Background of poly(*p*-phenylene terephthalamide) (PPTA)

Different methods for the synthesis of PPTA:

- Interfacial polycondensation method
- Vapor polycondensation method
- Direct polycondensation method
- Low temperature method

Reactant mixing

Isolation of oxygen and water

Rapid and strongly exothermic

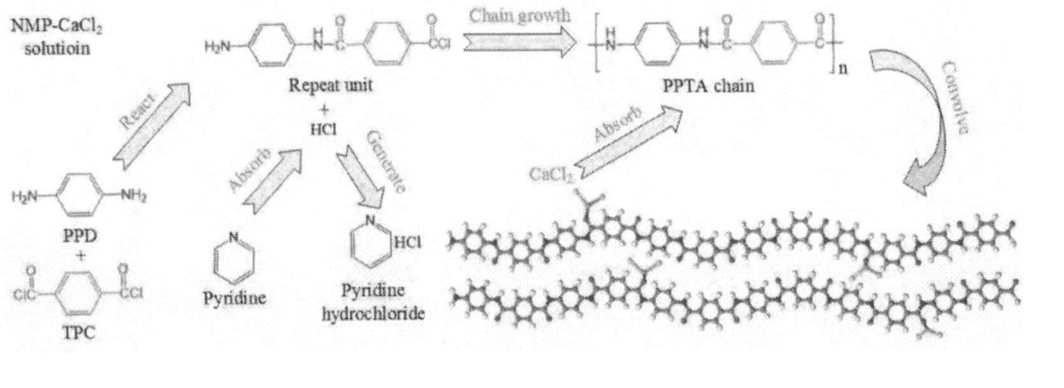

Experimental design

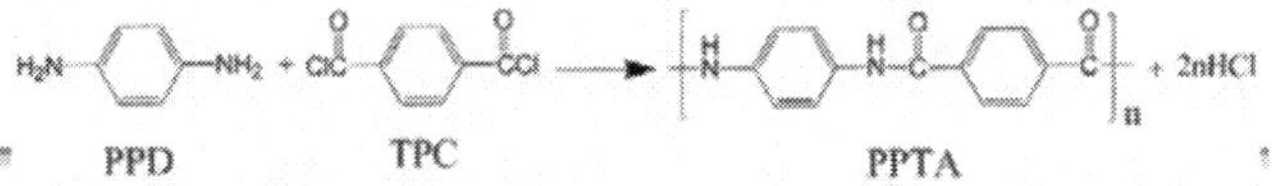

Non-aqueous suspension polycondensation method

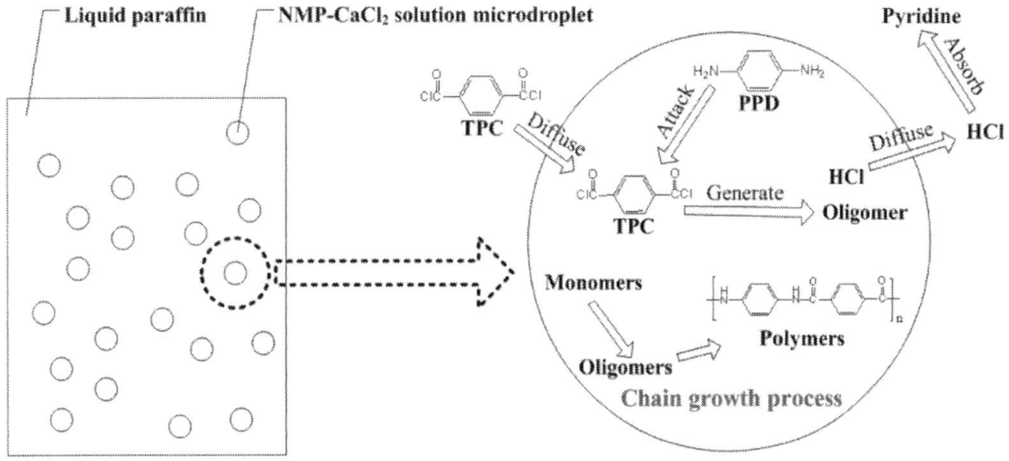

Surrounded by non-reactive solvent

Easy removal of HCl

Temperature and viscosity

Outline

6

- **Background Introduction**
- **Experimental section**
- **Conclusions**

Non-aqueous suspension

7

NMP microdroplets suspended in liquid paraffin under microscope

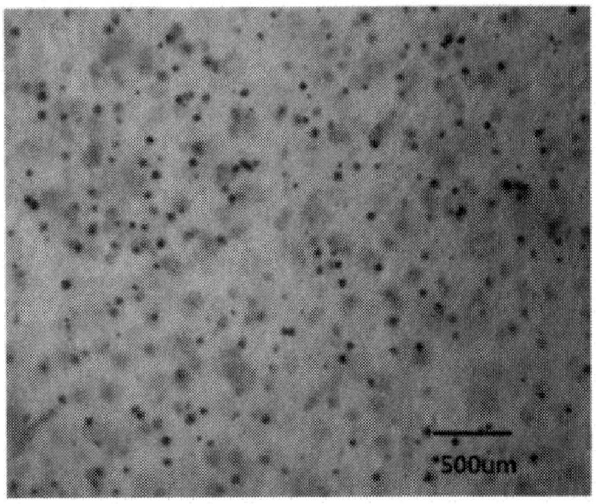

$$\overline{d} = k\frac{D_V \cdot R \cdot v_d \cdot \varepsilon}{D_S \cdot N \cdot v_m \cdot C_s}$$

NMP solution was dispersed in liquid paraffin under the strong shearing force from the emulsifying machine, forming a non-aqueous suspension with droplet size in dozens of micrometers.

Arshady, R., Colloid. Polym. Sci., 1992, 270(8): 717-732

Characterization

8

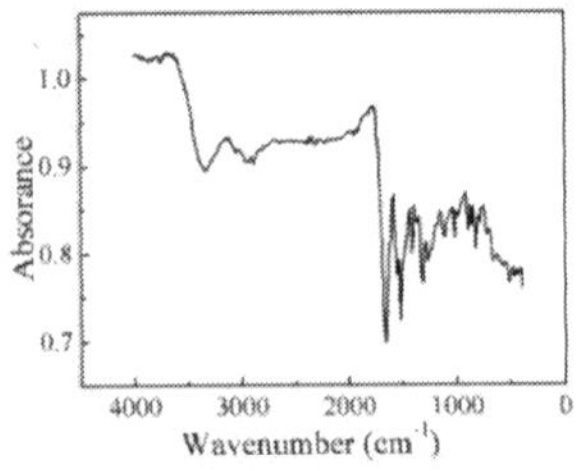

IR

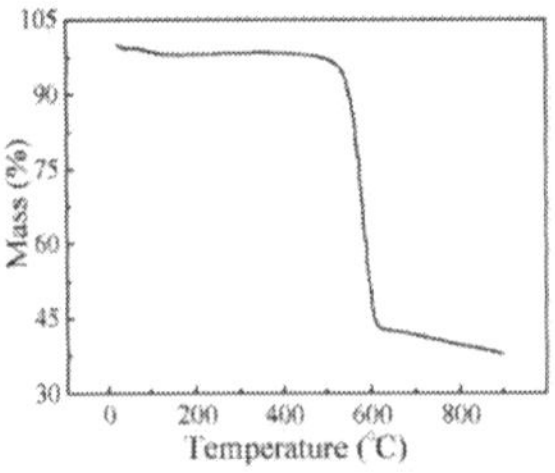

TGA

Sample No.	N	C	H	O
Theoretical	11.76	70.59	4.20	13.45
1	11.32	68.92	4.28	15.48
2	11.40	69.23	4.26	15.11
3	11.41	69.14	4.27	15.18

EA

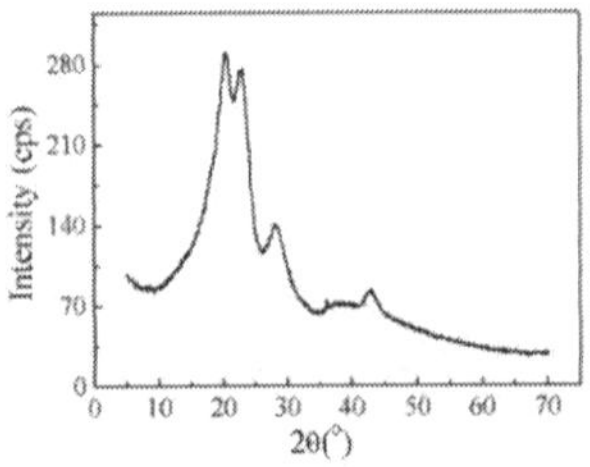

XRD

(a) (b)

SEM

Effects of different variables on the M_W of PPTA 9

Phase ratio

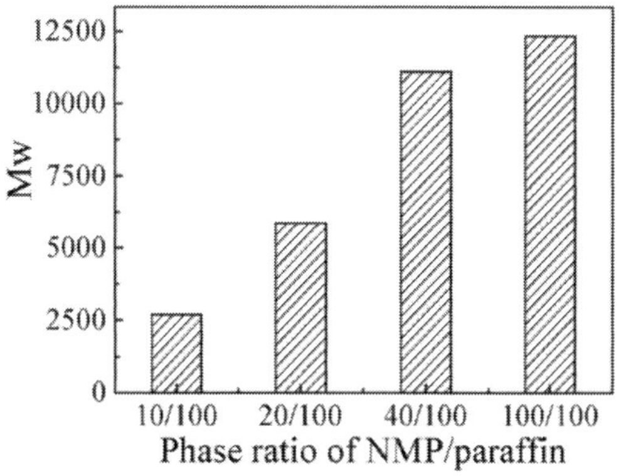

Reaction time

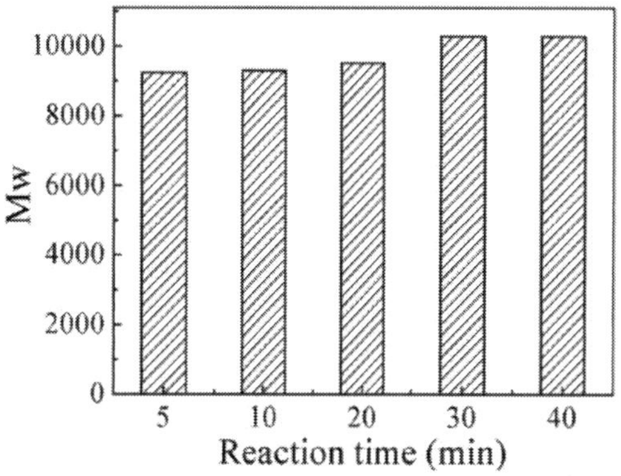

Small → Unbalance of monomer molar ratio

Large → Microdroplets coalesce quickly

Fast reaction

Effects of different variables on the M_W of PPTA 10

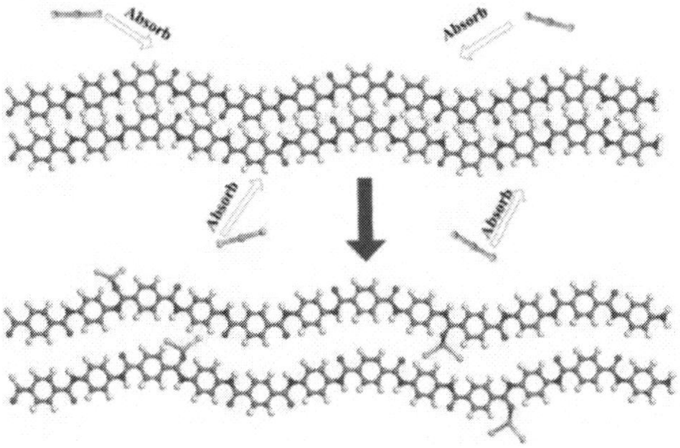

Complexation between calcium ions and carbonyl groups destroys the strong hydrogen bonding between PPTA chains

$CaCl_2$ concentratioin

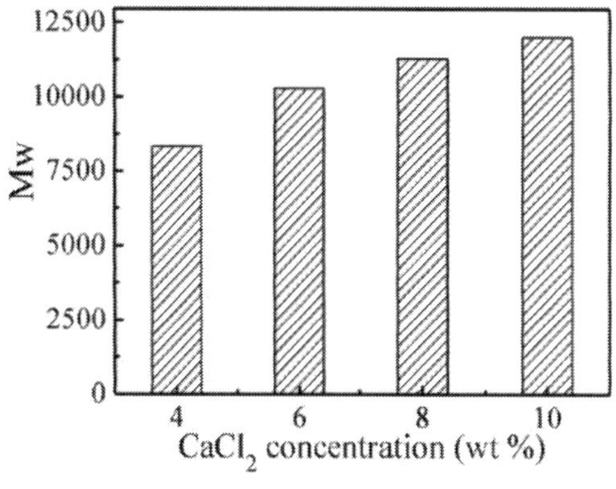

Effects of different variables on the M_W of PPTA

11

Monomer concentration

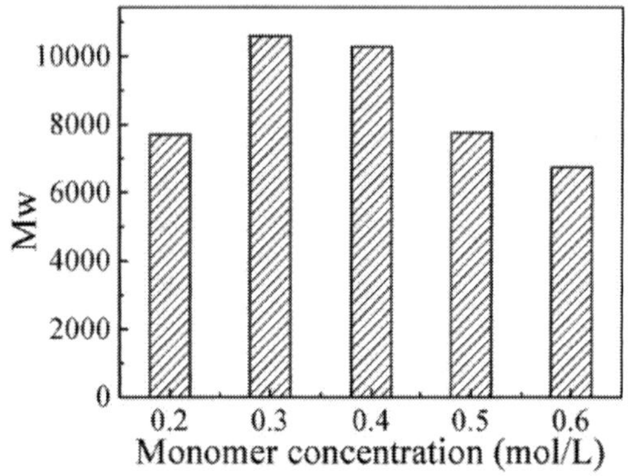

Medium monomer concentration range is better

Monomer molar ratio

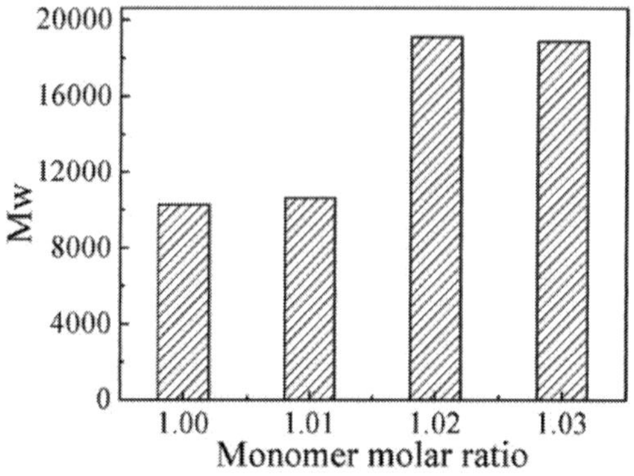

$$\overline{X}_n = \frac{1+r}{1+r-2rP}$$

About 2 percent more TPC

Effects of different variables on the M_W of PPTA

12

Temperature

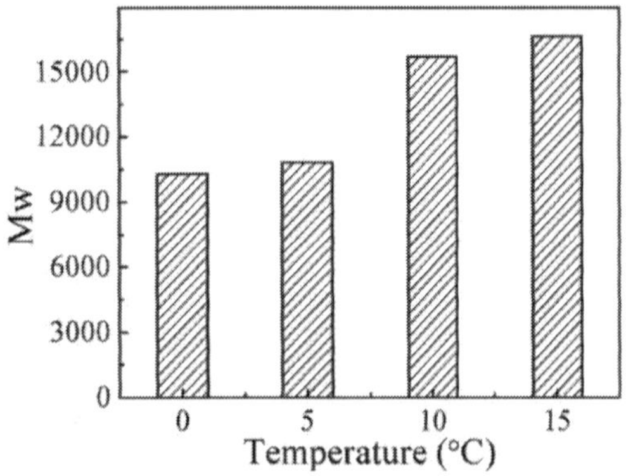

↑ Solubility of monomers and polymers

↓ Viscosity of the NMP phase

Stirring rate

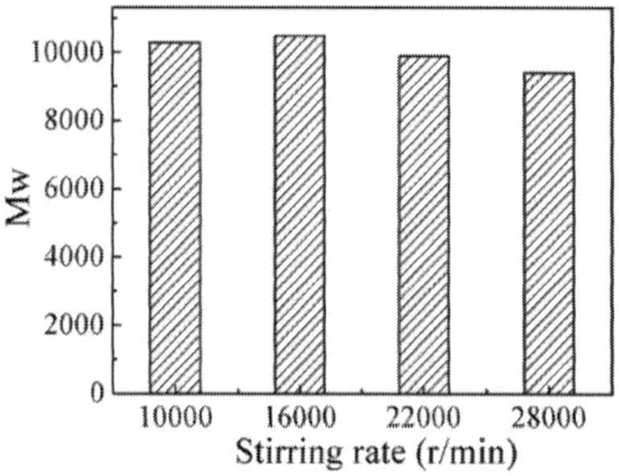

Overall effect is not huge

$$\overline{d} = k\frac{D_V \cdot R \cdot v_d \cdot \varepsilon}{D_S \cdot N \cdot v_m \cdot C_s}$$

Preparationof PPTA with a microreactor

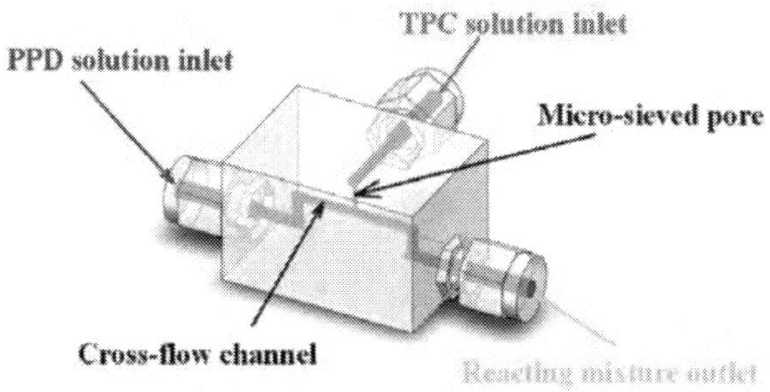

T-shaped microreactor

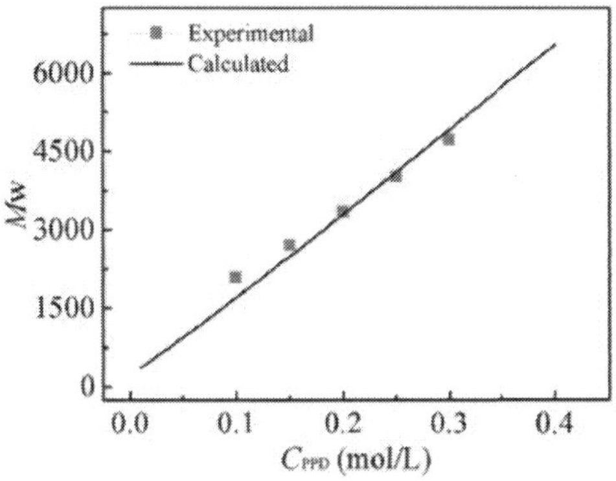

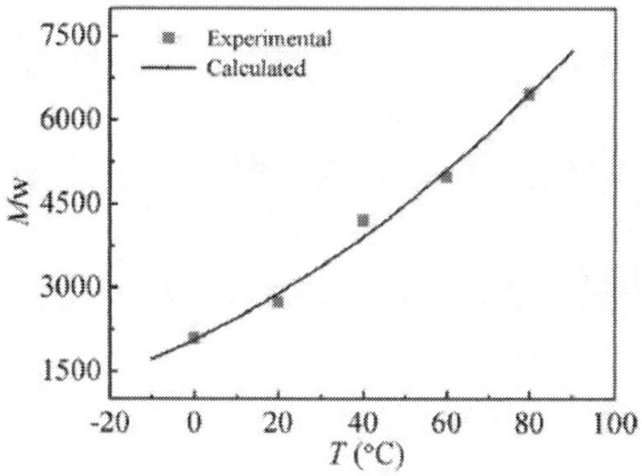

Outline

- **Background Introduction**
- **Research Contents**
- **Conclusions**

Conclusion

- The polycondensation between PPD and TPC was proposed in a non-aqueous suspension system to synthesize PPTA.
- Each of the NMP microdroplets suspended in paraffin liquid serves as a microreactor, in which the polycondensation takes places.
- Application of this new approach leads to less strict demand for removal of oxygen and water, easy removal of HCl, well-controlled temperature and low system viscosity
- The molecular weight can be adjusted easily by changing the following reaction parameters: phase ratio of NMP/paraffin, reaction time, $CaCl_2$ concentration, monomer concentration, monomer molar ratio, temperature and stirring speed.

Thank you for your listening!

Any question will be highly appreciated!

379180 Mathematical Model for the Bulk Polymerization of Styrene Using Multifunctional Initiators

Abstract:

A detailed mathematical model for bulk polymerization of styrene (St) using multifunctional initiators –both linear and cyclic- was developed. The model is based on a kinetic scheme that considers chemical and thermal initiation by sequential decomposition, propagation, transfer to monomer, diffusion-controlled termination by combination and re-initiation reactions and is used to predict the evolution of the chemical species and the molecular structure of the obtained polymeric species. The model calculates the evolution of total or average variables (e.g, total radicals, total polymer, molecular weight distributon moments) as well as the evolution of the detailed complete polymer molecular weight distribution, with polymer species characterized by chain length and number of undecomposed labile groups. The mathematical model was adjusted and validated using experimental data obtained from bulk polymerizations of styrene using various peroxide-type multifunctional initiators: diethyl ketone triperoxide (DEKTP, ctrifunctional cyclic), pinacolone diperoxide (DPP, bifunctional cyclic), 1,1-bis(tert-butylperoxy)cyclohexane (L331M80, bifunctional linear) and benzoyl peroxide (BPO, monofunctional linear) at reaction temperatures of 110-130 °C, where initiator decomposition is mostly sequential. The model is able to very adequately predict experimental polymerization rates and complete molecular weight distributions by adjusting the initiator decomposition constants. The model is used to theoretically evaluate initiator structure and functionality, as well as reaction conditions in a bulk St polymerization process. For a given initiator, the model predicts the existence of optimum temperature and initial initiator concentration resulting in high productivity and increased product quality, which was experimentally validated.

390814 Propylene Polymerization with Metallocene/Methylaluminoxane Catalysts: Mechanisms, Modeling and Simulation

Abstract:

AIChE-2014

Abstract

Polymers and co-polymers of propylene find extensive industrial and commercial applications like engineering plastics, elastomers and synthetic rubbers. Transition metal compounds are generally used as catalysts in polymerization of propylene, Ziegler Natta catalyst being the first successful kind among them. These catalysts are the derivatives of transition metals such as titanium ($TiCl_4$, $TiCl_3$), vanadium (VCl_4, $VOCl_3$), etc. and are generally employed with co-catalysts which are organometallic compounds, such as $Al(CH_3)_3$, $Al(C_2H_5)_3$, etc.

Ziegler-Natta catalysts are heterogeneous catalysts with multiple sites. Metallocene/methylaluminoxane complexes are the catalysts used next to Ziegler Natta catalysts in the polymer industry recently. Metallocene catalyzed propylene polymerization has recently attracted research interest since these catalysts allow the production of tailored macromolecules with properties those can be accurately designed. A broad spectrum of properties and applications of the polypropylene can be attained with metallocenes due to their single types of sites.

Two major types of metallocene catalysts are used in propylene polymerization –first type being bis-metallocene complexes of transition metals, (Cp_2MtCl_2) and the second type based on complexes containing one cyclopentadienyl group like [C_5H_4-$SiMe_2$-(tert-Bu)N]$TiCl_2$-MAO and [C_5Me_4-$SiMe_2$-(tert-Bu)N]$TiCl_2$-MAO, etc.
Homogeneity and single site characteristics of metallocenes enable to correlate metallocene structures with polymer properties such as molecular weight, stereochemical microstructure, behavior and mechanical properties. Understanding reaction mechanisms and kinetics of metallocene catalyzed polymerization reactions is very important to design/identify a catalyst for promising product properties or to modify the existing Ziegler-Natta based technology with metallocenes.

This article deals with the presents state of the art on the reaction mechanisms and kinetics involved in Ziegler-Natta and metallocene catalyzed propylene polymerization. Further, an approach of microscopic modeling and simulation of propylene polymerization through different metallocene catalyst systems is presented and discussed using a novel *Natural Logarithmic Differential Evolution* optimization algorithm.

References:

1. Kissin Y., In: Alkene Polymerization Reactions with Transition Metal Catalysts:Studies in surface science and catalysis, Elsevier publications, vol.173.

2. Ochoteco E., Vecino M., Montes M. and Cal J. C., Chemical Engineering Science 56: p. 4169–4179 (2001).

3. Song F., Hannant M. D., Cannon R. D. and Bochmann M., Macromol. Symp. 213: p. 173-185 (2004).

4. W. Kaminsky and Laban A., Applied Catalysis A: General 222: p. 47–612 (2001).

5. Zhang J., Wang X. and Jin G., Coordination Chemistry Reviews 250: p. 95–109 (2006).

367865 Simulation of Urethane Foaming Processes

Abstract:

Urethane formulations typically consist of two to five co-monomers and from four to eight additives including catalysts and surfactants. These thermoset polymers form in a process of rapidly changing physical and chemical properties. The complexity of this process has resulted in an industry that has been dominated by expertise embedded in people and formulation development that can accurately be described as an art. This creates a particularly difficult situation for advances in sustainability, especially use of soy-based polyols, because of how the momentum of proven formulations dominate the industry.

This paper is on the simulation of the complex urethane forming processes. A simulation program for urethane-forming reactions has been developed to assist in adopting soy-based resins in formulations. This paper describes Matlab program used to simulate urethane-forming reactions including the ability to simulate the reactions, viscosity, and gas formation in formulations containing up to three different polyols.

373748 Mechanical and Thermal Properties of Cross-Linked Phenolic Resins Using Molecular Dynamics

Abstract:

To gain insight into the design of materials, it is valuable to understand how the chemical make-up at the nano-scale can influence the thermal and mechanical bulk properties. An atomistic computational study allows us to manipulate the structural make-up of individual phenolic chains as well as generate various cross-linked (or cured) systems. In this study, molecular dynamics simulations of bulk phenolic systems were performed with the software LAMMPS. An all-atom force field was chosen to investigate how the strength and thermal conductivity of the phenolic material varies as a function of the degree of cross-linking and chemical make-up of the phenolic chains. Small-scale mechanical tests were performed to compute various moduli for the phenolic systems above and below the glass transition at varied degrees of cross-linking. The thermal conductivity was obtained using the Green-Kubo approach for the virgin phenolic system as well as the strained systems.

375126 Directed Differentiation in 3D Gels *Via* Photoreversible Protein Patterning

Abstract:

Synthetic hydrogels have emerged as a unique class of biomaterials that enable stem cells to be cultured in three-dimensions within near-physiological, synthetic microenvironments. Recent strategies have been developed that permit bioepitopes (*e.g.*, peptides, full-length proteins) to be introduced at any point in time and space to affect cell function spatiotemporally within user-defined subvolumes of the bulk material. While these techniques have been successfully utilized to direct a variety of basic cellular functions, advanced platforms that permit biological cues to be both *introduced* and subsequently *removed* would be beneficial in recapitulating the dynamic abundance of signaling biomolecules in the native, temporally-variable niche and in modulating complex cellular behavior. In this work, we demonstrate that the combination of two bioorthogonal light-based chemistries provides for the reversible immobilization of protein cues spatially within a hydrogel. The highlighted approach enables precise control over 4D biochemical functionalization of a synthetic polymer network in response to user-defined photonic stimuli. Results further illustrate the versatility of such dynamic biomacromolecular signal presentation in directing 4D progenitor cell fate.

378113 Dual-Stage Growth Factor Release within 3D Protein-Engineered Hydrogel Niches Promotes Adipogenesis

Abstract:

The engineering of biomimetic microenvironments from hydrogels is an emerging strategy to achieve lineage-specific differentiation *in vitro*. In addition to recapitulating critical matrix cues found in the native three-dimensional (3D) niche, the hydrogel can also be designed to deliver soluble factors that are present within the native inductive microenvironment. We demonstrate a versatile materials approach for the dual-stage delivery of multiple soluble factors within a 3D hydrogel to induce adipogenesis. We use a Mixing-Induced Two-Component Hydrogel (MITCH) embedded with alginate microgels to deliver two pro-adipogenic soluble factors, fibroblast growth factor 1 (FGF-1) and bone morphogenetic protein 4 (BMP-4) with two distinct delivery profiles. We show that dual-stage delivery of FGF-1 and BMP-4 to multipotent, human adipose-derived stromal cells (hADSCs) significantly increases lipid accumulation compared with the simultaneous delivery of both growth factors together. Furthermore, dual-stage growth factor delivery within a 3D hydrogel resulted in substantially more lipid accumulation compared to identical delivery profiles in 2D cultures. Gene expression analysis shows upregulation of key adipogenic markers indicative of brown-like adipocytes. These data suggest that dual-stage release of FGF-1 and BMP-4 within 3D microenvironments can promote the *in vitro* development of mature adipocytes with insulin sensitivity.

387118 Identifying the Differentiation and Proliferation Stage of Single Hematopoietic Stem Cells Using Raman Microspectroscopy

Abstract:

The ability to monitor the differentiation and proliferation status of individual hematopoietic stem cells is critical to understanding the effects of various external factors, such as matrix properties and cytokines, on hematopoiesis. Raman microspectroscopy is a promising method for single cell identification and characterization. Unlike other invasive methods, Raman microscopy has the ability to detect changes in the composition of living cells without compromising the cells' morphology, proliferation, or pluripotency. However, identifying cells according to their Raman spectra is difficult due to a lack of spectral bands that are unique to a single cell line. Here, we report our ability to identify the differentiation stage of individual mouse hematopoietic stem cells and mature cells by multivariate analysis of their Raman spectra. Confocal Raman microscopy was used to obtain spectral fingerprints of single cells in various stages of hematopoiesis. A model of the cell type-specific variance in the Raman spectra was constructed using partial least-squares discriminant analysis (PLS-DA) and enabled accurately identifying cells of "unknown" differentiation stage, including closely-related long- and short-term hematopoietic stem cells with low prediction error (<10%). Our high identification accuracy cannot be attributed to spectral contributions from the polyacrylamide gel substrates, which varied in stiffness and thickness, and culture media. In addition, principal component analysis was used to identify the cell cycle stage of each single cell based on the principal component corresponding to DNA-related Raman peaks. Stem cells were found to be more proliferative than mature cells. Noteworthy, the Raman contributions that correlated to cell cycle phase did not significantly interfere with identification of cell differentiation stage. This method allows for non-invasive monitoring of stem cell differentiation over time with location specificity, and can be used to correlate external stimuli with stem cell fate decisions.

Electrospinning is an attractive technique for the creation of anisotropic fibrous meshes to guide development of organized connective tissues, but the small fiber diameters can inhibit cell infiltration and proliferation. In contrast, hydrogels permit the formation of cellularized 3D scaffolds, but lack contact guidance features and anisotropic tensile properties of connective tissues. Our central hypothesis is that an electrospun fiber/hydrogel composite can support proliferation of mesenchymal stem cells (MSC) while providing topographical, chemical and mechanical cues to direct differentiation into different tissue phenotypes. To examine this, we developed a family of pseudo-3D microenvironments consisting of a thin layer of electrospun fibers within a collagen/MSC phase. In particular, we varied the orientation of electrospun polycaprolactone (PCL) fibers and incorporated hydroxyapatite (HA) nanoparticles into the fibers to determine the impact of fiber alignment and chemistry on MSC morphology, proliferation, and expression of the fibroblastic and osteoblastic phenotypes.

Our preliminary studies show that MSCs migrate from the collagen phase to the fiber surfaces, and assume a spindle morphology (Figure 1) when presented with aligned polycaprolactone (PCL) fibers and a polygonal morphology when presented with randomly oriented PCL/HA fibers. We are currently examining the effects of fiber orientation and HA content on expression of the fibroblastic and osteoblastic phenotype. As a next step we will examine the effect of composite mechanical properties by replacing the PCL fibers with elastomeric polyurethane fibers. In our presentation, we will show how the topographical, chemical and mechanical properties of fiber/collagen composites can be tuned and how they can affect MSC morphology and phenotype.

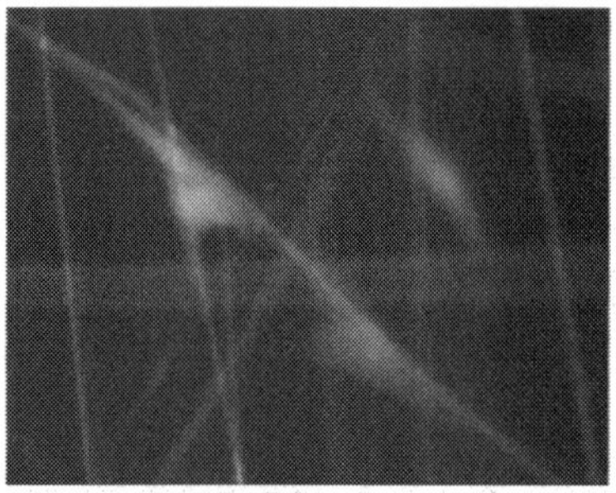

Figure 1. MSCs (green) in a 0.5 wt% collagen gel migrate to sparse electrospun fibers (red).

389353 Nanotemplated Polyelectrolyte Films As Porous Biomolecular Delivery Systems

Abstract:

Biomaterials capable of delivering controlled quantities of bioactive agents, while maintaining mechanical integrity, are needed for a variety of cell contacting applications. We describe here a nanotemplating strategy toward porous, polyelectrolyte-based thin films capable of controlled biomolecular loading and release. Films are formed via the layer-by-layer assembly of charged polymers and nanoparticles (NP), then chemically cross-linked to increase mechanical rigidity and stability, and finally exposed to tetrahydrofuran to dissolve the NP and create an intra-film porous network. We report here on the loading and release of the growth factor bone morphogenetic protein 2 (BMP-2), and the influence of BMP-2 loaded films on contacting murine C2C12 myoblasts. We observe nanotemplating to enable stable BMP-2 loading throughout the thickness of the film, and find the nanotemplated film to exhibit comparable cell adhesion, and enhanced cell differentiation, compared to a non-porous cross-linked film (where BMP-2 loading is mainly confined to the film surface).

389722 Plasma Proteins Alter the Vascular Wall Adhesion of Drug Carriers in a Material & Donor Specific Way

Abstract:

Introduction: The nanoscale plasma protein interaction with intravenously injected particulate carrier systems is known to modulate their organ distribution and clearance from the bloodstream. However, the role of this plasma protein interaction in prescribing the adhesion of carriers to the vascular wall remains relatively unknown.

Materials and Method: poly(lactide-co-glycolic-acid) (PLGA) micro and nanospheres were fabricated via the oil-in-water emulsion solvent evaporation method. Polystyrene and Silica microspheres were obtained commercially. All particles were conjugated, via carbodimide chemistry, with the sLe^{A}carbohydrate ligand that specifically binds to E-selectin on activated endothelial cells (ECs). The blood flow adhesion of the inflammation-targeted particles to the vascular wall was evaluated in a parallel plate flow chamber consisting of a monolayer of activated ECs at the bottom wall of the chamber.

Result and Discussion: We show that the adhesion of vascular-targeted PLGA spheres to endothelial cells is significantly inhibited in human blood flow, with up to 90% reduction in adhesion observed relative to adhesion in simple buffer flow depending on the particle size and the magnitude and pattern of blood flow. This reduced PLGA adhesion in blood flow is linked to the adsorption of certain high molecular weight plasma proteins on PLGA and is donor specific, where large reductions in particle adhesion in blood flow (>80% relative to buffer) is seen with ~60% of unique donor bloods while others exhibit moderate to no reductions. The depletion of high molecular weight immunoglobulins from plasma is shown to successfully restored PLGA vascular wall adhesion. The observed plasma protein effect on PLGA is likely due to material characteristic since the effect is not replicated with polystyrene or silica spheres. These particles effectively adhere to the endothelium at a higher level in blood over buffer flow. Finally, grafting of $sLe^{A}$to the PLGA surface via high density (brush conformation) of polyethylene glycol (PEG) chains did not eliminate the negative adhesion of PLGA particles to ECs in blood flow.

Conclusion: In this work, the adsorption of certain plasma proteins from human blood onto PLGA carriers were found to prevent these particles from effectively adhering to activated endothelial cells in *in vitro* assays, an effect that was not observed for particles of other material types. Our results also show that the extent of the negative adhesion effect of plasma proteins on PLGA particles is dependent on specific blood donors and the targeting ligand density but not the targeting ligand type. Overall, the presented data suggests that specific knowledge of the plasma protein composition across different humans may be critical to VTC design and their successful clinical use, i.e., highlighting the need for a shift toward personalized medicine in the design of targeted therapeutics. Alternatively, it is possible that with a detailed understanding of the specific proteins that affect particle vascular targeting, novel biomaterials can be designed to resist the adsorption of these proteins in order to achieve enhanced vascular targeting irrespective of the plasma composition of different individuals.

390053 Understanding and Manipulating the Interface Between Biomolecules and Carbon Nanotubes for the Next Generation of Bioanalytical Tools

Abstract:

Enzymes have potential applications in industrial catalysis, biosensing, drug delivery and decontamination, however with limited usage due to their low operational stability and yield loss in synthetic environments. Enzyme immobilization onto nano-sized solid supports has been proposed as an alternative to ensure enzyme stability, retention and recovery. However, the nanosupport has often been shown to affect the enzyme activity; a deep understand of the enzyme-nano interface reaction is thus needed if advances in the listed applications are to be made. Herein, we present a comprehensive and systematic study of the interface reactions between enzymes and nanosupports with different characteristics (i.e. surface area, charge and aspect-ratios). The interface reactions are discussed relative to the enzyme kinetics and retained activity at the nanosupports. Knowledge gained from this study can further be used to optimize enzyme-nanomaterial interactions in order to maintain optimal levels of enzyme activity while enhancing stability upon immobilization for use in a broad spectrum of applications.

390426 Visualization and Assay of γ-Secretase Enzyme Inhibition and Substrate Loading in Intact Lipid Bilayers Via Microsphere-Supported Biomembrane Systems

Abstract:

The intramembrane protease gamma-secretase is a current target of therapeutic intervention, with pivotal pathological functions within Alzheimer's disease and cancer. Our primary objective is to expand methods for *in vitro*studies of the intramembrane proteases (IMPs) with the development of a microsphere-supported biomembrane (proteolipobead) platform. Despite extensive studies, understanding of IMP regulation and catalysis has been hampered by membrane-associated enzymology. The precise reaction mechanisms of the intramembrane proteases remain to be elucidated and furthermore, rigorous analysis of the kinetics of interfacial biocatalysis in these systems has not yet been undertaken.

High throughput of screening of gamma-secretase drug candidates has been carried out in bulk assay systems using cell membrane fragments, solubilized enzymes and is underway in proteoliposomes. A critical barrier to further progress in the study and screening of gamma-secretase is that such bulk systems do not allow for the direct in situ quantification of enzyme, substrates, or inhibitors or their relative distributions within the structures under assay in native lipids. Due to these factors and the instability of these systems there are limitations in the scope of assays possible using solubilized systems and proteoliposomes in bulk solution. We have expanded in vitro models of gamma-secretase to include proteolipobeads with tunable lipid compositions (microsphere-supported biomembranes), using novel substrates such as Notch1-Sb1 (truncated Notch 1 with FLAG tag) to probe the enzyme. Using confocal and super-resolution imaging in tandem with flow cytometry and cryoTEM, we will present findings from i) direct measurements of substrate and gamma-secretase loading and in-membrane orientation within supported biomembranes, ii) the localization and quantification of enzyme:substrate and enzyme:inhibitor complexes within supported biomembranes, and iii) the introduction of new protein-orthogonal biomembrane tethering strategies that include membrane fusion.

382121 Green and Black Tea Polyphenols Mechanistically Inhibit Amyloid-β Aggregation in Alzheimer's Disease

Abstract:

Alzheimer's disease (AD) is currently the 6th leading cause of death and is the only disease among the top 10 that cannot be prevented, cured or treated. The amyloid cascade hypothesis states that the cause of the disease is the deposition of plaques comprised of insoluble amyloid-β (Aβ) fibrils in the brain. Naturally occurring Aβ monomer can aggregate via a nucleation dependent pathway in which Aβ monomer forms nuclei that conjoin to create soluble aggregates. These soluble aggregates are then lengthened through elongation and thickened through association to form insoluble fibrils. Research has shown that Aβ aggregates are neurotoxic. Consequently, inhibition of Aβ aggregation is one therapeutic strategy for AD. Epidemiological studies have correlated consumption of green tea, and to a lesser extent black tea, with a reduced incidence of AD. This study sought to explain this correlation by identifying the Aβ aggregation inhibitory capabilities of key polyphenol components in green and black tea.

Polyphenols studied include epicatechin, epigallocatechin and epigallocatechin gallate found in green tea and theaflavin and theaflavin monogallate found in black tea. Four different experiments were used to target unique steps along the aggregation pathway: a monomer aggregation assay to monitor the overall aggregation; an oligomerization assay to monitor the nucleation step; an association assay to monitor the late stage lateral binding of soluble aggregates; and an elongation assay to monitor the late stage lengthening of soluble aggregates. Monomer aggregation is induced via agitation in aqueous Tris-HCl buffer and is monitored using intensity of dynamic light scattering (DLS) and dot blot analysis with aggregate conformation specific antibody LOC. Oligomerization is initiated via the dilution of DMSO-solubilized Aβ in PBS and is monitored via SDS-PAGE and Western blotting to observe changes in oligomer size. Association and elongation are initiated by incubating SEC-purified Aβ soluble aggregates with salt or protein monomer, respectively, and are monitored for increases in aggregate size using DLS.

All catechin compounds exhibited moderate to slight inhibition of the overall aggregation, while the theaflavin compounds exhibited moderate to complete inhibition. To gain further insight into the process of inhibition, the mechanistic steps of aggregation, oligomerization, association and elongation, were examined individually. Theaflavins but not catechins were found to inhibit oligomerization. Conversely, catechins were more effective inhibitors of association than theaflavins. Both catechins and theaflavins showed similar inhibitory capabilities toward elongation. Together, these results demonstrate that while catechins and theaflavins can both inhibit Aβ aggregation, the mechanisms of action of these compounds is distinct.

Catechins and theaflavins show different inhibitory capabilities at varying mechanistic steps of the Aβ monomer aggregation pathway. Catechins affect only the later stages of aggregation, suggesting that the catechins may bind a specific structure present in aggregates. Conversely, theaflavins show inhibitory capabilities at every stage of the Aβ monomer aggregation pathway, alluding to a sequence specific recognition. Furthermore, better inhibitory capabilities, for both polyphenol categories, can be correlated with the number of gallate groups.

377164 Zwitterionic Fusion in Hydrogels and Spontaneous and Time-Inde Pendent Self-Healing Under Physiological Conditions

Abstract:

Self-healing materials can eliminate damage or fatigue during normal utilization, thus holding promise for enhanced lifetimes and enduring strength. They have been demonstrated by exploiting polymer-clay interactions, phase segregation, redox reactions, photo-reactive groups, and microvascular networks. However, up to now few materials have been reported to be effective in spontaneously repairing themselves under physiological conditions upon damage. For these few existing spontaneously self-healing materials, healing can only be achieved immediately after rupture occurs (i.e. in less than one minute) or at low pH values. Here, we report the first spontaneously healing material, driven by a completely new mechanism, *zwitterionic fusion*. It can repair at any lengths of separation time after damage (i.e., time-independent behavior) without any healing agents added or external energy input. As a result, two or more pieces of hydrogels can be re-connected easily after separation. In addition, this hydrogel demonstrates exceptional non-fouling behavior and efficient functionalization, which make them also suitable for biomedical-related applications.

388012 Biological Isolation of Protein from Shrimp Waste

Abstract:

Chitin is the second most abundant renewable natural biopolymer after cellulose. It is a linear polysaccharide composed of (1-4) - linked 2-acetamido- 2-deoxy-D-glucose units and has a chemical composition similar to that of cellulose. Chitosan, a derivative of chitin, is a linear polysaccharide of (1-4)-linked-D-glucosamine (GlcN) units. Chitin and chitosan are extensively employed in cosmetics, biotechnology, agriculture, medicine, food, and other industries. Chitin is commonly isolated from shrimp waste that is generated in large quantities by the shrimp processing industry. The first step in chitin isolation is removal of protein. The regular method for protein removal is the treatment of shrimp waste with the NaOH solution at high temperature. This method is energy demanding and environmentally unfriendly. Furthermore, harsh treatments of the waste material with a base at elevated temperatures may have a negative effect on the physical properties of the polymer as well. This research is focused on the development of alternative methods for protein removal from shrimp waste. The alternative techniques involve biological and physical methods for protein removal. The biological method of protein removal involves aerobic and anaerobic digestion of shrimp waste with *B. subtilis,* and mixed cultures from municipal sewage sludge. The physical methods for protein removal involve washing of the shrimp waste with water at different temperatures, sonication, and accelerated solvent extraction. Out of the methods investigated, the biological method shows promising results and will be discussed during the presentation. The aforementioned methods for protein removal from shrimp waste provide an alternative for chitin extraction in more environmentally friendly matter.

386214 Bactericidal Hydrogels Via Photointiated Thiol-Acrylate Interfacial Polymerization

Abstract:

Rapid and precise pathogen deactivation is of crucial importance in the medical, food, and biodefense industries. Antimicrobial peptides (AMPs) are a tremendous natural resource for developing decontamination surfaces owing to their innate broad spectrum activity. Thus far, AMP immobilization has been predominantly restricted to inorganic solid supports such as gold or glass despite recognized limitations, such as nonspecific biomolecular adsorption, labor-intensive peptide immobilization, and singular peptide orientation upon immobilization. Alternatively, the exchange of gold or glass for poly(ethylene glycol) (PEG) hydrogel supports would eliminate nonspecific binding in addition to broadening the scope of immobilization techniques and end applications. Therefore, the primary objective of this work was to establish a novel platform for the immobilization of Cecropin A (CPA) to the surfaces of PEG hydrogels via photoinitiated thiol-acrylate interfacial polymerizations. Using a 1:1 stoichiometric ratio of thiol:acrylate in the interfacial monomer formulation allowed coatings to be formed under ambient conditions without complete thiol conversion. The unreacted thiol was then conjugated with maleimide moieties within CPA sequences over a range of surface densities and configurations. Surface density was controlled by the weight fraction of PEG thiol and PEG acrylate in the interfacial monomer solution, while CPA surface configuration was altered by the type of PEG thiol that was incorporated. Specifically, the molecular weight of 4-arm PEG thiol (5 kDa, 10 kDa, and 20 kDa) was varied to investigate the influence of linker length on antibacterial activity, and the degree of thiol functionality (bifunctional 5 kDa, 4-arm 10 kDa, and 8-arm 20 kDa) was varied to control the arrangement of CPA on the surface. Propidium idodide/Syto 9 staining was used to quantify the activity of CPA against *E. coli* K235.

380042 Monodisperse Gold Nanoparticles Supported By Rice Husk Silica for Heterogeneous Catalysis Applications

Abstract:

Gold nanoparticles (Au NPs) based catalysts have received more and more attentions in the past two decades because of their unique catalytic properties in many important industrial processes. Both the experimental findings and theoretical predictions have demonstrated that the size of Au NPs plays a critical role in governing the catalytic activity. Smaller Au NPs typically exhibit higher catalytic activity. Herein, we report a facile method to synthesize Au NPs based heterogeneous catalyst using silica from rice husks (RHs) as a support. Although silica supported Au NPs have been extensively studied, the silica supports are typically synthesized from tetraethyl orthosilicate (TEOS) via a sol-gel process. Notably, in this study, silica from RHs was first used as the catalyst support. The silica nanoparticles obtained by calcining HCl treated RHs, which possess a rough surface, was first modified by (3-aminopropyl)triethoxysilane (APTES), which was designed to play a dual role: one was to absorb Au precursor ($AuCl_4^-$) to silica surface and the other was to stabilize the resultant Au NPs obtained by reducing $AuCl_4^-$ using sodium borohydride ($NaBH_4$). Characterizations of the nanostructures revealed that the Au NPs were formed with narrow size distribution around 2-5 nm, which was very critical for essential catalytic activities. The RH silica supported Au NPs exhibited excellent catalytic activity for the reduction of 4-nitrophenol by $NaBH_4$.

Multifunctional Polyampholyte Polymers as a Tissue Engineering Platform

Matthew T. Bernards,[1,2*] Tianyi Liu,[1] and Hayder Al-Naseri[1]
Chemical Engineering[1] and Bioengineering[2], University of Missouri, Columbia MO

There is significant interest in the development of a stealth polymeric scaffold that is adaptable as a tissue regeneration platform. Such a platform must be resistant to nonspecific protein adsorption, capable of protein conjugation, and have controllable mechanical properties. Our recent work has demonstrated that polyampholyte polymer hydrogels have all three of these beneficial properties. Polyampholyte polymers are a subclass of zwitterionic polymers which are composed of mixtures of positively charged and negatively charged monomer subunits. They are resistant to nonspecific protein adsorption and cell attachment in their base state, they are capable of covalently attaching adhesion proteins in controlled patterns and facilitating targeted cell attachment when activated, and they have controllable mechanical properties based on the selection of the monomer subunits. In this presentation each of these properties will be demonstrated and their adaptability for multiple tissues will be discussed. Furthermore, we have begun to demonstrate the use of these hydrogels as a platform for multiple tissue regeneration applications and these systems will also be presented.

363043 Crosslinking of Extracellular Matrix Scaffolds Derived from Pluripotent Stem Cell Aggregates for Neural Differentiation

Abstract:

Pluripotent stem cells (PSCs), including embryonic stem cells (ESCs) and induced pluripotent stem cells (iPSCs), emerge as promising sources for tissues engineering, drug screening, and disease modeling due to their unique proliferation ability and differentiation potential. At various developmental stages, PSCs and their progeny secrete large amounts of extracellular matrices (ECM) which could interact with regulatory growth factors to modulate the microenvironment of stem cells. Hence, ECMs derived from different types of PSC aggregates have broad signaling capacities and can be used as scaffolds to mediate cellular differentiation. However, the fast degradation of PSC-derived ECMs impedes their wide applications for in vitro cell expansion and in vivo transplantation. To address this issue, this study investigated the effect of crosslinking on the ECMs derived from PSCs and their regulatory capacity on the proliferation and differentiation of the reseeded PSC-derived neural progenitor cells (NPCs). To create different biological cues, ECMs from undifferentiated aggregates, spontaneous embryoid bodies, or PSC-derived NPC aggregates were decellularized. The derived ECM scaffolds were crosslinked using genipin or glutataldehyde to enhance the stability. The scaffolds were characterized by scanning electron microscopy and atomic force microscopy for topography and elasticity. PSC-derived NPC aggregates were reseeded on different ECM scaffolds and the cellular compositions of neural progenitors, neurons, and glial cells were characterized. Human induced pluripotent stem cells were also evaluated for the expressions of pluripotent markers and neural differentiation markers on different ECM scaffolds. The results indicated that various PSC-derived ECM scaffolds affected stem cell expansion and neural differentiation through intrinsic biological cues and biophysical properties and have potential for in vitro applications and in vivo delivery.

References:

1. Sebastien Sart, Teng Ma, Yan Li. Extracellular matrices decellularized from embryonic stem cells maintained their structure and signaling specificity. Tissue Eng Part A. 2014, 20: 54-66.

2. Sébastien Sart, Yijun Liu, Teng Ma, Yan Li. Microenvironment regulation of pluripotent stem cell-derived neural progenitor aggregates by human mesenchymal stem cell secretome. Tissue Eng Part A 2014, In Press. doi:10.1089/ten.TEA.2013.0437.

AUTHOR INDEX

AUTHOR INDEX

AUTHOR INDEX

AUTHOR INDEX

AUTHOR INDEX

AUTHOR INDEX

AUTHOR INDEX

AUTHOR INDEX

AUTHOR INDEX

AUTHOR INDEX

AUTHOR INDEX

AUTHOR INDEX

AUTHOR INDEX

AUTHOR INDEX

AUTHOR INDEX

AUTHOR INDEX

AIChE
120 Wall Street, FL 23
New York, NY 10005-4020

ISBN 978-1-5108-1264-2

CURRAN ASSOCIATES INC.
proceedings
.com